Biotechnology in Agriculture Series

General Editor: Gabrielle J. Persley, Science Adviser, The Agricultural Research Group, Environmental and Sustainable Development, The World Bank, Washington DC, USA.

For a number of years, biotechnology has held out the prospect for major advances in agricultural production, but only recently have the results of this new revolution started to reach application in the field. The potential for further rapid developments is, however, immense.

The aim of this book series is to review advances and current knowledge in key areas of biotechnology as applied to crop and animal production, forestry and food science. Some titles focus on individual crop species or groups of species, others on specific goals such as plant protection or animal health, with yet others addressing particular methodologies such as tissue culture, transformation or immunoassay. In some cases, relevant molecular and cell biology and genetics are also covered. Issues of relevance to both industrialized and developing countries are addressed, and social, economic and legal implications are also considered. Most titles are written for research workers in the biological sciences and agriculture, but some are also useful as textbooks for senior-level students in these disciplines.

Titles Available:

1: Beyond Mendel's Garden: Biotechnology in the Service of World Agriculture *G.J. Persley*
2: Agricultural Biotechnology: Opportunities for International Development *Edited by G.J. Persley*
3: The Molecular and Cellular Biology of the Potato *Edited by M.E. Vayda and W.D. Park*
4: Advanced Methods in Plant Breeding and Biotechnology *Edited by D.R. Murray*
5: Barley: Genetics, Biochemistry, Molecular Biology and Biotechnology *Edited by P.R. Shewry*
6: Rice Biotechnology *Edited by G.S. Khush and G.H. Toenniessen*
7: Plant Genetic Manipulation for Crop Protection *Edited by A. Gatehouse, V. Hilder and D. Boulter*
8: Biotechnology of Perennial Fruit Crops *Edited by F.A. Hammerschlag and R.E. Litz*
9: Bioconversion of Forest and Agricultural Plant Residues *Edited by J.N. Saddler*
10: Peas: Genetics, Molecular Biology and Biotechnology *Edited by R. Casey and D.R. Davies*
11: Laboratory Production of Cattle Embryos *I. Gordon*

LABORATORY PRODUCTION OF CATTLE EMBRYOS

Ian Gordon
BSc, MA, PhD, DipAgric (Cantab), NDA

Professor Emeritus
Department of Animal Science and Production
University College
Dublin
Ireland

CAB INTERNATIONAL

CAB INTERNATIONAL
Wallingford
Oxon OX10 8DE
UK

Tel: Wallingford (01491) 832111
Telex: 847964 (COMAGG G)
E-mail: cabi@cabi.org
Fax: (01491) 833508

A catalogue entry for this book is available from the British Library.

ISBN 0 85198 928 4

Typeset by Colset Pte Ltd, Singapore
Printed and bound in the UK at the University Press, Cambridge

CONTENTS

Preface

The world of animal production in Ireland and the UK has undergone radical changes in the past 40 years, since the author took his first steps in research towards the control of reproduction in farm animals. This monograph is devoted to a technology which is in process of development but which will probably not reach the farm on any scale before the end of the century. It is the technology which will take over from artificial insemination (AI) as the farmer's main tool in producing calves in the numbers and of the quality that he/she requires.

Cattle embryo transfer (ET), as introduced into commercial practice from 1970 onwards, has enabled cows to produce calves of a genetic quality and composition quite different from that of the surrogate mother herself. It also enables cows, under suitable conditions, to produce two calves rather than one. In appropriate farming systems, such breeding innovations can be the basis of improved animal efficiency. In Ireland, the production of twins rather than singles from suitable suckler and dairy cows may be a valuable means of providing calves for beef rearing in appropriate numbers and of acceptable quality.

Elsewhere, as well as the production of twins from beef suckler cows, it is possible that high-quality single beef calves may be produced by ET in Holstein milking cows. These are among the reasons why attention became focused in Dublin 30 years ago on researching low-cost methods of producing cattle embryos.

In comparison with the technique of AI in cattle, which has been around in one form or other for the past 100 years, *in vitro* fertilization (IVF), as a technique with relevance to livestock production, is a very recent innovation. A quarter-century ago, researchers could still be found arguing about accepting evidence that mammalian IVF had actually been achieved. As recently as a few years ago, many well-respected researchers would have considered it improbable that large numbers of cattle embryos would soon be produced from oocytes recovered from ovaries collected indiscriminately at the local abattoir. None the less, from a handful of laboratories conducting research on *in vitro* maturation and fertilization of cattle oocytes 5 years ago, the number currently active in producing cattle embryos for research and other purposes must now be numbered as well in excess of 100.

The starting-point for Dublin interest in laboratory-produced cattle embryos arose from a suggestion, in a 1962 paper dealing with the induction of twinning in beef cattle, that transfer of a single embryo to the non-pregnant uterine horn, a few days after breeding the cow, might have merit as an approach to twin production. The need for a plentiful supply of inexpensive beef embryos to explore such an approach subsequently led, in 1966, to the regular collection of cattle ovaries from a local abattoir close to the research laboratory. In cattle, early attempts to mature and fertilize cattle oocytes recovered from slaughterhouse ovaries were made in 1968 by Joe Sreenan. However, almost 20 years were to pass before the first set of beef cattle twins, produced by entirely *in vitro* laboratory procedures, were born on the university farm.

Before that time, studies in the early 1980s by Bob Moor and colleagues at Cambridge with sheep had provided crucial evidence on the supportive role of follicular cells during oocyte maturation. This acted as the spur to increased efforts in Dublin from 1984 onwards. By the end of 1987, largely as a result of the efforts of graduate research students, *in vitro* production of cattle embryos was a reality and a new Irish biotechnology company (Ovamass Ltd) was established in an effort to commercialize the production process.

The present work, which is based on more than 3000 reports, marks an attempt to cover the literature dealing with the laboratory production of cattle embryos and the way in which such embryos might be used in commercial applications and in research. It is clear from the massive volume of data reported in the literature by workers such as Chris Polge and Kehuan Lu that cattle embryos can now be produced in large numbers and in a consistent, predictable way. They can, as shown by Stan Leibo and others, survive freeze–thawing virtually the same as *in vivo*-produced embryos and after transfer can give rise to normal pregnancy rates.

All this is documented in the monograph; data presented in tables are largely drawn from the work of graduate students working with the author in Ireland. Probably the most important point to keep in mind is that embryo production technology can provide the necessary launching pad for many other developments in cattle breeding, particularly the farm application of large-scale embryo sexing and cloning techniques. It is also highly relevant to many areas of research in molecular biology and to future developments in recombinant deoxyribonucleic acid (DNA) technology.

Certainly, there is considerable scope for much further research to improve the technology of cattle embryo production; in most cattle IVF laboratories the production of blastocysts plateaus out at 20–40% of inseminated oocytes despite numerous variations on the basic technique. The fact that the majority of oocytes after fertilization do not develop into usable embryos is a measure of the ground yet to be covered and one of the reasons for compiling this monograph.

Some amount of space in the book is devoted to the potential use of the embryos in livestock farming. No apology is made for that; thought of practical applications was the reason for starting IVF work in the first place. In this, as a disciple of John Hammond, the author has always been committed to seeing science widely applied for the common good. The relentless growth of the human population and the need to sustain reliable food supplies at reasonable cost makes it essential to maintain a highly motivated and technically based livestock industry. In Ireland, as in many other countries, the infrastructure of the commercial AI services would readily permit exploitation of the new technology dealt with in this book. It is a matter of ensuring that the new embryo production technology is integrated into livestock production systems in a way acceptable to farmers and to society at large.

Research in laboratory embryo production to date has mainly focused on cattle, due to the fact that ET technology is much further developed commercially in that species than among the other farm animals. However, what can be achieved in cattle today should be equally possible in most of the other farm species tomorrow.

Finally, there is increasing interest in several countries in the use of IVF as an adjunct to conventional cattle ET. The superovulatory response of donor cows is notoriously variable and there are many valuable animals from which no embryo can be recovered, for one reason or another. The transvaginal aspiration of ovarian follicles, using real-time ultrasonics, may be the way to deal with such poor responders, but this is an approach which requires the necessary knowledge and expertise in IVF technology.

It is hoped that this monograph will be of some practical use to those who are about to establish an IVF laboratory. Towards that end, there are several appendices dealing with methods used in the Dublin laboratory for the production and use of embryos.

ACKNOWLEDGEMENTS

This book had its origin some years ago in a laboratory manual which the author compiled for graduate students pursuing studies in the IVF area of reproductive physiology. Thanks are due, first and foremost, to those graduate students who have worked with me at University College, Dublin over the last 30 years. Almost all the information in tables and many of the illustrations are drawn from the works they submitted for higher degrees. The author also wishes to express grateful thanks to editors and publishers who gave permission to reproduce illustrations and tables from books and original articles. Acknowledgements are also due to the staff of CAB International for all their help in the preparation of the work.

The final preparation of the text was carried out after retirement, in the tranquil surrounds of Lyons Research Farm. The author greatly appreciates the office and facilities provided by the University to enable the work to be completed.

Finally, a word of appreciation and thanks is due to a supportive wife, who cheerfully organized affairs to allow the writing to be done.

1 Developments in *In Vitro* Fertilization (IVF) Technology

1.1 Historical Aspects

More than a century ago, there were those who recognized that advancement of knowlege of mammalian fertilization would be greatly aided if the process could be observed *in vitro*. This meant developing conditions similar to those within the oviduct that would permit the egg and sperm to be maintained in a fully functional state.

1.1.1 First IVF reports

Back in the 1940s, much of the knowledge of fertilization was based on the study of sea urchin eggs, although early experiments to provide information on mammals were reported by Gregory Pincus and colleagues in the previous decade. A much better understanding of the basic mechanisms of mammalian fertilization came in the 1950s, with the recognition of the phenomenon of sperm capacitation (Austin, 1951; Chang, 1951) and the first comprehensive demonstration of IVF in the rabbit, resulting in the birth of young in 1959 (Chang, 1968; Chang *et al.*, 1977). For the first time, a repeatable technique for fertilizing mammalian eggs *in vitro* had been reported, using sperm recovered from the rabbit uterus 12 hours after mating.

Prior to that time, in work going back as far as 1878, studies with rabbit and other laboratory animals had failed to demonstrate that egg cleavage was the result of sperm penetration rather than activation of the egg by some other factor. Extensive experiments were performed by Gregory Pincus and associates during the 1930s in which they treated rabbit eggs with sperm and maintained them in culture. In some instances, the birth of young was recorded after the transfer of eggs treated with sperm to the oviducts of rabbits in which ovulation had been induced. However, such results were not regarded as conclusive because the experimental procedures did not exclude other possibilities of fertilization.

The successful fertilization of mouse eggs by sperm recovered from the uterus was first reported by Whittingham in 1968. It was first found by Chang and associates in the hamster and subsequently in the mouse that epididymal sperm cells could be prepared for IVF (artificially capacitated) without the conditioning required in the female tract. *In vitro* capacitation of sperm recovered from epididymides was achieved in mice and rats by the early 1970s.

1.1.2 Cattle IVF

One of the earliest attempts to fertilize artificially matured cattle oocytes *in vitro* was that of Sreenan (1970), who used bull sperm preincubated in a medium containing the enzyme α-amylase. The first genuine success in the fertilization *in vitro* of an artificially matured bovine oocyte is that recorded by Iritani and Niwa (1977) in Japan. The first calf resulting from IVF of an ovulated oocyte was born a few years later (9 June 1981). This was the result of the pioneering efforts of Brackett *et al.* (1982) in the USA. Two other IVF

Fig. 1.1. First cattle twins born after transfer of embryos produced by totally *in vitro* methods, reported in papers by Lu *et al.* (1987a, 1988c). The twins were displayed at various agricultural shows in the Republic and in Northern Ireland during 1988 to generate farmer interest in the use of IVMFC-derived embryos in cattle twinning.

calves were born in Georgia in 1983 (Brackett *et al.*, 1984). In Canada, using a laparoscopic technique to recover secondary oocytes, Lambert *et al.* (1983) succeeded in producing six IVF calves. This was done by recovering oocytes close to the time of ovulation, fertilizing them *in vitro* and using the rabbit oviduct as an *in vivo* embryo culture system.

The first calves born after IVF of artificially matured oocytes were those reported by Hanada *et al.* (1986) from Japan; in this work, embryos were cultured to the blastocyst stage in the rabbit oviduct and subjected to freeze–thawing prior to transfer. One of the first pregnancies produced by totally *in vitro* procedures (*in vitro* maturation (IVM) and *in vitro* culture (IVC) of the early embryo) was that reported from Dublin by Lu *et al.* (1987a); this resulted in the birth of the twins shown in Fig. 1.1.

1.2 Cambridge Contributions

The development of IVF/embryo transfer (ET) technology in mammals is associated with many valuable pioneering contributions from workers at Cambridge in the past century.

1.2.1 Days of Heape, Marshall and Hammond

Walter Heape's name will for ever be associated with the first successful egg transfer in the rabbit. However, quite apart from that, Heape had an important formative influence on the new science of reproductive biology in his years at Cambridge (Biggers, 1991; Heap, 1992). The first edition of Marshall's classic *Physiology of Reproduction* was dedicated to Walter Heape when it was published in 1910. It was Heape who, in 1905, had first proposed that the gonads were dependent for their activity and control on a generative ferment produced elsewhere in the body in minute quantities. This suggestion was subsequently taken up and developed by Marshall and Hammond in their studies on factors influencing embryo mortality at the School of Agriculture in Cambridge. In his book *The Breeding Industry*, published in 1906, Walter Heape also drew attention to various factors affecting reproductive efficiency in farm animals.

As observed by Asdell (1969), the first comprehensive treatise on reproduction was Aristotle's *Generation of Animals* and more than 2000 years were to pass before the next work, Marshall's *Physiology of Reproduction*, appeared in 1910. Up to that time, it was not uncommon for physiology textbooks to relegate discussion of the reproductive organs to a few final pages. In his treatise, Marshall presented all the relevant information on the breeding season, fertilization and other matters relating to reproduction. Although most discussion was with mammals, Marshall did not hesitate to draw from his knowledge of the reproductive processes in invertebrates and other forms of animal life.

The appearance of the expanded second edition of his book in 1922 acted as an important stimulus to research activities pursued in many laboratories of reproductive physiology between the two world wars. With Marshall as the biologist and John Hammond as the agricultural scientist, the Cambridge School of Agriculture became a renowned international centre for reproductive research.

1.2.2 The Animal Research Station

The Animal Research Station, 307 Huntingdon Road, Cambridge, was founded by John Hammond as part of the Cambridge School of Agriculture in 1932; after the Second World War it was established as a unit of the Agricultural Research Council (ARC). The primary objective of this ARC unit was to investigate problems related to ET in cattle. Among the many graduate students to work in Cambridge was Chang, who was to make several outstanding contributions to the development of IVF technology.

Chang came to Cambridge to work under John Hammond and Arthur Walton in 1939. In 1945, he had completed his PhD and several years of postdoctoral work, much of it focused on developing IVF/ET procedures in the rabbit. After the war, he moved to the Worcester Foundation of Experimental Biology at Shrewsbury, Massachusetts. With Marden in Cambridge and Chang in the USA, the first transatlantic shipment of fertilized rabbit eggs was accomplished in the early 1950s (Chang and Marden, 1954). Among the Australian graduates working with Hammond around that time was Des Dowling, who examined various aspects of embryo transfer in rabbits (Dowling, 1949). An important finding in Cambridge in 1949 arose from the work of John Hammond Jr, who was the first to consistently obtain *in vitro* embryo development in mice using a simple, defined medium. This was the first of many studies examining the *in vitro* requirements of mouse embryos. Other landmark work at that time was reported by Bob Averill who showed that sheep embryos could be stored in the oviduct of the rabbit (Averill *et al.*, 1955).

It was at the Animal Research Station in the late 1960s that Tim Rowson and colleagues first demonstrated that acceptable pregnancy rates could be achieved after embryo transfer in the cow. This was the main stimulus to the commercial development of cattle ET operations worldwide.

In the 1970s and early 1980s at the Animal Research Station, the work of Bob Moor and associates with sheep oocytes underlined the extremely important principle that normal embryonic development depends on successful maturation of the egg cell. In a series of important contributions, Moor and coworkers stressed the importance of the interactions between the oocyte and surrounding follicular cells. It was clearly shown that follicular somatic cells have both supportive and regulatory roles; it was suggested that a relatively simple IVM procedure should be possible for cattle (Moor *et al.* 1984).

In the early 1970s at the Animal Research Station, Ian Wilmut and Steen Willadsen were at the forefront in developing effective freezing procedures for cattle and sheep embryos. Alan Trounson and Steen Willadsen made many valuable contributions to knowledge in the micromanipulation of embryos and, by 1986, the first reports of successful cloning in sheep by nuclear transfer had appeared (Willadsen, 1986).

1.2.3 Cambridge, Babraham and beyond

In 1976, the Animal Research Station was incorporated into the ARC Institute of Animal Physiology, upon the retirement of its then director, Professor Thaddeus Mann. Among the many achievements of the station in the late 1970s and early 1980s was the birth of lambs and piglets by IVF techniques (Cheng *et al.*, 1986). The station continued to operate at Huntingdon Road under Tim Rowson and later Chris Polge until 1986, when it was finally closed by the Agricultural and Food Research Council (AFRC) as part of a restructuring programme. The station staff moved across Cambridge into new laboratories at the AFRC Physiology Institute in Babraham, in what now constitutes the Department of Molecular Embryology.

However, the Animal Research Station currently serves as the headquarters of Animal Biotechnology Cambridge (ABC), a company which has Chris Polge as its scientific director and has been devoted to research in several areas of biotechnology and reproductive technology, including the production of calves after separation of X- and Y-chromosome bearing sperm and IVF (Cran *et al.*, 1993). In 1988, ABC Ltd launched its system for cattle embryo production, the main commercial objective being to provide singleton high-quality beef embryos for the Holstein/Friesian dairy cow. This was done on the understanding that 7/8 beef calves produced from laboratory-fertilized embryos would have carcass values substantially higher than those of Continental × Holstein/Friesian stock. A year later, ABC Ltd took over the Irish company, Ovamass Ltd. Until 1992, ABC Ltd continued the production of beef embryos in Ireland to service the various cattle artificial insemination (AI) stations which had expressed interest in providing their clients with a twinning service.

What started as the Institute of Animal Physiology at Babraham after the Second World War became the Cambridge Research Station in 1986. The current focus of the station's work is towards advancing knowledge of the molecular, cellular and systemic processes in the animal body. Particular attention is paid to the mechanisms controlling development, fertility, growth, behaviour, health and welfare of animals (Cross, 1988). In line with increasing emphasis on molecular biology, what was the Agricultural and Food Research Council, responsible for the Cambridge Research Station, has become the Biotechnology and Biological Sciences Research Council. The activities of Bob Moor and colleagues in the Cambridge station's Department of Molecular Embryology continue to be especially relevant to all aspects of *in vitro* embryo production.

The Cambridge scene would not be complete without recalling that Bob Edwards, located in the university's physiology laboratory, was actively collaborating with Chris Polge and others at the Animal Research Station in the mid-1960s, in maturing the oocytes of various mammals, including cattle, sheep and pigs (Edwards, 1965, 1966). Although his agricultural interests are less well-known than his pioneering efforts in the development of human IVF technology, he contributed in no small measure to the formidable body of expertise located in Cambridge at that time.

Further afield, Ben Brackett in Georgia, Neal First in Wisconsin and Bob Foote in Cornell, together with their many students and associates, are especially well recognized for having pioneered many valuable developments in cattle IVF, in cloning and in genetic engineering (Brackett & Oliphant, 1975; Brackett *et al.*, 1980a, b, 1987; Brackett and Bousquet, 1984; First *et al.*, 1986; First, 1987, 1989, 1990a, b; Foote, 1987, 1993).

In Japan, the first reported success in the fertilization *in vitro* of artificially matured cattle oocytes was that of Akira Iritani and associates in 1977. It was in that country, a few years later in 1986, that Hanada and colleagues were to report the birth of the first calves derived from IVM/IVF, using the rabbit oviduct as the *in vivo* culture system. In Canada, Sirard, Lambert and colleagues at Quebec pioneered the use of laparoscopy to recover *in vivo*-matured oocytes from the ovaries of donor cows. After capacitation of fresh semen with high ionic strength (HIS) medium and IVF the rabbit oviduct was employed as the *in vivo* culture system to obtain embryos for transfer. In Denmark, the work of Torben Greve and coworkers has also been a valuable contribution towards the development of effective cattle IVF technology (Greve *et al.*, 1984, 1989a, b; Greve and Madison, 1991).

1.3 Dublin Contributions

In the Irish Republic, the beef industry is of major importance to the economy of the country. More than 40% of current foreign earnings come from agricultural exports and beef and dairy products are the main contributors to such exports. An important competitive advantage that Ireland holds over most other European Union (EU) countries is its potential to produce beef and dairy products cheaply from grassland in an environmentally sensitive way. The challenge facing Irish livestock farmers is to produce top quality beef from cattle systems seen to be welfare- and consumer-orientated.

The biological and economic efficiency of the country's beef and dairy cattle is limited in various ways by their reproductive performance. Since the mid-1960s, a research programme in reproductive physiology has been followed in University College, Dublin, aimed at developing new possibilities in reproduction control and breeding improvement. The efficiency of beef production and the quality of the carcasses that are produced in Ireland have been limited by the fact that most calves have come from the dairy herd. The ability to produce Continental bull calves from dairy cows, using sexed beef embryos, could be of particular commercial interest.

1.3.1 Early work in cattle

Some of the earliest studies exclusively devoted to bovine oocytes recovered from slaughterhouse ovaries were those of Sreenan (1968, 1969), who reported

Fig. 1.2. Twins produced by single embryo transfer to the contralateral horn of the uterus of the bred recipient cow. Research in cattle twinning by embryo transfer was initiated in University College, Dublin, in 1966. The aim was to develop a commercially feasible method that could be applied to beef and dairy cattle under the particular farming conditions that obtain in the Irish Republic. By the mid-1970s, several sets of twins had been born on the university farm by the method described above.

nuclear maturation occurring after a period of 24 hours in culture. As noted in later sections, the primary oocyte is arrested in prenatal life in prophase of the first meiotic division. Resumption of nuclear maturation (in the follicle destined for ovulation) only occurs under the influence of the preovulatory release of luteinizing hormone (LH), some 24 hours prior to ovulation.

Around the same time, an effective procedure, using the rabbit oviduct as an *in vivo* culture system, to permit the continued cleavage of the early cattle embryo to the blastocyst stage of development was reported (Sreenan and Scanlon, 1968; Sreenan *et al.*, 1968). There were also early attempts by Sreenan (1970) to fertilize artificially matured bovine oocytes *in vitro* and *in vivo* (in the sheep oviduct). A few years later, an effective procedure for cattle twinning by non-surgical transfer was reported (Boland *et al.*, 1975; Gordon, 1975); this was applied in a limited experimental scale, using embryos recovered from superovulated donor cattle (see Fig. 1.2).

1.3.2 *Twins for increased cattle efficiency*

Improving the biological and economic efficiency of production systems is of obvious concern to those undertaking research and development work in the beef industry. Lifetime efficiency of the beef animal is closely related to reproductive rate and increasing the twinning rate (normally 2–3%) could be one means of markedly increasing animal efficiency. Various studies suggest that twinning could increase the biological and economic efficiency of beef production by 20–25%.

In the EU, many of the calves reared for beef come from a decreasing dairy cow population. In Ireland, where most beef cattle originate in the dairy herd, induced twinning in even 10% of that herd could have a marked effect on calf numbers. In 1989, the Irish Department of Agriculture and Food introduced a temporary grant scheme to encourage farmers to use modern twinning technology to increase the national calf crop. For the EU as a whole, the production of twins rather than singles from some proportion of the cattle population could be one means of avoiding a calf deficit in some future time.

1.3.3 *The need for low-cost embryos*

For twinning by embryo transfer to be adopted commercially in Ireland, a low-cost supply of good-quality beef embryos is required. Although twinning by embryo transfer was shown to be technically feasible by the mid 1970s (see Gordon, 1975), the embryos recovered after superovulation of donor beef cattle were too costly to form the basis of any large-scale application of the

Fig. 1.3. IVF calves born in autumn 1987 in Dublin after IVM and IVF with culture in the sheep oviduct. Eighteen calves were born alive to 13 recipient cows after non-surgical transfer of embryos; this represented an embryo survival rate of 59%.

technology. The requirement to produce low-cost embryos by an alternative approach became all the more important.

As a result of expanding knowledge of essential factors influencing cytoplasmic maturation of oocytes (Staigmiller and Moor, 1984a, b) and factors affecting sperm capacitation in cattle (Parrish *et al.*, 1984, 1986a, b), a system for maturing and fertilizing cattle oocytes *in vitro* was developed in Dublin

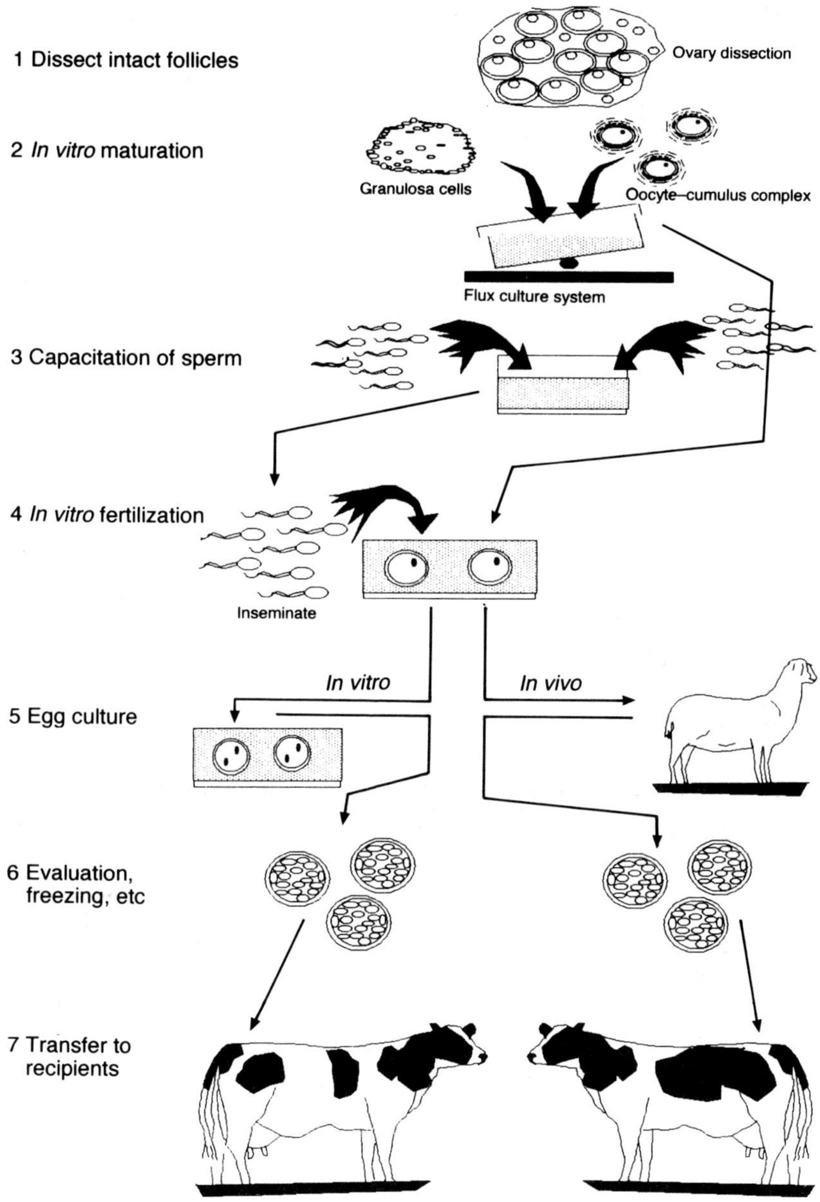

Fig. 1.4. Diagrammatic representation of cattle IVF system (from Lu, 1990).

(Lu *et al.*, 1987b). In the early work, an *in vivo* culture period in the sheep oviduct was employed to bring the embryos to the blastocyst stage for transfer (see Fig. 1.3). Late in 1987, the first calves were born from a totally *in vitro* production system, using the bovine oviductal cell coculture method (Lu *et al.*, 1987b, 1988a). A diagrammatic representation of the cattle IVF system as developed by Lu (1990) and associates is provided in Fig. 1.4.

1.3.4 Commercializing the embryo production process

Towards the end of 1987, a commercial company (Ovamass Ltd) was established to produce beef cattle embryos for use in cattle twinning in Ireland. The technical efficiency of the *in vitro* embryo production system, as developed in Ireland and employed by Ovamass, is dealt with in a paper by Lu and Polge (1992). This covers production over a 2-year period, from September 1989 to August 1991, in which 874,133 oocytes were collected from cattle ovaries obtained from abattoirs (12.2 per ovary). The average number of blastocysts produced per slaughter animal was 6.2 (see Table 1.1).

Table 1.1. Efficiency of IVF system expressed per ovary (from Lu and Polge, 1992).

	Oocytes			Embryos (blastocysts)	
Period	Collected	IVM	IVF	Total	Freezable
1/9/89–31/8/90	12.6	10.7	10.0	2.9	2.2
1/9/90–31/8/91	11.8	10.5	9.8	3.2	2.4
Total	12.2	10.6	9.9	3.1	2.3

For commercial success, it is essential that embryos for twinning transfer can be handled in much the same way as frozen bull semen. Research by Ovamass Ltd led to the development of a technique whereby frozen embryos could be thawed and a direct transfer made to the recipient animal (Bourke, 1992). This direct transfer method has resulted in acceptable pregnancy rates (Lu and Polge, 1992). By 1992, 130 inseminators from Irish AI stations had been trained to transfer beef embryos for twinning.

Cattle twinning by ET had been researched for almost 20 years in Dublin prior to the establishment of Ovamass Ltd in 1987. Results with *in vivo*-produced embryos from superovulated beef heifers suggested that good pregnancy rates (60%) and acceptable twinning rates (50%) could be achieved using bred recipient cattle (see Gordon, 1983). The time of establishing Ovamass Ltd in 1987 coincided with a period in Irish farming when calves for beef rearing had become both scarce and costly. It was therefore viewed as particularly opportune to initiate a commercial venture directed towards increasing the country's beef calf population.

1.3.5 Embryo production yields achieved

As a result of developments in oocyte recovery procedures, it is now possible to produce much larger numbers of embryos per slaughter animal. Data from the Dublin IVF laboratory now show oocyte yields of 88 per dam and blastocyst yields of 15.4 per dam (Carolan *et al.*, 1992a, 1993). Elswhere, it may be mentioned, there have been reports showing as many as 23 blastocysts per slaughter animal (Hamano and Kuwayama, 1993). Such embryo yields are certainly comparable to what may be achieved using current conventional superovulation procedures in cattle (Goode and Donaldson, 1990; Armstrong, 1991, 1993). As shown in Fig. 1.5, which refers to work conducted by Ovamass in the 1989–91 period, the proportion of oocytes that yielded freezable quality embryos at the end of the production process was relatively low; such data indicate the considerable scope there is for improving the existing technology.

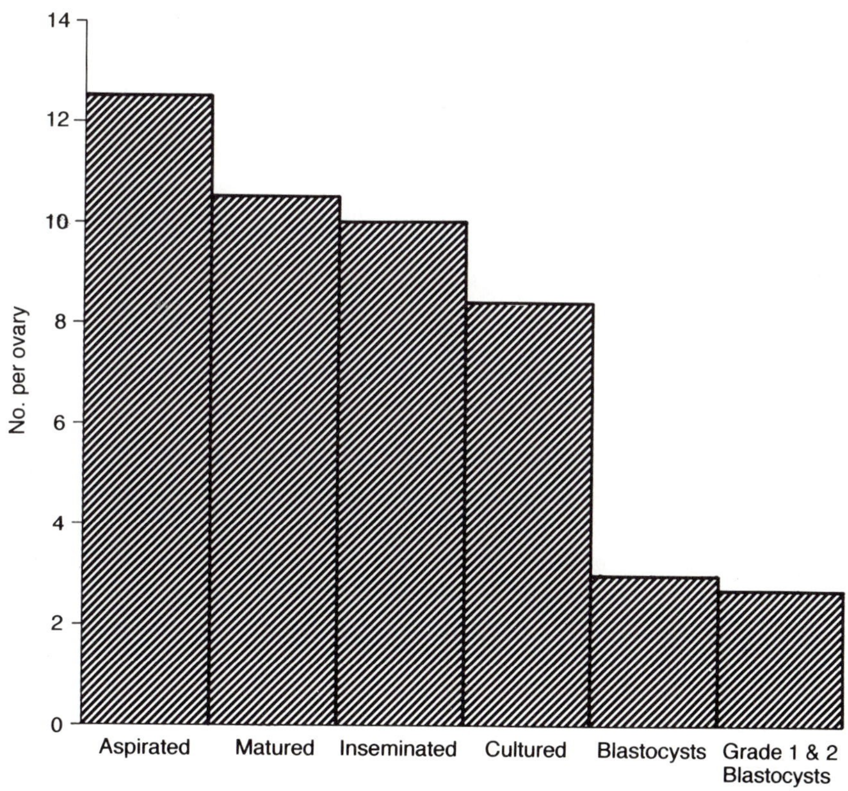

Fig. 1.5. *In vitro* production of cattle embryos in the Ovamass laboratories in 1989–91, expressed as yield of oocytes and embryos per ovary.

1.4 Developments in Embryo Transfer (ET) Technology

1.4.1 The pre-1970 period

In England, after the Second World War, the ARC Unit of Animal Reproduction, located at 307 Huntingdon Road, Cambridge, with John Hammond as director, had as its primary objective the development of ET procedures for use in cattle (see Gordon, 1983). Much of the early embryo recovery and transfer work was with sheep; as well as being easier to handle they had the important advantage of being less costly than cattle. By the late 1960s, researchers led by Tim Rowson were able to report that acceptable pregnancy rates could be achieved, using surgical intervention, in cattle. This provided much of the stimulus to commercial concerns to develop ET technology. In Ireland, Sreenan and Beehan (1974) were among the earliest to confirm the effectiveness of the transfer procedures.

1.4.2 Developments from 1970 onwards

Commercial cattle ET started in North America and elsewhere during the early 1970s, primarily as a means of multiplying the number of young produced by exotic breeds of beef cattle. In the early years, embryo recoveries usually involved midventral laparotomy with the donor under halothane anaesthesia. As noted by Hasler (1992), ET was not widely used with dairy cattle in the early 1970s because the udder of milking cows made midventral laparotomies rather difficult. However, by the mid-1970s, workers in several laboratories had developed non-surgical embryo recovery procedures to the point where they matched the results of surgical intervention.

Around the same time, non-surgical transfer techniques, most based on the Cassou inseminating 'gun' (Instruments de Medecine Veterinaire (IMV)), were also being developed. The availability of non-surgical recovery and transfer procedures opened the way to exploitation of cattle ET on the farm. An effective freeze–thaw method, which permitted embryos to be shipped to the furthest parts of the globe, completed the crucial requirements for commercial exploitation of cattle ET at that time.

After an initial explosion of interest in cattle ET in the USA and Canada in the 1970s (see Seidel, 1984), much of the uptake of the technique in the 1980s was in western Europe and in certain of the eastern European countries (see Van Lieshout, 1987; Wagner, 1987; Thibier, 1992). Early stages in the use of cattle ET in China in the 1980s is recorded by Hasler (1989).

A recent report by the International Embryo Transfer Society (IETS) (Anon, 1993) shows that about 300,000 cattle embryos were transferred worldwide (see Table 1.2).

1.4.3 Commercial advantages of cattle ET

The improvement of the genetic quality of cattle has traditionally been met by the pedigree breeder and in the past 40–50 years by way of AI, using

Table 1.2. Overall bovine ET activity in 1992 (from Anon., 1993).

Continents	Flushes	Numbers of Transferable embryos	Numbers of Embryos transferred
Africa[a]	1,837	10,660	9,707
North America[b]	34,413	–	126,587
South America[c]	5,009	21,164	20,245
Asia[d]	10,368	4,071[e]	28,829
Europe	22,699	109,559	96,300
Oceania[f]	330	2,297	986
Total	74,656	147,751	282,654

[a]Morocco, Republic of South Africa and Zimbabwe.
[b]Canada and 1991 data for USA.
[c]Argentina and Brazil.
[d]Korea, Japan and Thailand plus 1991 data from the People's Republic of China and India.
[e]The Japanese figure for this column was not reported.
[f]Partial report from New Zealand only.

progeny-tested and performance-tested bulls. The advent of an effective freezing technique was a particularly valuable milestone in the development of cattle ET technology. It meant that embryos could now be used in international trade, procedures having been agreed between countries to ensure that the transfer of embryos would not result in the transmission of pathogenic agents.

The commercial advantages of cattle ET include: (i) facilitating genetic improvement in the cattle industry by obtaining a larger number of calves from cattle of high genetic quality; (ii) enabling embryos to be moved from country to country in the frozen state, thereby reducing the need for long-distance cattle movements, with their attendant welfare problems; (iii) permitting high quality breeding stock to be available for sale in much larger numbers than was previously possible; and (iv) exploiting developments in reproductive technology, such as embryo sexing and embryo splitting. It should also be remembered that ET technology is the means by which advances in techniques such as IVF, large-scale cloning technology and genetic engineering can be exploited. Genetic screening by way of deoxyribonucleic acid (DNA) probes/polymerase chain reaction (PCR) technology is likely to be applied to embryos and to assist in the selection of the most appropriate genotypes for transfer (Hasler, 1991). Eventually, a point may be reached at which ET rather than AI may be regarded as the method best fitted to advance the quality of farm livestock.

1.4.4 *Current status of cattle ET technology*

Although the basic procedures (superovulation, embryo recovery, storage/freezing and transfer) employed in cattle ET are now well established (Mapletoft, 1985), there is considerable room for improvement in various areas of ET technology. As pointed out by Hasler (1992), the outcome of superovulation treatment has shown little improvement in the past 10–15 years. Although authors differ in their estimates, there is ample evidence in many instances to show that no more than five to seven embryos of transferable quality are recovered per flushed donor. The variability in response to a standard superovulation treatment is notorious. Marked improvements in the current pregnancy rates with fresh (65–80%) and frozen (55–70%) embryo transfers in suitable recipients are not likely (Hasler, 1992).

Although cattle ET may offer access to the highest-quality genetics at a fraction of the cost of importing a live animal, commercial (*in vivo*) ET does not, as currently practised, involve dealing with embryos on farms in the way which is commonplace with bull semen and AI. It is estimated by the IETS that the number of cows eligible as donors for commercial ET is less than 1% (of the total breeding cow population) in developed countries and less than 0.1% in the developing countries.

Future developments in commercial ET technology will almost certainly include the recovery of oocytes by ultrasound guided aspiration with subsequent IVF/IVC. In this, the need is for well-tested and repeatable IVF technology which can be applied in a cost-effective way. This technology could certainly permit embryos to be made available from those donors which, for one reason or other, will not provide them by conventional superovulation and flushing procedures.

1.5 Laboratory-produced Embryos in Cattle Production Systems

Effective techniques for the large-scale laboratory production of cattle embryos constitute only a first step towards getting this new form of ET technology usefully and extensively applied in normal farming practice. For ET to be provided along the same lines as AI, it will obviously be necessary to have clear guidelines on many aspects of transfer technology, including the most appropriate non-surgical transfer procedures and methods of dealing with recipient cattle. Other considerations, such as the most suitable pregnancy testing, feeding and management methods to employ with the pregnant recipient animal, have also to be addressed.

The methods employed in getting laboratory-produced embryos used in farming are likely to vary according to whether recipients are dairy cows or beef suckler animals. AI has been extensively employed around the world in dairy cattle but used much less in beef animals. The initial application of laboratory-produced embryos might be expected to follow a similar route because the established AI services are obviously geared to dealing with dairy

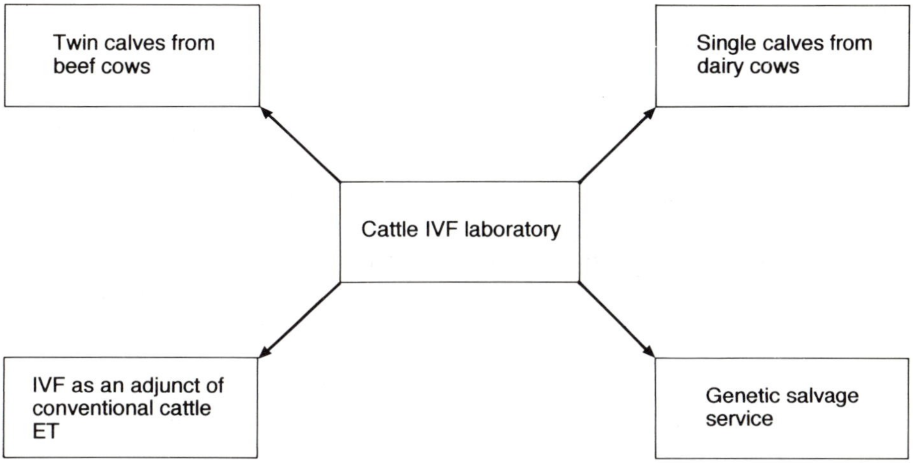

Fig. 1.6. Potential uses of IVF technology in cattle.

animals rather than beef. Some of the potential uses of cattle IVF are set out in Fig. 1.6.

1.5.1 Beef calves from high-yielding dairy cattle

It is estimated that about two-thirds of beef production in the UK is based upon calves produced from the dairy herd (Simm *et al.*, 1990; Reynolds and Street, 1991; Mayne and McEvoy, 1993a). However, the suitability of such calves for beef production may be questionable, given the trend towards greater use of the more extreme dairy-type blood lines in the dairy herd. There has been an increasing percentage of Holstein blood in the Friesian–Holstein population in recent years, primarily the result of semen and embryo importations into the UK from Canada and the USA. Similar trends are also increasingly evident in the Irish Republic. In 1991, the proven bull stud in Irish cattle AI stations contained 49% Holstein blood compared with 13% in 1983. In young bulls in progeny tests, Holstein blood has increased from 15% to 61%. It is inevitable that such changes will influence the quality of calves reared for beef.

Research in Northern Ireland has shown a 23% improvement in the efficiency of lean meat production between Friesian steers and Continental × Friesian steers (Mayne *et al.*, 1991). However, current trends both in the Irish Republic and the UK are towards greater efficiency in the dairy herds and this is likely to mean an increasing number of unsuitable calves for beef rearing. As the dairy herd becomes more extreme, the beef industry is likely to suffer increasingly from a lack of suitable calves for rearing.

One means of ensuring high-quality beef from extreme dairy cows is by producing embryos derived from good-quality beef cattle oocytes and sired using semen from suitable bulls (e.g. 7/8 Continental crossbreds). Such embryos could then be employed in single transfers to those dairy cows that are destined to produce calves for beef rather than dairy replacements. The use of embryos from half-bred Continental beef dams, by increasing the proportion of beef genes in calves produced from the dairy herd to at least 75%, could result in a useful improvement in the beef-producing potential of the dairy herd.

With the advent of semen sexing to ensure the birth of bull calves, IVF embryos could have even greater appeal. According to some estimates, sexing embryos for beef rearing could increase saleable meat production by about 29% (Reynolds and Street, 1991). However, the cost of the IVF beef ET would have to be in line with that of a beef insemination for it to have genuine commercial appeal. There is no future for techniques in beef rearing which do not have a clear cost–benefit advantage.

The production of twins from dairy cows, under suitable conditions of feed and management, may occasionally be a feasible option. The genetic composition of beef calves may be an important consideration in this. Single, high-quality Continental calves may be a possibility with milking animals, but twins may not, because of the total fetal load involved. If twin beef calves are to be produced, then *in vitro* matured, fertilized and cultured (IVMFC) embryos based on Hereford or Angus genetics may be more appropriate. However, this would have to be in market conditions where the lighter carcass weights would be treated preferentially. Weight restrictions on beef carcasses, as currently proposed by the EU, will impose a 340 kg limit on cattle carcasses qualifying for intervention support as from summer 1994. Such changes in favour of lighter beef carcasses may influence decisions on the farm as to the most appropriate beef calves to produce.

1.5.2 Twins from beef suckler cows

The EU dairy herd produces the majority of the calves that are reared for beef. The milking cow population has been contracting since 1984 with the imposition of milk quotas and, although there has been an increase in suckler cow numbers in most countries in recent years, the overall effect is a decrease in the size of the EC cattle herd. In such a context, the availability of commercially acceptable procedures for the production of twin calves in beef cattle systems may have some relevance. Twinning in beef cattle by ET can increase the output of calves per cow, without any detrimental effect on efficiency of feed use or carcass quality of the animals reared (Patterson, 1993b). There are probably marginal regions in the EU which would be well suited to beef production based on suckler cattle systems if the economic efficiency of the enterprise could be suitably improved; twins may be one way of achieving this.

Using laboratory produced embryos, it will certainly be possible, in due course, to predetermine the sex of the calves born after twinning transfers. Beef producers prefer steers rather than heifers for rearing, primarily because heifers are generally 10% less efficient in feed conversion, gain 10% less weight than steers and may run the risk of becoming pregnant. The biological efficiency of the suckler cattle system could be markedly enhanced by the application of an effective sexing procedure. With continued refinements, the sperm separation procedure already reported from Cambridge (Cran, 1992; Cran *et al.*, 1993, 1994; Miller *et al.*, 1993) is likely to be a ready means of ensuring the birth of bull calves.

Such biologically efficient cattle systems may well be in keeping with good environmental management, and so regarded as a desirable part of the EU's rural development policy.

Elsewhere in the world, there are likely to be various opportunities for producing twins rather than singles from beef cattle. In the USA, where beef consumption per capita is almost double that in the EC, there is strong interest in methods of increasing economic efficiency of the beef production process (Anderson *et al.*, 1982). In North America, there are pure-bred Hereford and Angus herds of beef cattle in abundance, and undoubtedly opportunities exist for exploiting a low-cost twinning procedure, based on laboratory-produced embryos (Gordon, I., 1988, 1989a, b). It may even be feasible, thinking in terms of current developments in do-it-yourself (DIY) ET in North America, for twin induction to be eventually conducted by appropriately trained stockmen in the same way as AI is currently carried out.

In all of this, it should be remembered that twin calf production is, as yet, a relatively untouched area of cattle husbandry. The most appropriate feeding and management methods for use under various farming conditions have yet to be fully determined. There is an obvious need to keep welfare considerations in mind in the development of such new production systems.

1.5.3 Calves from valuable donor cows

One possibility of making embryo production technology available to pedigree cattle farmers, large and small, is by using the ovaries of the genetically valuable cow after its death. As noted by Allen (1991a), in the UK there would be something like 750,000 cows culled each year from dairy herds and a corresponding number of replacement heifers brought in. The value of a laboratory-produced embryo, whether from dairy or beef animals, will be determined by the animal's pedigree, the IVF costs and the charge arising from the transfer of the embryo. The IVF technology would only come into play upon the slaughter of the cow after the end of its productive life. The best available dairy or beef sires would be employed in providing semen for the IVF.

The use of pedigree cattle ovaries to produce embryos is likely to have increasing appeal as it becomes possible to produce much greater numbers of embryos per animal ovary. Already, blastocyst yields averaging more than 20 per dam have been reported by some Japanese laboratories (Hamano and Kuwayama, 1993).

1.5.4 Using IVF as an aid to conventional cattle ET

Conventional cattle ET technology is based on the superovulation of high-quality donor cattle (dairy and beef) and the subsequent recovery of embryos by flushing the uterus 1 week after breeding to the bull of choice. Such technology inevitably involves repeated handlings of donors in the administration of hormones and in the breeding/flushing procedures.

It is widely recognized that superovulation treatments for conventional embryo production in donor cows have certain disadvantages, which include the marked variation in superovulatory response and the differences among donors in the number and quality of embryos recovered. There is also the fact that the calving interval of the donor, when this is a high-yielding dairy cow, may be extended by an average of 2 months (Bak *et al.*, 1989), adding to the general cost of applying ET.

An alternative to conventional superovulation procedures would be of considerable interest if it could provide a predictable supply of embryos. Recent advances in IVF technology make it possible for the repeated recovery of primary oocytes from the live cow to be linked to laboratory embryo production procedures in a way that could be commercially viable.

Recent calculations by workers in The Netherlands suggest that it may be realistic to think in terms of recovering in excess of 100 oocytes annually by way of transvaginal ultrasound-guided aspiration of vesicular follicles (Kruip *et al.*, 1991, 1993). Elsewhere, Ryan, D.P., *et al.* (1990a) have reported that oocytes recovered from cows in the early months of pregnancy can be used to produce embryos for transfer, after IVF and culture to the blastocyst stage. These authors note that the use of genetically superior pregnant cows for the laboratory production of embryos should maintain the normal calving interval and improve the genetic quality of the herd. There may even be the possibility of using semen sexed by flow cytometry (Cran *et al.*, 1993) in the planned production of bulls or heifers.

As well as the recovery of oocytes by aspiration in cyclic and pregnant animals, there would be other occasions (e.g. in prepubertal heifers) when it would be possible to produce embryos from high genetic quality cows in a way not normally available using conventional superovulation and embryo recovery regimens. It would even be possible to arrange to have embryos sired by several different bulls if the fertilization process is *in vitro* rather than in the cow oviduct.

1.5.5 Animal health and welfare considerations

Good health and freedom from physical and psychological stress are of paramount importance to good animal welfare on the farm. The welfare implications of novel techniques involving biotechnology and embryo manipulation must be kept constantly under review. It must be clearly recognized by the applied animal scientist that there are increasing numbers of people in the developed countries who are expressing a profound aversion to industrialized agriculture. Although the target of much of this emotion is the poultry and pig scene, all livestock producers must respond in a positive way to the challenge posed by these emerging views. Non-surgical embryo transfer need be no more stressful than many routine husbandry operations where cattle have to be handled, treated, dosed or injected for welfare reasons. It is important that public perception of the transfer technique should recognize this fact.

The technique of non-surgical transfer, which would be the usual means of establishing pregnancies in cows with laboratory-produced embryos, is not likely, as mentioned above, to raise questions of welfare different from those associated with routine AI. In this regard, note can be taken of studies such as those of Macaulay *et al.* (1986), in which evidence is provided indicating no additional stress with AI when compared with natural mating or with no mating at time of oestrus. In conventional cattle ET, the marginal commercial advantage to be gained from surgical rather than non-surgical transfer makes it difficult to justify the continued use of such intervention.

The capacity of the recipient animal to deal with the fetus is likely to be a much more controversial consideration with regard to the public perception of cattle ET. Appropriate procedures for the selection of suitable recipients, together with adequate attention to feeding and management during pregnancy, must always be an integral part of any application of IVF/ET technology on the farm. This will become all the more important as ET techniques, previously restricted to high-value pedigree cattle, become increasingly commonplace in beef and dairy herds with the availability of IVMFC-derived sexed embryos.

1.6 *In Vitro* Fertilization in Other Farm Animals

It is useful to keep in mind that the new embryo production technology developed for cattle is likely, with appropriate modifications, to have direct relevance to similar possibilities in sheep, goats, pigs, horses, buffaloes and deer. Such embryo technology is likely to be the foundation on which other developments (e.g. large-scale cloning, sexing and genetic engineering) can be built. It should be noted, however, that even in zebu cattle (*Bos indicus*) work in Brazil by Oliveira and associates has shown that there can be marked differences in the effectiveness of IVM/IVF/IVC procedures compared with that observed in European cattle (Oliveira *et al.*, 1991a, b, 1992, 1994). Quite apart from the use of IVF technology in basic research (see Fig. 1.7), the eventual commercial impact of *in vitro* production procedures, when applied to the full range of farm species, is likely to be considerable. In developing countries, techniques applicable to zebu cattle, buffaloes, goats and sheep may have a particular value and relevance.

1.6.1 Sheep and goats

There are many countries around the world in which sheep and goats feature prominently in the livestock industry. New forms of reproductive technology may well be employed to advantage in these species. The production of ruminant embryos *in vitro* has been reviewed by Trounson (1992). Sheep were the first of the farm animals to be used in IVF studies by Dauzier and Thibault in France during the 1950s. However, the first reports of lambs being born from oocytes matured and fertilized *in vitro* came much later. Cheng *et al.*

Fig. 1.7. A plentiful supply of cattle embryos for research purposes. Using slaughterhouse ovaries, it is now possible to produce large numbers of blastocysts using the technology described in this monograph (from Lonergan, 1992).

(1986), using a variety of ovine sperm capacitation procedures, reported the birth of lambs after transfer of *in vitro*-produced zygotes. These results were subsequently confirmed by Crozet *et al.* (1987a); in their studies, the early sheep embryos were generally cultured in rabbit oviducts prior to transfer to the sheep. The birth of normal lambs was also reported from former Czechoslovakia after the transfer of *in vitro*-produced zygotes (Slavik and Fulka,1991) or two- to four-cell embryos (Slavik *et al.*, 1992).

Work at Cambridge by Gandolfi and Moor (1987) showed that the presence of somatic cells increased the rate of cleavage and development of sheep zygotes in culture. The authors reported that sheep zygotes cocultured for 6 days with oviductal cells would, in many instances, develop to the morula/blastocyst stage, whereas those cultured without any cells did not proceed beyond the 16-cell stage. The successful coculture of IVM/IVF sheep embryos with oviductal cells was also reported by Fukui *et al.* (1988a).

Other studies at Cambridge, reported by Fukui *et al.* (1988b), made it clear that individual rams can affect not only the IVF rate but also the ability of zygotes to cleave after fertilization. These authors concluded that selection of rams with high fertilization ability is likely to be a key factor in sheep IVF work.

In China, a report by Bou Shorgan *et al.* (1990) showed that IVM/IVF sheep embryos could develop to the blastocyst stage in culture without oviductal cells: 11 lambs were born as evidence of the viability of such embryos.

Studies reported by Czlonkowska *et al.* (1991) in Poland showed that normal lambs could be born after IVM, IVF and *in vitro* coculture of early-stage embryos with oviductal cells. In New Zealand, Pugh *et al.* (1991b) were among the first to demonstrate that lambs can be produced after IVM and IVF using frozen–thawed sperm and oocytes collected during the non-breeding season; such results showed year-round production of IVM/IVF sheep embryos to be possible. In Australia, Holm *et al.* (1991a) and in Canada, Kelk *et al.* (1992) also reported the *in vitro* production of sheep blastocysts, with oviductal cells generally being employed in the IVC system. The Canadian workers also recorded pregnancies and lambs in the five recipient ewes employed in testing the viability of the embryos.

Although a Japanese paper (Hanada, 1985a) reported the birth of a kid after IVF of ovulated goat oocytes in 1985, the first goat kid produced by IVM/IVF appears to be that arising from the studies of Crozet *et al.* (1993) in France; they used the sheep oviduct for the early culture of the caprine embryos. At this time, there would seem to be no report of births arising from totally *in vitro*-produced goat embryos.

Elsewhere, goat IVF research has been reported from India (Chauhan and Anand, 1991), Spain (Martino *et al.*, 1992, 1993; Mogas *et al.*, 1992, 1993; Palomo *et al.*, 1992) and the USA (Pritchard *et al.*, 1990, 1991, 1992).

1.6.2 Pigs

In the pig, although births were reported from Cambridge after IVF of oocytes matured *in vivo* (Cheng *et al.*, 1986), it was the work of Mattioli *et al.* (1989) in Italy that first demonstrated that pig oocytes matured and fertilized *in vitro* can undergo normal embryonic development. These authors reported blastocyst formation, establishment of pregnancies and the birth of live piglets. Previously, the same group had shown that if pig oocytes are matured with the whole wall of the everted follicle, they become both capable of forming normal male pronuclei and more fertilizable (Mattioli *et al.*, 1988a, b).

Elsewhere, various aspects of pig IVF technology have been reported from Japan (Yoshida *et al.*, 1989, 1990, 1992a, b, 1993a, b; Yoshida and Ishigaki, 1991), China (Jinyi *et al.*, 1990; Wang, W.H., *et al.*, 1991; Wang, Z.K., *et al.*, 1992; Xia and Qin, 1992), Hungary (Vajta *et al.*, 1991), Canada (Zheng and Sirard, 1992), Mexico (Befancourt *et al.*, 1993) and England (Ding *et al.*, 1988, 1991, 1992a, b; Ding and Foxcroft, 1992); some of these studies have resulted in the birth of live piglets after IVM/IVF/IVC (Yoshida *et al.*, 1993a). Polyspermy after IVF is well recognized as being a major problem in the *in vitro* production of pig embryos. A report by Coy *et al.* (1993a) suggests that gamete coincubation time may be an important factor in this.

1.6.3 Horses

Research by Palmer *et al.* (1990, 1991) in Nouzilly resulted in the first successful pregnancy and live birth after IVF of an *in vivo*-matured oocyte. As yet, IVF of an artificially matured oocyte has not been reported in this species.

Elsewhere, some novel American experiments have demonstrated the ability of equine oocytes transferred to a mature Graafian follicle to undergo maturation, ovulation, fertilization and development to the blastocyst stage (Hinrichs and DiGiorgio, 1990, 1991).

Other aspects of equine IVF have been dealt with in reports from France (Okolski *et al.*, 1990, 1991) and the USA (Campo *et al.*, 1990; Ball and Miller, 1991, 1992; Freeman *et al.*, 1991; Hinrichs, 1991).

1.6.4 Buffaloes

There could be considerable interest in developing the technology of IVF embryo production for buffaloes. Buffaloes are among the most important domestic ruminants in more than 40 countries, mostly those in tropical and subtropical regions. In some developed countries, such as Bulgaria, Italy and the USA, buffaloes have been introduced for the production of special products such as cheese and low-fat buffalo beef (Hinkovski, 1990).

Although non-surgical procedures for recovering embryos from superovulated donor buffaloes have been available for some years, such methods suffer from the problem of low superovulatory responses and may be difficult and expensive to carry out (Madan, 1990; Misra *et al.*, 1990). It is generally accepted that the application of cattle ET directly to buffaloes has met with limited success. The poorer reproductive efficiency has been attributed to the formation of relatively smaller numbers of primordial follicles and to higher levels of follicular atresia.

Research in buffalo IVF technology has been reported from India (Totey *et al.*, 1991a, b, 1992, 1993a, b), Thailand (Chuangsoongneon and Kamonpatana, 1991), Italy (Bacci *et al.*, 1991), Taiwan (Lu and Hsu, 1991; Pavasuthipaisit *et al.*, 1992) and Malaysia (Jainudeen *et al.*, 1993). In Japan, Suzuki *et al.* (1991b) reported the production of embryos by IVM/IVF/IVC: a year later, Suzuki *et al.* (1992) reported the birth of the first buffalo calf produced by such procedures. The first IVM/IVF/IVC buffalo calf was born in China in 1993 (H.S. Jiang, personal communication). Workers with buffaloes continue to express optimism that, with the appropriate IVF system, ovaries from abattoirs may provide an abundant and easily accessible source of buffalo embryos for breed improvement and other purposes. However, until the emergence of satisfactory techniques for recovering much higher numbers of oocytes from mammalian ovaries, the commercial application of IVF technology in this species is likely to limited.

1.6.5 Deer

In certain countries, deer farming has developed over the past 20 years to the point where it has become an accepted part of the animal production scene. Reports on the extent of deer farming in New Zealand and Australia (Fennessy and Taylor, 1989; Jordan, 1992) and the UK (Fletcher, 1989) have been published. It may not always be appreciated that farmed deer are known to be very susceptible to tuberculosis (*Mycobacterium bovis*) and that appropriate control measures, both against the spread of *M. bovis* to cattle and to prevent human infection may sometimes need to considered (Fanning and Edwards, 1991). According to Fukui *et al.* (1991a), the development of effective IVF techniques in deer could be valuable in the *in vitro* hybridization of different deer species which do not readily hybridize naturally (e.g. Père David's and red deer). These authors, working at Ruakura in New Zealand, report on the successful maturation and fertilization of red deer oocytes, with some proportion cleaving to the two- to eight-cell stage. As yet, however, no birth of young after IVF has been reported for this species.

1.7 Human *In Vitro* Fertilization

There is much common ground between procedures that are used in IVF/ET in farm animals and the techniques that are part of assisted reproduction in humans (Betteridge and Rieger, 1993). For that reason, it is worth spending a moment looking at some areas of mutual interest.

1.7.1 Historical aspects

Since the birth of Louise Brown in 1978, human IVF has undergone considerable development; tens of thousands of oocyte retrievals are now made annually, resulting in the birth of thousands of children worldwide (Lee, 1991; Winston and Handyside, 1993). The application of this IVF technology, coupled to the fact that the average success remains low (in the UK, the live birth rate from each treatment cycle started in 1992 was 12.5%), means that a considerable amount of research is conducted in many areas that are directly relevant to laboratory embryo production in farm animals.

1.7.2 Establishment of pregnancy by embryo transfer

Although the advent of assisted reproductive techniques has led to a revolution in the treatment of human infertility, it is the poor viability of embryos, with consequent implantation failure, which is seen as a major factor limiting the successful establishment of pregnancies. In human IVF, it is well accepted that success is greater when more than one embryo is transferred simultaneously. The same would be true in a normally monotocous species, such as the cow, where pregnancy rates can be directly related to the number of embryos transferred. In ruminants, there is a growing body of evidence suggesting that failure in embryo–maternal signalling for pregnancy recognition is a significant source of embryo mortality. Ways are now available by which the incidence of embryo mortality in cattle and sheep may be reduced by the use of agents that enhance maternal recognition. The importance of effective communication between embryo and maternal endometrium (by cytokines/growth factors) is likely to be as great in humans as in ruminants.

1.7.3 Controlled superovulation

Basic mechanisms involved in the primate menstrual cycle are much the same as those in the bovine oestrous cycle (see Section 2.2). During each menstrual cycle, a group of follicles begin a well-characterized pattern of growth and development. Such groups of synchronously growing follicles are known as cohorts. One member of a cohort is able to continue development to become the preovulatory follicle. The follicle destined for ovulation is believed to be functionally, as well as morphologically, dominant since it inhibits the development of other competing follicles in both ovaries (Hodgen, 1990; Hillier, 1993).

It is believed that the dominant follicle proceeds through to the preovulatory stage because it has been activated by a certain threshold level of follicle stimulating hormone (FSH), which occurs during the early follicular phase of the menstrual cycle. Once the ovulatory follicle has been activated in this way, it can apparently continue to develop in the presence of subthreshold levels of FSH. By secreting oestradiol and inhibin, with their consequent negative feedback effects on FSH, the dominant follicle ensures that the level of this gonadotrophin falls below the threshold level, thereby hastening the regression of other follicles with ovulatory potential. In the treatment of anovulatory women, where the aim is to induce a single ovulation, there may be a possibility of producing, by recombinant DNA technology, isoforms of FSH which can help to ensure the development of the one dominant follicle (Baird, 1993).

The fact that the normal human menstrual cycle is characterized by unifollicular ovulation, whereas the needs of human IVF are for multiple follicle development, means that there has been a proliferation of superovulation regimens, most of these being based on forms of FSH extracted from the urine of postmenopausal women. The common feature of such regimens is amplification and/or prolongation of the intercycle increase in FSH, thereby encouraging maturation of several rather than one vesicular follicle. This is achieved by increasing or extending release of endogenous FSH (using clomiphene, gonadotrophin-releasing hormone (Gn-RH)) or by administration of exogenous gonadotrophins; the latter is the approach usually favoured, because dose can be regulated according to follicular response.

The mechanism of Gn-RH action is assumed to be via the down-regulation of pituitary gonadotrophins, resulting in follicular regression through removal of gonadotrophic (presumably FSH) support. Exogenous FSH is then effective in stimulating uniform follicle growth because of the absence of a dominant follicle. Pretreatment with Gn-RH analogues also means that a premature surge of LH can be avoided and this permits greater flexibility in the timing of human chorionic gonadotrophin (HCG) (for ovulation induction) and oocyte recovery.

In the first decade of human IVF activity, ovarian stimulation was usually by way of using human menopausal gonadotrophin (HMG), and the efficacy and safety of urinary gonadotrophins have been amply demonstrated by the birth of thousands of healthy babies. In 1992, the first report of an established pregnancy and subsequent birth after stimulation with a new biosynthetic FSH preparation (Org 32489) was made (Devroey *et al.*, 1993). It is believed by some that the future of gonadotrophin therapy is likely to lie in the application of recombinant human FSH of high biochemical purity and specific bioactivity (Mannaerts, 1992).

According to Franks (1992), there is an increasingly important place in human IVF programmes for controlled superovulation, in which stimulation is limited to two to four follicles. The author suggests that such controlled superovulation may become possible by the use of pure preparations of FSH and LH, which are now becoming available, as noted above, by way of recombinant DNA technology. Where the aim is to induce a single ovulation (in the treatment of anovulation), pure LH may be of particular use (Hillier, 1993); the gonadotrophin may promote the terminal maturation of a single preovulatory follicle and simultaneously arrest the development of multiple less mature follicles that would otherwise occur in response to treatment with FSH.

In cattle research, there is currently considerable research activity in elucidating the hormonal mechanisms regulating multiple ovulation. On the one hand, attempts are made to restrict the number of ovulations to no more than two (e.g. for induced twinning); on the other, there are those interested in promoting a much larger superovulatory response for conventional cattle ET purposes. Much new information about follicular dynamics in the cow has emerged since the advent of real-time ultrasonics for daily ovary scanning.

1.7.4 In vitro *maturation*

Although the laboratory production of cattle embryos is almost invariably based on the artificial maturation of the follicular oocyte, in human IVF oocyte maturation normally occurs *in vivo*, with recovery of the oocyte a few hours prior to the estimated time of ovulation. However, with modern techniques and the aid of new ultrasound imaging, vaginal retrieval of oocytes can be successfully undertaken in non-stimulated ovaries (Cha *et al.*, 1991; Dandekar *et al.*, 1991; Kumar *et al.*, 1991). Although a pregnancy was reported by Cha *et al.* (1991), there are those who believe that much more research on the maturation of human oocytes is required before the prospects of widespread clinical application of this procedure can be fully evaluated (Gosden *et al.*, 1993).

In Australia, however, Trounson *et al.* (1994) have demonstrated that human IVM is feasible; these authors suggest that the procedure is likely to be adopted as a normal procedure for the treatment of infertile women. The workers showed that immature oocytes recovered by ultrasound-guided aspiration from women with polycystic ovaries, can be matured-fertilized and developed in conditions similar to those used for bovine embryo production; pregnancy was established when embryos were returned to the patient's uterus.

It was noted some years ago (McLaren, 1988) that a reliable, safe and effective method of inducing maturation of immature oocytes *in vitro* might be the means of revolutionizing human IVF. Similar views have been expressed in more recent times by Abate *et al.* (1992), who observe that human IVF based on recovery and maturation of the primary oocyte could represent a significant improvement on the grounds of its economic, psychological and practical implications.

1.7.5 *Early embryo culture*

In cattle, embryo transfer is routinely performed at the 7-day old blastocyst stage of development. In contrast, transfer in human IVF is usually performed on the second day after oocyte retrieval, when embryos are at the two- to four-cell stage. As observed by Bolton (1992), this tradition has arisen despite the fact that, *in vivo*, the embryo would not reach the uterus until 2–3 days later, by which time it would be at the morula/blastocyst stage. The origin of the current technique apparently lies in the fear that prolonged exposure to sub-optimal culture conditions may compromise embryo viability to a greater degree than its exposure to environmental conditions that are possibly sub-optimal in the uterus.

However, techniques of coculture in dealing with cattle embryos support the view that this need not be so. A common coculture method employed with cattle embryos is one involving the use of granulosa and cumulus cell monolayers. The viability of embryos is similar to that coming from oviductal monolayers, which have been employed in the coculture of human embryos (Bongso *et al.*, 1990, 1991a, b, 1992; Bangso and Fong, 1991; Yeung *et al.*, 1992). Granulosa cells may have merit in coculture with human embryos in forming a particularly safe autologous system. There is some evidence suggesting that such a system may improve human embryo development *in vitro* (Plachot *et al.*, 1993).

Lopata (1992) in Australia suggests that excellent approaches towards refining human embryo culture are now available and that the transfer of blastocysts rather then two- to four-cell eggs may help in optimizing pregnancy success rates. As shown by Dokras *et al.* (1993), there would be a need to have effective procedures for evaluating the quality of blastocysts if a change is made to a later time of transfer. The effect of the primary nutrients, pyruvate and glucose, on development to the blastocyst stage, was examined by Conaghan *et al.* (1992a, b, 1993); pyruvate was found to be an essential nutrient, but omitting glucose in the early cleavage stages assisted in overcoming the development block at the four- to eight-cell stage and significantly increased the number of trophoectoderm cells in the resultant blastocysts.

1.7.6 *Cryopreservation of embryos and oocytes*

Cattle embryos were among the first mammalian embryos to be successfully frozen and preserved in liquid nitrogen storage units (Wilmut and Rowson, 1973). Countless thousands of normal healthy calves have been born world-wide as testimony to the value of cryopreservation as part of modern cattle ET technology. However, almost all such freezing is with blastocysts rather than early embryos, which are much more sensitive to reduced temperatures and to cryopreservation. There are certain situations in which embryo freezing is justified in human IVF, but it would seem appropriate to perform this at the optimal stage of development if the maximum benefit is to be derived from the technique.

It would seem that considerable effort would be justified in developing effective techniques for the cryopreservation of human oocytes (Marrs and Quinn, 1990). Certain of the ethical difficulties in embryo freezing would be overcome if it were possible to store the secondary oocyte successfully. Surplus oocytes recovered from the superovulated infertile IVF patient could be held until treatment was completed by a confirmed pregnancy. However, as in cattle, the chemosensitivity and cryosensitivity of oocytes during the process of cryopreservation continue to pose considerable difficulties, although the first human birth after fertilization of a freeze–thawed oocyte was reported several years ago (Chen, 1986). It is likely that studies with cattle and other mammalian oocytes will greatly assist in the development of effective procedures for use in human IVF.

1.7.7 Gamete micromanipulation

As noted by Winston and Handyside (1993), failure of fertilization is a frequent outcome of routine IVF treatment for male infertility. In cattle, it has been shown in Japan that bovine oocytes, matured *in vitro*, can develop to the blastocyst stage after being fertilized by the intracytoplasmic injection of an immotile 'dead' sperm (Goto *et al.*, 1990a, c, 1991, 1992a, b). Normal healthy calves were produced by such a procedure. Elsewhere, a bull which was sterile when used in natural service produced embryos when its sperm were injected into the ooplasm of the oocyte (Yang and Anderson, 1992). In situations in which bull sperm may be available in limited numbers (e.g. sperm sexed by flow cytometry), there may be particular justification for using such microassisted fertilization procedures, if their efficiency can be improved. In humans, sperm microinjection appears to be capable of achieving fertilization rates much superior to those recorded so far in cattle.

Recent reports from some human IVF units show fertilization rates in excess of 50%, with subsequent cleavage and the birth of normal healthy babies, after intracytoplasmic sperm injection (Palermo *et al.*, 1992b; Baker *et al.*, 1993; Van Steirteghem *et al.*, 1993a, b); the use of such microassisted fertilization is likely to be extremely valuable in dealing with severe forms of male infertility. This new form of reproductive technology, as well as other micromanipulation methods to enhance fertilization, including zona thinning, zona drilling and subzonal sperm insemination, has an important part to play in making IVF procedures more effective under some conditions.

1.8 Chapter Contents

The information in this monograph is primarily aimed at meeting the needs of technical and scientific personnel interested or engaged in laboratory embryo production in cattle. The work also seeks to provide for the general reader interested in mammalian reproductive biology. There is also the important question of making use of the embryos in research and commercial practice. Apart from information directly drawn from experience in the Dublin IVF laboratory, the book makes reference to more than 3000 reports that have appeared in the scientific literature, most of them in the past 5 years.

1.8.1 Essential steps in cattle embryo production

The subject matter covered in the following seven chapters includes the recovery and maturation of the primary cattle oocyte, the capacitation of bull sperm, the fertilization of the secondary oocyte and the subsequent culture of the zygote and embryo to the blastocyst stage of development. The temporary storage of the embryo in the 'fresh' state or its long-term cryopreservation in a liquid nitrogen container is discussed. Then, there is the matter of utilizing the embryos by way of non-surgical transfer to recipient cattle and the care and management of such recipients through the ensuing pregnancy.

The final chapter is devoted to the use of oocytes and embryos in research and commercial practice, particularly in areas such as large-scale cloning, embryo sexing and the application of recombinant DNA technology in cattle. Such areas of endeavour could not be readily contemplated but for the developments that have already occurred in the laboratory production of embryos. For that reason, an attempt has been made to be comprehensive in the coverage of reports, although the text does not seek to cater for those researching primarily in these areas.

1.8.2 Commercial interest in embryo production technology

Commercialization of cattle embryo production technology inevitably means that each of the technical steps (IVM/IVF/IVC/freeze-storage/transfer) in the process is constantly under review with the aim of improving the product to meet the needs of the market-place. Effective freeze–thaw procedures are particularly important for the optimal commercial exploitation of the embryo production technology. Although there are well-established procedures for the cryopreservation of *in vivo*-produced cattle embryos, there is inevitably much less information available about the viability of freeze–thawed cattle embryos derived from IVF and IVC.

However, results reported by Leibo and Loskutoff (1993) showed that the survival rate of *in vitro*-derived embryos (produced by the method of Yang, N.S., *et al.*, 1990) cooled in ethylene glycol was virtually identical to that previously reported for *in vivo* embryos cooled in glycerol, the traditional cryoprotectant employed in freezing. The same authors emphasize the relevance of the embryo production techniques to the results that can be anticipated after freezing. There is also extensive evidence that high pregnancy rates can be achieved in cattle after the non-surgical transfer of fresh or frozen *in vitro*-produced embryos (Polge *et al.*, 1991). In view of the fact that it is the frozen embryo which is the main commercial product, particularly in international commerce, there is clearly much interest in developing optimal freeze–thaw methods.

1.8.3 The search for more oocytes

Although the bovine ovary contains thousands of immature egg cells, relatively few can be recovered by current procedures. However, in mice and rats, IVC systems for growing and maturing oocytes are being developed (Eppig and Schroeder, 1986; Eppig *et al.*, 1988, 1990; Gandolfi *et al.*, 1988a; Daniel *et al.*, 1989; Nayudu *et al.*, 1990, 1992; Qvist *et al.*, 1990; Eppig, 1991a; Merriman *et al.*, 1993). There are even reports of normal ovulation from individual mouse follicles grown in culture (Boland *et al.*, 1993). The field has been well reviewed by Gosden *et al.* (1993); they draw attention to the successful work of Eppig and colleagues in culturing mouse cumulus–oocyte complexes (COCs), recovered from preantral follicles, on collagen-coated membranes. Early attempts to use a similar approach with pig oocytes has been less successful (Hirao *et al.*, 1993). However, there is clearly considerable scope for examining the possibility of developing appropriate culture systems specifically for cattle and other farm animals.

2 Oocyte Recovery and Maturation

2.1 Introduction

2.1.1 Historical

In most mammalian species, some hours before rupture of the follicle and ovulation, the fully grown oocyte resumes meiosis, progressing from prophase of the first meiotic division to metaphase of the second division. This complex biological process, which transforms the primary oocyte into an unfertilized egg, is known as oocyte maturation (see reviews by McGaughey, 1983; Thibault *et al.*, 1987; Moor, 1988; Motlik, 1989; Siracusa *et al.*, 1990; Tornell *et al.*, 1991). The trigger for the resumption of meiosis *in vivo* is the preovulatory surge of luteinizing hormone (LH). It is believed that LH is the most important gonadotrophin *in vivo* for inducing ovulation and oocyte maturation.

The first obvious sign of resumed meiosis under the microscope is the dissolution of the nuclear membrane, termed germinal vesicle breakdown (GVBD). This is illustrated in Fig. 2.1. Extrusion of the first polar body (PB) and formation of the second meiotic spindle subsequently occur as the oocyte matures and the surrounding cumulus cells undergo expansion (see Fig. 2.2).

Studies on the *in vitro* maturation (IVM) of mammalian oocytes were first reported some 58 years ago, when Pincus and Enzmann (1935) observed that rabbit primary oocytes resumed meiosis spontaneously when they were liberated from vesicular follicles and placed in a suitable tissue culture medium. They recovered the primary oocytes from the large follicles of unmated rabbits and used sterile rabbit serum as their culture medium. Using similar methods, Pincus and Saunders (1939) cultured human follicular oocytes and reported that about 50% of these resumed meiosis and that some 30% proceeded to metaphase II with formation of the first PB.

A detailed study of the maturation of rabbit oocytes was carried out by

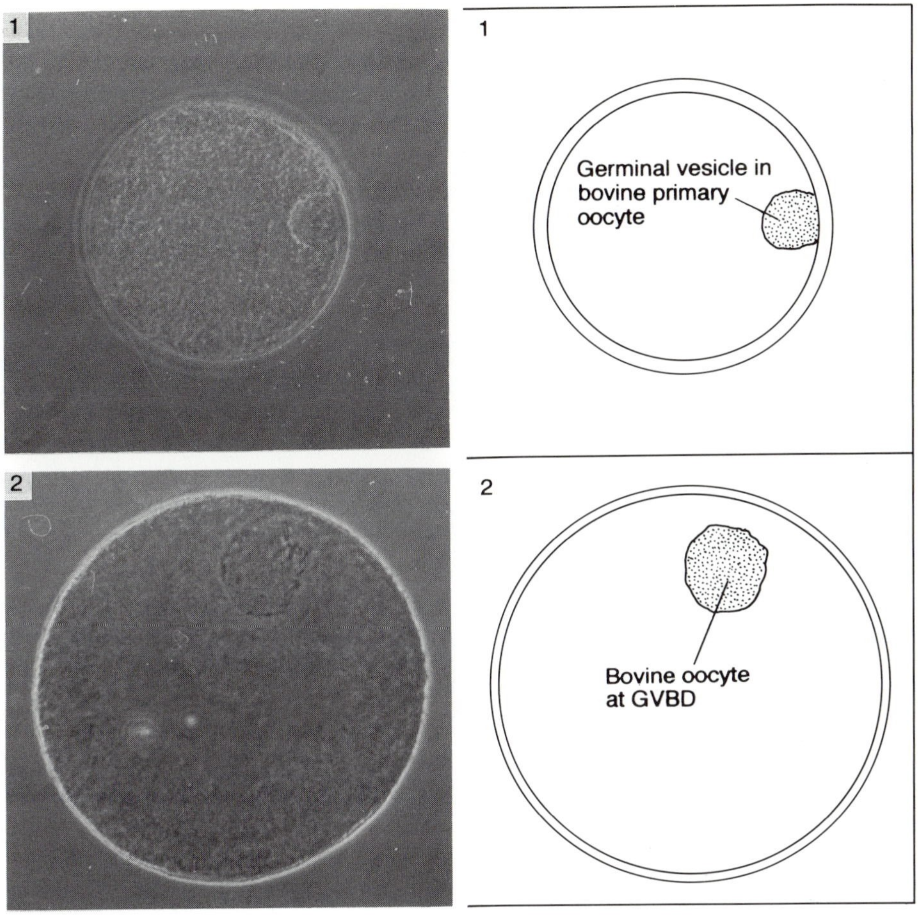

Fig. 2.1. Germinal vesicle-stage bovine oocyte (1) and germinal vesicle breakdown (GVBD) (2).

Chang (1955). He observed that maturation was apparently unaffected by the addition of pituitary extracts to the culture medium, by follicle size or by the animal's reproductive status.

In the 1960s, Edwards (1962, 1965, 1966) observed that such nuclear maturation of oocytes removed from the follicular environment was a common phenomenon among several other species, including cattle. His work in Cambridge at that time did much to stimulate interest in various of the critical factors involved in bringing mammalian eggs to maturity in a culture medium in the laboratory. He was also one of the first to culture oocytes within their follicles *in vitro* as well as liberated oocytes and concluded that an inhibitory factor was present in the follicle which disappears on release of the oocyte.

Other work in the late 1960s and early 1970s with cattle oocytes included attempts in Dublin to determine the chronology of maturational changes *in*

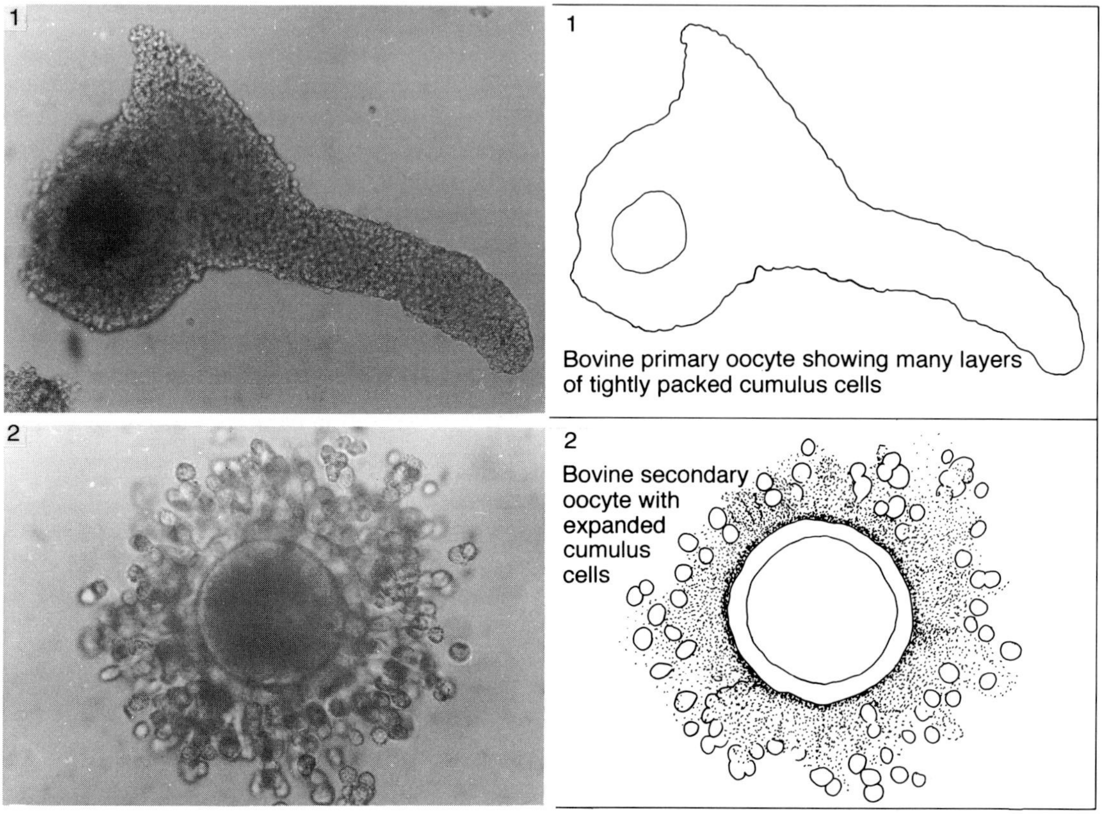

Fig. 2.2. Before (1) and after (2) *in vitro* maturation of the bovine oocyte.

vitro (Sreenan, 1968, 1969), observations on oocyte maturation at the chromosomal level (Jagiello *et al.*, 1974) and examination of the biochemical changes that occur during maturation of the ruminant oocyte (Crosby *et al.*, 1981; Crosby and Moor, 1984, 1985).

2.1.2 Nuclear and cytoplasmic maturation

For general information on many features of mammalian oocyte maturation, there are a number of excellent reviews available (Leibfried-Rutledge *et al.*, 1987, 1989a, b; Moor, 1988; Wassarman, 1988b; Siracusa *et al.*, 1990; Tornell *et al.*, 1991; Mattioli, 1992; Moor *et al.*, 1992b). Although nuclear changes can be clearly followed in the maturing oocyte, other, much less evident, changes affect the oocyte's cytoplasm and surface. There are also interactions between the oocyte and the surrounding follicular cells. Understanding the complex signalling mechanisms that operate between the follicular cells and the oocyte has long presented a challenge to researchers.

Although it has been recognized for many years that oocytes will resume meiosis spontaneously in a suitable culture medium, it has also been known that, after apparently normal nuclear maturation, they may be quite incapable of giving rise to a normal embryo. Reports for cattle oocytes (Sreenan, 1968, 1970) and other ruminant oocytes for many years clearly showed that these matured eggs had an extremely low capacity for fertilization and normal embryonic development. Artificially matured oocytes were apparently characterized by an almost total inability to support even the earliest phases of embryonic life.

In order to appreciate just how poor such embryonic development was, it is possible to cite an analysis of published data (comprising more than 1500 secondary oocytes from mice, rabbits, sheep and cattle) carried out by Cambridge workers (Moor and Warnes, 1978). This showed that less than 1% of *in vitro*-matured oocytes were capable of normal embryonic development when transferred to recipients. As the evidence gradually accumulated, it was evident that the concept of oocyte maturation must be widened to include all those events that permit the oocyte to express its developmental potential after fertilization and should not be limited merely to nuclear events or the ability of the oocyte to be fertilized (Thibault, 1977; Thibault *et al.*, 1987). It is now firmly established that the acquisition of biological competence by the bovine oocyte involves numerous structural and biochemical changes within the cell as part of cytoplasmic maturation. It is also clear that positive support from follicular cells (granulosa/cumulus) is essential during IVM to confer the requisite developmental competence to the oocyte (Staigmiller and Moor, 1984a, b).

2.1.3 *Work with* in vivo-*matured oocytes*

As shown later in this chapter, an understanding of IVM of the cattle oocyte requires knowledge of the numerous events that occur in the live animal as well as information derived from work at the laboratory bench. Although most efforts at maturing bovine oocytes have been with those released from the follicle into an appropriate culture medium, there have been those who have used *in vivo*-matured eggs. It is important to elucidate the nature of the many changes that take place in the follicle destined for ovulation during the final 24 h before ovulation in the cow (Loos *et al.*, 1991b). More precise information on the involvement of hormones, growth factors and cytokines in the *in vivo* maturation of the oocyte in the preovulatory follicle may provide vital clues as to what is required in an optimal maturation medium. There have also been those who have studied how the oocyte matures within the intact follicle cultured *in vitro*.

2.1.4 *Crucial considerations in maturing the oocyte artificially*

Procedures currently employed in artificial maturation of cattle oocytes are still in their infancy. There is a constant need to emphasize the fact that effective oocyte maturation is at the very foundation of embryo production. It is not only *in vitro* fertilization (IVF) embryo production which is at stake: the fields of large-scale cloning and genetic engineering in cattle are also dependent on having high-grade secondary oocytes and embryos available in large numbers.

The maturation procedure developed in Dublin (Lu *et al.*, 1987b, 1988a) was initially based on the dissected follicle, with subsequent recovery of an intact cumulus–oocyte complex (COC), additional cumulus/granulosa cells, trace amounts of antibiotic and oestrous cow serum (OCS) in commercially available culture medium (M-199); maturation was by way of a non-static (flux) culture system operating at 39°C in a 5% carbon dioxide atmosphere at maximum humidity. In evaluating the effectiveness of the maturation process, there is an obvious need to examine morphological features within the oocyte and changes which occur in the cumulus cells outside the oocyte.

During IVM, extensive redistribution of intracellular organelles occurs; mitochondria migrate to occupy a perinuclear location and the cortical granules migrate outward to lie just beneath the vitelline membrane (Cran, 1987, 1989; Cran and Esper, 1990). At all times, there is the need to realize that the crucial test of successful maturation is the ability of the oocyte to undergo normal embryonic development after sperm penetration and fertilization.

2.2 The Bovine Oestrous Cycle and Associated Events

2.2.1 *Oestrus and the oestrous cycle*

The development of effective IVM and IVF procedures in cattle requires accurate information on all aspects of follicular development in the live cow as well as an understanding of the various endocrine events associated with the ovarian cycle.

Oestrus is the period of sexual receptivity in the cow and is characterized by the animal being willing to be mounted by other cattle, both male and female (see Phillips, 1993). The bovine oestrous cycle averages 21 days in cows and 20 in heifers, with a normal range of 18–24 days. The duration of oestrus will vary according to breed, management and a variety of environmental factors: 12–16 h may be taken as an average duration, with a range of 3–28 h (Allrich, 1993). Ovulation usually occurs some 10–12 h after the end of the heat period. This would agree with reported timings of the preovulatory LH peak, occurring in the early hours of heat (Dobson *et al.*, 1973; Bernard *et al.*, 1983), with ovulation following after a further 24 h.

The different endocrine and hormonal interactions that occur during the cycle, involving gonadotrophin-releasing hormone (Gn-RH), gonadotrophins (follicle-stimulating hormone (FSH), LH, ovarian steroid/peptide hormones (oestradiol, progesterone, inhibin) and prostaglandin $F_{2\alpha}$ ($PGF_{2\alpha}$) of uterine origin are shown diagrammatically in Fig. 2.3.

During pro-oestrus, gonadotrophins induce final maturation of the preovulatory follicle, resulting in increased secretion of oestradiol. Oestradiol, in the relative absence of progesterone, acts on receptors in the brain to induce

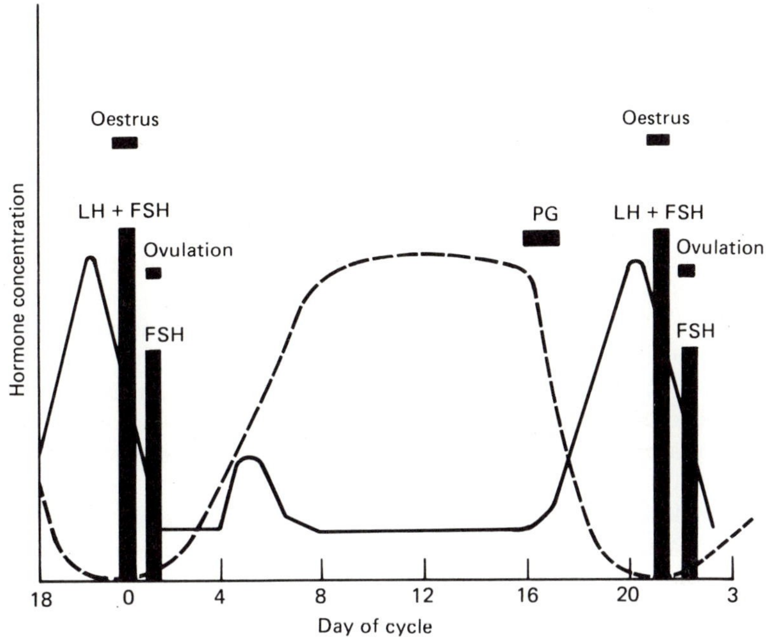

Fig. 2.3. Diagrammatic representation of changes in hormone levels during the bovine oestrous cycle. Solid line represents oestradiol concentration; dashed line represents progesterone concentration (after Peters, 1985).

sexual receptivity. The LH surge brings about follicle rupture and the formation of the corpus luteum (CL); this structure is known to contain two morphologically and functionally distinct steroidogenic cell types, the large and small luteal cells. The morphology of these two cell types during the oestrous cycle and pregnancy have been described in the literature. Large cells are characterized by having a large spherical nucleus, a highly folded plasma membrane, stacks of rough endoplasmic reticulum and protein-containing secretory granules. Small cells have an irregularly shaped nucleus and large numbers of cytoplasmic lipid droplets. The small, theca-derived, cells are more numerous than the large cells, which are at least partly derived from granulosa cells (Hansel *et al.*, 1990). It has been suggested that the two hormones which principally regulate luteal function (LH and $PGF_{2\alpha}$) act on the two cell types through different second-messenger pathways (Wiltbank *et al.*, 1990). LH appears to stimulate small luteal cells through the cyclic adenosine monophosphate (cAMP)/protein kinase A second messenger system. $PGF_{2\alpha}$ apparently acts directly on the large luteal cells by way of the calcium/protein kinase C second messenger system.

There has been considerable progress, in recent years, in understanding the endocrine mechanisms that control the pattern and timing of uterine secretion of $PGF_{2\alpha}$ in the cow's oestrous cycle (Silvia *et al.*, 1991). It should be remembered that uterine prostaglandins induce luteal regression in a relatively small group of mammals, which includes ruminants, horses, pigs and guinea-pigs. In these species, uterine secretion of $PGF_{2\alpha}$ can be stimulated by oxytocin during the oestrous cycle. It is believed that uterine prostaglandin and luteal oxytocin comprise a positive feedback loop. Oxytocin can stimulate uterine secretion of prostaglandin and prostaglandin can stimulate secretion of oxytocin from the CL.

2.2.2 *Expanding knowledge of follicular dynamics*

A sound understanding of the processes involved in the growth and differentiation of vesicular follicles destined for ovulation in the cow is essential for those working in the artificial maturation of bovine oocytes. It was first suggested by Rajakoski (1960), more than 30 years ago, that follicular development in the heifer occurs in two waves. The first wave started on day 3 and ended at midcycle with development and subsequent regression of a single ovulatory-size follicle. The second wave became evident after midcycle and terminated with the ovulation of a follicle. In The Netherlands, Kruip (1982) concluded from his studies that at least two different groups of large follicles appeared in the ovaries of the cow during the oestrous cycle.

In more recent times, ultrasonographic observations have clearly confirmed that ovarian follicular development during the oestrous cycle is characterized by two or three consecutive follicular waves per cycle (Pierson and Ginther, 1987a; Fortune *et al.*, 1988; Sirois and Fortune, 1990). Similar follicular waves are also evident in prepubertal (Evans, A.C.O., *et al.*, 1992, 1993; Hopper *et al.*, 1993) and pregnant animals. Evidence provided by Pierson and Ginther (1987b) shows that the CL exerts a positive intraovarian effect; a greater number of small vesicular follicles (2–3 mm diameter) has been observed in the ovary bearing the CL than in the contralateral ovary during the oestrous cycle and early pregnancy.

There may be marked individual variations in follicular dynamics among cattle, with as few as one to as many as four waves of follicular growth occurring within an oestrous cycle. Each wave involves the recruitment of a dominant follicle (DF), which continues to grow while suppressing the growth of other follicles larger than 4 mm in diameter (Savio *et al.*, 1993). It is clear that follicular growth in cattle is under both endocrine and paracrine control, and the local factors involved include steroids, growth factors, cytokines and other regulatory molecules (Driancourt, 1991).

These factors can act on ovarian cells either directly or indirectly by modulating the response to circulating gonadotrophins. As observed by Hunter *et al.* (1992), it is possible that, at particular stages of folliculogenesis, the different components of the endocrine and paracrine systems vary in importance. In the early stages of vesicular follicle development, when the follicle is poorly vascularized and steroidogenesis is minimal, locally produced paracrine and autocrine factors may have a more important effect on cell division than endocrine factors. With increasing follicular development and vascularization, endocrine control becomes more important. Evidence has been presented by Findlay (1993) of a local regulatory role for inhibin, activin and follistatin in the control of folliculogenesis.

The growth and maturation of the preovulatory bovine follicle proceeds in distinct phases which include recruitment, selection and dominance. Follicular dynamics in dairy cattle was followed by Lucy *et al.* (1991) by daily scanning and other means. Evidence was found that diet and energy balance may induce specific changes in the secretion of insulin-like growth factors (IGFs) and their binding proteins, leading to altered function of the DF.

In order to set much of this in context, it should be emphasized that events occurring in the preovulatory follicle are of obvious interest in devising appropriate protocols for use in the *in vitro* culture (IVC) of the immature oocyte.

2.2.3 Gonadotrophin release

The preovulatory surge of LH, which occurs in the early hours of oestrus and about 1 day ahead of follicle rupture, is responsible for initiating the nuclear and cytoplasmic maturation of the primary oocyte that is present in the follicle destined for ovulation. In attempting the artificial maturation of the oocyte, it has usually been a matter of simulating certain of the critical events that occur in the live cow in course of the final 24 h period between LH release and follicle rupture. The cow is a monovular species, which implies that generally only one oocyte reaches maturity at the end of each oestrous cycle. It seems probable that the endocrine and paracrine factors controlling the final maturation and ovulation of the single follicle in the cow are more complex than those in sheep, where multiple ovulations (two to three) are commonplace.

In considering the endocrinology of the cow's oestrous cycle, it should be noted that pulsatile or episodic secretion is the common mode of secretion for most of the hormones involved. In such secretion, small amounts of hormone are released at different time intervals. In terms of LH activity, high-amplitude, low-frequency pulses occur during the luteal phase of the oestrous cycle, resulting in a low average plasma LH level. During the preovulatory phase, this changes to a low-amplitude, high-frequency pattern, resulting in higher plasma concentrations of LH.

Evidence supports the view that the pulsatile pattern of LH secretion is a consequence of pulsatile Gn-RH secretion. During the luteal phase, the LH frequency is about one pulse every 4 hours: in the preovulatory phase, and after luteolysis, this is likely to increase to one per hour or greater. Those seeking to measure the preovulatory LH surge take blood samples at intervals of 5–15 minutes in recognition of this increased pulse frequency. As progesterone secretion declines, LH pulse frequency increases. This greater LH frequency stimulates oestradiol secretion from ovarian follicles and oestradiol elicits the preovulatory LH surge that induces ovulation and initiates luteinization.

Knowledge of the factors influencing FSH secretion has been less certain (Price, 1991). It is generally accepted that Gn-RH is responsible for stimulating FSH secretion, although the effects are less evident than for LH. It is believed that Gn-RH plays a relatively minor role in the regulation of FSH levels: this may be reflected in the pattern of FSH secretion, which is not pulsatile.

The regulation of FSH release is believed to be exercised by the ovaries and involves an interaction between oestradiol and inhibin. In sheep, and perhaps in cattle, it is evident that oestradiol is involved in the control of short-term fluctuations in FSH plasma concentrations while inhibin regulates the longer term levels (Price, 1991). Whether oestradiol and inhibin also differentially regulate the biological activity of FSH remains uncertain.

The relationship between ovarian follicular dynamics and FSH concentrations during the cow's oestrous cycle is dealt with in a study reported by Hamilton *et al.* (1992). They record FSH concentration increasing before a follicular wave and then returning to the basal level by 16 h after the wave. The authors conclude that initiation of follicular waves during the bovine oestrous cycle appears to follow a transient increase in the secretion of FSH. The important role of FSH during follicular dominance is also dealt with in studies reported by Turzillo and Fortune (1993).

2.2.4 *The dominant follicle*

Knowledge of the process of folliculogenesis in cattle has increased considerably in the recent decade. Follicular development is now known to be a well-organized series of events (see review by Adams, 1994). During the oestrous cycle, there are two or three waves of follicular development and the end of the cycle is characterized by the emergence of a DF, an event which coincides with the marked decline in the progesterone concentration as the CL regresses (Bergfelt *et al.*, 1991; Ko *et al.*, 1991). Follicular waves consist of cohorts of follicles that are 5 mm or greater in diameter, which appear approximately every 7 days. Oestrous cycles with three waves are longer, and have longer luteal phases, than cycles with two waves, which suggests that the number of waves per cycle is determined by the time of luteal regression.

The pattern of follicular turnover and development in prepubertal cattle is dealt with by Evans, A.C.O., *et al.* (1992). They found that 9-month-old prepubertal Hereford heifers showed patterns of follicular waves consistent with postpubertal wave dynamics. They observed significant transient increases in FSH level occurring prior to the initiation of a new wave of follicles and speculated that such increases were important in initiating each new wave.

In contrast to the two or three waves of follicular development per ovarian cycle in cattle and sheep, pigs and humans apparently show only a single wave of follicular development during the preovulatory phase (Driancourt, 1991). Two processes are believed to be involved in the growth and development of the follicle destined for ovulation in the cow. The first process is that of follicle recruitment, which results in the development of a group of follicles (cohort) from which the preovulatory follicle emerges. The second process is follicle selection, whereby one follicle becomes dominant and continues towards ovulation while the others regress.

The process by which the ovulatory follicle is selected in the cow remains obscure. According to Adams *et al.* (1992a, b, c), in a study dealing with factors regulating follicular waves in heifer cattle, the time of selection of a DF of a particular wave (defined here as the time of divergence in growth profiles of subordinate follicles vs. the DF) is first apparent within 2 days of the detected emergence of a wave. The time of follicle selection appears to coincide with a significant fall in FSH concentration. The diameters of the largest and second largest follicle show the least difference 5 days before ovulation and thereafter diverge. This divergence is thought to represent selection of the DF. All other large follicles (>7 mm diameter) start undergoing atresia at the approximate time that the follicle destined to ovulate becomes the DF.

The factors that lead to the dominance of one follicle in the cow and the mechanisms that suppress the growth of the subordinate follicles are not well understood. There is support for the view that a decline in FSH level is a component in the selection of the DF (Adams *et al.*, 1992a; Stock and Fortune, 1993; Stock *et al.*, 1993). Evidence supports the view that the DF is functionally (not merely morphologically) dominant since it inhibits the development of other competing follicles in both ovaries. It appears that the DF produces a range of ovulatory control factors to ensure that only it will progress through to the point of ovulation (Fortune *et al.*, 1990; Bergfelt *et al.*, 1991; Ko *et al.*, 1991). It is suggested by Badinga *et al.* (1991) that increased aromatase activity by the DF at the time it establishes its dominance may play a critical role in the suppression of subordinate follicles.

Reports by Hillier (1989) and Williams and Hodgen (1989), dealing with concepts of gonadotrophin-dependent granulosa cell function in the human preovulatory follicle, emphasize the roles of FSH and LH in regulating oestrogen biosynthesis. They suggest that the induction by FSH of LH-responsive granulosa cell aromatase activity may be central to the mechanism whereby one oestrogen-secretory follicle develops and ovulates in the human menstrual cycle. They note that at the start of the cycle, when the FSH level rises, the ovaries contain several follicles with varying potentials for FSH-dependent development. Each follicle is believed to require a different amount of stimulation by FSH for it to undergo preovulatory development.

The follicle which eventually becomes dominant is thought to be the one whose granulosa cells most rapidly acquire high levels of aromatase and LH receptors in response to the intercycle FSH rise (i.e. the one with the lowest FSH threshold). During the midfollicular phase of the menstrual cycle, oestradiol secretion by the follicle begins to increase and the steroid negatively feeds back to decrease pituitary FSH release. This results in a progressive decrease in the FSH concentration, which limits FSH development of other follicles with a relatively high threshold. By such means, it is thought that only one follicle becomes fully mature, being protected against the fall in circulating FSH by the high responsiveness of its granulosa cells to both FSH and LH.

The regulatory roles of inhibin and activin in human folliculogenesis are also dealt with by Hillier (1991). Research suggests that inhibin and related proteins, such as activin, may have a local regulatory role at or near their sites of formation in the ovaries. It is suggested that during the first half of the follicular phase of the menstrual cycle, before follicle selection occurs, interfollicular differences in granulosa cell production of activin may contribute to the establishment of a variable threshold response to FSH. The follicle whose granulosa cells are producing the greatest amounts of activin would be expected to be most responsive to FSH. Activin may also act simultaneously to reduce thecal androgen synthesis within this follicle, at a time when its requirement for an aromatase substrate is still minimal.

By the midfollicular phase, in the follicle responding most rapidly and extensively to FSH, granulosa cell aromatase activity, inhibin production and LH-receptor expression rise to critical levels. Since aromatase activity and inhibin production are coupled to the LH receptor, this follicle is selected and becomes destined to ovulate.

During the second half of the follicular phase of the menstrual cycle, LH-responsive granulosa cell inhibin production in the preovulatory follicle continues to increase, as does aromatase activity and oestrogen secretion. The inhibin produced acts locally to enhance LH (and IGFs)-stimulated androgen synthesis in the theca interna, creating a paracrine positive feedback loop through which granulosa cells secure the increased supply of aromatase substrate necessary to sustain preovulatory rates of oestrogen secretion.

In the cow, where two or three waves of follicles develop each cycle, there are additional considerations. During the luteal phase, because the hormonal environment cannot support final growth and development, DFs undergo regression, losing first functional and subsequently morphological dominance (Sirois and Fortune, 1990). Such loss of functional dominance would permit the recruitment of a new follicular wave. It is believed that the emergence of the DF of the first follicular wave during the oestrous cycle occurs, not only because of appropriate FSH support (Turzillo and Fortune, 1993), but also because of the absence of any inhibitory influences, as a result of the ovulation of the preceding DF.

Evidence for the role of steroids, inhibin and other peptides in the growth and regression of bovine follicles has been provided by several authors (Findlay, 1990, 1993; Martin *et al.*, 1991; Adams *et al.*, 1992a; Ireland *et al.*, 1992; Bo *et al.*, 1993; Guilbault *et al.*, 1993a, b; Stock and Fortune, 1993). It is apparent that the attainment and loss of follicular dominance is closely related to certain crucial changes in the hormonal environment. Studies in follicle hormone production in sheep suggest that inhibin does not play an important role in follicle selection in that species (Mann *et al.*, 1992a, b).

Changes in peripheral progesterone concentrations appear to be particularly critical. Adams *et al.* (1992c) have shown that the greater development of the DF of the first wave, compared with subsequent anovulatory waves, is a result of the lower progesterone levels that operate during its growth phase. It is also clear that exposure to low circulating levels of progesterone at the end of the growing phase results in continued growth and prolonged maintenance of the DF (Adams *et al.*, 1992c; Savio *et al.*, 1993; Stock and Fortune, 1993).

The evidence of Adams *et al.* (1992a, b, c), Stock and Fortune (1993) and Savio *et al.* (1993) strongly supports the view that negative feedback effects on LH pulse frequency is a mechanism by which changes in plasma progesterone influence follicle growth and regression. The demise of the non-ovulatory DF during the bovine oestrous cycle occurs by way of the negative feedback effects of luteal progesterone, which maintains low LH pulse frequency and oestradiol production.

If a subluteal concentration of progesterone is maintained artificially, prolonged dominance of the DF may ensue, and such prolongation can apparently compromise fertility. The cause of the reduced fertility associated with the extended lifespan of the DF is not known (Savio *et al.*, 1993; Stock and Fortune, 1993). If the normal lifespan of an active DF in the cow (about 9 days) is extended, it may influence oocyte quality adversely; that is one possibility. It may, on the other hand, be a question of an endocrine imbalance arising from prolonged DF activity (oestrogen production?) which adversely affects the uterine environment.

2.2.5 Ultrasound imaging

It has been recognized for some time that follicular development in the ovaries of the cow is a dynamic process. Traditionally, changes in vesicular follicle populations have been determined by slicing excised ovaries taken at different stages of the oestrous cycle or by monitoring via rectal palpation.

In recent years, the advent of real-time ultrasonic scanning of the reproductive organs per rectum has removed the disadvantages associated with previous methods and provides a valuable new approach to the study of follicular dynamics in the cow (Pierson and Ginther, 1986, 1987a, 1988; Omran *et al.*, 1988; Pierson *et al.*, 1988; Greve and Purwantara, 1992, 1993). It is now the basis of an extremely accurate technique for the estimation of the vesicular follicle population within the limits of resolution imposed by the scanning device; follicles as small as 2–3 mm can be visualized, measured and sequentially monitored. According to Huhtinen *et al.* (1992), real-time ultrasonic examination has become the method of choice for monitoring follicle development.

The ultrasound instruments employed in examining the ovaries are B-mode, real-time scanners. B-mode refers to brightness modality, wherein the ultrasonic imaging is a two-dimensional display of dots, the brightness of which is proportional to the amplitude of the returning echoes. Real-time imaging refers to the moving display which is continuously presented on the monitor screen. Two types of real-time, B-mode ultrasound instruments are marketed, linear-array scanners and sector scanners. In the linear array scanner, sound waves are emitted perpendicular to the transducer along the row of crystals. For transrectal examinations in the cow, it is important to select equipment with a durable, atraumatic probe designed for intrarectal insertion (Pierson *et al.*, 1988). For the aspiration of ovarian follicles, a sector scanner equipped with a vaginal transducer is employed (Pieterse *et al.*, 1988a, b, 1991a, b, c; Pieterse, 1990).

2.2.6 Intraovarian events and their regulation

The identification of intraovarian factors and an understanding of their role in the follicular maturation and ovulatory processes would do much to shed light on key factors influencing folliculogenesis in the cow. There is, for example, a growing body of evidence on the important role of the IGF system and other growth factors in folliculogenesis (May *et al.*, 1991; Armstrong and Xia, 1993; Monget *et al.*, 1993); this has implications for the employment of grcwth hormone in influencing the ovarian follicle population in cattle (Gong *et al.*, 1990, 1991a, b, 1992, 1993) and perhaps in humans (Carlsson *et al.*, 1992).

In terms of cytokine activity, there is evidence suggesting the existence of a complete intraovarian interleukin-1 (IL-1) system replete with ligands, receptor and receptor antagonist. In view of the fact that IL-1 is an established mediator of inflammation and ovulation may constitute an inflammatory-like reaction, Kokia *et al.* (1992) suggest that IL-1 may play an intermediary role in the process of ovulation. There is strong support for the view that IL-1 may be the centre-piece of an intraovarian regulatory loop concerned with the promotion of the preovulatory cascade.

2.2.7 *Other endocrine events*

The fact that the peptide hormone oxytocin is produced by luteal cells in the bovine ovary is now well established (see Peters, 1985). It is apparent that the formation of oxytocin receptors in the cow's endometrium is essential for the synthesis of $PGF_{2\alpha}$. The formation of these receptors is apparently dependent on oestradiol of follicular origin.

The wave of follicular growth that occurs between days 12 and 15 of the bovine oestrous cycle would seem to be the source of this oestradiol. Circulating oxytocin of luteal origin binds to oxytocin receptors in the endometrium of the uterus and initiates prostaglandin synthesis and release. The prostaglandin acts locally on the CL to reduce progesterone production and the release of more oxytocin, which stimulates the further release of prostaglandin, thereby establishing a positive feedback loop between the

Table 2.1. Appearance of corpus luteum at different stages of the bovine oestrous cycle (based on Ireland *et al.*, 1979).

Stages of oestrous cycle	*N*	Appearance of corpus luteum (CL)	Estimated range of days spanned by each stage of cycle
I	30	Surface of CL red, follicle recently ruptured (0.5–1.5 cm in diameter)	1–4
II	42	Point of rupture of follicle covered over, apex of bisected CL red or brown (1.6–2 cm), vasculature visible on surface of CL	5–10
III	52	Bisected CL is orange or yellow and has no red or brown colour remaining, vasculature prominent on surface of CL (2 cm)	11–17
IV	22	Vasculature absent from surface, large follicle present	18–20

ovary and the uterus. This brings a fall in progesterone concentration to baseline value in about 24 h.

2.2.8 Determining stage of cycle in slaughterhouse cattle

In the collection of ovaries from slaughtered cyclic cattle for oocyte maturation, it is sometimes necessary to estimate the stage of the cycle. This can be done with considerable accuracy on the basis of CL appearance (colour, size, vasculature), as described by Ireland *et al.* (1979, 1980); descriptions are in Table 2.1. Such information may also be useful when collecting cattle oviducts for the preparation of epithelial cell monolayers. The appearance of the CL at the midcycle stage is shown in Fig. 2.4.

2.3 Prenatal Development of the Bovine Ovary

The series of events that result in the female germ cell (oocyte) being able to perform its ultimate function of interacting with the sperm develops over a long period, extending from the stage of oogonia differentiation in the fetal ovary to the final maturation of the oocyte just before its release at ovulation. The process of differentiation and maturation of the oocyte can be arranged into a prenatal and postnatal phase, each consisting of periods of enhanced activity followed by periods of quiescence. Various aspects of oogenesis and folliculogenesis in the prenatal heifer and the postnatal cow are covered in several papers and reviews (Baker and Franchi, 1967; Baker, 1971, 1982; Baker and Hunter, 1978; Byskov, 1982; Monniaux *et al.*, 1983; Russe, 1983; Mariana *et al.*, 1991).

It has long been recognized that very few of the primary oocytes present in the bovine ovaries develop to the point of becoming secondary oocytes and being ovulated. It has been estimated that there may be more than 200,000 oocytes in primordial follicles in the ovaries of the heifer calf at birth, although no more than 300 of these are likely to reach the ovulatory stage (Erickson, 1966a).

2.3.1 Migration of primordial germ cells

Special cells, the primordial germ cells, are believed to be responsible for initiating the processes that result in the invasion of the gonad and the differentiation of the ovary. The cells can be recognized because of their large size and round shape and their high alkaline phosphatase activity also makes it possible to identify them by their histochemical characteristics (Zamboni, 1972). The primordial germ cells, which are the sole source of adult germ cells, apparently arise in the yolk sac of the developing embryo and migrate, by amoeboid movements in mesenteries, or by way of the bloodstream, to the undifferentiated gonad (Baker, 1971).

With the arrival of the germ cells, the epithelium of the gonadal ridge begins to thicken and proliferate actively. This results in the gonadal ridge

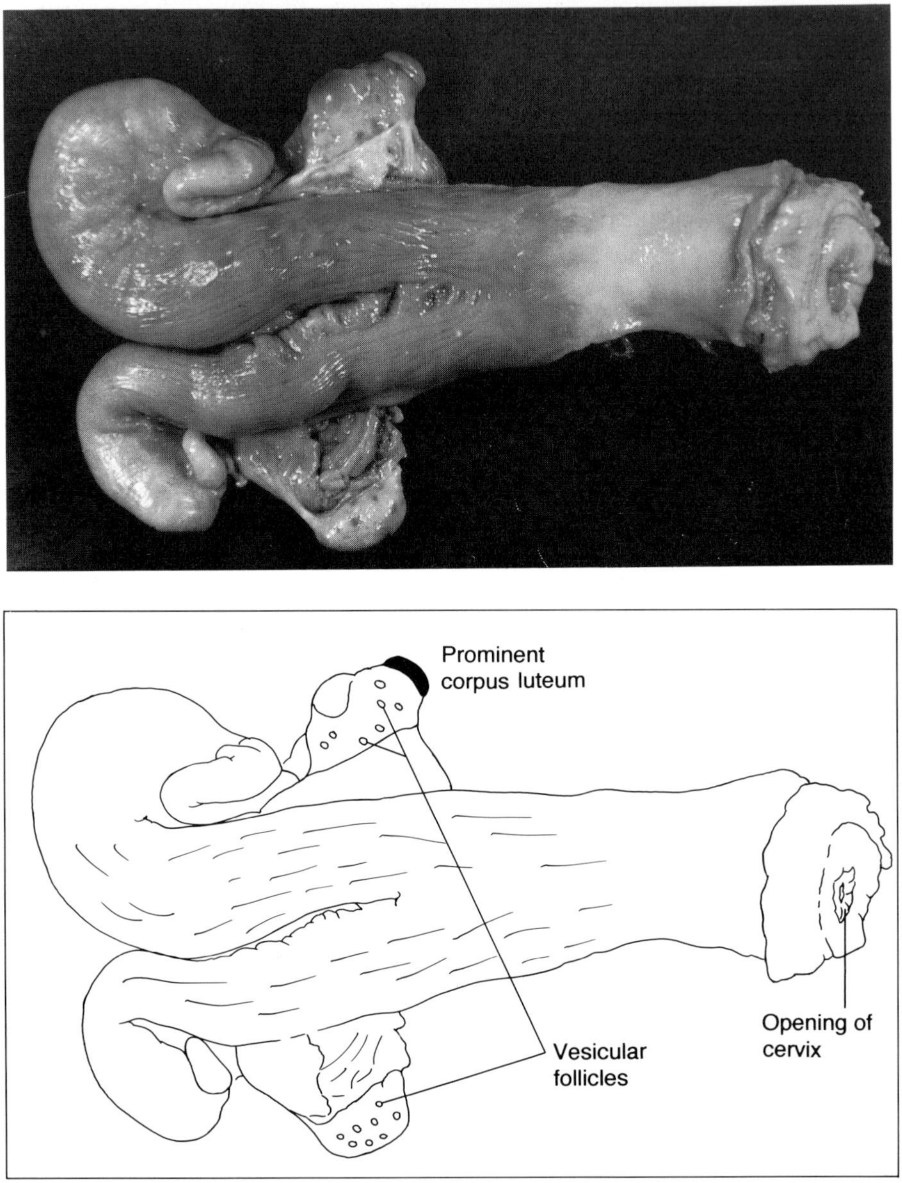

Fig. 2.4. Estimating the stage of the oestrous cycle on the basis of corpus luteum evaluation.

becoming visible in the bovine embryo at about 28 days of age and differentiated at 41 days into an ovary possessing a well-defined cortex by 45 days (Marion and Gier, 1971). According to Lavoir *et al.* (1992), between days 35 and 130 of gestation, germ cells undergo mitotic divisions before entering meiosis around day 75–80. During this period, the number of viable germ cells may reach as many as 2 million per animal, of which 95% degenerate by the time the heifer calf is born (Erickson, 1966b).

The primordial germ cells that move into the cortical region of the presumptive gonad, together with supporting epithelial cells, form what are known as the cortical sex chords. After a process marked by cell growth and by the redistribution of cytoplasmic organelles, primordial germ cells within the ovary multiply actively and are transformed into oogonia. At the end of mitotic proliferation, oogonia enter meiotic prophase and differentiate into primary oocytes. In giving rise to oocytes, oogonia undergo the two cell divisions of meiosis in order to reduce their diploid chromosome and deoxyribonucleic acid (DNA) complement ($2n$) to the haploid state (n). On entering prophase of the first meiotic division, the oogonium becomes known as a primary oocyte: after completion of the first meiotic division and entry into the second, it becomes a secondary oocyte.

There is interest in obtaining primordial germ cells as a possible alternative to embryonic stem (ES) cells for use in nuclear transfer and genetic engineering (Lavoir *et al.*, 1992). Such germ cells are mitotically dividing, and may be derived from the same cells of the embryo's inner cell mass as the stem cells. As observed by Gosden (1992a, b) and Gosden and Telfer (1992), there may be great potential in the exploitation of ovarian germ cell manipulation and transplantation in future developments in reproductive technology.

2.3.2 Formation of oogonia

After a period of mitotic replication, a proportion of the oogonia enter the preleptotene and leptotene stage of the first meiotic prophase. Once meiotic prophase commences, there is no endocrine requirement for continued meiotic progression. It is during preleptotene that the final DNA replication takes place in preparation for meiosis. It is this synthetic activity which signals the transformation of oogonia into primary oocytes. Such oocytes then progress through leptotene and zygotene. During zygotene, homologous chromosomes pair and synapse to form what appears to be single chromosomes, but in actual fact they are bivalents composed of four chromatids. Although part of the growth of the primary oocyte consists of an enlargement of the nucleus, by far the greatest contribution comes from an increase in the number and distribution of cytoplasmic organelles (Baker, 1971).

The oocytes then enter pachytene, which is characterized by crossing over and recombination between chromatids. By the end of prenatal life, oocytes have passed through diplotene and entered the dictyate stage, where they remain until stimulated to resume meiosis prior to ovulation. Cessation of mitotic activity in oogonia occurs at about day 150. The proportion of germ cells which have reached the pachytene and diplotene stages of meiosis increases markedly between day 170 and the time of birth, and the prediplotene stages disappear around this time (Baker and Hunter, 1978).

In the mammals that have been studied, the first germ cells to enter meiosis are situated at the inner part of the ovarian cortex. As soon as these oocytes reach the diplotene stage, they become surrounded by somatic cells and begin to grow and eventually to form follicles. Such a growth pattern has been recognized for many years, but still little is known about the mechanisms that control it. Byskov *et al.* (1992) tested the theory, using mice, that the number of follicles which begin to grow is inversely related to the number of non-growing follicles; their results showed that such follicles exert growth inhibition on each other when placed close together.

2.3.3 *The primordial follicle*

The follicle is the functional unit of the mammalian ovary and has a dual role: it provides the microenvironment for oocyte growth and maturation and is responsible for hormone (steroid/peptide) production. There are three basic types of follicles, as described by Erickson (1966a, b): (i) primordial; (ii) growing, in which oocyte growth and increase in follicular cell numbers and layers is occurring; and (iii) vesicular, in which there is a fully grown oocyte with granulosa cells and a layer of differentiated thecal cells.

Primordial follicles occupy the peripheral areas of the ovarian cortex in cattle. These follicles are separated from one another by compactly organized stroma consisting of fibroblasts, blood vessels and bundles of collagen and reticular fibres. Primordial follicles constantly move from the non-growing pool to enter a growth phase. An interaction between primordial follicles and the surrounding stroma is clearly involved. Each primordial follicle consists of a centrally located oocyte with a single layer of squamous or flattened granulosa cells, demarcated by an all-encompassing basal lamina which separates it from the surrounding fibrous stroma. Hirshfield (1991) has reported evidence suggesting that thecal cells may be present in primordial or primary follicles. If this is true, then all components of the follicle may be present at the very start of follicle growth.

The regulatory factors involved in the origin, differentiation and accumulation of various organelles in the cytoplasm of the primordial oocyte remain to be determined. The limited biochemical data available on oocytes in primordial follicles indicate that there is some amount of protein synthesis during this non-growing phase. The biochemical factors responsible for the selective growth of some follicles, while others remain quiescent, remain unknown. Nutrient materials required for the maintenance of primordial follicles are believed to enter the oocyte by way of a process of membrane diffusion, by pinocytosis, by active transport or through irregular gaps between adjacent follicle cells.

Some authors distinguish between primordial and primary follicles. Hulshof *et al.* (1992) state that primordial and primary bovine follicles cannot be clearly distinguished by their diameter (30–60 μm) but do show different morphological features. In primordial follicles, oocytes are surrounded by one layer of four to eight flattened granulosa cells; in primary follicles they are surrounded by 11–20 cubical granulosa cells, organized in one layer.

In pigs, Greenwald and Moor (1989) describe two types of primordial follicle: a peripheral zone of clustered oocytes with an incomplete investment of pregranulosa cells and a deeper region in which individual follicles are surrounded by a separate layer of pregranulosa cells. It seems likely that the latter follicles represent a further step in differentiation and are the forerunners of primary follicles. In the report of Carroll *et al.* (1990a), primary follicles in the mouse ovary are described as being surrounded by one or two layers of granulosa cells.

Formation of the zona pellucida

According to Baker (1971), the zona pellucida (ZP) first appears in oocytes that have a cuboidal layer of granulosa cells around them. The ZP is the relatively thick extracellular coat that is formed by and surrounds the bovine primary oocyte, increasing in width as the oocyte increases in diameter. The ZP first appears as islands of fibrillar material situated in spaces between adjacent granulosa cells and the oocyte surface. As oocyte growth continues, the ZP becomes a denser and thicker meshwork of interconnected filaments completely surrounding the oocyte and largely separating it from the granulosa cells. Contact between the oocyte and the granulosa cells is maintained via junction complexes formed between oocyte microvilli and granulosa cell extensions that penetrate the ZP.

In structure, the ZP is composed of three glycoproteins, termed ZP1, ZP2 and ZP3, with molecular weights in the region of 200,000, 120,000 and 80,000 (Wassarman, 1988a). The ZP is an open network of fibrils which permits free passage of large molecules. Of the three glycoproteins, ZP3 appears, at least in the mouse, to have a key role to play in the binding of sperm to the surface of the ZP and in the initiation of the acrosome reaction.

The ZP contains sperm receptors that mediate sperm–oocyte interaction as a preliminary to fertilization; after sperm penetration, the ZP also participates in the zona reaction, an important defence mechanism against polyspermy.

The onset of meiotic prophase, the nuclear stage characteristic of the primary oocyte, occurs between days 75 and 80 of prenatal life (Erickson, 1966b). From this time until day 150, oogonia continue to proliferate by mitosis, and they coexist within the ovarian cortex with primary oocytes at the leptotene, zygotene, pachytene and diplotene stages of the first meiotic prophase (Baker and Hunter, 1978).

Techniques are available by which primordial follicles can be isolated. In pigs, Greenwald and Moor (1989) removed the ovarian cortex, which contains the primordial follicles, and incubated fractions in a collagenase solution. After incubation, the suspension was filtered and a pure population of primordial follicles isolated. Such procedures may be valuable in examining factors involved in follicle development in the cow. Studies reported by Carroll *et al.* (1990a) have demonstrated the possibility of freeze-storing primary mouse follicles. Several workers have described culture systems capable of allowing growth and maturation of preantral follicles in species, including hamsters (Roy and Greenwald, 1989), mice (Torrance *et al.*, 1989; Carroll *et al.*, 1990a, 1991a), cattle (Nuttinck *et al.*, 1993) and humans (Roy and Treacy, 1993). Further refinements along these lines may eventually open the way to preserving and growing the primary follicles of cattle, so that literally thousands of oocytes, rather than the current 40 or so per ovary, can be obtained.

2.3.4 *Development of growing follicles*

The growing follicle category includes secondary follicles, which can be distinguished from primary follicles on the basis of possessing several layers of cubical granulosa cells and greater oocyte size (Hulshof *et al.*, 1992). The start of follicle and oocyte growth is apparently regulated within the ovary, being a function of the size of the pool of non-growing follicles. The growth of the oocyte and its surrounding follicular cells is coordinated and the oocyte progresses through a series of well-defined morphological stages. Factors involved in the regulation of growth in preantral follicles are still largely unknown. It is generally held that the earliest stages of follicle growth are independent of circulating gonadotrophins, but there is some evidence that such growth is regulated by both FSH and high local concentrations of oestrogen, although the source of oestrogen is unknown.

2.3.5 *Vesicular follicles*

Erickson (1966b) noted that vesicular follicles are first seen in calf ovaries at 250 days of gestation; by day 270, in some instances, the ovary may show numerous vesicular follicles (Russe, 1983). According to Monniaux *et al.* (1983), the antrum first appears in bovine follicles ranging in diameter from 0.115 mm (when 10% possess an antrum) to 0.280 mm (when some 90% are antral). The vesicular follicle consists of: (i) the primary oocyte, enclosed within the ZP, containing a large nucleus and prominent nucleolus; (ii) a complex somatic component including inner layers of columnar cells, which form specialized intercellular junctions with the oolemma, a number of layers of granulosa cells, and an outer thecal layer in communication with blood vessels; and (iii) a large fluid-filled cavity (antrum). Follicular fluid (FF) is a unique body fluid containing a large number of biochemical components derived from blood and other local secretions. Throughout the entire growth phase of the oocyte and follicle, the oocyte remains arrested in dictyate of the first meiotic prophase.

The fact that vesicular follicles are present in the calf *in utero* raises the possibility of maturing and fertilizing oocytes and markedly reducing the generation interval. Some of the breeding implications of this are discussed by Georges and Massey (1991).

2.3.6 *Follicular atresia in the fetal ovary*

The growth of the bovine ovarian follicle is continuous from the primordial stage to the terminal stage, culminating in ovulation. Such growth can be stopped at any point by the degenerative process known as atresia. Biochemical and morphological changes are associated with atresia, and it represents a process or processes by which oocytes are lost from the ovary other than by ovulation. The earliest stages of atresia are difficult to diagnose histologically, particularly in primordial follicles. A physiological role for atresia has yet to be determined. In practical terms, the atretic process can be at its most striking in the fetal ovary. According to the evidence of Erickson (1966b), the number of germ cells in the fetal ovaries of the heifer calf increased to a peak value of 2,739,000 at day 110 of prenatal life, but had fallen to 150,000 by the time of birth.

In the postnatal ovary, the atretic process continues. It has been estimated, for example, that the infantile human ovary contains about 400,000 oocytes; in the 30 years of reproductive life about 400 oocytes may mature and be ovulated. On this basis, only about one oocyte in 1000 leaves the ovary by ovulation; the other 999 are lost in the process of atresia.

2.4 The Bovine Ovary in Postnatal Life

2.4.1 The prepubertal animal

Vesicular follicles are present in bovine ovaries at all stages from birth of the heifer calf through to old age. According to Baker and Hunter (1978), little is known of the changes occurring in the ovary between birth and puberty. As mentioned above, there is evidence that vesicular follicles are present in the heifer calf's ovary before birth (Mariana *et al.*, 1991). Aspiration of oocytes from the ovaries of 2–3-month-old calves was reported by Dahlhausen *et al.* (1981), but no attempt was made to fertilize them. In the studies of Duby and Robl (1987), calves were superovulated and oocytes aspirated for IVF after gonadotrophin treatment; the authors suggested that calves may provide a reliable source of oocytes in large numbers.

In France, Mhawi *et al.* (1991) dealt with various aspects of follicular development in calves shortly after birth. In Japan, Kajihara *et al.* (1991a) obtained a yield of 8% blastocysts with oocytes recovered from calves and processed through their IVM/IVF/IVC system; an oocyte from a 4-month-old Holstein calf resulted in a live birth after transfer of a frozen–thawed embryo.

Studies in superovulation (see Armstrong, 1991, 1993; Stubbings *et al.*, 1993) have shown that follicle response of calf ovaries to FSH stimulation increased progressively from 3 to 9 weeks of age and that oocytes recovered laparoscopically from these follicles are capable of producing embryos and calves after IVF. Whether follicle growth and regression occurs in waves in such young calves is unknown. There is clear evidence, however, as reported by Evans, A.C.O., *et al.* (1992, 1993) and Hopper *et al.* (1993), that follicular waves and vesicular follicles of all sizes occur in the ovaries of prepubertal heifers.

2.4.2 Vesicular follicle population

There are a number of reports in the literature which set out the number of vesicular follicles recorded in bovine ovaries. A summary of such data, listed

Table 2.2. Numbers of vesicular follicles in the ovaries of the cow as shown in the literature.

Authors	Mean number of vesicular follicles
Erickson (1966b)	47 per pair of ovaries (1 mm and above)
Scanlon (1969)	49 per pair of ovaries (1 mm and above)
Ireland *et al.* (1979)	40 per pair of ovaries (1 mm and above)
Leibfried and First (1979)	50 per pair of ovaries (1 mm and above)
Matton *et al.* (1981)	49 per pair of ovaries (1 mm and above)
McNatty *et al.* (1984)	52 per pair of ovaries (2 mm and above)
Pierson and Ginther (1987b)	42 per pair of ovaries (2 mm and above)
Prado *et al.* (1989)	38–41 per pair of ovaries (3–4 mm and above)
Marion *et al.* (1968)	382 per pair of ovaries (0.4 mm and above)
Dufour and Roy (1985)	142–228 per ovary (0.16 mm and above)
Maurasse *et al.* (1985)	324 per pair of ovaries (0.14 mm and above)
Lussier *et al.* (1987)	127–490 per ovary (0.13 mm and above)
Le Van Ty *et al.* (1989)	233 per pair of ovaries (0.16 mm and above)

in chronological order, is provided in Table 2.2. Many of the reports summarized deal only with those vesicular follicles which were readily visible on the surface of the ovary (≥ 1 mm diameter). However, it is clear from Table 2.2 that there are much larger numbers of follicles if all those from the start of vesiculation are included. A number of these are below the ovary surface and can only be observed when the ovarian cortex is exposed. Vesiculation (i.e. formation of an antrum), as given in the report by Marion *et al.* (1968), started in bovine follicles of 0.5 mm diamater. Later workers, however, have shown vesiculation starting in follicles as small as 0.12 to 0.16 mm (Monniaux *et al.*, 1983; Dufour and Roy, 1985; Maurasse *et al.*, 1985; Lussier *et al.*, 1987); such information is clearly useful in considering ways and means of increasing the number of oocytes that can be recovered from vesicular follicles.

2.4.3 *Follicle development*

A review by Spicer and Echternkamp (1986) dealt with factors known at that time to influence ovarian follicular growth, function and turnover in the cow. The evidence of Lussier *et al.* (1987) showed that antrum formation in this species may occur over a certain range of follicle diameters (0.14–0.28 mm). Early growth of follicles can be attributed to an increase in the number of granulosa cells and an increase in the surface of the granulosa layer (see Fig. 2.5).

According to Marion *et al.* (1968), the stratum granulosum at the time of vesiculation is two to seven cells thick and the theca is a thin layer of orientated stroma cells surrounding the follicle. A membrana propria is clearly

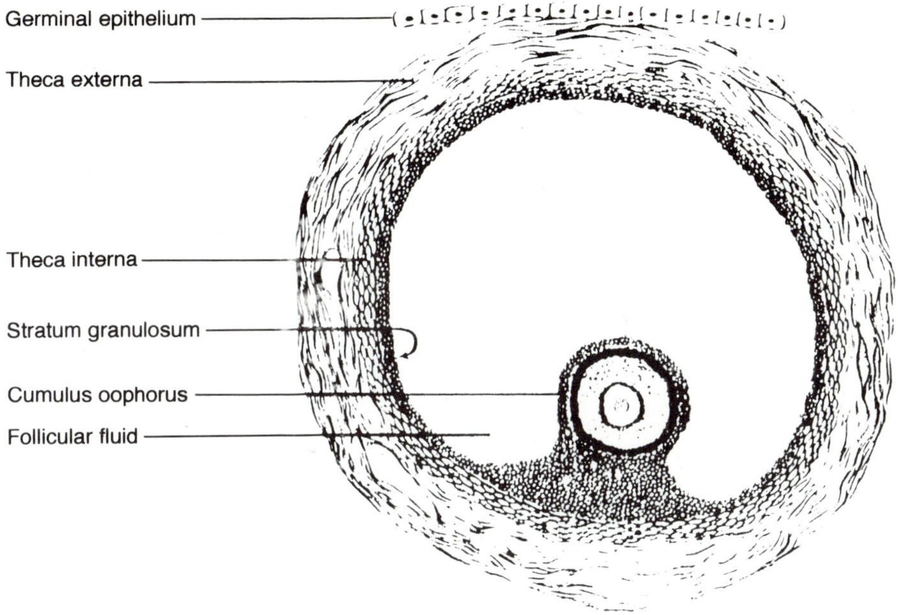

Fig. 2.5. A large mature vesicular follicle and surrounding tissue in bovine ovary (after Hafez and Gordon, 1962).

evident at the primary follicle stage and continuous thereafter. In follicles measuring 1–2 mm, the oocyte becomes surrounded by an increasingly thick layer of cumulus oophorus, several cells thick in a 2 mm follicle. In most instances, the cumulus oophorus and stratum granulosum are broadly continuous. The thecal layer consists of glandular interna cells surrounded by fibrous externa cells. Five to six days are considered sufficient for a small (2 mm) vesicular follicle to reach preovulatory size (Savio *et al.*, 1993).

During the growth phase, the granulosa cells differentiate into two subpopulations organized as pseudostratified epithelial mural granulosa cells in contact with the basement membrane that surrounds the follicle, and cumulus cells enclosing the oocyte and coupled by gap junctions to both the oocyte and the surrounding mural granulosa cells. These two cell types are known to differ in their distribution of receptors and in steroidogenic characteristics. In the preovulatory follicle, they also differ markedly in their response to gonadotrophins, which stimulate cumulus cells to produce and secrete hyaluronic acid but which have no such effect on granulosa cells.

At the start of oestrus, the follicle destined for ovulation is about 10 mm in diameter. By this stage, the cumulus oophorus has increased to eight to ten cells thick and the hilus is usually a slender stalk. The stratum granulosum has increased in thickness until it exceeds 100 μm. Within 24 h, follicle diameter increases to ovulation size, 16–18 mm. During this period of rapid preovulatory expansion, mitoses in both granulosa and theca interna cells continue at the same level as previously, which is quite inadequate to maintain

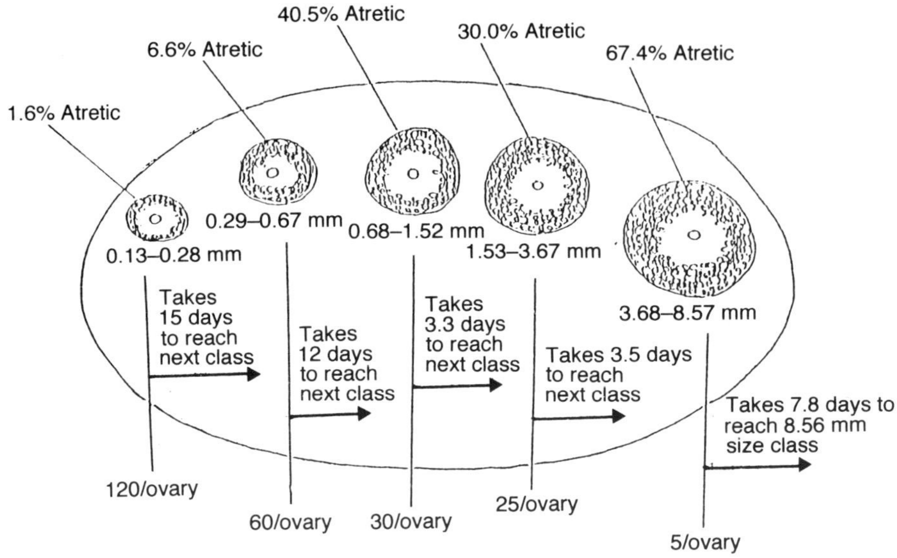

Fig. 2.6. Follicle size categories in the bovine ovary and the incidence of atresia (based on data from Lussier *et al.*, 1987).

the thickness of these layers (Marion *et al.*, 1968). The preovulatory surge of gonadotrophin, in the hours preceding ovulation, stimulates cumulus cells to undergo expansion and the mural granulosa cells to become luteal cells. Follicular atresia may begin at any stage of vesicular follicle development.

It has been estimated that the vesicular follicle takes two cycle lengths (42 days) to develop from vesiculation to the preovulatory stage and about a week for the vesicular follicle to grow from 0.67 mm diameter to 3.67 mm (see Fig. 2.6).

It should be noted that vesicular follicles must reach a certain size before they can be expected to yield meiotically competent oocytes (i.e. oocytes capable of resuming meiosis). Recognition of this in the cow was first recorded by Motlik and Fulka (1986), who observed that follicles below 1–2 mm diameter are unlikely to contain meiotically competent oocytes. This has subsequently been confirmed in several other reports (Kauffold *et al.*, 1990; Lonergan, 1990).

Thecal cells

In the development of the vesicular follicle, factors regulating the structural and functional differentiation of the thecal cells at the cellular and molecular level are poorly understood. Thecal cells are known to secrete transforming growth factor alpha (TGF-α) and TGF-β, and the cells are known to have both autocrine and paracrine functions. The most abundant steroid products of mature thecal interna cells are androgens produced from the catabolism of cholesterol. Enzymatic conversion of cholesterol to androgens in the thecal cells results in dehydroepiandrosterone (DHEA), which is then metabolized to androstenedione.

The major source of this cholesterol in the cow is believed to be the internalization of blood-borne lipoproteins as the thecal cells of the growing vesicular follicle become highly vascularized. It is possible that LH provides the main stimulus for the various steroidogenic activities of the thecal cells. During follicle growth, thecal cells are believed to respond to LH primarily by activation of enzyme systems which permit the cells to synthesize androgens.

There is evidence that the theca interna cells in cattle are a source of oestradiol, although the quantities produced, relative to androstenedione, are very small. In some species, which include cattle, thecal cells have been found to have an active aromatase enzyme system. The ability of bovine thecal cells to secrete androgens precedes acquisition of the aromatase system by granulosa cells. Throughout the oestrous cycle, 20–60 vesicular follicles may produce androgens, whereas only a few such follicles develop the ability to transform androgens into oestrogens. There is evidence that acquisition of the aromatase system, which is regulated by FSH, may be a key factor in permitting the selection and emergence of the dominant follicle.

Basement lamina

The basement lamina or lamina propria is the acellular layer that is present between the thecal and granulosa cell layers. The basement lamina of the growing follicle forms a bipartite structure composed of a homogeneous acellular layer upon which the peripheral layer of granulosa cells rests and an outer region of bundles of collagen fibres. A high rate of transfer of molecules from the theca to the outer cells of the avascular stratum granulosum is possible because of the arrangement of arterioles, capillaries and venules around the basement lamina.

A blood–follicle barrier does not appear to exist, and the selective permeability of the basement membrane is apparently under the control of gonadotrophins and oestrogens. In vesicular follicles approaching ovulation, the basal lamina starts to break down, leading to invasion of the stratum granulosum by thecal cells and capillaries.

2.4.4 Oocyte growth

Some of the structures that undergo change during growth of the oocyte are noted in Fig. 2.7.

Nucleus and nucleolus

The nucleus, or germinal vesicle (GV), of growing oocytes increases in diameter during the growth phase; during the period of nuclear enlargement, chromosomes remain as highly diffuse bivalents.

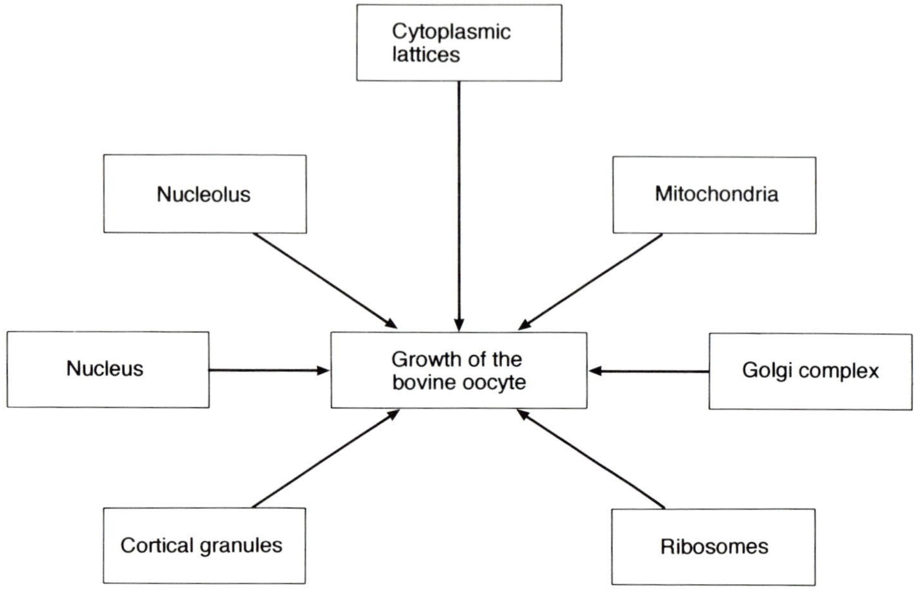

Fig. 2.7. Components that undergo change during oocyte growth.

Although rodent oocytes apparently reach their full size before antrum formation (see review by Thibault *et al.*, 1987), in the cow the oocyte is still growing when the follicle reaches the antral stage. At the time the follicular antrum is fully differentiated, the follicle is about 0.4–0.8 mm in diameter and the mean diameter of the oocyte is about 110 μm (excluding the ZP). As shown in the report of Motlik and Fulka (1986), at this stage of growth, the nucleolus of the oocyte possesses vacuolated and heterogeneous structures composed of fibrillogranular elements with discrete fibrillar centres. The nucleolus plays a key role in relation to transcription as it is the site of ribosomal ribonucleic acid (rRNA) synthesis. The fibrillar components are believed to be the sites of rRNA transcription whereas the granules represent the mature rRNA organized in preribosomal particles. The high ^{3}H-uridine incorporation in nucleoplasm and nucleolus, as well as the localization of the nucleolus, organizing region (NOR) inside the vacuolated nucleolus, indicate that the growing bovine oocyte is engaged in rRNA and homogeneous nucleolar (hn) RNA synthesis (see Table 2.3).

During further follicular growth, bovine oocytes gradually reach their full size; their nucleoli become progressively formed by a compact electron-dense fibrillar material and fibrillogranular vacuolated areas are localized at the nucleolar periphery. As shown in the studies of Motlik and Fulka (1986), nucleolar compaction coincides with achievement of full size of the bovine oocyte. Nucleolar compaction is accompanied by the fragmentation of the single or two large initial nucleoli, to form a few small compact nucleoli, 2–3 μm in diameter. A large fibrillar centre is associated with each small

Table 2.3. Relationship between the size of the bovine oocyte, its RNA synthetic activity, morphology and meiotic competence (from Motlik and Fulka, 1986).

Follicular category (mm)	Mean oocyte diameter (μm)	Type of nucleolus[a]	RNA synthetic activity[b]		% of oocytes in		
			Nucleoplasm	Nucleolus	GVBD	MI	MII
0.5–1.6	110–120	VFG	+++	+++	37	12	0
1.8–3.0	126	LV	++	+	93	9	84
4.0–8.0	135	C	+	–	88	9	79

[a]Type of nucleolus: VFG – vacuolated, fibrillogranular; LV – nucleolus with large central vacuoles surrounded with fibrillogranular material; C – exclusively compact.
[b]^{3}H-uridine labelling after 15–20 min pulse.

Table 2.4. Relationship between follicle size and oocyte diameter in cattle (from Lonergan, 1990).

Follicle diameter (mm)	Number of oocytes examined	Oocyte diameter ($x \pm$ s.d.)
0.1–0.5	21	91.0 ± 12.5[a]
0.5–1.0	55	112.3 ± 11.8[b]
1.0–1.5	27	124.5 ± 9.2[c]
1.5–2.0	7	131.1 ± 12.2[c,d]
2.0–3.0	41	133.8 ± 10.3[d]
3.0–4.0	54	130.0 ± 9.7[d]
4.0–5.0	35	130.2 ± 7.1[d]
5.0–6.0	20	130.0 ± 11.9[d]

a, b, c, d – Values with different superscripts differ significantly ($P < 0.05$).

nucleolus (Crozet *et al.*, 1986; Crozet, 1989).

The oocyte continues to grow, as follicle diameter increases up to about 2 mm (see Table 2.4). This is taken from the studies of Lonergan (1990). In the same work, evidence was found that meiotic competence of oocytes was low in those recovered from follicles less than 1.5–2.00 mm in diameter (Table 2.5).

Some of the reasons for the absence of meiotic competence in growing ungulate oocytes have been identified. In pigs, growing oocytes (90 μm in diameter) are unable to resume meiosis; Christmann *et al.* (1993) have shown that such oocytes have sufficient cell cycle components for G2 to M transition, but lack ability to convert these components to active M-phase promoting factor (MPF) until growth is almost completed.

Table 2.5. Maturation competence of bovine oocytes from different follicle sizes after *in vitro* culture (from Lonergan, 1990).

Follicle diameter (mm)	% oocytes post-GVBD		% reaching metaphase II	
0.1–0.5	–	(0/4)[a]	–	(0/4)[a]
0.5–1.0	45.16	(14/31)[b]	16.12	(5/31)[b]
1.0–1.5	69.23	(9/13)[c]	15.38	(2/13)[b]
1.5–2.0	100	(7/7)[d]	57.14	(4/7)[c]
2.0–3.0	97.96	(48/49)[d]	87.75	(43/49)[d]
3.0–4.0	95.76	(113/118)[d]	88.13	(104/118)[d]
4.0–5.0	97.43	(38/39)[d]	84.61	(33/39)[d]
5.0–6.0	100	(24/24)[d]	87.5	(21/24)[d]

a, b, c, d – Values in the same column with different superscripts differ significantly ($P < 0.05$).

Mitochondria

Oocyte growth is accompanied not only by a substantial increase in the number of mitochondria present, but also by marked changes in their ultrastructure (Fleming and Saacke, 1972; Kruip *et al.*, 1983). During the growth phase, mitochondria are closely associated with smooth endoplasmic reticulum (SER); they are present in increasing numbers and are transformed from the elongated shape to the round and oval shape. Fully grown oocytes contain round or oval, highly vacuolated mitochondria that have arched and concentrically arranged crista (Zamboni, 1972). During oocyte maturation, the mitochondria cluster intensively in the cortical regions and are associated with lipid droplets and elements of SER. At the end of maturation, the clusters of mitochondria disaggregate and become randomly distributed in the central region of the oocyte.

Golgi complex

As in the case of mitochondria, the Golgi complex undergoes marked ultrastructural changes during oocyte growth, as shown by Golgi complex activity (Zamboni, 1972). During the early stages of oocyte growth, Golgi complexes become more active, as shown by lamellae that are spaced further apart, by the appearance of vacuoles and by the proximity of numerous lipid vesicles. From the mid to late stages of oocyte growth, Golgi complexes exhibit increased numbers of very swollen, stacked lamellae that are associated with numerous vacuoles, granules, coated vesicles and lipid vesicles. These changes are consistent with increased participation of the Golgi complexes in the processing and concentration of secretory products and cortical granule formation during oocyte growth.

Cortical granules

Cortical granules are not evident in the oocytes of primordial follicles and their formation can be regarded as a feature of oocyte growth and differentiation (Kruip *et al.*, 1983; Cran, 1987, 1989; Cran and Esper, 1990). These granules fuse with the vitelline membrane at fertilization and, by releasing their contents into the perivitelline space, alter the functional properties of the ZP. Cortical granules are small, spherical, membrane-bound organelles, which are thought to resemble lysosomes. The first evidence of cortical granules within the oocyte is usually found at the onset of follicular growth. Oocyte growth is accompanied by the formation and accumulation of increasing numbers of the granules. Although it is believed that cortical granules are derived from the Golgi complex, there is evidence suggesting that there is a contribution from multivesicular bodies as well as from granular endoplasmic reticulum. Migration of cortical granules to the plasma membrane is a common feature of the final stages of oocyte maturation.

Ribosomes and cytoplasmic lattices

During the growth phase of the oocyte, the number of ribosomes present in the cytoplasm increases while the ribosome density decreases. The number of ribosomes present in polysomes also increases during oocyte growth, while the number of single polysomes decreases dramatically. The changes in ribosome population during oocyte growth is believed to be related to changes in rates of protein synthesis during this period. Cytoplasmic lattices are lattice-like structures (plaques, lamellae, fibrillar arrays), which first appear in the early stages of oocyte growth and increase markedly in number throughout the growth period. The function of this extremely abundant cytoplasmic component is not known.

Biochemical aspects of oocyte growth

Many biochemical changes are known to occur during the growth of the oocyte. Carroll and Whittingham (1993) examined Ca^{2+} oscillations in mouse oocytes at different stages of growth and after meiotic resumption was stimulated *in vivo*. They showed that the mechanisms of Ca^{2+} homeostasis in oocytes change during growth and maturation. During their growth, primary oocytes are known to synthesize a variety of proteins. The general pattern of protein synthesis is believed to remain constant throughout oocyte growth, but there are changes in the relative rate of synthesis of specific proteins.

It has been shown that the reprogramming of protein synthesis after GVBD is essential for ungulate oocytes to complete meiotic maturation and pronuclear development. Transformation of the sperm nucleus to a male pronucleus requires protein synthesis extending at least until the early metaphase II stage (Ding *et al.*, 1992b).

The specific proteins synthesized during oocyte growth include mitochondrial proteins, ribosomal proteins, ZP proteins, glycoproteins, histones, tubulin, actin, calmodulin, lactate dehydrogenase, creatine kinase, glucose-6-phosphate dehydrogenase (G6PD) and other proteins involved in fertilization (Wassarman, 1988a, b). Structural proteins and enzymes are synthesized and stored throughout oocyte growth, but the turnover of nascent protein in growing oocytes occurs at a relatively low rate, partly accounting for the large stores inherited by unfertilized eggs. Some of the proteins synthesized by growing oocytes, such as those in the ZP and the cortical granules, are essential for fertilization and early embryonic development.

The RNA content of oocytes is known to increase markedly during their growth phase, with RNA accumulation exhibiting biphasic kinetics relative to oocyte volume. During the early and midgrowth stages, changes in nucleolar ultrastructure and in levels of RNA polymerase activity are consistent with high rates of transcription of ribosomal RNA. As the compact areas of the nucleolus progressively increase, the synthesis of RNA decreases.

2.4.5 Endocrine events during follicle growth and development

Gonadotrophins

FSH is known to be of crucial importance in the recruitment and selection of DFs in the cow, as a result of its ability to stimulate aromatase activity in the granulosa cell, which converts the follicular microenvironment from androgenic to oestrogenic. In the preovulatory bovine follicle, the FSH level in FF is higher than in luteal-phase follicles before the plasma LH peak and shows a peak at about 4 h after the LH plasma surge (Dieleman and Bevers, 1987). The high FF concentration of FSH at time of ovulation may be associated with the assumed role of this gonadotrophin in ovulatory-related processes (such as cumulus cell expansion and mucification).

The LH concentration in FF is believed to decrease markedly as follicle size increases. This may reflect some utilization of the fluid LH by follicular cells, particularly granulosa cells, which are bathed in the FF. In the preovulatory follicles, the levels of LH are not different from those in the luteal-phase follicles but show a peak in the FF at about 4 h after the peak in the peripheral blood (Dieleman and Bevers, 1987). This lag may be due to the occupation of free LH receptors in the follicular wall since the inward flow of small proteins such as gonadotrophins may not be impeded by the blood–follicle barrier.

Although the maximum of the LH peak in the FF is about one seventh of that in the peripheral blood, the LH level in the FF is higher than the plasma LH level from 7 h to 20 h after the LH plasma peak, when luteinization of the granulosa cells occurs. The tailing of the LH surge in FF is believed to be due to the gradual release of LH from the receptors in the follicle wall. It is well accepted that LH provides the stimulus *in vivo* that brings about the resumption of meiosis.

The prolactin level in the peripheral blood during the luteal phase of the bovine oestrous cycle is higher than that during the period of preovulatory follicular development (Dieleman and Bevers, 1987). The level of prolactin in FF does not, however, show any significant change throughout preovulatory follicular development; there is also no difference in prolactin level between preovulatory and luteal phase follicles. Although prolactin is believed to play a beneficial role in the process of cytoplasmic maturation in mammalian oocytes, there is no clear view as to how it exerts its effect.

Oestradiol and progesterone

In non-atretic vesicular follicles, the oestradiol level in FF rises as follicle size increases and becomes dominant in the preovulatory follicle. The plasma progesterone concentration, on the other hand, does not change in any clear way and remains very low during follicular development (Kruip and Dieleman, 1985; Dieleman and Bevers, 1987; Dieleman and Blankenstein, 1987). After the preovulatory LH surge, the oestradiol level starts to fall, accompanied by an increase in the progesterone level. In the preovulatory follicle, oestradiol ceases to be synthesized by granulosa cells and is replaced by progesterone just prior to ovulation (Moor and Seamark, 1986).

It is believed that oestradiol plays an important role in the process of follicular development and ovulation. The steroid sensitizes the granulosa cells to respond to gonadotrophins and stimulates the proliferation and differentiation of the granulosa cells. It also influences the hypothalamus–pituitary axis in producing a small increase in the LH level and a higher frequency of its pulsatile release, eventually resulting in the preovulatory surge. The physiological role of progesterone in the follicle itself is not well understood. By the time oestradiol concentrations start to decrease and progesterone levels to increase, events indirectly attributable to the LH surge, the meiotic maturation of the oocyte is under way. The maturation of the oocyte is likely to be influenced by the change in the oestradiol/progesterone balance within the follicle.

Androstenedione and testosterone

It is believed that small vesicular follicles are androgenic, regardless of whether they are destined to ovulate or become atretic. As follicle size increases, the concentrations of androstenedione and testosterone in FF decrease gradually in both non-atretic and atretic follicles (Kruip and Dieleman, 1985). Small follicles in the late luteal phase and follicular phases of the cycle contain appreciably more testosterone in the FF than similar-sized follicles in the early luteal stage. The preovulatory follicles contain significantly greater concentrations of androstenedione and testosterone in the FF than the luteal-phase follicles (Dieleman and Bevers, 1987). After the preovulatory LH surge, the testosterone levels in the FF continue decreasing but the androstenedione levels increase again. Androstenedione appears to be the major form of the androgen in the FF, with a level which is more than ten times that of testosterone.

It is evident that androgens synthesized by thecal cells are obligatory substrates for oestrogen biosynthesis and may play a role in follicular atresia. Their role in oocyte maturation remains uncertain.

2.4.6 Induction of multiple ovulations in the cow

For those engagaed in IVF research, there may occasionally be a need to obtain *in vivo*-matured oocytes for comparative studies with artificially matured oocytes; for that and other reasons, a knowledge of superovulatory treatments and information on the normality of oocytes shed after such treatment is relevant to the present discussion. Of commercial relevance is the fact that superovulation techniques and embryo transfer enable the lifetime reproductive output of the cow to be increased tenfold.

The stimulation of cattle to induce additional ovulations (i.e. superovulation) has been the subject of much research during the past 50 years and has always been an important consideration in the development of commercially acceptable embrgo transfer (ET) technology in cattle. Descriptions of superovulation techniques go back to the days of Smith and Engle in 1927, who recorded a fourfold increase in ovulation rates, after the intramuscular implantation of anterior pituitary tissue in mice and rats.

A few years later, Cole and Hart in the USA demonstrated the ability of the serum of pregnant mares to induce multiple ovulations in rats, establishing the basis for what eventually became the most widely used gonadotrophin treatment for superovulation. Pregnant mare serum gonadotrophin (PMSG), a glycoprotein hormone present in the blood of the mare between days 40 and 130 of gestation, is unique among gonadotrophins in possessing both FSH and LH biological activities within the one molecule. Much of the early research on cattle superovulation involved the use of PMSG (see Gordon, 1975).

Although superovulation has been a key element of cattle ET, and despite much research over the years in this area, there has been little significant improvement in the effectiveness of the technique in recent years (Hahn, 1992, 1993; Hasler, 1992). Indeed, there are those who would regard the current practice of superovulation to be an art that has not advanced appreciably since it was first carried out by Smith and Engle more than 60 years ago (Armstrong, 1991, 1993). Highly variable quantitative and qualitative superovulatory responses are well recognized as a major constraint to the application of cattle ET technology (Hahn, 1992).

Animal variability is regarded as at least as important a factor determining superovulatory response as the particular hormone preparation employed. There is, however, much evidence supporting the view that a higher and more consistent response can be obtained with FSH preparations than with PMSG. In recent years, the most widely used commercial preparations have been Folltropin (Vetrapharm Inc.) and FSH-P (Burns Biotech), both prepared from pig pituitaries. The superovulatory responses which can be induced with such preparations are well established, although questions remain about the extent to which embryo quality may be influenced by the composition and dose level of the preparation (Mapletoft *et al.*, 1988; Mapletoft and Pierson, 1993).

In the USA, workers at six locations compared treatment with a *h*orse *a*nterior *p*ituitary (HAP) extract with a single batch of porcine FSH (pFSH); in this, Staigmiller *et al.* (1992) found HAP to be an acceptable alternative to pFSH for cattle superovulation. There was, however, no reduction in the variability in superovulatory response, which was one of the main hopes in carrying out the study with HAP. There was a wide variation in superovulatory response among the six locations, for both HAP and pFSH treatments, which is probably an indication of the subtle nature of factors influencing the ovulatory response of cattle to exogenous gonadotrophins. All locations used the same hormone batches, dosages, procedures and protocols, and yet the uncontrolled differences resulted in two- to fourfold response differences between the two FSH preparations.

In using FSH, much attention has focused on the amount of LH contained within the particular preparation employed (Lindsell *et al.*, 1986). It has been believed that the superovulatory response is influenced by the relative amounts of FSH and LH in the product (the FSH : LH ratio). All commercially available FSH preparations are pituitary extracts from farm animals (usually pigs) and contain variable amounts of LH. The outcome of superovulation treatments involving FSH preparations that are reputedly less contaminated than usual with LH has been reported (Donaldson and Ward, 1985, 1987; Donaldson *et al.*, 1986; Tribulo *et al.*, 1991); such preparations appeared to have a favourable effect on embryo quality.

Studies reported by Herrier *et al.* (1991a, b) have shown that the use of an FSH preparation with a defined FSH : LH ratio can eliminate one important source of variability. A review by Mapletoft and Pierson (1993) suggests that the maximum acceptable level of LH contamination of an FSH preparation is between 15 and 20% of the original LH content of the extract. The recent availability of bovine FSH (bFSH) of high biochemical purity, produced by recombinant DNA technology, should guarantee batch-to-batch consistency and the absence of protein contaminants, including other gonadotrophins, when it is marketed for commercial use (Looney and Bondioli, 1988). This bFSH has been shown to possess high biological activity and, within the dose ranges evaluated by Bellows *et al.* (1991) in beef cattle, induced an excellent superovulatory response.

The most favourable time at which superovulatory treatment can be initiated in the oestrous cycle of heifers and cows is coincident with an interphase period between the day of maximum diameter of the first DF and emergence of the second DF (days 9–13; day 0 = day of oestrus). Such treatment is followed by $PGF_{2\alpha}$ 48 h later to induce luteal regression, oestrus and ovulation. According to Mapletoft and Pierson (1993), the length of a cow's cycle may provide a clue as to the most appropriate time to start the superovulation treatment. Cattle with 21–23-day cycles should be started on day 9; animals with an 18–20-day cycle should start on day 10.

It is clear, from several reports, that there is a decreased superovulatory response when FSH treatment is initiated in the presence of a DF. In Canada, Guilbault (1991) used ultrasonics to monitor development and showed that the presence of a DF at the time of initiating gonadotrophin treatment reduced superovulatory response by 40–50%. Huhtinen *et al.* (1992) in Finland showed that, in the absence of a dominant follicle in the ovaries at the time of gonadotrophin stimulation, cows produced twice the number of transferable embryos compared with the animals treated with gonadotrophin in the presence of a DF. It would clearly be valuable to have an accurate method of recognizing the presence of a DF or a method of inhibiting its development before gonadotrophin treatment is applied. In Canada, Nasser *et al.* (1993) also emphasize the importance of the follicular wave status of the donor animal in determining its superovulatory response. In Germany, Bungartz and Niemann (1993) used ultrasonics to accurately diagnose the presence of a DF, but this would not seem to be a commercially feasible practice.

Superovulation treatment regimes aimed at modifying postovulatory FSH levels in the donor cow have been reported by several workers. Treatment is designed to stimulate follicular growth at the start of the cycle. Although some studies (Touati *et al.*, 1991) have suggested a favourable response, the general view now is that such gonadotrophin priming treatment reduces rather than increases the ovulatory response (Guilbault *et al.*, 1990, 1992; Calder *et al.*, 1991; Gray, B.W., *et al.*, 1991, 1992; Calder and Rajamahendran, 1992). It appears that priming with FSH at the start of the cycle, while it does induce increased follicular development, results in a higher proportion of anovulatory follicles. The cause of this is not well understood.

The effect of dose schedules and route of administration of the FSH preparation has been reported by several groups (Bo *et al.*, 1991; Purwantara *et al.*, 1991; Hockley *et al.*, 1992). Single injections can be used rather than the conventional eight doses spread over 4 days; such simplifications may be particularly important in dealing with beef or zebu cattle, where the restraint of animals for multiple injections may be an unwelcome source of stress (see Mapletoft and Pierson, 1993).

Pituitary down-regulation by means of Gn-RH in ovarian stimulation regimes in human IVF programmes has been widely employed for some years. Such treatment is aimed at eliminating the DF and enabling uniform follicle growth in response to FSH treatment. In the USA, Savio *et al.* (1991) attempted to influence the superovulatory response of cattle treated with a potent Gn-RH analogue on day 3; the treatment decreased rather than increased response. The extent to which pituitary down-regulation by Gn-RH may favourably influence superovulatory responses in cattle has yet to be adequately examined.

2.4.7 *Recombinant bovine somatotrophin (rBST)*

Growth hormone (somatotrophin) is secreted by the anterior pituitary gland. Its most obvious action in the growing animal is to induce linear growth of the long bones. In this, growth hormone acts on the epiphyseal growth plates of the long bones of the young animal. It is believed that most of these effects on growth are indirect and are mediated via IGF-I, which is secreted mainly by the liver in response to growth hormone. Recombinant bovine somatotrophin (rBST) has become available in recent years and this has increased the understanding of the way in which the hormone acts in the reproductive processes (Webb *et al.*, 1992, 1994).

There is every reason to believe that growth hormone plays an important role in reproductive biology (Christman and Halme, 1992). There are even those who take the view that growth hormone can legitimately be classified as a gonadotrophin on the basis of the evidence currently available. Whatever gonadotrophic action the hormone may possess, it is likely to depend on the concurrent presence of conventional gonadotrophins (FSH/LH). For that reason, the term 'cogonadotrophin' may be appropriate (Katz *et al.*, 1993). The availability of rBST has led to its use as a cotreatment with FSH preparations. Although some workers have found that this may enhance the superovulatory response in cattle (Rieger *et al.*, 1991a; Kuehner *et al.*, 1993a), others record no difference (Gray *et al.*, 1993). Attempts to evaluate the pituitary content of somatotrophin in cattle, using clonidine, have been reported (Kuehner *et al.*, 1993b).

It is now clear that rBST stimulates the synthesis of IGF-I, not only in the liver, but in various other tissues. In the ovary, for instance, IGF-I is found in high concentrations in FF, particularly in the DF. Studies reported by Spicer and Geisert (1992) have shown that small bovine vesicular follicles contain significantly less IGF-I than medium–large follicles. Whether this was due to increased local biosynthesis or increased diffusion of IGF-I from serum is not clear. There is a large body of evidence suggesting the existence of an intraovarian IGF system complete with ligands, receptors and binding proteins (Katz *et al.*, 1993). Given that rBST can stimulate ovarian IGF-I and that IGF-I stimulates granulosa cell function, it is reasonable to assume that the hormone may influence superovulatory responses.

The observations of Gong *et al.* (1990, 1991a, b) showed that treatment with rBST led to a twofold increase in the population of vesicular follicles 2–5 mm in diameter in heifers. The authors concluded that rBST probably acted via increased peripheral IGF-I concentrations, although in this and in subsequent work (Gong *et al.*, 1992, 1993) a direct effect of BST at the ovarian level could not be excluded. Information reported elsewhere (Langhout and Spicer, 1990; Langhout *et al.*, 1991) lent some support for the view that rBST may have a direct effect on ovarian activity in cattle. These authors showed that the granulosa cells of small and large follicles increased protein synthesis in response to the hormone. Elsewhere, functional growth hormone receptors have been shown to be present in human granulosa cells (Carlsson *et al.*, 1992), which supports the view that the hormone has a direct action in the ovary.

In human follicular stimulation programmes, there is evidence that growth hormone supplementation can enhance ovarian response to gonadotrophins in some studies (Ibrahim *et al.*, 1991; Lee and Ng, 1991; Volpe *et al.*, 1991, 1992) but not in others (Younis *et al.*, 1992). The evidence from several studies has been dealt with in a review by Katz *et al.* (1993). In France, Hugues *et al.* (1991) examined the use of growth hormone-releasing hormone (GH-RH) in women undergoing superovulatory response in IVF programmes. Their data led them to conclude that activation of growth hormone release by GH-RH enhanced the hormonal ovarian response to human menopausal gonadotrophin (HMG) and that such treatment may have a value in improving follicular maturation regimes.

2.4.8 Characteristics of preovulatory follicles and oocytes after superovulation

The cytological characteristics of oocytes recovered from cattle after superovulation treatment were reported almost 30 years ago (Hafez and Ishibashi, 1964); no adverse effect on the pattern of nuclear maturation was found. It has become evident, however, from many subsequent reports, that the superovulation treatments employed in cattle may compromise the quality of the oocytes released at time of ovulation. As observed by Foote and Ellington (1988), superovulating cattle can cause abnormal follicular steroidogenesis, premature maturation and ovulation of oocytes, deviant systemic hormone profiles and other changes. Despite such problems, it has to be noted that thousands of oocytes go on to produce normal healthy calves each year. FF was examined in preovulatory bovine follicles in the studies of Wise *et al.* (1986); endocrinological or ionic differences were not noted between follicles harvested from control and gonadotrophin-stimulated animals.

There are studies in which more than one third of oocytes aspirated from superovulated cattle in the preovulatory period have been found to display abnormal nuclear maturation (Hyttel *et al.*, 1991a, b). There may also be adverse effects operating on sperm transport and survival arising from the superovulation treatment. Loos *et al.* (1991a) examined follicular levels of progesterone and oestradiol in superovulated cattle and concluded that oocyte and follicle development were out of phase in some instances, which could explain some of the defects that are believed to occur in oocytes. In The Netherlands, Dieleman and Bevers (1993) recorded preovulatory follicles containing more than a fourfold higher oestradiol concentration in superovulated cows than in non-stimulated animals; such an inappropriate endocrine environment may influence oocyte quality.

The characteristics of preovulatory follicles in the normal cow have been described by Staigmiller and England (1982) in terms of steroid production and follicle dimensions. The characteristics of preovulatory follicles and oocytes after different superovulation treatments is dealt with by Laurincik *et al.* (1993a). These authors show that certain treatments (FSH-P/Gn-RH) can reduce the incidence of follicular atresia, as measured by opaqueness of follicles and various morphological characteristics associated with the oocyte. The first chronological picture of an ovulation wave in superovulated cattle was provided by Laurincik *et al.* (1993b); they showed that ovulation occurred over an 8 h period, with most ovulations occurring in the first 4 h.

One problem with current superovulation regimes lies in the fact that the usual preovulatory surge of LH may be absent in some cows. For that reason, some investigators have used Gn-RH in an effort to improve the ovulatory response (Laurincik *et al.*, 1991a; Petr *et al.*, 1991; Wubishet *et al.*, 1991). This usually takes the form of a dose of Gn-RH administered on the day of oestrus; this may not necessarily result in LH release.

2.4.9 Follicular atresia in the cow

Bovine ovaries acquire their lifetime quota of oocytes before birth, when ovarian ageing begins. As noted, in dealing with the prenatal ovary, the store of primordial follicles starts to decrease almost as soon as it is established. According to Erickson (1966a), who examined cows from birth to 20 years of age, numbers of primordial follicles remain stable until about the fourth to sixth year and then decline thereafter until the near zero point is reached at 20 years.

In humans, according to Faddy *et al.* (1992), it is apparent that the age-dependent loss of follicles can no longer be regarded as a simple programme running at a constant rate from birth to menopause. These authors show that the rate of follicle disappearance increases with age and that it more than doubles when numbers fall to the critical figure of 25,000 at 37.5 years old. It is known that the FSH level rises after 40 years, presumably as a consequence of decreasing negative feedback control by ovarian steroids/peptides, and it has been suggested that FSH may increase the rate of recruitment of primordial follicles. There is certainly evidence supporting the view that the development of human preantral follicles depends very much on FSH (Roy and Treacy, 1993).

Although atresia is the ultimate fate of 99.9% of bovine follicles, the mechanisims leading to follicular degeneration remain unclear. One possibility, mentioned by Faddy *et al.* (1992), is that FSH may be involved in programmed cell death in the human ovarian follicle. Indeed, ovarian follicular atresia may serve as a particularly useful model in elucidating the mechanisms involved in physiological cell death.

Many papers have been devoted to what differentiates a healthy from an atretic ovarian follicle (Murdoch, 1992; Monniaux *et al.*, 1993). On the macroscopic level, there is evidence that growing vesicular follicles are more likely to protrude from the ovarian surface whereas those undergoing regression have a greater proportion of their surface embedded within the ovary (Spicer *et al.*, 1987). One of the earliest signs of atresia in vesicular follicles is a degenerative change affecting the granulosa cells, in which their aromatase activity is lost. There may also be a decrease in androstenedione synthesis by thecal cells during atresia. A paper by Price *et al.* (1993), dealing with changes in inhibin-β levels in follicles, indicated that this peptide was not involved in the initiation of atresia.

In The Netherlands, Van den Hurk and Dijkstra (1992) have reported the presence of atypical granulosa cells in vesicular bovine follicles. These cells differed from normal granulosa cells both histologically and ultrastructurally; their presence in follicles coincided with the occurrence of degenerative changes in the cytoplasm of adjacent granulosa cells in non-atretic follicles.

It is not uncommon for the oocyte of a large atretic follicle to undergo a process termed pseudomaturation (Murdoch, 1992). In this, there can be a breakdown of the GV and a concurrent loss of cell-to-cell contact among GV and between cumulus cells and the oocyte.

2.5 Vesicular Follicles and Their Oocytes: From Birth to Old Age

2.5.1 Calves and prepubertal heifers

Vesicular follicles are present in bovine ovaries at all stages from birth to old age (Bedirian and Baker, 1975). Relatively few studies deal with the numbers and quality of oocytes recovered from cattle at different ages and in different reproductive states.

Aspiration of oocytes from the ovaries of 2.5-month-old calves and their subsequent *in vitro* maturation were reported some years ago (Dahlhausen *et al.*, 1981); attempts to fertilize them (IVF) were not made. In more recent times, Duby and Robl (1987) superovulated heifer calves and aspirated oocytes for IVF after exogenous LH treatment. On the basis of their results, they were able to recommend calves as a reliable source of oocytes in large numbers. This may be a commercial consideration, should calf ovaries be employed as a primary source of oocytes. In those EU countries where veal production involves the slaughter of young calves, including heifer calves (The Netherlands, Belgium, France), laboratory embryo production may be based on ovaries recovered from calves rather than maiden heifers.

In Japan, Kajihara *et al.* (1991a) used oocytes from slaughterhouse calves in IVM/IVF/IVC and succeeded in producing offspring from a 4-month Holstein calf. In Canada, Armstrong *et al.* (1991a, 1992) reported using calves as sources of oocytes for embryo production. As with Irvine *et al.* (1993) in Australia, the Canadian workers recovered oocytes by laparoscopic aspiration of gonadotrophin-stimulated calf ovaries. The normality of oocytes from calves did appear to be satisfactory and that could open the way to producing embryos from calves as a means of markedly shortening the generation interval. This could have useful implications for the acceleration of genetic gain in cattle.

In Germany, Torner *et al.* (1992) have shown that oocytes recovered from 4-month-old calves were capable of cleavage after IVM/IVF; these authors record an average of 120 oocytes per ovary, of which 36 possessed a compact cumulus covering. Large numbers of oocytes per ovary in slaughterhouse prepubertal cattle were also recorded by Thonon *et al.* (1993).

A comparison in which calf and cow slaughterhouse ovaries were used as sources of oocytes (Palma *et al.*, 1993a) led to the conclusion that calves were less efficient in an IVF programme. Further work in Germany reported by Palma (1994) has indicated that supplementation of the IVM medium with a higher level of FSH (20 μg rather than 10 μg ml^{-1}) led to a higher embryo yield. In France, Revel *et al.* (1993) found that the rate of blastocyst formation from 3-month-old calf oocytes was significantly lower than that from cow oocytes. In Canada, Levesque and Sirard (1994), using 40-day-old calves, found similar evidence. In an earlier paper dealing with pigs, Pinkert *et al.* (1989) had also reported that zygotes from prepubertal gilts did not possess the same *in vitro* development potential as zygotes from mature animals. In Dublin, a limited amount of information was obtained to show that no difference may be expected in IVF outcome using oocytes from cyclic and prepubertal heifers (Lu, 1990).

2.5.2 *Age, oestrous cycle and pregnancy*

Bovine oocytes used in embryo production may come from cattle with diverse reproductive backgrounds. Lonergan (1990) found clear evidence in Dublin that cow ovaries may provide lower numbers of oocytes than are generally recovered from maiden heifers. Elsewhere, Moreno *et al.* (1992) reported that the number of oocytes recovered was significantly greater in heifers than in cows; they also noted that heifer ovaries without a CL gave the greatest yield of grade 1 oocytes. Studies on the effect of age on blastocyst yield after IVM/IVF/IVC by Mermillod *et al.* (1992a) in Belgium showed a higher yield from younger cows (1–3 years) than older animals. The extent of follicular atresia in heifers and cows was the subject of a report by Silvan *et al.* (1993).

There have occasionally been those reporting evidence of higher-quality oocytes coming from cows rather than heifers (Zhang *et al.*, 1991a). It may be a question of the nutritional status of animals rather than age and parity. It would not be expected that cull cows coming from the dairy herd would be in such good body condition as heifers specifically prepared for the beef market.

Oestrous cycle stage

Various authors have recorded follicle numbers during different phases of the bovine oestrous cycle (Leibfried and First, 1979; Kubajak *et al.*, 1984) and according to whether follicles are located ipsilateral or contralateral to the cycle's CL (Pierson and Ginther, 1987b). Although there are minor cyclical variations, it is apparent that the CL can exert a mild but positive intraovarian effect; a greater number of small vesicular follicles (2–3 mm diameter) may be found in the ovary bearing the CL than in the contralateral ovary during the oestrous cycle and in early pregnancy. Studies by Lonergan (1990) in Dublin showed that the number of vesicular follicles in the 2–6 mm diameter size category did not differ significantly during any of the cycle phases examined (Table 2.6).

In terms of oocyte quality, other work in Ireland reported by Tan and Lu (1990) examined the effect of cycle stage on both cleavage rate (eggs cleaving to at least the two-cell stage by 48 h after IVF) and the yield of blastocysts. No clear evidence was found (see Table 2.7) to suggest an effect of cycle.

In former Czechoslovakia, papers by Machatkova *et al.* (1991a, b, 1993) have recorded evidence of a higher developmental potential in oocytes

Table 2.6. Effect of stage of oestrous cycle on the number of vesicular follicles visible on the surface of the bovine ovary (from Lonergan, 1990).

Stage of oestrous cycle (day)	No. of pairs of ovaries	Ipsilateral				Contralateral			
		Total	< 2 mm	2–6 mm	> 6 mm	Total	< 2 mm	2–6 mm	> 6 mm
I (1–4)	11	$^{bc}21.18^{1}$	$^{ac}3.91^{1}$	$^{a}17.18^{1}$	$^{a}0.09^{1}$	$^{bc}21.54^{1}$	$^{ac}3.18^{1}$	$^{a}18.0^{1}$	$^{a}0.36^{1}$
II (5–10)	65	$^{ac}20.15^{1}$	$^{a}2.65^{1}$	$^{a}16.72^{1}$	$^{b}0.78^{1}$	$^{ac}20.23^{1}$	$^{a}3.0^{1}$	$^{a}16.54^{1}$	$^{b}0.69^{1}$
III (11–17)	68	$^{b}23.16^{1}$	$^{ac}3.52^{1}$	$^{a}18.87^{1}$	$^{b}0.77^{1}$	$^{b}23.60^{1}$	$^{bc}4.41^{1}$	$^{a}18.32^{1}$	$^{b}0.87^{1}$
IV (18–20)	43	$^{b}24.46^{1}$	$^{bc}4.28^{1}$	$^{a}19.39^{1}$	$^{b}0.79^{1}$	$^{bc}20.46^{2}$	$^{ac}3.72^{1}$	$^{a}15.88^{2}$	$^{b}0.86^{1}$

a, b, c – Figures within the same column not sharing the same superscript differ significantly ($P < 0.05$).
1, 2 – Figures not sharing the same superscript differ significantly ($P < 0.05$) between ipsilateral and contralateral ovaries.

Table 2.7. Embryo yield from oocytes recovered at different stages of the oestrous cycle (from Tan and Lu, 1990).

Phase	Oocytes insem.	Cleavage (%)	BL/insem. oocytes (%)
ELP	469	365 (77.8)	101/469 (21.5)
LP	599	445 (74.3)	132/599 (22.0)
LLP	612	480 (78.4)	150/612 (24.5)

BL = blastocyst; ELP = early luteal phase; LP = luteal phase; LLP = late luteal phase.

recovered from cattle in the 14–16 days stage of the oestrous cycle than at days 7–9 or at unknown stages. These authors suggest that this effect may be related to the relative proportions of dominant follicles found in the different cycle phases.

In Hungary, Vajta *et al.* (1992) conducted a study to determine if the oocytes from pregnant cattle are capable of producing the usual yield of embryos after IVM/IVF/IVC; they showed that pregnant cows were as acceptable in this regard as cyclic animals. Similar findings have been reported elsewhere (Behboodi *et al.*, 1991, 1992; Sivakumaran *et al.*, 1991; Neugebauer, 1992) and some authors have even reported significantly greater yields of oocytes from pregnant cows than from non-pregnant animals (Moreno *et al.*, 1993). For those commercial IVF concerns contemplating the recovery of oocytes by vaginal ultrasound-guided aspiration of follicles in the ovaries of early pregnant cattle, this may be of particular interest.

2.5.3 Breed, nutrition and individual animal variability

In Ireland, working with beef animals, Lonergan (1990) found that, while Continental crossbreds (Charollais, Limousin, Simmental) had somewhat lower numbers of surface vesicular follicles than the traditional Hereford crossbreds, there was a remarkable uniformity in follicle numbers among the various breed types examined.

The effect of nutrition on the ovarian follicular population of beef heifers is dealt with in a report by Maurasse *et al.* (1985); they found that a high-energy diet accelerated the growth rate and turnover of follicles. In post-partum beef cattle, it has been shown that animals maintained on low-plane nutrition possessed fewer vesicular follicles than those on a high plane in the early weeks after calving (Prado *et al.*, 1989). In general, however, young beef cattle prepared for slaughter would have a good nutritional status; this should ensure that follicle populations are not limited by feeding deficiencies. Short-term fasting of beef cattle has been shown to significantly decrease plasma IGF-I levels (Spicer *et al.*, 1992).

There are those who have reported on the effect of dietary protein on follicle growth and embryo development in superovulated, non-lactating Holstein cattle (Garcia-Bojalil *et al.*, 1991). It was found that feeding diets that differed widely in crude protein content failed to alter ovarian follicular development and subsequent embryo yields.

Individual animal variability

There is evidence suggesting that the yield of embryos after IVM and IVF can be markedly influenced by the individual animal (Goto *et al.*, 1990c; Berg and Brem, 1991; Mermillod *et al.*, 1992b, d). High individual variations in oocyte number and in the yield of blastocysts after IVM/IVF/IVC may explain experimental variations in cattle embryo production. According to Berg and Brem (1991), differences in embryo yield after IVF are likely to be comparable to those observed after conventional superovulation treatments. In Japan, however, Funahashi *et al.* (1991) concluded that the oocytes aspirated from the follicles of different donor heifers had the same developmental capacity and that the production of calves from any individual donor animal should be possible. In Dublin, Carolan *et al.* (1993), in a study of individual animal variability, were able to produce blastocysts from each of the 12 heifers that were used.

2.5.4 Gonadotrophin treatment prior to slaughter

It may, under certain circumstances, be possible to think in terms of recovering oocytes from cattle that have undergone superovulation treatment immediately prior to the time of slaughter. In former Czechoslovakia, Pavlok *et al.* (1989) used cattle COCs recovered from cows stimulated 24 h previously with 2000 iu PMSG; whether that influenced results is unclear, in the absence of controls. Cattle given FSH preparations some 48–72 h ahead of slaughter may possess ovaries containing a number of follicles that have commenced their final growth phase and which may be capable of yielding oocytes of higher than average quality. A study reported by Lu *et al.* (1991) compared oocytes recovered from slaughterhouse ovaries 2 days after the commencement of superovulation treatment with those from untreated animals; a significantly higher yield of embryos was produced by gonadotrophin-treated animals.

Elsewhere, Jablonka *et al.* (1991), working with cyclic sheep, showed that FSH treatment resulted in increased numbers of medium and large follicles after 48 h. In Dublin, Lonergan *et al.* (1993a, 1994) showed that FSH treatment prior to slaughter in cattle led to a greater proportion of larger follicles and a higher number of good-quality oocytes for IVM. If follicle size is to be regarded as an important determinant of oocyte quality (see Section 2.7.3 below), then techniques for increasing the number of larger-sized follicles may be one means of increasing embryo yield per ovary. The administration of prostaglandin (Prosolvin; Intervet) has also been shown to result in an increase in the number of large follicles (>6 mm) in the cow ovary, but apparently is associated with atresia in such follicles (Hurk and Dijkstra, 1992).

2.6 Recovering the Primary Oocyte

2.6.1 Aspiration techniques: old and new

Recovery of bovine oocytes by aspiration of vesicular follicles, using an appropriate pipette or syringe and needle, has been the method most commonly employed (Sreenan, 1968; Hunter *et al.*, 1972; Leibfried and First, 1979; Mayr *et al.*, 1986; Xu *et al.*, 1987a, 1988a). One of the difficulties associated with the aspiration approach lies in the fact that oocytes may only be retrieved from some 30–60% of the punctured follicles. Katska (1984) recorded a recovery rate of 43% with aspiration and 100% using follicle dissection. Of those oocytes recovered by aspiration, 45% were classified as morphologically normal, whereas 63% were in this category when recovered from isolated follicles. Such data imply that good-quality oocytes may be less readily obtained by aspiration than by follicle dissection. Leibfried-Rutledge *et al.*

Table 2.8. Total number and proportion of good-quality oocytes recovered from bovine ovaries.

Authors	Total no. per ovary	Acceptable quality (%)
Aspiration		
Katska and Smorag (1984)	10.2	45.0
Iwasaki *et al.* (1987)	9.4	56.4
Suss *et al.* (1988a)	11.5	46.5
Berg and Brem (1989)	15.0	63.0
Preinberg *et al.* (1989)	16.0	79.8
Sato *et al.* (1990)	10.0	80.0
Vergos (1990)	14.2	46.1
Lonergan (1992)	9.7	31.3
Lu and Polge (1992)	12.2	–
Follicle dissection		
Katska (1984)	–	63.2
Vergos (1990)	16.3	68.2
Lonergan (1992)	17.2	60.7
Ovary slicing after aspiration		
Iwasaki *et al.* (1987)	13.0	29.7
Sato *et al.* (1990)	16.0	76.9
Takagi *et al.* (1992)	32.1	–
Ovary slicing		
Suss and Madison (1983)	20–30	–
Xu *et al.* (1992a)	55.5	70.3
Hamano and Kuwayama (1993)	31.6	–
Carolan *et al.* (1994)	44.2	–

(1985) recorded a 60% recovery rate, which gave an average yield of five to eight acceptable-quality oocytes per ovary pair. Elsewhere, Suss *et al.* (1988a) reported an average of 23 oocytes per ovary pair, with 47% (10.6 per ovary pair) of these showing an acceptable COC. In Dublin, several studies have shown that only 50–60% of oocytes may be recovered from punctured follicles by aspiration in contrast to a figure of >95% when follicle dissection is employed (Lonergan, 1990; Vergos, 1990).

Data dealing with the number of oocytes recovered per ovary, using the different methods described in the literature, are in Table 2.8.

When transvaginal, ultrasound-guided follicle aspiration is employed in live cattle, there may be a need to refine evacuation techniques so that the punctured follicle is flushed with medium after the initial recovery attempt (Kruip *et al.*, 1991); oocyte recovery presumably can fail because the immature egg remains firmly bound to follicular cells.

The advantage of follicle aspiration is in terms of speed of operation, which may be particularly important in commercial embryo production (see Fig. 2.8). In comparison with follicle dissection, it is possible to recover oocytes three times faster than by dissection. Nevertheless, it is also important to take note of quality considerations; there is no great merit in speed of operations in recovering oocytes unless it is associated with increased embryo yield.

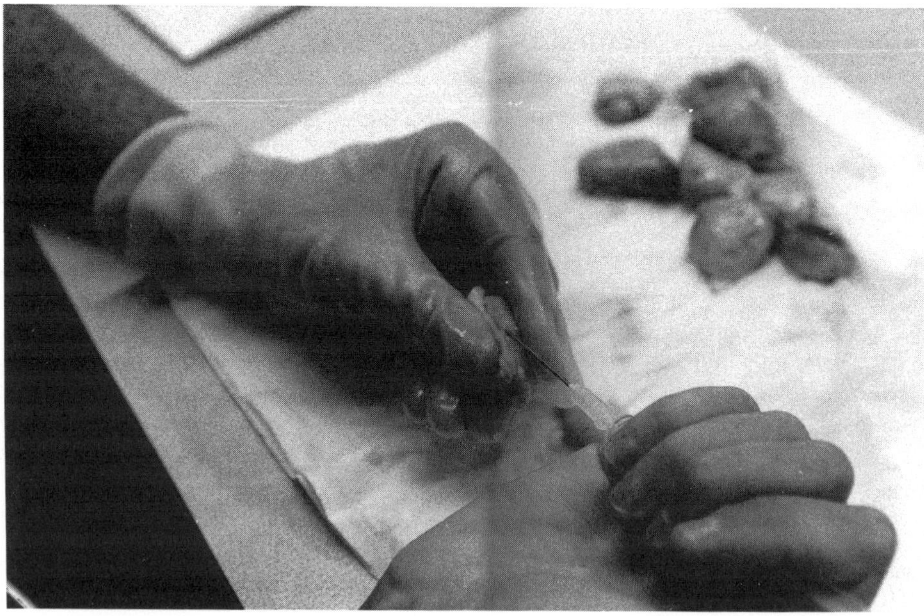

Fig. 2.8. Oocyte recovery by aspiration, the most widely used of the recovery procedures for cattle and other farm animals. In this instance, vesicular follicles (2–8 mm diameter) are being aspirated using a 2 ml syringe and an 18 G needle.

2.6.2 Dissecting the intact follicle

Dissection of the intact vesicular follicle (2–8 mm diameter) and its subsequent carefully controlled rupture were described by Staigmiller and Moor (1984a, b) as an important element in their IVM technique employed in sheep. This method was used in Dublin (see Fig. 2.9) as a means of recovering morphologically acceptable oocytes with the least disruption to the surrounding cumulus cells (Lu *et al.*, 1987a, b, 1988a). An advantage of follicle dissection, as the oocyte recovery method, lies in being able to identify non-atretic follicles. In sheep, Moor *et al.* (1978) described the morphological characteristics which permit such follicles to be identified. These include features such as a uniformly bright translucent appearance, evidence of extensive vascularization and possession of a regular stratum granulosum layer within the follicle. Atretic follicles, on the other hand, may be expected to show a dull, grey, opaque appearance and to be poorly vascularized.

Several reports deal with comparisons between aspiration and follicle dissection in cattle (Bottcher *et al.*, 1989; Lonergan, 1990; Vergos, 1990; Jiang, H.S. *et al.*, 1991) and in sheep (Wahid *et al.*, 1992). Such work generally supports the view that significantly greater yields of oocytes in the highest-quality grades may be obtained by follicle dissection rather than aspiration. Part of the explanation may lie in the fact that aspiration of the oocyte can result in

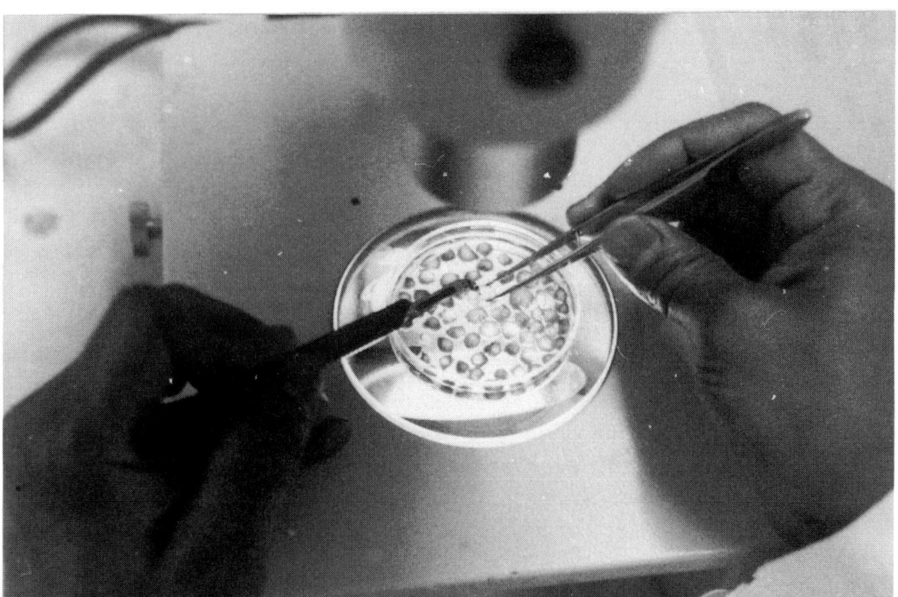

Fig. 2.9. Oocyte recovery by follicle dissection and rupture. This was the method originally used by Lu (1990) in recovering bovine oocytes for maturation and fertilization studies. The cumulus–oocyte complex is liberated by rupturing the intact follicle in a sterile plastic dish containing the dissection medium.

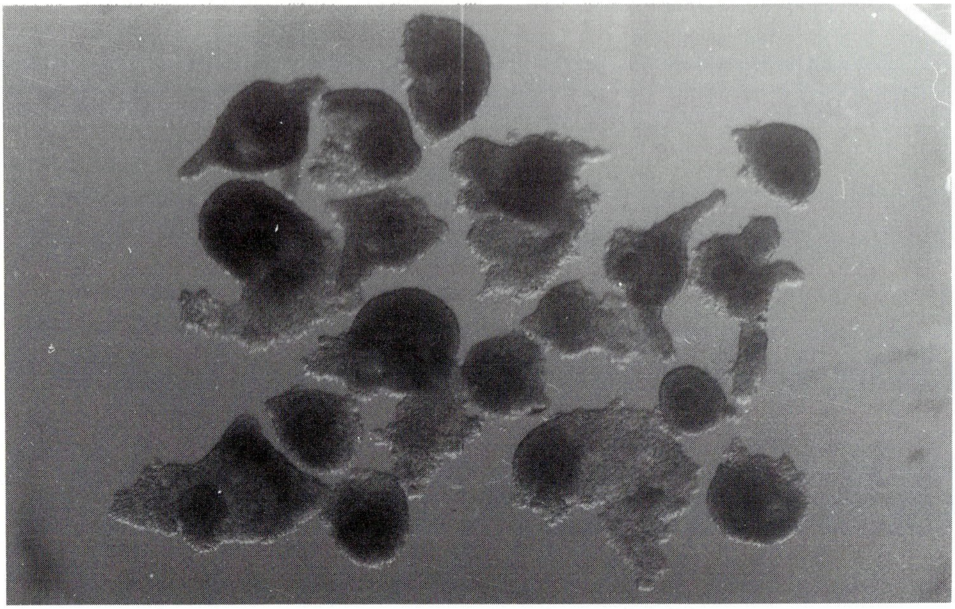

Fig. 2.10. Primary cattle cumulus–oocyte complexes after recovery showing good (5 layers) cumulus cell attachment.

greater disruption of surrounding cumulus cells. There is also the possibility that aspiration does not always succeed in retrieving the highest-quality oocytes; this may be due to the cumulus oophorus being firmly attached to the stratum granulosum.

Data for the large-scale use of aspiration in embryo production in Ireland are provided by Lu and Polge (1992), who recorded an average of 12.2 oocytes per ovary from some 0.75 million slaughterhouse ovaries.

2.6.3 *Ovary slicing techniques*

Various workers have reported on oocyte recovery techniques involving the slicing, cutting or dissection of the bovine ovary in an effort to obtain a good yield of immature eggs for maturation. The slicing techniques have been either directly applied to ovaries or applied after completion of aspiration. Data in Table 2.9 are from studies in sheep by Wahid (1993) in which a comparison was made between ovary slicing and aspiration.

One of the first reports of slicing in cattle was that of Suss and Madison (1983), who described a procedure by which they recovered 20–30 oocytes per ovary. More recently, slicing procedures have been employed in Dublin with both cattle (Carolan *et al.*, 1993, 1994) and sheep (see Table 2.9). In Canada, Xu *et al.* (1992a, b) reported oocyte yields averaging 55 per animal for the two animals involved. In Japan, Hamano and Kuwayama (1993) used a

Table 2.9. The effect of collection methods (aspiration versus slicing) on the recovery of sheep follicular oocytes* (from Wahid, 1993).

Method of collection	No. of ovaries	Classification of oocyte** A + B	C	D	E	Total
Aspiration	149	179	328	95	6	608
%		29.5[a]	54.0[a]	15.6[a]	1.0[a]	100
Mean oocyte/ovary		1.2	2.2	0.6	0.04	4.1
Slicing	85	571	449	196	79	1295
%		44.1[b]	34.7[b]	15.1[a]	6.1[b]	100
Mean oocyte/ovary		6.8	5.2	2.3	0.9	15.2
Overall	234	750	777	291	85	1903
%		39.4	40.8	15.3	4.5	100
Mean oocyte/ovary		3.2	3.3	1.2	0.3	8.1

*Oocytes were obtained from follicles of $\leqslant$ 6 mm in diameter.
**Oocyte classification: A, cumulus–oocyte complexes (COCs) with > 5 layers of cumulus cells; B, COCs with 3–5 layers of cumulus cells; C, COCs with 1–3 layers of cumulus cells; D, denuded oocytes; E, oocytes with expanded cumulus cells or atretic.
a, b – Values in the same column with different superscripts were significantly different at $P < 0.05$ (χ^2).

cutting technique to recover an average of 32 oocytes per ovary. This was a threefold increase on the numbers recovered by aspiration. In both sheep (Kay and Frylinck, 1992) and goats (Mogas *et al.*, 1992) substantially larger numbers of oocytes have been retrieved by way of surface slicing procedures than by aspiration.

A comparison between oocyte recovery by aspiration and by ovary slicing was made by Carolan *et al.* (1993, 1994). In this, there was a threefold increase in oocyte yield using the slicing method (see Table 2.10). In accepting oocytes for maturation, the normal criteria are based on the number and compactness of the surrounding cumulus cell layers (see Fig. 2.10). However, when the surface dissection method was employed, oocytes were measured to determine whether a number may not yet have attained full growth and be meiotically incompetent. However, no significant difference was evident between those obtained by ovary slicing and those obtained by routine aspiration (see Fig. 2.11).

The use of M-199 as the preferred holding medium for oocytes recovered by slicing, rather than phosphate-buffered saline (PBS), was also demonstrated by these workers. This would be in agreement with work reported by Keller *et al.* (1993), who found that short-term exposure of bovine oocytes to suboptimal Ca^{2+} and Mg^{2+} concentrations (such as in PBS) could be

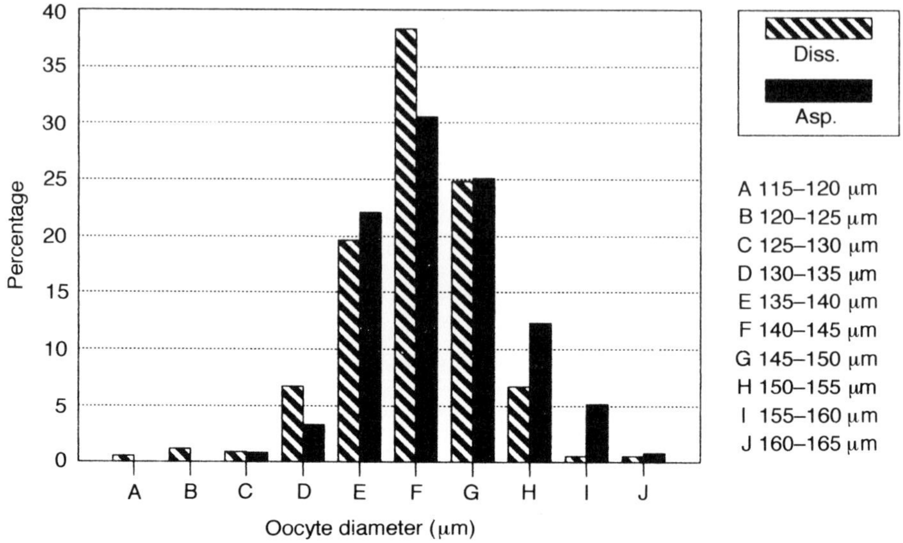

Fig. 2.11. Effect of recovery method on the diameters of good-quality oocytes obtained (from Carolan *et al.*, 1994).

Table 2.10. The number and type of oocytes obtained by aspiration or dissection of the surface of bovine ovaries (from Carolan *et al.*, 1994).

	Oocyte classification*						
Treatment	A	B	C	D	E	F	Total
Aspiration**	4.3 ± 0.9^a (30.9)	2.5 ± 0.6 $(18.0)^x$	1.7 ± 0.1^a (12.2)	3.1 ± 0.9 $(22.3)^x$	1.4 ± 0.5^a $(10.1)^x$	0.7 ± 0.2 (5.0)	13.9 ± 2.3^a (100)
Dissection***	17.1 ± 1.9^b (38.7)	3.1 ± 0.4 $(7.0)^y$	7.5 ± 1.4^b (17.0)	5.1 ± 1.5 $(11.5)^y$	9.8 ± 1.8^b $(22.2)^y$	1.7 ± 0.3 (3.8)	44.2 ± 4.4^b (100)

*Oocyte classification: A, COCs with an even cytoplasm and four or more cumulus cell layers; B, COCs with an even cytoplasm and less than four cumulus cell layers; C, COCs with a granular cytoplasm; D, denuded oocytes with an even cytoplasm; E, denuded oocytes with a granular cytoplasm; F, oocytes with an expanded cytoplasm.
**Three replicates (of 20 ovaries).
***Seven replicates (of three ovaries).
Results are given as the no. of oocytes in each class recovered per ovary ± SEM and the proportions of the total in that class (%).
a, b/x, y – Values in the same column with different superscripts differ significantly ($P < 0.05$) (t-test) (χ^2).

detrimental. The results from slicing ovaries from 12 cows and using almost all the the available oocytes in maturation is given in a report by Carolan *et al.* (1993). These data are in Table 2.11 and show that an average of 15.4 blastocysts were obtained per animal. Of interest in this is the fact that oocyte recovery by slicing rather than aspiration can result in marked increases in

Table 2.11. Developmental capacity of unselected bovine oocytes recovered from individual pairs by surface dissection (from Carolan *et al.*, 1993).

Cow no.	IVM	IVF	Cleavage (%) Rate	Cleavage (%) Index	Blastocyst yield (%) Day 7	Blastocyst yield (%) Day 8	Blastocyst yield (%) Day 9	Blastocysts per dam
1	60	57	78.9	93.3	20.0	24.4	33.3	15
2	38	29	82.8	83.3	16.7	20.8	20.8	5
3	70	65	75.4	83.7	22.4	30.6	30.6	15
4	65	53	71.7	86.8	28.9	34.2	36.8	14
5	45	42	71.4	80.0	33.3	40.0	53.3	16
6	119	102	76.5	87.2	24.4	34.6	39.7	31
7	61	60	65.0	51.3	2.6	7.7	12.8	5
8	60	50	68.0	55.8		2.9	05.8	2
9	150	127	64.6	75.6	19.5	23.2	26.8	22
10	78	74	73.0	59.3	14.8	25.9	31.5	17
11	132	118	74.6	76.1	22.7	29.5	36.4	32
12	31	28	75.0	52.4	19.0	28.6	52.3	11
Total	909	805	72.3	75.4	19.4	26.1	31.7	185
Average	75.8	67.1			9.4	12.6	15.4	15.4

blastocyst yield, after IVM/IVF/IVC. The slicing takes about three times longer per ovary than aspiration, but this is compensated for in the threefold increase in yield. In terms of work load, slicing can provide the same number of oocytes per unit of time as aspiration, but with much fewer ovaries. In those countries or circumstances where bovine ovaries are in short supply, the slicing method may have much to recommend it.

Slicing after aspiration

The slicing of ovaries after preliminary aspiration of follicles is dealt with in several reports (Iwasaki *et al.*, 1987; Arlotto *et al.*, 1990; Sato *et al.*, 1990; Takagi *et al.*, 1992). There would appear to be no merit, in terms of oocytes recovered or their quality, in combining the two procedures.

2.6.4 *Other mass harvesting procedures*

Over the years, various workers have attempted to recover large numbers of oocytes from ovaries. A procedure involving trypsin digestion of bovine ovarian tissue, to obtain oocytes for maturation, was described in the USA by Strickland *et al.* (1976); they recorded an average of 221 oocytes/ovary. The method involved trypsin digestion of ovarian tissue for 1 hour, with subsequent filtering of material through a 200 μm screen. A subsequent paper (Strickland *et al.*, 1977) reported that oocytes survived this extraction procedure. Mass harvesting procedures have been employed in other species. In pigs, Oikawa (1978) described a method in which pig ovaries were first minced and the ovarian tissue subsequently filtered; some 300–400 oocytes per ovary were recovered. Similar numbers were reported by Homa *et al.* (1986, 1988), who were able to separate a population of large, meiotically competent oocytes from smaller, meiotically incompetent oocytes. In rabbits, a mechanical and enzymatic approach to recovering oocytes is described by Fukunari *et al.* (1989).

2.6.5 Ovary storage: temperatures and time-limits

In the recovery of oocytes for maturation, the question of the time interval between animal slaughter and oocyte recovery from the ovaries is an important consideration. There is also the matter of defining temperature limits which should be observed in storing ovaries before they are processed in the laboratory. In general, workers dealing with cattle oocytes recover them within 1–2 h of animal slaughter; ovaries are usually stored at about 30°C. (Sato *et al.*, 1977; Shioya *et al.*, 1988a; Schleger *et al.*, 1990; Risopatron *et al.*, 1991; Sekine *et al.*, 1992). There has been the suggestion that oocytes falling below 30°C may not complete nuclear maturation (First and Parrish, 1987a). In The Netherlands, Kruip and Vernooy (1982) concluded that a collection and working temperature of 18–21°C was probably the most appropriate. In pigs, Rosenkranz (1993) reported results suggesting that a transport temperature of 20°C rather than above (30–37°C) improved the outcome of IVM.

Much work has been carried out in Ireland in defining time and temperature limits for ovary storage (Yang, N.S., *et al.*, 1990; Vergos, 1990; Lonergan, 1992); some of the relevant data are provided in Table 2.12.

It is evident that slaughterhouse ovaries can be stored for 8 h or even beyond, provided this is done at a reduced temperature (24°C). A paper by Solano *et al.* (1994) reports on preserving cattle ovaries at 4°C for 12–24 h and records that cleavage rate was little affected; however, the critical parameter is embryo yield rather than cleavage. In terms of operator comfort in the IVF laboratory, a temperature lower than 30–37°C is clearly desirable.

Table 2.12. Effect of ovary storage conditions (time and temperature) on the outcome of IVF and early embryo development (from Yang, N.S., *et al.*, 1990).

Storage hours	Cleavage rate (%)			% blastocysts		
	37°C	24°C	4°C	37°C	24°C	4°C
0	77.3	82.5	37.8	19.9	31.4	3.7
4	69.4	80.3	26.2	10.6	23.0	0
8	8.0	78.5	23.0	3.4	27.5	0.8
16	0	62.9	28.0	0	12.6	0
24	0	71.2	28.4	0	14.7	1.6

As a general principle, it can also be assumed that animal death is rapidly followed by degenerative changes in the ovaries and such changes may be much more serious at body temperatures (30–37°C) than when the gonads are cooled to a lower temperature (20–25°C).

Reports on ovary storage for other mammals include that of Johnston *et al.* (1989), who showed that cat ovaries could be held at 4°C for 2–6 h; oocytes were still capable of undergoing maturation and fertilization. In mice, Schroeder *et al.* (1991) recorded the birth of young after oocytes had been recovered from ovaries held for several hours after the death of the animals.

It is evident that there is reasonable latitude in the handling and storage of bovine ovaries, but the general principle of rapid processing should always be observed.

2.6.6 *Oocyte recovery from the live cow*

An increasing number of reports in recent years deal with oocyte recovery from the live animal. The earliest work in this field was by North American researchers. This was directed at the recovery of *in vivo*-matured (secondary) oocytes recovered by an endoscope inserted through the skin of the right paralumbar fossa (Lambert *et al.*, 1983, 1986; Sirard and Lambert, 1985, 1986; Sirard *et al.*, 1985a, b; Schellander *et al.*, 1989a, c). The donor animals were superovulated and endoscopy performed under local anaesthesia near the expected time of ovulation. It should be noted that secondary oocytes, surrounded by expanded cumulus cells, should be easier to retrieve than primary oocytes, which are likely to be well attached to follicular cells. The oocytes were fertilized *in vitro* (usually with sperm capacitated by high ionic strength medium) and resultant embryos cultured in the rabbit oviduct for 4–5 days before transfer to recipient cattle.

A method of aspirating follicles by the aid of ultrasound scanning was used by Callesen *et al.* (1987), the authors claiming the method had advantages over the laparoscopic procedures then in use. The ovaries were retracted by rectal manipulation and placed under the skin of the sacroischiadic region where the transducer was located; follicles were visualized and punctured and oocytes were recovered.

At Cornell, Ellington *et al.* (1990a) described a surgical procedure for obtaining cattle zygotes from Holstein heifers at 40–48 h after the start of oestrus. At Louisiana State University, Ryan, D.P., *et al.* (1990a) and Meintjes *et al.* (1993) collected oocytes from gonadotrophin-treated cattle during the first trimester of pregnancy; they suggested that this might be an alternative for producing embryos from valuable donors. The timing of laparoscopic aspiration of preovulatory follicles in superovulated cattle is dealt with by Laurincik *et al.* (1991b), the authors drawing attention to the difficulties in accurately monitoring follicle development.

In the USA, Fayrer-Hosken and Caudle (1991) reported a highly effective laparoscopic technique whereby they recovered oocytes from 79% of the follicles aspirated in superovulated cows. The authors conclude that, by allowing visualization and manipulation of the genital tract, the laparoscope can permit successful, rapid and minimally invasive treatment of several forms of cattle infertility. In France, Maurel (1991) demonstrated the value of a rapid LH assay in enabling precise determination of ovulation in the cow to be made. This may be useful in attempting recovery of the mature oocyte.

The repeated recovery of primary bovine oocytes, at relatively short time intervals, by transvaginal ultrasound-guided aspiration techniques has been an active research area, particularly in Canada (Stubbings *et al.*, 1988a, 1990a; Walton *et al.*, 1993), the USA (Gibbons *et al.*, 1994), Germany (Rath, 1993; Simon *et al.*, 1993; Wiebke *et al.*, 1993), UK (Scott *et al.*, 1994) and The Netherlands (Pieterse *et al.*, 1988a, b, 1991a, b, c; Kruip *et al.*, 1990, 1991, 1993; Vos *et al.*, 1990; Van der Schans *et al.*, 1991; Baltussen *et al.*, 1992; Boni *et al.*, 1992, 1993a, b; Roelofsen *et al.*, 1993).

Oocyte recovery by transvaginal laparoscopy

An alternative approach for transvaginal oocyte collection, involving laparoscopy, has been described by Reichenbach *et al.* (1993a, 1994) in Germany; these workers conclude that transvaginal laparoscopy is a useful alternative to the ultrasound-guided procedure.

Oocyte yields possible from live animals

In some early reports of ultrasound-guided oocyte recoveries from The Netherlands, it was calculated that such follicle aspiration methods could result in an annual yield of 130 oocytes per animal and an output of 30 embryos. Workers in that country suggest that IVM/IVF using oocytes collected twice weekly from the same live animal over a period of several months may be a useful alternative to the use of slaughterhouse ovaries in large-scale embryo production. In Germany, Rath (1993) confirms that repeated aspirations can be performed at least twice weekly for several months without any need for hormonal stimulation. Attempts have been made to increase oocyte recovery rate in twice-weekly aspiration schedules by the use of PMSG or vaccination against inhibin (Fry *et al.*, 1994). Gibbons *et al.* (1994) have confirmed previous reports showing that greater numbers of transferable embryos can be produced from twice- rather than once-weekly aspirations.

2.6.7 Oocyte maturation inhibitors

In the live cow, competent oocytes do not resume meiosis prior to the preovulatory LH surge, and only then in DF which are much larger than the 1–2 mm follicles at which such competence is first attained.

There may be occasions in practice when it would be useful to maintain meiotic arrest in the oocyte liberated from the follicle. If, for example, an IVF system was being employed in conjunction with aspiration of primary oocytes from cattle ovaries under field conditions, there may be a need to hold such oocytes at the GV stage until they reach the IVF laboratory. It has already been noted that oocytes have the unusual ability to undergo maturation spontaneously when released from follicles and cultured *in vitro* (Pincus and Enzmann, 1935). Follicle-enclosed oocytes, on the other hand, do not resume meiosis, *in vivo* or *in vitro*, in the absence of stimulation by gonadotrophin. It follows, therefore, that there is an inhibitory agent(s) within the vesicular follicle (>2 mm diameter) responsible for maintaining the oocyte in the GV state. When primary bovine oocytes are cocultured with part of the follicle wall, it has been observed that a majority of oocytes remain arrested (Loos *et al.*, 1992a; Richard and Sirard, 1993); the implication is that a strong inhibitory factor is secreted from the follicle wall.

Studies reported by Sirard and Coenen (1993a) evaluated the short-term (24 h) effect of the coculture of bovine follicle hemisections (including theca and granulosa cells) with COCs on the resumption of meiosis; it was found to be possible to maintain meiotic arrest for at least 24 h by such coculture. It was suggested that a factor may have been released from the follicular wall and that its presence in the culture medium inhibited meiosis.

Early studies by Chang in the 1950s suggested that FF may contain an oocyte maturation inhibitor (OMI). It is known that pig FF (pFF) can inhibit the IVM of pig COCs. Sreenan (1969) recorded limited nuclear maturation in bovine oocytes cultured in bovine FF (bFF). Sirard and First (1988) reported similar findings, and also showed that bFF inhibited maturation of hamster oocytes. Results reported by Petr *et al.* (1989) show that porcine cumulus cells release a factor capable of inhibiting GVBD in both pig and cattle oocytes. Intact sheets of porcine mural granulosa cells are able to prevent the resumption of meiosis in bovine oocytes (Kalous *et al.*, 1993); similarly, intact sheets of mural granulosa cells will maintain meiotic suppression of horse oocytes (Hinrichs and Schmidt, 1993).

The concept that meiotic arrest at the GV stage in the bovine oocyte is sustained by a heat-stable polypeptide factor(s) in bFF was supported in work by Ayoub and Hunter (1990). This and other evidence does suggest that an OMI exists that is not species specific, is produced by somatic cells of the follicle, is released into the FF and is responsible for the maintenance of meiotic arrest in the oocyte (Motlik *et al.*, 1991).

The results of Sirard and Bilodeau (1990a, b) showed that live granulosa cells are capable of maintaining bovine oocytes in the GV stage for at least 24 h. However, according to these workers, granulosa cells in limited numbers do not inhibit resumption of meiosis, especially if the IVM contains OCS or added gonadotrophins. The role of bovine granulosa cells is thought to be one of transmitting inhibitory signals until an acute LH stimulation changes their signal to neutral or to stimulatory for the cumulus.

When cumulus cells are added to oocytes, in the early period after their liberation from vesicular follicles (earlier than about 45 min), they are capable of inhibiting maturation (Sirard and First, 1988; Sirard and Lambert, 1988). Elsewhere, workers have noted an inhibitory effect of granulosa cells on oocyte maturation (Alm *et al.*, 1990; Konig *et al.*, 1990). In Dublin, Vergos (1990) examined the effect of delaying the addition of cumulus/granulosa cells to the IVM medium until at least 1 hour beyond the time of oocyte liberation from the follicle (by which time oocytes would have lost the ability to respond

Table 2.13. The effect of granulosa cell origin and time of addition to the maturation medium on the *in vitro* maturation of primary bovine oocytes[a] (from Vergos, 1990).

Follicle size (mm)	Treatment	Oocytes: In culture	Oocytes: With polar body	Oocytes: Cumulus cell expansion (%)
Control[b]		83	44 (53.0)[c]	0
2–5	A[d]	91	74 (81.3)[e]	100
	B[f]	88	71 (80.6)[e]	100
8–10	A	84	68 (80.9)[e]	100
	B	97	76 (78.3)[e]	100
Overall		443	333 (75.2)	

[a]Data from five replicates.
[b]No granulosa addition.
[c]$P < 0.01$.
[d]Routine addition.
[e]$P > 0.05$.
[f]Granulosa cell addition with 1 h delay.

to inhibitory factors). In the normal IVM routine, in which additional cumulus/granulosa cells are employed, there was no evidence of an inhibitory effect on oocyte maturation (see Table 2.13).

Motlik and Kubelka (1990) draw attention to the fact that recent studies have revealed striking similarities between cell cycle factors controlling mitosis and meiosis. If mitotic and meiotic cycles are regulated by identical factors, which seems likely, the authors suggest that interphase-specific substances from granulosa cells might act as a negative signal, passing from granulosa cells into oocytes and preventing the activation of the molecules controlling the cell cycle. After the preovulatory gonadotrophin surge, when LH is known to bind to its granulosa cell receptors, there is probably an acute response of granulosa cells to the gonadotrophin, resulting in their final differentiation. Such activity may release the oocyte from inhibition, as well as acting upon the granulosa cells.

There are those who believe that the resumption of meiosis in the bovine oocyte *in vivo* is the result of a stimulatory factor in the bFF becoming active, in the follicle destined for ovulation, rather than due to removal of an inhibitory factor (Romero *et al.*, 1990); these workers found that bFF obtained after the preovulatory LH surge in the cow stimulated a higher percentage of oocytes to mature than bFF collected before LH release.

Biochemical factors influencing meiotic arrest in cattle and pig oocytes have been dealt with in several papers (Liehman *et al.*, 1986a, b; Gotze *et al.*, 1990; Sanbuissho *et al.*, 1990a; Sirard, 1990; Aktas *et al.*, 1991a, b; Eyestone and Boer, 1991; Homa, 1991; Jung *et al.*, 1991; Shim *et al.*, 1992; Sirard *et al.*, 1992).

2.7 Indicators of Follicle and Oocyte Quality

2.7.1 Atretic and non-atretic follicles

The ability to identify good-quality follicles and oocytes is of considerable importance. There are a number of rules to be observed in the selection process. A scheme by which ruminant ovarian follicles can be classified was described by Moor *et al.* (1978); this scheme, with minor modifications, has been employed by other workers (Kruip and Dieleman, 1982; McNatty *et al.*, 1984). The types of criteria used are shown in Table 2.14.

The proportion of vesicular follicles in the cow ovary recorded as atretic has differed in various published reports (Choudary *et al.*, 1968; Marion *et al.*, 1968; Dufour and Roy, 1985; Maurasse *et al.*, 1985; Lussier *et al.*, 1987). In general terms, the usual assumption is that 50% or so vesicular follicles may be atretic. One of the more recent papers, however, gives a figure of 233 vesicular follicles for the cow, with almost 90% of these being non-atretic (Le Van Ty *et al.*, 1989).

The morphological features of non-atretic follicles in sheep were described by Moor *et al.* (1978). Non-atretic follicles have a bright translucent appearance, whereas atretic follicles can be expected to show a dull, grey, opaque appearance. The macroscopic appearance of bovine follicles, dissected free of ovarian tissue and transilluminated, was found to be highly correlated with histological indications of atresia in cattle (Kruip and Dieleman, 1982, 1985). The presence of thecal cells in the membrana granulosa probably reflects an early stage of atresia in the bovine follicle (Van Den Hurk *et al.*, 1991). The relationship between morphological features of the bovine follicle and the developmental capacity of the oocyte retrieved is dealt with in a paper by Wurth and Kruip (1992).

According to Grimes and Ireland (1985, 1987), the appearance of the surface of large vesicular follicles (≥ 5 mm diameter) can probably be used as some indicator of oocyte quality. Subsequently, Grimes *et al.* (1987) showed that opaque vesicular follicles are probably the result of the growth of stromal tissue over the surface of the follicle as they undergo atresia. They demonstrated that oocytes recovered from opaque follicles are more likely to be strongly atretic (42%) than those from clear (21%) or intermediate (23%) follicles. In Russia, Smirnova and Galieva (1989) proposed a classification scheme for follicles which could be employed in selecting oocytes for IVM. Results from these and other studies indicate that macroscopic criteria are likely to provide a reasonable means of identifying non-atretic follicles in terms of morphology and functional ability.

Hormonal and biochemical factors may be used in identifying atretic and non-atretic follicles. In Canada, Stubbings *et al.* (1990b) assessed various methods of selecting oocytes on the basis of follicular steroid levels; abnormal follicular steroid concentrations were associated with abnormal oocytes. In

Table 2.14. Criteria employed in the classification of bovine follicles.

Moor *et al.* (1978)	Kruip and Dieleman (1982)	McNatty *et al.* (1984)
Non-atretic Uniformly bright, translucent appearance, extensive vascularization, regular granulosa layer	*Non-atretic* Uniformly bright, appearance, very firm vascularization, regular granulosa layer, no free-floating particles in the follicular fluid	*Non-atretic* Visible thecal capillaries when viewed at ×10 magnification under a dissecting microscope, no debris in follicular fluid, 75% of expected maximum number of granulosa cells for a given follicle size, healthy oocyte, pink to red theca interna
Intermediate Some loss of translucency, slightly greyish appearance	*Light-atretic* Some loss of translucency, slightly greyish appearance, some very small free-floating particles in the follicle cavity	*Intermediate* Same as for non-atretic except theca interna is white
Atretic Dull grey opaque appearance, poorly vascularized, showing empty blood vessels, detachment of membrana granulosa, free bodies in the antral cavity	*Atretic* Dull greyish appearance blood vessels either empty or irregularly filled with clotted blood, partial detachment of membrana granulosa and mass of free-floating globules in antral cavity	*Atretic* All the follicles other than those listed above
	Heavy-atretic Dark, often spotted appearance and very dark cumulus mass	

Germany, Mukhopadhyay *et al.* (1991) reported that a high bFF prorenin level could be a useful indicator of follicular atresia in cattle ovaries. In Canada, Hazeleger *et al.* (1993) examined the use of bFF parameters to select developmentally competent bovine oocytes; they concluded that morphological criteria and bFF progesterone concentration may be useful in such selection.

2.7.2 *Oocyte morphology: classification schemes*

Selection of bovine oocytes for IVM on the basis of visual assessment of morphological features was first examined by Leibfried and First (1979). Since that time, many reports have proposed classification schemes, based on the compactness and quantity of surrounding follicular cells and other morphological features (Suss and Madison, 1983; Katska and Smorag, 1984; Xu *et al.*, 1986a, b; Shioya *et al.*, 1988b; Tan *et al.*, 1988; Jinghe *et al.*, 1989; Loos *et al.*, 1989, 1991a; Schellander *et al.*, 1989d, e; Smirnova and Galieva, 1989; Younis *et al.*, 1989a; Kim and Park, 1990; Thonabulsombat *et al.*, 1990; Yang and Lu, 1990; Dominko and First, 1991; Oikawa *et al.*, 1991; Hazeleger and Stubbings, 1992; Madison *et al.*, 1992). In pigs, evidence has been presented by Meinecke and Meinecke-Tillman (1993) showing that the signal(s) triggering nuclear progression depends on events in the cumulus layer of the COC, rather than within the oocyte itself. For many reasons, it is now clear that

Table 2.15. Oocyte quality as determined by morphological examination.

Category	COC
Criteria as employed by Loos et al. *(1989)*	
1	Compact multilayered cumulus investment; homogeneous ooplasm; total COC light and transparent
2	Compact multilayered cumulus investment; homogeneous ooplasm but with a coarse appearance and a darker zone at the periphery of the oocyte; total COC slightly darker and less transparent
3	Less compact cumulus investment; ooplasm irregular with dark clusters; total COC darker than 1 or 2 above
4	Expanded cumulus investment; cumulus cells scattered in dark clumps in a jelly matrix; ooplasm irregular with dark clusters; total COC dark and irregular
Criteria as employed by Younis et al. *(1989b)*	
Selected	Homogeneous-appearing ooplasm and compact cumulus cells tightly adherent to the zona pellucida
Unselected	Other categories with incomplete cumulus complements and heterogeneous ooplasm
Criteria as employed by Loos et al. *(1991a)*	
1	Compact multilayered cumulus investment; homogeneous ooplasm; total COC light and transparent
2	As 1 but with coarse appearance and a darker zone at the periphery of oocyte
3	Total COC darker than 1 or 2; less compact cumulus; ooplasm irregular with dark clusters
4	Expanded cumulus investment

Table 2.16. Effect of oocyte type used in IVM on subsequent fertilization rate and embryo yield (from Lonergan, 1992).

	Oocyte type				
	Denuded	Expanded	2–3 layers	4–5 layers	Compact
CR	61.9%[a] (26/42)	65.8%[a] (25/38)	77.8%[a] (63/81)	90.7%[b] (117/129)	90.2%[b] (111/123)
CI	57.7%[a] (15/26)	65.0%[ab] (16/25)	30.2% [c] (19/63)	75.2%[b] (88/117)	80.2%[b] (89/111)
EY	Nil (0/26)	Nil (0/25)	9.5%[a] (6/63)	17.1%[a] (20/117)	31.2%[b] (34/109)

a, b, c – Values in the same row with different superscripts differ significantly ($P < 0.05$). Data from three replicates.
CR, cleavage rate; CI, cleavage index; EY, embryo yield.

an oocyte classification scheme with a good predictive value is of considerable importance in an embryo production system. Some of the schemes suggested by authors for COC classification are shown in Table 2.15.

In Dublin, Lonergan *et al.* (1992) established a clear relationship between oocyte morphology and embryo yield after IVM/IVF (see Table 2.16). Oocytes can be broadly categorized into those which are acceptable (see Fig. 2.12) and those that are not (see Fig. 2.13). Loos *et al.* (1989) distinguished four categories of bovine COCs as used in IVM studies based on the compactness and transparency of the cumulus investment and homogeneity and transparency of the ooplasm. Subsequently, Loos *et al.* (1992a) described the morphology of these classified oocytes after IVM and compared it with that of oocytes collected after *in vivo* maturation. The authors concluded that there can be considerable morphological variability among oocytes capable of normal development. After *in vivo* maturation, secondary oocytes formed a homogeneous group in their morphology whereas *in vitro*-matured oocytes formed a heterogeneous group.

Laurincik *et al.* (1992) separated oocytes into two groups: (i) those in which a dark rim of corona cells were visible around the ZP; and (ii) those in which the corona displayed the same density as the cumulus cell mass. The authors concluded that COCs with a dark rim of corona cells mature at a higher rate and maintain unexpanded cumulus cell characteristics and efficient oocyte–corona coupling longer than complexes with even density of the cumulus mass. According to this, the presence of a dark rim of corona cells may be used as a criterion for selection of oocytes for IVM.

Oocyte cytoplasm granulation

Although it is often stated that bovine COCs showing heterogeneous ooplasm should not be selected for IVM (Leibfried and First, 1979; Greve and Madison, 1991; Brackett and Zuelke, 1993), results from work in Dublin suggests that

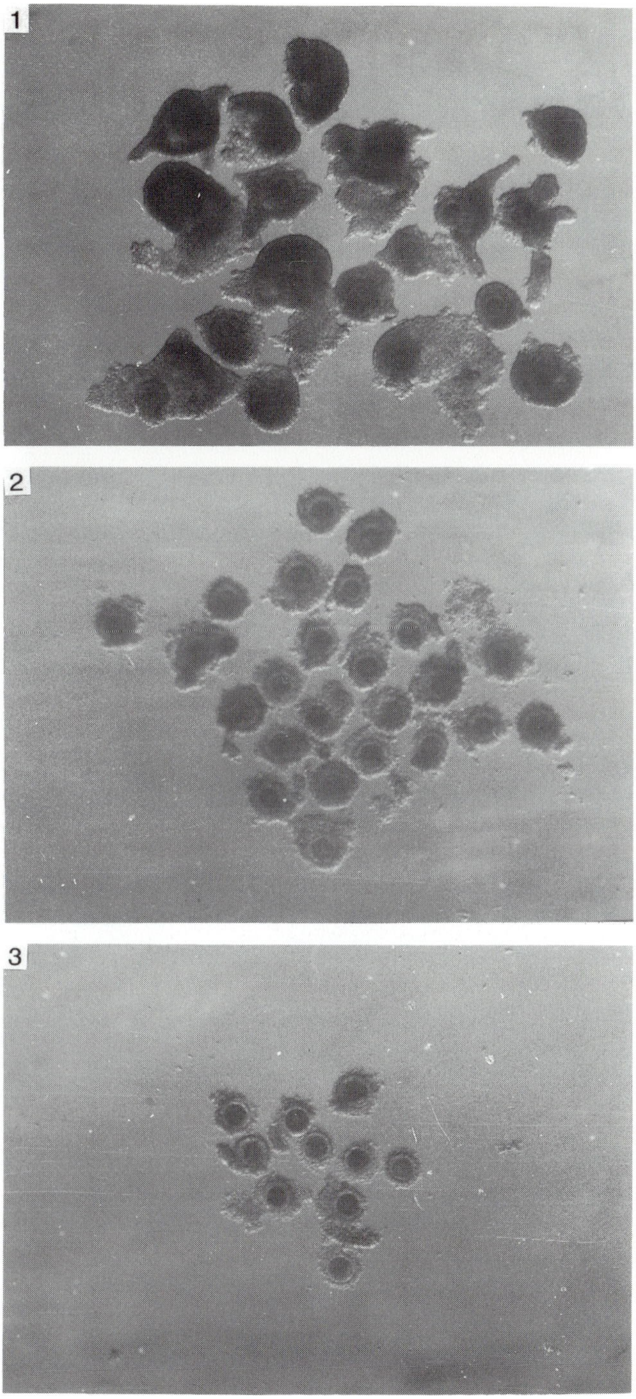

Fig. 2.12. Acceptable-quality bovine primary oocytes. (1) Oocytes showing many tight layers of cumulus cells; (2) oocytes showing three to four cumulus cell layers; (3) oocytes showing two to three cell layers.

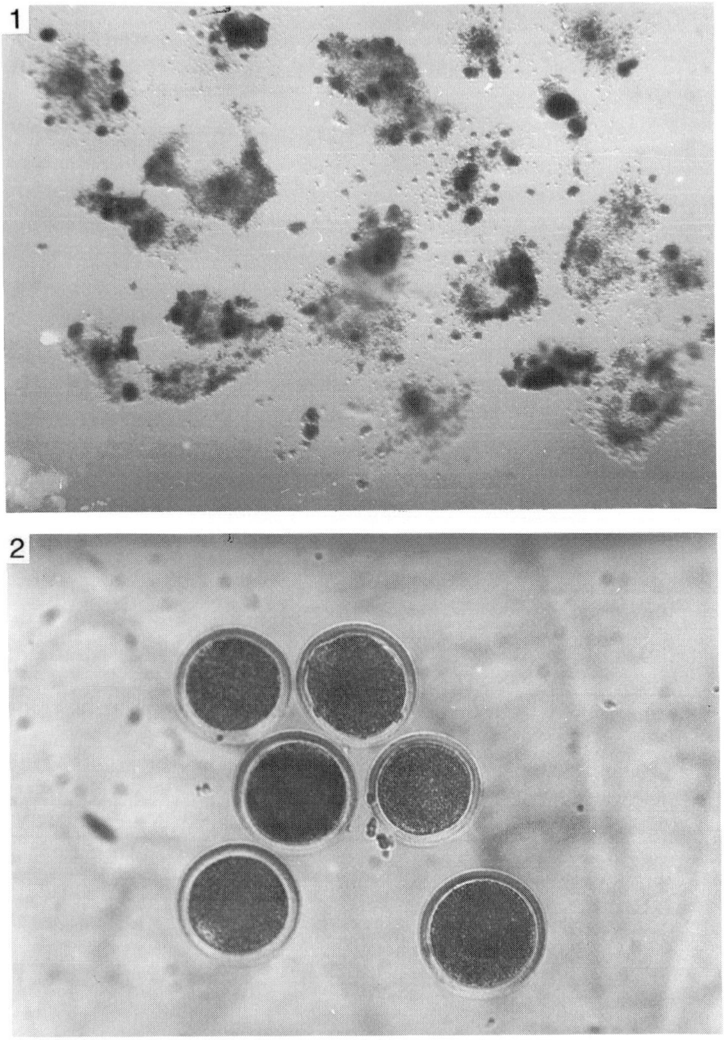

Fig. 2.13. Unacceptable-quality bovine primary oocytes. (1) Bovine oocytes with expanded cumulus cells, with the cells scattered in dark clumps in a jelly-like matrix; (2) denuded oocytes.

there may be no clear basis for COCs with granulated cytoplasm being excluded from IVM (Monaghan, 1993).

In the USA, Arlotto *et al.* (1992) reported a correlation between oocyte size and time taken to mature artificially. There was also evidence that larger than average oocytes resulted in higher embryo yields.

Ultrastructural features

Various authors have examined the ultrastructure of oocytes to determine whether this may provide guidance on their quality. Fleming and Saacke

(1972) were among the first to describe the ultrastructural characteristics of bovine oocytes recovered from large vesicular follicles. Loos and Kruip (1988) showed that the cumulus cell process endings in the bovine oocyte differed between atretic and non-atretic follicles, but it was not clear whether this was associated with a different metabolic coupling or not. The intimate association between cumulus cells and the oocyte was described by Tan *et al.* (1988) and Jinghe *et al.* (1989) in electron microscopy studies. Stains specific for DNA were employed by Schellander *et al.* (1990a) to reveal nuclear structures in cattle oocytes. In Colorado, Van Blerkom *et al.* (1990) carried out an analysis of the cytoplasmic distribution of mitochondria, lipid droplets and vesicles in more than 5000 GV-stage bovine oocytes. Among other things, they found evidence that acquisition of developmental competence for IVF and the ability of the embryo to develop subsequently were related to the organization of the cytoplasm at the GV stage.

2.7.3 Follicle size and oocyte quality

The size range of vesicular follicles used for oocyte recovery in cattle IVM/IVF/IVC studies is generally 2–6 mm. The size at which bovine vesicular follicles can be expected to contain meiotically competent oocytes is now known with some certainty (Motlik and Fulka, 1986; Lonergan, 1990; Tan and Lu, 1990). Elsewhere, other investigators showed that oocytes derived from 1–2 mm follicles resulted in significantly reduced maturation and fertilization rates when compared with larger follicles (Fuhrer *et al.*, 1989; Schellander *et al.*, 1989b; Lucas-Hahn and Niemann, 1991b; Pavlok *et al.*, 1992, 1993).

The acquisition of meiotic competence in oocytes varies with species. In rodents, oocytes have reached full size and have attained meiotic competence by the time follicles develop an antrum; in cattle the oocytes are still growing when follicles reach the antral stage. According to Thibault *et al.* (1987), the growing oocyte becomes progressively capable of resuming meiosis; meiotic competence appears when the oocyte is about 80% the size of the fully grown oocyte.

In Ireland, there is ample evidence showing a clear relationship between follicle size and oocyte quality (Barnes *et al.*, 1990; Tan and Lu, 1990; Lonergan *et al.*, 1991b; McCaffrey *et al.*, 1992a, b; Monaghan, 1993). Oocytes recovered from follicles less than 2 mm diameter have shown significantly lower cleavage rates and embryo yields than follicles above that size. Oocytes from follicles larger than 6 mm have given significantly higher embryo yields than those taken from the normal 2–6 mm range. Data in Table 2.17 are taken from work by Lonergan (1992). Other work in Dublin by Monaghan (1993) has recorded almost a threefold increase in blastocyst yield from follicles > 8 mm in comparison with follicles in the normal range.

The indications are that large follicles (> 6 mm) provide the oocyte with a microenvironment which improves its quality. Some indication of qualitative change is evident in Table 2.18, taken from Lonergan (1992), which shows that a significantly higher percentage of oocytes taken from > 6 mm follicles were in the highest-quality grade.

Table 2.17. Effect of follicle size on embryo yield following IVM/IVF/IVC (from Lonergan, 1992).

Follicle size (mm)	Cleavage		Blastocyst yield
	Rate	Index	
2–6	90.8% (99/109)	86.9% (86/99)	34.3%[a] (34/99)
>6	92.9% (91/98)	82.4% (75/91)	65.9%[b] (60/91)
χ^2	NS	NS	$P < 0.001$

a, b – Values in the same column with different superscripts differ significantly ($P < 0.05$).

Table 2.18. Effect of follicle size on the type of oocyte obtained for IVM (from Lonergan, 1992).

Follicle size (mm)	Oocytes recovered	Denuded	Expanded	2–3 layers	3–4 layers	Many layers
2–6	87.7%[a] (186/212)	3.8%[a] (7/186)	14.5%[a] (27/186)	12.9%[a] (24/186)	22.0%[a] (41/186)	46.8%[a] (87/186)
>6	77.0%[a] (104/135)	1.9%[a] (2/104)	11.6%[a] (12/104)	6.7%[a] (7/104)	9.6%[b] (10/104)	70.2%[b] (73/104)
χ^2	$P < 0.01$	NS	NS	NS	$P < 0.01$	$P < 0.001$

a, b – Values in the same column with different superscripts differ significantly ($P < 0.05$).

Fine-structure analysis of the nuclear and nucleolar morphology of early IVF-derived embryos by Pavlok *et al.* (1992) showed that oocytes for IVM should be selected from follicles of a diameter above 2 mm.

It has been suggested that no single IVM culture system is likely to provide ideal conditions for oocytes taken from every category of bovine follicle (Kruip *et al.*, 1979). In the live cow, vesicular follicles are growing over a period of several days from the small or medium size (2–6 mm) to the preovulatory size (10–20 mm); in IVM, the events in nuclear and cytoplasmic maturation are compressed into a mere 24 h. There may be need to take account of follicle size, as a source of variation in IVF outcome, on the basis of oocytes having been exposed to different cellular and endocrine influences. The intrafollicular environment comprises an ever-changing environment of steroid and peptide hormones, growth factors, cytokines and other molecules, each of which may act alone or in concert with one or more of the other factors in influencing oocyte and follicle development.

Lonergan (1992) was able to relate morphological features of the oocyte to follicle size (see Table 2.18); significantly more high-grade oocytes were obtained from large follicles than from smaller follicles (70% vs. 47%). It is known that in porcine follicles rapidly growing *in vivo*, both thecal and granulosa cell layers increase in thickness and complexity, with a prominent capillary bed evolving immediately outside the basement membrane (King, G., *et al.*, 1991). It is only to be expected that the highest-grade oocytes come from the larger follicles, but they must be non-atretic as well as large.

In Denmark, Assey *et al.* (1992, 1994) compared the follicular environment and structure of oocytes originating from the DF with those of the two largest subordinate follicles examined a few days after ovulation. The DF was characterized by an oestrogen-dominated environment, healthy cumulus cells junctionally coupled to the oocyte, and other morphological features. Subordinate follicles, on the other hand, were characterized by a progesterone-dominated environment, usually with degenerate cumulus cells, meiotic activation and other features of atresia. In human IVF studies, androstenedione and progesterone concentrations in preovulatory FF have been measured by T. Kobayashi *et al.* (1991); follicles yielding oocytes that cleaved after IVF showed a shift in steroidogenesis from oestrogen to progesterone dominance and declining aromatase activity.

Elsewhere, Sirard *et al.* (1992) have drawn attention to the fact that only about one third of bovine oocytes selected for IVM/IVF/IVC result in blastocysts; this, they take as indicative of poor quality in most oocytes selected for maturation. The authors note that bovine oocytes are usually drawn from follicles which may be several days away from possible ovulation. Dramatic changes in oocyte nuclei, especially in nucleoli, are known to occur as the bovine follicle grows from 1 to 20 mm. Such changes may have a crucial effect on the developmental potential of the oocytes. It is known that a very stable form of RNA accumulates in the oocyte and that it is translated during maturation, fertilization and early embryonic development; such RNA accumulation may be influenced by the nature of follicle growth.

In Canada, Blondin (1993) attempted to determine the developmental capacity of individual bovine oocytes in relation to the diameter and level of atresia of follicles; he found that increasing follicle diameter had a greater influence than either oocyte morphology or atretic status. In cloning studies, Barnes *et al.* (1993a) records that oocytes derived from 1–5 and 6–8 mm follicles result in 21% and 42% development to the morula/blastocyst stage, respectively, after nuclear transfer.

In Denmark, oocytes from DFs collected in the final 2 days before the LH surge in the cow present features which indicate that the primary oocyte only then achieves the competence to undergo final maturation (Hyttel *et al.*, 1993). The implication here is that bovine oocytes recovered from follicles other than the preovulatory ones have yet to undergo certain prematurational events, the nature of which is currently unknown. For such reasons, it may be necessary to think in terms of exposing primary oocytes to an appropriate pre-IVM treatment to bring them to the level of developmental competence normally found in oocytes within preovulatory follicles. Barnes *et al.* (1993a) suggest that future IVM protocols that inhibit the resumption of meiosis for 2–3 days as well as altering gonadotrophin and steroid parameters may be the way forward with oocytes from the smaller follicles (1–5 mm). Elements that may be useful in examining an appropriate pre-IVM treatment for oocytes are illustrated in Fig. 2.14.

It has been observed in mice that progression of the earliest stages of embryo development depends upon resources produced during oocyte development (Eppig, 1992; Eppig *et al.*, 1992a). Evidence supports the view that mouse oocytes sequentially acquire the capacity to develop to more advanced stages of early embryonic development and that this acquisition is correlated with advancing follicular development. According to this, mouse oocytes in preantral follicles are not competent to undergo GVBD but this competence is acquired shortly after antrum formation. Oocytes from early antral follicles that undergo GVBD, however, do not complete the first meiotic division but become arrested at metaphase I. As murine follicles enlarge, an increasing proportion of oocytes isolated from them are capable of completing development to metaphase II, where meiosis is arrested until fertilization.

In mice, it is known that metaphase II oocytes can be fertilized *in vitro* and will develop to the two-cell stage, regardless of the stage of antral follicle development from which the oocytes are isolated. However, the ability of two-cell embryos to develop to the blastocyst stage is clearly restricted when the oocytes are isolated from the earlier stages of antral follicle development. It would appear that murine oocytes from early antral follicles are embryo-

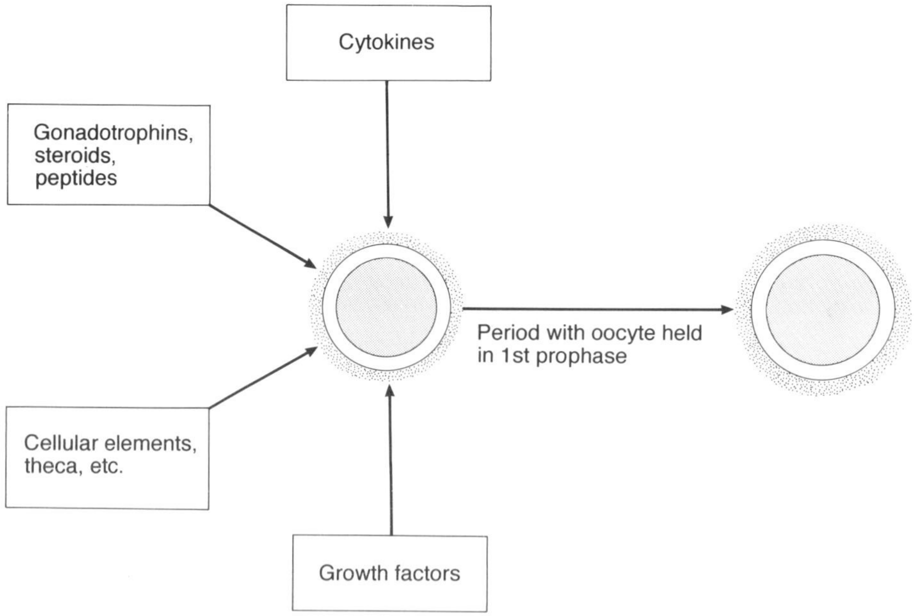

Fig. 2.14. A pre-IVM oocyte treatment for improving oocyte quality.

restricted because they either lack, or are unable to produce, maternal factors essential for appropriate activation of the zygotic genome. Translated into cattle terms, it may be speculated that a proportion of oocytes are unable to progress beyond an early stage of development because they lack similar maternal factors. A pre-IVM treatment to provide these oocytes with the maternal factors required for early embryonic development may be a possibility worth exploring.

2.7.4 Oocyte yield per ovary: what seems possible?

Yet to be fully explored and exploited are the large numbers of oocytes that are present in preantral and small (<1 mm diameter) vesicular follicles in the cow. At birth, the bovine ovary contains many thousands of primordial follicles. Each primordial follicle has the potential to mature, secrete oestrogen and ovulate, but only a small fraction of them will do so. The only primary oocytes currently acceptable for use in IVM are those located in the vesicular follicles that are 2 mm or larger in diameter. The challenge is to develop methods whereby follicles can be grown *in vitro* from the primary stage through to the preovulatory stage. There would be obvious attractions in developing methods whereby bovine oocytes present in early vesicular follicles (<1 mm diameter) could be grown to full size and to meiotic competence *in vitro*.

In rats, a report by Gandolfi *et al.* (1988a) dealt with a culture method suitable for growing isolated ovarian follicles from the preantral to the vesicular stage under controlled conditions. An oocyte culture system in mice, using preantral and antral follicles, was described by Eppig *et al.* (1988); the method involved putting the immature oocytes through a 12-day culture period, in which they were maintained in meiotic arrest. In Canada, Daniel *et al.* (1989) developed a culture system which supported the growth and maturation of oocytes from immature rats; they demonstrated that follicle-enclosed, midgrowth rat oocytes (40–45 μm diameter) could be grown to full size (70–80 μm diameter) in culture, with 60–70% acquiring meiotic competence. There was evidence that such oocytes could be matured and fertilized.

Further reports in mice (Eppig and Schroeder, 1989; Eppig *et al.*, 1990; Eppig, 1991a) demonstrated for the first time that oocytes with full developmental capacity may be produced *in vitro* from oocytes isolated before they have completed their growth phase or acquired meiotic competence. In Japan, Hirao *et al.* (1990) cultured oocyte–granulosa cell complexes, isolated from 10-day-old mice, in the presence of a chemical inhibitor of meiosis; the oocytes grew and when transferred to inhibitor-free medium resumed meiosis. A report by Qvist *et al.* (1990) showed that primary mouse follicles could be grown *in vitro* to the preovulatory stage.

The growth and acquisition of meiotic competence by oocytes from fresh and frozen–thawed primary follicles collected from 10-day-old mice was studied by Carroll *et al.* (1991b); follicles were cultured in collagen gels for 12 days, during which time mean oocyte diameter increased from 48 to 60 μm. The function of the intact follicular unit in mice cultured *in vitro* was also studied by Hartshorne and Barlow (1992). A system of individual follicle culture was used by Nayudu *et al.* (1992) in mice and marmoset monkeys to study follicular development and oocyte function. Preantral mice follicles were cultured by Nayudu and Osborn (1992), who showed that the follicle is a developmental unit, not requiring a blood vascular or neural system for growth and development from preantral to periovulatory stage.

According to these various groups, continued improvements in such culture systems are likely to come from an increased understanding of the development and function of granulosa cells in preantral and early antral follicles, oocyte–granulosa cell interactions and oocyte metabolism. Endocrine factors involved in the regulation of growth in preantral follicles are still unclear.

Some workers have attempted to study aspects of carbohydrate metabolism in the developing mouse follicle, using culture systems which support development from primary to preovulatory stages (Boland *et al.*, 1991, 1993). Culture conditions necessary for bovine preantral follicle viability are dealt with by Hulshof *et al.* (1993); they found that the addition of pyruvate to their culture medium (minimal essential medium, MEM) was essential for follicle viability.

Methods for the mechanical and enzymatic isolation of bovine preantral follicles have been described by various workers (Figueiredo *et al.*, 1992, 1993; Taha and Schellander, 1992; Nuttinck *et al.*, 1993). Procedures for the isolation of preantral bovine follicles, as employed by Figueiredo *et al.* (1993), are shown in diagrammatic form in Fig. 2.15.

There is need to elucidate the factors responsible for development of such follicles through to the vesicular stage. The isolation and long-term culture of human preantral follicles have been reported by Roy and Treacy (1993), using modifications of a procedure previously employed successfully in hamsters (Roy and Greenwald, 1989). The authors showed that, contrary to the view that preantral follicles do not require FSH support for growth, the development of human preantral follicles depends very much on the gonadotrophin. They also reported the differentiation of preantral into antral follicles over a 6-day period in their agar-based three-dimensional culture system.

In the possible utilization of oocytes in the thousands of follicles populating the bovine ovary, there is the development of culture systems for bringing the oocyte to full size and to meiotic competence *in vitro*. There may be another approach towards utilizing the oocytes, which does not involve growing the follicle and its oocyte. It may be possible simply to make use of the nucleus of the oocyte.

In Italy, Barboni *et al.* (1992) have demonstrated that when growing bovine oocytes (40–60 μm diameter) are fused (electrofusion) with fully grown enucleated oocytes, the reconstituted oocytes are capable of undergoing nuclear maturation and of sustaining male pronucleus formation after IVF. The possibility may exist, therefore, that GVs obtained from meiotically incompetent oocytes can be activated by the cytoplasm of fully grown oocytes. It may be possible to employ a form of nuclear transfer, using nuclei from very immature oocytes to produce secondary oocytes from a particularly

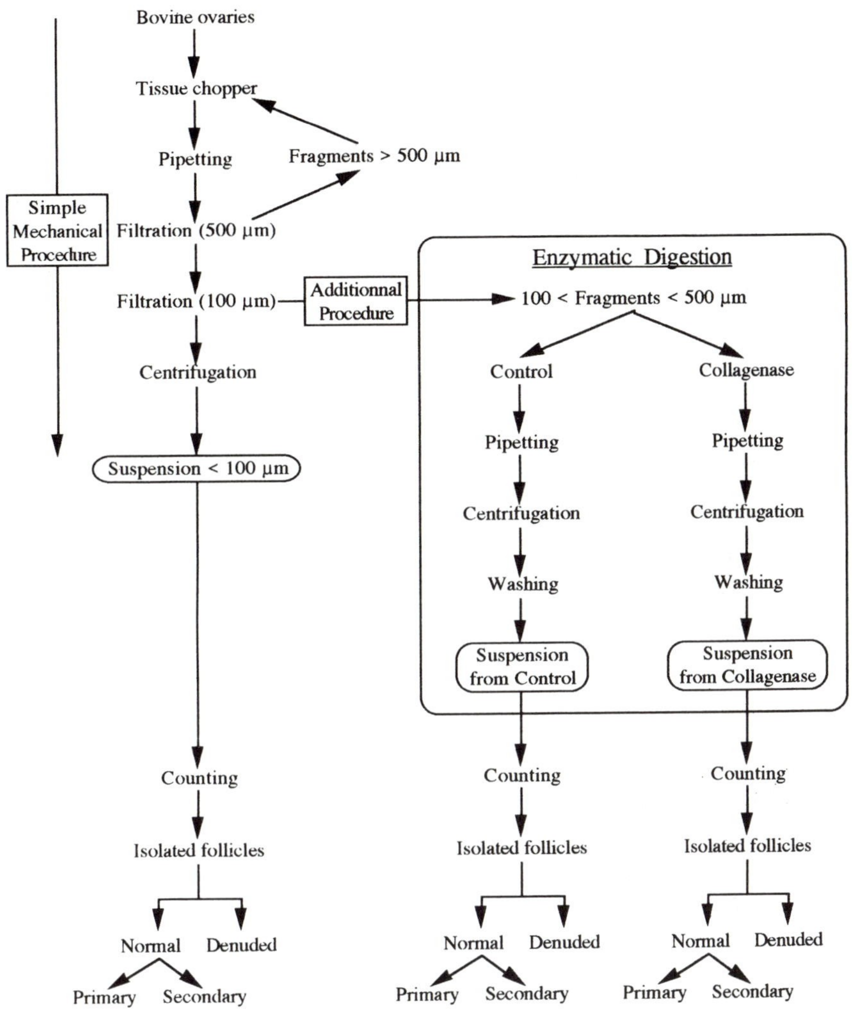

Fig. 2.15. Mechanical and enzymatic procedures for the isolation of preantral bovine follicles (from Figueiredo *et al.*, 1993).

desirable genotype, in a novel approach to producing embryos of specific genotypes.

2.8 Oocyte Maturation *In Vivo*

2.8.1 Germinal vesicle breakdown (GVBD)

During maturation, the nucleus of the bovine oocyte, which had entered prophase of the first meiotic division at an earlier stage, undergoes reduction division. In reduction division, the number of chromosomes in the oocyte is reduced to the haploid number (n = 30 chromosomes). The nuclear maturation stages of mammalian oocytes are well documented in the literature. At the molecular level, it is recognized that meiosis in the oocyte involves many interactions between cell cycle molecules and their target substrates in the nucleus and cytoplasm of the cell (Moor *et al.*, 1992b).

The process of nuclear maturation begins with the recession of the nuclear membrane (GVBD), the disappearance of the nucleoli and the condensation of the chromosomes. Oocytes with a relatively slow rate of maturation (cattle, sheep and pigs) require active protein synthesis for the resumption of meiosis (Hunter and Moor, 1986, 1987) whereas in rodents and rabbits this is apparently not necessary (Motlik *et al.*, 1990; Kastrop *et al.*, 1991a, b). It appears that fully grown rodent and rabbit oocytes possess all the proteins essential for chromatin condensation and nuclear membrane breakdown. Cattle, sheep and pig oocytes, on the other hand, are equipped before resumption of meiosis for chromatin condensation only; proteins required for GVBD

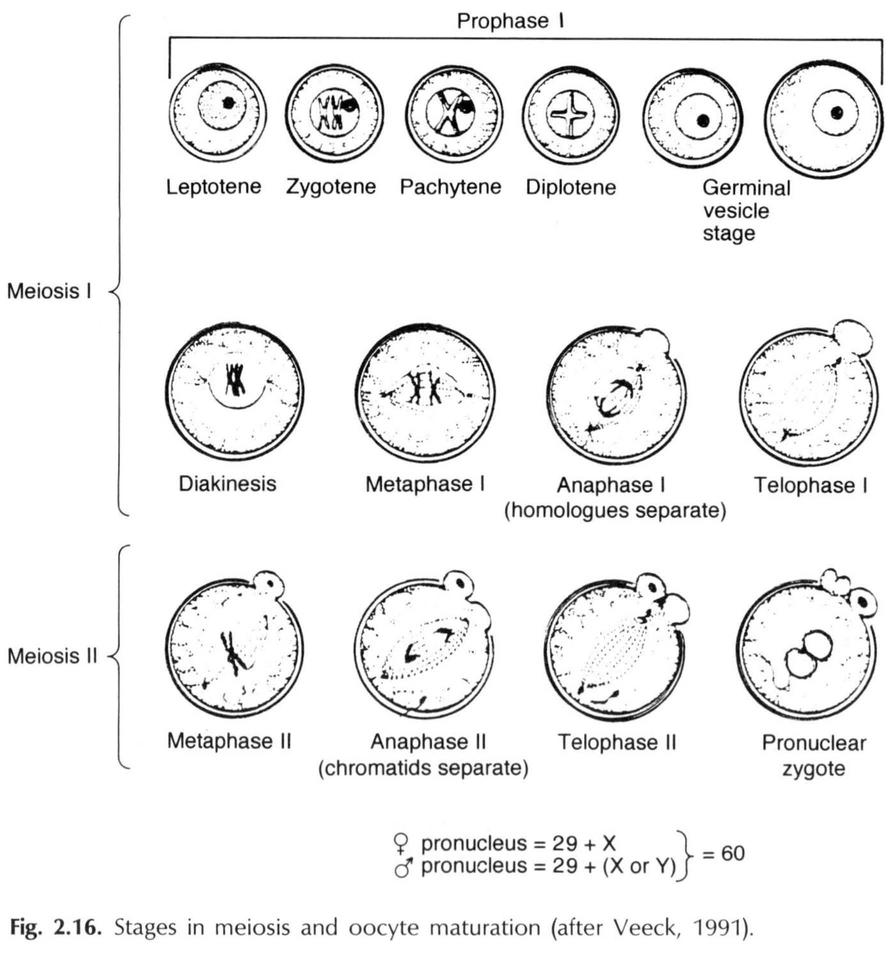

Fig. 2.16. Stages in meiosis and oocyte maturation (after Veeck, 1991).

are synthesized at a specific time after the resumption of meiosis (Simon *et al.*, 1989).

With the occurrence of GVBD, the chromosomes condense into a compact form. The centrosome (a specialized area of dense cytoplasm) then divides into two centrioles around which asters appear. These asters move apart and a spindle is formed between them. The chromosomes, in diploid pairs, are set free in the cytoplasm and become arranged on the equatorial plate of the spindle (metaphase I). The primary oocyte now undergoes two meiotic divisions. In the first division, two daughter cells are formed, but one of these contains almost all the cytoplasm; the second, much smaller, cell is the first polar body (see Fig. 2.16). In addition to chromosomes, the first polar body contains a variety of organelles, including mitochondria, ribosomes and cortical granules.

The progression from telophase of the first meiotic division to extrusion of the first polar body ensures that one group of chromosomes is expelled into the perivitelline space while the other group remains in the cytoplasm of the oocyte. The egg may now be termed a secondary oocyte and the vitelline chromosomes revert to spindle formation, the second meiotic spindle. The second meiotic division begins and proceeds up to metaphase II, but is not completed unless sperm penetration occurs or the oocyte is otherwise activated. Completion of the second meiotic division is accompanied by extrusion of the second polar body into the perivitelline space. The oocyte at this stage contains just the haploid number of chromosomes.

In the live cow, GVBD is observed to occur 4–8 h after the preovulatory LH peak (Kruip *et al.*, 1983). In superovulated cattle, GVBD has been observed to occur 6–12 h after the start of maturation (Xu *et al.*, 1986b). There have been suggestions that nuclear maturation is more rapid *in vitro* than *in vivo* (King *et al.*, 1986). According to Loos *et al.* (1991c), in a study of single-ovulating cows, GVBD occurs at 8 h and metaphase II is reached at 19 h.

2.8.2 Biochemical and physiological events during maturation

Nuclear maturation *in vivo* takes about 24 h in the cow; the same time interval holds for the sheep. It is recognized that oocyte maturation involves two phases: the first is an inductive phase, up to the occurrence of GVBD, in which the oocyte undergoes few structural or synthetic changes but appears to be undergoing reprogramming by the somatic elements in the follicle (cumulus cells, etc.). In the final 18 h synthetic phase, most components of the oocyte undergo reorganization but it appears unlikely that cumulus cells have the same crucial influence on the oocyte (see Fig. 2.17).

The work of Moor (1978) indicated that the high concentration of oestradiol found in the ovine preovulatory follicle during the first 6–8 h after the LH surge is more essential for cytoplasmic maturation than for the resumption of meiosis. The inductive phase of maturation is one in which the cumulus cells are thought to act in a particularly crucial supportive role. Eventually, more precise information will become available on the nature of events that occur during maturation and how these influence subsequent development of

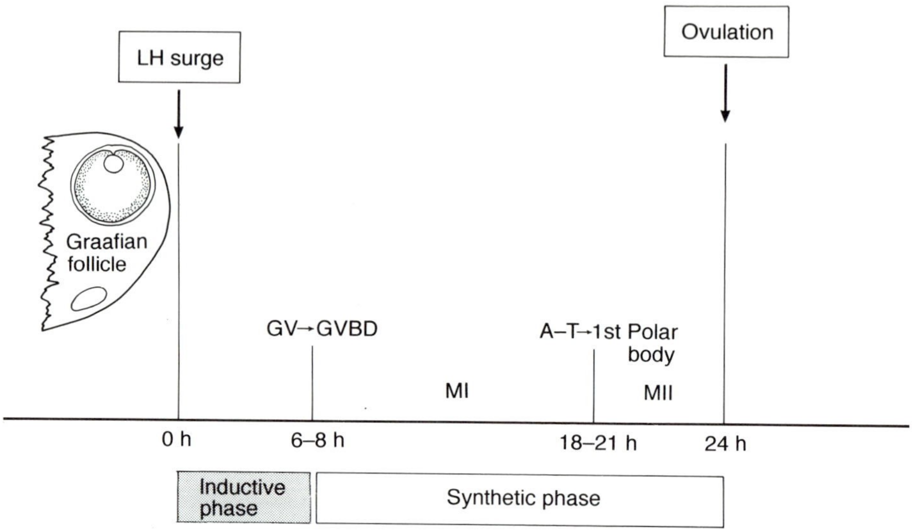

Fig. 2.17. Chronology of events during oocyte maturation in the cow. GV, germinal vesicle oocyte; GVBD, germinal vesicle breakdown; MI, metaphase I; A–T, anaphase–telophase; MII, metaphase II.

the oocyte. One such report is that of Rieger and Loskutoff (1994) who have studied changes in the metabolic activity of bovine oocytes during the course of maturation *in vitro*; results suggest that oxidative metabolism increases and is the major site of cellular energy production during the maturation period.

In The Netherlands, the biochemical aspects of bovine oocyte maturation have been the subject of several reports (Kastrop *et al.*, 1988, 1990a, b; Kastrop, 1990). There is evidence showing that certain of the proteins synthesized during maturation persist throughout the early development of the embryo (Moor and Gandolfi, 1987). Oocyte maturation is marked by the completion of intracellular organization and the formation of the second metaphase plate. It is not until the storage of messenger RNA (mRNA) is exhausted in early embryonic life that the oocyte's own programming ceases to be a dominant directive force in development.

After meiotic maturation is resumed in the bovine oocyte, RNA synthesis is barely detectable. However, the rate of RNA synthesis was found to be significantly higher at the start of maturation when the oocyte is coated by a well-developed, multilayered cumulus oophorus than if denuded (Kozikova and Sirotkin, 1992); the indications are that cumulus cells may have a regulatory role in RNA synthesis. Cessation of rRNA synthesis is apparently associated with dissolution of the nucleolus upon mixing of the nucleoplasm and cytoplasm. It is believed that about 20% of maternal RNA is degraded, with as much as one half of the polyadenylated RNA accumulated during growth being lost during meiotic maturation.

Although RNA synthesis decreases markedly after the bovine oocyte resumes meiosis, the synthesis of mRNA is seen to be obligatory for GVBD in cattle and other ungulates. A specific inhibitor of RNA polymerase II, α-amanitin, reduces the maturation rate in cattle and sheep COCs (Hunter and Moor, 1987), although it is without effect on denuded oocytes. It may be that cumulus cells are essential for transporting the inhibitor into the oocyte. Evidence reported by Rabahi *et al.* (1992, 1993) in France suggests that during the preovulatory period in the cow, the gonadotrophin surge may act directly on the COC. However, it is likely that endocrine factors first reach the granulosa cells located closer to the blood-vessels and mural granulosa cells are known to possess more LH receptors than cumulus cells. It is possible that granulosa cells secrete, in response to the gonadotrophin surge, compounds able to modify COC metabolism.

2.8.3 Morphological features of matured oocytes

A number of workers have reported on the morphology of bovine oocytes after *in vivo* maturation (Sirard and Lambert, 1985, 1986; Lambert *et al.*, 1986); the ideal morphology is described as an oocyte with an expanded and clear cumulus cell mass. After IVF, such oocytes gave higher cleavage rates than oocytes without cumulus cells or possessing a compact cumulus. Ultrastructural changes recorded in ovulated bovine oocytes by Hyttel *et al.* (1986a, b) included the formation and enlargement of the perivitelline space and the rearrangement of mitochondrial clusters and vesicles which occurred about 15 h after the LH peak and coincident with a substantial increase in progesterone dominance in the FF. Schellander *et al.* (1989a) assessed quality in oocytes recovered after *in vivo* maturation as excellent when the ooplasm was evenly granulated, the perivitelline space small and the cumulus cells fully expanded without evidence of degenerate cells.

2.9 *In Vitro* Maturation (IVM) Culture Systems

2.9.1 Culturing the intact follicle

The culture of intact ovarian follicles has been reported by several groups; the objective may have been oocyte maturation or a study of endocrine events in the follicle. In the USA, Rush *et al.* (1973) reported studies with small bovine follicles (1–2 mm diameter) cultured for 60 h on stainless steel grills suspended over hormone-supplemented medium; no evidence of oocyte maturation was obtained. In Cambridge, Moor and Trounson (1977) found that 40–50% of sheep oocytes matured within intact follicles *in vitro* subsequently gave rise to normal embryos and young; this was in contrast to their consistent failure at that time to effectively mature oocytes liberated from follicles. Intact follicle culture was also used in several studies (Warnes *et al.*, 1977; Moor, 1978; Moor *et al.*, 1978) at Cambridge to examine steroidogenesis and protein synthesis in sheep oocytes.

A comparison between intra- and extrafollicular maturation of cattle oocytes was made by Fukui *et al.* (1987b), the workers concluding at that time that the culture of intact follicles provided a more suitable environment for the oocyte. Elsewhere, workers used intact follicle culture to study steroidogenesis in calves (Peluso and Hirschel, 1987, 1988) and in adult cattle and sheep (Skyer *et al.*, 1987; Terqui *et al.*, 1988a, b; Kruip and Dieleman, 1989). The use of follicle-enclosed oocytes has also allowed various hormonal effects to be examined in other species (Yoshimura *et al.*, 1990, 1991, 1992).

Although the intrafollicular maturation of the bovine oocyte may appear to have an advantage over conventional IVM culture systems, intact follicle culture is generally regarded as tedious, time-consuming and requiring considerable skill; for such reasons, it cannot be regarded as suitable for large-scale embryo production. With intact follicle culture, there is the need for gonadotrophin supplementation of the culture medium and for longer periods of culture compared with those employed with the liberated oocyte. Extrafollicular maturation, unlike intrafollicular, can obviously be monitored visually from GVBD to metaphase II and extrusion of the first polar body.

For those working with horses, there is an unusual way of making use of the intact follicle in the live animal. This involves the transfer of primary oocytes into the preovulatory follicle which is about to ovulate (Hinrichs and DiGiorgio, 1990, 1991). Oocytes were transferred into the lumen of the preovulatory follicle via a needle placed through the flank. Twenty transfers were performed, each of 7–18 oocytes, with the mares inseminated afterwards and later flushed. The authors suggest that this method might be a useful means of obtaining large numbers of mature oocytes, fertilized oocytes or early embryos.

2.9.2 *Simple and complex IVM media*

The culture medium employed in IVM not only can affect the proportion of bovine oocytes that reach metaphase II and become capable of undergoing fertilization, but can also influence subsequent embryonic development (Bavister *et al.*, 1992). The suitability of seven commercially available culture media for IVM in cattle was examined by Rose and Bavister (1992); they showed that the choice of medium could markedly affect subsequent embryo development.

The culture media employed in the extrafollicular maturation of the bovine oocyte can be broadly divided into simple and complex. Simple media are usually bicarbonate-buffered systems containing basic physiological saline with pyruvate, lactate and glucose. The main differences between various types of simple media lie in differences in their ion concentration and in the concentrations of the energy sources. The media are generally supplemented with serum or albumin with trace amounts of antibiotics added (penicillin, streptomycin, gentamicin). Complex media contain, in addition to the basic components of simple media, amino acids, vitamins, purines and other substances, mainly at levels at which they are found in serum. Fixed nitrogen is present as free amino acids.

In some studies, an attempt has been made to establish whether all stages of *in vitro* embryo production (maturation, fertilization, embryo development) could be carried out in the one medium (Waterman *et al.*, 1991); RF10 (modified F-10) medium containing 0.6% bovine serum albumin (BSA) was used with some success.

One of the most widely employed complex media for IVM of bovine oocytes is tissue culture medium 199 (M-199). It consists of Earle's salts, buffered with *N*-(2-hydroxyethyl)-piperazine-*N'*-(2-ethanesulphonic acid) (HEPES) and sodium bicarbonate and supplemented with pyruvate, lactate, amino acids, vitamins, purines and other substances found in serum. M-199 has been the standard medium employed in IVM of cattle and sheep oocytes in Ireland (Lu and Gordon 1987; Lu and Polge, 1992). Elsewhere, Xu *et al.* (1987b, 1992a, c) in Denmark and Canada have employed the alternative complex Ham's F-10 medium for IVM.

Most cattle IVF laboratories routinely use M-199 as the basic IVM medium, and there have been very few reports suggesting that other media may be more appropriate. In one comparison of complex media for IVM, Hawk and Wall (1993) concluded that, under their conditions, F-10 is superior to M-199 and B2. However, in this work, B2 was found to be more effective for *in vitro* culture. The F-10 medium consists of a Krebs Ringer bicarbonate buffer system supplemented with pyruvate, lactate, amino acids, vitamins, trace elements and other substances found in serum, such as thymidine and hypoxanthine. Penicillin and streptomycin are also added, as are albumin and serum. In Germany, Behrens (1992) has reported employing a modified Menezo-B2 medium in maturation studies.

The effect of ethylenediaminetetraacetic acid (EDTA) (0.1 mM) and pyruvate (0.25 mM) in cattle IVM culture medium (M-199 with supplements) on subsequent developmental ability of oocytes was reported by Chung and Seidel (1993); there was some indication that EDTA may be beneficial but pyruvate was without effect.

The main point to be made about IVM media, whether simple or complex, is the need to work towards defining all constituents on the basis of a proved requirement. For example, both M-199 and Ham's F-10 contain the purine, hypoxanthine, which has been implicated in the meiotic arrest of mice and monkey oocytes (Eppig *et al.*, 1985; Warikoo, 1988). In fact, hypoxanthine-supplemented medium can be employed to deliberately inhibit the resumption of meiosis in the oocytes of some mammals. The currently employed IVM media almost invariably contain bovine serum or cellular constituents, which means many unknown substances are present and the concentrations of numerous constituents cannot be accurately defined.

The fact that current media employed in cattle IVM were initially designed for general cell culture means that they may contain factors which adversely affect oocyte maturation. With that in mind, Sirard and Coenen (1993b) prepared a new IVM medium for cattle oocytes, based on measured bFF constituents. This bovine synthetic FF was used to evaluate various products on meiotic resumption.

2.9.3 *Buffering systems and osmolarity*

The buffering system employed in IVM media will depend on whether the medium is exposed to air or to a carbon dioxide-enriched atmosphere. The advantage of HEPES- or phosphate-buffered media for short-term work with oocytes and embryos is that they do not require a carbon dioxide-controlled gas phase to maintain a relatively constant pH. The IVM will normally be isotonic with the natural tissue fluids (FF/seminal plasma) which are in contact with the gametes of the living animal. Osmolarity is generally arranged to be between 275 and 285 mosm, which is considered to be the optimum range.

2.9.4 *Water quality considerations*

Water is the major constituent of any IVM medium. The use of ultrapure water, free from contaminants, is crucial for its preparation (Bavister, 1992a, b). There are several basic groups of contaminants to be found in water. These include dissolved ionized and non-ionized solids and gases, particulate matter, microbials and pyrogens. The principal methods for purifying water are glass distillation, deionization, filtration, reverse osmosis, adsorption and ultrafiltration. Rigorous quality-control testing should be carried out at regular intervals to ensure that water purity is acceptable. Use of a 0.2 μm filter should remove all bacteria, since the smallest known bacterium is about 0.3 μm. Optimum results have been claimed using the Millipore reverse osmosis (RO) and Milli-Q (MQ) systems. However, high purity ion-exchange systems such as the MQ can be regarded as entirely acceptable when the culture medium contains protein (blood serum/BSA); protein helps to fix heavy metal ions and reduce their adverse effects. Japanese workers used the hatching rate of mouse embryos as an appropriate indicator of their Millipore MQ system (Fukuda *et al.*, 1987). According to Wisconsin workers, quality-control testing of water and media can be effectively conducted using the hamster sperm bioassay technique (Rinehart *et al.*, 1988).

The most reliable physical measurement of water quality for *in vitro* culture appears to be the electrical conductivity, which should be as close to 0.06 $\mu S\,ml^{-1}$ as possible (Fukuda *et al.*, 1987).

2.9.5 *Static and flux culture systems*

Cambridge workers (Staigmiller and Moor, 1984a, b; Cheng, 1985; Cheng *et al.*, 1986), dealing with sheep, reported findings which provided a firm basis for a culture system permitting a substantial increase in the percentage of extrafollicular oocytes maturing to metaphase II *in vitro*. They showed, under the culture conditions employed, that gentle agitation (flux system) of the IVM medium prevented the attachment and subsequent differentiation of cumulus cells; the resultant maintenance of the correct state of somatic cell differentiation during culture was believed to markedly improve the quality of oocytes. It is necessary to specify the culture conditions used; this involved a 2 ml volume of IVM medium, containing 20 COCs and additional cumulus/granulosa cells.

Early studies in Dublin with bovine oocytes (Lu and Gordon 1987, 1988a, b) was based on the Cambridge IVM culture system as described by Staigmiller and Moor (1984a); gentle and regular agitation was provided throughout the 24 h IVM period. Elsewhere, Gradl *et al.* (1988) with cattle oocytes and Braun (1988) with sheep oocytes also report an advantage of a non-static over a static IVM culture system. The main factors influencing the need or otherwise for a flux culture system are likely to include (i) volume of IVM medium; (ii) number of COCs per unit volume; and (iii) the addition of cumulus/granulosa cells.

Despite the suggested virtues of a flux system for cattle IVM, many reports in the literature record successful maturation using non-static culture systems. For the most part, these usually involve IVM in microdroplets (e.g. 10–20 COCs in 50–100 μl volumes). The use of microdroplets under oil is a system which has certain advantages, including the prevention of or reduction in the evaporation of water, protection from microbial contamination, attenuation of temperature and gas fluctuations and ease of examination during culture. The lightweight paraffin oil originally employed by most workers has been replaced in most instances by silicone oil (liquid polydimethylsiloxane); such oil is believed to be inert and to be relatively free of factors which may prove toxic to oocytes and embryos.

In some reports, however, it has been shown that steroids such as oestradiol may diffuse out of the microdroplets into the surrounding oil (Miller and Pursel, 1987). This may be a factor influencing results obtained in cattle IVM (Xu *et al.*, 1987b).

It may simply be a question of the IVM system preventing the differentiation of granulosa/cumulus cells and this may be done in various ways, including culture in microdroplets without agitation.

2.9.6 *Effect of maturation time*

During IVM, GVBD occurs from 3 h to more than 12 h of culture. Although some discrepancies exist in regard to the time of abstriction of the first polar body (Hyttel *et al.*, 1989c), it is usually recorded as occurring 18–24 h after the start of maturation (Sreenan, 1968, 1970; Jagiello *et al.*, 1974; Motlik *et al.*, 1985; Greve *et al.*, 1986; Hyttel *et al.*, 1986c; King *et al.*, 1986; Xu *et al.*, 1986b).

In the studies of Sirard *et al.* (1989) on the timing of nuclear events during IVM, the GV was evident from 0 to 6.6 h, GVBD occurred at 6.6–8.0 h, chromatin condensation at 8–10.3 h, metaphase I at 10.3–15.4 h, anaphase I at 15.4–16.6 h, telophase I at 16.6–18.0 h and metaphase II at 18.0–24.0 h.

Xu *et al.* (1986b) found that the IVM culture period required for GVBD and abstriction of the first polar body was related to the thickness and compactness of cumulus investments, with an approximately 6 h delay in heavily compacted COCs. Similarly, Spiropoulos and Long (1989) found that partially denuded oocytes, totally denuded oocytes and oocytes with expanded cumulus cells at the start of IVM progressed to metaphase II faster than compact COCs.

In Canada, Semple *et al.* (1993) have reported that bovine oocytes

Table 2.19. The effect of maturation time on maturation rate and subsequent embryo development (from Monaghan, 1993).

Treatment	Meta II %	Cleavage rate %	Cleavage index %	Blastocyst yield % Day 7	Blastocyst yield % Day 8	Blastocyst yield % Day 9
18 h	81.3 (61/75)	73.6 (304/413)	70.0 (213/304)	8.9[a] (27/304)	16.1 (49/304)	18.8[a] (57/304)
24 h	79.7 (51/64)	76.0 (301/396)	70.4 (212/301)	14.3[b] (43/301)	23.3 (70/301)	25.9[b] (78/301)

a, b – Values in the same column with different superscripts differ significantly ($P < 0.05$).

achieve developmental competency within 14 h of commencing IVM; their data also suggest that early fertilization can lead to significantly higher yields of blastocysts. In Russia, Prokofiev *et al.* (1992) concluded, on the basis of their embryo yields, that a wide range in the duration of oocyte is possible (18–24 h). In Ireland, Monaghan *et al.* (1993) found a 24 h maturation period to be superior to 18 h, which was in agreement with data obtained previously in Ireland (Lu, 1990). Details are in Table 2.19. It is evident that, while there was no significant difference between treatments in terms of maturation and cleavage rates, there was a significant difference in final blastocyst yield in favour of the 24 h maturation period. In contrast, data reported elsewhere have either shown no difference in blastocyst yield between 18 and 24 h IVM periods (Prokofiev *et al.*, 1992) or have favoured earlier (14–18 h vs. 24 h) maturation periods (Semple *et al.*, 1993).

2.9.7 Other considerations

Antibiotic cover

An ideal antibiotic for oocyte culture should have broad-spectrum anti-bacterial activity as well as freedom from toxicity. The use of trace amounts of penicillin (50–100 iu ml^{-1}) and streptomycin (50–100 $\mu g\ ml^{-1}$) in flushing fluids and culture media has been routine practice in cattle ET for many years. In the IVM medium, antibiotics are generally included to provide cover against the growth and proliferation of microorganisms during the period of culture. The concentration of antibiotics is known to be non-toxic, but their inclusion in the medium must obviously be allied to rigorous standards of hygiene in the IVF laboratory. In this regard, the use of laminar flow cabinets and the filtration and sterilization of all components of the culture medium are part of the essential precautions which must be taken in any cattle IVF laboratory.

There has been at least one report, uncommon though it is, of certain batches of penicillin–streptomycin proving to be spermicidal and possibly ovicidal (Sertich *et al.*, 1988). The use of gentamicin sulphate (50 μg ml^{-1}) is an alternative to penicillin/streptomycin (Younis *et al.*, 1989a; Lonergan, 1992); this antibiotic can tolerate autoclaving and high temperatures and possesses biological and biochemical properties which make it particularly useful. However, there is at least one report (Mizoshita *et al.*, 1993) containing results suggesting that gentamicin should not be used in cattle embryo production.

Patterns of energy metabolism during IVM

IVM of primary bovine oocytes is known to cause changes in protein synthesis and phosphorylation. In Canada, Rieger and Loskutoff (1993, 1994) examined patterns of energy metabolism in bovine oocytes during maturation; their preliminary results suggest that oxidative metabolism is the major source of energy for such oocytes.

Light environment

As part of routine maturation, fertilization and early culture of cattle embryos, they are inevitably exposed to varying amounts of either daylight or artificial light. There would be the suspicion that lesions caused by light exposure might, in certain instances, contribute to embryo failure. A study by Bedford and Dobrenis (1989) in rabbits would suggest that it is probably unnecessary to compromise the efficiency of IVM/IVF procedures in order to minimize light exposure of oocytes and early embryos. In Belgium, Barlow *et al.* (1992), working with mice, showed that light exposure is not a limiting factor in experiments and studies involving the micromanipulation of oocytes. On principle, however, oocytes should not be exposed to light for any longer than is necessary. Maturation and other events normally occur under conditions of darkness in the cow's reproductive tract.

Glutathione

The composition of the IVM medium has a definite impact on the ability of the pig oocyte to form the male pronucleus after sperm penetration. Yoshida (1993) has found that glutathione (GSH) synthesis occurs throughout the IVM of pig oocytes and is an important cytoplasmic factor for regulating sperm nuclear decondensation and male pronucleus formation after sperm penetration in pig oocytes. The Waymouth MB 725/1 medium they employed in IVM contains several substrates of GSH, which may help in explaining its effectiveness in supporting the functional maturation of pig oocytes.

Heparin

Heparin sulphate and dermatan sulphate are glycosaminoglycans (GAGs) found in FF. Fenton and Ax (1990) and Fenton *et al.* (1993) showed that, although these GAGs would stimulate cumulus cell expansion in bovine oocytes and increase the production of hyaluronic acid, they did not enhance fertilizability of the oocytes. Elsewhere, Olson *et al.* (1990) reported reduced cleavage rates when cattle oocytes were matured in medium containing heparin.

Vitamins

The concentration of some vitamins in cattle follicles has been determined. Chew *et al.* (1984) recorded vitamin A levels and, on that basis, it may be argued that the concentration present in current IVM media may be too low. Schweigert *et al.* (1987) suggest that media for bovine oocyte maturation contain either none or unphysiologically low levels of vitamin A. These authors suggest that 25–30 μg 100 ml^{-1} of the vitamin should be present to avoid a lack of this factor adversely affecting essential protein synthesis. It has also been noted that this vitamin may be one of the factors influencing growth and development of the DF in cattle (Schweigert *et al.*, 1988; Schweigert and Zucker, 1988).

2.10 Temperatures and Gas Phases

2.10.1 Core temperature of the cow

Until well into the 1980s, research studies in cattle IVM were almost invariably carried out at a temperature of 37°C. However, in the last decade, it has come to be appreciated that cattle actually have an average rectal temperature of 38–39.3°C, with a core temperature which is somewhat higher. Lenz *et al.* (1983b) were among the first to examine the effect of temperature on cattle IVM and IVF. Results indicated that maturation was not affected in the range 35–39°C, but IVF proved to be temperature-dependent, the fertilization rate being higher at 39°C than at 35, 37 or 41°C.

The authors conclude that it is not surprising that IVF events are optimal at 39°C, if this is the true core temperature of the cow. In pigs, Eng *et al.* (1986) produced compelling evidence to show that optimal IVM and IVF rates occur when oocytes are cultured at the core temperature of that species (39°C). At 37°C, the percentage of oocytes matured was 37% as compared with 71% at 39°C; a matter of 2°C could double the maturation rate.

There may be those who observe that vesicular follicles in the ovary may not necessarily be at the core temperature of the animal. In rabbits, for example, follicle temperature can be some 3°C lower than the animal's core temperature. In IVM studies in cattle, Katska and Smorag (1985) concluded that there may be a relatively high level of tolerance of meiosis to the culture temperature. It was not until a temperature some 5°C below that of body temperature was reached that there was an obvious slowing down in the maturation rate. These workers suggest that 35°C should be regarded as the minimal maturation temperature. Ultrastructural aspects of bovine oocytes cultured at temperatures ranging from 35 to 39°C have been reported by Morstin and Katska (1986).

2.10.2 Optimum temperatures for maturation, fertilization and early embryo culture

Work in Ireland has closely examined the effect of temperature on maturation, fertilization and early culture of bovine embryos (Wang, W.L. *et al.*, 1991). Results from the three studies involved are detailed in Table 2.20.

In experiment I, oocytes were matured in a humidified atmosphere of 5% carbon dioxide in air at 36, 37, 38, 39 and 40°C and then fertilized and cultured at 39°C. In experiment II, oocytes were matured at 39°C, fertilized at 36, 37, 38, 39 and 40°C and then cultured as early embryos at 39°C. In experiment III, oocytes were matured and fertilized at 39°C and then cultured as early embryos at 36, 37, 38, 39 and 40°C. It is evident from the data presented that the optimum temperature for IVM is 38–39°C; for IVF and IVC, the optimum temperature is 39°C. The detrimental effect of using the 40°C temperature is clearly demonstrated, particularly in fertilization and early embryo development. These data point up the need for precise temperature control, with high-quality equipment and laboratory expertise.

2.10.3 Temperature control during and after IVM

It may be important to ensure that bovine oocytes are not exposed to reduced temperatures (e.g. below 20°C) once they have commenced IVM and during the period of IVF. Cambridge studies with sheep oocytes provided clear

Table 2.20. The effect of temperature on the *in vitro* maturation, fertilization and development of bovine oocytes (from Wang, W.L., *et al.* 1991).

Exp.	Temp. °C	Oocyte insem. or embryos cult.	Cleavage (%)	BL on Day 7 (%)	Total BL (%)	BL hatched (%)
	36	433	364 (84.1)	98 (22.6)[a]	146 (33.7)[a]	104 (71.2)
	37	408	341 (83.7)	96 (23.5)[a]	138 (33.8)[a]	94 (68.1)
I	38	308	267 (86.6)	103 (33.4)[b]	135 (43.8)[b]	109 (80.7)
	39	418	359 (85.8)	116 (27.8)[ab]	166 (39.7)[ab]	128 (77.1)
	40	417	349 (83.7)	96 (23.0)[a]	141 (33.8)[a]	109 (77.3)
	36	634	500 (78.9)[a]	45 (7.1)[b]	159 (25.1)[b]	94 (59.1)[a]
	37	597	508 (85.1)[b]	96 (16.1)[c]	216 (36.2)[c]	157 (72.7)[bc]
II	38	584	487 (83.4)[b]	111 (19.0)[c]	213 (36.5)[c]	165 (77.5))[c]
	39	581	496 (85.4)[a]	170 (29.3)[d]	230 (39.6)[c]	188 (81.7)[c]
	40	569	435 (76.5)[a]	19 (3.3)[a]	49 (8.6)[a]	39 (61.2)[a]
	36	228		7 (3.1)[a]	67 (29.4)[b]	51 (76.1)[b]
	37	393		39 (9.9)[b]	145 (36.9)[bc]	111 (76.6)[b]
III	38	472		79 (16.7)[c]	202 (42.8)[cd]	152 (75.2)[b]
	39	507		156 (30.8)[d]	247 (48.7)[d]	200 (80.9)[b]
	40	413		23 (5.6)[a]	77 (18.6)[a]	40 (51.9)[a]

a, b, c, d – $P < 0.05$ within experiments and columns.
BL = blastocyst.

Table 2.21. Effect of cooling bovine secondary oocytes to 18–22°C on IVF and subsequent embryonic development (from Lonergan, 1992).

	Cleavage rate	Cleavage index	Embryo yield
Normal	82.9% (73/88)	63.0% (46/73)	21.9% (16/73)
Cooled	84.2% (86/114)	65.6% (63/96)	17.6% (16/91)

$P > 0.05$.

evidence of induced chromosomal abnormalities in oocytes cooled to 20°C at all stages of meiosis through to metaphase II (Moor and Crosby, 1985). This may point to the need for close temperature control in the laboratory handling of bovine oocytes after IVM and during the IVF process. In practical terms, once bovine oocytes emerge from IVM, there may be a particular need for them to be maintained at 30°C or above until they reach an embryonic stage at which they will tolerate lower temperatures. In Dublin, however, Lonergan *et al.* (1992) assessed the effect of cooling bovine oocytes to room temperature (18–22°C) before IVF on subsequent cleavage rate and embryo yield; no adverse effect was recorded (see Table 2.21).

2.10.4 Effect of gas atmosphere

Work in Texas by Azambuja *et al.* (1993) examined the effect of gas phase on the IMV process. A comparison was made between the conventional 5% CO_2 in air gas phase and one supplying 5% CO_2, 5% O_2 and 90% N_2. More embryos developed to the morula or blastocyst stage when the conventional gas phase was employed. In Madison, Pinyopummintr and Bavister (1994) concluded that low oxygen concentration (5–10%) is detrimental for both maturation and fertilization of bovine oocytes. It has been shown, in pigs, that primary oocytes will undergo nuclear and cytoplasmic maturation in HEPES-buffered media, supplemented with gonadotrophins and incubated in air (Hagen *et al.*, 1991).

2.11 Bovine Serum and Other Protein Sources

2.11.1 Bovine serum albumin (BSA)

It is common practice to include macromolecules from various sources in IVM culture media; bovine serum, or occasionally BSA, is generally employed for that purpose. The studies of Leibfried-Rutledge *et al.* (1986a) and Sanbuissho and Threlfall (1988) found that fetal calf serum (FCS) was superior to BSA as a protein supplement in IVM medium (as shown by cumulus cell viability and expansion). Similar findings with sheep oocytes were reported by Braun (1988), who found that FCS but not BSA was able to support FSH-induced cumulus expansion.

BSA can be regarded as a variable component of a maturation medium, since it has been recognized for some time that the commercial preparation is a chemically impure product. For that reason, any IVM medium containing commercial BSA cannot be regarded as being chemically defined (Kane, 1978, 1987). For example, four lots of Fraction V BSA were tested by Bavister and McKiernan (1991) for their ability to support two-cell hamster embryos in culture. Results showed that such preparations can produce highly variable effects on cultured embryos and cells, ranging from highly inhibitory to highly stimulatory.

For some time, it has been recognized that BSA was probably contaminated with some low molecular weight compound (Kane, 1978, 1987). Attempts to isolate and identify this contaminant eventually proved successful (Kane, 1990); it proved to be citrate (Gray, C.W., *et al.*, 1992). It has been demonstrated that citrate as an additive to culture medium can enhance early embryonic development in rabbit blastocysts. The growth-promoting effect of citrate on rabbit blastocysts is discussed by Kane and Fahy (1993); among various possibilities, it could be involved in chelating toxic metal ions or in promoting the uptake of ions such as Ca^{2+} or Fe^{2+} by the embryo in chelated form.

2.11.2 Sources of bovine serum

Bovine serum, in the form of FCS or OCS, has been employed as the main protein source in bovine IVM studies for some years. Culture media for oocyte maturation are almost always supplemented with some source of fixed nitrogen. Although many substances have been tested, BSA and various forms of bovine serum (see Fig. 2.18) have been those most frequently employed.

In Dublin, the studies of Lu *et al.* (1987b; Lu and Gordon, 1987) showed that OCS had a significant and marked effect, compared with FCS, on the

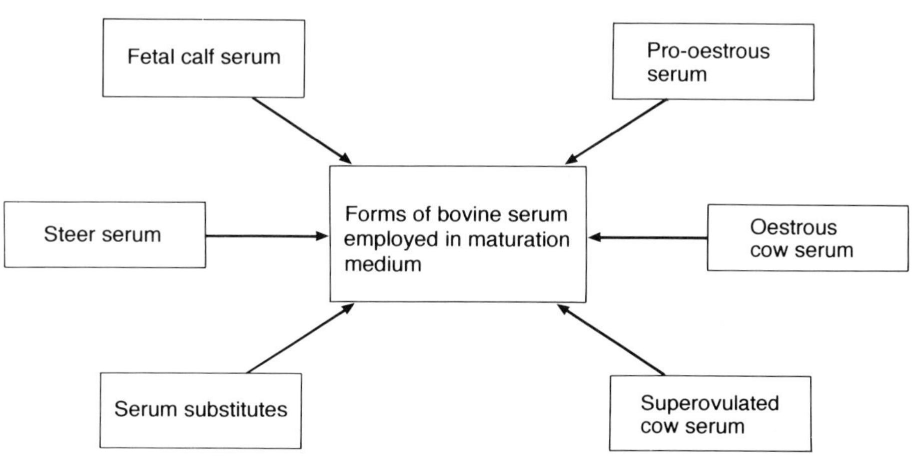

Fig. 2.18. Forms of bovine serum employed in cattle IVM studies.

Table 2.22. Comparison of oestrous cow serum and steer serum in the IVM medium (from Lonergan, 1992).

	Cleavage rate	Cleavage index	Embryo yield
OCS	82.9% (87/105)	74.7% (65/87)	15.5% (13/84)
SS	87.8% (108/123)	75.9% (82/108)	17.6% (19/108)

$P > 0.05$.

percentage of secondary oocytes that were fertilized and which cleaved during the first 2 days after insemination. A number of workers elsewhere have used OCS in their IVM studies in cattle (Sanbuissho and Threlfall, 1985, 1989; Xu *et al.*, 1987a; Sanbuissho, 1989; Yoon *et al.*, 1989; Chen-Lu and Lu, 1990; Kim, C.I., *et al.*, 1990; Kim, C.K., *et al.*, 1990; Schellander *et al.*, 1990b).

In the USA, studies by Younis *et al.* (1989a) suggested that pro-oestrous cow serum (collected on the day prior to oestrus) may be more effective in IVM than OCS. Comparisons of sera in Ireland by Lonergan (1992) suggest that steer serum (SS) in IVM medium is as effective as OCS (see Table 2.22).

For commercial large-scale embryo production, the recovery and processing of SS is much easier to organize than the collection of serum from animals in oestrus or at defined stages of their oestrous cycle.

2.11.3 Constituents of bovine serum

As noted by Kane (1978, 1987) the effectiveness of bovine serum in IVM and IVF may vary markedly from one batch to the next. Data on the amino acid content of bovine serum is provided by Fallon *et al.* (1988). It is clear that this constituent can show wide variations (see Table 2.23); similar variability may presumably be expected with hormones, growth factors, cytokines, vitamins and the many other substances found in bovine serum.

In using OCS rather than other forms of serum in Dublin (Lu *et al.*, 1987b), it was assumed that concentrations of steroids/gonadotrophins in that particular preparation differed markedly from serum collected at other stages of the oestrous cycle or from other categories of cattle. Workers who have quantified the changes in peripheral plasma hormone concentrations during the bovine oestrous cycle have recorded that oestradiol levels can increase threefold from a basal value of 5.0 pg ml^{-1} to a peak value of 15.0 pg ml^{-1} on the day of oestrus (Dobson *et al.*, 1973). The peak concentration of LH also occurs around the time when oestradiol is at its highest level.

It was thought that such changes in hormone concentrations might go some way in explaining why OCS might genuinely differ in its effectiveness in IVM from other forms of bovine sera. In the study reported by Younis *et al.* (1989a), suggesting that pro-oestrous serum (collected on the day prior to heat) may be more effective than OCS, it was found, on analysis, to contain high levels of LH and prolactin. In Dublin, Vergos (1990) compared bovine serum collected in the early hours of oestrus with that obtained at random

Table 2.23. Amino acid concentrations in sera from 25 clinically normal cattle (from Fallon *et al.*, 1988).

Amino acid	Mean (μmol l^{-1})	Range[a]	Coefficient of variation (%) Survey	Method
Alanine	266	130–500	26.9	1.9
Arginine	246	80–620	36.1	19.6
Asparagine[b]	37	16–78	34.9	6.3
Citrulline	107	24–320	63.7	20.3
Glutamic acid[c]	524	369–722	17.5	4.1
Glycine	453	280–690	21.0	2.0
Histidine	70	20–160	55.7	40.0
Hydroxyproline	37	10–100	49.7	4.4
Isoleucine	116	60–200	26.4	3.5
Leucine	147	86–240	23.2	4.0
Lysine	133	70–230	28.4	7.3
Methionine	18	10–28	25.6	8.4
Ornithine	138	73–240	27.6	4.8
Phenylalanine	63	40–94	20.0	2.9
Proline	85	56–124	20.0	2.4
Serine	131	73–221	23.8	2.5
Threonine	78	32–170	31.7	1.3
Tryptophan	30	16–52	27.1	6.6
Tyrosine	95	43–190	32.3	7.5
Valine	257	140–440	25.9	1.6
Branched chain/aromatic amino acids	3.4	2.1–5.2	21.9	3.9

[a]Mean ± 2 s.d. (logarithmically transformed).
[b]Includes aspartic acid.
[c]Includes glutamine.

throughout the oestrous period: no difference was evident in the effectiveness of the two forms of sera in IVM.

In Austria, Peli and Schellander (1993) studied the effect of several batches of OCS containing different oestradiol and LH concentrations; they showed that sera with high LH or oestradiol levels are not a prerequisite for obtaining satisfactory cleavage rates and embryo yields.

As well as that, it is clear that SS can replace OCS in IVM most effectively and it is only to be expected that hormone concentrations would differ substantially between OCS and SS preparations.

Quite apart from the requirement of bovine serum as a source of fixed nitrogen, its inclusion in the maturation medium may have other values. In preserving the fertilizability of matured mouse oocytes, for example, Downs *et al.* (1986) suggested that it may be important to include serum in the IVM medium to prevent hardening of the ZP, which could adversely affect fertilization. Subsequently, Schroeder *et al.* (1990) reported that fetuin, a major glycoprotein constituent of FCS, can prevent hardening of the ZP during IVM; the authors suggest that fetuin acts by preventing the action of proteolytic enzymes originating from precociously released cortical granules. Identification of fetuin as a substance that prevents zona hardening may be useful in characterizing those factors that influence cytoplasmic maturation. Elsewhere, workers with rat (Zhang, X., *et al.*, 1991) and mice (Yamazaki and Ishibashi, 1990; George *et al.*, 1992) oocytes also showed that FCS would prevent zona hardening and improve the fertilizing capacity of oocytes.

Serum levels employed in IVM media

The level of bovine serum employed in many of the Dublin IVM studies has been 20% (Lu *et al.*, 1987b). Minimal levels of OCS in an IVM medium have been examined by various authors. Vergos (1990) recorded that at least 1% of OCS was required for IVM, but there was no clear merit in exceeding a 10% concentration. The routine in Dublin has continued to employ 10–20% serum in recent years (Lonergan, 1992; Monaghan, 1993).

2.11.4 Heat treatment and preparation of serum

The deactivation of bovine serum (a temperature of 56°C for 30 min) prior to use for supplementation of media for IVM/IVF/IVC is a routinely practised procedure. However, little information is available in the literature to suggest that this is an absolute necessity for successful culture. Bavister *et al.* (1992) employed serum which was not heat-inactivated in their studies; these authors failed to find any definitive evidence in the literature showing that such heat treatment was necessary. Further than that, they suggest that the rationale for heat-treating serum (destruction of complement factors) may be faulty and that it may have an adverse rather than a beneficial effect. In human IVF, Imoedemhe *et al.* (1992c) carried out studies which led them to the conclusion that serum deactivation by heating is not necessary.

On the preparation of serum, there is information showing that steroids such as progesterone can be very unstable in whole blood collected from cattle (Breuel *et al.*, 1988). The processing of serum after it is collected may be a consideration in the preparation of IVM media. It is possible that differing procedures may give rise to variable results, arising from changes occurring in hormone concentrations during the collection, handling and storage of whole blood. For the IVF laboratory processing its own bovine serum, the aim must be to ensure that the optimal and repeatable procedure is always employed. There is also the question of effects arising from microfiltration of serum.

Microfiltration is the usual method of sterilizing tissue culture fluid. Inhibition of cell-line growth has been recorded after the addition to the culture medium of serum filter-sterilized through a membrane filter, which was possibly due to the leaching of toxic factors from the filters themselves. Harrison *et al.* (1990), in a study of embryotoxicity of Micropore filters used in medium sterilization, showed the importance of discarding the initial filtrate, or preflushing the filters.

2.12 Bovine Follicular Fluid (bFF)

2.12.1 bFF as a protein source

In the different stages of vesicular follicle development in cattle, primary oocytes are bathed in extracellular fluid, which is conventionally termed 'follicular fluid'. FF is a serum transudate modified by follicular metabolic activities, containing specific constituents such as steroids and glycoproteins synthesized by the cells of the follicle wall. During follicular growth, equilibrium is established between serum and FF. The metabolite concentrations in the two compartments are similar and are, in turn, similar to those in blood serum. FF contains several components of major physiological significance and buffers the internal environment of the follicle against the influence of external conditions; it maintains the oocyte in a state awaiting signals to resume meiosis and progress towards ovulation. Follicular development is assumed to be regulated by circulating pituitary gonadotrophins, the release of which is controlled by feedback mechanisms mediated by ovarian steroids and peptides.

Use was made of bFF as an IVM medium and an embryo storage medium by Sreenan (1969); it was found to inhibit bovine oocyte maturation in its undiluted form. Bevers *et al.* (1992), in The Netherlands, similarly reported that meiosis in bovine oocytes was delayed in 100% bFF. Shi (1991) in Dublin found that inclusion of bFF at a high level (40%) in the IVM medium had an adverse effect. Leibfried and First (1980) and Sirard and First (1988) reported that bFF significantly reduced the proportion of bovine oocytes resuming meiosis. Sirard and Bilodeau (1990b) examined the effect of bFF in IVM in the presence or absence of granulosa cells. The presence of bFF reduced the number of granulosa cells needed to maintain meiotic inhibition. Bottcher *et al.* (1990) reported that bFF from small vesicular follicles could inhibit meiosis in cattle oocytes. The inhibitory effect on meoisis of bFF, particularly that from small follicles, was also shown in a study reported by Ayoub and Hunter (1993a).

In Dublin, Lonergan (1992) found that supplementation of the IVM medium by bFF at the 10–20% level favoured subsequent embryonic development in cattle. He found, as did Shi (1991), bFF to be an effective substitute for SS at the 10–20% (v/v) level (see Table 2.24).

In the USA, Romero *et al.* (1991) concluded from their studies with cattle IVM that 20% bFF, obtained under specific conditions, enhanced oocyte quality. In Germany, Bottcher *et al.* (1992) record that a 20% level of bFF from oestrous cows increased embryo yield over that found with FCS or bFF from non-oestrous cows. In Japan, Iritani *et al.* (1992a) also record evidence of a favourable effect from including bFF in IVM medium at the 20–30% level. In Uruguay, bFF at the 30% level increased fertilization rate and embryo yield in comparison with 10% OCS (Larocca *et al.*, 1993). The implication in these various reports is that certain factors in bFF may favourably influence oocyte quality.

2.12.2 Composition of bFF

FF is known to contain the putative oocyte maturation inhibitor (OMI) which is believed to be responsible for the maintenance of meiotic arrest in ungulate oocytes (Jagiello *et al.*, 1977; Sato *et al.*, 1977; Sirard and First, 1988). However, a follicular factor has also been shown to facilitate the completion of cytoplasmic maturation in bovine (Fukui *et al.*, 1987a) and porcine (Mattioli *et al.*, 1988b; Naito *et al.*, 1988, 1990) oocytes. In pigs, several authors have noted that male pronucleus formation rate may be significantly enhanced when oocytes are cultured in pFF.

Yoshida *et al.* (1992a, b) examined the effect of pFF on the maturation of porcine oocytes and their subsequent developmental capacity *in vitro*. These authors found that pFF contains a substance, with a molecular weight between 10,000 and 200,000, which improves the rate of cumulus expansion, nuclear maturation, fertilization and embryonic development. Such reports emphasize the dual role of FF, in maintaining meiotic arrest on the one hand and in facilitating the completion of oocyte maturation on the other. In The Netherlands, Bevers *et al.* (1992) reported that meiosis was delayed in bovine oocytes incubated in 100% bFF.

Some information is available on external factors influencing the composition of bFF. In the USA, Spicer *et al.* (1991) examined the concentrations of IGF-I, oestradiol and progesterone in the follicles of cattle maintained on different dietary regimes. Oestradiol level in oestrogen-active (preovulatory) follicles was twofold greater in animals on high feed levels as compared with cattle on low levels.

A study reported by Homa and Brown (1992) determined the fatty-acid composition of bFF in small and large developing bovine follicles and examined the effect of a range of saturated and unsaturated fatty acids on spontaneous breakdown of GVs in oocytes. They found that linoleic acid significantly inhibited GVBD at relatively low dose levels. There are indications that the fatty-acid environment of immature bovine oocytes in small follicles is more conducive to maintaining meiotic arrest than that of developing follicles.

2.13 Serum-free and Chemically Defined Media

2.13.1 Use of serum substitutes

There is a need in cattle IVM to work continuously towards chemically defining all the constituents in the culture medium. Whether it is a question of energy substrates, nutrients or salts, the IVM medium should be formulated on the basis of an oocyte's proved requirement for particular constituents. The use of a serum-free medium for maturation would provide a convenient system for studying with much greater precision the mechanisms that regulate events in the bovine oocyte.

Several commercial products are available as serum substitutes for use in cell culture; one such product is Ultroser G. This substance consists of a combination of six main groups of compounds: growth factors, adhesive factors, mineral trace elements, hormones, binding proteins and vitamins. Although no information on specific components is available, the product is stated to have a low protein concentration (one third that of FCS), a constant composition and an absence of microbial contamination. Reports of the use of the material in cattle IVM are limited, although it has been used in embryo culture.

Ectors *et al.* (1989a, b) successfully matured cattle oocytes in Menezo B2 medium supplemented with 10% Ultroser G (plus oestradiol/LH/FSH). In Dublin, Lonergan (1992) also used the substitute successfully in cattle IVM without hormone supplementation at concentrations of 1–4%.

Work in Ireland also indicated that OCS may be replaced in the IVM medium by Ultroser HY. The biological activity of the reconstitued Ultroser solutions is fivefold that of FCS. A report by Flood *et al.* (1992) notes that fetal bovine serum (FBS) is routinely employed in cattle IVM/IVF/IVC systems but can be a costly supplement. These workers used a less expensive alternative in the form of a commercial product 'Fetalclone' (approximately half the cost) and report that this supplement compared favourably with FBS.

Organic substances of large molecular weight have been employed as serum substitutes. A study reported by Saeki *et al.* (1990a, b, 1991a) employed polyvinyl-pyrrolidone (PVP) at a 0.3% concentration as a substitute. In the absence of hormones (oestradiol/LH/FSH) in the IVM medium, there was no yield of blastocysts. In Dublin, however, Lonergan (1992) found that PVP could effectively replace serum in the absence of hormone supplementation; in this work, he employed additional granulosa cells in the IVM medium, which may explain the ability to produce blastocysts. Studies by Monaghan (1993) employed gonadotrophin and oestradiol at various concentrations in serum-free IVM medium supplemented with 0.3% PVP; at certain dose levels, it was possible to obtain blastocyst yields comparable to those achieved with the standard protocol.

2.13.2 *Serum-free media*

The maturation and fertilization of mouse oocytes in serum-free media was examined by Serta *et al.* (1991), who showed that the oocytes could be matured and fertilized in the absence of serum or FF. They employed a modified Ham's F-10 with BSA to provide a serum-free medium and suggest that the previously reported 'zona hardening' was not due to a lack of serum. It was thought that the integrity of sperm zona receptors may be impaired in suboptimal culture conditions.

Eppig *et al.* (1992b) have provided further evidence that the ZPs of mouse oocytes cultured in serum-free medium were much harder than freshly isolated GV-stage oocytes. Such hardening could be prevented by the addition of fetuin. There was, however, evidence of unknown serum-borne factors that have a beneficial effect during oocyte maturation. A serum-free medium pro-

Table 2.24. Use of bovine follicular fluid (bFF) from follicles of different sizes as a substitute for steer serum (SS) in a cattle IVM medium (from Lonergan, 1992).

Treatment	Cleavage rate	Cleavage index	Embryo yield
10% SS	86.3% (145/168)[a]	79.3% (115/145)[a]	28.9% (42/145)[ab]
20% SS	89.0% (84/194)[a]	72.9% (62/85)[a]	30.7% (23/75)[a]
10% bFF (2–6 mm)	91.5% (140/153)[a]	75.7% (106/140)[a]	21.5% (28/130)[a]
20% bFF (2–6 mm)	88.8% (96/108)[a]	72.9% (70/96)[a]	32.3% (20/62)[ab]
10% bFF (>6 mm)	84.5% (98/116)[a]	77.6% (76/98)[a]	32.6% (29/89)[ab]
20% bFF (>6 mm)	85.6% (107/125)[a]	82.2% (88/107)[a]	33.9% (36/106)[b]

a, b – Values in the same column with different superscripts differ significantly ($P < 0.05$).

ved less effective than one containing serum, in terms of fertilization and early embryonic development.

As noted in an earlier section (2.11.3) fetuin is the major glycoprotein in FCS and is thought to be the factor responsible for preventing zona hardening in mouse oocytes during IVM in the absence of serum. Work by Fontes *et al.* (1992b) investigated the effect of fetuin in serum-free IVM medium used with cattle oocytes. In this study, blastocyst yield was lower with fetuin than with FCS. This led the authors to conclude that there may be unknown positive factors present in FCS which are important for early embryonic development.

The factors responsible for effective cattle IVM cannot be clearly identified and defined when serum and other such substances are included in the medium. It is highly desirable that the maturation medium should be chemically defined if the role of hormone supplements and growth factor additives is to be elucidated. It is also necessary to use a defined medium in any study of oocyte metabolism. Downs (1992), working with mouse oocytes, altered the type and concentration of energy substrates in the IVM medium and showed that this can markedly influence oocyte maturation. Several authors, as noted above, have reported successful oocyte maturation in cattle in serum-free medium (Larson and Parks, 1990; Saeki *et al.*, 1990b, c, 1991a; Vergos, 1990; Coskun *et al.*, 1991; Lonergan, 1992; Monaghan, 1993).

2.14 Somatic Cell Support

2.14.1 The cumulus–oocyte complex

It is evident from many reports in the literature that gonadotrophins, steroids and cellular factors all interact to provide essential support for the oocyte dur-

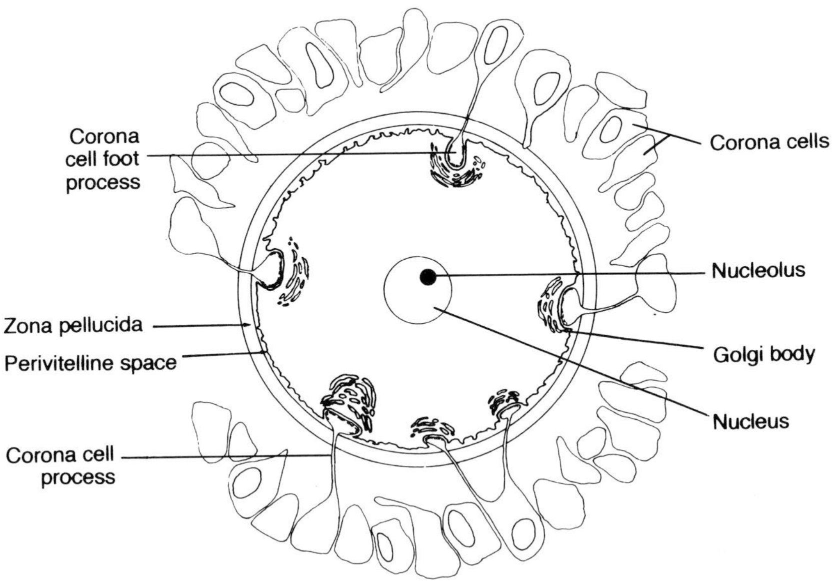

Fig. 2.19. Diagram of ultrastructure of various components of the growing bovine oocyte and of a portion of its zona pellucida associated with corona cells showing various organelles (from Thibault *et al.*, 1987).

ing maturation in the live cow. The preovulatory follicle can be viewed as a structure in which millions of somatic cells contribute towards the welfare of one germ cell, the oocyte. For some years, it has been appreciated that the oocyte is intimately associated with adjacent cumulus cells. Those cells immediately adjacent to the oocyte (the corona radiata cells) have long cytoplasmic extensions which penetrate the ZP and terminate in bulbous swellings closely associated with the oocyte membrane. The importance of these processes and gap junctions is further enhanced by the known impermeability of the oocyte membrane to various metabolites of low molecular weight (e.g. choline, uridine and inositol). The entry of such molecules is believed to be exclusively via the cumulus cell processes and gap junctions. The various components of the primary bovine oocyte and surrounding follicular cells are illustrated in Fig. 2.19.

The metabolic cooperation between oocyte and cumulus cells serves an important nutritive role during maturation. It is known that the cumulus cells remain in contact, via gap junctions, during the entire maturation period (Sutovsky *et al.*, 1993). It is possible that an important role of follicular cells is to compensate for deficiencies in the oocyte's membrane, which is poorly equipped to meet the metabolic requirements of the germ cell. The spherical bovine oocyte has the minimum possible surface area/volume ratio of any cell shape. It is also known that carrier systems in the oolemma are often either totally absent or poorly developed (Moor, 1988).

It is also clear that cumulus cells are linked to one another by gap junctions and that these junctions permit the passage of small molecules from one cell to another. The oocyte and the entire mass of surrounding cumulus/granulosa cells in the follicle can be regarded as a structural and functional syncytium (Thibault *et al.*, 1987; Suss *et al.*, 1988b). The bovine oocyte depends for its maturation on this fact and any undue disturbance of the intimate connections between cumulus cells and the dependent oocyte is likely to adversely affect its subsequent developmental capacity.

In France, Rabahi *et al.* (1991) investigated the *in vivo* qualitative and quantitative changes in protein synthesis by granulosa and cumulus cells of preovulatory bovine follicles relative to the LH surge and the resumption of meiosis in the oocyte. Cumulus cells are considered to be a subtype of granulosa cells and are believed to retain some of the physiological properties of granulosa cells, including gonadotrophin binding and steroid secretion. According to Rabahi *et al.* (1991), various morphological, cytochemical and histochemical differences are recognized to exist between cumulus and granulosa cells which would suggest functional specialization of these cells. A further paper by Rabahi *et al.* (1993) showed that paracrine control of bovine COC metabolism by preovulatory granulosa cells could be exerted not only by way of intercellular contacts but also via substances secreted in the FF.

It is well established that FSH and LH are the major regulators of granulosa cell differentiation. The action of these gonadotrophins is modulated by other peptide hormones, neurotransmitters, steroids, growth factors, cytokines and other factors produced locally by the ovary. Studies by Luck *et al.* (1990) have shown that bovine granulosa cells taken from large vesicular follicles and cultured *in vitro* secrete decreasing amounts of oestradiol and immunoreactive inhibin and increasing amounts of oxytocin and progesterone.

Cumulus cells are linked to granulosa cells, which in turn, are closely associated with the theca interna cells, even though physically they are separated by the basement membrane. It is possible, despite this membrane, that substances produced by thecal cells find their way into the follicular cells and subsequently into the oocyte. There is certainly clear evidence showing that thecal cells can provide support for the oocyte during maturation, even though the means by which their secretory products reach the cell is unclear.

2.14.2 Use of additional cumulus/granulosa cells

One of the firm principles upon which bovine oocyte maturation as practised in Dublin is based is that the IVM culture system provides additional cumulus/granulosa cell (3–5 million cells ml^{-1}) support for the oocyte (Lu *et al.*, 1987b). The source of cumulus/granulosa cells (e.g. follicle size) used in the IVM medium does not appear to be a crucial consideration (Vergos, 1990; Carolan *et al.*, 1992b; Suh *et al.*, 1993). The available evidence strongly supports the view that coculture of the bovine oocyte with live cumulus cells *in vitro* greatly enhances their capacity to undergo normal fertilization and embryonic development.

The first crucial and comprehensive studies showing that cumulus cells play an obligatory role in the process of mammalian oocyte maturation were those of Staigmiller and Moor (1984a) in sheep. They clearly demonstrated that the addition of supplementary cumulus cells to the sheep COC enabled the oocyte to undergo full maturation and normal embryonic development. They also used cumulus cells from pig follicles in sheep IVM to show that cumulus cells supporting maturation need not necessarily be species-specific. Elsewhere and at a later date, Saeki *et al.* (1990c) showed evidence of enhanced quality in bovine oocytes cultured with rabbit preovulatory follicle cells.

Prior to the report of Staigmiller and Moor (1984a), studies by Fulka and Motlik (1980) in coculturing rabbit oocytes with cumulus cells led them to conclude that such coculture favourably influenced subsequent embryonic development. In rats, a study by Wert and Larsen (1989) supported the view that the cumulus oophorus (the mass of cumulus cells immediately surrounding the oocyte) acts as a bridge between the membrana granulosa and the oocyte, allowing the transfer of metabolites and signal molecules from the larger population of granulosa cells to the oocyte.

Beneficial effects of cumulus cells in oocyte maturation have also been reported by other workers in cattle (Fukui and Sakuma, 1980; Critser *et al.*, 1986a, b; Lutterbach *et al.*, 1987; Faundez *et al.*, 1988; Chikamatsu *et al.*, 1989; Fukui and Ono, 1989; King *et al.*, 1990; Fassi-Fihri *et al.*, 1991; Mochizuki *et al.*, 1991a, b; Cox *et al.*, 1993; Suh *et al.*, 1993), pigs (Park *et al.*, 1990a), sheep (Wahid, 1993) and rats (Vanderhyden, 1989; Vanderhyden and Armstrong, 1989). Studies reported by Chian and Niwa (1994) have shown that the presence of intact cumulus cells for at least 12 h is necessary for normal cytoplasmic maturation of cattle oocytes. Where necessary, it has proved possible to preserve granulosa cells by freezing (Chupin *et al.*, 1992) and to use them subsequently in IVM.

It should be noted that there is increasing evidence that meiotic maturation involves a two-way interaction between cumulus cells and the bovine oocyte. Results reported by Allworth and Albertini (1992, 1993), for example, strongly suggest that bovine oocyte IVM involves modification of the cumulus cell cytoskeleton that may regulate the onset, progression and completion of oocyte maturation.

2.14.3 *Theca cell support*

Although IVM systems for use in cattle, as reported by many workers, take note of the possible need to add cumulus or granulosa cells (or a combination of both) to the maturation medium, reports of the addition of thecal cells have seldom been reported. Thecal cells are known to be the major source of androgens, which are obligatory substrates for oestrogen biosynthesis by granulosa cells, and also to secrete growth factors that promote granulosa cell proliferation and play an important role in the regulation of follicular development in the adult bovine ovary. Thecal tissue has also been shown to provide optimum conditions both for maturation and subsequent development of porcine oocytes (Ding *et al.*, 1988; Mattioli *et al.*, 1988a, b, 1989).

Studies by Shi (1991) examined the effect of thecal cells on cattle IVM; it was found that, whether in suspension or in the form of a monolayer or

Table 2.25. Cleavage and embryonic development of bovine oocytes matured *in vitro* in a culture system supplemented with thecal cells, granulosa cells or their conditioned medium (from Shi, 1991).

Treatment	Replicates	Cumulus expansion	Oocytes insem.	Cleavage (%)	% blastocysts developed
1 CM	7	–	442	391 (88.5)[a]	37.1 (164/442)[ab]
2 CM + GC	10	+	549	469 (85.4)[ab]	35.9 (197/549)[b]
3 CM + GCCM	4	+	295	240 (81.4)[b]	33.9 (61/180)[b]
4 CM + GM	4	+	306	275 (89.9)[a]	41.2 (73/177)[ab]
5 CM + TC	4	+	295	257 (87.1)[ab]	44.1 (130/295)[a]
6 CM + TCCM	4	+	595	509 (85.5)[ab]	39.8 (237/595)[ab]
7 CM + TM	7	+	407	361 (88.7)[a]	42.6 (124/291)[ab]

a, b – Within columns, values with different superscripts are significantly different ($P < 0.05$).
CM, control culture medium; GC, granulosa cells; GCCM, granulosa cell-conditioned medium; GM, granulosa cell monolayer; TC, theca cells; TCCM, thecal cell-conditioned medium; TM, thecal cell monolayer.

Table 2.26. Developmental capacity of bovine oocytes matured in IVM medium with different proportions of TCCM (from Shi, 1991).

Treatments	Oocytes inseminated	Cleavage (%)	Blastocysts developed (%)
Control	357	295 (82.6)[a]	125 (35.0)[a]
25% TCCM*	299	257 (86.0)[ab]	109 (36.5)[a]
50% TCCM	299	271 (90.6)[b]	134 (44.8)[b]
100% TCCM	326	279 (85.6)[ab]	108 (33.2)[a]

*25% TCCM, thecal cell-conditioned medium replacing 25% by volume of standard IVM medium.
The data shown in the table are from six replicates.
a, b – Within columns, values with different superscripts are significantly different ($P < 0.05$).

when used in the preparation of a conditioned medium, these cells apparently enhanced the acquisition of developmental competence (see Table 2.25). Thecal cells added to the IVM medium in comparable numbers to granulosa cells (5 million ml^{-1}) resulted in a significantly higher blastocyst yield (44.1% vs. 35.9%). The same worker (Shi, 1991) produced a thecal cell-conditioned medium (TCCM) and showed that, at certain concentrations, it could significantly enhance the developmental capacity of bovine oocytes (see Table 2.26). However, when the proportion of conditioned medium used in IVM was increased to 100%, the frequency of oocytes developing to the blastocyst stage decreased significantly. In Russia, Malenko (1990) cocultured bovine oocytes with either granulosa cells or with both granulosa and theca interna cells; the combination resulted in higher fertilization and development rates.

The effect of gonadotrophins and exogenous steroids on hormone production by bovine thecal cells *in vitro* was examined by Wrathall and Knight (1993); they found evidence that LH and oestradiol stimulated the production of androstenedione by these cells.

2.14.4 Theca–granulosa cell interactions

The mechanism by which the growth and differentiation of granulosa and thecal cells is controlled remains uncertain. However, there is ample evidence to show the importance of granulosa–theca cell interaction. Theca–granulosa cell interaction was the subject of a study by Roberts and Skinner (1990); their results suggest the existence of a feedback loop in which androgens produced by thecal cells are used as a substrate for granulosa cell aromatization to oestrogens, which, in turn, may stimulate thecal cell production of androgen. In Japan, Kotsuji *et al.* (1990) developed a culture system in which bovine granulosa and theca cells were allowed to attach to opposite sides of a collagen membrane. They showed that communication between the two cell types resulted in reciprocal modulation of morphology, proliferation and function.

Subsequently, Kotsuji *et al.* (1992) studied the role of granulosa and theca cells in cattle IVM and concluded that this produced a response characteristic of the growing follicle. There is evidence that growth factors such as TGF-α and TGF-β may be secreted by thecal cells and may also modulate their growth and function (Roberts and Skinner, 1991). Lobb and Dorrington (1992) have suggested that the relative concentrations of TGF-α and TGF-β may determine the rates of proliferation of granulosa cells during follicular development. In sheep, Fricke *et al.* (1992) have reported evidence showing that fibroblast growth factors secreted by granulosa cells may regulate thecal cell proliferation *in vivo*.

2.14.5 Preparation and use of conditioned medium

Vergos (1990) in Dublin attempted to devise an IVM medium which would eliminate the necessity to work with granulosa cells on the day of oocyte maturation. To this end, the standard IVM medium was prepared in 2 ml aliquots, including 20% OCS and granulosa cells, and incubated for 1 day. The medium was then centrifuged and the supernatant either used fresh or after storage at −20°C. Data in Table 2.27 show that both forms of the conditioned medium resulted in cleavage rates and embryo yields comparable to those found with the standard IVM procedure.

In cattle IVM, as noted above, work in Ireland by Shi (1991) examined the use of conditioned medium prepared from thecal cells on maturation and subsequent development of the oocyte. It is much more difficult to prepare live thecal cells than live granulosa/cumulus cells; for that reason, TCCM may be more acceptable in any large-scale production of bovine embryos. The same author found that the effectiveness of TCCM varied with the culture stage of thecal cells; TCCM from day 3–4 of the culture was found to be the most potent (see Table 2.28).

Table 2.27. Use of a non-cellular preconditioned medium for the maturation of the bovine oocyte[a] (from Vergos, 1990).

Treatment of oocytes during IVM	In modified Tyrode's medium				In conditioned medium derived from OCS–BOEC		
	After 48 h from IVF			At 72 h:			
	Cleavage rate (%)	Cleavage index (%)	4-cell (%)	8-cell (%)	Morulae (%)	Blastocyst (%)	Total (%)
CM[b]	36/68 (52.9)[c]	21/36 (58.3)[d]	–	–	1 (1.7)	Nil (0)	1/56 (1.7)[c]
Standard method[e]	62/74 (83.8)[d]	42/68 (69.3)[d]	35/42 (83.3)[d]	31/35 (88.6)[d]	5 (62.5)	3 (37.5)	8/31 (25.8)[d]
Fresh C-MM[f]	80/99 (80.8)[d]	49/80 (61.2)[d]	35/49 (71.4)[d]	33/35 (94.8)[d]	4 (40.0)	6 (60.0)	10/33 (30.3)[d]
Frozen C-MM[g]	28/32 (87.5)[d]	19/28 (67.8)[d]	11/19 (57.8)[c]	10/11 (90.9)[d]	1 (50.0)	1 (50.0)	2/10 (20.0)[d]

[a] Data from three replicates.
[b] Data from five replicates.
[c] Significant at $P < 0.05$ level.
[d] Non-significant.
[e] Standard IVM procedure.
[f] Fresh conditioned maturation medium.
[g] Frozen conditioned maturation medium.

Table 2.28. Developmental capacity of bovine oocytes matured in TCCM from different culture stages (from Shi, 1991).

Culture stage	Oocytes insem.	Cleavage (%)	Blastocysts developed (%)
Day 0–2	313	265 (84.7)[ab]	108 (34.5)[a]
Day 3–4	310	273 (88.1)[ab]	128 (41.3)[a]
Day 5–6	312	275 (88.1)[b]	101 (32.4)[b]
Day 7–8	296	252 (85.1)[ab]	108 (36.5)[ab]

The data shown in the table are from five replicates.
ab – Within columns, values with different superscripts are significantly different ($P < 0.05$).

2.14.6 *Culturing zona-free and artificially denuded oocytes*

There is evidence suggesting that cumulus cells can be removed mechanically from the bovine COC and yet normal maturation be retained (Dominko and First, 1991); this appears to be possible if the denuded oocyte is cultured with conditioned IVM medium. Work in Ireland by Shi (1991) also involved the denudation of bovine oocytes and their subsequent culture in conventional IVM medium. It was found that removal of cumulus cells prior to oocyte maturation resulted in a significant reduction in the developmental capacity of oocytes (see Table 2.29); this reduction was proportional to the degree of cumulus cell removal. The addition of supplementary cumulus cells was unable to reverse the decrease in developmental capacity due to loss of the attached cumulus cells. Whether this was due to a failure of live cumulus cells to produce metabolites and substrates essential for the oocyte is unknown. In the USA, however, Hawk *et al.* (1992) have demonstrated that bovine oocytes selected for good quality can be stripped of almost all surrounding cells before IVM and still retain their developmental potential.

Table 2.29. Effect of cumulus cells on *in vitro* maturation of bovine oocytes (from Jiang, 1991).

Cumulus loss	Extra cumulus	Oocytes	Cleavage (%)	Blastocysts (%)	Hatched (%)
Intact	–	192	164 (85.4)[a]	80 (41.7)[a]	66 (82.5)[ab]
IVF*	–	195	157 (80.5)[a]	88 (45.1)[a]	75 (85.2)[ab]
Most	–	190	158 (83.2)[a]	56 (29.5)[b]	49 (87.5)[ab]
Most	+	181	141 (77.9)[ab]	56 (30.9)[b]	47 (83.9)[ab]
Almost	–	157	131 (83.4)[a]	31 (19.7)[c]	24 (77.4)[ab]
Almost	–	160	114 (71.3)[b]	32 (20.0)[c]	25 (78.1)[ab]

*The cumulus cells of oocytes were removed almost completely at the time oocyte maturation was completed.
a, b, c – Within columns, values with different superscripts are significantly different ($P < 0.05$).

In Canada, Xu *et al.* (1990) have demonstrated that even the zona-free bovine oocyte is capable of being matured, fertilized and subsequently developed to the blastocyst stage of development. Zona-free and cumulus-enclosed bovine oocytes were cultured *in vitro* in a study by Bilodeau *et al.* (1993) on cAMP and the resumption of meiosis; increases in cAMP in cumulus cells may activate a stimulatory signal for maturation and such a signal may overcome the inhibitory effect of a high content of cAMP in the oocyte itself.

Use of cumulus cell monolayer

In the buffalo, attempts have been made to culture denuded oocytes on a cumulus cell monolayer (Pawshe and Totey, 1993); a 6-day-old monolayer was found to support IVM at the highest rate.

2.14.7 *Influence of oocyte on somatic cells*

Although the importance of cumulus/granulosa cells for oocyte development has been recognized for some years, the potential influence of the oocyte on the function of these somatic cells and their responses to stimulation by endocrine, paracrine and autocrine factors has been appreciated much more recently (Buccione *et al.*, 1990; Eppig, 1991b; Eppig *et al.*, 1993). It now appears that oocytes, at least in some species, can promote the organization of the follicle, the proliferation of granulosa cells and the differentiation and function of cumulus cells. In mice, it has been shown that this cumulus expansion enabling factor is secreted by oocytes cultured *in vitro* (Eppig *et al.*, 1993). A diagrammatic representation of possible interactions between oocyte and cumulus cells is provided in Fig. 2.20.

2.15 Hormone Supplementation of the IVM Medium

Bovine oocyte maturation *in vitro* has been investigated by numerous research groups in recent years. Procedures for IVM have differed (Armstrong *et al.*,

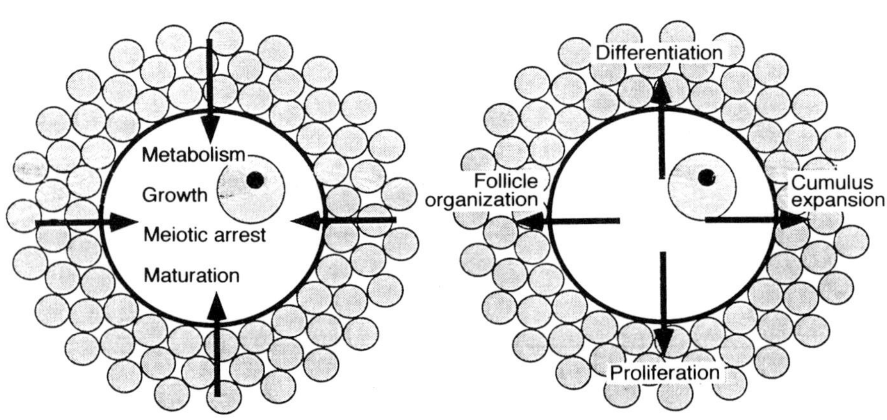

Fig. 2.20. Oocyte–cumulus cell interactions (based on Eppig, 1991b).

1991b; Bavister *et al.*, 1992; Brackett and Zuelke, 1993) and the use of hormone supplements in the medium has been controversial, although the majority of workers currently employ them (Fukushima and Fukui, 1985; Fukui and Ono, 1988; Huang *et al.*, 1989; Sirard *et al.*, 1989; Younis *et al.*, 1989b; Saeki *et al.*, 1990b, c, d, 1991b; Fukushima *et al.*, 1991; Luvoni and Oliva, 1991). Goto and Iritani (1992), reviewing evidence on the hormone supplementation of IVM media, suggest that the hormone content of serum employed in supplementing such media should be taken into account. Supplementation of IVM medium (M-199 + 10% OCS) with FSH and/or LH in studies by Olson *et al.* (1991) showed no beneficial effect on embryo yield but markedly enhanced cumulus cell expansion.

In Dublin, it has been found that bovine oocytes matured in serum-free, hormone-free medium can produce blastocysts after IVF/IVC but that the yield of embryos is enhanced by the addition of gonadotrophin and oestradiol (Lonergan, 1992; Monaghan, 1993).

In Dublin, early work indicated that exogenous hormones may not be required in the IVM system employed (Lu and Gordon, 1987; Lu *et al.*, 1987b, 1988a; Lu, 1990); this has remained true in subsequent large-scale embryo production (0.76 million oocytes) carried out by Ovamass in Ireland over a 2-year period (Lu and Polge, 1992). In sheep, there is the evidence provided by Galli and Moor (1991) showing that oocytes undergoing IVM in the absence of exogenous gonadotrophins have much greater developmental capacity subsequently. An examination of different oocyte maturation and sperm capacitation procedures by X. Yang *et al.* (1993a) showed that IVM treatment in the absence of hormone supplementation could be effective when certain sperm treatment procedures were followed. In Australia, Trounson *et al.* (1994) observe that neither gonadotrophins (FSH/LH) nor growth factors (epidermal growth factor (EGF)/TGF-β) are essential for the production of cattle embryos.

2.15.1 Gonadotrophins (LH, FSH and prolactin) and thyroid-stimulating hormone (TSH)

Luteinizing hormone (LH)

The LH concentration in bFF has been observed to decrease significantly as follicle size increases. This is taken as reflecting some utilization of the LH in the bFF by follicular cells, particularly granulosa cells, which are bathed in that fluid. In the preovulatory follicle, the level of LH is no different from that in the luteal-phase follicle, but shows a peak in the bFF some 4 h after the peak in the peripheral blood (Dieleman and Bevers, 1987). This lag may be due to the occupation of free LH receptors in the follicle wall, since the inward flow of small molecules such as gonadotrophins may not be impeded by the blood–follicle barrier. Brackett *et al.* (1989) showed that the use of a highly purified LH preparation of bovine origin at a certain level ($100\,\mu g\,ml^{-1}$) in their IVM medium significantly enhanced oocyte quality (shown by increased embryo yield after IVF/IVC). Zuelke and Brackett (1990) also demonstrated the beneficial effect of LH in a serum-free medium.

Subsequently, studies by the same workers (Zuelke and Brackett, 1992, 1993) and an earlier report by Zuelke *et al.* (1991) sought to elucidate the mechanisms by which LH influences maturation of the bovine oocyte. Evidence was found that LH may alter calcium distribution within the ooplasm and that the gonadotrophin promotes increased glycolysis, combined with increased mitochondrial glucose oxidation, within cumulus-cell-enclosed bovine oocytes. It was also evident that LH exposure resulted in increased glutamine metabolism within the oocyte.

Follicle-stimulating hormone (FSH)

FSH is believed to be of crucial importance in the recruitment and selection of the DF in cattle, due to its ability to stimulate aromatase activity in the granulosa cell, which converts the follicular microenvironment from androgenic to oestrogenic. There appears to be no significant change in the level of FSH in the bFF of the cow in relation to follicle size. In the preovulatory follicle, the FSH level in bFF is higher than in luteal-phase follicles, before the plasma LH peak, and shows a peak at about 4 h after the LH plasma surge.

From about 6 h after the LH plasma peak, the pattern of FSH concentration in the FF shows a gradual decrease until about 20 h, after which it increases again (Dieleman and Bevers, 1987). The high bFF concentration of FSH at the time of ovulation is possibly correlated with the supposed stimulatory role of the gonadotrophin for ovulatory processes such as mucification of cumulus cells and activation of plasmin. It is well known that FSH stimulates the ovarian secretion of oestradiol and inhibin in the live animal, but their production by cultured granulosa cells *in vitro* appears to decline with time and is not responsive to FSH. In sheep, a report by Campbell and Scaramuzzi (1991) showed that FSH can induce granulosa cells from small follicles to produce both oestradiol and inhibin, in the presence of insulin; in large follicles, however, large doses of insulin and FSH were required.

In view of the known importance of FSH in follicle development in the cow, the gonadotrophin is commonly employed in IVM media in an attempt to make the maturation process more effective. Much of such work suggested that FSH has a beneficial effect and that the presence of the gonadotrophin in the medium enhances expansion of cumulus cells surrounding the oocyte, which in turn enhances sperm capacitation and the fertilization process. However, measurements of meiotic status, fertilization and cleavage rates may be inadequate indices of physiological maturation. Full nuclear and cytoplasmic maturation of the bovine oocyte can only be demonstrated by the formation of normal embryos. A study by Eyestone and Boer (1993) sought to identify specific developmental stages influenced by FSH (in a chemically defined cattle IVM medium); it was concluded that FSH enhanced early embryo development rather than meiotic maturation or fertilization.

Prolactin (PRL)

The PRL level in the peripheral blood of the cow during the luteal phase of the oestrous cycle is higher than that during the period of preovulatory follicle development (Dieleman and Bevers, 1987). However, the concentration of PRL in the FF does not show any significant change throughout preovulatory development, nor are differences evident between preovulatory and luteal-phase follicles. However, relative to follicle size, the PRL concentration in bFF rises as follicle size increases.

While there has been little compelling evidence that PRL plays a beneficial role in the cytoplasmic maturation of the bovine oocyte, it may have some involvement. In fact, the studies of Yoshimura *et al.* (1989) with rabbits suggested that the addition of PRL to their maturation medium, in conjunction with LH, FSH and oestradiol, increased the percentage of oocytes subsequently developing to the morula/blastocyst stage in a dose-dependent manner. This report did indicate a possible contribution of PRL to the acquisition of normal developmental competence by the mammalian oocyte. In cattle, however, Saeki *et al.* (1991a) reported that the addition of PRL to the IVM medium failed to enhance the developmental capacity of oocytes.

Thyroid-stimulating hormone (TSH)

The α subunit of the TSH molecule is identical to the subunits of LH, FSH and human chorionic gonadotrophin (HCG). Some workers have examined the use of this hormone for its gonadotrophic properties in oocyte maturation. Working with mouse oocytes, De Silva and Pearl (1990) reported evidence indicating that TSH possessed gonadotrophic properties in bringing about cumulus cell expansion. Data showing a beneficial effect of TSH during cattle IVM on subsequent embryo development were reported by Younis and Brackett (1992); they suggest the need for further studies to investigate possible follicular receptors and their affinity to TSH.

2.15.2 *Ovarian proteins and peptides (inhibin, activin and oxytocin)*

Inhibins and activins are dimeric proteins comprising two of three subunits, which were initially purified from gonadal fluids based upon their abilities to stimulate or inhibit FSH production by cultured pituitary cells. A very considerable volume of literature is already building up on the role of inhibin and activin in follicular development in mammals. In rats, for example, Woodruff *et al.* (1990) reported how inhibin and activin locally regulate folliculogenesis; inhibin increased granulosa cell proliferation, while activin decreased cell proliferation. The inhibin content of the preovulatory bovine follicle is dealt with in several reports (Guilbault *et al.*, 1991; Juengel *et al.*, 1991). Evidence for the production of inhibin by granulosa cells in culture has been reported by Wrathall and Knight (1991). Other studies at Reading include the use of novel assays specific for dimeric inhibin (Beard *et al.*, 1991; Knight and Muttukrishna, 1993). Analysis of bFF samples from bovine follicles of differing sizes (3–20 mm diameter) showed evidence of a significant increase in dimeric inhibin concentration with increasing follicular volume. Much of this type of information is likely to be valuable at some point in elucidating the role of hormonal factors in bovine oocyte maturation.

Oxytocin

The nonapeptide, oxytocin, was originally detected in the magnocellular neurones of the hypothalamus. In more recent times, this peptide has been found in the ovaries of cattle and other farm ungulates. In cattle and sheep, it appears that luteal oxytocin is synthesized and stored in the large luteal cells, which are believed to be derived from the granulosa cells. It is believed that ovarian capacity to synthesize the hormone varies markedly at different stages of the ruminant oestrous cycle (Wathes *et al.*, 1992a, b). It is also known that oxytocin is present in the FF of preovulatory cattle, sheep and goat follicles. Studies by Voss and Fortune (1991, 1992) have shown that follicular oxytocin is predominantly secreted by the granulosa cells in cattle amd that these cells show a greatly increased capacity to secrete the peptide late in the follicular phase, after the LH surge. Among the hormones to which the maturing oocyte is exposed in the FF of the cow, oxytocin is clearly one; it remains to be seen whether it has any function in oocyte maturation.

2.15.3 Steroids (oestradiol, and progesterone)

As noted elsewhere (Section 2.4.5), oestradiol plays an important role in follicle development. The steroid sensitizes the granulosa cells to respond to gonadotrophins and stimulates the proliferation and differentiation of granulosa cells. The role of progesterone in the follicle is less well defined, although it is clear that the balance between oestradiol and progesterone changes in the preovulatory follicle during the period when maturation of the bovine oocyte occurs. As to the effect of including oestradiol in the cattle IVM medium, particularly when an alternative source of the hormone (e.g. OCS) is not employed, the general consensus appears to be that it should be included at a level of 1 $\mu g\ ml^{-1}$ (Hensleigh and Hunter, 1983a b; Xu *et al.*, 1986a; First and Parrish, 1987a; Fayrer-Hosken *et al.*, 1988; Garcia *et al.*, 1988; Sirard *et al.*, 1988a, b; Yoshimura *et al.*, 1989; Wiemer *et al.*, 1991a, b). There is some evidence that at higher dose levels, oestradiol may have an adverse effect on spindle formation and polar body extrusion (Kruip *et al.*, 1988). A study of the effect of certain steroids (oestradiol, progesterone, androstenedione) on the primary bovine oocyte suggested that they may alter protein uptake and incorporation (Pontbriand *et al.*, 1989).

2.15.4 Other hormones

Insulin

The effect of insulin on various aspects of IVM and IVF of bovine oocytes has been examined. Although work by Stubbings (1989) with insulin in IVM was apparently without effect, a study reported by Zhang *et al.* (1991a) showed that cumulus-enclosed bovine oocytes significantly improved cumulus cell expansion scores and fertilization rates after exposure to insulin-supplemented M-199 medium. It was also observed that the number of cells per embryo was significantly higher when insulin (1 μg ml^{-1}) was employed. In a further report (Zhang *et al.*, 1991c), the same workers showed that insulin added to the IVM medium significantly increased the level of maturation, fertilization and cleavage of pig oocytes, compared with those matured *in vitro* without insulin.

Growth hormone (bovine somatotrophin)

Little information is available on the effect of BST on ovarian function and even less on oocyte maturation. It is known that growth hormone increases IGF-I secretion by granulosa cells *in vitro*. In former Czechoslovakia, Sirotkin and Nitray (1992) record that BST treatment during IVM had no marked influence on the resumption of meiosis, but significantly delayed its completion in a dose-related manner. The subsequent developmental capacity of oocytes was not affected. The effect of pig growth hormone in the IVM medium on cytoplasmic maturation of porcine oocytes was examined by Hagen and Graboski (1990), who found evidence that the hormone accelerated the normal course of maturation.

2.15.5 Controlling time sequence of hormonal events in IVM

In the live cow, as the DF develops to the preovulatory stage, oestradiol predominates; finally, this steroid ceases to be synthesized by the follicle and is replaced by progesterone just prior to ovulation. For such endocrinological changes to be fully simulated *in vitro*, it may be necessary to think in terms of deliberately changing the steroid environment from oestradiol-dominant to progesterone-dominant in the course of cattle IVM treatment. In The Netherlands, Kruip *et al.* (1988) concluded that high concentrations of oestradiol may have a disorganizing effect on the meiotic spindle of the bovine oocyte and that the steroid should not be present for more than 10 h after the onset of IVM. These authors suggest that, in adding hormones to IVM medium, consideration be given to bringing the time sequence of endocrine events more in line with what is known to occur in the cow's preovulatory follicle.

Studies in Canada (Stubbings *et al.*, 1988a, b; Stubbings, 1989) examined the effect of adding FSH/LH to the IVM medium 6 h after the onset of maturation, with oestradiol and insulin being present throughout the entire 24 h culture period; higher fertilization rates and development to morulae (81% vs. 60% and 36% vs. 10%, respectively) were obtained when gonadotrophins were added at 6 h. In pigs, Funahashi and Day (1993a) also took account of the fact that cytoplasmic maturation probably requires a strictly regulated sequence of hormonal changes during maturation. They showed that transferring pig COCs from hormone-supplemented IVM medium, at 20 h after culture onset, enhanced the cytoplasmic maturation of oocytes.

2.16 Growth Factors and Cytokines in the IVM Medium

2.16.1 *Paracrine and autocrine regulation of ovarian events*

The meiotic maturation of mammalian oocytes is known to be regulated by gonadotrophins in the live animal. The endocrine factor that is generally accepted as regulating the induction of bovine oocyte maturation *in vivo* is the preovulatory surge of LH. It is clear that LH release is but one of many factors involved in a complex sequence of events occurring in the preovulatory follicle in the hours preceding ovulation. It is also evident that oocyte maturation is regulated by growth factors and cytokines. It is likely that such substances will attract increasing research attention in the years ahead.

It is now well recognized that there is a cell-to-cell communication system operating within the bovine ovary; the terms paracrine and autocrine regulation have been introduced to cover such communications. Paracrine communication involves local diffusion of regulators (growth factors/cytokines) from producer to target cells; this may include more than one cell type. On the other hand, autocrine communication postulates that a regulatory agent acts on the surface receptor of its cell of origin. In this, a single cell serves as a site of regulator production and reception as well as action. Growth factors and cytokines may be viewed as local regulators involved in subtle coordination of cellular proliferation and differentiation. It seems likely that the bovine oocyte, in the period immediately preceding ovulation, is exposed to an ever-changing intrafollicular environment of hormones, growth factors and cytokines, each of which may be acting alone, or more probably in concert with other molecules, in affecting its cytoplasmic and nuclear maturation.

2.16.2 *Use of growth factors*

Among the growth factors currently implicated in the modulation of oocyte maturation are EGF, TGF-α, TGF-β, IGF-I and IGF-II. EGF is a single-chain polypeptide with a molecular weight of 6045 daltons, and is known to be a potent mitogen for many cells, including granulosa cells. In a study of EGF and its receptor in the human ovary, Maruo *et al.* (1993) found that the expression of EGF and EGF receptor increased as the size of the oocyte increased; such findings imply a possible role for EGF in modulating oocyte maturation.

A large body of evidence now suggests the existence of an intraovarian IGF system, complete with ligands, receptors and binding proteins (Adashi, 1990; Adashi *et al.*, 1991). It appears that this intraovarian IGF system is largely concerned with amplification of gonadotrophin action in the various events involved in follicular growth and differentiation. IGF-I is a 70-residue single-chain polypeptide with about 45% sequence homology with insulin.

In vitro studies have established that granulosa cells can secrete IGF-I and that this growth factor is stimulatory to granulosa cell mitosis and steroidogenesis (Adashi *et al.*, 1991). Studies reported by Spicer and Geisert (1992) in cattle showed that small ovarian follicles contained significantly less IGF-I than medium-sized or large follicles. Whether this was due to increased local biosynthesis or increased diffusion of IGF-I from serum was not clear. Production of other growth factors is known to be in close proximity to the follicle; EGF-like activity is evident in pFF and both TGF-α and TGF-β have been identified in the thecal cells of bovine ovaries (Lobb and Dorrington, 1992).

The effect of growth factors on oocyte maturation has been examined in cattle by several investigators. Sanbuissho *et al.* (1990b) demonstrated that EGF is capable of promoting bovine oocyte maturation *in vitro*. This same factor was employed in cattle IVM by Takagi *et al.* (1991) without any evident effect on oocyte quality. The binding of EGF to granulosa cells was examined by Rose *et al.* (1991). Illera *et al.* (1992a, b; Illera and Petters, 1993) examined the effect of EGF on bovine IVM and reported that the growth factor had a significant effect on IVM when cumulus-enclosed oocytes were used, but not with denuded oocytes. In Georgia, Harper and Brackett (1991, 1992) have reported on the action of EGF and IGF-I in IVM; a beneficial effect was evident when these growth factors were combined with gonadotrophins.

In Germany, the addition of 50 ng ml^{-1} of IGF-I to the IVM medium improved the subsequent yield of embryos (Herrler *et al.*, 1992). Studies by Jiang (1991) in Ireland evaluated the effect of TGF-α, TGF-β and IGF-II in the maturation of bovine oocytes. In this work, evidence was found that inclusion of these three growth factors significantly improved the quality of oocytes (see Table 2.30). The same worker also found some evidence of a favourable effect of EGF in the cattle IVM medium. It should be noted that EGF is structurally and functionally related to TGF-α. It may be speculated that some of these growth factors are among the molecules present in bFF or TCCM which may be responsible for the favourable effects observed after the substances are added to the cattle IVM medium (Roberts and Schlitz, 1991). There is evidence in mice that TGF-α plays a role in meiotic maturation and that its effect is

Table 2.30. Effect of growth factors on *in vitro* maturation of bovine oocytes and their subsequent development (from Jiang, 1991).

				Embryos developed to				
				Blastocyst				Hatched blastocysts
TGF-α	TGF-β	IGF-II	Cleavage (%)	D7	D8	D9	Total (%)	(%)
+	+	+	162/187 (86.6)	89	14	4	107 (57.2)[c]	90 (84.1)
+	+	–	153/179 (85.5)	79	14	6	99 (55.3)[bc]	85 (85.9)
+	–	+	159/182 (87.4)	82	17	3	103 (56.6)[c]	87 (84.5)
+	–	–	148/176 (84.1)	80	12	5	97 (55.1)[bc]	81 (83.5)
–	+	+	155/185 (83.8)	75	10	9	94 (50.8)[abc]	78 (83.0)
–	+	–	157/189 (84.7)	68	11	6	85 (45.0)[ab]	70 (82.4)
–	–	+	150/178 (84.3)	68	13	3	84 (47.2)[abc]	71 (84.5)
–	–	–	161/188 (85.6)	61	11	10	82 (43.6)[a]	69 (84.2)

a, b, c – $P < 0.05$.

mediated by the surrounding cumulus cells (Downs, 1989; Brucker *et al.*, 1991). In rats, Feng *et al.* (1991a) have shown that TGF-β significantly influenced oocyte maturation. A maturation-promoting effect of EGF was reported by Das *et al.* (1991a, b) in mice; the same authors subsequently tested EGF in IVM of human oocyes and obtained promising results in their maturation. In The Netherlands, Loos *et al.* (1993) found TGF-α and EGF to be without effect in their studies of bovine oocyte maturation. Harper and Brackett (1993), on the other hand, suggest that the presence of EGF in serum may be one of the undetermined components contributing to enhanced oocyte maturation.

Elsewhere, in Spain, Lorenzo *et al.* (1992, 1993) and workers in the USA (Park and Lin, 1993) found some evidence of a favourable effect of EGF in their cattle oocyte maturation media. The effect of EGF in pigs was examined by Illera *et al.* (1992a), Arellano *et al.* (1993) and Coskun and Lin (1992), who found that the growth factor enhanced oocyte maturation *in vitro* in the absence of hormones or FF. In all of this, it is essential that an appropriate chemically defined IVM system is available for use in the elucidation of growth factor effects.

2.16.3 *Use of cytokines*

Although the concept of stress and corticosteroids as negative influences on immune function is well recognized, it is now evident that the cells of the immune system can be markedly influenced by the endocrine system. Arkins and Kelley (1992), in reviewing the neuroendocrine control of immune competence, concluded that growth hormone and, probably, PRL, play critical roles in the development and function of immune tissues. The physiological mediator of many growth hormone actions, IGF-I, is known to have a profound effect on a number of immune events. It is also clear that IGF-I itself is produced by some immune cells, in response to growth hormone stimulation, and may act in an autocrine manner. The neuroendocrine and immune systems can now be viewed as active partners participating in the maintenance of the animal's defences.

Cytokines are small regulatory peptides or glycoproteins, with molecular weights ranging from 6000 to 60,000, which are synthesized and secreted by activated immune and mesenchymal cells (Ben-Rafael and Orvieto, 1993). Cytokines are believed to act generally in a paracrine or autocrine manner rather than like hormones, on distant target cells. In Israel, Zolti *et al.* (1991) studied the secretory products of the human COC; they found IL-1, IL-6 and colony-stimulating factor 1 (CSF-1) to be secreted. It is possible that cytokines originating from the oocyte have a role in preparing the maternal immune or endocrine systems for subsequent events in fertilization and early pregnancy (Ben-Rafael and Orvieto, 1993).

The cells of the immune system include macrophages, monocytes, T lymphocytes and B lymphocytes. Activated macrophages engulf and process antigen and produce several cytokines, one of the most important of these being IL-1. This is a 17-kilodalton polypeptide which appears to play a role in the developing follicle by targeting both the granulosa and the theca interstitial cells as its sites of action (Hurwitz *et al.*, 1991). Resident ovarian macrophages have long been recognized as potential *in situ* regulators of ovarian function; it appears that these cells are gonadotrophin-dependent and probably subject to cyclical hormonal regulation.

Little information is available on the role of cytokines in follicle growth and development in cattle. One report by Grizzle *et al.* (1991) has shown that human recombinant IL-1β may have an effect on the maturation and differentiation of the growing follicle.

It has been observed that ovarian follicle rupture, followed by oocyte extrusion, probably involves mechanisms common to a typical inflammatory response. Such a reaction is known to involve the production of IL-1 by monocytes and macrophages. There is evidence to suggest that macrophages may possess gonadotrophin receptors and are activated by gonadotrophin. Results presented by Sjogren *et al.* (1991) are compatible with the view that IL-1α has a potential regulatory role in the granulosa–luteal transition that occurs with the formation of the CL.

2.17 Evaluating the Maturation Process

2.17.1 Stages in nuclear maturation

During bovine oocyte maturation, chromatin structures within the immature egg proceed through a series of morphological arrangements, starting at prophase of the first meiotic division and continuing to metaphase of the second division. Several workers have described the changes which occur during IVM. Early work in Dublin showed that about 80% of bovine oocytes had matured by 24 h after the onset of culture (Sreenan, 1968, 1969); in that study, the nuclear membrane disappeared after 5–6 h (GVBD) and metaphase I was reached after about 12 h. More recent work in Dublin by Lu (1990) shows much the same result (see Table 2.31), with more than 90% of oocytes completing maturation within these time limits. Such findings are in agreement with those of Fukushima and Fukui (1985), who cultured bovine oocytes *in vivo* in the rabbit oviduct. In the data of Lu (1990), the peak period of oocytes showing metaphase II was 24 h. Reports in the literature show a wide range of times for the occurrence of the peak period for metaphase II, varying from 20 h (Fulka and Motlik, 1980), 21 h (Jagiello *et al.*, 1974), 20–24 h (Suss and Wuthrich, 1985), 24 h (Xu *et al.*, 1986b; Sirard *et al.*, 1988a), 24–30 h (Sato *et al.*, 1977) to 31 h (Edwards, 1965). Such differences may have been due to difficulties inherent in the culture conditions and recording procedures. The mature bovine oocyte is shown in Fig. 2.21. Spiropoulos and Long (1989) recorded most bovine oocytes at metaphase II after 19–24 h in IVM culture and observed that partially denuded or wholly denuded oocytes progressed to metaphase II more rapidly than compact COC.

In the Dublin work, an effect on nuclear maturation was evident according to whether or not a hormone supplement was employed in the IVM

Table 2.31. Effect of hormones (gonadotrophins + oestradiol) added to the culture medium on nuclear maturation of the bovine oocyte (from Lu, 1990).

Cult. period (h)	Hormone additives	No. of oocytes exam.	Stages of nuclear meiosis (%)							
			CV	GVBD	D	Pre-MI	MI	AI	TI	MII
4	+	55	14.5	80.0	5.5	–	–	–	–	–
	–	55	0.9	87.3	3.6	–	–	–	–	–
8	+	50	–	22.0	40.0	36.0	–	–	–	2.0
	–	52	–	13.5	48.1	38.4	–	–	–	–
12	+	58	–	–	–	48.3	50.0	1.7	–	–
	–	58	–	–	–	56.9	43.1	–	–	–
16	+	60	–	–	5.0	–	70.0	25.0	–	–
	–	57	–	–	1.8	1.8	42.1	50.9	3.5	–
20	+	66	–	–	3.0	–	3.0	–	75.8	18.2
	–	65	–	–	–	–	–	–	81.5	18.5
24	+	75	–	–	–	–	1.3	–	13.3	85.3
	–	75	–	–	1.3	–	–	–	9.3	89.3

Data are from five replicates.
AI, anaphase I; D, diakinesis; GV, germinal vesicle; GVBD, germinal vesicle breakdown; MI, metaphase I; MII, metaphase II; TI, telophase I.

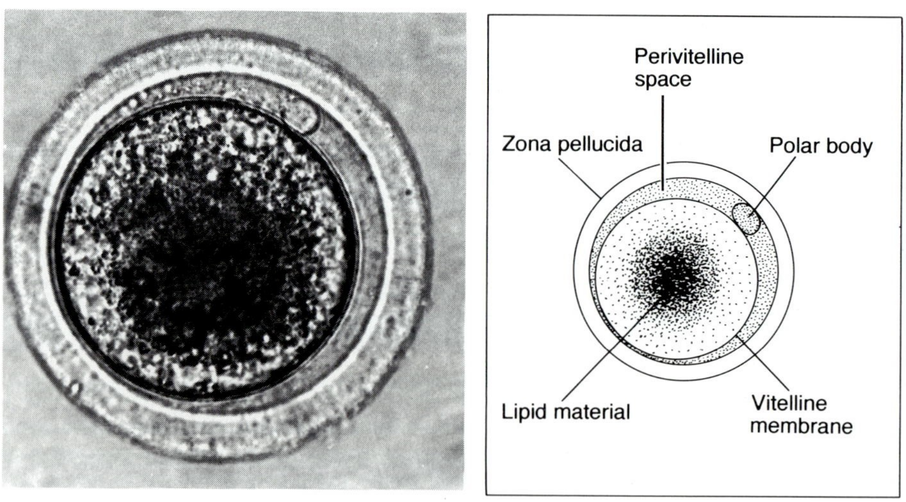

Fig. 2.21. The matured bovine oocyte.

medium; a delay was evident in the resumption of meiosis when oocytes were cultured with the hormones (Lu, 1990). A similar delay in the resumption of meiosis was reported in bovine oocytes by Suss *et al.* (1988a), this delay being related to the increase in cAMP level measured in the oocytes. cAMP has been shown to inhibit the resumption of meiosis in pig oocytes (Rice and

McGaughey, 1981). It may be that the addition of hormones to the IVM medium induces a rise in cAMP in the bovine oocytes, resulting in a transient delay in the resumption of meiosis.

2.17.2 Cumulus cell expansion

Maturation of the liberated bovine oocyte *in vitro* involves well-defined changes in the cumulus cells which surround it. As reported by Hensleigh and Hunter (1983c), the cumulus cells expand to form a spherical mass in three dimensions and the COCs appear to 'float' in the culture dish. A classification scheme employed by Hunter and Moor (1987) for assessment of bovine oocytes after IVM is in Table 2.32. This is based on the degree of cumulus expansion. It is believed that cumulus cells are stimulated by gonadotrophin (FSH) and growth factor (EGF) to produce and secrete hyaluronic acid, which disperses the cells, a process known as expansion or mucification. Cells of the membrana granulosa do not undergo expansion but become luteal cells in the subsequent formation of the CL.

In Dublin, Lu and Gordon (1987) showed that expansion of cumulus cells in cattle IVM was significantly affected by the employment of additional cumulus/granulosa cells. This may imply production by such cells of a factor, such as EGF, which plays a role in cumulus expansion.

In most human IVF programmes, oocyte maturation is assessed by the degree of cellular expansion and mucification of the associated cumulus cells. As pointed out by Osborn and Moor (1985), although such an assessment allows obviously atretic oocytes to be discarded, it does not indicate whether the oocytes are completely mature, nor can the readiness of the oocyte to undergo fertilization be ascertained. The same authors also note that cumulus expansion may sometimes occur independently of oocyte maturation.

Cumulus cell–oocyte coupling in pigs was examined by Motlik *et al.* (1986), who record that the corona radiata cells only expand towards the end

Table 2.32. Classification scheme for assessment of oocytes after maturation (based on the assessment of the degree of cumulus cell expansion) (from Hunter and Moor, 1987).

Grade 1: Full cumulus cell expansion
This is characterized by the extremely adhesive nature of the cumulus cells and the enlargement of the cumulus cell mass to at least $\times$ 3 diameters ($>$ 300 μm) away from the zona pellucida

Grade 2: Moderate cumulus cell expansion
This is where the observed expansion of the cumulus cells would be of the order of $\times$ 2 diameters ($>$ 200 μm) away from the zona pellucida

Grade 3: Slight expansion of cumulus cells
This category would be characterized by cumulus cells being tightly adherent to the zona pellucida

of the maturation process; for much of maturation, oocyte and corona radiata remain closely coupled. In the report of Mattioli *et al.* (1988a), again in pigs, it is shown that cumulus cell expansion proceeds from the superficial layers to the deeper ones, with the corona radiata layer being reached only at the end of maturation.

It is known that cumulus cell expansion during IVM involves radical changes in cellular shape and in interactions between follicle cells and the oocyte (Allworth and Albertini, 1992, 1993); in the course of cattle oocyte maturation, there are complex rearrangements of the cytoskeleton of cumulus cells and the cortical ooplasm.

The application of quantitative videographic techniques in work with sheep, rat and mouse oocytes provided evidence that the degree of expansion observed in IVM medium lacking substrates of hyaluronic acid synthesis may be less than the expansion which occurs *in vivo*; supplementing the IVM medium with glucosamine permitted expansion similar to that found *in vivo* (Chen *et al.*, 1989, 1990a, b).

There is now clear evidence in some species that the proliferation, differentiation and function of mouse granulosa cells are dependent upon the oocyte (Buccione *et al.*, 1990; Vanderhyden *et al.*, 1990, 1991; Vanderhyden, 1993). There is also information indicating that a factor(s) secreted by mouse and rat oocytes is necessary for cumulus cells to undergo expansion. In pigs, however, oocytes can secrete a cumulus expansion enabling factor, but expansion of pig COCs does not depend on this factor (Singh *et al.*, 1993a, b; Vanderhyden, 1993); this suggests that regulation of cumulus expansion in the pig COC is different from that in rodents.

2.17.3 Morphological assessment and staining methods

After *in vitro* maturation of the bovine oocyte, the cell is surrounded by an expanded cumulus, there are changes in the dimensions of the perivitelline space, expulsion of the first PB into that space and the establishment of the second metaphase spindle tangential to the vitelline surface. Suss and Madison (1983) noted that the PB could only be detected in about 50% of oocytes with a chromosomal configuration of metaphase II, such results implying that absence of the PB in an oocyte did not necessarily indicate a failure in nuclear maturation. Many other morphological and ultrastructural features have been described by workers, and are valuable in assessing the normality or otherwise of the maturation process.

Kruip *et al.* (1983) are among those who have described changes occurring in the oocyte of the live cow in organelle structure and location in relation to the time of the preovulatory LH surge. One of the first visible effects of the LH surge is the formation of the perivitelline space, with the simultaneous loss of intercellular contact between the cytoplasmic processes of the corona radiata cells and the vitellus. During maturation, there is a migration of vesicles, mitochondria and lipid droplets to the centre of the oocyte, resulting in an almost organelle-free cortical region characterized by the presence of 'nests' of SER and cortical granules beneath the vitelline membrane. Nucleolar changes at the ultrastructural level *in vivo* during oocyte maturation in the cow have been described by Crozet *et al.* (1986). The ultrastructure of bovine oocytes during IVM has been reported by Hyttel *et al.* (1987a) in Denmark and Loos *et al.* (1992b) in The Netherlands. Such studies are valuable in establishing standards against which IVM systems can be evaluated.

The normal technique used in Dublin for evaluating nuclear maturation has involved fixing the naked oocyte in ethanol : acetic acid (3 : 1) for at least 24 h and then staining with 1% lacmoid in 45% acetic acid solution (Lu, 1990). A paper by Byun *et al.* (1991), however, described a rapid staining method, using Tyrode's solution, acetoethanol, basic fuschin and acetoglycerol; this has been employed in assessing nuclear configuration both in cattle (Fair, 1992) and in human oocytes (Gordon *et al.*, 1992). Elsewhere, use has been made of a DNA-specific stain (Hoechst 33342) for visualizing the chromatin of live oocytes under ultraviolet (UV) light (Viuff *et al.*, 1991); however, the stain/UV light appeared to compromise subsequent fertilization.

In any discussion of evaluating bovine oocyte maturation, it must be emphasized that normality cannot be confirmed by any morphological examination. Rose and Bavister (1992) suggest that the developmental capacity of embryos derived from IVM/IVF oocytes would be a much more definitive way of assessing the normality of maturation, given that IVM/IVF embryos are capable of developing to the blastocyst *in vitro*. Such views date back to the findings of Moor and Trounson (1977), who suggested that the only valid guide to complete oocyte maturation was the capacity of the oocyte to develop after fertilization to at least the blastocyst stage of development. The same authors stressed the importance of not using earlier cleavage stages as a measure of normal maturation, noting that many abnormalities induced during culture are not detectable until the blastocyst stage.

2.18 Maturation of Oocytes in Other Farm Animals

The amount of information on oocyte maturation in farm species other than the cow is limited. It is clear that maturation conditions appropriate for cattle oocytes are not necessarily as effective in the other farm species; each IVM culture system must be developed on the basis of providing oocytes capable of producing viable embryos and young at term.

2.18.1 Sheep and goats

Work in Dublin in the late 1960s by Crosby (1969) showed that most sheep oocytes reach metaphase II after about 24 h of *in vitro* maturation (see Table 2.33). Staigmiller and Moor (1984a, b) were the first to show that, under appropriate IVM culture conditions, sheep oocytes are capable of acquiring full developmental competence. Oocyte recovery was by follicle dissection and they employed M-199 supplemented with 10% FCS, gonadotrophins, oestradiol and additional granulosa cells; oocytes were cultured at 37°C for 24 h in a culture system providing gentle agitation. Developmental competence

Table 2.33. The progression of nuclear maturation in sheep oocytes matured *in vitro* (from Crosby, 1969).

	Culture period (h)														
Maturation stages	2	4	6	8	10	12	14	16	18	20	22	24	25–27	28–30	48	Totals
Vesicular nucleus	11	7	3	1	14	1	5	1	2	1	4	6	14	2	–	72
Diakinesis chromatin	1	3	5	5	5	3	2	0	2	0	2	0	3	0	–	31
Prometaphase	–	–	8	3	1	1	1	3	4	4	9	5	8	0	–	47
Metaphase plate	–	–	6	8	8	11	21	14	0	4	4	14	35	11	3	139
Late anaphase	–	–	–	–	–	–	2	2	9	1	0	2	7	–	–	23
Telophase	–	–	–	–	–	–	1	3	7	7	3	4	5	–	–	30
Prometaphase II + polar body	–	–	–	–	–	–	–	–	5	–	1	0	0	–	–	6
Metaphase II + polar body	–	–	–	–	–	–	–	–	2	–	5	7	40	11	3	68
All stages	12	10	22	17	28	16	32	23	31	17	28	38	112	24	6	416

was assessed by transfer of oocytes to inseminated recipient ewes. Other work in Cambridge used M-199, containing 10% FCS, in sheep IVM (Agrawal and Polge, 1988); 53% of oocytes reached metaphase II in the study. In France, Szollosi *et al.* (1988) described an *in vitro* system which apparently induced full meiotic maturation in 98% of follicular sheep oocytes recovered from follicles 2–6 mm in diameter; the IVM medium contained FSH, LH ($10\,\mu g\,ml^{-1}$ of each), oestradiol ($1\,\mu g\,ml^{-1}$) and additional mural granulosa cells. The gonadotrophin requirements for normal oocyte maturation *in vivo* in sheep were examined by G.Evans *et al.* (1992) in Australia; these workers concluded that LH pulses are not essential for oocyte maturation or early embryonic development in sheep.

In the collection and storage of sheep ovaries after slaughter, Moodie and Graham (1989) showed that a transport temperature of 22°C was the most suitable of those investigated (5, 22 and 37°C). Wahid (1993) showed that aspiration of follicles *in situ* was a less effective oocyte recovery procedure than ovary slicing (see Table 2.8). Earlier work in Dublin examined several methods of surface-slicing sheep ovaries (Mehmood *et al.*, 1992); such methods resulted in a marked increase in oocyte yield (14–36 per ovary) in comparison with aspiration. The IVM/IVF of sheep oocytes collected by follicle aspiration using a laparoscopic procedure is reported in a paper by Baldassarre *et al.* (1994) from Argentina; data showed such oocytes to be as capable of maturation and fertilization as abattoir-derived oocytes.

In the IVM of sheep oocytes, Wahid *et al.* (1991) showed the value of morphological evaluation of oocytes prior to maturation. Subsequently, the same workers reported that high rates of nuclear maturation can be achieved in the absence of hormone supplementation (Wahid *et al.*, 1992).

Elsewhere, Galli and Moor (1991) had earlier drawn attention to results with sheep indicating that the addition of steroids during maturation enhanced neither cell cycle progression nor developmental potential of IVM ovine oocytes. The same workers reported that sheep oocytes undergoing IVM in the absence of gonadotrophins have much higher viability than many of those cultured in the presence of gonadotrophins. In Australia, O'Brien *et al.* (1994) found no beneficial effects from using hormones (FSH + LH + oestradiol) in their sheep oocyte IVM medium. In New Zealand, Pugh *et al.* (1991b) found evidence of higher cleavage rates with oocytes originating from FSH-prestimulated donors than from oocytes recovered from untreated sheep.

In goats, in which the maturation period is, as with sheep, about 24 h, Younis *et al.* (1990, 1991) found that the addition of FSH or LH to the IVM medium (serum-supplemented M-199) resulted in a significant improvement in maturation rate. Cytogenetic studies of goat oocytes after IVM were reported by Deb and Goswami (1990). De Smedt *et al.* (1991, 1992) matured caprine COCs in IVM medium supplemented with hormones and granulosa cells. They found that most oocytes from follicles of 2–6 mm diameter achieved meiotic maturation (86%) whereas those from follicles of 1–2 mm diameter did not (24%). In Spain, working with prepubertal goat oocytes, Palomo *et al.* (1992) examined different IVM culture systems. As with cattle and sheep, protein synthesis was required for GVBD in goats (Le Gal *et al.*, 1992).

A study by Mogas *et al.* (1992) examined ovary slicing in comparison with aspiration and follicle dissection in recovering oocytes from prepubertal goats. Substantially higher numbers of oocytes were recovered by slicing than by other procedures (92.3 vs. 1.04–2.39). The effect of these recovery methods on IVM was dealt with by Martino *et al.* (1992). Other work in Spain (Mogas *et al.*, 1993) demonstrated a beneficial effect on cleavage rate when hormones (FSH/LH/oestradiol) were added to the goat IVM medium; increasing the LH concentration was without effect. In a further paper from the same laboratory, it is suggested that granulosa cells from adult goat ovaries may confer developmental competence on oocytes recovered from prepubertal ovaries (Martino *et al.*, 1993).

2.18.2 Pigs

Studies with the ovaries of slaughtered gilts in former Yugoslavia, reported by Stancic *et al.* (1992), show 7.9 and 45.3 oocytes per ovary for aspiration and dissection procedures, respectively. Protein synthesis and metabolic cooperation between follicle cells and oocytes during oocyte development in the pig, from an early stage, have been reported (Lazzari and Moor, 1990; Lazzari *et al.*, 1991; Morbeck *et al.*, 1991). The crucial importance of the interaction between somatic and germinal cells in oocyte maturation in pigs has been well emphasized (Mattioli, 1989, 1992; Mattioli *et al.*, 1990, 1991a, b, 1992; Park, S.B., *et al.*, 1991; Nagai, 1994). In this species, it has been shown in early work that cumulus–oocyte interaction is easily lost *in vitro* and that only cultures involving the presence of abundant follicle wall components were able to create the conditions for competent oocyte maturation (Mattioli *et al.*, 1988a, b).

A study by Mattioli *et al.* (1989) suggested that porcine oocytes apparently depend on progesterone both for the maintenance of cell coupling with their associated cumulus cells and for subsequent embryonic viability. The rate of male pronucleus formation is enhanced by the addition of granulosa cells to the IVM medium (Park *et al.*, 1990a). In Germany, Eroglu and Meinecke (1990) have argued that failure of IVM pig oocytes to undergo normal fertilization and development may be due to inadequate changes in the synthesis of specific polypeptides during maturation. Studies reported by Miller, K.F., *et al.* (1991) have shown that the level of immunoreactive inhibin in FF is related to meiotic stage; however, the role of the peptide, if any, in maturation remains unclear.

The duration of meiotic maturation is substantially longer in pigs (42–44 h) than in cattle, sheep and goats (24 h). The importance of the maintenance of intercellular coupling between cumulus cells and the oocyte was emphasized by Moor *et al.* (1990), who concluded that the contribution made by follicle cells to cytoplasmic maturation is fully evident only when gonadotrophins are present in the IVM medium and when a non-static culture system is used. The addition of LH and FSH to the IVM medium was found to accelerate and facilitate meiotic maturation in several studies (Galeati *et al.*, 1990; Musse, 1990; Mattioli *et al.*, 1991a); LH selectively improved cytoplasmic maturation. Evidence from the work of Wu *et al.* (1991a, b) suggested that porcine oocyte quality was higher in those recovered from large follicles than from small (range 1–5 mm diameter). In Illinois, Shyu *et al.* (1990) have described a classification system suitable for porcine oocyte maturation, based on the chromatin arrangement.

An IVM medium containing a high concentration of cysteine (such as Waymouth MB 752/1) was found to enhance the ability of pig oocytes to form the male pronucleus (Yoshida and Ishigaki, 1991; Yoshida *et al.*, 1992a, b, 1993a, b). It is believed that the addition of cysteine to the IVM medium may be an important factor in the synthesis of glutathione by the pig oocyte and in enhancing formation of the pronucleus (PN) after sperm penetration. The microinjection of glutathione directly into pig oocytes has been shown to increase male PN (MPN) formation (Naito and Toyoda, 1992); the indications are that failure to form the MPN may be partly due to a low intracellular glutathione concentration.

Results from a study by Ding and Foxcroft (1992) in Canada confirmed that follicular shells apparently secrete factors that support cytoplasmic, if not nuclear, maturation. They also showed that MPN formation rate, in sperm-penetrated oocytes, was correlated to diameter and shell weight of cocultured follicles; large follicles were more supportive of cytoplasmic maturation than small ones. Elsewhere, Funahashi and Day (1993a, b) showed that the removal of hormone supplements from IVM medium at 20 h after culture enhanced cytoplasmic maturation and cumulus expansion. They also found that a serum-supplemented IVM medium reduced the ability of pig oocytes to form an MPN; serum supplementation apparently promoted rapid progression of meiotic maturation. The authors suggest that serum in the IVM medium may not allow pig oocytes sufficient opportunity for intercellular communication with cumulus cells, which may influence MPN formation after sperm penetration. In Missouri, Lee, S.J., *et al.* (1992) have shown that the maturation rate of pig oocytes was increased by supplementation of the IVM medium (M-199 + hormones) with 20% dialysed pFF. In Cambridge, Jung *et al.* (1992) have documented various biochemical changes in the pig oocyte during IVM. The use of brilliant cresyl blue (BCB) as a measure of the X-chromosome-linked enzyme, G6PD, was reported by Ericsson *et al.* (1993) as being a possible way of screening pig oocytes for IVM.

Changes in follicular endocrinology during the final maturation of pig oocytes *in vivo* have been described by Xie *et al.* (1990c). In the laboratory, the studies of Reed *et al.* (1991) and Illera and Petters (1993) evaluated the addition of EGF + IGF-I to IVM media in the presence or absence of FF and hormones. Addition of the growth factors adversely affected oocyte quality in the presence of hormones and pFF but had a favourable effect when oocytes were cultured without gonadotrophins or pFF. In Canada, Singh and Armstrong (1992) and Singh *et al.* (1993b) examined the effects of combinations of EGF, TGF-β and FSH, LH, oestradiol and androstenedione on maturation and cumulus expansion in pigs; the authors concluded that growth factors are likely to interact with gonadotrophins and steroids and should not be viewed in isolation. Other studies reported by Singh and Armstrong (1994) support the view that EGF of follicular origin may play a role in the regulation of follicle development and oocyte maturation in the pig; evidence was also given about a possible role of EGF in regulating polyspermy in this species.

2.18.3 Horses

It is, as yet, uncertain how long the primary horse oocyte takes to mature *in vivo*. Unlike farm ruminants and pigs, in the mare there is no discrete preovulatory LH surge preceding release of the oocyte. It is known, however, that ovulation will occur about 36 h after HCG administration in mares with mature follicles; oocytes collected at that time are in metaphase II. In Poland, Okolski and Sionina (1992) reported coculture of oocytes with theca interna cells to be most successful in IVM; the same authors found that most horse oocytes reached full nuclear maturation after 36 h of IVM culture.

Hinrichs (1991) drew attention to the fact that only three reports at that time were available on successful maturation and fertilization of horse oocytes recovered from slaughterhouse specimens; in each case, the COCs had been recovered by scraping follicles to obtain cumulus-intact oocytes. In a subsequent report, Hinrichs *et al.* (1993) attempted to define meiotic configuration of horse oocytes after different IVM periods. They found that the time course for maturation of horse oocytes was relatively long, taking 32 h or possibly longer. In Poland, Mlodawska *et al.* (1991) observed metaphase I after about 20–24 h and metaphase II after about 40–44 h in a majority of their cultured oocytes. In Japan, the studies of Choi *et al.* (1993a, b) suggested that an IVM culture period of 24–32 h may be appropriate. In Georgia, Willis *et al.* (1991a, b, 1994) have examined various factors relating to oocyte maturation and have shown that 15 h of culture (M-199 + 15% mare serum) can result in significant numbers of oocytes reaching metaphase II (47%); these oocytes are apparently mature with respect to nuclear configuration and cortical granule migration. The implication from such work is that equine oocytes cultured for 15 h may be appropriate candidates for IVF. Methods for collecting and maturing equine oocytes *in vitro* are detailed in a paper by Shabpareh *et al.* (1993).

In the collection of equine oocytes, a report by Meintjes *et al.* (1994) has shown repeated transvaginal ultrasound-guided oocyte recovery to be a safe and effective procedure (15.2 oocytes recovered per mare), which could be applied during the first half of pregnancy; this approach made it possible to harvest more oocytes from early-gestating mares than is normally anticipated from cyclic mares. In Colorado, Dippert *et al.* (1994) have shown that elimination of the DF and aspiration of all follicles 3 days later improved both oocyte recovery and maturation.

2.18.4 Buffaloes and deer

A limited amount of information is available on IVM and IVF of buffalo oocytes (see review by Madan *et al.*, 1994). In Japan, Suzuki *et al.* (1991b) used M-199, supplemented with 5% oestrous buffalo serum in attempts to mature oocytes; embryo yields were low, but transfers to 14 buffalo recipients resulted in one pregnancy confirmed at 90 days. It is clear that yield of oocytes per ovary is low in this species compared with cattle. This has been attributed to the low number of primordial follicles in the buffalo (Totey *et al.*, 1991a, b). The ultrastructure of buffalo oocytes during maturation is dealt with in a report by Boni *et al.* (1991) in Italy. These workers report that cumulus cells are in contact with the oocyte plasma membrane via cytoplasmic processes that penetrate the ZP, as described in the other farm ungulates.

Studies in buffalo IVM have shown that addition of hormones can result in a significant increase in maturation rate (Totey *et al.*, 1991a, b, 1992). On the other hand, Madan *et al.* (1994) routinely culture oocytes in M-199 with 10% oestrous buffalo serum in the absence of hormones. As shown by Singh and Majumdar (1992), the buffalo oocyte matures after 24 h of culture. A comparison of various media and supplements used in IVM by Totey *et al.* (1993a) suggested that Ham's F-10 medium, supplemented with FCS and hormones (FSH/LH/oestradiol) was the most suitable culture system for use in buffaloes. In an effort to improve the oocyte yield per buffalo ovary, Pawshe and Totey (1993) cultured denuded oocytes on a cumulus cell monolayer; they found that the 6-day-old monolayer would support maturation.

In New Zealand, a preliminary study on the effects of culture duration and time of gonadotrophin addition on the maturation of red deer (*Cervus elaphus*) oocytes was made by Fukui *et al.* (1991a). Ovaries were recovered after slaughter and vesicular follicles aspirated and matured in M-199 supplemented with 10% FCS and additional granulosa cells. These authors report acceptable maturation rates (61–77%) when oocytes were cultured for 20–28 h. The cleavage rate was increased when gonadotrophin was added at 6 h after the onset of IVM rather than at 0 h.

3 Capacitating Sperm

3.1 Introduction

3.1.1 Historical

The production of sperm by the bull (spermatogenesis) is a lengthy process, occurring in the testes, whereby stem cell spermatogonia at the base of seminiferous tubules divide by mitosis to maintain their own numbers and to cyclically produce primary spermatocytes which undergo meiosis to produce haploid spermatids, which differentiate into sperm released at the tubular lumen. The bull's sperm is a small, highly condensed cell which is only about one twenty-thousandth the volume of the bovine oocyte.

Fertilization of the bovine oocyte involves a sequence of events in which the sperm: (i) is motile (to reach the oocyte and move through the zona pellucida (ZP)); (ii) has the ability to undergo capacitation and express the acrosome reaction (AR); (iii) has the capacity to bind to the ZP and vitelline membrane by acquiring the correct binding proteins during maturation and exposing these binding sites to the oocyte at the appropriate time; and (iv) is able to fuse with the oolemma and be incorporated into the oocyte. Sperm acquire many of these capabilities as they pass through the epididymis (see reviews by Rogers, 1978; Yanagimachi, 1990; Zaneveld *et al.*, 1991; Roldan and Gomendio, 1992).

The phenomenon of capacitation was first recognized and described independently by Austin (1951) and Chang (1951). They demonstrated that mammalian sperm finally become competent to penetrate the secondary oocyte only when they have spent some time in the female tract. Of historical interest in cattle *in vitro* fertilization (IVF) are the reports of Iritani and Niwa (1977) and Iritani (1980), who examined the possibility of capacitating bull sperm in the isolated bovine oviduct and uterus and in the reproductive tract of the live rabbit.

3.1.2 *The capacitation process*

Capacitation is now clearly recognized as a process involving the sperm in a complex series of biochemical and physiological reactions. It is believed that the initial stages of capacitation involve the removal and alteration of components derived from the seminiferous tubules, epididymis, vas deferens and seminal plasma which are adsorbed by the sperm plasma membrane. During their passage from the testis through the epididymis, sperm are modified to yield mature fertile cells for storage in the tail of the epididymis until released at ejaculation or otherwise voided in the urine.

One important aspect of epididymal maturation is modification of sperm surface properties. Sperm as they leave the testis and enter the head (caput) of the epididymis are not capable of penetrating oocytes. It appears that additional maturational changes must occur in the caput and/or corpus epididymidis before the capacity to penetrate is acquired. Before being able to fertilize the oocyte, however, the bull sperm must migrate through the reproductive tract by negotiating various barriers. In preparation for this journey, the surface of the sperm is protected by glycoprotein coats secreted by the epididymis, and additional coats are added to the sperm surface when the gametes are exposed to seminal plasma at ejaculation. It is believed that an important part of the capacitation process is the gradual removal or alteration of these protective coats from the sperm surface, especially in the region of the acrosome (Yanagimachi, 1990). The removal or alteration of these coats would permit exposure of receptor sites, allowing sperm to interact specifically with oocyte receptors.

Difficulties have arisen in adequately defining the term 'capacitation', for the process involves several events, some of which are known to be reversible. It was observed by Chang in the 1950s that the fertilizing ability of capacitated rabbit sperm could be reversibly inhibited by exposure to seminal plasma. Reversal of bull sperm capacitation is dealt with in a report by Alfonso and Hunter (1991); they found that the decapacitation activity of bovine seminal plasma on capacitated sperm was concentration dependent and was only evident when more than 50% of the medium was seminal plasma. In Israel, Rubinstein and Breitbart (1991) suggest that spermine is the major decapacitating factor in ram seminal plasma; they view the role of spermine as preventing premature capacitation and the AR.

In dealing with the phenomenon of capacitation, some investigators include the AR in it, while others do not. The definition used here is that favoured by Furuya *et al.* (1992), who took the view that capacitation is a sequence of biochemical changes that enables the sperm to undergo the AR in response to the ZP, which appears to be the physiological inducer of that reaction. It should be added that Ca^{2+} entry is regarded as essential to induce the AR in bull sperm, although the precise mechanism of calcium entry and the involvement of other ions associated with calcium entry is not clearly understood (Ruknudin, 1989; Kim and Babcock, 1991).

3.1.3 Hyperactivation

The changing motility patterns shown by sperm during capacitation have been described by several workers. Hyperactivation involves a radical transformation in motility, when sperm show a vigorous, asymmetrical flagellum-beating pattern. Hyperactive movement is also characterized by sharply curved, asymmetric tail beats and frequent changes in direction. Various advantages have been proposed for hyperactivation, including an increased ability to remain free of the oviductal epithelium, increased manoeuvrability in the convoluted lumen of the fallopian tube and increased ability to move through viscous fluids.

Evidence for a possible advantage of hyperactivated mouse sperm in remaining free of the oviductal wall is provided by DeMott and Suarez (1991, 1992). They characterized the pattern of progress of individual sperm and found that most were stuck to the wall for most of the time. Progress up the oviduct to the site of fertilization occurred intermittently as sperm release from the wall, swim a short distance and reattach. The authors concluded that

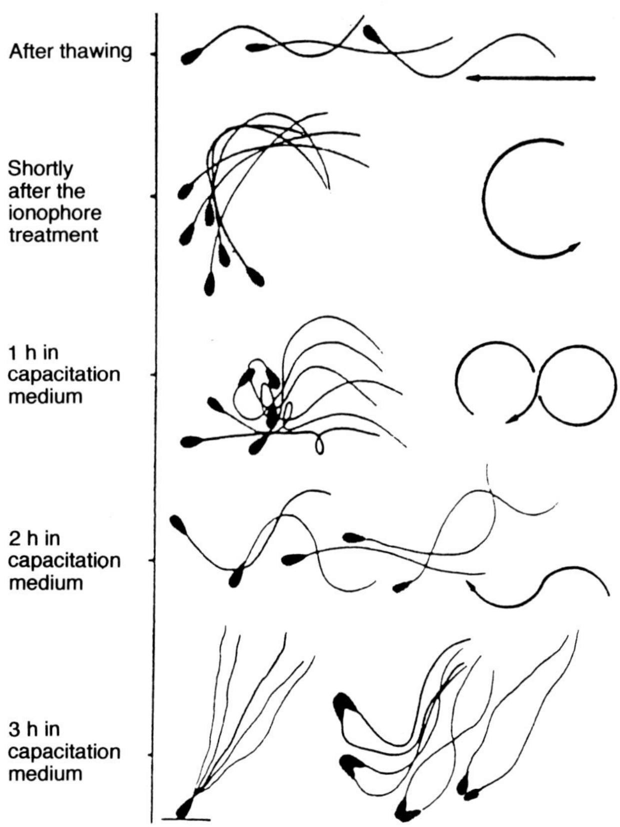

Fig. 3.1. Changing aspects of movement patterns of bull sperm during *in vitro* capacitation (from Tateno and Mikamo, 1987).

sperm broke free from the oviductal wall by using hyperactivated tail beats and that such movement plays an important role in moving sperm towards the site of fertilization. In the cow, Pollard *et al.* (1992) suggest that hyperactivation of bull sperm in the oviduct mediates their release from the epithelium, migration from the isthmus to the ampulla, retention at the site of fertilization and the acquisition of high fertilizing capacity.

The motility patterns shown by bull sperm after capacitation are described by Tateno and Mikamo (1987); according to these authors, it was possible to determine whether sperm were adequately capacitated (using ionophore A23187) by studying their movement patterns (see Fig. 3.1). The sperm moved progressively after thawing but after capacitation they stopped progressive movement and started to circle around themselves by beating the entire length of the tails. After about 1 h, many sperm showed vigorous movement of the neck and the tail, i.e. hyperactive movement. After a further hour, most sperm began to move progressively by a flexible beating of the tail and, later still, the head of the sperm became sticky.

Although hyperactivation involves such a radical change in the motility pattern of the bull sperm, it is of interest to note that the development of the hyperactivated flagellar pattern is possibly not so much a maturational process as a return to a pre-existing potential shown by epididymal sperm (Suarez, 1988).

3.1.4 Factors involved in the acrosome reaction

As to the AR itself, this probably represents a sequel to the preceding changes covered under the term capacitation; the AR involves multiple fusions between the outer acrosomal and overlying plasma membranes. These membranes, like those of somatic cells, are composed of lipid layers. In the dispersal of the acrosomal contents during the AR, the outer acrosomal membrane and the overlying plasma membrane fuse to form a number of vesicles (see Fig. 3.2).

The AR is known to be triggered in the presence of extracellular Ca^{2+}. In the absence of extracellular Ca^{2+} sperm of all the farm mammals fail to initiate the AR. A study by Storey *et al.* (1992) led them to conclude that the AR induced in mouse sperm by the mouse ZP proceeds through a series of reactions, which include a transient rise in intracellular Ca^{2+} in the sperm head. The acrosomal status of mammalian sperm can be determined by various procedures (see review by Cross and Meizel, 1989).

The mechanisms by which monovalent and divalent cations may participate in the induction of the AR have been discussed (Bavister, 1986; Bhattacharyya *et al.*, 1986). It is believed that the levels and/or ratios of monovalent cations found *in vivo* may be important in regulating sperm capacitation, AR and fertilizing ability. There is evidence that the Na^+/K^+ ion ratio increases steadily from a low level in the epididymal plasma to the oviductal fluid in the ampullary region of the fallopian tube. Some of the biochemical changes in bull sperm during capacitation *in vitro* have been reported by White *et al.* (1992); they recorded no significant change in oxygen

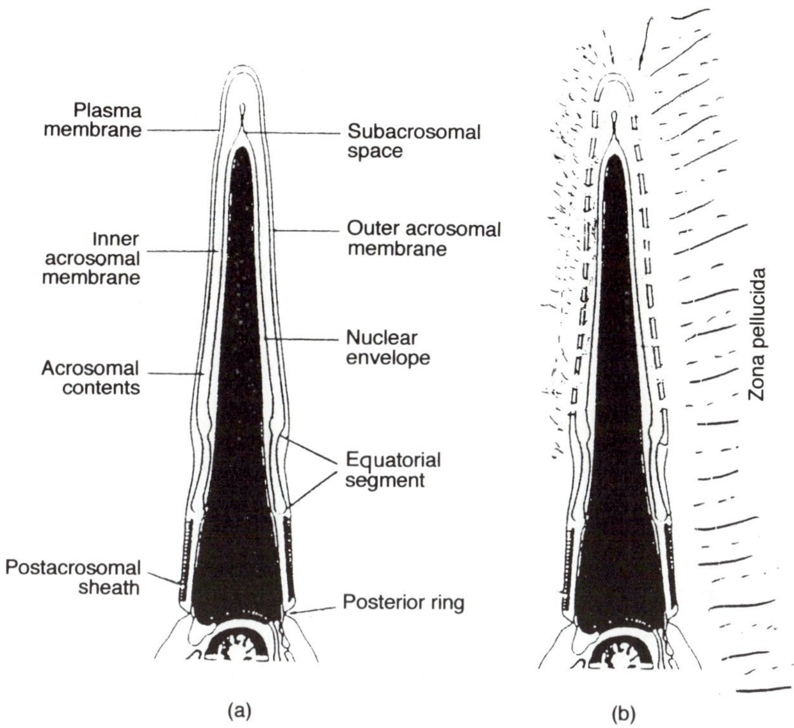

Fig. 3.2. Schematic drawing of the bovine acrosome reaction. (a) Intact sperm; (b) sperm during fusion of the plasma and outer acrosomal membranes.

uptake or cyclic adenosine monophosphate (cAMP) and adenosine triphosphate (ATP) levels specifically associated with the capacitation process.

The findings of Crozet (1984) indicate that the AR of capacitated bull sperm occurs at the surface of the ZP in cattle, as it does in sheep (Crozet and Dumont, 1984). In the matter of bull sperm having the ability to penetrate the secondary oocyte, it is believed that sperm binding sites appear on the surface of the ZP at an early stage of follicle development.

The blocking effect of carbohydrates on human fertilization has been examined in studies reported by Mori *et al.* (1993). These authors suggest that, besides being an energy source, D-fructose in semen may have another importtant role as one of the factors involved in maintaining sperm in an uncapacitated state and for preventing the AR before and after ejaculation.

3.1.5 Artificial induction of capacitation

For IVF, it is clearly essential to have a means of artificially capacitating sperm; over the past 50 years, much time and effort have been devoted to this problem. Part of the artificial capacitating procedure involves the removal (by thorough washing) of seminal proteins and the other substances that coat the sperm membrane of ejaculated semen. Some workers have sought to establish the role of substances in seminal plasma by using epididymal bull sperm which have not been coated with accessory gland proteins. It is clear that differences in the capacitating ability of bull sperm do exist. Epididymal bull sperm can be capacitated in a simple salt solution (Ball *et al.*, 1983a, b) whereas ejaculated sperm do not capacitate well unless specific capacitating methods and agents are used. Presumably, this is the result of the decapacitation factors present in bull seminal plasma. A capacitation procedure previously employed in preparing epididymal bull sperm for IVF was modified to use ejaculated semen in the report of Critser *et al.* (1984).

3.2 Capacitation in the Cow

In cattle, the time required for capacitation of ejaculated sperm *in vivo* has been estimated as about 6 h, based on the first penetration of oocytes after artificial insemination (AI) (see Hunter, 1985). However, factors in the female cattle tract involved in capacitation and their origins remain largely unknown. According to a report by Murase *et al.* (1992), in the natural mating process, cervical mucus in the cow may select motile sperm and probably plays a role in the induction of sperm capacitation and the AR.

3.2.1 Sperm transport in the reproductive tract

A knowledge of the mechanisms involved in sperm migration through the cow's reproductive tract is necessary in order to understand how capacitation is achieved in that species. A comprehensive review of sperm migration in laboratory species has been provided by Katz *et al.* (1989). It is clear, from many reports, that the bovine oviduct is not simply a tube through which sperm must pass to reach the oocyte. As a result of their high motility, sperm pass through the uterotubal junction (UTJ) and enter a reservoir in the caudal isthmus. The formation of this reservoir may be influenced by the thick, tenacious mucus and the binding of the sperm cells to the mucosal surface (Suarez, 1992). The migration of bull sperm through the fallopian tube appears to be carefully regulated, so that relatively few sperm reach the site of fertilization (ampullary section of the oviduct) at any one time (Hunter, R.H.F., *et al.*, 1991). According to Suarez (1992), it is possible that the release of sperm from this reservoir may require them to become capacitated and hyperactivated; this may enable sperm to pull free of the mucosal surface and move on through the oviduct.

There is some suggestion that sperm transport may differ between the two oviducts of the cow according to the site of ovulation. It appears that the oviduct ipsilateral to the ovulating ovary of the cow may exhibit greater muscular and secretory activity than the contralateral tube (see Ellington, 1991).

On the other hand, Shalgi *et al.* (1992) suggest that hamster sperm require a progressive linear type of motility pattern to pass through the UTJ and only show hyperactivation after entering the oviduct.

3.2.2 Role of oviductal secretory cells

There is ample evidence to suggest that the bovine oviduct may play a key role in sperm capacitation by providing an environment that facilitates both this process and fertilization. Certainly, there is much information now available to show that the bovine oviduct does not function simply as a site for sperm retention, but probably participates in the physiological maintenance of sperm fertilizing capacity and consequently contributes to the fertilization of the oocyte (Pollard *et al.*, 1991a). There is evidence that bovine oviductal cells secrete factors that are capable of supporting and maintaining bull sperm viability and motility for up to 30 h *in vitro* (Ijaz *et al.*, 1994). It is known that secretory epithelial cells of the ungulate oviduct produce a wide array of glycoproteins, and that this production is mediated by the hormones of the oestrous cycle (Erickson-Lawrence *et al.*, 1989). The volume and composition of oviductal fluid are known to vary throughout the oestrous cycle (Killian *et al.*, 1987). In terms of growth factors, there is evidence from studies in the human that insulin-like growth factors (IGFs) may act as mediators of steroid action in the oviduct (Giudice *et al.*, 1992).

Prior to fertilization, sperm and oocytes are exposed to the oviductal fluid, which contains secretions from these secretory cells as well as serum transudate. Oviductal fluid recovered from different regions of the bovine oviduct and at different stages of the oestrous cycle can differentially affect sperm fertilizing ability (Grippo *et al.*, 1992; Killian and Grippo, 1992). These workers found that bull sperm incubated in ampullary oviductal fluid collected close to oestrus/ovulation penetrated more oocytes than those incubated in other fluid samples. Other reports also show evidence that media supplemented with oestrus-associated oviductal protein can enhance bovine sperm capacitation and fertilization (Xu and King, 1990b; King, R.S., *et al.*, 1992).

The fact that glycine is present in high concentration in oviductal fluid led Marquant-Le Guienne *et al.* (1993a) to use that amino acid in their artificial capacitation medium; they found a marked improvement in fertilization rate and almost twice as many embryos were produced using the glycine medium compared with control.

High-density lipoprotein (HDL) has been detected in bovine oviductal fluid and its possible participation in the capacitation process has been suggested (Ehrenwald *et al.*, 1990); the level of HDL is elevated during the follicular phase of the oestrous cycle. The lipid content of oviductal fluid in the cow was examined by Henault and Killian (1993a, b); concentrations of lipids, cholesterol and glycerides were highest in the ampulla and least in the isthmus. Such differences may reflect changes in the luminal environment affecting the sperm membrane.

Patterns of oviductal motility are known to vary during the cycle. Bennett *et al.* (1988) also noted that there was a much higher level of motility in the oviduct ipsilateral to the active ovary. The regional morphology of the bovine oviduct was examined by Eriksen *et al.* (1993), who found evidence of local differences in the microenvironment which may be relevant to the sequential progression of fertilization and early development of the embryo.

Recent advances in tissue culture techniques have made possible the culture of monolayers of bovine oviductal epithelial cells (BOEC) *in vitro* (Ellington *et al.*, 1989a, 1991a); these authors suggest that the BOEC coculture system offers an excellent opportunity for studying capacitation, the AR and the fertilization process in cattle. In their studies, they found that bull sperm interacted with the BOEC in a manner very similar to that seen *in vivo* and that stages in capacitation were stimulated by this interaction. It is also now clear that the actual physical contact of sperm with oviductal epithelial cells *in vitro* is sufficient to trigger production of certain proteins (Ellington *et al.*, 1989b, 1991b, 1993a; Anderson and Killian, 1992); these authors suggest that, in addition to maintaining sperm viability, such proteins may help in preventing premature AR, inhibit complement-mediated antibody formation against sperm cells and enhance gamete recognition.

The association of protein components of bovine oviductal fluid with the surface of bovine sperm was examined by McNutt *et al.* (1992); they suggest that oviductal proteins that are enriched on the sperm surface may modulate sperm function in preparation for interaction with the oocyte. A study reported by Wegner and Killian (1992) sought to determine the significance of the synthesis of an oestrus-associated protein found in bovine oviductal fluid; they suggest that the association of this protein with the ZP of the bovine oocyte may modify the ZP in such a way as to favourably influence zona–sperm interaction.

In France, Guyader *et al.* (1989) and Guyader and Chupin (1991) also suggested that the incubation of bull sperm on BOEC monolayers may be an efficient method of capacitating sperm and one that closely resembles the *in vivo* process.

A review by Hunter (1991) considers the role of oviductal macromolecules in the interaction between sperm and oocyte in the live animal. There is some evidence that glycoproteins elaborated by the porcine oviduct bind to the ZP of newly ovulated pig oocytes and thereafter are partially hydrolysed by sperm proteases at fertilization (Nagai and Moor, 1990); such glycoproteins may act to minimize the risk of simultaneous penetration of the zona by several sperm.

In other studies, Raychoudhury and Suarez (1991) found that porcine sperm attached to their oviductal explants remained motile for 2 days or longer, when compared with unbound sperm. They suggest that high oestradiol levels at the onset of oestrus in the sow may play a role in binding sperm to the oviductal epithelium; the fall in oestradiol level prior to ovulation may be responsible for the release of these sperm from the isthmic reservoir. Perhaps a similar hormonal sequence may operate in the cow.

3.2.3 Glycosaminoglycans in the reproductive tract

Glycosaminoglycans (GAGs) are reputed to be among the most intricate structures in living organisms; they include heparin, heparan sulphate, chondroitin sulphate, dermatan sulphate, hyaluronic acid and keratan sulphate. The most familiar member of the GAGs is heparin, discovered as far back as 1916, whose action in the prevention of blood clotting is widely recognized. Techniques currently available permit the physiochemical and biological properties of GAGs to be examined in detail.

GAGs which have been identified as effective inducers of capacitation *in vitro* are known to be abundant in the follicles of the cow and to be released by the ovulating follicle into the ampullary region of the oviduct at time of ovulation. A view expressed by Falcone *et al.* (1991), on the basis of their studies in the human, was that follicular fluid released at ovulation may enhance the fertilizing potential of sperm. In pigs, Hansen *et al.* (1991) also suggest that follicular fluid released into the oviduct may play some role in sperm capacitation. As well as being present in follicular fluid, GAGs are found throughout the cattle reproductive tract and it is believed that their production is modulated by the action of the ovarian steroid hormones. Heparin was the most potent GAG of several examined by Handrow *et al.* (1982). The studies of Lee and Ax (1984) showed that the concentrations of GAGs in the bovine reproductive tract decrease significantly anterior to the cervix; the authors concluded that such changes may be biologically important to sperm. They recorded increased concentrations of GAGs at time of oestrus and decreased levels during the luteal phase of the cycle. Similar changes have been noted in sheep (Lee *et al.*, 1986).

Evidence exists to show that whereas low doses of GAGs can induce capacitation of bovine sperm, high doses may prevent it; in seminal plasma, there can be GAGs at concentrations exceeding 1 mg ml^{-1} (Lee *et al.*, 1985) and this may be a factor in explaining why high doses of GAGs or seminal plasma can inhibit capacitation (Lee *et al.*, 1983; Miller and Ax, 1986).

According to Herz *et al.* (1985), there may be something unique in the concentration of GAGs in the ampulla of the oviduct ipsilateral to the ovulating ovary of cow after oestrus which may result in a high percentage of capacitated sperm. In cattle, essentially a monovular species, it may not be surprising to find that the side of the reproductive tract ipsilateral to the ovulating follicle may be more effective in promoting capacitation than the opposite, contralateral side. Parrish *et al.* (1989a) collected oviductal fluid from bovine oviducts and evaluated it for its effect on capacitation of bull sperm *in vitro*; results indicated that a heparin-like GAG in the fluid may be a capacitating agent in the live animal.

It is now clear that uterine, oviductal and follicular fluids recovered from the cow are capable of capacitating bull sperm *in vitro* and that all these fluids contain GAGs. The effects of various GAGs have been examined most thoroughly using bovine sperm. In a review article, Miller and Ax (1990) suggest that the mechanism of heparin-induced capacitation involves binding to sperm and altering plasma membrane proteins, thereby activating ion channels to increase cytosolic calcium and pH.

3.3 Available Capacitation Procedures

Artificial capacitation procedures are aimed at simulating the sequence of events which normally occurs in the cow's reproductive tract. It has proved to be more difficult to capacitate sperm artificially in some species than in others. Methods used in dealing with the sperm of farm animals have usually been based on modifications of techniques employed in laboratory species, these having been studied much more intensively. Sperm from laboratory species (rabbits, mice, hamsters) can be capacitated by several forms of manipulation involving various types of media. The function of such manipulations is to remove from sperm the surface proteins which are otherwise inhibiting capacitation.

3.3.1 Historical

Since the early description of the need for sperm capacitation before fertilization (Austin, 1951; Chang, 1951), much research activity has been devoted to the development of methods to artificially induce capacitation and the subsequent AR *in vitro*. The first authentic demonstration of mammalian fertilization *in vitro* was that recorded by Chang in the late 1950s in rabbits, using sperm recovered from mated does. Subsequently, it was found that epididymal sperm cells could be prepared to fertilize (capacitated) without having to undergo preparation in the female tract.

In Dublin, attempts were made to fertilize *in vitro*-matured cattle oocytes, using bull sperm exposed to a variety of procedures, including exposure to amylases (Sreenan, 1970). In the USA, Brackett and colleagues in the early 1970s used immunological methods to show that the capacitation process involves removal or alteration of seminal plasma components that normally coat the surface of the rabbit sperm. Work with mice, and subsequently with rabbit sperm, suggested that capacitation could be achieved in media of

Table 3.1. Approaches to the artificial capacitation of bovine sperm *in vitro*.

Method	Author(s)
High-ionic-strength medium (HIS)	Brackett *et al.* (1982)
Bovine follicular fluid	Fukui *et al.* (1983b)
Standard-ionic-strength medium	Iritani *et al.* (1984)
Heparin	Parrish *et al.* (1984)
Elevated pH	Cheng (1985)
Ionophore A23187	Hanada (1985b)
Liposomes	Graham *et al.* (1986)
Percoll gradient/hypotaurine	Utsumi *et al.* (1988)
Oviductal cell monolayer	Guyader *et al.* (1989)
TEST yolk	Ijaz and Hunter (1989c)
Caffeine	Niwa *et al.* (1988)
Oviductal fluid	Parrish *et al.* (1989a)

elevated ionic strength (Brackett and Oliphant, 1975). Some of the various approaches to the artificial capacitation of bull sperm are set out in Table 3.1.

3.3.2 Modifying osmolarity and pH of media

Although the IVF of mammalian oocytes was first recorded in the literature more than 30 years ago, it is only in relatively recent times that IVF using artificially capacitated bull sperm has been demonstrated and verified by the birth of calves (Brackett *et al.*, 1982; Sirard *et al.*, 1986a). The technique employed to capacitate bull sperm by these workers involved the use of a hypertonic medium (380 mosm), as previously employed with success in capacitating rabbit sperm (Brackett and Oliphant, 1975).

Later studies in the USA, however, questioned the requirement for such hypertonic treatment in inducing capacitation (Bondioli and Wright, 1983b); it was shown that treatment of ejaculated bull sperm with a chemically defined medium of normal osmolarity (290 mosm) was equally effective, assessed on the basis of sperm penetration of cattle oocytes.

In Japan, Iritani *et al.* (1984) confirmed that ejaculated bull sperm could be artificially capacitated by preincubation in an isotonic, chemically defined medium (290 mosm). In their work, however, prolonged incubation of the bull semen was employed as a preliminary treatment. Bull semen was collected on the day prior to IVF and sperm incubated for 14–18 h at 20°C, washed once and then preincubated for 8 h in the isotonic medium. It was believed that the early stages of capacitation, such as loosening of coating factors and changes in the plasma membrane of the sperm, were initiated during storage of the semen at 20°C. Further work along similar lines was reported by Iritani (1985).

Although it is probable that prolonged incubation of bull sperm (18–24 h) can allow surface proteins to be loosened by that means, it was noted by Wheeler and Seidel (1986) that such lengthy incubation periods markedly reduced sperm viability; clearly, for IVF purposes, there is little merit in having capacitated sperm that are immotile.

However, whether using high-ionic-strength (HIS) or normal medium, the fertilization rates achieved with ejaculated bovine sperm have usually been low, with considerable bull-to-bull variation being evident. The HIS capacitation treatment that resulted in the birth of the first IVF calf showed 26–35% penetration of cattle oocytes. Sirard *et al.* (1986a) saw the use of the HIS method as probably being limited to certain bulls (those that responded well); they did not regard the procedure as suitable for general application.

Importance of pH

In sheep and pigs, the importance of pH values used in capacitation procedures and IVF itself has been recognized for some time. In pigs, Cheng (1985) found that a high penetration rate (>89%) could be achieved when porcine sperm were capacitated at 37°C in medium held at pH 7.8 prior to being added to fertilization medium held at pH 7.4. and 39°C. In that study, it was

Table 3.2. Effect of pH of capacitation and fertilization medium on the outcome of *in vitro* fertilization of bovine oocytes matured *in vitro* (from Lu, 1990).

pH of capacitation		pH of fertilization medium			
		7.0	7.4	7.8	Total
7.0	Oocytes exam.	44	35	40	119
	% fertilized[a]	43.2^{b}	57.1^{bc}	72.5^{c}	57.1
	% cleaved	4.5^{A}	37.1^{B}	35.0^{B}	24.4
7.4	Oocyctes exam.	40	32	34	106
	% fertilized[a]	42.5^{d}	68.8^{e}	91.2^{f}	66.0
	% cleaved	22.5^{C}	9.4^{C}	70.6^{D}	34.0
7.8	Oocytes exam.	44	48	48	140
	% fertilized[a]	54.4^{g}	70.8^{gh}	79.2^{h}	68.6
	% cleaved	22.7	35.4	29.2	29.3
Total	Oocytes exam.	128	115	122	365
	% fertilized[a]	46.9^{E}	66.1^{FG}	80.3^{G}	64.1
	% cleaved	16.4^{H}	28.7^{IJ}	42.6^{J}	29.0

[a]No. of fertilized oocytes including no. of cleaved oocytes.
The percentage of fertilization or cleavage of treatment groups with different superscripts within columns is statistically different: $P < 0.05$ for b–h and $P < 0.01$ for A–J, respectively.

evident that the optimum pH for fertilization was different from that for sperm capacitation. In sheep, the same author also found that the optimal pH for capacitation was 7.8 and for fertilization it was 7.4. In cattle, on the other hand, Lu and Gordon (1987) recorded the optimal pH value for the capacitation medium as 7.4 and for the fertilization medium it was 7.8 (see Table 3.2). These pH values are still employed in current Dublin IVF procedures (Carolan *et al.*, 1994; Lonergan *et al.*, 1994).

In Minnesota, Ijaz and Hunter (1989b) concluded that the most effective system for capacitating bull sperm was a pH 7.6 wash followed by capacitation in pH 7.6 medium; they found that media pH hastened capacitation, but was not a capacitating agent.

3.3.3 Bovine serum albumin and follicular fluid

Although information on many of the molecular events involved in sperm capacitation is limited, the requirement for serum albumin in supporting sperm capacitation *in vitro* has been demonstrated in various laboratory species. Exposure of bull sperm to bovine serum albumin (BSA) in an appropriate physiological salt solution can result in capacitation *in vitro* (Goodeaux, 1979) and BSA is routinely included in culture media employed in IVF. A report by Dow and Bavister (1989) suggested that serum albumin may play a key role in removing cholesterol and/or zinc from sperm because this protein has considerable binding capacity for both molecules. It is believed that cholesterol, which stabilizes cell membranes, is removed from the sperm plasma membrane during capacitation; zinc ions, which also stabilize membranes may also be removed in the early stages of capacitation.

Bovine follicular fluid

Fukui *et al.* (1983a) examined various procedures for capacitating frozen–thawed bull sperm and the subsequent use of such sperm in IVF; they found evidence that porcine follicular fluid (pFF) could favourably influence capacitation. A sperm penetration rate of 56% was reported by Sugawara *et al.* (1984), using frozen–thawed bull sperm preincubated in media containing bovine follicular fluid (bFF). As to the compounds in bFF responsible for the capacitating effect, Grimek *et al.* (1984) did isolate low molecular weight proteoglycans from bovine follicles. Proteoglycans are composed of a protein core with numerous heteropolysaccharide side-chains. These side-chains are mainly composed of GAGs. Various GAGs from bFF were examined by Miller *et al.* (1987) and it is clear that bFF contains many compounds capable of capacitating bull sperm; for that reason, it has been used in bull sperm capacitating medium (Ayoub and Hunter, 1991; Iqbal and Hunter, 1991, 1992a, b).

3.3.4 Use of calcium ionophore (A23187)

The calcium ionophore, A23187, has been successfully employed, mainly by Japanese workers, to bypass the early stages of capacitation of bull sperm by directly increasing Ca^{2+} content of the cell and inducing the AR (Hanada, 1985b). Various authors have shown the importance of an influx of extracellular Ca^{2+} into the sperm in the capacitation process (Ruknudin *et al.*, 1988; Ruknudin, 1989). The calcium ionophore A23187 has been employed to achieve this influx (Byrd, 1981; Takahashi and Hanada, 1984; Bird and Houghton, 1988; Fukuda *et al.*, 1988). There is some evidence that the Ca^{2+} influx is the result of calcium entering a non-mitochondrial compartment as a consequence of the equilibration of the ion across the mitochondrial and plasma membranes of the sperm (Simpson and White, 1987). It is also known that ionophore treatment results in increased respiratory activity in bull sperm.

In Galway, a comparison of A23187 and P12 liposomes in capacitating bull sperm by Bird and Houghton (1988) led them to conclude that the ionophore was the more effective. A further study by Bird *et al.* (1989) showed that the ionophore treatment resulted in hyperactivation and a functional AR in bovine sperm, enabling them to penetrate zona-free hamster oocytes.

Workers in the USA compared treatment of bull sperm with A23187 and with heparin (Jiang, S., *et al.*, 1991; Jiang *et al.*, 1992); it was suggested that the simplicity of using the ionophore and the higher yield of embryos was an advantage of using that agent. The ionophore was also employed as the capacitating agent in studies reported from China (Jiang, J.Y., *et al.*, 1991).

3.3.5 Electropermeabilization

Capacitation is known to be associated with an increase in sperm membrane fluidity and permeability; the final stage of the process is the irreversible AR. As noted earlier, a crucial factor in the initiation of the AR is an influx of Ca^{2+}; many methods for inducing the AR are based on achieving permeabilization of the cell membrane in the presence of calcium. Work in Galway has shown that it is possible to induce the AR in human sperm by electropermeabilization (Tomkins and Houghton, 1988). Despite the efficiency of the technique, it was found to result in a significant decline in motility. The advent of the technique of Baekonization led Sweeney and Tomkins (1993a, b) to investigate the value of this method in the induction of the AR in bull sperm. They found that it was possible for the AR to occur without loss of sperm viability or motility. Baekonization involves the use of a sophisticated square wave generator in which the electrical pulses are discharged through the sample without the need for direct contact with the electrodes. As yet, there does not appear to be any report on the use of this novel AR-inducing procedure in cattle IVF.

3.3.6 Other methods

TEST–yolk buffer

TEST–yolk buffer has been employed in capacitating bull sperm (Ijaz and Hunter, 1989a, c, 1992; Ijaz *et al.*, 1989). It is believed that the capacitating activity resides in the TES and Tris molecules. The Minnesota workers concluded that storage of bull sperm in TEST–yolk buffer for 4–48 h would result in capacitation. A fertilization rate of 87% was recorded, using ejaculated bull semen diluted 1:10 in TEST–yolk buffer, cooled to 4°C and stored for 8 h (Ijaz and Hunter, 1989c). Further data along similar lines were provided in the report of Ijaz and Hunter (1992).

Calcium-free Tyrode's

Other work reported by researchers in Minnesota included a study in which bull sperm were washed in BSA–saline (pH 7.6) and then incubated for 4 h in Ca^{2+}-free Tyrode's medium (pH 7.6); this resulted in capacitation and sperm penetration of oocytes (Ijaz and Hunter, 1989b).

Platelet-activating factor (PAF)

The role of platelet-activating factor (PAF) in sperm function has been examined by Parks and Hough (1990); there was a marked increase in the percentage of acrosome-reacted sperm and a corresponding decrease in motility. These effects were observed over a very narrow concentration range; it was suggested that the response to PAF was the result of a non-physiological perturbation of membranes in the acrosomal region.

Liposomes

Graham *et al.* (1986) showed that phosphatidylcholine phospholipids having fatty acid chains of 12 carbons or less were effective in inducing bull sperm to undergo the AR. Incubation of liposomes made with phosphatidylcholine (PC) containing fatty acid chains of either ten (PC10) or 12 (PC12) carbons resulted in greater than 90% of sperm exhibiting the AR within 15 min. However, PC10 rapidly destroyed sperm motility, whereas with PC12 liposomes sperm remained motile for several hours. Increased penetration rates of zona-free hamster oocytes were obtained using bull sperm treated with these liposomes (Graham and Foote, 1987a, b; Graham *et al.*, 1987; Wheeler and Seidel, 1988a, b).

A report by Graham *et al.* (1987) extended these findings with liposomes to ram and stallion sperm; Kim *et al.* (1989) reported on their use in capacitating rabbit sperm. As with other approaches to capacitation, bull-to-bull variation is evident in liposome-treated fresh bovine sperm (Graham and Foote, 1987a, b); certain bulls required less PC12 to induce the AR than did others.

Amino acids and catecholamines

Beta amino acids, such as the sulphonated amino acid taurine and its precursor, hypotaurine, are known to be present in seminal and oviductal fluids. Studies with hamster sperm have shown hypotaurine to be about three times more potent than taurine. The work of Leibfried and Bavister (1980, 1982) showed that IVF of hamster oocytes can occur, even at very low sperm concentrations, in a chemically defined medium with BSA and a catecholamine such as epinephrine, but only when taurine has been added to sustain sperm motility. Leibfried and Bavister (1982) also showed that prior exposure of hamster sperm to hypotaurine was a prerequisite for the action of epinephrine in the induction of capacitation.

There is evidence in mammals such as the hamster indicating that catecholamines are involved in the fertilization process (Leibfried and Bavister, 1982). In mice, in the presence of catecholamines, sperm more readily effect fertilization (Stanger, 1983). In cattle, the addition of hypotaurine and epinephrine to media proved to be favourable in the IVF studies reported by Ball *et al.* (1983b). It should also be noted that these compounds may influence motility characteristics of sperm.

Chelating agents

Andrews and Bavister (1988, 1989) demonstrated that hamster sperm can be chemically capacitated *in vitro* with D-penicillamine, which is a divalent cation (zinc) chelator. The chelating action refers to the reversal of metal or magnesium inhibition of adenosine triphosphatase (ATPase). It is believed that sperm may encounter a reduced gradient of zinc concentration as they progress from the starting point in the caput epididymidis of the male to the site of fertilization in the oviduct of the female. Zinc is known to have a number of biological functions in somatic cells, including the inhibition of electron transport in the mitochondrial respiratory chain, resulting in an overall reduction in respiratory output. Whether it is possible that zinc inhibits sperm capacitation by maintaining them in a metabolically quiescent state, as they are during epididymal storage, is not clear.

Progesterone

In human sperm, the ovarian steroid progesterone has been shown to increase the passage of Ca^{2+} through the sperm plasma membrane and to increase the number of acrosome-reacted sperm. In the USA, Sueldo *et al.* (1992) reported that progesterone ($1\,\mu g\,ml^{-1}$) used in human IVF was associated with increased ZP binding, oocyte penetration and hyperactivated motility and the AR. Elsewhere in that country, Meizel & Turner (1993) reported that human sperm polyamine biosynthesis is necessary for the progesterone-initiated rapid increase in Ca^{2+} and the subsequent acrosome reaction. In London, it was also shown by Dasgupta *et al.* (1993) that progesterone could significantly accelerate capacitation and the acrosome reaction. The effect of progesterone on bovine sperm could be of interest, in view of the sequence of endocrine events after ovulation in the cow, where oestrogen decreases and progesterone increases.

3.4 Heparin and Other Glycosaminoglycans

3.4.1 Dose levels and duration of treatment

Of the various methods that have been examined and evaluated in capacitating bovine sperm for IVF, that involving the use of the GAG heparin is probably the one receiving the greatest research attention. It was developed as an *in vitro* capacitation and IVF system by workers at Madison. The first report was that of Parrish *et al.* (1984), who employed a pre-IVF heparin treatment ($10\,\mu g\,ml^{-1}$); this resulted in a marked increase in fertilization rate. The method as initially reported involved a 4 h incubation period with freshly collected bull sperm and 0.5 h or less with frozen–thawed semen. Since that time, many reports have been made by the Madison group dealing with the effect of heparin in modulating sperm physiology.

It is believed that the results of these studies support the view that capacitation of bull sperm by heparin probably reflects the *in vivo* mechanism (Parrish *et al.*, 1985a, b, 1988, 1989a, b; Handrow *et al.*, 1986, 1989). The effect of heparin level on capacitation in specific bulls is dealt with in reports from the UK (Lancaster *et al.*, 1990), Korea (Chung *et al.*, 1991) and Belgium (Van Soom *et al.*, 1991; Vlaenderen *et al.*, 1991). In Japan, Fukui *et al.* (1990) record that heparin dosage and incubation period for sperm capacitation are important factors affecting bovine IVF and subsequent embryo development.

The heparin-binding properties of sperm have been examined (Marks and

Ax, 1985; Miller *et al.*, 1987; Miller and Ax, 1988, 1989a, b, 1990; Nass and Ax, 1989). It is known that heparin-binding proteins are present in seminal plasma and that ejaculated sperm have more heparin-binding sites than epididymal sperm. Evidence suggests that heparin-binding proteins play a role in fertilization by attaching to the sperm surface, enabling heparin-like GAGs in the female tract to induce capacitation. Studies in Canada show that heparin induces changes in the calmodulin (CaM)-binding properties of sperm proteins and induces a reduction in CaM concentrations during capacitation (Leclerc *et al.*, 1989a, b, 1992).

Heparin should be regarded as a capacitating agent rather than an AR-inducing compound; it is also known that capacitation by heparin is temperature- and Ca^{2+}-dependent. Various biochemical studies of the action of heparin in bovine sperm capacitation have been reported by the Madison group (Uguz and Parrish, 1993).

Reviewing *in vitro* capacitation methods, First and Parrish (1988) concluded at that time that heparin treatment was probably the most effective for cattle. It appeared that any level of fertilization or polyspermy could be achieved in the cow, by adjusting either sperm concentration or heparin dose. In addition to heparin studies, other GAGs, such as chondroitin sulphate have been shown to be capable of inducing capacitation in bull sperm (Lenz *et al.*, 1983a, b); however, none of these compounds has been shown to have the effectiveness of heparin.

In Japan, Niwa and Ohgoda (1988) reported a synergistic effect of 20 $\mu g\ ml^{-1}$ heparin and 10 nM caffeine in their capacitation treatment of frozen–thawed bull sperm.

In Ireland, Lu (1990) found that heparin was consistently effective in markedly increasing fertilization rates over control levels; the fertilization rate was further improved as heparin dose increased (see Table 3.3). It was evident that the optimum dose of the agent (100 $\mu g\ ml^{-1}$) was tenfold that recommended by Parrish *et al.* (1986b). A preincubation period of 15 min was found

Table 3.3. Effect of heparin dose on pronucleus formation, polyspermy and the incidence of parthenogenetic activation of bovine oocytes fertilized *in vitro* (from Lu, 1990).

Dose of heparin ($\mu g\ ml^{-1}$)	No. of oocytes				
	Examined	Fertilized (%)	2 PN[a] (%)	Polyspermy (%)	1 PN[b] (%)
0	92	5 (5.4)[c]	2 (2.2)[g]	0 (0)	10 (10.4)
10	105	53 (50.5)[d]	45 (42.9)[h]	0 (0)	5 (4.8)
50	109	75 (68.8)[e]	55 (50.5)[hi]	3 (4.0)	8 (7.3)
100	115	96 (83.5)[f]	69 (60.0)[i]	8 (8.3)	6 (5.2)

[a] 2 PN – two pronuclei.
[b] 1 PN – one pronucleus, defined as parthenogenetic activation.
c, d, e, f, g, h, i – Percentages in the same column with different superscripts differ statistically ($P < 0.05$).

to be satisfactory (Lu and Gordon, 1988). Similar results with heparin dose levels up to 100 μg ml^{-1} were reported by Fukui *et al.* (1990) in Japan.

3.4.2 Glucose in the fertilization medium

Heparin treatments, as employed in Ireland (Lu *et al.*, 1987b, 1988a), contained a low concentration of glucose (5 mM) in the capacitation and fertilization media. This was in contrast to several other reports, suggesting that heparin treatment should be carried out in the absence of glucose (Parrish *et al.*, 1985a, b, c; Susko-Parrish and Parrish, 1988). Leibfried-Rutledge *et al.* (1989a) and Handrow *et al.* (1989) also reported that capacitation of bull sperm by heparin is inhibited by a 5 mM concentration of glucose. It was believed that such inhibition is related to events in the metabolism of the sperm. In the presence of glucose, sperm appear to metabolize it in preference to pyruvate or lactate. If glucose is the sole source of energy in the medium, the initiation of the AR is retarded. This is taken as support for the view that glucose is metabolized to lactate or pyruvate until the intracellular concentration reaches the level at which the sperm switches from glucose metabolism to pyruvate/lactate metabolism and becomes capable of undergoing the AR.

According to this view, it is possible that sperm capacitation in the uterine/tubal environment of the female tract may be regulated by the energy compounds available to them; in the presence of glucose, fructose or other seminal sugars, the AR of sperm may be retarded until such time as they are in an environment rich in lactate and pyruvate.

3.4.3 Heparin in the fertilization medium

The standard IVF procedure, as used in Dublin, involved a preincubation period of 15 min, during which sperm, in a Ca^{2+}-free modified Tyrode/albumin/sodium lactate/sodium pyruvate (TALP) medium, were exposed to heparin (Lu *et al.*, 1987a). Sperm were then added in a small volume of capacitation medium (5 μl) to droplets of fertilization medium (45 μl). An alternative procedure was examined by Shi (1991), Lonergan (1992) and Fair (1992), in which heparin was added directly to the fertilization medium, without the sperm undergoing any pretreatment. Many different dose levels were examined and it was clear that this procedure gave results entirely comparable to the earlier, preincubation method. Data from the work of Shi (1991) are in Table 3.4.

Shi (1991) found that although the fertilization rate of heparin-treated groups was significantly higher than controls, there was also evidence of a significant increase in embryo yield when the heparin concentration equalled or exceeded 18 μg ml^{-1}. There were indications that heparin not only may influence the fertilization process but may also affect subsequent embryo development. In Japanese work, there has also been evidence suggesting that heparin may influence embryo development as well as sperm capacitation (Kurosaka *et al.*, 1990).

Elsewhere, Guyader *et al.* (1990) examined the effect of heparin concentra-

Table 3.4. Effect of heparin concentration in the fertilization medium on the cleavage rate of bovine oocytes and their subsequent embryonic development (from Shi, 1991).

Heparin conc. ($\mu g\ ml^{-1}$)	Oocytes insem.	Cleavage (%)	% zygotes beyond the 2-cell stage	Blastocysts developed (%)	% blastocysts hatched
0	454	334 $(73.6)^{c}$	57.6 $(133/231)^{b}$	112 $(24.7)^{b}$	76.8 (53/69)
2	365	305 $(83.6)^{b}$	85.7 $(144/168)^{a}$	77 $(21.1)^{b}$	72.2 (26/36)
6	464	394 $(84.9)^{b}$	85.0 $(226/266)^{a}$	116 $(25.0)^{b}$	82.5 (52/63)
18	481	412 $(85.7)^{ab}$	89.8 $(237/264)^{a}$	174 $(36.2)^{a}$	75.0 (75/100)
54	491	438 $(89.2)^{a}$	87.9 $(255/290)^{a}$	195 $(39.7)^{a}$	86.8 (92/106)
162	432	365 $(84.5)^{b}$	87.4 $(229/262)^{a}$	166 $(38.4)^{a}$	80.0 (76/95)
486	92	83 $(90.2)^{ab}$	85.5 $(71/83)^{a}$	34 $(37.0)^{a}$	76.5 (26/34)

Data shown in the table from seven replicates.
a, b, c – Within a column, values with different superscripts are significantly different ($P < 0.05$).

tions in the fertilization medium (range: 0.05–2.0 $\mu g\ ml^{-1}$) on subsequent embryonic development; the highest embryo yield was when IVF occurred in the presence of 1 $\mu g\ ml^{-1}$ of heparin. Other reports on a fertilization medium containing varying concentrations of heparin in cattle IVF came from this same group in France (Marquant-Le Guienne *et al.*, 1990a, b, c).

3.5 Use of Fresh or Frozen Semen

3.5.1 Fresh semen

Various workers have employed both fresh and frozen bull semen in their cattle IVF studies (Bondioli and Wright, 1983b; Critser *et al.*, 1984; Lengwinat *et al.*, 1990). Wheeler and Seidel (1986) concluded from their work that fresh semen requires a longer capacitation period than frozen. In Madison, freshly ejaculated sperm had similarly required a longer heparin treatment period than frozen in the work of Parrish *et al.* (1984). In New Zealand, Seaton *et al.* (1991) found that fresh sperm gave better penetration rates than frozen–thawed sperm and, in Germany, Lengwinat *et al.* (1990) similarly record a significantly higher fertilization rate. However, frozen–thawed bull semen is likely to deteriorate more rapidly than fresh. One problem in using fresh bull sperm in IVF is likely to be the unpredictable quality of semen. With frozen–thawed bull sperm, they have at least passed through an initial screening before freezing.

3.5.2 Frozen semen

The majority of dairy cows in many countries around the world are currently bred by AI, usually with frozen semen. The freezing of bull semen in the ministraw (0.25 ml capacity; IMV) is in common use in the cattle AI industry and it is likely to be much more feasible for an IVF laboratory to obtain supplies of frozen semen rather than keep a bull for the provision of fresh ejaculates. The freezing of semen in cattle has been the subject of much research over the years, since the first calves born to frozen semen arrived in the early 1950s. The recommended glycerol levels for bovine semen cryopreservation have ranged from 15% to the usual 7–8% employed by most semen freezing units. The addition of cryoprotectant to semen is usually carried out at 5°C and an equilibration period of 2 h used (Salhab and Merilan, 1991). Despite the 40 and more years that have passed since Chris Polge's first success in freezing bull semen, probably no more than 50% of sperm currently survive the process; there remains the need for a more complete understanding of the factors influencing cell preservation by low-temperature storage.

In Dublin, in early cattle IVF work, ejaculates from several bulls were pooled after thawing (Lu *et al.*, 1987b). The use of pooled semen is a well-accepted method of minimizing male variability in AI in cattle, sheep and pigs. For commercial applications, however, it is generally necessary to employ clearly identified individual bulls. Extensive work in the production of cattle embryos for commercial use, as reported by Lu and Polge (1992), employed semen from individual bulls.

3.5.3 *Semen diluents*

The diluent employed in the freezing of bull semen could be a factor influencing capacitation by the use of heparin, according to First and Parrish (1987a, b); this may need to be examined in the light of the various diluents in commercial use in freezing (egg-yolk- or milk-based diluents). In Norway, Lonergan *et al.* (1993b), on the basis of *in vitro* maturation (IVM)/IVF/*in vitro* culture (IVC) studies, have reported that trildyl may be safely employed as an alternative to the regular milk-based diluent used in that country. As to the general vitality of frozen–thawed bull sperm, in comparison with those freshly collected, it is believed that semen dilution and freeze–thaw procedures do not substantially impair their energy metabolism (Ferrandi *et al.*, 1988); for that and other reasons, such sperm are likely to be acceptable for IVF work.

In the thawing of bull semen, Barth and Bowman (1988) report on the appropriate practical procedure which should be followed. This includes assessing sperm quality on the basis of the percentage of progressively motile sperm, rate of progression and the percentage of sperm showing normal morphology. Post-thaw viability was optimum at a thawing temperature of 35°C. To minimize semen contamination, straws should be carefully wiped with 70% ethanol prior to cutting and releasing their contents.

3.6 Bulls as a Source of Variability in IVF

Almost all authors reporting on cattle IVF have commented on the considerable variability that exists among different bulls in the ability of their

Fig. 3.3. The bull as a factor in cattle IVF. There is much to support the view that the bull may be an important source of variation in the results obtained in cattle IVF. Recent studies in separating X- and Y-chromosome-bearing sperm by flow cytometry have all involved the use of freshly collected semen. For many reasons, it would be much more convenient to use frozen–thawed rather than fresh semen. This aspect is likely to receive much research attention.

sperm to become capacitated. As the efficiency of cattle IVF systems has improved over the past decade, the importance of minimizing or eliminating the bull as an important source of variability has become all the greater. The advent of sexing by sperm separation has meant that fresh semen rather than frozen–thawed semen has had to be employed (see Fig. 3.3) As well as the practical interest, there are excellent scientific reasons for examining the fundamental causes of male variability.

3.6.1 High- and low-fertility bulls

In Canada, using ovulated oocytes in their cattle IVF studies, Lambert *et al.* (1984) employed HIS medium to capacitate bull sperm; with five different bulls, they recorded fertilization rates varying from 14 to 46%. These workers regarded individual bull variation as one of the most important factors affecting sperm preparation by the HIS method. Other workers, using other capacitation methods, recorded similar bull differences in their IVF studies (Hanada, 1985b; Aoyagi *et al.*, 1988; Niwa and Ohgoda, 1988).

Work in Minnesota, reported by Miller and Hunter (1987) appears to be an exception; they failed to find evidence of significant variation among 29 AI bulls in their capability to achieve IVF. Marked male variation has been

Table 3.5. Effect of individual bulls on the cleavage rate of bovine oocytes and their subsequent development (from Shi, 1991).

AI station	Bulls	Replicates	Oocytes insem.	% cleavage[a] (mean ± s.d.)	% blastocysts developed[b] (mean ± s.d.)
A	01	7	275	83.6 ± 5.41	20.9 ± 6.35
	02	8	334	89.1 ± 4.31	30.5 ± 7.95
	03	4	213	76.2 ± 10.8	19.6 ± 5.21
	04	4	203	81.8 ± 7.91	19.5 ± 1.45
	05	4	167	85.4 ± 3.31	10.9 ± 3.76
	06	4	133	77.4 ± 6.93	18.3 ± 13.1
	07	4	133	74.2 ± 11.9	10.6 ± 4.02
	Overall	35	1157	82.1 ± 8.36	20.1 ± 9.34
B	08	3	137	87.0 ± 9.10	23.8 ± 11.3
	09	6	256	85.6 ± 5.54	24.9 ± 7.57
	10	3	130	74.8 ± 4.28	16.5 ± 6.60
	11	3	137	89.8 ± 4.56	18.4 ± 3.61
	Overall	15	660	84.5 ± 7.49	21.7 ± 7.75

[a] $Fc = 3.12 > F_{0.05} = 2.78$, $P < 0.05$.
[b] $Fb = 3.70 > F_{0.05} = 2.78$, $P < 0.05$.

reported in work with sheep, both *in vitro* (Fukui *et al.*, 1988b) and in natural matings (Kalchreuter and Horneber, 1992). It is clear, taking into account the available evidence, that male variability is likely to be a general feature of farm animal IVF.

Workers at Madison took bulls of high fertility (non-return rate (NRR)% = 78%) and of low fertility (NRR% = 66%) and examined the outcome of IVF when sperm from these two groups were employed. The yield of IVF embryos recorded by Hillery *et al.* (1990) for the high-fertility bulls (32%) was almost double that recorded for those in the low-fertility group (18%). In Ireland, Shi *et al.* (1990) showed that although non-significant differences in cleavage rate were evident among bulls, the embryo yields could differ markedly and substantially (see Table 3.5). This supported the findings in other reports showing that bull differences may be evident in their contribution not only to the fertilization process but also to subsequent embryonic development (Eyestone and First, 1989a; Marquant-Le Guienne *et al.*, 1990a, b, c; Barandi *et al.*, 1993).

3.6.2 *Cause of bull variability*

Iritani *et al.* (1986) suggested that individual bull variability may be related to the stage of season, age of animal, ejaculate sperm quality and other factors. The data of further studies in Ireland by Shi *et al.* (1991) suggested, however, that the variation arising from individual bulls is greater than that arising from bull ejaculates in their contribution to fertilization and embryonic development (see Table 3.6). The differences between bulls appear to be dependent on their individual characteristics, rather than the result of a random error in semen sampling.

Although differences in embryo yield due to bulls have been clearly demonstrated by results in Ireland and elsewhere, how sperm exert control over such embryonic development is not clear. Shi (1991) advanced three possible explanations of the phenomenon. Firstly, seminal plasma may have been a source of variation in the sperm used in IVF; when epididymal sperm have been used, male differences are apparently not evident (Goto *et al.*, 1989). It is known that seminal plasma contains decapacitation factors; variation in the level of such factors may affect ejaculated sperm. Secondly, synthetic activity in oocytes is induced by sperm penetration (First and Parrish, 1987a) and sperm may differ in their ability to stimulate such activity. Thirdly, sperm may differ in the time taken for them to capacitate and this may affect subsequent embryonic development of the oocyte after IVF.

Elsewhere, Taft *et al.* (1992) have recorded evidence that sperm from subfertile bulls (used in AI) may undergo the AR and die prematurely. If this

Table 3.6. Effect of season on cleavage and embryonic development of bovine oocytes fertilized *in vitro* with semen samples from different bull ejaculates (from Shi, 1991).

Bull	Ejaculate	Replicates	Oocytes insem.	% cleavage	% blastocysts developed
01	Spring	5	222	85.1 ± 2.34	38.1 ± 9.20
	Summer	5	210	82.5 ± 3.93	33.3 ± 2.70
	Autumn	5	214	82.4 ± 6.27	33.1 ± 10.9
	Winter	5	211	84.4 ± 5.05	33.0 ± 8.47
	Overall		857	83.7 ± 4.45	34.4 ± 8.04
02	Spring	5	178	83.2 ± 5.49	31.9 ± 11.6
	Summer	5	166	83.8 ± 5.40	42.0 ± 7.10
	Autumn	5	194	84.8 ± 7.28	27.9 ± 4.02
	Winter	5	176	84.6 ± 5.51	34.8 ± 7.77
	Overall		714	84.1 ± 5.52	34.2 ± 9.11
03	Spring	4	178	89.3 ± 2.88	36.2 ± 8.30
	Summer	4	190	81.7 ± 7.49	30.9 ± 11.0
	Autumn	4	158	82.9 ± 5.49	33.7 ± 14.8
	Winter	4	137	83.7 ± 2.32	32.0 ± 9.59
	Overall		663	84.4 ± 5.38	33.2 ± 10.2
04	Spring	4	154	82.5 ± 5.97	46.1 ± 5.80
	Summer	4	157	88.6 ± 2.38	48.5 ± 5.90
	Autumn	4	154	87.1 ± 11.0	45.7 ± 9.50
	Winter	4	158	85.5 ± 3.34	50.1 ± 4.48
	Overall		623	85.9 ± 6.35	47.6 ± 6.28

Values shown in the table are mean ± s.d.

should occur in the tract of the cow, then the sperm population available for fertilization may be reduced, with a consequent adverse effect on fertilization. A review by Parrish *et al.* (1992) draws attention to the fact that events such as fertilization are likely to be influenced by cell cycle regulation. Bulls may influence events as a result of the mechanism by which the sperm activates the oocyte and triggers synthesis of compounds needed for the cell cycle. It might even be that the sperm cell itself contributes something that is required for regulating the zygotic cell cycle. As pointed out by Parrish *et al.* (1992), it is important for gamete and embryo biologists to be aware of recent developments in understanding cell cycle regulation.

In The Netherlands, Kruip *et al.* (1992) concluded that the commonly used parameters for evaluating sperm quality are probably inadequate for an accurate prediction of bull-fertility. Like others, they found that bulls tested by IVF had significantly different fertilizing capabilities *in vitro*. In Canada, Kroetsch and Stubbings (1992) found evidence of innate fertility differences between bulls used in IVF but concluded that sire effects on IVF can be rendered insignificant by using high insemination doses. The use of pooled semen (from individual bulls of high genetic merit) may be a means of reducing bull-factor variability in an IVF system (Aurich and Hahn, 1993); however, this may often not be commercially feasible.

3.7 Assessing Efficiency of Capacitation Procedures

3.7.1 Staining techniques

Staining techniques have been developed to determine the occurrence of the AR in the sperm of various mammalian species. In this, it is necessary to distinguish between sperm that have undergone a true AR from those that have undergone a false AR. The triple-staining technique, as described by Didion and Graves (1986a,b) and Didion *et al.* (1987), is claimed to be capable of doing this. Several other workers have reported staining techniques for evaluating the AR in bull sperm (Aalseth and Saacke, 1986; Yang *et al.*, 1989; Ooba *et al.*, 1991; Den Dass *et al.*, 1992; Way and Henault, 1992).

3.7.2 Oocyte penetration tests

The AR of the bovine sperm is the ultimate proof of capacitation and ability of the sperm to penetrate the secondary oocyte. In the cow, as in the ewe, the evidence indicates that the AR only occurs at the surface of the ZP. The penetration of the ZP is therefore used in some tests as the end point of capacitation.

A zona penetration assay for use with bull sperm was described by Wheeler and Seidel (1987), using 'dead' cattle oocytes. In this assay, it was possible to observe sperm directly in the perivitelline space; the method did not, however, evaluate the interaction of the sperm with the vitelline membrane. In hamsters, Boatman *et al.* (1988) developed salt storage procedures which enabled them to perform hamster sperm capacitation bioassays on a large scale. Elsewhere, however, Chian *et al.* (1991) reported on difficulties in assessing bull sperm capacitation and the AR using salt-stored bovine oocytes. In India, Sharma *et al.* (1991) used preserved buffalo oocytes in fertility tests.

The use of hamster oocytes, usually zona-free oocytes, has been described by several workers as a possible method of quantifying the effectiveness of capacitation procedures (Graham *et al.*, 1986, 1987; Graham and Foote, 1987a; Eaglesome and Miller, 1989; Schellander *et al.*, 1989b). This is based on the principle that bull sperm cannot penetrate the vitelline membrane of zona-free hamster oocytes without first undergoing an AR. In human sperm evaluation, this zona-free hamster oocyte test has been widely used. The score is usually expressed as the percentage of oocytes penetrated; a score in the range of 15–100% is taken as normal and 0–14% as below normal.

In human IVF, there is evidence of men who are consistently unable to fertilize human oocytes *in vitro*. In London, Pampiglione *et al.* (1993) effectively used an AR test, stimulated by calcium ionophore A23187, to predict such cases of fertilization failure.

3.7.3 Sperm–zona binding

It is clear, from many reports, that both sperm characteristics and oocyte quality may play a role in the efficiency of the IVF process. A paper by Mahadevan *et al.* (1987) on human IVF examined the number of sperm bound to the ZP. Their findings support the view that the number of sperm bound to the ZP is functionally important. The motility of sperm and the number of motile sperm used to inseminate oocytes were significantly correlated with the number of sperm bound to the ZP.

In cattle, Fazeli *et al.* (1991, 1992) attempted to establish a sperm–zona binding assay for bull sperm. Their results showed differences among bulls in the relative numbers of sperm bound to the ZP of cattle oocytes. Whether such differences in sperm–zona binding is related to the quality of bull semen remains uncertain. Sperm–zona binding has been shown to be greater in ejaculated bull sperm than in epididymal sperm (Ellington *et al.*, 1993b).

3.8 Capacitation Procedures in Other Farm Animals

3.8.1 Sheep, goats and pigs

The methods of capacitation that are highly effective in cattle may not be as successful in other species. Work with sheep and pigs by Cheng *et al.* (1986) led them to conclude that there are basic differences between these two species in their requirements for capacitation. In Cambridge, Fukui *et al.* (1987a) examined the effect of various factors on ram sperm preparation. Although heparin has been shown to accelerate the AR in bovine sperm suspension at a dose level of 10 μg ml^{-1}, no such effect was found by Williams *et al.* (1991) with ram semen; the authors concluded that this was probably the result of species differences or the need for higher heparin levels.

In former Czechoslovakia, however, Slavik *et al.* (1990) had earlier successfully employed heparin to capacitate ram sperm, and used them to penetrate bovine oocytes. The effect of freeze–thawing on *in vitro* capacitation of ram sperm has been dealt with in some reports (Artiga *et al.*, 1993; Garde *et al.*, 1993). Freshly ejaculated ram semen was found to require a longer capacitation period than frozen semen. It was also clear that frozen–thawed semen loses its fertilizing capacity *in vitro* more rapidly than freshly ejaculated sperm.

Although the ionophore-induced AR is believed by some to be similar to the normal *in vivo* reaction in capacitated sperm, Watson *et al.* (1991) did find clear ultrastructural differences in ram sperm; they concluded that the ionophore-induced reaction is not the same as the natural event.

In goats, De Smedt *et al.* (1991, 1992) capacitated ejaculated caprine sperm in modified defined medium (DM) as used for ram semen. Use of such sperm in the IVF of ovulated and *in vitro*-matured oocytes resulted in fertilization rates of 85 and 83%, respectively.

In pigs, sperm preparation and capacitation as employed by Mattioli *et al.* (1989) involved prolonged incubation (24–48 h) in a commercial semen diluent, with subsequent washing and use of Percoll layers to provide a highly motile fraction. This was a modification of procedures previously reported by Cheng *et al.* (1986). In Korea, Park *et al.* (1990b) suggested that storage of boar semen at 20°C for 10–12 h and preincubation at a high sperm concentration was a suitable method of inducing capacitation of sperm in that species. Combination treatments with ionophore A23187, heparin and caffeine have been examined by several workers (Wang, W.L., *et al.*, 1991; Lei *et al.*, 1992). No capacitation procedure comparable to the relatively simple heparin procedure used in cattle is currently available for use in pigs.

3.8.2 Horses and buffaloes

In horses, there are a limited number of reports dealing with the use of ionophore A23187-treated sperm in IVF attempts (Bezard *et al.*, 1989; Blue *et al.*, 1989). In one such report, both lysophosphatidylserine (LPS) and the calcium ionophore were found to be capable of capacitating equine sperm and inducing the AR. In a report from Newmarket, Zhang *et al.* (1990) also recorded using A23187 in horse IVF. They found that stallion sperm are particularly sensitive to the ionophore and concluded that it is probably necessary to determine the appropriate dose level on an individual basis. In the USA, Farlin *et al.* (1992) used heparin in a comparison which showed that incubation with 10 μg ml^{-1} was sufficient to induce capacitation in bull sperm but that a tenfold increase in dose level was required for equine sperm. Elsewhere in the USA, Fayrer-Hosken *et al.* (1992) has shown that heparin treatment enhances recognition of the equine ZP.

Relative to cattle, buffalo sperm appear to have poor fertilizing capacity and low viability if the semen is refrigerated or frozen with liquid nitrogen. In buffalo IVF, varying doses of heparin have been examined in some studies (Totey *et al.*, 1993b); it appears that heparin can capacitate buffalo sperm in a dose-dependent manner. As with the other farm animals, considerable variability exists among buffalo bulls in the fertilizing capacity of sperm.

4 *In Vitro* Fertilization

4.1 Introduction

Fertilization is a complex process, which results in the union of two gametes, the restoration of the somatic chromosome number and the start of the development of a new individual. Successful cattle *in vitro* fertilization (IVF) requires appropriate preparation of both sperm and oocyte, as well as culture conditions that are favourable to the metabolic activity of the male and female gametes (Brackett, 1981, 1983, 1992; Sirard, 1989a, b; Xu and King, 1990a). The basic fertilization process starts when a bull sperm first attaches loosely, and then binds tenaciously, to receptors on the zona pellucida (ZP) of the cow's secondary oocyte. The sperm, which has a large number of zona-binding proteins on its surface, binds to sperm receptors on the oocyte; such binding is usually species-specific, so that bull sperm should normally only be capable of fertilizing cattle oocytes. Many of the basic events of mammalian fertilization have been clarified and the molecular mechanisms underlying these events identified (reviews by Wassarman, 1988b, 1994).

4.1.1 In vitro *maturation and fertilization: early reports*

Even as recently as 1985, no more than ten IVF calves had been reported in the literature, all these births occurring in North America (Brackett *et al.*, 1982, 1984; Sirard *et al.*, 1986a); on each occasion, the calf was the result of fertilization *in vitro* of an ovulated or *in vivo*-matured oocyte.

The first report of successful *in vitro* maturation (IVM) and IVF of cattle oocytes was that of Iritani and Niwa (1977) in Japan but the birth of calves was not reported until the work of Hanada *et al.* (1986), using calcium ionophore for sperm capacitation; three calves were born, representing 21% of the embryos transferred to recipient animals. The IVF system described by

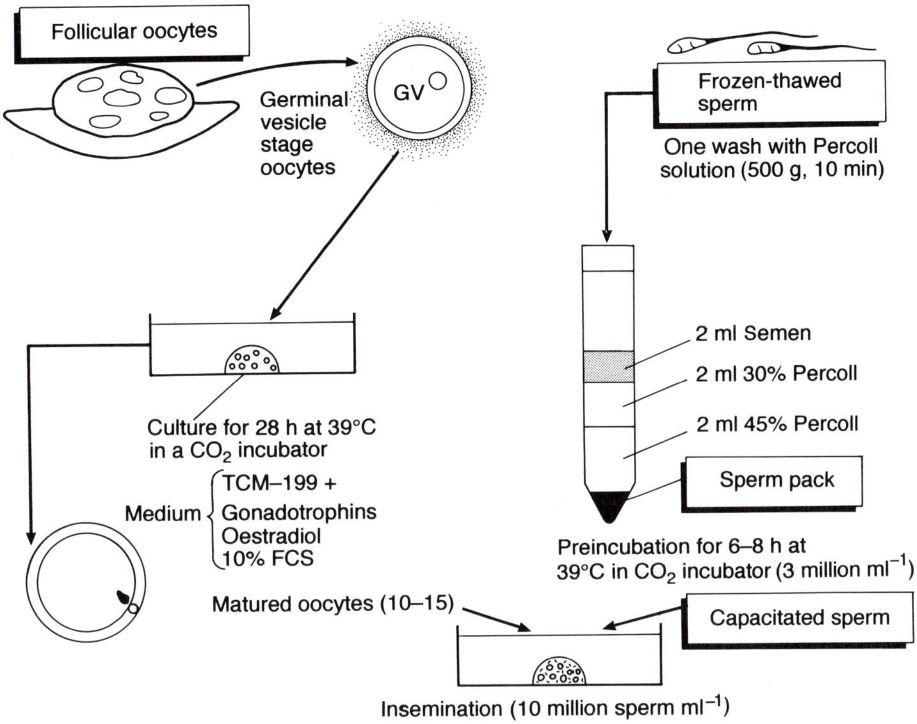

Fig. 4.1. *In vitro* fertilization system (Kyoto University) for cattle (from Iritani *et al.*, 1988).

Iritani and associates at Kyoto University (Iritani *et al.*, 1988) is presented in schematic form in Fig. 4.1.

Possibly the largest group of IVM/IVF calves born around that time came from the work of Lu *et al.* (1987b, 1988b). In this, 18 calves were born alive to 13 recipients after non-surgical transfer of embryos; this represented an embryo survival rate of 59%. The report of Lu *et al.* (1987b) not only set out information on the methodology employed in the production of IVM/IVF embryos but also provided the first convincing evidence of normal pregnancy rates with this type of embryo.

The main factors held to be responsible for the success of Lu *et al.* (1987a) at that time included: (i) the use of follicle dissection and subsequent follicle rupture as the means of recovering high-quality cumulus–oocyte complexes (COCs); (ii) the provision of additional granulosa cells (GCs) (3–5 million ml^{-1}); (iii) the use of oestrous cow serum and the absence of hormones; and (iv) the use of a non-static culture system. There was also evidence (see Table 3.2) showing the importance of pH values employed in the capacitation of sperm (7.4) and in the fertilization medium (7.8).

4.1.2 *Chapter contents*

Although there has been a remarkable increase in knowledge of the various events involved in the fertilization process, information is still far from complete and many of the biochemical processes underlying the process remain unknown. Research studies are increasingly concerned with events at the molecular level, particularly those involving the binding of the sperm head to the ZP. The present chapter deals with information on events involved in the natural process of fertilization in the cow and subsequently with the various steps that are essential in preparing sperm and oocytes for fertilization *in vitro*. An important point, emerging in much of the recent evidence, is the fact that the oviduct clearly plays an important role in providing for the needs of the sperm, as well as the needs of the oocyte. Advances in IVF techniques, using bovine oocytes (BOs) recovered from slaughterhouse cattle, have made it possible not only to provide large numbers of embryos for practical purposes and research but also to study many features of the fertilization process in the cow.

4.2 Fertilization in the Cow

4.2.1 *Fertilization in the cow's oviduct*

Fertilization in the cow is a complex process that normally occurs within the controlled and unseen environment of the oviduct. An understanding of sperm transport in the cow and factors affecting the number of sperm that reach the site of fertilization in the ampulla is provided in a number of reports (Hunter, 1985; Hunter, R.H.F., *et al.*, 1991). It is evident from these that there is probably a synchronization of sperm movement and ovulation in the cow, which has the effect of bringing relatively small numbers of sperm from a reservoir in the caudal isthmus to the ampullary region of the oviduct. In nature, the cow is served by the bull during the period of oestrus and sperm would normally be present in the reproductive tract several hours prior to the occurrence of ovulation (which occurs about 10–12 h after the end of the heat period). The lifespan of sperm in the cow's oviduct is believed to be of the order of 24–48 h, where freshly ejaculated semen is involved, but nearer to 12–24 h when frozen–thawed semen is employed in artificial insemination (AI).

4.2.2 *Lifespan of the secondary oocyte*

Although not as yet precisely defined, it seems probable that the unfertilized cow oocyte has a remarkably short viable lifespan (6–12 h) after it is released from the ruptured follicle. As pointed out by Hunter (1985, 1989), the microtubules of the meiotic spindle of the ovulated cow oocyte would be expected to become disorganized within several hours, with pairs of microtubules escaping laterally from the spindle apparatus and the consequent loss of chromosomes from the metaphase plate.

The cortical granules, which confer the block to polyspermy in the cow after releasing their contents into the perivitelline space of the activated oocyte, migrate from Golgi regions within the oocyte to take up their position just below the plasma membrane immediately prior to ovulation. The granules are believed to remain in position for several hours, until ageing of the oocyte sets in, when they begin to swell and wander away from the surface of the ooplasm. In this dispersed state, the release of the contents of the cortical granules into the perivitelline space cannot follow the activation of the oocyte and polyspermy is liable to occur. It is clear that the BO is designed to be fertilized very soon after ovulation and there is likely to be very little latitude in terms of its lifespan if sperm are not already in the reproductive tract at the time of follicle rupture. For cattle IVF, this probably means achieving fertilization of the secondary oocyte as soon as possible after nuclear and cytoplasmic maturation is complete.

4.2.3 Dispersion of cumulus cells

In considering the interaction of sperm and oocyte in the cow's oviduct, it is known that the oocyte is ovulated with a covering of expanded cumulus cells; these cells are apparently dispersed in the ampullary region of the oviduct within a few hours. The absence of a corona radiata on tubal ova was commented on by Laing (1945) in slaughter studies with inseminated cows. Lorton and First (1979) observed that cattle oocytes recovered from the oviducts within 2–3 h of ovulation are usually completely devoid of cumulus cells. Such observations are in general agreement with experiences in Dublin. Other data, in this instance from sheep, also show cumulus cells to be lost within a few hours of ovulation (Thompson and Wales, 1987). It is believed that most of the expanded cumulus cells are lost within 2 h but it appears that the corona radiata cells immediately surrounding the oocyte may take a further 2 h before they disperse.

In the normal *in vivo* fertilization process, therefore, bull sperm in the ampulla would be presented with the naked secondary oocyte. It is believed that individual sperm in the oviduct become capacitated over a period of several hours, so that sperm capable of penetrating the oocyte are available for a similar time interval.

4.2.4 Fertilization rate in cattle

Estimates of fertilization rates in cattle were reviewed by Diskin (1987). For heifers, the rates shown in the literature vary from 75% to 97%, with an overall mean of 88%; for cows, rates vary from 83% to 100%, with an overall mean of 90%. In Ireland, Diskin and Sreenan (1980) recorded a fertilization rate of 90% in beef heifers, using AI with frozen–thawed semen. The same workers concluded that fertilization failure accounts for only about 10% of overall reproductive failure in beef heifers while embryo death accounts for more than 30%. Available evidence suggests that fertilization failure probably accounts for about 10–12% of conception failure and shows little difference between heifers and cows or between the use of frozen–thawed semen and the bull in natural service. Diskin (1987) suggests, however, that published fertilization rates probably overestimate the actual rate achieved at farm level. In practice, for example, there may be up to 10% of cows that are inseminated without having shown a genuine heat period; obviously, fertilization cannot occur when the animal is not in genuine oestrus.

Although it is clear that a high fertilization rate is the norm in the cow, the causes of failure, when it does occur, are not well understood. It is unlikely to be a question of the individual cow or the timing of insemination alone; the quality of sperm is probably an important factor. Research in cattle IVF may go some way towards explaining some causes of fertilization failure, now that oocyte–sperm interaction can be studied in the laboratory.

4.2.5 Fertilization in superovulated cattle

It should be mentioned that the fertilization rates described above may not always hold good for superovulated animals, where several oocytes are likely to be released from the ovaries. It is not uncommon to recover embryos at different stages of development from superovulated donors. This may be the result of oocytes being fertilized over a period of several hours rather than at the one time. As a general rule, up to 50% of embryos recovered from superovulated cattle are likely to be unacceptable for subsequent transfer (Gordon, 1975); part of this problem may lie in the different times at which ovulation and consequently penetration and fertilization by sperm occur. This should be a problem avoided in cattle IVF, where it can be arranged for insemination to be carried out at the optimal time.

An article by Hyttel *et al.* (1991b), dealing with superovulated animals that had all shown a preovulatory luteinizing hormone (LH) surge, provides an insight into the structural, functional and chronological aspects of fertilization as it occurs *in vivo* in superovulated cows. About 25% of the oocytes were unfertilized, possibly the result of impaired sperm transport; limited numbers of sperm attached to the ZP (accessory sperm) also suggested defects in sperm transport. Cattle with deviant LH patterns, i.e. those in which the LH surge is premature, delayed or non-existent, may be expected to show much higher rates of fertilization failure.

4.3 Preparing Sperm for *In Vitro* Fertilization

4.3.1 Assessing quality of frozen–thawed semen

Using semen obtained from cattle AI stations implies that sperm must have shown a certain standard of motility for them to be accepted for freezing in the first instance. There is still, however, the need to check for satisfactory sperm motility and other characteristics after the thawing of frozen semen. It is as well to remember that the classical supravital stains are not appropriate for assessing the viability of sperm because of the presence of glycerol, which interferes with differential (live/dead) staining. A fluorescent staining system applicable to frozen–thawed bull semen was developed by Ericsson *et al.*

(1989); this did provide information on the viability and potential fertilizing capacity of such sperm. However, such specialist semen evaluation technology would seldom be available.

The use of computer-assisted motility analysis (CASMA) of frozen–thawed bull sperm is reported by Tuli *et al.* (1992); under some conditions, the use of a computerized semen analyser may help to eliminate subjective human error. The main disadvantage of such technology, as observed by Tuli *et al.* (1992), is in the high cost of equipment; a low-cost instrument, suitable for performing the essential functions of counting and assessing sperm motility, would be welcome. Apart from motility assessment, computer-assisted analysers can be of value in research when monitoring movement changes of bovine sperm during capacitation (Leidl *et al.*, 1993).

The diluent employed in bull semen preservation is a factor to keep in mind. Although the freezing of bull sperm has a long history, even with current techniques 50% or more of sperm may fail to survive the process. In the USA, Karabinus *et al.* (1991) showed differential diluent effects on post-thaw bull semen quality. It is evident that altering the composition of the semen diluent or the sequence of adding diluent components may influence the post-thaw quality of frozen–thawed bull sperm. It is likely that some proportion of those bull sperm that do survive freeze–thawing may not necessarily be fully functional. In all of this, it is as well to remember that freezing bull sperm remains a relatively inefficient procedure. Advances in an understanding of the general principles of cryobiology, applicable to the freezing of both bovine sperm and embryos, can be expected in the years ahead.

4.3.2 Swim-up procedures

Motility is an essential feature of bull sperm if they are to progress through the female reproductive tract and effect fertilization. Penetration of the sperm head through the ZP presumably requires particularly vigorous propulsion (hyperactive motility) by the sperm tail. The swim-up technique has become a well-accepted step in obtaining a highly motile sperm sample for use in cattle IVF. On occasions, attempts have been made to correlate swim-up findings with the fertility of bulls used in AI (Lathrop and Foote, 1986).

As currently employed in cattle IVF in Dublin, the swim-up technique usually involves taking the thawed semen sample and overlaying it with an appropriate volume of a suitable medium. This medium has generally been a calcium-free-Tyrode/albumin/sodium lactate/sodium pyruvate (TALP) preparation and is used both in the swim-up technique and in the subsequent washing of sperm. The semen sample is left in an incubator at 39°C for 30–60 min, during which time the progressively motile sperm in the semen swim up into the TALP. Although the swim-up procedure can result in a highly motile sperm sample, it can suffer from the disadvantage of a relatively low yield of sperm, which may be a practical consideration in using very expensive semen straws. In human IVF studies, Purvis and Egdetveit (1993) showed that leaning the swim-up tubes at 30°, rather than standing them vertically, increased sperm yield by 50–100%. Even the form of medium used in swim-up may be a consideration. In France, Marquant-Le Guienne *et al* (1993a) showed that the addition of 5 mM glycine to their TALP medium significantly influenced subsequent cleavage rate and embryo yield after IVF.

It was Parrish *et al.* (1984) who first reported a significant effect on the fertilization rate in cattle IVF using the swim-up technique. Fertilization frequency increased from 46% to 59% and swim-up consistently yielded samples showing highly motile sperm. These findings were confirmed in a subsequent report from the same laboratory (Parrish *et al.*, 1985b) and from studies reported from Georgia by Keefer *et al.* (1985). In practice, however, there may be instances in which the swim-up procedure is omitted prior to washing the sperm. In New Zealand, Catt (1990) dispensed with the swim-up procedure and showed that a larger number of oocytes could be fertilized with the available sperm. The need to use the swim-up may be greater when semen quality and sperm motility are poor.

Various modifications of the swim-up procedure are reported in human IVF literature. According to Ing *et al.* (1991), for example, a swim-down procedure was shorter, simpler and provided a significantly better recovery of sperm with sustained high motility than did the swim-up method.

4.3.3 Glass-wool filtration procedures

In the 1950s, it was shown that diluted bull semen, when filtered through a layer of small glass beads, resulted in dead sperm being retained by the filter while the live ones passed through. In more recent times, Daya *et al.* (1987) used glass-bead column separation for human semen; they found it to be a reliable and consistent method for separating motile sperm from poor-quality semen. Filtration of thawed human sperm samples through glass-bead columns was reported by Lechtzin *et al.* (1991); they found the method to be significantly more effective than the swim-up procedure for obtaining motile sperm.

The use of glass wool, rather than beads, is dealt with in other reports. In Chicago, Rana *et al.* (1989) report data suggesting that filtration of human sperm through a glass-wool column yields a sperm population with an increased oocyte-penetrating capacity. Other work in that city showed that combining treatment of human sperm with TEST–yolk buffer and glass-wool filtration would induce a synergistic effect (Holmgren and Jeyendran, 1993); this was demonstrated by a significantly enhanced oocyte penetration rate.

Glass-wool filtration and swim-up separation procedures were compared by Stubbings and Wosik (1991) in Canada; they used frozen–thawed bull semen to examine the effect on fertilization and embryo development rates. Comparable results were found but filtration with glass wool only required 3 min to complete, which made for a much faster sperm handling time in the laboratory.

4.3.4 Use of BSA/Percoll density gradients

Data reported by White *et al.* (1984) suggested that a population of sperm with a high frequency of viable, motile, morphologically normal bull sperm may be isolated using a discontinuous bovine serum albumin (BSA) gradient. In pigs, a highly motile population, with characteristics indicative of high fertility was isolated by way of a similar discontinuous BSA gradient (Estienne *et al.*, 1988); the procedure yielded a fraction containing more than 90% progressively motile sperm.

In Percoll density gradients, sperm are separated by equilibrium, when sperm density is equal to that of the gradient. It is used on the assumption that sperm with good nuclear morphology are denser than inferior sperm and are selected in the higher-density fraction of the Percoll gradient. In cattle IVF, the method is employed in the IVF system used by Iritani in Japan (see Fig. 4.1). Utsumi *et al.* (1988) used 30% and 40% Percoll in modified Tyrode's (m-Tyrode's) solution as their separation method. In a later paper, Utsumi *et al.* (1991) used 30% and 45% solutions. Sperm selection by Percoll gradient has also been employed by Belgian (Ectors *et al.*, 1989a, b) and French (Chupin, 1989) workers. In Madison, Rosenkrans *et al.* (1993) employed a 90% and 45% Percoll density gradient in isolating a motile sperm fraction in their studies.

Percoll density gradients have been compared with swim-up procedures in separating poor-quality human sperm; Lannou and Blanchard (1988) and Sapienza *et al.* (1993) found Percoll separation resulted in sperm of higher quality than those selected by swim-up. On the basis of their experience, Moohan and Lindsay (1993) suggest that the discontinuous Percoll density gradient technique should be adopted as the method of choice for the separation of sperm for all human-assisted reproductive techniques. In AI with frozen-thawed human sperm, the beneficial effects of Percoll treatment have also been recorded (Burkle *et al.*, 1993).

Sperm separation and sexing

It should be mentioned that there may be some possibility of sperm separation methods having an influence on the sex of the sperm in the sample. Work with human sperm, for example, has led to claims that it is possible to isolate fractions rich in Y-bearing sperm by layering over columns of BSA (Ericsson *et al.*, 1973). In more recent times, in 100 consecutive live births after IVF, a preponderance of male infants (a sex ratio of 1.83 as opposed to the expected 1.06) was reported by Thatcher *et al.* (1989); these authors speculate that their sperm separation procedure (swim-up in Ham's medium) may have inadvertently led to the preselection of Y-bearing sperm. As yet, there does not appear to be evidence in cattle IVF that sperm preparation procedures influence the sex ratio.

4.3.5 Use of hyaluronic acid

Reports from work in Sweden have described the use of hyaluronic acid (a member of the proteoglycan family) in isolating a motile fraction of bull sperm (Shamsuddin *et al.*, 1992b; Shamsuddin and Rodriguez-Martinez, 1993). Results from such work suggest that hyaluronic acid may be employed to improve the IVF results where semen is showing low post-thaw motility from a particular bull.

The conventional treatment of sperm, involving swim-up and washing, can be modified by completely omitting washing and centrifugation and allowing sperm to migrate through a hyaluronic acid preparation to the IVF medium (Larsson *et al.*, 1993).

4.3.6 *Sperm washing*

The washing of bull sperm to remove seminal plasma proteins rapidly and effectively prior to use of the male gametes in IVF is well recognized as essential. Bull semen used in IVF is likely to be frozen–thawed; this means that washing must also remove constituents of the cryopreservation medium, which may include factors such as egg yolk and glycerol. Washing the bull sperm by centrifugal sedimentation and resuspension in fresh medium is generally held to be the most rapid and effective method of removing such unwanted materials. In the sperm preparation method employed in cattle IVF in Dublin (Lu *et al.*, 1987b; Lu, 1990), motile sperm cells, recovered after swim-up into Ca^{2+}-free TALP medium (Ijaz and Hunter, 1989d) are centrifuged twice at 500 *g* for 10 min.

In human IVF, there is evidence that centrifugation *per se* induces sublethal damage in sperm (Alvarez *et al.*, 1993): sublethal damage is characterized by the delayed appearance of deleterious effects, usually in the form of decreased sperm motility over time. There may be a case for re-evaluating some of the centrifugation procedures employed in cattle IVF.

4.3.7 *Sperm storage after swim-up and washing*

In normal cattle IVF systems, matured oocytes are inseminated with sperm shortly after preparation. A study by Chen *et al.* (1992) demonstrated that high cleavage rates and embryo yields could be achieved when IVM oocytes were inseminated with sperm stored at 25°C for up to 8 h (see Table 4.1). In Dublin, Fair (1992) also recorded similar evidence for semen stored for 4 h at

Table 4.1. The effect of time of storage of frozen–thawed bull sperm on the *in vitro* fertilization of bovine oocytes (from Chen *et al.*, 1992).

Time (h)	No. insem.	No. cleav. (%)	Blastocysts (%)	Hatched (%)
0	237	187 (79.0)[a]	77 (32.5)[a]	59 (76.6)
8	286	237 (82.9)[a]	117 (40.9)[b]	94 (80.3)
12	218	185 (84.9)[a]	83 (38.1)[ab]	67 (80.7)
24	310	235 (75.8)[a]	107 (34.5)[ab]	86 (80.4)
48	214	129 (59.0)[b]	68 (31.8)[a]	54 (79.4)

a, b – $P < 0.05$.

25°C. In the IVF laboratory, it may be useful to have such latitude in the utilization of bull sperm.

4.3.8 Sperm doses

In vivo fertilization in the oviduct is likely to occur in conditions when the sperm : oocyte ratio is close to 1 (Hunter, 1993). In cattle IVF, the commonly used sperm doses vary from 0.5 to 5.0 million ml^{-1}. Expressed in terms of sperm : oocyte ratios, these would vary from 10,000 to 20,000 : 1 (Brackett *et al.*, 1982; Bousquet *et al.*, 1984; Lambert *et al.*, 1986; Parrish *et al.*, 1986b; Lu *et al.*, 1987b), although in some reports, the ratio has been reduced to as low as 500 : 1 (First and Parrish, 1987b). In human IVF, as few as 10,000 sperm ml^{-1} fertilization medium have been used successfully (Trounson, 1986b; Yates and De Kretser, 1987); this would be one hundredth the dose that is routinely employed in cattle IVF in Dublin (1 million ml^{-1}). In Denmark, the effect of sperm concentration on cattle IVF was examined by Madison *et al.* (1991b); the use of 1 million sperm ml^{-1} was recommended.

In Dublin, Vergos (1990) attempted to determine the optimum sperm density and volumes to employ in cattle IVF. Results in Table 4.2 suggest that

Table 4.2. The effect of volume and concentration of sperm dose on the *in vitro* fertilization of *in vitro*-matured bovine oocytes[a] (from Vergos, 1990).

Group	Treatment			*In vitro* fertilization					
	Volume (μl)	Sperm concentration ($\times 10^6$ sperm ml^{-1})	Sperm/oocyte ratio	No.	Rate (%)	Index (%)	Abnormal (%)	Activated (%)
A	1	0.29	2,800	61	31 $(50.8)^b$	17 $(54.8)^c$	2 $(6.4)^{bc}$	–
B	2	0.58	5,600	58	44 $(75.8)^c$	26 $(59.1)^c$	3 $(6.8)^c$	–
C	5	1.3	14,000	75	58 $(77.3)^c$	35 $(60.3)^c$	4 $(6.8)^c$	–
D	10	2.5	28,000	67	51 $(76.1)^c$	28 $(54.9)^c$	7 $(13.7)^b$	–
E	20	4.2	56,000	71	53 $(74.6)^c$	19 $(35.8)^b$	15 $(28.3)^b$	–
Overall				332	237 (71.4)	125 (52.7)	31 (13.0)	–
Dead sperm				40	0	0	0	7 $(17.5)^c$
No sperm				51	–	–		4 $(7.8)^c$

[a] Data from five replicates.
b, c – Values with different superscripts within the same column are significantly different ($P < 0.05$); values with same superscripts within the column are not significantly different ($P > 0.05$).

a sperm density of 0.58–1.3 million sperm ml^{-1} in volumes of 2–5 μl (added to 45 μl fertilization droplets, containing ten oocytes).

Sperm doses and naked oocytes

In human IVF, there is apparently no requirement for the oocyte to have cumulus cells attached to it for fertilization to occur (Trounson, 1986b). Work in Dublin by Kinis (1990) would appear to support this finding in cattle. He examined the effect of reducing the sperm dose according to whether artificially matured cattle oocytes were 'cleaned' of cumulus cells or were exposed to sperm in the normal way (partial mechanical stripping of cumulus cells). Results in Table 4.3 show no adverse effect of the citrate cleaning of oocytes on cleavage rate and embryo yield using IVF at the normal sperm dose (1 million ml^{-1}).

Studies by Kinis (1990) also examined the effect of citrate cleaning on the incidence of parthenogenetic activation and polyspermy; as shown in Table 4.4, there was no evidence of any adverse effect.

4.4 Enhancing Sperm Motility

It is clearly important to have highly motile bull sperm available for IVF. This may be achieved by applying various procedures for isolating motile samples (swim-up techniques, Percoll density gradients, etc.). There are also a number of chemical agents which may be employed to stimulate motility of bull sperm and to maintain motility.

Table 4.3. Effect of cumulus removal by the use of sodium citrate on the outcome of *in vitro* fertilization (from Kinis, 1990).

Types of oocytes	No. of oocytes inseminated	Cleav. rate[a] (%)	Cleav. index[b] (%)
Cumulus-free	1083	892 (82.3)	775 (86.8)
Cumulus-covered[c]	820	621 (75.7)	490 (78.9)

Treatment	No. of eggs transferred on monolayers	Total yield of usable embryos (%)[d]	Developmental stage reached: Morulae	Developmental stage reached: Blastocysts
Sodium citrate	687	290 (42.2)	199	91
Control	443	176 (39.7)	127	49

[a]Cleavage rate is the percentage of cleaved eggs.
[b]Cleavage index is the percentage of cells beyond the two-cell stage in relation to total number of cleaved eggs at 46–48 h post-insemination.
[c]Partial denudation by mechanical stripping by repeated pipetting.
[d]The eggs remained on oviductal monolayers for 6 days.

Table 4.4. Parthenogenetic activation and the incidence of polyspermy in sodium citrate-treated and intact bovine oocytes (from Kinis, 1990).

Treatment	No. of unfertilized oocytes	No. of replicates	Parthenogenetic cleavage (%)[a]
Sodium citrate	272	5	22 (8.0)[b]
Control	242	5	18 (7.4)[b]

Treatment	No. of zygotes with two or more pronuclei[c]	No. of replicates	Polyspermic ova (%)	3P	4P
Sodium citrate	80	4	5 (6.2)[d]	4	1
Control	72	4	6 (8.3)[d]	6	–

[a]Assessment of parthenogenetic activation was carried out at 48 h of culture.
[b]Non-significant difference.
[c]The ova were fixed 18 h post-insemination.
[d]Non-significant difference.

4.4.1 Use of caffeine and glutathione

Caffeine, a cyclic nucleotide phosphodiesterase inhibitor, has been employed as a motility-stimulating agent with bull sperm; it apparently achieves its effect by inhibiting phosphodiesterase, which results in an intracellular accumulation of cyclic adenosine monophosphate (cAMP). The effect of caffeine on human sperm motility has been widely documented; in Ireland, the transitory nature of this effect was demonstrated by Traub *et al.* (1982). The agent is known to markedly increase the motility and respiration of bull sperm (Garbers *et al.*, 1971; Ball and First, 1983; El-Gaafary *et al.*, 1990). In some reports, results suggest that caffeine may increase the motility of sperm in poor-quality ejaculates but may have little or no effect in high-quality ejaculates (Critser *et al.*, 1984).

In cattle IVF, treatment of frozen–thawed bull sperm with caffeine has been reported to result in increasing penetration rates (Ohgoda *et al.*, 1987). In mice, the findings of Pomeroy *et al.* (1988) showed that caffeine enhanced fertilization by increasing sperm motility; there were indications that the agent might enhance the acrosome reaction (AR), binding of sperm to the ZP and fusion of sperm with the vitelline membrane.

The post-thaw motility of buffalo sperm has been increased by 25–100% by caffeine treatment (Gehlaut and Srivastava, 1987). Fattouh and Abdou (1991) recommended the addition of 2 mM of caffeine to thawed buffalo semen to enhance the motility rate. The stimulating effect of caffeine on bull sperm penetrating ability in cattle IVF studies was reported by Niwa *et al.* (1988).

In Ireland, Bird *et al.* (1989) showed that caffeine had a stimulating effect on bull sperm at a concentration of 2.5 mM; at 5.0 mM, however, there was a decrease in motility. In Japan, Niwa and Ohgoda (1988) reported a synergistic effect of caffeine (5 mM) and heparin (10 μg ml^{-1}) in inducing capacitation and possibly the AR in frozen–thawed bull sperm, resulting in increased penetration rates *in vitro* of artificially matured oocytes. Later studies by Niwa *et al.* (1992) used caffeine and heparin in the fertilization medium, with and without glucose. A combination of 5 mM caffeine and heparin was employed in cattle IVF by Q.Z. Yang *et al.* (1992) in China; results from this combination were better than those after using heparin alone or ionophore A23187. In human IVF, Imoedemhe *et al.* (1992b) reported that caffeine at concentrations greater than AR mM can adversely affect the fertilizing capability of sperm and the cleavage of embryos derived from such sperm.

Glutathione

There is some evidence that glutathione (GSH) at the 5 mM concentration can increase the motility and fertility of frozen–thawed bull sperm (Slaweta and Laskowska, 1987); GSH is a tripeptide widespread in living organisms and may possibly exert its effect by counteracting the peroxidation effects which may otherwise develop after the death of sperm.

4.4.2 Sperm motility stimulating mixture (PHE)

The presence of a mixture of penicillamine, hypotaurine and epinephrine (PHE) in IVF media has been found to increase significantly sperm motility and the frequency of oocyte penetration (Leibfried and Bavister, 1982; First and Parrish, 1987a; Greve *et al.*, 1987; Boatman *et al.*, 1988; Susko-Parrish *et al.*, 1990; Miller, G.F., *et al.*, 1992, 1994). Others, however, record no favourable effect from the use of PHE (Long *et al.*, 1993a). In Dublin, Vergos (1990) examined the use of the PHE combination and found it to be effective in significantly increasing cleavage rates (see Table 4.5).

In the USA, Susko-Parrish *et al.* (1990) concluded that some of the variability which has been found in fertilization rate among investigators in cattle IVF may be due to the presence or absence of PHE in the fertilization medium. It is not clear what contribution each of the three ingredients present in PHE make to the overall effect of the mixture. In Germany, studies have re-evaluated the role of hypotaurine and epinephrine in the IVF system; there was no effect of these agents on blastocyst production in IVF medium containing 10 μg ml^{-1} heparin (Palma *et al.*, 1993c).

Careful preparation and storage of PHE are essential. In the mixture, hypotaurine is reasonably stable but catecholamines in solution, such as epinephrine, readily oxidize to cytotoxic adenochromes; this change is acceler-

Table 4.5. The effect of sperm motility factors on the *in vitro* fertilization of bovine oocytes matured *in vitro*[a] (from Vergos, 1990).

	In vitro fertilization						
		Cleavage				Abnormal blastomere	
Treatment[b]	No.	Rate	%	Index	%	formation	%
Control	323	227	70.3[c]	119	52.4	15	6.6[c]
Caffeine	304	229	75.3[c]	138	60.3[c]	17	7.4[c]
PHE	367	321	87.5[d]	243	75.7[d]	13	4.0[c]
Overall	994	777	78.1	500	64.3	45	5.7

[a]Data from nine replicates.
[b]Control: no additives. Caffeine, 5 mM. PHE: penicillamine, 2 mM; hypotaurine, 10 mM; epinephrine, 0.1 mM.
c, d – Values with different superscripts within the same column are significantly different ($P < 0.01$).

ated by exposure to light, oxidizing agents or alkaline pH, and is inhibited by penicillamine (Bavister, 1989).

4.4.3 Use of 2-CL-A, 2-DXA and PAF as motility stimulators

2-Chloroadenosine (2-CL-A) and 2-deoxyadenosine (2-DXA)

There is evidence showing that adenosine and its analogues (e.g. 2-chloroadenosine (2-CL-A)) stimulate cAMP production and are capable of increasing the motility of bull sperm by a mechanism that does not involve plasma membrane receptors (Hoskins *et al.*, 1990). A study by Archibong *et al.* (1992) evaluated the effect of 2-CL-A on human sperm; they showed that the agent stimulated forward motion but not the percentage of motile cells in freshly ejaculated human sperm. The same authors refer to data showing a stimulatory effect of 2-CL-A on bull sperm flagellar beat. Elsewhere, Imoedemhe *et al.* (1992a) found that another analogue of adenosine, 2-deoxyadenosine (2-DXA), enhanced fertilization rate in some asthenospermic patients, without any adverse effect being evident on embryonic development. There may be occasions when these agents may be of value in dealing with low-motility bull sperm.

In human IVF studies, Lacham-Kaplan and Trounson (1993) explored the use of 2-DXA and the phosphodiesterase inhibitor, pentoxifylline, separately and together, on the developmental capacity of mouse and human embryos, in order to clarify the possible detrimental effect of these agents. The authors conclude that 2-DXA is a potent inhibitor of embryos and oocytes should not be exposed to it.

Platelet-activating factor (PAF)

Platelet-activating factor (PAF; 1-*O*-alkyl-2-acetyl-*sn*-glycero-3-phosphorylcholine) is a potent phospholipid secreted by various cells and which is known to play a role in reproduction. It has been shown to increase sperm motility and enhance fertilization rates in some species (e.g. rabbit) but the mechanism whereby the growth factor interacts with gametes remains unclear (Roudebush *et al.*, 1993).

4.4.4 Use of pentoxifylline and theophylline

Pentoxifylline is a phosphodiesterase inhibitor which is known to stimulate human sperm *in vitro*, especially in samples from oligoasthenozoospermic patients (Yovich, 1993). Kay *et al.* (1993) in Glasgow found that pentoxifylline stimulates the development of hyperactive motility in washed human sperm incubated under capacitating conditions. The stimulatory effect was maintained for a period of 1 h after washing to remove the agent. According to Moohan *et al.* (1993), maximum response is found at a concentration of 2 mM pentoxifylline. Although Lacham-Kaplan and Trounson (1993) concluded that pentoxifylline could be employed in IVF, it is regarded as important to wash the agent thoroughly out of the sperm sample used for insemination to avoid any adverse effect on fertilization and early embryonic development (Tournaye *et al.*, 1993; Yovich, 1993).

Theophylline is another phosphodiesterase inhibitor which can significantly increase motility in bull sperm *in vitro* (Fayed and Hattab, 1991).

4.4.5 Use of superoxide dismutase mimics

There is evidence showing that cyclic nitroxide compounds with superoxide dismutase-like activity may prevent oxygen damage in tissue-cultured cells exposed to oxidative challenge. In the USA, Lindemann and Kanous (1991) screened two of these superoxide dismutase mimics (4-hydroxy-2,2,6,6-tetramethylpiperidine-1-oxyl (TEMPOL) and 2,2,6,6-tetramethylpiperidine-1-oxyl (TEMPO)) and found that they prolong the motility and extend the life of bull sperm. It is believed that the superoxide anion is extremely toxic to the motile apparatus of bull sperm and these compounds protect the cells. They may therefore have value as additives in the *in vitro* processing of sperm.

4.4.6 Cell-to-cell contact

It should be kept in mind that bull sperm motility may be enhanced merely by contact with cells or with medium which has previously been in contact with cells (preconditioned medium), as shown in the report of Sirard *et al.* (1986b). As observed by Pollard *et al.* (1991a), oviductal cells in coculture maintain both the motility and fertilizing capacity of bull sperm. This may be part of the oviductal role in influencing gametes prior to fertilization and providing for the needs of the early developing embryo subsequently.

4.5 Fertilizing *In Vivo*-matured Bovine Oocytes

4.5.1 Sources of in vivo *oocytes*

Two general approaches have been taken towards achieving IVF in cattle, one involving the use of oocytes recovered from the cow's reproductive tract after having matured (*in vivo*) in the animal's own ovary and the other involving oocytes matured artificially *in vitro*. It was using *in vivo*-matured oocytes, recovered by surgical intervention, that resulted in the first recorded IVF birth in cattle (Brackett *et al.*, 1982). However, Canadian researchers have been among those foremost in employing the laparoscopic approach to recover *in vivo*-matured oocytes from the preovulatory follicles of superovulated cattle close to the time of ovulation, without this compromising the fertility of the donors (Lambert *et al.*, 1983, 1986; Sirard *et al.*, 1985b, 1986c); using laparoscopy under local anaesthesia, these workers were able to achieve oocyte recovery rates of 80–90% in the aspirated preovulatory follicles.

In Germany, Kauffold *et al.* (1986) recovered secondary BOs after flushing the excised oviducts; a fertilization rate of 31% was recorded in 665 oocytes inseminated with sperm capacitated by high-ionic-strength (HIS) medium. In a subsequent report from this group, a fertilization rate of 34% was recorded (Kauffold *et al.*, 1988). It is clear that fertilization rates may be relatively low, even with oocytes recovered after considerable cost and difficulty.

4.5.2 Oocyte quality and IVF success rate

Sirard and Lambert (1986) were among those who stressed that the quality of oocytes recovered from donor cattle is crucial for the success of IVF. The oocyte, according to these authors, should present an ideal morphology, which means showing an expanded and clear cumulus cell mass around the oocyte. If they had that, then 50% of oocytes could be expected to reach the two-cell stage after IVF; where the recovered oocyte showed a compact corona radiata layer of cells, no more than 6% could be expected to cleave. It was concluded by Lambert *et al.* (1986) that variability in cattle IVF results may be attributable to defects in either oocyte quality or sperm quality.

In the Canadian work, considerable donor variation was clearly evident in the quality of the *in vivo*-matured oocytes. It need not follow that *in vivo* maturation, in superovulated donors, necessarily results in high-quality oocytes. Some proportion of oocytes from donors may possess an abnormal or suboptimal pattern of nuclear maturation which is not compatible with normal IVF (Greve *et al.*, 1986).

4.6 Preparing *In Vitro*-matured Oocytes for IVF

4.6.1 Removal of cumulus cells

In an effort to make the surface of the *in vitro*-matured cattle oocyte more accessible to sperm in IVF, workers have attempted to remove some or all of the cumulus cell layers, either by mechanical stripping in suitably sized micropipettes (Lu *et al.*, 1987b, 1988a; Cox, 1991), by the use of enzyme preparations, such as hyaluronidase (Xu *et al.*, 1987a; Park *et al.*, 1989), by chemical agents, such as sodium citrate (Kinis, 1989b, 1990; see Fig. 4.2 and Section 4.3.8) and by procedures such as vortexing (Parrish *et al.*, 1988).

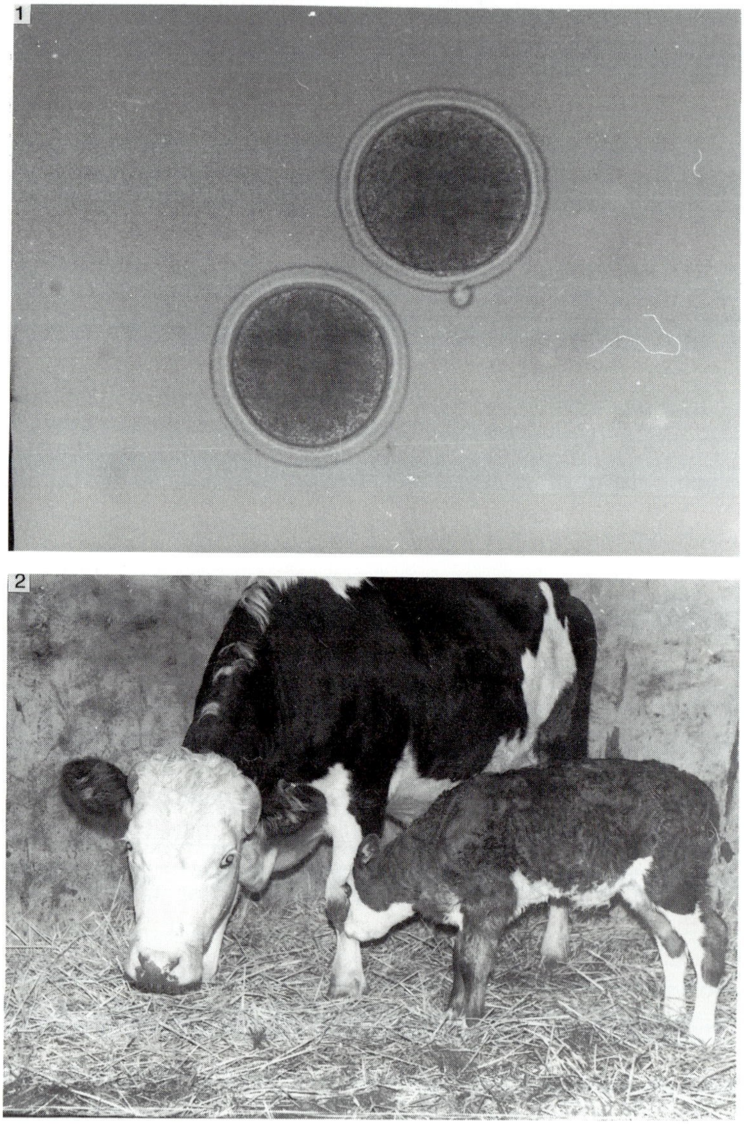

Fig. 4.2. Cleaning the secondary bovine oocyte with sodium citrate solution. Much work was carried out by Kinis (1990) and others in Dublin in cleaning the matured bovine oocyte with a sodium citrate solution (1). The use of this 'cleaning' process was examined at all stages through to embryo transfer and the birth of calves, as shown in (2).

What is described as a simple and rapid method of removing cumulus cells from BOs was reported by Stubbings *et al.* (1987); they used a supernatant prepared by repeatedly freeze–thawing bull sperm diluted in phosphate-buffered saline (PBS) and subsequently centrifuging it. It was believed that the extract contained lytic enzymes derived from the sperm acrosome; the oocytes were not used in IVF.

Clearly, it is undesirable that oocyte cleaning treatments should reduce ability to be fertilized or compromise subsequent embryonic development. In Germany, Bottcher *et al.* (1990) found that removing cumulus cells resulted in a considerable increase in the incidence of defective ZPs and ooplasm abnormalities. On the other hand, Cox (1991), who mechanically removed surrounding cumulus cells from artificially matured cattle oocytes, found no evidence of a treatment effect. Unlike the situation *in vivo*, where attached cumulus cells are rapidly eroded from the ovulated oocyte, dispersion of cumulus and corona cells after IVM does not appear to occur readily. According to Park *et al.* (1989), cumulus cells are likely to be firmly attached to IVM oocytes even 22–24 h after IVF.

Younis and Brackett (1991) examined how the presence or absence of cumulus cells around the oocyte may contribute towards the variability observed in cattle IVF results. Their findings agreed with those of Fukui (1990), who recorded that cumulus cells are necessary at the time of IVF to maximize the incidence of the AR. As early as the report of Ball *et al.* (1983b), the presence of some cumulus cells around the oocyte was recorded as beneficial. However, these authors did consider that denuded matured oocytes may have been damaged by exposure to hyaluronidase and/or mechanical pipetting. There is evidence, for example, that many hyaluronidase preparations are contaminated with trypsin; G.F. Miller *et al.* (1991) reported a decrease in cleavage rate of BOs exposed to this agent. They were also able to demonstrate that this problem could be overcome by the addition of trypsin inhibitor to the cumulus cell removal medium. Behalova and Greve (1993) have shown that sperm penetration rate was similar in denuded and cumulus-enclosed BOs; however, there was a significant increase in polyspermy in the denuded oocytes. The denudation technique in this instance did employ hyaluronidase. In Beltsville, Hawk *et al.* (1992) demonstrated that detaching most cumulus cells from oocytes before IVF increased fertilization rate markedly.

Looking at other farm species, in pigs, oocytes surrounded by expanded cumulus cells have exhibited a higher penetration rate and a greater incidence of male pronuclear formation than have denuded oocytes (Kikuchi *et al.*, 1993).

It may be the particular methods employed in denuding oocytes that are important. The denudation procedure may be at fault in some instances, as shown by G.F. Miller *et al.* (1991). An extensive examination by workers in Dublin (Kinis *et al.*, 1989b; Kinis, 1990) on using sodium citrate as the denuding agent has shown no adverse effect on cleavage rate and embryo yield. In this, a 3% sodium citrate solution is employed over a 5 min period, with gentle agitation to disperse the attached cumulus cells. Some of the potential advantages of using a simple, effective BO cleaning procedure are listed in Table 4.6. Cleaning the oocyte may have particular relevance for cloning

Table 4.6. Citrate cleaning of bovine oocytes.

Potential advantages of citrate cleaning of bovine oocytes	Examination of the cleaned oocyte permits easier assessment of quality
1. Potentially less risk of damage to the zona pellucida, which may have implications for freeze–thawing of embryos subsequently. **2.** Speed and reproducibility of oocyte preparation is enhanced. **3.** For research purposes, cumulus cell presence does not confound the interpretation of monolayer (IVC) studies. **4.** May result in a lower rejection rate of oocytes prior to insemination; may be 5% rejection rather than 10–15% with the partial stripping procedure. **5.** The citrate cleaning technique is well suited for scaling-up procedures in cattle IVF, allowing less skilled personnel and possibly lower numbers of sperm per insemination. One operator can clean more than 500 oocytes in about 15 min.	**1.** Permits the identification of polar bodies. **2.** Allows assessment of granulation of ooplasm to be made more easily. **3.** Can determine the extent of the perivitelline space. **4.** Can detect vacuolation of the ooplasm more accurately. **5.** Can apply computer enhancement techniques which are capable of measuring oocyte size, volume, colour and other features. **6.** No interference from cumulus cells in carrying out enzyme analyses or other procedures with the zygote or embryo. **7.** For sperm injection purposes (SUZI or ICSI) it is necessary to have a clean oocyte. **8.** Application of invasive procedures for embryo sexing is simplified by the absence of attached cumulus cells.

by nuclear transfer, sperm injection and other forms of gamete micromanipulation. The evidence from the live animal strongly suggests that BOs in the oviduct are likely to be naked rather than cumulus-covered at time of fertilization.

4.6.2 Temperature sensitivity of the secondary oocyte

In the live animal, the secondary BO has a limited lifespan in the oviduct; however, sperm are normally awaiting its release and fertilization occurs within a few hours of ovulation. The cattle IVF system must also ensure that sperm meet with the oocyte shortly after its maturation and under the appropriate temperature conditions. It is known that secondary oocytes can be very sensitive to reductions in temperature (Moor and Crosby, 1985); sheep oocytes cooled below 29°C for 3 h exhibited various chromosomal defects. In human IVF, it has also been shown that transient cooling of the secondary oocyte to room temperature can result in irreversible disruption of the meiotic spindle (Pickering *et al.*, 1990); such findings have obvious implications for oocyte handling procedures.

For such reasons, there should be careful control of the temperature environment of the secondary BO. However, results from studies in Dublin (see Table 2.18) suggest that manipulation of the oocyte and IVF can be carried out at normal IVF laboratory temperatures (25–30°C) rather than requiring higher temperatures (>30°C), which are uncomfortable for the laboratory personnel (Lonergan *et al.*, 1991b).

4.6.3 *Effect of maturation time*

There is ample evidence to show that, *in vivo*, the oocyte maturation period is about 24 h in the cow. The routine procedure in Dublin has been to mature BOs artificially for that same length of time (Lu *et al.*, 1987a, b, 1988a, b, c). In the live cow, ageing of the mature oocyte prior to fertilization is likely to compromise its subsequent embryonic development (Hunter, 1989). At Madison, Susko-Parrish *et al.* (1991) examined the influence of ageing in cattle oocytes produced by an IVM system; the time at which sperm penetrated the oocytes was not affected by the age factor, but there was evidence of reduced embryonic development in oocytes aged 32 h.

There is some flexibility in the duration of maturation periods. Russian and other investigators found that an 18 h IVM period yielded similar results (cleavage rate and embryo yield) to 24 h. In Dublin, Monaghan *et al.* (1993) found no significant difference in cleavage rate between 18 and 24 h IVM periods, but blastocyst yield was higher at the 24 h period (see Table 2.19); this did not agree with data recorded elsewhere (Prokofiev *et al.*, 1992; Semple *et al.*, 1993). It is still not entirely clear whether a 24 h period is to be regarded as the most appropriate for cattle oocytes.

4.7 Using the Oviduct in Fertilization Attempts

4.7.1 *Xenogenous fertilization*

Some early studies examined xenogenous fertilization as a possible method of obtaining bovine embryos. Sreenan (1968, 1970) used sheep inseminated with bull semen and transferred artificially matured BOs to the oviducts and recovered them 2–3 days later. Some apparently normal cleaved embryos were recovered, but it is believed that this was probably the result of parthenogenetic activation. However, the cow oviduct was employed in the first successful fertilization *in vivo* of cattle oocytes matured *in vitro* (Newcomb *et al.*, 1978); normality of embryos was confirmed by the birth of twin calves. Although the yield of embryos was low (3%), the Cambridge work was important in being the first to show that some artificially matured cattle oocytes were capable of being fertilized, producing embryos and developing through to full term.

Other studies in which cattle oocytes have been introduced into oviducts or uterus for fertilization include those using the pig reproductive tract (Bedirian *et al.*, 1975), the pig and rabbit tract (Hirst *et al.*, 1981; Dooley *et al.* 1983; Fukui *et al.*, 1983a; Lazar *et al.*, 1992) and the rat uterus (Tsujii, 1988). A study reported by Myers *et al.* (1992) compared the developmental capacity of *in vitro*-matured BOs fertilized either *in vivo* (in the cow oviduct) or *in vitro*; surprisingly, perhaps, the authors found greater development occurred in the oocytes fertilized *in vitro*.

4.7.2 *Fertilization in the mouse oviduct*

Selgrath and Ebert (1990) chose to examine fertilization of pig oocytes in a xenogenous environment, the oviduct of the mouse. Their studies showed that it was possible to fertilize such oocytes in the mouse and they suggested that this approach might be an alternative to IVF. In Dublin, Sharif (1992) was able to show that *in vitro*-matured BOs can be fertilized and can subsequently develop to the blastocyst stage in the isolated mouse oviduct system (Table 4.7). The oviducts were maintained at two temperatures (37 and 39°C) and a comparison was made with the normal IVF routine. Although there was no temperature effect evident in the isolated oviducts, a significantly higher fertilization rate occurred at 39°C than at 37°C in the normal IVF process. The results achieved by this worker did not suggest that prefertilization exposure of the BO to the mouse oviductal environment enhanced penetration and subsequent fertilization; work with hamsters, however, does suggest that favourable oviductal modulation of sperm and oocyte interactions may occur (Rubin *et al.*, 1993).

4.7.3 *Modified GIFT techniques for the cow*

There have been some reports of workers recovering *in vivo*-matured oocytes and attempting to fertilize them subsequently in the oviducts of the cow. In Canada, Shea *et al.* (1983) suggested that infertility in some valuable cows might be overcome by transferring their oocytes into the oviducts of other (host) animals for fertilization. These authors recovered ovulated oocytes surgically and transferred 1100 into inseminated host cows for fertilization; 41% were recovered subsequently, but only 21% of these showed evidence of fertilization.

A modified gamete intrafallopian transfer (GIFT) and fertilization technique has been reported for cattle by Fayrer-Hosken and Stroud (1989) and Fayrer-Hosken and Caudle (1991). The plan was for laparoscopically collected cattle oocytes to be transferred into the oviduct of an inseminated recipient, resulting in fertilization *in vivo* and subsequent embryo development. The authors suggest that such a technology might enable multiple embryos to be obtained from hitherto infertile cows.

4.8 *In Vitro* Fertilization Culture Systems

Many factors are now recognized as contributing towards the successful interaction of BO and sperm in cattle IVF. In terms of the culture systems in which cattle IVF is carried out, there are various possibilities. In Dublin, the procedure employed by Lu *et al.* (1987b, 1988a) is based on the use of microdroplets of modified TALP (50 μl) on a fertilization plate under mineral oil. The overlying oil serves the dual function of minimizing evaporation and preventing the entry of microorganisms.

4.8.1 *The fertilization medium*

Fertilization in the cow normally takes place in a closely controlled environment within the confines of the ampulla of the fallopian tube. In setting up an appropriate system for cattle IVF, it is clear that the medium employed must be capable of providing the secondary oocyte and the capacitated sperm with the conditions which will permit sperm penetration to occur readily. The fertilization medium employed in Dublin (Lu *et al.*, 1987b, 1988a) is based on the TALP formulation as originally described by Bavister and Yanagimachi (1977) in their studies with hamsters.

The TALP medium is an m-Tyrode's preparation containing 25 mM sodium bicarbonate and BSA (0.6%). This TA (Tyrode/albumin) medium is further modified by varying amounts of the energy sources, sodium lactate and sodium pyruvate (LP). Optimal concentrations of these latter constituents appear to be 10 mM lactate and 1.0 mM pyruvate in the presence of 5.6 mM glucose; the osmolarity is held at 285–295 mosm. As mentioned earlier (Section 3.3.2), in studies reported by Lu and Gordon (1987), a pH value of 7.8 was found to be optimal for the TALP fertilization medium (see Table 3.2).

When TALP is used as the fertilization medium, BSA is employed as the protein source. Saeki *et al.* (1993) examined the effect of BSA and heparin in the IVF medium on fertilization and subsequent development; they showed that IVM BOs can be fertilized in a BSA-free medium with a low incidence of polyspermy. Studies reported by Tajik *et al.* (1993) examined how different protein supplements in BO fertilization can influence sperm penetration in bovine IVF; it was found that protein supplementation was not required when oocytes were enclosed in cumulus cells but was essential if they were cumulus-free.

Attention has been drawn to evidence that, in the bovine oviduct, the Ca^{2+} concentration is around 3.1 mM, whereas it is only 2.0 mM in the TALP–HEPES fertilization medium (Hyttel *et al.*, 1989a). In sheep IVF, there are those who recommend using high Ca^{2+} concentrations in the fertilization medium (Huneau and Crozet, 1988).

4.8.2 *Water quality*

In preparing the cattle IVF system, it is essential to keep in mind considerations that apply to the preparation and storage of the fertilization medium. In preparing media, glassware should preferably be silicone-coated to prevent adsorption and contamination problems. As with the IVM medium, water quality in the IVF system must be closely monitored and carefully controlled. There are numerous articles relating to quality control of such media in human IVF and how they may be monitored by one bioassay or other.

A sensitive hamster sperm motility assay was used by Stewart-Savage and Bavister (1988); they found evidence of deterioration in medium quality being correlated with an increase in pH during storage, probably the result of a loss of carbon dioxide and bicarbonate. The hamster assay is more sensitive than that using rat or mouse sperm (Reed and Petters, 1991). Elsewhere, the sperm motility assay has proved of value in testing the toxicity of plastic ware, filtration systems, aspiration needles and other pieces of laboratory equipment that may come in contact with oocytes, sperm and embryos (Rinehart *et al.*, 1988); the sperm motility assay is believed to be superior to the mouse embryo bioassay in differentiating the water quality aspects of IVF.

4.8.3 Temperature and other environmental factors

It has been recognized for some time that fertilization of cattle oocytes is a temperature-dependent process (Lenz *et al.*, 1983a); available evidence would seem to support the view that a temperature of 39°C is not only optimal for BO maturation but for sperm penetration as well (see Table 4.7). This would certainly be compatible with known events in the live animal. In Cambridge, working with pigs, Cheng *et al.* (1986) found that the most important factor influencing successful penetration of porcine sperm into oocytes was the temperature of incubation; less than 1% of oocytes were fertilized at 37°C but a high sperm penetration rate (89%) was achieved when sperm and oocytes were incubated together at 39°C. Marked temperature dependence of ARs in ram sperm is evident in the report of Watson *et al.* (1991); increasing numbers of sperm showed the AR as temperature increased by as little as 2°C. In goats, the AR reaction occurred more readiliy at 39.5°C than at 37°C in some of the studies reported by Kusunoki *et al.* (1989).

The effect of temperature on cattle IVF is also dealt with by Iqbal and Hunter (1992b) in Minnesota; using heparin as the capacitating agent, they record a significantly higher fertilization rate at 39°C than at 37°C (77% vs. 46%).

Table 4.7. Effect of two culture temperatures on IVF of *in vitro*-matured bovine oocytes in the isolated mouse oviduct system (IMOS)[a] (from Sharif, 1992).

T (°C)	Culture systems	No. of oviducts	No. of oocytes	Cell stage 1	2	>2	CR (%)[b]	CI (%)[c]
37°C	IMOS	24	110	49	24	37	55.7[d]	60.5[d]
	Droplets	–	150	59	28	63	60.8[d]	69.2[de]
39°C	IMOS	22	120	58	25	37	51.7[d]	59.7[d]
	Droplets	–	150	35	29	86	76.7[e]	74.8[e]

[a] Data are pooled from three replicates.
[b] CR: cleavage rate (no. of cleaved oocytes/no. of oocytes inseminated).
[c] CI: cleavage index (no. of oocytes cleaved beyond two-cell stage/total no. cleaved).
d, e – Values with different superscripts in the same column differed significantly ($P < 0.05$).

Light

In the course of cattle IVF, oocytes and sperm are inevitably exposed to either daylight or artificial light for variable periods; there would be those who express concern as to the possible adverse effects of light. Work with rabbit oocytes suggests that it is probably unnecessary to compromise the efficiency of IVF procedures in order to minimize light exposure. It appears, from Dublin experience, that BOs, sperm and early embryos are reasonably tolerant of the lighting conditions that normally occur in an IVF laboratory.

Gas atmosphere

It is necessary to incubate gametes and embryos in a rigorously controlled atmosphere of either 5% carbon dioxide in air or 5% carbon dioxide, 5% oxygen and 90% nitrogen, using maximum humidity to prevent medium evaporation. Away from the incubator, and in the preparation of sperm (swim-up, washing) and oocytes (cumulus cell partial removal), there are several instances when media are exposed to air. These operations are typically carried out in a laminar flow hood, where the temperature, humidity and gas phase may be quite different from those in the incubator. Although the modified TALP media employed in such preparations are buffered with HEPES, it is desirable that procedures on the laboratory bench are dealt with as rapidly as possible.

4.8.4 Granulosa and oviductal cells in the fertilization medium

Workers in cattle IVF have exposed oocytes to sperm individually or in groups of varying numbers (e.g. 5–20); it is generally agreed that IVF may be more effective when oocytes are exposed to sperm in groups rather than singly (Hensleigh and Hunter, 1983c). The implication is that oocytes and associated cells may produce and secrete substances promoting fertilization.

In the live cow, fertilization is believed to occur in the presence of relatively few sperm in the fallopian tube, but these sperm would obviously be in contact with the live cells of the oviduct at all times. It might also be mentioned that in the living animal, the ovulated oocyte is in constant contact with oviductal cells as from the instant of ovulation; there may be substances produced by oviductal cells which are especially important for sustaining and capacitating sperm (Pollard *et al.*, 1991b) as well as for development of the embryo after fertilization.

In Ireland, the addition of oviductal cells, in suspension, to the TALP fertilization medium has been observed to improve the quality of subsequent embryonic development (Chen-Lu *et al.*, 1991). Data are in Table 4.8. A study by Sivakumaran *et al.* (1991) in Canada showed a clear advantage of using oviductal cells in their IVF system. Elsewhere in that country, the effect of oviductal cells and heparin on bull sperm capacitation *in vitro* was examined by Xu and King (1990a) and King, R.S., *et al.* (1991); they also record a beneficial effect of the oviductal cells. The presence of oviduct epithelial cells in the IVF medium, as well as the amino acids glycine and alanine, was also reported to have a beneficial effect in the work of Moore and Bondioli (1993). Studies by Sirard *et al.* (1986b) and Ijaz and Sirard (1993) provided evidence that GCs and oviductal cells secrete factor(s) which support bull sperm

Table 4.8. Effect of oviductal cell suspensions on *in vitro* fertilization and development of bovine oocytes (from Chen-Lu *et al.*, 1991).

Treat.	Cleavage (%)	Treat.	Blastocysts (%)	Frozen (%)	Hatched (%)
A	84.0	A1	44.9 (184/410)	45.7^a	80.0
	(1511/1798)	A2	47.6 (524/1101)	45.0^a	82.7
B	81.5	B1	47.4 (349/737)	36.4^b	84.6
	(1449/1778)	B2	52.5 (369/701)	36.6^b	86.9

a, b – $P < 0.05$.

A, oviductal cell suspension added to IVF medium; B, control treatment; A1 and B1, coculture with granulosa cells; A2 and B2, coculture with granulosa and oviduct cells.

viability and motility for up to 30 h. In Japan, Choi *et al.* (1991) found no beneficial effect on fertilization from the presence of bovine oviductal cells in their IVF medium.

In horses, the coculture of stallion sperm with oviductal cells has resulted in a high percentage of such sperm showing evidence of capacitation (Ellington *et al.*, 1993c). There is also evidence in human IVF studies that ampullary cell coculture may be useful in improving fertilization rates and subsequent embryonic development (Bongso *et al.*, 1991a). In such systems, the insemination of oocytes is carried out in the presence of freshly detached human oviductal cells; such cells are used because sperm may bind to established monolayer cells, restricting their motility.

4.8.5 Endotoxins and quality control

In human IVF, there is some evidence that exposure of human gametes to media contaminated by endotoxin may result in a reduction in the pregnancy rate. In cattle, studies in Denmark suggest that endotoxin may induce polyspermy but has no adverse effect on embryonic development (Madison *et al.*, 1991a, b). The source of the endotoxin may be the BSA employed in the preparation of IVF media. The bacterial toxin known as endotoxin is a biologically active lipid derived from the outer membrane of Gram-negative bacteria. Endotoxic fragments are released from bacterial cells after their death, either by natural autolysis or by artificial disruption. Treatment of water and media by conventional methods of heat or chemical sterilization or filtration neither alters the properties of endotoxin nor removes the endotoxic fragments.

Studies in mice (Randall *et al.*, 1991) and rats (Miyake *et al.*, 1991) have provided some information on possible detrimental effects of certain BSA sources and actions of endotoxins.

In human IVF, the presence of > 1 ng bacterial endotoxins ml^{-1} of medium has resulted in significantly lower fertilization and pregnancy rates. In the USA, Zarutskie *et al.* (1992) showed that high-quality media and strict measures to control bacteria are essential in keeping endotoxin at a non-significant level. Dumoulin *et al.* (1991) examined five types of commercially available BSA and showed that these contained endotoxins in varying concentrations. It was considered possible that embryos cultured in a medium containing such endotoxins might elicit an immune response after transfer which might compromise their survival.

IVF quality control considerations

The efficiency of IVF procedures depends on appropriate quality control and there are several excellent reviews dealing with this (Boone and Shapiro, 1990; Schiewe *et al.*, 1990a). The use of rodents (mice, hamsters) in IVF quality control has been reviewed by Leibo (1990). Quality control may need to be wide-ranging. In human IVF work, for example, the repainting of a hospital corridor coincided with a particularly unsuccessful phase of clinical IVF activity. Whether this really constituted a genuine problem or not, it is essential at all times to be aware of any factor that may possibly militate against the effectiveness of the IVF system.

4.9 Interaction of Sperm and Oocyte

4.9.1 Sperm–oocyte recognition mechanisms

In the most recent decade, evidence has been accumulating to show that fertilization in farm and other mammals is mainly controlled by interactions between sperm and components of the oocyte. In the mouse, it is known that the sperm AR is induced after binding to one of the ZP glycoproteins, ZP3. After binding to the sperm receptors, ZP3 appears to initiate the sequence of events leading to acrosomal exocytosis. Some of the biochemical events which may play a regulatory role in the process of the ZP-induced AR in the mouse sperm are dealt with in a report by Furuya *et al.* (1992).

Increasing attention is also being devoted to the molecular basis of mammalian gamete recognition (Miller, D.J., *et al.*, 1992; Tesarik, 1992; Green, 1993). The presence of sperm-specific antibodies, which may be recognized by a receptor on the plasma membrane of the BO (the oolemma) and play a role in sperm–oocyte interaction, has been discussed by Kellerman and Hunter (1992); they concluded that a bovine oolemmal receptor does exist that binds bovine immunoglobulin G (IgG). In pigs, daily variation in ejaculates can influence the number of sperm that bind to the ZP, but Fazeli *et al.* (1992) found no evidence for this in cattle.

4.9.2 Early events in fertilization in vitro

The mechanisms of oocyte penetration have been well documented in laboratory species (Fraser, 1984; Wassarman, 1988a, b, 1994) but the literature describing these events in farm animals is much more limited. Knowledge of the early development of the BO after fertilization, including the time at which the male and female pronuclei (PNs) form and when the first and subsequent cleavages of the early embryo occur is clearly essential to an understanding of the normality or otherwise of zygotes/embryos produced by IVF.

Fertilization involves activation of the oocyte by the sperm. Without this stimulus, the oocyte would be unable to form PNs and become a zygote. A striking feature of activation in the cow is that the vitellus shrinks in volume, expelling fluid into the perivitelline space. At the same time, the sperm head in the vitellus swells and acquires the consistency of a gel, losing its characteristic shape. The final structure, which resembles the nucleus of a somatic cell much more closely than it does a sperm head, is termed the male PN.

An illustration of certain of the cellular interactions that occur around the time of fertilization is provided in Fig. 4.3.

Since the first report of successful IVF in the cow (Brackett *et al.*, 1982), a number of investigators have examined details of fertilization and associated events in the cow, both in the living animal itself (Hyttel *et al.*, 1987a, b, c, 1988a, b, 1991a) and under *in vitro* culture (IVC) conditions (Hyttel *et al.*, 1989a, b, c). Using *in vitro*-matured bovine oocytes, Xu and Greve (1987, 1988) presented a detailed analysis of early events during IVF (sperm penetration, PN development, activation of the second meiotic division, female chromosome decondensation and PN formation). Other studies in Denmark demonstrated that BO maturation and fertilization, as they occur *in vivo* in the cow, constitute a series of well-synchronized nuclear and cytoplasmic

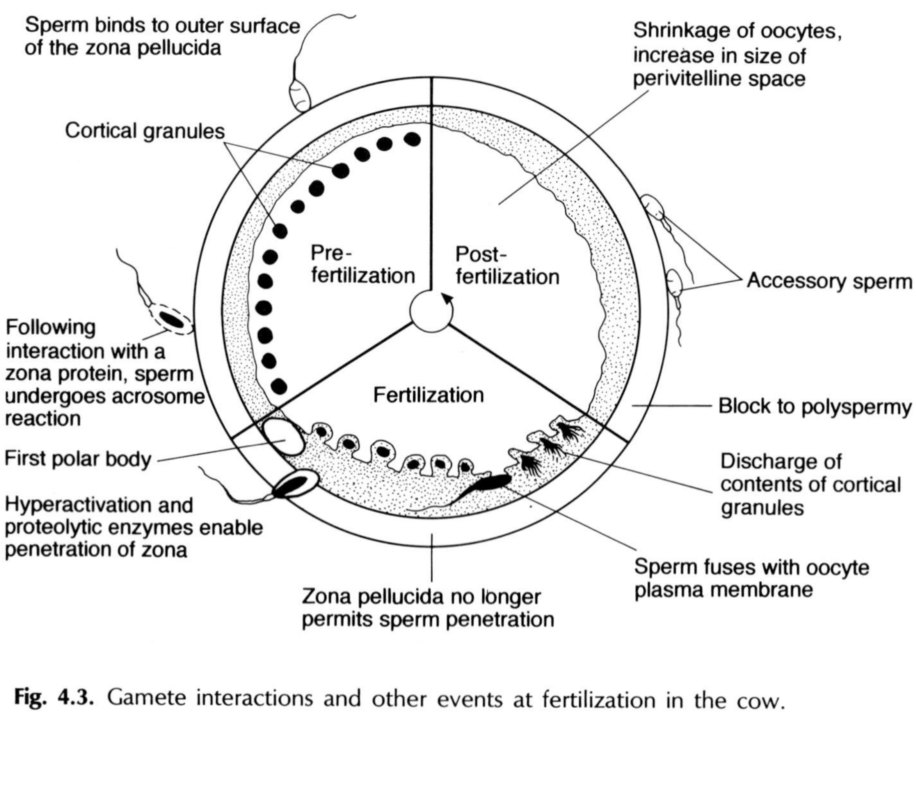

Fig. 4.3. Gamete interactions and other events at fertilization in the cow.

Table 4.9. Description of developmental stages of male and female pronuclei (PN) (from Jiang, 1991).

Stage	Male	Female
PN-1	The complete sperm in the ooplasm, head stains heavily, no sign of decondensation is seen, tail is still attached to head	Chromosomes are contracted to a dot or second meiotic division has just started (anaphase–telophase II)
PN-2	Sperm head starts to decondense, enlarges in size, stains less, and tail is detached	Second polar body is in the process of extrusion, or second meiotic division is completed
PN-3	Sperm head is further decondensed, nuclear envelope starts to form, mid-place is found in the vicinity of the head	Chromosome decondensation occurs, nuclear envelope starts to form, second polar body is located close to the PN
PN-4	Chromosome decondensation is completed, the more or less spherical PNs are surrounded by a complete envelope	
PN-5	PNs enlarge in size, the PN contents stain evenly, second polar body and the sperm tail can still be located	
PN-6	PNs reach maximum size, the distance between the two is very small, sperm tail is still visible	

changes (Hyttel *et al.*, 1988a, b); under certain *in vitro* conditions, however, some well-defined deviations may occur (Hyttel *et al.*, 1989c). In Japan, Saeki *et al.* (1991a, b) determined the time sequences of sperm penetration and formation of male and female PNs in BOs inseminated *in vitro*; they concluded that their culture conditions were suitable for the normal development of PNs.

In Ireland, an analysis of penetrated IVM BOs at various times after *in vitro* insemination by Jiang (1991) led to the definition of the six stages of early PN development set out in Table 4.9. The time required for sperm penetration of IVM oocytes (with attached cumulus and corona cells) was less than 4 h, similar to the report on bovine *in vivo* fertilization by Hyttel *et al.* (1988a, b). Sperm head decondensation occurred within 1–2 h of oocyte penetration and the male PN developed after a further 3–5 h (see Figs 4.4 and 4.5).

In the same study and using the same IVM/IVF system as described by Lu *et al.* (1987b), the incidence of polyspermy was recorded by Jiang (1991) as 10–12% (see Table 4.10). Such polyspermy frequencies are comparable to those recorded in previous cattle IVF studies in Ireland (Kinis, 1990; see Table 4.4) and elsewhere (Parrish *et al.*, 1986b). Other abnormalities observed included polygyny, asynchronous development of male and female PNs and preactivation of cytokinesis. Asynchrony of PN development appeared as a

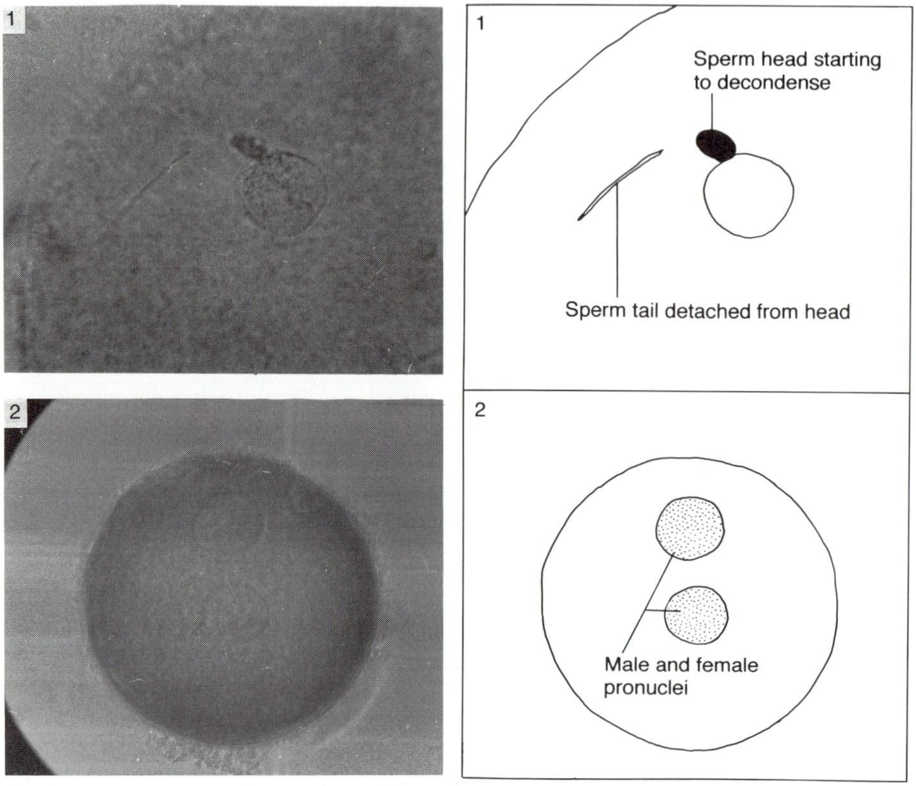

Fig. 4.4. Decondensation of sperm head (1) and formation of male and female pronuclei (2).

Table 4.10. examination of *in vitro* fertilization of *in vitro*-matured bovine oocytes with cumulus and corona cells (from Jiang, 1991).

	Time of examination (hours after insemination)								
	4	6	8	10	12	16	20	24	28
Ova exam.	118	179	145	135	102	133	251	202	166
Monospermy	27	81	95	94	70	111	203	157	128
%	22.9	45.3	65.5	70.2	68.6	83.5	80.9	77.7	77.1
Polyspermy			2	11	10	15	32	25	20
%			1.4	8.2	9.8	11.3	12.8	12.4	12.1
Asynchronous ova		2	7	12	11	14	19	17	12
%		1.1	4.8	9.0	10.8	10.5	7.6	8.4	7.2
Penetration %	22.9	45.3	66.9	78.3	81.4	94.7	93.6	90.0	89.8

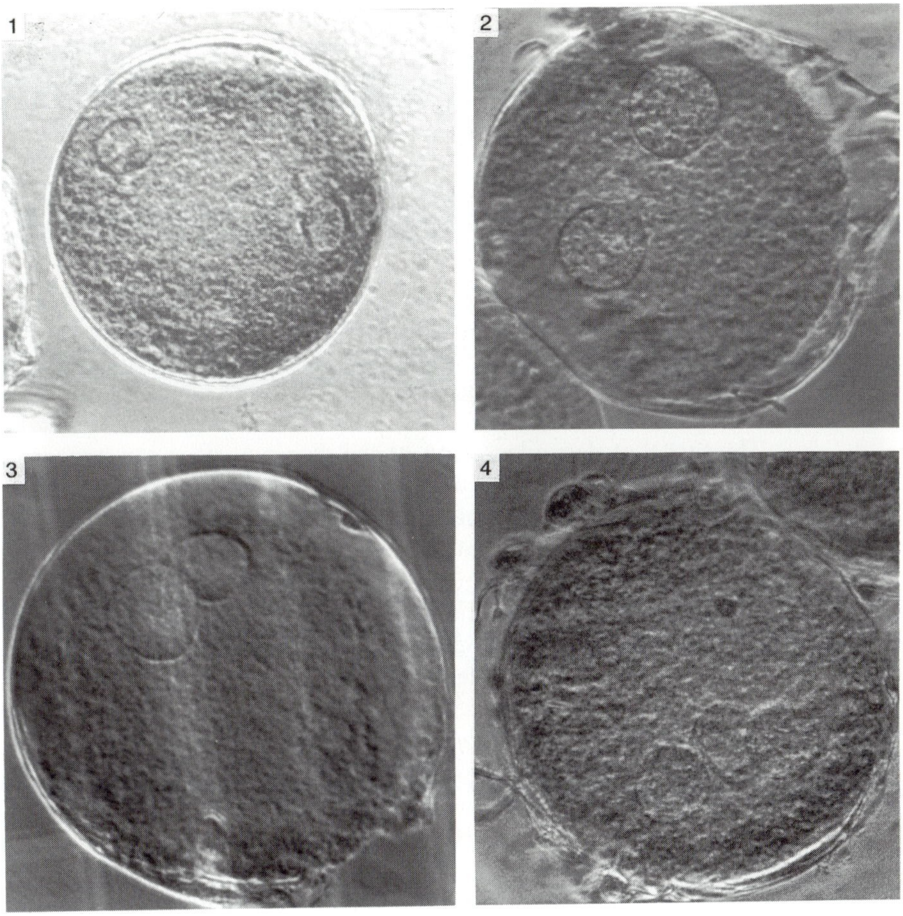

Fig. 4.5. Pronuclei in the bovine oocyte: normal and abnormal. (1) Pronuclei at stage PN-3; (2) pronuclei at stage PN-6; (3) pronuclei about to fuse; (4) abnormality – three pronuclei evident.

prominent abnormality, but was less serious than recorded in the work of Xu and Greve. Polygyny is shown by the presence of more than two PNs, or two PNs and one decondensing sperm head. The occurrence of this defect, according to Xu and Greve (1987, 1988), is probably due to the failure of the first or second polar body to be extruded, or from parthenogenetic activation. Preactivation of cytokinesis comprises symmetrical division of the oocyte ooplasm, similar to normal cytokinesis; PNs may be found in one or each of the two cellular components (Xu and Greve, 1988). In China, Sun *et al.* (1994) used electron microscopy to observe the timing and ultrastructure of sperm penetration, the AR, PN formation and other features of IVF of BOs.

Biochemical changes that occur as a result of sperm penetration include changes in the pattern of intracellular Ca^{2+}. It is known that such changes in Ca^{2+} levels at fertilization are involved in the induction of cortical granule exocytosis and the resultant block to polyspermy that occurs in the cow. Sperm-induced transient rises in Ca^{2+} have been shown to be propagated through the BO in the form of a wave (Sun *et al.*, 1993).

4.9.3 *Sperm penetration of maturing oocytes*

It has been reported for several mammals that the ZP of the immature oocyte at the germinal vesicle (GV) stage is penetrable by sperm; decondensing sperm heads, on the other hand, are rarely observed in such oocytes. It is evident that bull sperm can penetrate a maturing BO at metaphase I (Niwa *et al.*, 1991); the same has been reported in pigs (Nagai *et al.*, 1988; Kikuchi and Nagai, 1993). The studies of Niwa *et al.* (1991) showed that BOs at the GV stages, as well as those undergoing maturation, are equally penetrable by bull sperm *in vitro*.

The fact that decondensation of the sperm nucleus can apparently occur at all stages is in contrast to reports for most other mammalian species. Maturation of BOs is not required for sperm penetration into the vitellus or for sperm nuclear decondensation under the *in vitro* conditions employed by workers in Japan (Niwa *et al.*, 1991; Abeydeera and Niwa, 1992; Chian *et al.*, 1992a). These authors also showed that there is no block to polyspermy at the level of the ZP or the vitelline membrane. The indications are that there is a failure of cortical granule exocytosis in the immature BO.

4.9.4 *Crossing the interspecific sperm barrier*

It is generally held that the mammalian oocyte is protected against foreign sperm penetration by a barrier, which operates at the level of the ZP, the vitelline membrane or both. An IVF culture system can be useful in examining events that are normally difficult or impossible to study under natural conditions in farm animals. In former Czechoslovakia, Slavik *et al.* (1990) demonstrated that ram sperm can penetrate IVM BOs as readily as bull sperm. A subsequent report (Slavik and Fulka, 1992) showed that the ZP of the BO does not represent a barrier against penetration by goat sperm. The penetration of BOs by ram sperm *in vitro* was also reported in Canada by Kelk *et al.* (1991); however, when cows were inseminated with ram semen, no evidence of fertilization was obtained. The authors conclude that fertilization of BOs by ram sperm may be an *in vitro* phenomenon. This apparent *in vivo* block to fertilization may be a useful model for studying certain forms of fertilization failure.

4.10 Postinsemination Treatment of Oocytes

4.10.1 *Effect of sperm exposure time*

In the report of Lu *et al.* (1987b), IVM BOs were incubated with sperm for 18–20 h after *in vitro* insemination. In the live cow, the ovulated oocyte is believed to encounter very few free-swimming sperm at the site of fertilization; blocks to polyspermy would usually have time to be expressed before another sperm reaches the oocyte. With IVF, clearly much larger sperm numbers are in the vicinity of the BO; with increasingly efficient capacitation procedures, the possibility exists of several sperm penetrating the ZP and vitelline membrane. The time interval during which sperm and oocytes are coincubated in the fertilization medium is a further consideration. In some IVF systems, a significantly higher incidence of polyspermy has been shown for oocytes cultured with sperm for 24 h than for those cultured for 8 h (Chian *et al.*, 1992a).

Bull sperm used for IVF have the potential to release hydrolytic enzymes in the small volume of fertilization medium (50 μl) used to contain the gametes. It has been suggested that these enzymes may have a deleterious effect on the

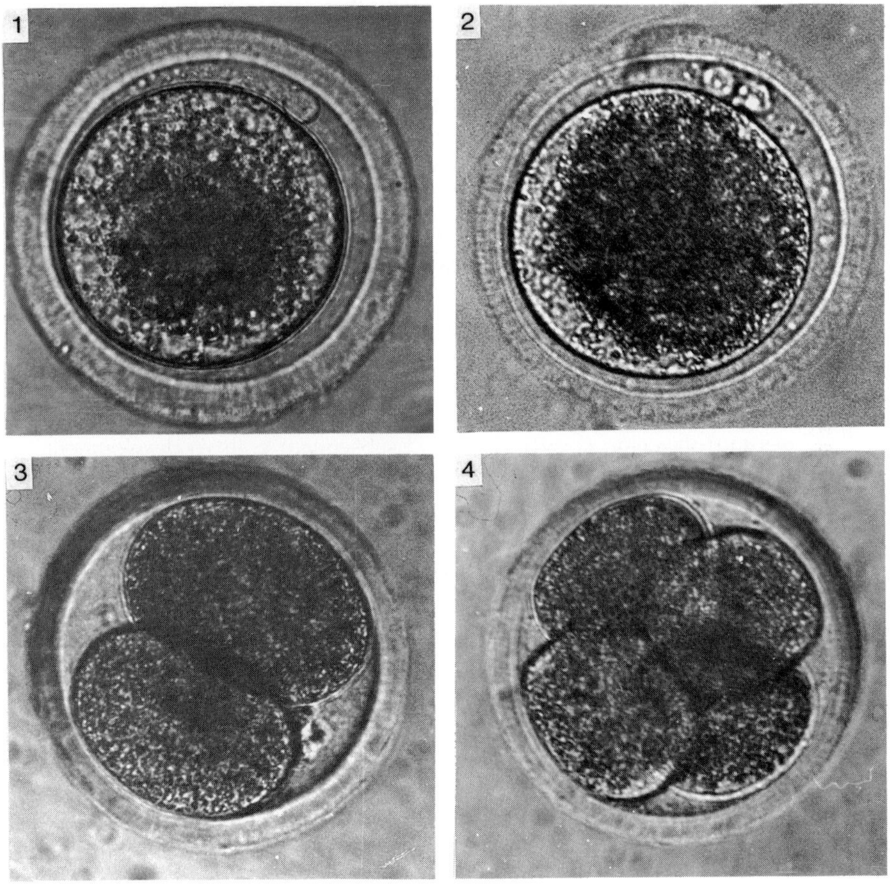

Fig. 4.6. Polar bodies and cleavage at 48 hpi as evidence of fertilization. (1) Oocyte showing the first polar body in the perivitelline space; (2) two polar bodies evident in the perivitelline space; (3) first cleavage occurs somewhere around 30 hpi; (4) by 48 hpi the majority of embryos should have reached the four-cell stage and beyond.

developmental capacity of oocytes. Rehman *et al.* (1993) sought to determine the optimal sperm/oocyte incubation time for achieving maximum fertilization rates after IVM/IVF; the highest fertilization rate and embryo yield resulted when oocytes were incubated with sperm for 24 h.

4.10.2 Transfer of presumptive zygote to embryo culture medium

In the original protocol put forward in Dublin for cattle embryo production, presumptive zygotes were placed in the fallopian tubes of sheep some 18–20 h after insemination (Lu *et al.*, 1987b); at this time, first cleavage of embryos had not occurred. In the live cow, according to Laing (1949), first cleavage occurs about 30 h post-insemination (hpi); *in vitro*, it has been recorded to occur at 26–32 hpi (Barnes *et al.*, 1987). When a totally *in vitro* production system was subsequently developed (Lu *et al.*, 1988a), presumptive zygotes were examined at 44–48 hpi for evidence of cleavage (see Fig. 4.6); they were then moved to the *in vitro* embryo culture (IVC) system.

Further studies in Ireland by Wang (1991) examined the effect of time of transfer from the TALP fertilization medium to a GC monolayer, as shown by cleavage rate and subsequent embryonic development. Presumptive zygotes/embryos were transferred to the GC monolayer at 7, 24, 44 and 72 hpi. As shown in Table 4.11, transfer to the IVC culture system at 7 hpi resulted in a lower cleavage rate, presumably because a proportion of the oocytes were still not fertilized at that stage. It was concluded that transfer at 24 hpi was preferable to the other options.

In agreement with these findings, Lonergan (1992) found transfer at 24 and 48 hpi to be more beneficial to further embryonic development than transfer

Table 4.11. The effect of time of transfer from fertilization medium to the granulosa cell monolayer on cleavage rate and subsequent embryo development (from Wang, 1991).

	Transfer time (h)*			
	7	24	44	72
No. inseminated	384	755	1039	499
No. cleaved	290	603	845	406
%	75.5[a]	79.9[a,b]	81.3[b]	81.4[b]
No. embryos	290	603	845	406
No. blastocysts	158	338	375	158
%	54.5[c]	56.1[c]	44.4[d]	38.9[d]
No. blastocysts	57	174	244	99
Hatched blastocysts	48	147	187	67
%	84.6[e]	81.6[e]	76.6[e,f]	67.6[f]

*Total of ten trials; transfer from fertilization medium to culture medium with granulosa cell monolayers.
a, b – $P < 0.05$.
c, d – $P < 0.01$.
e, f – $P < 0.05$.

Table 4.12. Effect of interval from insemination to transfer to granulosa cell monolayer on cleavage and blastocyst yield (from Lonergan, 1992).

	Cleavage (%)		Blastocyst yield (%)		
hpi	Rate	Index	Day 7	Day 8	Day 9
24	77.3 (422/546)	76.5^{a} (323/422)	11.7^{a} (29/249)	20.1^{a} (50/249)	22.1^{a} (55/249)
48	74.0 (432/584)	62.0^{b} (268/432)	6.0^{b} (17/282)	15.3^{a} (43/282)	20.0^{a} (56/282)
72	75.3 (399/530)	66.92^{b} (267/399)	0.7^{c} (2/300)	2.0^{b} (6/300)	3.3^{b} (10/300)

a, b, c – Values in the same row with different superscripts differ significantly.

at later stages; there was a substantial decrease in embryo yield when transfer was delayed to 72 hpi (see Table 4.12). It was also evident in this work that embryos transferred at 24 hpi showed a significantly higher day-7 blastocyst yield than those transferred later. It was concluded that transfer at 24 hpi was the preferred option.

4.11 Gamete Micromanipulation

There are four main approaches to gamete micromanipulation which may assist fertilization under conditions of reduced sperm number or when sperm motility is compromised or even non-existent: (i) zona thinning; (ii) zona drilling, in which a hole is introduced into the zona, which facilitates access of sperm to the oocyte surface; (iii) subzonal sperm insertion (SUZI), in which one or more sperm are placed under the ZP, after which they fuse spontaneously with the oocyte; and (iv) intracytoplasmic sperm injection (ICSI), in which a sperm is microinjected into the cytoplasm of the oocyte. The various possibilities are illustrated in Fig. 4.7.

Apart from the potential practical utility of such manipulations, they can also be a valuable means of investigating events involved in normal sperm–oocyte interactions. Factors influencing the success of sperm injection in mouse, hamster, rabbit and cattle oocytes were the subject of research several years ago (Markert, 1983; Keefer and Brackett, 1987; Clarke and Johnson, 1988; Lassalle and Testart, 1988; Martin *et al.*, 1988; Keefer *et al.*, 1990). However, microassisted fertilization should probably best be regarded as an encouraging but ill-defined set of procedures that may be useful in human and animal IVF in certain circumstances (see reviews by Iritani *et al.*, 1988, 1992a; Keefer, 1990; Iritani, 1991; Gordon, 1992; Ng *et al.*, 1992a, b; Gwatkin, 1993).

According to Fishel *et al.* (1993), more than 300 babies had been born at that time from microassisted fertilization procedures carried out in human IVF clinics. The long-term implications of such procedures, however, do need to

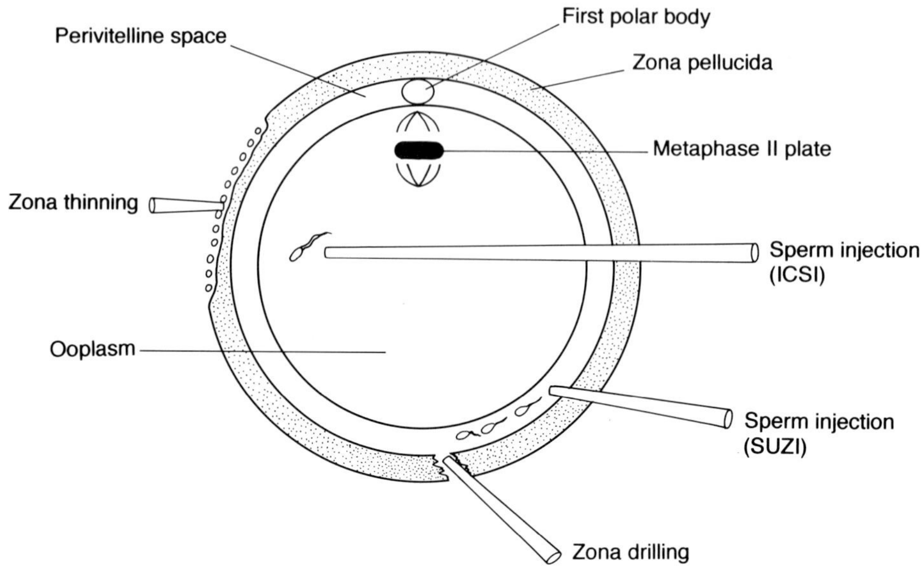

Fig. 4.7 Four micromanipulation approaches to effecting fertilization with limited sperm numbers.

be taken into account (Tucker *et al.*, 1993; Vines, 1993); for that reason, there is probably much of value in gathering and evaluating data from farm animal models as a background to applications in humans.

4.11.1 Zona thinning procedures

There may be occasions when it would be useful to achieve fertilization of IVM BOs with lower than usual sperm doses. Zona thinning by enzymatic digestion has not apparently been reported in cattle IVF. Zona thinning as one approach to human IVF patients with a history of fertilization problems was suggested by Kiessling *et al.* (1988); they used a mild trypsinization treatment. The study involved eight oocytes, two of which were fertilized but did not continue development. The efficacy of zona thinning, using the mouse as the model, was reported by Mordel *et al.* (1992). This work demonstrated zona thinning by partial digestion with pronase to be safe and capable of enhancing fertilization at very low sperm concentrations. They suggested that such a procedure may be of value in IVF with men producing sperm of reduced penetrating ability.

As discussed above (Section 4.9.1), sperm–ZP recognition and binding is the first stage in the interaction between the gametes. The binding is mediated by sperm receptors present in the ZP (ZP3) and protein present in the sperm plasma membrane. The fact that fertilization occurred, in the studies of Mordel *et al.* (1992), despite a digested zona surface indicates that ZP3 probably exists at a sufficient concentration to elicit the sperm–oocyte response throughout the entire thickness of the ZP of the mouse. The ability of ZP3 to function as a sperm receptor is attributable to the presence of certain oligosaccharides (Miller, D.J., *et al.*, 1992); pronase, which digests protein, does not affect carbohydrate structure. The data of Mordel *et al.* (1992) suggest that pronase treatment may be used to enhance fertilization efficiency at very low sperm concentrations; ZP3 continues to act effectively under such conditions, as both sperm receptor and inducer of the AR.

Zona-free oocytes

Under ordinary *in vivo* and *in vitro* conditions, the ZP of the cow oocyte is regarded as essential for normal fertilization. It induces the AR of bull sperm and serves to block polyspermic fertilization. However, studies with IVM BOs, using pronase, have clearly shown that the ZP is not indispensable for IVM and IVF (Xu *et al.*, 1990, 1991a); although the percentage of ZP-free zygotes that became blastocysts was lower than usual, the embryos were capable of giving rise to pregnancies on transfer to recipient cattle. A similar result has been shown in mice (Naito *et al.*, 1992); normal mice were produced from oocytes fertilized and developed in the absence of ZPs.

4.11.2 Zona drilling and partial zona dissection

Zona drilling covers the use of micromanipulation to introduce a gap in the ZP either mechanically, by laser ablation or by localized application of a zona solvent from a microneedle (Gordon, J.W., 1988, 1990b; Gordon *et al.*, 1988). Partial zona dissection (PZD) is a mechanical technique which creates a breach in the zona, rather than a drill hole.

The earliest reports of drilling in mouse oocytes involved introducing a gap in the ZP through which sperm could gain access to the oocyte membrane (Depypere *et al.*, 1988, 1991; Gordon and Laufer, 1988); live young were born after using the technique. In Canada, Vanderhyden *et al.* (1989) used acid Tyrode to drill holes in the ZPs of rat oocytes. Wazzan *et al.* (1990), working with mice, showed zona drilling to be useful in enhancing fertilization when epididymal sperm were used. The suggestion was that such a technique may be used in men with proximal epididymal obstruction where only caput sperm are available. Sperm from long-term vasectomized mice were also used successfully in fertilizing oocytes after PZD in the report of Barlow *et al.* (1991).

Extremely promising results with zona drilling in the mouse were reported by Edirisinghe *et al.* (1991); drilled oocytes showed much higher fertilization rates than zona-intact oocytes (61% vs. 6%) at very low sperm concentrations (10,000 sperm ml^{-1}). They were able to use fewer sperm numbers by a factors of 10–50 times less than those required for zona-intact oocytes. These authors report more than 97% of zygotes developing to the blastocyst stage without the induction of parthenogenetic activation. Keeping the hole in the zona small by carefully controlling the amount of acidic medium is mentioned as a possible reason for their high level of success. A paper by Cohen and Feldberg (1991) suggested that the ability of drilled mouse oocytes to hatch may be influenced by the process; they showed that these oocytes may be rescued by performing an additional and more aggressive opening elsewhere on the zona.

In Dublin, studies on zona drilling of IVM BOs have demonstrated an increase in fertilization rate under certain conditions of low sperm number (Kinis *et al.*, 1990c, f, 1991a). In this work, acid Tyrode (pH = 2.5) was employed as the drilling agent.

In human IVF, Cohen *et al.* (1988) found zona drilling to be unsuitable and used mechanical force to open the zona (zona cutting); they reported monospermic fertilization with normal cleavage and subsequent pregnancies. Zona drilling and zona cutting were used in a human IVF programme reported by Payne *et al.* (1991). It was found that although both drilling and cutting increased the fertilization rate using suboptimal semen, no pregnancies resulted. A review by Cohen (1991a) draws attention to some of the advantages and disadvantages of breaching the ZP. It is now evident that, as currently practised, zona drilling may be of limited effectiveness in human IVF; for human embryos with a zona thickness greater than 15 μm, zona drilling may actually reduce implantation rates. Zona drilling can result in the embryo attempting to hatch through a thickened zona in an hour-glass pattern, with its consequent entrapment.

Results in mice now show that hatching occurs as a result of lysis of the zona by enzymes elaborated by the trophectoderm cells of the blastocyst (Gordon and Dapunt, 1993a). Zona drilling may interfere with the effectiveness of this lytic process and result in a reduced hatching rate. PZD may have several adverse effects on human oocytes, according to Terriou *et al.* (1993); these include: (i) possible toxicity of hyaluronidase in removing cumulus cells; (ii) damage during manipulation; (iii) exposing the oocyte to harmful variations in temperature and pH during manipulation; and (iv) breaching the oocyte may permit entry of debris or metabolites which decrease fertilization rate/embryo quality.

Results reported by Levron *et al.* (1993) also suggest that a major determining factor for the relatively low fertilization rate after PZD may be related to failure of the sperm to enter the perivitelline space in a substantial proportion of manipulated human oocytes.

Use of laser beams

Several reports are available on the use of laser (an acronym for light amplification by stimulated emission of radiation) technology to achieve accurate incisions in the ZP, mainly working with mouse and hamster oocytes (Blanchet *et al.*, 1992; Coddington *et al.*, 1992; Feichtinger *et al.*, 1992a; Neev *et al.*, 1992, 1993; Tadir *et al.*, 1992, 1993; El-Danasouri *et al.*, 1993; Laufer *et al.*, 1993). In these species, laser drilling enhanced the efficiency of IVF at low sperm concentrations without adversely affecting subsequent embryo development. Neev *et al.* (1992) also quantified laser parameters on human oocytes. Progress in laser technology in recent years has resulted in the introduction of many lasers for biomedical studies and applications. The potential advantages of the laser as a micromanipulating tool include its simplicity, accuracy, small effective spot size (in the range of 0.5–1.0 μm) and manoeuvrability without any mechanical handling.

4.11.3 Subzonal sperm insertion (SUZI)

The birth of young reported by Mann (1988) in Australia was one of the earliest studies to demonstrate that SUZI (also called under zona insemination (UZI), sperm microinsemination (SMI) and microinsemination by sperm transfer (MIST)) is a feasible means of achieving fertilization and normal development during pregnancy in the mouse. The microinjection of sperm into the perivitelline space, or SUZI, has subsequently been employed in human IVF as a treatment in certain instances of male infertility (see review by Fishel *et al.*, 1993a, b). A number of children have been born after the application of this form of micromanipulation.

It is clear that it is possible to obtain fertilization and subsequent normal embryonic development by way of this approach (Bongso *et al.*, 1989; Wittemer *et al.*, 1991; Krzyminska *et al.*, 1992; Hedges and Saunders, 1993; Lippi *et al.*, 1993; Tucker *et al.*, 1993); the incidence of polyspermy appears to be influenced by the number of sperm injected. In France, polyspermy occurred when SUZI was performed with more than ten injected sperm (Lassalle *et al.*, 1987). According to Sakkas *et al.* (1992), the injection of four sperm was optimal in enhancing the fertilization process without increasing the risk of polyspermy. Palermo *et al.* (1992b) and Sjogren *et al.* (1992) injected up to three sperm and it was believed that the low incidence of polyspermy was related to this low number. The report of Alikani *et al.* (1992) showed that more than half of the motile sperm from men with normal semen profiles can fuse with the oolemma, once placed in the perivitelline space. There are some results suggesting that subzonally inseminated human oocytes cocultured with cumulus cells have an enhanced fertilization rate (Kobayashi *et al.*, 1993).

According to Garrisi *et al.* (1993), SUZI offers the advantage in human IVF that motile sperm can be harvested from very poor semen samples and guaranteed access to the oolemma and that certainty of direct contact between sperm and oocyte is achieved. However, they stress that success with SUZI depends on a high degree of technical skill and rigorous maintenance of normal pH and environmental control for the oocyte. It is highly inappropriate to perform SUZI in human patients who could achieve equivalent fertilization rates by means of routine IVF. According to Imoedemhe and Sigue (1993), human patients with previous failed fertilization in routine IVF practice should be considered for SUZI. It is possible to use epididymal sperm in SUZI, as reported by Soong *et al.* (1993). It is also possible to use SUZI and transfer oocytes to the fallopian tube (Imoedemhe *et al.*, 1993).

It has been found, in both mice and rabbits, that the processing of sperm to initiate capacitation and the AR is a prerequisite for achieving high fertilization rates using the SUZI technique (Minhas *et al.*, 1991; Palermo and Van Steirteghem, 1991; Palermo *et al.*, 1992a). The beneficial effects of PAF in the treatment of rabbit sperm before SUZI is reported by Minhas *et al.* (1993). In work with human oocytes, Parinaud *et al.* (1992) attempted to provide human capacitated sperm for SUZI by exposing them to a high concentration of progesterone; they found that such treatment was effective in inducing capacitation of sperm without their motility and viability being impaired. Some workers have employed hyperosmotic sugar solutions (sucrose/trehalose) to dehydrate oocytes and increase the volume of the perivitelline space to facilitate sperm deposition (Schwiderski *et al.*, 1990b; Yang and Foote, 1990; Yang, N.S., *et al.*, 1990).

A limited number of reports dealing with SUZI in cattle have appeared in Germany (Schwiderski *et al.*, 1990b), the USA (Yang, K., *et al.*, 1990; Heuwieser *et al.*, 1991) and Thailand (Pavasuthipaisit *et al.*, 1994); the technique was regarded as useful for studying sperm–oocyte interactions, but fertilization rates were generally low.

A glimpse into the future may have been provided in papers by Kobayashi *et al.* (1992a, b), who developed a computer-controlled micromanipulation system for the subzonal insemination of a single capacitated mouse sperm.

In discussing the need or otherwise for using capacitated sperm in SUZI, it might be recalled that zona-free BOs have been fertilized (Xu *et al.*, 1990, 1991a) and clearly such sperm did not undergo the normal physiological ZP-induced AR.

4.11.4 Intracytoplasmic sperm injection (ICSI)

The advantages of ICSI (also called direct insemination of cytoplasm of oocyte (DISCO), direct intracytoplasmic insemination (DICI), intracytoplasmic insemination (ICI) and microinsemination into cytoplasm (MIMIC)) are that any sperm type (sperm head, immotile sperm, dead sperm, grossly defective sperm) can be injected and that both capacitation and the AR may be unnecessary for fertilization (Yang *et al.*, 1988; Hoshi *et al.*, 1992; Iritani *et al.*, 1992b; Gwatkin, 1993; Hosoi and Iritani, 1993; Lundin *et al.*, 1993). There are even reports of using round spermatids, injected into hamster oocytes, as a means of achieving fertilization (Ogura and Yanagimachi, 1993).

ICSI in cattle

One of the more dramatic reports of recent years dealing with gamete micromanipulation in farm animals was that of Goto and colleagues in Japan (Goto *et al.*, 1990a, b, c, 1992a, b; Goto, 1993), who clearly demonstrated that BOs matured *in vitro* can undergo cleavage and embryonic development after the injection of 'dead' bull sperm into the cytoplasm of the oocyte. This work resulted in the birth of a normal healthy calf in the summer of 1990. Other work in Japan, as reported by Iritani *et al.* (1992b) and Hosoi and Iritani (1993), has shown that acceptable fertilization and cleavage rates may be achieved in rabbits by the injection of a single, uncapacitated sperm into the ooplasm.

When a bull sperm encounters an oocyte in the oviduct of the live cow, an intricate surface reaction is triggered which results in the activation of the oocyte. It is this activation which leads to completion of the second meiotic division, formation of the female PN, release of cortical granule contents and the development of the zona reaction and the vitelline block. Results reported by Keefer *et al.* (1990) indicated that, in contrast to results in rabbits and hamsters, BOs are not sufficiently stimulated by the sperm injection procedure to complete meiosis. In the work of Goto *et al.* (1990a, b, c) and subsequently, BOs were deliberately activated by a 10 min period of exposure to calcium ionophore (A23187). In the studies of Heuwieser *et al.* (1992a, b), a 5 min exposure to the ionophore (10 μM) was employed. Such activation methods were based on results reported by Ware *et al.* (1989), who examined the effect of various dose levels of A23187 on BO activation. Ethanol (7% for 7 min) has also been employed to activate oocytes in the work reported by Pavasuthipaisit *et al.* (1994) in Thailand.

Resistance of sperm head

Changes that occur in the bull sperm as it moves through the epididymis renders the nucleus very resistant to physical and chemical disruption. The same is true of human sperm (Yanagida *et al.*, 1991). The resistance of the bull sperm nucleus to low temperature is clear from the countless millions of calves that have been born after AI with frozen–thawed semen. On the other hand, even after 30 min exposure to a temperature of 90°C, human sperm are capable of developing into pronuclei after injection into hamster oocytes. In Hawaii, Katayose *et al.* (1992) showed that human sperm remained capable of decondensing and developing into PNs even after 20 days storage in ethanol/methanol. The association of deoxyribonucleic acid (DNA) with the sperm-specific basic proteins, protamines, serves to condense chromatin and renders the sperm nucleus structurally stable and genetically inactive. In farm species, such as the cow, the sperm nucleus is further stabilized (becomes resistant to physical and chemical disruption) by an extensive cross-linking of protamines by disulphide bonds, which occurs during transit of the sperm through the epididymis.

Mammalian sperm DNA is the most tightly compacted eukaryotic DNA, being at least sixfold more highly condensed than DNA in mitotic chromosomes (see review by Ward and Coffey, 1991). It is known that human and bull sperm nuclei can be freeze-dried and stored for months and still be capable of developing into PNs.

Of practical interest are the results of workers at Cornell in cattle showing higher fertilization rates with sperm injection than with SUZI (Heuwieser *et al.*, 1992a, b) and succeeding in achieving fertilization with sperm from a bull suffering from the 'stump tail' infertility defect (Yang and Anderson, 1992). Other studies at Cornell compared normal cattle IVF with sperm injection, using semen from bulls with a proved record in artificial breeding. Almost no oocytes were fertilized by some bulls using IVF, whereas sperm injection was effective in achieving a reasonable level of fertilization (Heuwieser *et al.*, 1992a, b). Such results make it clear that sperm injection can be superior to IVF in certain instances of bull infertility or subfertility. In Australia, Catt *et al.* (1994) have also recorded superior results using ICSI rather than SUZI in work with cattle oocytes; they record that sperm damage (scoring the sperm tail with a pipette) enhanced the ability of the oocyte to complete the normal fertilization process. Whether this was the result of sperm damage releasing components which signal the oocyte to activate is unclear.

However, the real practical significance of ICSI in cattle may eventually be in terms of devising computer-controlled fertilization systems which make optimal use of sperm, particularly sperm in low numbers that may have been 'sexed' by flow cytometry or other techniques. As observed by Fishel *et al.* (1993), eventually microassisted fertilization may become possible without the need for microtools, using laser technology to manipulate sperm into the oocyte. In the meantime, there is much of fundamental interest in developing the sperm injection approach to fertilization.

In cattle, as in other species, a clearer understanding of the activation process may be the means of improving ICSI technology. In particular, a procedure in which activation of the oocyte is consistently initiated by ICSI, as appears to be the case in human oocytes, should be carefully examined. The use of ICSI in the mouse seems to be less successful than in humans (Ron-El *et al.*, 1993); it may be a matter of species variation.

ICSI in human IVF

In human IVF studies in Belgium, fertilization rates have been substantially higher using ICSI than after SUZI (Palermo *et al.*, 1992b; Joris *et al.*, 1993; Vankelecom *et al.*, 1993; Van Steirteghem *et al.*, 1993a, b, c). Other groups reporting >50% of oocytes fertilizing and cleaving normally after ICSI include Baker *et al.* (1993) in Australia and Lundin *et al.* (1993) in Sweden. A comparison between SUZI and ICSI is provided in Table 4.13, taken from studies reported by Van Steirteghem *et al.* (1993a). It is possible to use the ICSI technique in instances of extreme oligospermia in the male (Vankelecom *et al.*, 1993) or in cases of acrosomeless sperm (Lundin *et al.*, 1993). The injection of a sperm into the ooplasm appears to be sufficient stimulus to activate the oocyte, without the benefit of membrane fusion. PN formation is apparently complete in most oocytes by 8 h after ICSI (Nagy *et al.*, 1993a). The fertilizability of 1-day-old non-fertilized metaphase II oocytes obtained

Table 4.13. Fertilization and embryo development after SUZI and ICSI (from Van Steirteghem *et al.*, 1993a).

	SUZI	ICSI	*P*
Metaphase II oocytes injected	3287	3944	
Intact oocytes	3008 (92%)	3401 (86%)	0.001
Two-PN oocytes	500 (17%)	1873 (55%)	0.001
Transferable embryos	400 (80%)	1412 (75%)	0.05

after standard human IVF procedures has been examined by Nagy *et al.* (1993b); it is evident that such oocytes can be fertilized by the ICSI procedure.

According to Edwards and Van Steirteghem (1993), a sufficient volume of calcium-containing culture medium may be injected during ICSI to invoke calcium-induced calcium release, which initiates many of the processes of oocyte activation. The medium employed in ICSI by the workers in Belgium contains 5 mM calcium; with the sperm, a volume of about 1 pl of this calcium-rich medium is injected deep into the oocyte, at some distance from the polar body. It may be this step which triggers oocyte activation. In various species, including the cow, it is known that sperm penetration triggers a series of Ca^{2+} oscillations, or repetitive waves of Ca^{2+} release, which occur for several hours (Homa *et al.*, 1993). In Utah, White *et al.* (1993) examined the mechanism of BO activation; there was evidence that inositol-1,4,5-trisphosphate (IP3) can induce Ca^{2+} transients similar to those reported at fertilization.

It has been possible to make direct measurements of intracellular Ca^{2+} changes in human oocytes in response to activation by sperm (Taylor *et al.*, 1992b, 1993); calcium transients have been recorded every 10–25 minutes for up to 6 h after IVF. It is also known that fertilization initiates a series of changes in the pattern of protein synthesis in the BO (Smith *et al.*, 1989); such fertilization proteins may be involved in the maternal programming of early development of the embryo up to the time of transition to control by the embryo's genome.

4.12 Assessing Efficiency of the IVF Procedure

In assessing the efficiency of IVF procedures and in evaluating the normality or otherwise of the embryos that emerge from the production system, there is the obvious need to first specify what is to be regarded as normal, as shown in the live cow; this provides the background against which to compare IVM/IVF zygotes/embryos with their naturally occurring counterparts.

4.12.1 Criteria for assessing fertilization

The penetration of the bull sperm into the oocyte initiates a series of events involving the nuclear and cytoplasmic components of sperm and oocyte. Shortly after the sperm penetrates through the ZP, the non-membrane-bound sperm decondenses within the oocyte cytoplasm. Subsequently, a nuclear envelope forms and swelling of the male PN is evident. Simultaneously, meiosis resumes in the oocyte with the extrusion of the second polar body and formation of a nuclear envelope around the decondensing female chromatin. Formation of male and female PNs occurs simultaneously and within a few hours of sperm penetration. In contrast to what is possible with the oocytes of some species (e.g. mouse, human), in cattle, due to the presence of lipid vesicles and other pigments, the PNs are not visible, other than when special techniques are employed (differential interference contrast microscopy, staining, centrifugation).

4.12.2 *Staining procedures*

The BO has an opaque cytoplasm, which requires prolonged fixation and staining procedures (24–48 h) to reveal nuclear status and other features in detail. The development of a rapid staining procedure by Byun *et al.* (1991) in Korea has permitted much more rapid evaluation of events in oocyte maturation and fertilization to be made. The rapid staining method takes 40 min and involves the use of acid Tyrode's solution, acetoethanol, basic fuschin and acetoglycerol. With this technique, maturation and fertilization of the BO *in vitro* can be analysed in a much shorter time period than was previously possible.

The release of cortical granules and the completion of meiosis are two of the earliest responses observed during the process of BO activation induced either by the entry of the fertilizing sperm or by the action of some other activating agent. The contents released from the cortical granules modify the properties of the oocyte plasma membrane and the overlying ZP to prevent the entry of further sperm into the oocyte, the so-called vitelline block and zona reaction. A paper by Lee and Ahuja (1991) describes an immunocytochemical technique which may be employed to study the kinetics of the release of the cortical granule contents.

Changes in intracellular Ca^{2+} concentration, using the fluorescent calcium indicator Fura-2AM, have been recorded by Mizuno *et al.* (1994) in BOs immediately after sperm–oocyte fusion until the four-cell stage; they showed that the magnitude of the first Ca^{2+} transient was greater than subsequent repetitive transients, which were observed to occur up to the PN stage.

4.12.3 *The presence of accessory sperm*

After sperm entry into the vitellus of the BO, a reaction in the ZP (zona reaction) results in the failure of additional sperm to achieve zona penetration. Sperm firmly bound to the ZP, after penetration of the first sperm, are termed 'accessory sperm'. In the live cow, the number of such accessory sperm has been shown to be positively associated with fertility in cattle. In a study in which cattle were inseminated with sperm encapsulated in protamine sulphate microcapsules, Munkittrick *et al.* (1992) found evidence that high accessory sperm numbers per embryo were significantly related to bovine embryo quality. DeJarnette *et al.* (1990, 1992) and Nadir *et al.* (1992, 1993) are other workers to report a positive relationship between accessory sperm number and embryo quality in the cow.

4.12.4 *Cleavage rate as a measure of IVF efficiency*

One of the standard evaluation procedures employed in cattle IVF is the percentage of inseminated oocytes which have cleaved when examined 2 days later; this measure is termed the cleavage rate. In the live animal, information on the times at which first cleavage occurs is limited; it is given as about 30 hpi by Laing (1949). In Dublin, a further measure of IVF effectiveness is the percentage of cleaved embryos which have progressed beyond the two-cell stage. In the live cow, the evidence shows the four-cell stage to be reached at about 42 h (see Fig. 4.8). In IVF studies in Dublin, the term cleavage index is used to cover the percentage of cleaved embryos that are beyond the two-cell stage. There is good evidence that this index, under appropriate embryo culture conditions, may be an accurate predictor of blastocyst yield (Vergos, 1990; Sharif, 1992; Lonergan, 1992; Lonergan *et al.*, 1992). Some of this information is provided in Table 4.14, taken from the studies of Lonergan (1992).

The normality of a bovine embryo is likely to be a matter of development in relation to age, as well as the morphological features that it shows. With

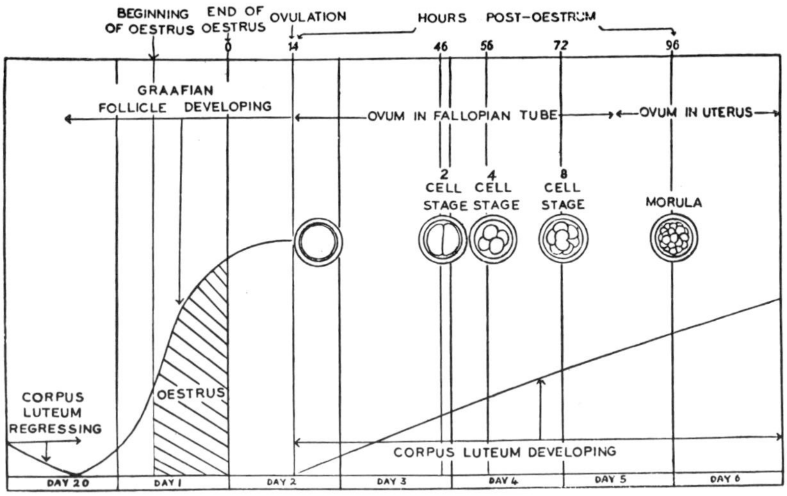

Fig. 4.8. Diagrammatic representation of events at and about oestrus in the cow (Laing, 1949).

Table 4.14. Effect of cell stage at 48 hpi on blastocyst yield after IVM/IVF/IVC (from Lonergan, 1992).

	Blastocyst yield (%)		
	Day 7	Day 8	Day 9
One-cell	Nil (0/100)	Nil (0/100)	Nil (0/100)
Two-cell	Nil (0/152)	Nil (0/152)	Nil (0/152)
Four-cell	12.8 (40/312)[a]	22.1 (69/312)[a]	24.7 (77/312)[a]
> Four-cell	39.8 (45/113)[b]	47.8 (54/113)[b]	52.2 (59/113)[b]

a, b – Values in the same column with different superscripts differ significantly ($P < 0.001$).

IVF, where most emphasis is upon IVC from the zygote through to the blastocyst stage, many opportunities are available for setting up appropriate standards, using photographic or videographic techniques.

4.12.5 *Indicators of fertilization*

In human IVF, some thought has been given to agents that may serve as indicators of fertilization. In the normal course of events, detectable amounts of human chorionic gonadotrophin (HCG) are not found until after implantation. There is some evidence, however, that an assay for PAF may be a means of detecting the human conceptus prior to implantation. Fertilization has been found to increase the concentration of PAF fourfold over unfertilized oocytes (Nakatsuka *et al.*, 1992).

4.13 Fertilization Abnormalities

The two most commonly described fertilization abnormalities in cattle IVF are polyspermy and parthenogenesis.

4.13.1 *Incidence and detection of polyspermy*

The effect of different sperm capacitation treatments on the outcome of cattle IVF was examined by Chikamatsu *et al.* (1989), who recorded a markedly higher incidence of polyspermy (33%) when bull sperm were capacitated using caffeine and calcium ionophore (A23187) than when using a heparin procedure (8%). This relatively low incidence of polyspermy after heparin treatment would be in general agreement with the experience in Ireland. In the standard Dublin IVF protocol, as described by Lu *et al.* (1987b), the incidence of polyspermy has been recorded as 10% in the studies of Jiang (1991).

Other work, this time by Fair (1992), using the same protocol, also indicated a low rate (<5%) of polyspermy. Although some workers have expressed reservations about employing high heparin dose levels, because of the risk of a high incidence of polyspermy (First and Parrish, 1988), such heparin levels have been employed in Dublin without problems. The effects of polyspermic fertilization on development within the first cycle have been described by Long *et al.* (1993b); these authors record that 55% of zygotes in their study had been fertilized by more than one sperm.

4.13.2 *Incidence and detection of parthenogenesis*

Parthenogenesis is defined as the development of the oocyte without the intervention of the male gamete. In terms of nomenclature, parthenogenome is the name given to a parthenogenetic embryo (see Graham, 1974). The early stages of parthenogenesis can be artificially induced or may occur spontaneously in a wide variety of mammals; in general, the activated oocyte seldom cleaves more than twice before ceasing development (Graham, 1974; Whittingham, 1980). Adult parthenogenomes have apparently never been recorded in mammals, although they do occur in some vertebrates. It has been established by Surani and colleagues that, in mammals, parthenogenesis is not compatible with normal development due to the absence of paternally imprinted genes. Among farm animals, parthenogenetic strains of turkeys are known to exist.

In human IVF, there is evidence that oocytes can be parthenogenetically activated; there is evidence that acid Tyrode's solution, ethanol, the protein synthesis inhibitor puromycin and the calcium ionophore A23187 can induce activation. It has been suggested that the use of activated human oocytes for cytogenetic research may contribute to progress in preimplantation genetics (De Sutter *et al.*, 1992). In cattle IVF, activation of oocytes and development to the blastocyst stage has been recorded (see Table 4.15 from the studies of Monaghan, 1993).

A study by Tada and Takagi (1992) in Japan examined early development and X chromosome inactivation in ethanol-induced mouse parthenogenomes. Although some embryos did progress, abnormal development was evident in every parthenogenome as early as 5 days after activation. The failure of X-inactivation in the trophectoderm apparently contributed to the defective development of these parthenogenomes. The absence of a paternal genome in parthenogenetic mice appeared to limit their embryonic development (Henery and Kaufman,1992); there is also evidence in that species showing that diploid parthenogenomes have a developmental advantage over haploid embryos and possess a greater cell number.

Apparently normal cleavage of cattle embryos to the eight-cell stage of division, after IVM and transfer to the sheep oviduct, was described by

Table 4.15. The effect of ethanol and electrical activation of secondary bovine oocytes on their subsequent development *in vitro* (from Monaghan, 1993).

Treatment	Cleavage rate (%)	Blastocyst yield (%)		
		Day 7	Day 8	Day 9
Inseminated control	79.0 (173/219)	9.8^{a} (17/173)	17.9^{a} (31/173)	24.3^{a} (42/173)
Ethanol activated	28.6^{b} (92/322)	0^{b} (0/92)	1.1^{b} (1/92)	3.3^{b} (3/92)
Electroactivated	23^{b} (35/152)	0^{ab} (0/35)	5.7^{ab} (2/35)	5.7^{b} (2/35)
Untreated control	8.5^{c} (27/316)	0^{ab} (0/27)	3.7^{ab} (1/27)	3.7^{b} (1/27)

a, b, c – Values in the same column with different superscripts differ significantly ($P < 0.05$).

Trounson *et al.* (1977); this was the result of parthenogenetic development. It is now evident that under certain conditions, parthenogenetic development of transfered BOs can occur quite commonly in the oviduct of the cow, ewe and rabbit. This may explain why work reported by authors in Denmark (Xu *et al.*, 1986a) has shown a higher percentage of cleaved cattle embryos coming from rabbit oviducts when no bovine sperm had been employed than when sperm were used (15% vs. 13%); it is clear that due caution must be exercised in assessing the normality of cattle embryos in the early cleavage stages. In the USA, a similar level of parthenogenetic development has been recorded (16%) in bovine embryos recovered from rabbit oviducts (Wall and Hawk, 1988). It appears probable that BOs exposed to reduced temperatures during handling and manipulations may be activated.

Parthenogenetic development of activated BOs to the hatched blastocyst has been reported from Ireland (Lee, S.H., *et al.*, 1992). Stages in this develop-

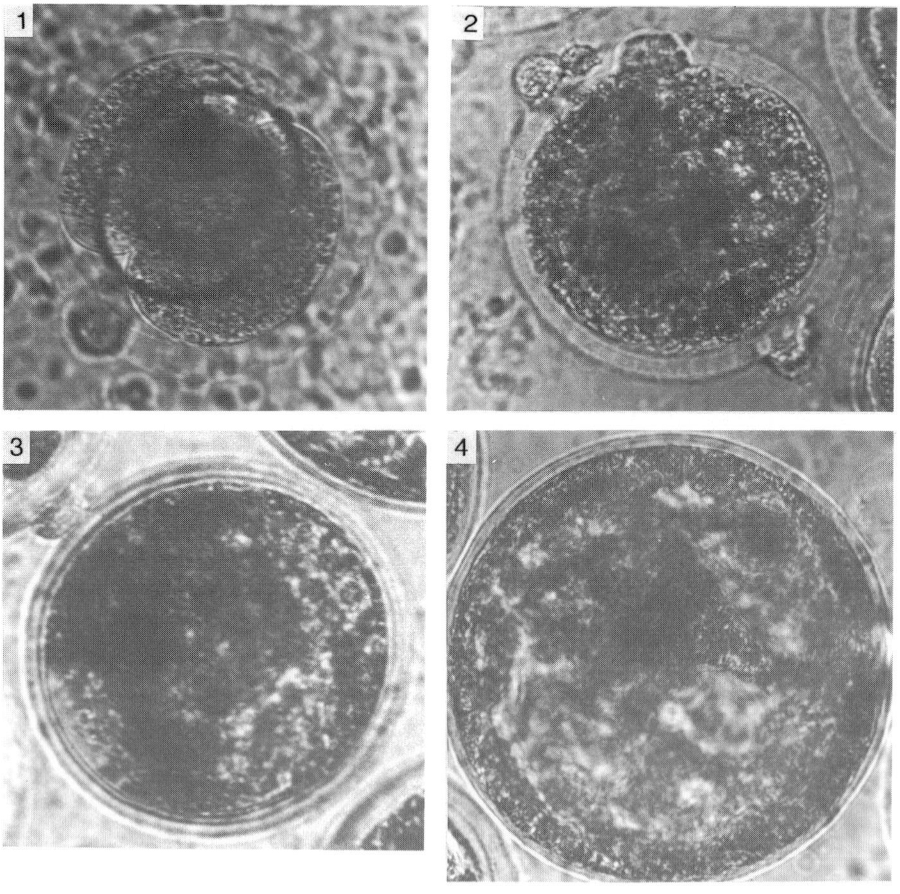

Fig. 4.9. Bovine blastocysts produced by ethanol activation (see Lee, S.H., *et al.*, 1992). (1) Two-cell embryo; (2) late morula; (3) blastocyst; (4) expanded blastocyst.

ment, up to the expanded blastocyst stage, are shown in Fig. 4.9. Other studies have shown similar results (Collas and Barnes, 1992; Fukui *et al.*, 1992; Minamihashi *et al.*, 1993; Plante and King, 1993).

Studies conducted by Ayoub and Hunter (1993a, b) examined a number of factors that may result in BO activation; they found that activation could readily occur under common IVC conditions. Their evidence certainly strengthens the view that early cleavage is not a sufficient measure of embryo quality. In Japan, Nagai (1992) used ethanol as the activating agent; 5% of the IVM BOs progressed to the eight-cell stage after 60 h of culture. In Utah, electrical stimulation was employed by Collas and Barnes (1992), who reported high rates of activation (69%) and 15% developing to the blastocyst stage; both activation and subsequent development were dependent on the presence of extracellular Ca^{2+}.

Artificial induction of activation

Controlled parthenogenetic activation of BOs may result in a better understanding of the mechanisms of spontaneous activation and suggest possible ways of preventing it, thereby improving the general efficiency of the cattle IVF system. Studies in Ireland have shown that electrical stimulation would induce activation of IVM bovine embryos, resulting in subsequent cleavage (Kinis, 1990; Monaghan, 1993); cleavage rate is usually markedly below that for oocytes exposed to sperm in the normal way. In the same laboratory, using a 7% ethanol activation procedure, S.H. Lee *et al.* (1992) recorded a cleavage rate of 16 to 34.5% in oocytes activated after IVM periods varying from 24 to 42 h. In some of the oocytes activated after a 30 h maturation period, development continued to the blastocyst and hatched blastocyst stage stage. As noted above (Table 4.15), studies by Monaghan (1993) have also showed that various stimuli can result in activation and development to the blastocyst stage. The production of parthenogenetic blastocysts after electrical stimulation (Pugh *et al.*, 1991a) or ethanol activation (Wahid, 1993) of *in vitro*-matured ovine oocytes has also been recorded; as with cattle, some of these sheep blastocysts progressed to the hatching stage of development (see Table 4.16). An earlier paper by Ware *et al.* (1989) had suggested that spontaneous activation of BOs is unlikely to occur until 26 h or more after the onset of IVM; they concluded that maturational age of the oocyte is the main determinant of activation.

In Canada, Yang and Smith (1993) assessed the development of IVM BOs after exposure to a single electric impulse; results indicated that parthenogenetically activated oocytes have a limited potential to develop beyond the eight-cell stage, with none of 194 stimulated oocytes reaching the hatched blastocyst stage. Elsewhere in Canada, Plante and King (1993) showed that aged BOs can self-activate in the absence of sperm and even reach the blastocyst stage, but much later than IVF embryos.

Table 4.16 Development of parthenogenetic ethanol-activated sheep oocytes matured *in vitro* (from Wahid, 1993).

Treatment (hour IVM)	No. of oocytes activated	No. of oocytes cleaved	% Cleaved	Blastocyst yield (%)
Control	87	3	3.4[a]	0 (0)
16	169	12	7.1[ac]	0 (0)
20	171	22	12.9[bc]	0 (0)
22	113	35	31.0[d]	1 (2.9)[a]
24	325	142	43.7[e]	4 (2.8)[a]
26	97	55	56.7[f]	1 (1.8)[a]
28	168	68	40.5[de]	1 (1.5)[a]
32	163	77	47.2[ef]	0 (0)
Total	1293	414	32.0	7 (1.7)

a, b, c, d, e, f – Values with different superscripts in the same column were significantly different at $P < 0.05$ (χ^2).
Data were obtained from two replicates (except for 24 h with four replicates).
Control: without activation, oocytes were removed after 24 h IVM.
Blastocyst: after 10 days IVC.

4.13.3 Chromosomal abnormalities in IVF embryos

In the live cow, a higher incidence of karyotypically abnormal embryos is found in the early days after fertilization, suggesting that such abnormalities tend to die in the early embryonic period and are not encountered in later stages. Methods of preparing chromosomes for the evaluation of the karyotypes of early bovine embryos have been described (King *et al.*, 1979). Working with *in vivo* cattle embryos, Gayerie de Abreu *et al.* (1984) recorded incidences of 9% and 6% abnormal karyotypes in the early-stage embryos from cows and heifers, respectively; all these abnormalities proved to be polyploids ($3n$ or $4n$) or haploids ($1n$). Several other authors have reported on chromosomal abnormalities in bovine embryos produced *in vivo* or by *in vitro* procedures (King, 1985; Iwasaki and Hamano, 1991; Iwasaki *et al.*, 1992).

4.14 A Laboratory Test for Predicting Bull Fertility

Semen used in AI is routinely subjected to motility and density evaluation after collection from the bull. Semen showing low initial sperm progressive motility or low sperm density is normally not accepted for use in AI. Although semen characteristics are often associated with fertility levels shown by bulls in the field, there are many instances in which poor fertility rates occur despite apparently satisfactory results after semen evaluations. The availability of a procedure in which bull fertility would be accurately predicted on the basis of results achieved in IVF would be useful. As suggested by Andersson *et al.* (1992), such a test could save expense and time in comparison with the need to inseminate cows and assess bulls on the basis of non-return rates.

A means of increasing the number of frozen semen straws from particularly valuable AI bulls could have important economic advantages in commercial AI applications. In the freezing of bull semen, it is common practice to dilute ejaculates to provide a final concentration of 100 million sperm ml^{-1} and to load doses of at least 25 million sperm into 0.25 ml straws. Assuming that only some 50% of sperm survive the freeze–thaw process, this provides 10 million motile sperm for each insemination dose. If it were possible to use substantially lower sperm doses, 5 million sperm per straw rather than 25 million, then high-quality bulls could be used more extensively. For such reasons, refinement of techniques for bull semen evaluation is an important area of semen physiology research.

In human IVF, there is interest in tests that may be valuable for predicting fertilization rates *in vitro* (Liu and Baker, 1992); it is clear that conventional semen analysis provides limited information on human sperm fertility.

4.14.1 Effect of bull on IVF outcome

Studies reported from Cambridge in sheep (Fukui *et al.*, 1988a) and from Madison in cattle (Leibfried-Rutledge *et al.*, 1986b; Hillery *et al.*, 1990) suggest that the selection of rams and bulls producing sperm with a high fertilizing ability may be an important factor in achieving successful and consistent IVF results. The Madison studies suggested the existence of bull effects on embryonic development which may not be related to sperm penetration rate or the apparent normality of fertilization.

On the basis of such evidence, First and Parrish (1987a, b) suggest that more accurate prediction of bull fertility might follow from IVF data, including information on early embryonic development. In Ireland, studies using the normal heparin treatment of sperm (Lu *et al.*, 1987b) have shown clear evidence of differences among bulls in cleavage rates and embryo yields (Shi, 1991); embryo yields in one study showed a threefold variation (from 10.6% to 30.5%) according to the bull employed in IVF.

The effect of bull ejaculates collected in different seasons on cleavage rate and embryo yield was also examined by Shi (1991); bull fertility was apparently independent of seasonal variation but significant differences in embryo yield were evident among the four bulls (see Table 3.6). In Japan, however, Otoi *et al.* (1993b) have reported evidence suggesting that embryo yield after IVM/IVF/IVC may be influenced by factors associated with ejaculates from the same bull; differences were even evident using semen straws prepared from the one ejaculate.

In France, a high correlation has been found between the IVF results for bulls and their field fertilization rates estimated on the basis of the non-return rate 60–90 days after insemination (Marquant-Le Guienne *et al.*, 1990a, b, c, 1992b; Marquant-Le Guienne and Humblot, 1991, 1992b). The highest

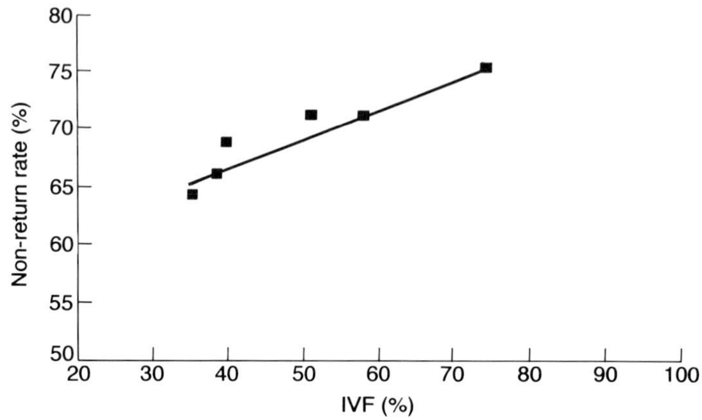

Fig. 4.10. Regression line between field non-return rates and IVF rates with 0.05 $\mu g\ ml^{-1}$ heparin. $y = 0.22x + 0.58$; $r = 0.83$; $P < 0.01$. (From Marquant-Le Guienne *et al.*, 1990a).

correlation with non-return rate was found with fertilization rate plus embryonic development to the blastocyst stage ($r = 0.73$). The French workers conclude that IVF data can be complementary to conventional semen evaluation procedures in the selection of young bulls with good reproductive potential (see Fig. 4.10 taken from Marquant-Le Guienne *et al.*, 1990a).

IVF technology was also used to evaluate the semen of different bulls in Italy (Lacalandra *et al.*, 1992); semen from eight bulls was examined and variations in fertilization rates and embryo yields were recorded after putting oocytes through their IVM/IVF/IVC system. In the USA, Mortimer *et al.* (1991) evaluated the *in vitro*-fertilizing ability of semen from young bulls showing sperm abnormalities (proximal droplet); a high incidence of proximal protoplasmic droplets severely compromised fertility *in vitro* but as the bulls matured the proportion of droplets decreased and IVF rates increased.

Although it is clear that genuine correlations are likely to exist between bull fertility and the outcome of IVF, it has yet to be established whether this is likely to be of value in identifying high-fertility bulls for use in commercial AI. In all of this, IVF data must come from a standard protocol as well as taking account of the various procedures used in bull semen processing. There are obviously many variables involved, both male and female.

4.14.2 Other forms of fertility testing

Methods that may have potential in the laboratory testing of bull fertility are outlined in Fig. 4.11.

Acrosome reaction assays

Some workers have used tests in which agents are employed to induce capacitation and the AR in bull sperm. Marks and Ax (1985), in the USA,

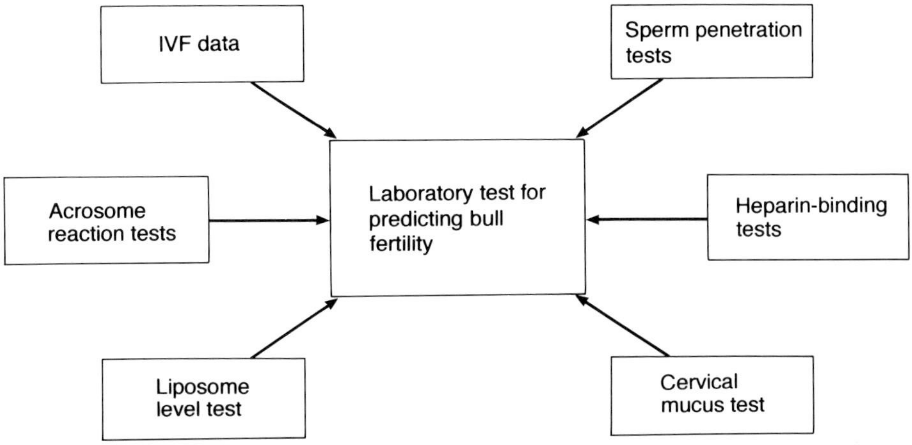

Fig. 4.11. Laboratory tests used in predicting bull fertility.

showed that heparin-binding sites on bovine sperm varied among bulls with different non-return rates and suggested that this might be the basis of a test for predicting bull fertiility. A laboratory assay using chondroitin sulphate was described by Lenz *et al.* (1986, 1988); the authors concluded that it could be used to predict the relative fertility of bulls. In Germany, Blottner *et al.* (1990) used a heparin test and found a correlation between non-return rates and the outcome of that test.

In the UK, an attempt was made by Whitfield and Parkinson (1992) to relate non-return rates to the induction of the AR (assessed on the basis of the loss of the apical ridge of the sperm head). This was based on previous reports by Madison workers that glycosaminoglycans, such as heparin, promote an AR in bull sperm (Handrow *et al.*, 1982). A significant correlation ($r = 0.86$) was obtained by Whitfield and Parkinson (1992) between the fertility predicted on the basis of their heparin-induced AR and the 90-day non-return rate. In view of evidence that heparin capacitates rather than induces the AR in bull sperm (see Section 3.4.1), it may be useful to have further information on this approach to fertility testing.

Sperm penetration assays

Zona-free hamster oocyte penetration tests have been employed as a means of estimating the fertility of different types of sperm, including bull (Davis *et al.*, 1987a, b) and ram (Hattingh and Kay, 1992) sperm. When this test is used with bull sperm, however, functional interpretation of the results is difficult because of gamete heterology. A heterospermic, competitive penetration assay, using zona-free BOs, was therefore used by Henault *et al.* (1992a, b); evidence was found that the assay, when read 3 h after insemination, is capable of detecting subtle differences in bull sperm fertility. Also in the USA, Truelson *et al.* (1992) developed an *in vitro* homologous oocyte penetration assay for bull semen and compared it with the zona-free hamster oocyte penetration assay. A previous report from Germany by Linnenbrink (1990), however, showed no significant correlation between sperm penetration rate (intact IVM BOs) and non-return rates.

Cervical mucus penetration tests

The bovine cervical mucus penetration test is apparently an effective tool in human andrology, being used in assessing sperm motility. A kit is available commercially (Penetrak, Serono Diagnostic) for using this approach in sperm testing. In Italy, Galli, A., *et al.* (1991) have reported on using the Penetrak kit with frozen–thawed bull sperm; there was no correlation between mucus penetration and fertility.

Tests involving liposomes

The relationship between the non-return rates of bulls and the percentage of motile and acrosome-reacted sperm, after treatment with different levels of P12 liposomes, has been reported by Davis *et al.* (1987b). Further work with liposomes was reported by Graham and Foote (1987b), who observed differences among bulls according to their tests. Fresh sperm from high-fertility bulls required less P12 liposome to induce the AR than sperm from low-fertility males; with frozen–thawed sperm, the opposite proved to be true. These authors note that, in cattle AI, bulls showing evidence of low fertility produce semen that freezes poorly; they suggest that this may be due to the fact that membrane damage is greater in the low-fertility bulls during freezing and less P12 is required to further destabilize the sperm membrane and initiate the AR.

Heparin-binding proteins

A study in Texas sought to determine the relationship of heparin-binding proteins (HBP) present in seminal plasma to bull fertility (Bellin *et al.*, 1993); there is the possibility that bull fertility may be determined by measuring the presence of certain HBPs on sperm membranes.

Assessing motility characteristics

A study reported by Bailey *et al.* (1992) examined the possibility that freezing bull sperm may alter normal Ca^{2+} regulation and that this may contribute to reduced fertility, perhaps by affecting motility characteristics; CASMA (Hamilton-Thorne HT-T2030 Motility Analyzer) was employed with frozen–thawed sperm but there was no correlation with field fertility.

4.15 *In Vitro* Fertilization in Other Farm Animals

4.15.1 Sheep and goats

Information on the production of sheep embryos from *in vitro*-matured and fertilized oocytes is limited. Early attempts at IVF in the USA involved various forms of sperm capacitation, including use of hypertonic medium (Bondioli and Wright, 1980, 1983a); low penetration rates ($<15\%$) achieved at the time were assumed to be due to capacitation failure. Studies in Cambridge resulted in improvements in the capacitation process and led to the birth of ten lambs (Cheng, 1985; Cheng *et al.*, 1986). In France, Crozet *et al.* (1987a, b) obtained high fertilization rates with both ovulated oocytes and IVM ovine oocytes; a normal lamb was produced from an ovulated oocyte.

A high incidence of polyspermy (27%) was recorded by Crozet (1986), the author drawing attention to the need to reduce the incidence. Increasing the Ca^{2+} concentration in the IVF medium to 10 mM stabilized the fertilization rate in studies involving several rams and ovulated sheep oocytes (Huneau and Crozet, 1988). Several other authors reported IVF results, using the same ram sperm capacitation procedures as Crozet *et al.* (1987a, b); sperm treatment involved the use of modified Brackett medium supplemented with sheep serum (Cognie *et al.*, 1989, 1991; Czlonkowska *et al.*, 1991; Pugh *et al.*, 1991a, b). In China, Shorgan *et al.* (1990) used ionophore A23187 to capacitate ram sperm and transferred embryos at various stages to 20 recipient ewes; ten ewes became pregnant and 11 lambs were born.

In former Czechoslovakia, Slavik and Fulka (1991) and Slavik *et al.* (1992) described the use of M-199 enriched with heparin as their sheep fertilization medium. In these studies, high fertilization and cleavage rates of IVM ovine oocytes were achieved, but transfer of early-stage embryos resulted in a low birth-rate in recipients (five lambs from 34 embryos). In Australia, Walker *et al.* (1994) concluded from their studies that the use of synthetic oviductal fluid (SOF) + 2% oestrous sheep serum for both swim-up and IVF was an acceptable procedure. Elsewhere, in the same country, Stojanov *et al.* (1994) have shown that ram sperm may maintain their fertilizing capacity during more than 2 weeks of chilled storage.

The birth of three kids in studies reported by Rao *et al.* (1984) provided evidence of their success in fertilizing goat oocytes in the rabbit oviduct. Reports by Younis *et al.* (1990, 1991), however, were among the first to document pregnancies after the transfer of goat embryos produced by IVM and IVF. They demonstrated that heparin was an effective agent in capacitating goat sperm; in contrast to results in cattle (Niwa and Ohgoda, 1988), the addition of caffeine (2 mM) to heparin did not enhance the outcome of goat IVF.

In India, studies in the maturation and fertilization of goat oocytes *in vitro* have been reported by Chauhan and Anand (1991) using several forms of IVM media; in Ham's F-12 medium, 70% of oocytes reached metaphase II, with 63% of these being penetrated.

In France, De Smedt *et al.* (1991, 1992) have reported studies in IVM/IVF in the goat. In this, goat COCs were matured in M-199, supplemented with

10% fetal calf serum (FCS), GCs and hormones (follicle-stimulating hormone (FSH)/LH/oestradiol). It was recorded that 86% of oocytes from follicles 2–6 mm in diameter progressed to metaphase II. Brackett's defined medium, containing 20% oestrous sheep serum, was employed in the capacitation of sperm. The authors recorded a cleavage rate of 58%; the incidence of polyspermy was 20%. The birth of a goat kid, after IVM/IVF and early embryo culture in the sheep oviduct, is reported by Crozet *et al.* (1993). In this work, goat sperm, separated by a swim-up procedure, were capacitated in Brackett's defined medium supplemented with 20% oestrous sheep serum.

4.15.2 Pigs

A review of some problems facing IVF of pig oocytes has been provided by Hunter (1990); the author suggests concentrating attention on the factors contributing to the oviductal microenvironment during fertilization *in vivo* in that species. Work in Cambridge (Cheng, 1985; Cheng *et al.*, 1986) resulted in the birth of piglets after the IVF of ovulated porcine oocytes. Early problems in the maturation of pig oocytes were eventually overcome by Mattioli *et al.* (1988a, b), who matured the oocytes with the whole wall of the everted follicle; this led to IVM/IVF embryo production and the subsequent birth of young. Studies in the fertilization of pig oocytes have been reported from several countries, including Korea (Park *et al.*, 1990b, 1991), Japan (Miyano *et al.*, 1990b; Yoshida *et al.*, 1990; 1993a), China (Jinyi *et al.*, 1990; Wang, Z.K., *et al.*, 1992), Hungary (Vajta *et al.*, 1991), the UK (Ding *et al.*, 1992a, b) and North America (Lee, S.J., *et al.*, 1992; Zheng and Sirard, 1992; Funahashi and Day, 1993a, b, c).

It is well recognized that one of the most acute problems in pig IVF is that of polyspermy. Cheng *et al.* (1986) found that oocytes exposed to pig sperm for more than 8 h showed a significantly increased incidence of polyspermy. Mattioli *et al.* (1989) found that removal of adherent sperm from the ZP 6–8 h after the start of gamete coincubation reduced multiple sperm penetration. Other workers have followed this practice (Nagai and Moor, 1990; Nagai *et al.*, 1990; Yoshida *et al.*, 1990), but polyspermy has remained a serious problem.

It would appear that a marked species difference exists between pigs and ruminants in the incidence of polyspermy. According to Coy *et al.* (1993a, b), this may be due to the blockage to polyspermy being physiologically slower in pigs than in cows or because exocytosis and dispersal of the cortical granule contents during IVF is more normal in cows than in pigs. The average number of sperm penetrating porcine oocytes matured *in vitro* was found by Miyano *et al.* (1990a) to be significantly higher than the number penetrating ovulated pig oocytes. Workers in Spain have reported using a gamete coincubation period of 4 h and suggest that this may be optimal for pig IVF (Coy *et al.*, 1993b).

A report by Funahashi and Day (1993b) showed that the prefertilization incubation of boar sperm in suitable concentrations of pig follicular fluid can reduce the incidence of polyspermy in a pig IVF system; these workers suggest that follicular fluid, released by the sow ovary at ovulation, may assist in preventing polyspermy *in vivo*. It is believed that a low level of follicular fluid in prefertilization media increased monospermic fertilization by induction of the AR and a consequent reduction in the number of sperm binding to the ZP. The properties of pig zona and of zona glycoproteins during IVF have been studied by Hatanaka *et al.* (1992). At Beltsville, Park and Pursel (1991) reported that multiple additions of sperm in IVF rather than a single addition can reduce the incidence of polyspermy.

A further problem in pig IVF relates to the quality of sperm that are available. The fertilizing ability of boar sperm stored in the Kiev diluent was examined by Hamano *et al.* (1990); fertilization rates decreased from 98–100% at day 0/1 to zero at day 5. Large variations in fertilization rate have been evident with frozen–thawed sperm (Nagai *et al.*, 1988). Using the zona-free hamster oocyte test, Clarke and Johnson (1987) showed that penetration by frozen–thawed boar sperm, separated by a swim-up procedure, was significantly decreased in comparison with penetration by sperm from fresh or liquid-stored ejaculates. Percoll gradient centrifugation is a procedure employed to isolate a highly motile sperm fraction from boar semen and young have been born after its use in IVM/IVF (Mattioli *et al.*, 1989). The effect of different sperm preparation methods was examined by Zheng *et al.* (1992); they froze boar semen in 0.5 ml straws, by what they describe as optimal techniques, and subsequently used fractionation techniques to isolate highly motile sperm.

The successful use of primary porcine oocytes in a homologous IVF assay is described in a paper by Martinez *et al.* (1993); these authors tested the fertilizing ability of boar sperm and suggested that such tests may be used in fertility evaluations. If confirmed, this test may be of interest to those operating commercial AI services. There is need for much further information on the changes in cryopreserved pig sperm, as in the report of Hashizume *et al.* (1990), who examined ultrastructural changes after pellet freezing.

4.15.3 Horses

Information on IVF in the mare is limited to studies conducted in recent years. In France, Palmer *et al.* (1990) recovered oocytes from preovulatory follicles, and incubated them in B2 medium, supplemented with 15% FCS for 6 h before IVF; sperm capacitation involved the use of the calcium ionophore A23187. This resulted in a fertilization rate of 26% and the birth of the first IVF foal in the summer of 1990 (Palmer *et al.*, 1991). Other reports on horse IVF have been provided by workers in Poland (Okolski *et al.*, 1990), and the USA (Campo *et al.*, 1990).

4.15.4 Zebu cattle and buffaloes

The results of studies in IVM/IVF in zebu cattle (*Bos indicus*) have been provided by Oliveira *et al.* (1991a, b, 1992, 1994) working in Brazil; the authors draw attention to the possible differences between the needs of zebu cattle and those of European breeds. It was found, for example, that higher heparin dose levels and higher sperm concentrations were required for zebus. The general conclusion was that IVF techniques developed for *Bos taurus* must be adapted to meet the specific requirements of zebu cattle.

The IVM of buffalo oocytes has been reported by Totey *et al.* (1991a, b, 1992) in India; they found frozen–thawed semen to be less effective than fresh semen. The same workers recorded a relatively low yield of morphologically acceptable oocytes per ovary (0.46 per ovary). It is believed that the low yield of oocytes may be the result of the low number of preantral follicles present in the buffalo ovary, in comparison with the cow. Fertilization rates have been lower than those reported in cattle IVF and the optimal capacitating treatments have yet to be satisfactorily formulated. In view of the limitations of oocyte recovery rate and buffalo semen characteristics, it may be better to await developments in technology that enable much larger numbers of oocytes to be available per ovary before placing undue emphasis on possible practical applications of IVF technology in this species.

Apart from studies in India, reports dealing with buffalo IVF have been provided by workers in several other countries that have buffalo populations; these include Thailand (Chuangsoongneon and Kamonpatana, 1991), Italy (Bacci *et al.*, 1991) and Taiwan (Lu and Hsu, 1991). A calf was born after transfer of an embryo produced by the IVM/IVF/IVC system of Suzuki *et al.* (1991b, 1992). The majority of these reports are characterized by low fertilization and embryo development rates.

5 Culturing the Early Embryo

5.1 Introduction

The culture of farm animal embryos *in vitro* requires an appropriate environment so that the early embryo can undergo several cleavage divisions to enable it to reach the blastocyst stage of development. Much progress has been made in providing for these needs over the past decade (review by Petters, 1992).

5.1.1 Historical

Although well over a century has passed since the first report of *in vitro* culture (IVC) of a mammalian embryo (Schenk, 1880), most activity in this field has been in the last 50 years, particularly with the emergence of effective tissue culture techniques (Brinster, 1968). Much of the work has been with rabbit and mouse embryos; Hammond Junior, working in Cambridge in 1949, was the first to successfully culture the mouse embryo through its early cleavage stages *in vitro*, using egg white in a physiological salt solution as the medium. The earliest reports of a successful procedure permitting the continued cleavage of the early cattle embryo to the blastocyst stage outside the cow came from Dublin (Sreenan and Scanlon, 1968; Sreenan *et al.*, 1968). In the same year, work in Cambridge confirmed that cattle embryos could be cultured to the hatching stage in the rabbit oviduct (Adams *et al.*, 1968).

5.1.2 In vivo *culture systems*

Since the late 1960s, workers have used a variety of animal systems, including mice, rabbits, sheep, cattle and even chicken eggs, for the *in vivo* culture of early bovine embryos. The use of the oviduct as an *in vivo* culture system helped to focus attention on the unique properties of this part of the female reproductive system. It clearly must be regarded as much more than just a tube carrying the embryo down into the uterus. All the evidence suggests that the fallopian tube plays a crucial role in the life of the gametes and in the early embryo in the days immediately after fertilization.

5.1.3 In vitro *culture systems*

The cells of the oviduct were used in the first successful coculture systems in sheep and cattle. The use of the oviductal monolayer culture system in sheep marked a particularly important milestone (Gandolfi *et al.*, 1986; Gandolfi and Moor, 1987). It subsequently became apparent that not only oviductal cells but a wide variety of other somatic cells are capable of providing an environment in which farm animal embryos can develop. There is good evidence to suggest that certain of the current coculture systems are highly effective in providing for the needs of cattle embryos. In Ireland, work by McCaffrey *et al.* (1990, 1991, 1992c) has demonstrated that a high proportion of *in vivo*-produced cattle zygotes can be brought to the blastocyst stage using the bovine oviductal epithelial cell (BOEC) monolayer culture system; not only that, but a high percentage of such blastocysts were of high-quality grade (see Table 5.1).

In looking towards the optimal embryo culture system, however, it is essential to think in terms of chemically defined media that are soundly based on a knowledge of embryo metabolism and on known embryo preferences for energy substrates and other essential nutrients. It is only by that means that it will be possible to provide for all those factors crucial for early embryonic development.

Table 5.1. Development of one- to four-cell bovine embryos on an oviductal cell monolayer (from McCaffrey *et al.*, 1990).

	Control	BOEC-0	BOEC-3
n	32	46	37
Initial cell stage of ova (M ± SE)	2.00 ± 0.16	2.19 ± 0.14	2.38 ± 0.18
No. ova ≥ 16-cell (%)	16 (50)	36 (78)	30 (81)
No. morulae/blastocysts (%)	0 (0)	36 (78)	30 (81)
No. of grade 1 and 2 ova (%)	0 (0)	31 (67)	26 (70)

BOEC-0, culture dishes seeded with bovine epithelial cells on day of ovum recovery; BOEC-3, culture dishes seeded 3 days prior to start of embryo culture.

5.1.4 *Chapter contents*

The present chapter examines the use of the various *in vivo* and *in vitro* systems that have been employed in culturing the early embryo, from the time of fertilization to the stage at which it is ready for transfer to a recipient animal or for storage (i.e. from zygote to blastocyst). In this, the quality of the IVC embryos must be entirely comparable to that of *in vivo*-produced embryos.

5.2 Early Embryo Development in the Bovine Oviduct

5.2.1 *The oviductal microenvironment*

The formation and function of oviduct fluid in mammals has been reviewed in a paper by Leese (1988). Ellington (1991) has dealt specifically with the anatomy and physiology of the bovine oviduct and covers some of the important events of reproduction which occur in it. As clearly shown in the excellent treatise of Hunter (1988), the oviduct of the cow provides the environment for the transport and capacitation of bull sperm and the transport and fertilization of the secondary oocyte. Several authors have reported on the secretions of the bovine oviduct, to determine the amounts produced and fluctuations in their rate of production during the cow's oestrous cycle. The recovery of cattle reproductive tracts from abattoirs provides a ready source of oviducts that can be useful in several areas of research (Fig. 5.1).

Studies to identify the nature of the products secreted by the oviducts could obviously assist in the development of appropriate media for early embryo culture. There is some evidence that the lipid environment of the oviductal lumen may affect the fertilization process and early cleavage of the embryo, by influencing membrane fluidity and the provision of energy sources (Killian *et al.*, 1989; Henault and Killian, 1991, 1993a, b); lipid synthesis was

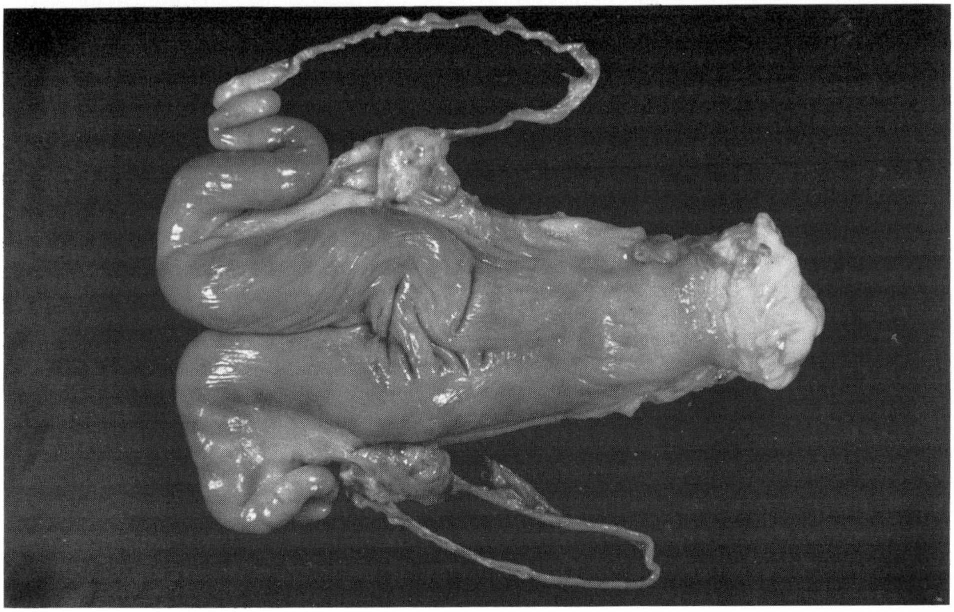

Fig. 5.1. Abattoir materials provide a ready supply of oviductal cells for setting up a BOEC monolayer culture system. In addition to providing oocytes for studies in maturation and fertilization, cattle tracts at the abattoir can be used as the source of oviductal cells for establishing a coculture system or for other research purposes.

greatest in the ampullary section of the fallopian tube during the follicular phase of the cycle.

There is growing evidence that the oviduct may provide the early embryo with compounds essential for normal development; these may include certain oviductal proteins and growth factors (Gandolfi *et al.*, 1993; Persson *et al.*, 1993; Viuff *et al.*, 1994). A novel protein, which first appeared in bovine oviductal fluid at oestrus, or day 1 of the cycle, and which remained in the fluid for 3–4 days has been reported (Gerena and Killian, 1988); the protein is presumed to be synthesized and secreted by oviductal cells. A study by Boice *et al.* (1990a, b) demonstrated that bovine oviductal epithelial secretory cells synthesize, and release at oestrus, oviduct-specific glycoproteins into the lumen of the fallopian tube; the evidence indicates that these proteins are oestrogen-dependent (Murray, 1990). These compounds are believed to be similar to oviductal polypeptides previously described by Geisert *et al.* (1987). Such reports strongly suggest that the bovine oviduct synthesizes and secretes unique glycoproteins; the possibility exists that these may affect fertilization and early embryonic development by their actions on the gametes. It remains for the precise biochemical nature and biological function of such oviductal glycoproteins to be clarified.

The view is taken by some workers that the early bovine embryo cannot survive on its own after fertilization but relies on somatic elements in the fallopian tube and uterus for its survival (Gandolfi and Moor, 1987, 1988; Gandolfi *et al.*, 1986, 1989a, b, c); they believe that this is reconcilable with evidence that oviductal cells secrete certain specialized proteins for a time period equivalent to that which the embryos spend in the tube. Evidence is available in sheep (Gandolfi *et al.*, 1988a, b, 1989a, b, c, 1990; Modina *et al.*, 1989) to show that a protein specifically secreted by the bovine oviductal cells at the time when embryos are passing through the oviduct is bound to the zona pellucida (ZP) and subsequently incorporated into the cytoplasm of the embryo. The Italian workers believe that the oviduct–embryo relationship is particularly close and that IVC systems which do not recognize this may not be fully capable of providing for the needs of high-quality embryos (Gandolfi, 1994). The method employed in Dublin in recovering oviductal epithelial cells for use in embryo studies is shown in Fig. 5.2.

In sheep, there are indications that growth factors, including insulin-like growth factor I (IGF-I), play a role in the development of oviductal embryos (Williams and Butler, 1991). Expression of the IGF-II gene by sheep oviductal cells has been demonstrated (Kincy *et al.*, 1992). There is evidence suggesting that platelet-derived growth factor (PDGF) is present in bovine oviductal fluid, probably associated with a carrier protein (Gandolfi *et al.*, 1991). There is also a report by Larson *et al.* (1991) suggesting that, during the fourth cell cycle of bovine embryo development, PDGF signals initiation of the first embryonically controlled development.

In pigs, there is evidence of the synthesis and secretion of three major glycoproteins during pro-oestrus, oestrus and met-oestrus; it appears that the ampulla and the isthmus of pig oviduct each secrete unique proteins and that such secretions are linked to endocrine events during the oestrous cycle (Buhi *et al.*, 1990).

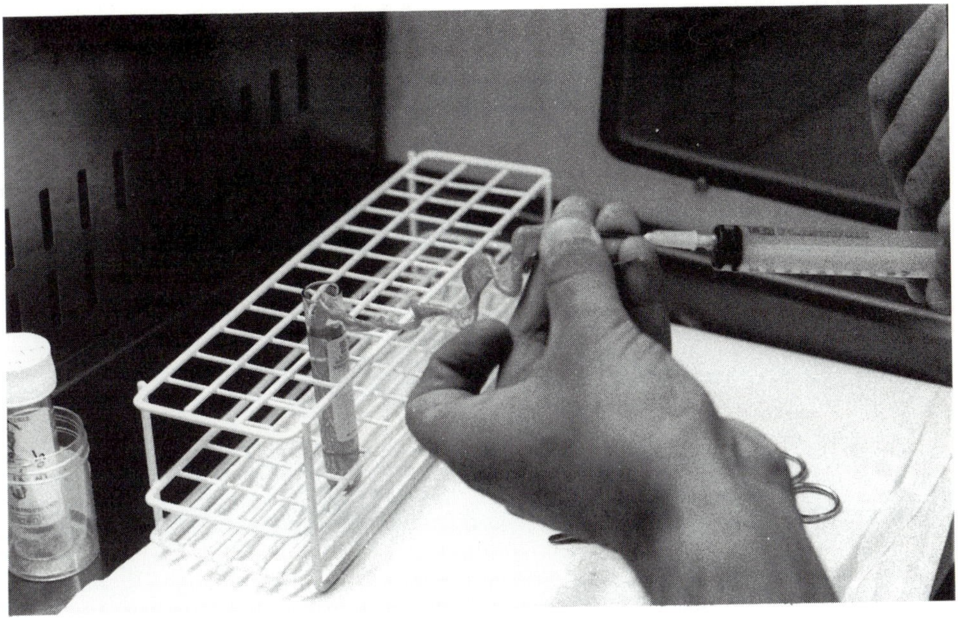

Fig. 5.2. Flushing the bovine oviduct for epithelial cells. Oviducts were excised from the cow tract and ligated at the ovarian and uterine extremities. They were then placed on ice in a Thermos flask and transported to the laboratory. The oviducts were then straightened and flushed with 5 ml of M-199 + 10% SS to obtain the epithelial cells.

5.2.2 Cleavage and compaction of the bovine embryo

The various stages in the early development of the bovine embryo, as recorded *in vivo*, are illustrated diagrammatically in Fig. 5.3. First cleavage has been recorded at about 30 h and second cleavage some 10–12 h later (Hamilton and Laing, 1946; Laing, 1949). In Belgium, working with *in vitro*-matured/*in vitro*-fertilized (IVM/IVF) cattle embryos, Van Soom *et al.* (1992) record peak concentrations of two-, four-, eight- and 16-cell stages emerging at 36, 42, 60 and 102 hours post-insemination (hpi).

An important stage in the development of the early mammalian embryo is the time at which cell compaction occurs. This will vary according to species. In the mouse, for example, it occurs as early as the eight-cell stage; in the cow it occurs at a later stage. Although one report suggested that compaction of the bovine embryo occurs at the 16-cell stage (Seidel, 1983), most workers agree that it occurs at the 32-cell stage or even beyond (Prather and First, 1988; Picard and Betteridge, 1989; Bondioli *et al.*, 1990a, b; Van Soom *et al.*, 1992). Under the microscope, compaction is evident by the tight formation of the blastomeres; the event is an important preliminary to the formation of the trophectoderm (TE), the inner cell mass (ICM) and the development of the blastocoele cavity. As shown by Van Soom *et al.* (1992), compaction usually occurs between days 5 and 6 after fertilization.

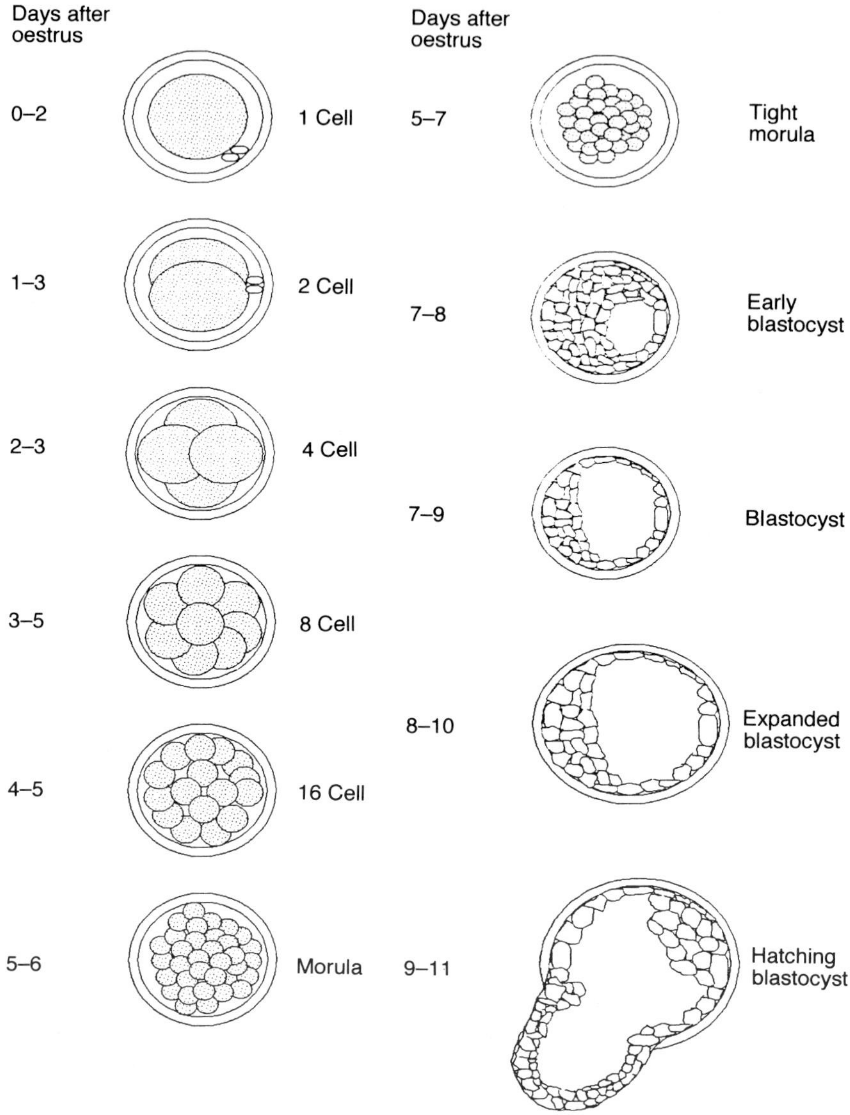

Fig. 5.3. Stages in the development of the bovine embryo.

Polarization of early blastomeres is believed to be the basis of compaction, and various workers have studied the mechanism of polarization. Koyama *et al.* (1993) examined the initiation of polarity in cattle IVM/IVF embryos and found it occurring at the 16-cell stage; polarity is evident in the distribution of surface microvilli on the blastomeres.

It should be mentioned that compaction is important to those concerned in cloning by nuclear transfer, for it marks the point at which it is no longer possible to separate the individual blastomeres that may be employed as donor cells in nuclear transplantation. Bondioli *et al.* (1990a, b), describing their cattle cloning research, report using donor embryos up to the 64-cell stage to provide blastomeres. It is also noteworthy that cattle embryos can undergo compaction even in the absence of the ZP (Xu *et al.*, 1991a).

It is believed that the cytoskeleton (microtubules and microfilaments) interacting with the plasma membranes are the major contributors to compaction (Gualtieri *et al.*, 1992). It is during the compaction period that the outer cells of the compact morula form tight junctions and the morula shows some expansion due to the action of the Na^+/K^+ pump. The Na^+/K^+ pump transports ions across the membranes into the centre of the ball of compact blastomeres and osmotic action then causes transport of fluid to the interior of the morula, resulting in some expansion of the embryo due to hydrostatic pressure.

5.2.3 *Transformation of morula to blastocyst*

Transformation of the morula to the blastocyst stage is regarded as an important step in early embryonic development; the process gives rise to two different cell types, those of the ICM and the trophoblast cells. The trophoblast cells, or TE, will develop into tissues of the fetal placenta and associated extraembryonic membranes, whereas the ICM cells will form the three germ layers of the developing embryo (endoderm, mesoderm and ectoderm). In some species, such as the pig, it is apparent that oestrogens are required for the transformation of the compacted morula to the cavitated blastocyst (Niemann and Elsaesser, 1986). Factors influencing oestradiol production by pig embryos were dealt with in a report by Hofig *et al.* (1990); the effect of dehydroepiandrosterone (DHEA), IGF-I and insulin in the culture medium was to increase oestrogen production. However, the endocrine factors involved in early embryonic development in cattle remain largely unknown. As observed by Watson (1992), the advent of techniques for the laboratory production of bovine and other farm embryos now enables a new source of material to be available for investigating blastocyst development.

Evidence for the requirement of embryos for specific energy substrates for transformation are provided in a report on mice (Martin and Leese, 1993); an obligatory requirement for glucose to support transition from morula to blastocyst was found in that species. The patterns of energy metabolism in IVM/IVF/IVC bovine embryos were determined in an earlier paper by Rieger *et al.* (1991b); the metabolism of glucose increased significantly between the eight- and 16-cell stages and continued to increase to the expanded blastocyst stage. Factors important for the normal development of ICM cells in rat and mouse embryos were examined by Hewitson and Leese (1992) and Brison *et al.* (1992, 1993); their data suggest that ICM cells develop in a microenvironment, the blastocoele fluid, which is well supplied with energy substrates. The trophoblastic cells apparently play a role in regulating the supply of these substrates to the ICM.

Changes in protein content of rabbit embryos as they progress from zygote to the peri-implantation blastocyst are reported by Morgan and Kane (1993); they found a significant increase in protein content in embryonic cells at the time of transformation of the morula to the blastocyst stage. For such reasons, it is not surprising to find that there is an amino acid requirement for the development of cleavage-stage rabbit embryos to blastocysts *in vitro*.

5.2.4 Embryonic mortality

Even in normal healthy cattle, a proportion of embryos which have passed through the fallopian tubes into the uterus fail to continue development, generally during the first 3 weeks of pregnancy (Diskin and Sreenan, 1980). Embryonic mortality has long been recognized as a major source of loss in breeding cows and numerous studies have reported on it. Some of the factors currently recognized to be involved in embryonic mortality in farm ruminants are set out in Fig. 5.4.

In cattle, it should be noted that attachment of the embryo to the uterine wall does not occur until about 3 weeks after conception, by which time considerable growth and the development of the embryo have taken place (Greenstein and Foley, 1958). Embryo recovery at specific intervals after breeding has been employed to determine the time at which embryonic mortality occurs. Sreenan and Diskin (1986) summarized many of the published reports on the extent and timing of embryonic mortality; from this, it appeared that the greatest single increase in the rate of embryo loss occurred between day 15 and day 18.

As observed by Betteridge and Loskutoff (1993), there is some suggestion that embryonic loss may be greater after embryo transfer (ET) than after

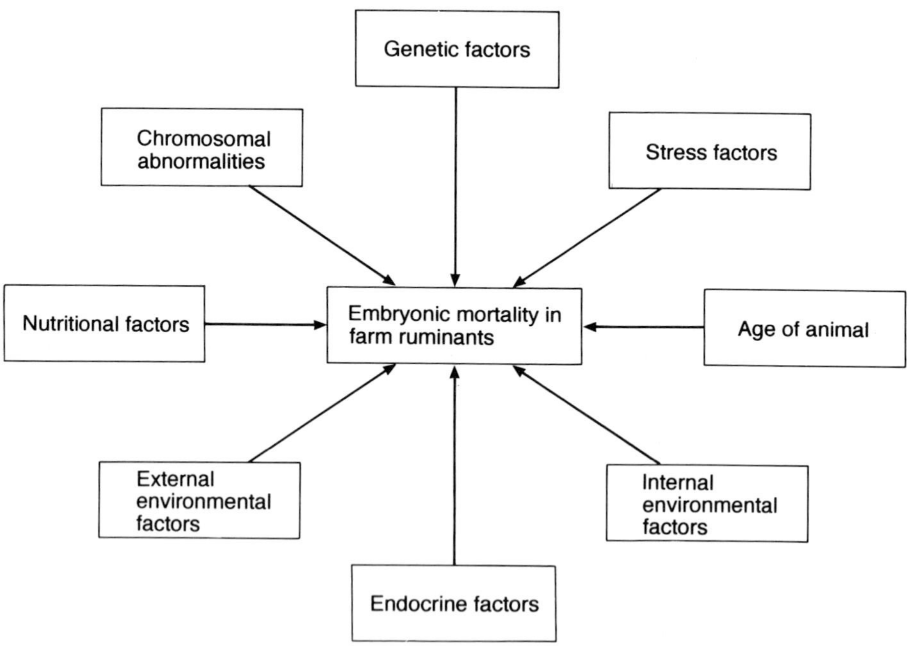

Fig. 5.4. Factors affecting embryo mortality in farm ruminants.

artificial insemination (AI). It would not be altogether surprising if this were true, given the insults inevitable to embryos in terms of their exposure to environmental factors quite foreign to the oviduct and uterus of the cow. Although most embryo losses after AI occur in the first 3 weeks of pregnancy, after ET there are some worrying indications of continuing loss in the period of placental establishment, which need to be evaluated. In establishing confidence in ET as a realistic alternative to AI or natural service in cattle production systems, there is a clear need for firm information on this question.

Embryonic development covers the period up to about 50 days of age in cattle; beyond that, fetal growth and development continue through to term. There is still a scarcity of information on the factors affecting the development of the embryo and the fetus; the influence of maternal factors needs to be clearly separated from those arising from the quality of the embryo itself. There is clearly scope for research which will help in explaining early embryo mortality in cattle as well as providing the standards which may be used in producing the laboratory cattle embryos that may be expected to have the optimum chance of survival in the recipient cow. Efforts to fully understand factors responsible for embryonic mortality must certainly include a study of the mechanisms involved in the maternal recognition of pregnancy (Bazer *et al.*, 1994).

In passing, it may be noted that early embryonic mortality in the pig has been directly related to the sequence of events at ovulation and early embryogenesis (Pope *et al.*, 1990; Xie *et al.*, 1990a, b). The pattern of oocyte and follicle development is one of the sources of disparity in the age of pig embryos; later-ovulating oocytes are the last to be fertilized and become the smaller blastocysts. Although normal, they will die preferentially as a result of an asynchronous and hostile uterine environment which has developed in response to the oestradiol secretions of their more advanced littermates (Pope, 1992).

5.3 *In Vivo* Culture Systems

It is evident, from many reports, that the oviducts of several species (including rabbits and sheep) can sustain early cattle embryos *in vivo* and yield morulae and blastocysts capable of establishing normal pregnancy rates in recipient animals. Although the practical employment of such *in vivo* culture systems is likely to be of historic interest only in the years ahead, they can be regarded as having contributed substantially to knowledge of factors influencing early embryo development.

5.3.1 The rabbit oviduct

Although early attempts in the USA to store cattle embryos in the rabbit oviduct produced negative results (Hafez and Sugie, 1963), later efforts in Ireland were successful (Sreenan and Scanlon, 1968; Sreenan *et al.*, 1968) when it became evident that bovine blood serum was probably an unsatisfactory medium for the recovery and transfer of cattle embryos to the rabbit. These observations were confirmed and extended by Cambridge workers (Adams *et al.*, 1968; Rowson *et al.*, 1969). According to Polge and Rowson (1975), it was inadvisable to store bovine embryos in the rabbit for more than 3 days or in the ewe for more than 4 days; storage beyond this period produced embryonic abnormalities. In Dublin, however, it was found that early-cleavage cattle embryos continue developing satisfactorily for several days, and that a high percentage are capable of developing into a normal conceptus on transfer to a recipient cow (Gordon, 1975, 1976; Boland, 1984).

In more recent times, Canadian and Japanese workers engaged in cattle IVF studies have frequently employed the rabbit oviduct as a temporary incubator (Hanada, 1985b; Sirard *et al.*, 1985a; Hanada *et al.*, 1986; Lambert *et al.*, 1986; Fukui and Ono, 1988; Ueda *et al.*, 1988). According to Sirard and Lambert (1986), there is some evidence suggesting that cleaved cattle embryos may be better retained in the oviduct than those that are not cleaved. In using the rabbit, the fallopian tube is usually ligated at the uterotubal junction (UTJ) to prevent the cattle embryos passing into the uterus.

The number of cattle embryos transferred per oviduct has usually been in the range 5–20; the expected rate of embryo recovery from the ligated tube appears to be about 70% (Boland, 1984). The doe may be treated or not treated with human chorionic gonadotrophin (HCG) to induce ovulation prior to the transfer of embryos. In the work of Fukui and Ono (1988), for example, a dose of 100 iu HCG was employed 2 days prior to introducing embryos into the oviduct. However, there is no clear evidence that the rabbit's reproductive status is important in determining the effectiveness of the *in vivo* culture period.

The period of embryo culture in the rabbit oviduct has usually been of the order of 3–7 days. It is not clear whether cattle IVF workers have always taken adequate account of the recorded adverse effects of prolonged culture in the rabbit oviduct. In studies with sheep embryos, for example, Cambridge workers recorded that the optimum storage period was about 3 days, with pregnancy rates declining after 5 days and reaching negligible levels after 7 days (Polge and Rowson, 1975). On the other hand, workers such as Hawk *et al.* (1988) have cultured cattle embryos from the zygote stage for periods of 7–9 days and recovered embryos at the morula or even the hatched blastocyst stage; such embryos were apparently able to survive after transfer to recipients. The explanation may lie in the age at which cattle embryos are introduced. At the zygote stage, they may survive in the rabbit oviduct for 7 days; introduced at the eight-cell stage, they may not survive more than 4 days.

In the USA, Hawk *et al.* (1989) examined the effect of leaving the rabbit oviduct open rather than ligating it in their studies with cattle embryos; this resulted in the loss of most embryos. It was believed that the embryos may have passed out through the cervix in those instances where the oviducts were left open.

5.3.2 The sheep oviduct

Sheep oviducts have been employed for many years as an *in vivo* culture system. In work reported almost 40 years ago in France, it was shown that sheep embryos retained in the ligated oviducts of the ewe for up to 8 days were capable of developing into normal healthy lambs on transfer to recipient animals (Wintenberger-Torres, 1956). In more recent times, it has been shown by several groups that the ligated sheep oviduct can provide an adequate environment, not only for sheep embryos but also for those from other farm species, including cattle (Fehilly *et al.*, 1984). In Ireland, work dates back to the early studies of Sreenan (1968, 1969), who attempted to fertilize IVM bovine oocytes in the oviducts of sheep previously inseminated with bull semen. In providing for the needs of cattle oocytes/embryos, it might be assumed that the closely related sheep has a possible advantage over the rabbit. However, in some regards, there appears to be little difference between the two species.

Those who have compared the sheep and rabbit oviducts in the culture of cattle embryos report the two *in vivo* culture systems as yielding similar results (Lutterbach *et al.*, 1987; Fukui *et al.*, 1989; Westhusin *et al.*, 1989). However, for transfer purposes, the ewe's greater size and consequent oviductal capacity mean it can probably accept many more embryos than the rabbit. The sheep has also been the subject in Ireland of much research in the control of oestrus and ovulation (see Gordon, 1983); this made it possible, with relative ease, to have animals available in all seasons of the year to take embryos at a clearly defined stage of cycle.

The appropriate technique for the artificial control of oestrus and ovulation in the sheep (cyclic or anoestrous) involves the intravaginal administration of a suitable progestagen (e.g. fluorogestone acetate (FGA), medroxyprogesterone acetate (MAP)) over a period of 12 days in conjunction with an injection of gonadotrophin (500–750 iu pregnant mare serum gonadotrophin (PMSG) by i.m. injection) at the time of pessary removal. The ewe can be expected to come into oestrus about 36 h after pessary withdrawal and the heat period has a duration of a further 36 h. This means that ovulation can be expected to occur some 60–70 h after pessary withdrawal.

The usual procedure in Dublin involved introducing presumptive bovine zygotes into the ligated ovine oviduct shortly after the occurrence of a controlled ovulation; this was on the assumption that this timing corresponded most closely to *in vivo* events in the cow. The recovery rate of cattle and sheep oocytes/embryos from the oviduct was found to be influenced by a number of factors. The recovery rate from ligated tubes could often be markedly lower than from non-ligated tubes (Quirke, 1968; Sreenan, 1969; O'Gorman, 1970; Vergos, 1990). Even the breed of ewe could significantly influence recovery rates; Crosby (1969) recorded a yield of 36.5% in Blackface Mountain and 46.2% in Galway sheep. It was speculated that such breed differences arose from stressors acting in the sheep; Blackface ewes, for example, were relatively unaccustomed to handling, in comparison to the more placid Galway.

In more recent times, the ewe oviduct system assumed considerable importance in the work in Dublin, by providing for the needs of presumptive cattle zygotes for a period of about a week (Lu *et al.*, 1987b). Recovery rates of 60–70% were achieved, using ligated oviducts and introducing zygotes, in groups of 10–50 per tube (Lu *et al.*, 1988a). With experience, it became evident that several hundred presumptive zygotes could be transferred to the tube without adversely affecting recovery rate (Gordon *et al.*, 1988); this procedure was subsequently applied in a semicommercial setting for the large-scale production of beef embryos (see Table 5.2). In this, of the 55,797 oocytes transferred, 31,469 (60%) were recovered after about a week and 23.5% of these had developed to the morula/blastocyst stage. The usual time of transfer was 20 h after insemination of the IVM oocytes; the embryos were exposed to tubal factors for almost a week. The yield of embryos, as a percentage of those oocytes that were assumed to have cleaved, was 39.2%. It is not unreasonable to assume that this *in vivo* culture system provided a type of microenvironment comparable to that found in the live cow. The estimated cleavage rate (60%) must be regarded as relatively low; this may well have been the result of fertilization failure. Other studies by Jiang *et al.* (1989) dealt with the transfer, at 20–24 h after insemination, of 1171 presumptive zygotes to the sheep oviduct; 629 (54%) of these were recovered and 27% had developed to the morula/blastocyst stage. Kinis (1990) and Vergos (1990) also showed embryo yields of similar magnitude (24–27%) using this type of culture system. In sheep, in contrast to the rabbit, non-ligation of the oviduct was not found to have an adverse affect on the subsequent recovery of embryos (Vergos, 1990).

In naturally ovulating sheep, there is evidence that nutritionally mediated shifts in endocrine status (e.g. low progesterone levels) may influence embryo quality. However, there is currently no evidence from studies in Ireland that clearly links ewe nutritional or endocrine status with the quality of cattle embryos recovered after *in vivo* culture.

In the USA, the Madison group (Eyestone *et al.*, 1985, 1987b) reported studies in which 254 early cleavage-stage (one- to two-cell) embryos were transferred to sheep (small numbers per oviduct); a 60% recovery rate was recorded. There was no difference in embryo recovery rate between intact and

Table 5.2. Summary of results showing yields of cattle embryos recovered after IVM/IVF and culture in the sheep oviduct.

Authors	Oocytes to ewe	Recovered from eve	Oocytes cleaved	Yield of embryos	Maturation details
Lu *et al.* (1987a)	202	149 (74%)	98 (66%)	4 (2.7%)	FCS + hormones
Lu *et al.* (1987c)	119	78 (66%)	55 (71%)	23 (29.5%)	FCS + hormones
Lu *et al.* (1987b)	222	143 (64%)	121 (85%)	86 (60.0%)	OCS ± hormones
Lu *et al.* (1988a)	668	420 (63%)	334 (80%)	191 (46.0%)	OCS – hormones
UCD/OVAMASS data	55,797	31,469 (56%)	18,895 (60%)	7401 (23.5%)	OCS – hormones

ovariectomized sheep, or between cyclic and anoestrous ewes. The general conclusion was that ovarian activity in the ewe is not essential for the normal development of the cattle embryos in the oviduct. Other North American studies employed sheep that were either pregnant or under the influence of progesterone throughout the entire *in vivo* culture period (Biery *et al.*, 1988); even under these conditions, recovery rate and embryo development were held to be satisfactory. In Canada, the transfer of cattle embryos to the ligated oviducts of cyclic ewes was followed by a 78% recovery rate 5 days later (Xu *et al.*, 1989). It may be noted that the requirements for early development of pig embryos are readily met by the sheep oviduct, as shown in studies reported by Prather *et al.* (1991c).

5.3.3 *The cow oviduct*

Although there have been isolated instances in which the cow oviduct has served as an *in vivo* culture system (Newcomb *et al.*, 1978; Xu *et al.*, 1987a; Peura and Aalto, 1989), it was clearly impracticable for this species to be employed on a routine basis. In the report of Peura and Aalto (1989) from Finland, early cattle embryos were transferred to the oviducts of recipient heifers and recovered non-surgically several days later; the authors reported 26/169 (17%) at the morula/blastocyst stage on recovery. The use of a laparoscopic oviductal transfer approach has been reported by Fayrer-Hosken *et al.* (1988, 1989); this technique could be of interest for those who wish to transfer and recover bovine embryos by non-surgical procedures. There is, however, no reason to believe that the bovine oviduct may have a particular merit over alternative oviductal culture systems. In France, Chupin *et al.* (1988) showed the capacity of the oviducts of prepubertal calves (3 months old) to sustain early bovine embryos; they concluded that no specific hormonal preparation was required before the oviductal cells became able to support embryonic development.

5.3.4 *The isolated mouse oviduct culture system*

Until it becomes clear what the mammalian oviduct provides for the early developing embryo, an isolated organ culture system may have advantages over some culture systems (oviductal/granulosa cell monolayers), which may not necessarily maintain cells in the same state of differentiation as in the live animal. In this regard, the isolated mouse oviduct culture system (IMOS) has been successfully used in the culture of mouse, rat, hamster and pig embryos from the one-cell stage to the morula/blastocyst stage (Papaioannou and Ebert, 1986; Minami *et al.*, 1988; Ebert and Papaioannou, 1989; Krisher *et al.*, 1989a, b). Methods involved in culturing cattle embryos in the IMOS, taken from Sharif (1992), are summarized in Fig. 5.5.

Studies by Sharif *et al.* (1991a, b, c) in Dublin were the first to report the development of early bovine embryos (three- to four-cell stage, produced by IVM/IVF) to the morula/blastocyst stage, using Chatot–Ziomek–Bavister (CZB) as the culture medium. It was not clear, in this work, why the isolated

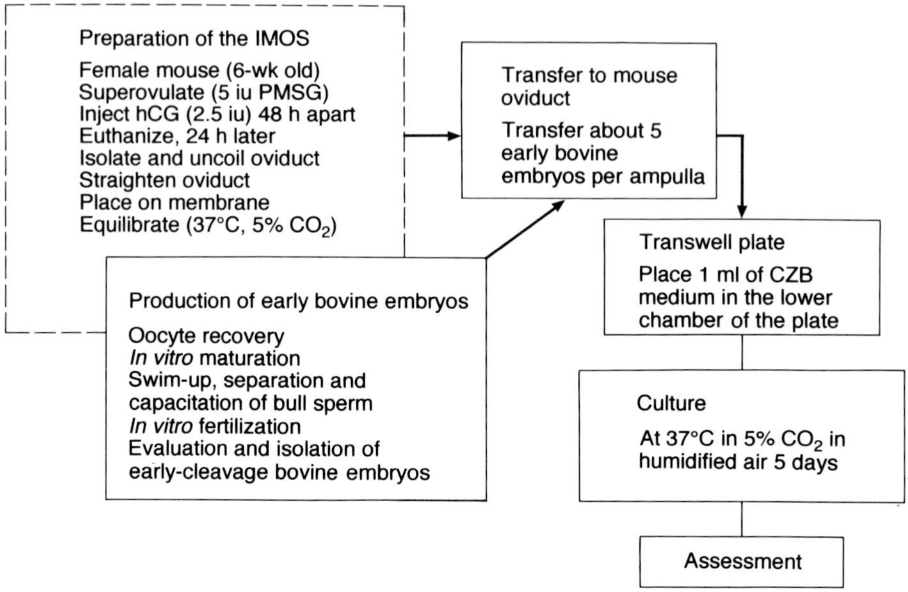

Fig. 5.5. A schematic representation of procedures involved in culturing bovine embryos in the isolated mouse oviduct system (from Sharif, 1992).

mouse oviduct when cultured with CZB medium supported embryo development but with media such as M-199 (with or without serum), Krebs Ringer bicarbonate (KRB) and modified Krebs Ringer medium did not. In comparing the composition of the media employed (see Table 5.3), the absence of glucose in CZB is apparent; although glucose is also absent in KRB, that medium is also lacking in most other energy substrates. The Dublin authors concluded that the mouse system was capable of giving embryo yields comparable to those found with *in vivo* culture systems (rabbit/sheep oviducts).

There were indications that early bovine embryos differed in their requirements from the porcine embryos. Other work in Dublin included a histological evaluation of the isolated mouse oviduct (Sharif *et al.*, 1992a); this showed that the isolated oviduct was well maintained throughout the culture period (about a week). Several other studies on factors influencing cattle embryo development in the isolated mouse oviduct were reported by Sharif *et al.* (1991d, 1992a, b, c, d, 1993). There was no evidence, however, that the mouse oviduct provided more favourably for the needs of the bovine zygote or embryo than alternative culture systems.

In the USA, Sparks *et al.* (1992a) examined the isolated mouse oviduct system as a means of supporting development of bovine embryos from the one-cell to the blastocycst stage in comparison with the BOEC monolayer culture system. Unlike the Dublin workers, their results led them to conclude that the explanted mouse oviduct was incapable of providing an adequate environment for the development of cattle zygotes. As noted above, it is

Table 5.3. Composition of medium M-199, KRB, mKRB and CZB

	Medium[a]			
Components	M-199[b]	KRB	mKRB	CZB
NaCl	8.000	6.721	6.721	4.769
KCl	0.400	0.356	0.356	0.360
$NaHCO_3$	0.350	2.100	2.106	2.110
KH_2PO_4	0.060	0.162	0.162	0.161
$MgSO_47H_2O$	–	0.293	0.294	0.294
Na_2HPO_4	0.048	–	–	–
$CaCl_2$	0.140	0.347	0.189	–
$CaCl_22H_2O$	–	–	–	0.249
Na pyruvate	–	–	–	0.029
BSA	varies	–	4.000	5.000
Na lactate	–	–	–	3.508
Glucose	1.000	–	1.000	–
EDTA	–	–	–	0.041
Glutamine	–	–	–	1.000

[a]Components are in g l^{-1} except for glutamine, which is added at a concentration of 1.0 mM. Penicillin and streptomycin were added at concentrations of 100 iu ml^{-1} and 50 μg ml^{-1}, respectively.
[b]M-199 also contains certain amino acids, vitamins and trace elements.

unclear why the mouse oviduct when cultured with CZB medium supports cattle embryo development but with a medium such as M-199 does not.

5.3.5 *The chick embryo culture system*

An unusual *in vivo* culture system was developed by workers in Louisiana, in which mice, goat and cattle embryos have been cultured, from the pronuclear to the blastocyst stage, in the amniotic cavity of the 4-day-old chicken embryo (Blakewood *et al.*, 1988, 1989a, b, 1990). The evidence indicated that cattle embryos developed in this system at rates comparable to those found in the cow itself. These workers suggested, on the basis of such studies, that the eggs of domestic chickens may provide an alternative to some of the other culture systems that are available. There is, however, a problem in keeping embryos for more than 3 days at one time in the chicken egg. If the intended culture period is 6–7 days, this means using two chicken eggs, which renders the procedure tedious and unsuited for large-scale use. The use of extracted chicken amniotic fluids for the culture of mouse embryos was not found to confer the same embryotrophic qualities as *in situ* culture in the chicken amnion (Blakewood and Zhang, 1993).

5.4 Metabolism of the Early Embryo and the Development Block

5.4.1 *Early embryo metabolism*

Much research has been conducted in attempting to determine the nature of the factors that affect the development of the early mammalian embryo (McKiernan and Bavister, 1990a, b; McKiernan *et al.*, 1991). Much of this work has been with mice, but it has become increasingly clear that the differences between species make it essential for methods to be developed specifically for each mammal. The metabolism of the early mammalian embryos is dealt with in a review by Leese (1991); this author draws attention to the effectiveness of the CZB culture medium (Chatot *et al.*, 1989) in permitting about 50% of one-cell embryos (in a blocking strain of mouse) to develop to the blastocyst stage. In the oviduct, the materials utilized by the early cleaving embryo are likely to change as it grows older; this has been shown in studies with mouse embryos, but is likely to be a general principle that applies to all farm mammals, including the cow. Early mouse and rat embryos are known to shift from predominantly utilizing pyruvate to glucose at the late morula and early blastocyst stages (Brison and Leese, 1991; Dufrasnes *et al.*, 1993). Studies by Peura (1990) in Finland failed to find an effect from the presence of two intermediates of the Krebs cycle (citrate and malate) and one precursor of acetyl-coenzyme A (CoA) (acetate) in the IVC medium on development of IVM/IVF-derived bovine embryos.

There are many studies suggesting that glucose may be detrimental to early embryo development, including work with cattle (Moore and Bondioli, 1990), sheep (Ledda *et al.*, 1991), hamster (Seshagiri, 1990; Seshagiri and Bavister, 1990a, b, 1991a, b) and human embryos (Conaghan *et al.*, 1992a, b, 1993). Studies in mice have shown that optimal development can occur during the first 48 h in media such as CZB, which contains glutamine but no glucose (Chatot *et al.*, 1989). The sugar does, however, become important at later stages. Work with inbred strains of mice has shown that glucose added to pyruvate/lactate medium at the time of compaction supports transition from morula to blastocyst (Brown and Whittingham, 1990); in the absence of glucose, mouse embryos are arrested at the morula stage. Glucose is not utilized as a significant energy source by early sheep embryos until after the 16-cell stage. Williams and Butler (1991) in Idaho measured the adenosine triphosphate (ATP) and adenosine diphosphate (ADP) content of early-stage sheep embryos. The authors suggest that the increase in glucose uptake by sheep embryos after the 16-cell stage may be due in part to a decrease in the ATP/ADP ratio. The ability of early sheep embryos (one to four cells) to develop *in vitro* in CZB medium was examined by McGinnis and Youngs (1992) in Idaho. Glucose supplementation of this medium was found to increase cytoplasmic fragmentation of embryos but did not inhibit cleavage. The effects of pyruvate on the development of early sheep embryos in a modified CZB medium, containing glutamine but lacking glucose and lactate, was examined by Youngs *et al.* (1992); the embryos developed to the blastocyst stage in the absence of pyruvate, lactate and glucose. In New

Zealand, studies with early sheep embryos show that glucose metabolism is not critical for development, but is beneficial at low concentrations (Thompson *et al.*, 1991, 1992a, c); higher concentrations of glucose, however, can inhibit development, possibly by inhibiting the tricarboxylic acid (TCA) cycle.

5.4.2 *The development block*

One obstacle to progress in developing embryo transfer technology in cattle in the 1960s and 1970s was the fact that cattle embryos at the one- to four-cell stages, cultured by conventional tissue culture techniques, rarely progressed beyond the 8–16-cell stage; on the other hand, embryos cultured at the 8–16-cell stage could generally be expected to undergo normal compaction, blastulation and hatching (Wright and Bondioli, 1981). It was this difficulty in early embryo cleavage that led to the development of the various *in vivo* culture systems described earlier (see Sections 5.3.1–5.3.5).

It is now evident that early embryonic development in mammals is driven by protein and messenger ribonucleic acid (mRNA) stored within the oocyte. Transcription during early cleavage stages results in quantitative and qualitative changes in protein synthesis, which bring about the maternal/embryonic transition in the control of transcription. This transition is now known to occur at the late four-cell to early eight-cell stage in the bovine embryo (King *et al.*, 1988b; Jones and First, 1990, 1991). In the mouse, on the other hand, transcriptional activity from the embryonic genome is initiated at the two-cell stage; in both cattle and sheep, it is evident that activation of the embryonic genome is delayed in comparison with rodents (Watson *et al.*, 1992a, b, 1993); patterns of gene expression within the early cattle embryo show that bovine development displays several characteristics which distinguish it from that of rodents.

Although the cells of the eight-cell cattle embryo apparently remain alive during the period of blocked development, they cannot be rescued once the block is initiated (Eyestone and First, 1986). In Madison, Eyestone and First (1991) used *in vivo*-produced embryos from superovulated cattle to determine whether one- to two-cell embryos cultured *in vitro* to the eight-cell stage would continue to develop if transferred to a permissive environment. They found the eight-cell-stage embryo to be essentially irreversible, their evidence suggesting the existence of a particularly sensitive phase during the four- to eight-cell stage. This is believed to reflect the fact that the block is caused by transcriptional failure. Results reported by Pollard *et al.* (1991a) suggested that the critical period(s) when BOEC act to mediate passage through the 8–16-cell developmental block are during the four- to eight-cell transition or at the eight-cell stage itself.

There are several ways in which it is possible to overcome the development block as it is now known to occur in mammalian embryos. Several authors have shown that specific compounds or changes in medium composition may be capable of permitting embryos to pass through the block stage. Barnes and Eyestone (1990) drew attention to evidence in mice and hamsters suggesting that inadequate energy generation may occur at the time of cleavage arrest and that the presence of glucose in the culture medium during early cleavage is inhibitory. In cattle, Larson, R.C., *et al.* (1989, 1990, 1992a, b), using a defined medium, reported evidence supporting the view that fibronectin or transforming growth factor beta (TGF-β) and basic fibroblast growth factor (bFGF) promote development of bovine embryos beyond the eight-cell 'block' *in vitro*.

In mice, the removal of glucose and the addition of glutamine can overcome the two-cell block in that species (Chatot *et al.*, 1989, 1990). Using the CZB medium, mice embryos (from blocking strains) have been cultured from the one-cell to the blastocyst stage; this medium is a modification of Brinster's ovum culture medium 2 (BMOC-2), containing an increased lactate/pyruvate ratio, 0.1 mM ethylenediaminetetraacetic acid (EDTA) and 1 mM glutamine. It is known that glutamine is utilized by cells in culture as an energy substrate; this may occur in the early developing mouse embryo. Glutamine is also known to be an important amino acid in the synthesis of purine and pyrimidine nucleotides and, at a time when the cells of the mouse embryo are starting to divide more rapidly, the amino acid may be required to enhance the nucleotide pool.

5.4.3 Activation of the bovine embryonic genome

A study of protein profiles by Barnes *et al.* (1987) has shown changes occurring at the early to mid-eight-cell stage in the bovine embryo, at a time of increased transcription of RNA; this was taken to indicate the onset of transcription and translation directed by the embryonic genome.

As observed by Crosby *et al.* (1988), early embryonic development in many mammalian species, including cattle, is dependent on products that are synthesized by the oocyte during maturation and early cleavage, although this maternal legacy involves a variety of macromolecules and includes the storage of mRNA, which provides the major template for protein synthesis during the early cleavage divisions of the embryo. Data in a report by Frei *et al.* (1989) indicate that, from the one-cell to the eight-cell stage of development of the bovine embryo, there is a decrease in the level of protein synthesis, with no major changes in the qualitative pattern of protein synthesis taking place during this time. At the 8–16-cell stage, these same workers found evidence of a change in the qualitative pattern of protein synthesis being established in the bovine embryo, with this pattern persisting up to the blastocyst stage.

One method of monitoring the activation of the bovine embryonic genome lies in a study of the nucleolus organizer regions (NORs) in early embryos. Evidence provided by King *et al.* (1985b) indicated that NORs in cattle embryos become transcriptionally active at about the eight-cell stage. Other work by King *et al.* (1988a) notes that the eight-cell stage in the bovine embryo is of much longer duration than the two-, four- or 16-cell stages. It is believed that, once RNA genes become transcriptionally active, their activity can be maintained *in vitro* in the absence of the factors presumed to be essential for their activation. Stages in the development of IVM, IVF and IVC (IVMFC)-derived embryos from two-cell to late morula are shown in Fig. 5.6.

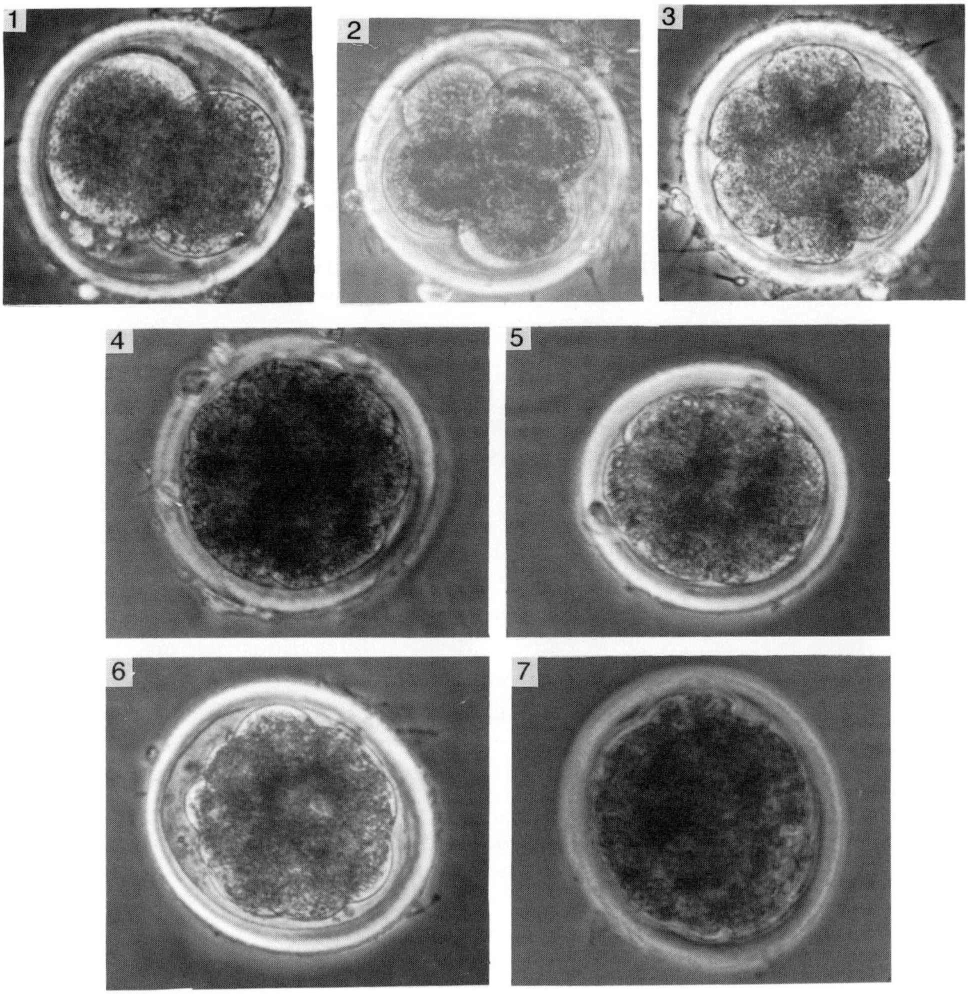

Fig. 5.6. From two-cell to late morula stage in 6 days. The bovine embryo reaches the two-cell stage at about 30 h after IVF (1); the four-cell (2) and eight-cell (3) stages are found at 48 h after IVF; by day 3 the embryo should have reached the 16-cell stage (4) and by day 4 it should be at the 16–32-cell stage (5); by day 5 it would be forming a compact morula (6) and by day 6 it can be expected to have reached the late morula stage (7).

As pointed out by King *et al.* (1989), the organelle that morphologically reflects changes in the metabolism or physiology of the cell is the nucleolus, which can be considered to be a good marker of the onset of embryonic transcription. It is a matter of observing the morphological features of the nucleoli to confirm the time at which activation of the embryonic genome occurs.

5.5 *In Vitro* Culture Systems

5.5.1 *Synthetic oviductal fluid (SOF) formulations*

The general experience, before the advent of coculture techniques, was that very limited development could be achieved with cattle embryos in conventional tissue culture media. This was recognized in studies reported as long ago as the early 1950s (Wintenberger *et al.*, 1953). An early exception to that general rule was the work of Tervit *et al.* (1972) in Cambridge, who were among the first to describe the successful culture of one- to eight-cell cattle embryos to the morula/blastocyst stage in a medium termed 'synthetic oviductal fluid' (SOF) and using a 5% rather than a 20% oxygen-in-air gas atmosphere (see Table 5.4). Limited information on further embryonic development and the establishment of pregnancy in a recipient animal was also provided. The medium was based on a previous biochemical analysis of sheep oviductal fluid. However, there are those who note that the composition of current mammalian embryo culture media bears little resemblance to the ionic and metabolite composition of oviductal fluid (Gardner and Sakkas, 1993).

Although SOF and low oxygen tension, or modifications of that system, have been employed in cattle and sheep by various authors, there has not always been general agreement that they provide more favourably for the needs of early embryos than alternative techniques. A few years after the report of Tervit *et al.* (1972), workers in the USA examined several culture

Table 5.4. Composition of synthetic oviductal fluid – original formulation (from Tervit *et al.*, 1972) and a modified SOF formulation (from Takahashi and First, 1992).

Original		Modified	
Constituent	Concentration	Component	Concentration
NaCl	107.70 mM	NaCl	107.70 mM
KCl	7.16 mM	KCl	7.16 mM
KH_2PO_4	1.71 mM	KH_2PO_4	1.19 mM
$MgCl_2$	0.49 mM	$CaCl_2$	1.71 mM
$CaCl_2$	1.71 mM	$MgCl_2$	0.49 mM
$NaHCO_3$	25.07 mM	$NaHCO_3$	25.07 mM
Na lactate	3.30 mM	Sodium lactate	3.30 mM
Na pyruvate	0.33 mM	Sodium pyruvate	0.30 mM
Glucose	1.50 mM	BSA[a]	3.0 mg ml^{-1}
BSA	32 mg ml^{-1}	Phenol red	1.3 μg ml^{-1}
Penicillin	100 iu ml^{-1}	Gentamicin sulfate	50.0 μg ml^{-1}
Streptomycin	50 μg ml^{-1}		
pH – 7.2 to 7.4 270 mosm			

[a]Crystallized and lyophilized; the osmolarity and pH of mSOF are 270–280 mosm and 7.2 to 7.3, respectively.

media and gas phases for their ability to sustain early cattle embryos produced *in vivo* (Wright *et al.*, 1976a, b). The SOF medium promoted no more than one cell division or less in eight-cell embryos; using Ham's F-10 medium, supplemented with 10–50% fetal calf serum (FCS), some embryos developed to the expanded blastocyst stage.

The studies of Thompson *et al.* (1989a) showed limited development of sheep embryos to the blastocyst stage in SOF; this work did not support the view that SOF with low oxygen tension may necessarily overcome the block condition in sheep. In other studies in New Zealand, Fukui *et al.* (1991b) found SOF, with 5% oxygen and either FCS or human serum, to be their preferred system. However, the same study showed that the success of a particular system (e.g. SOF and 5% oxygen) probably depends on the correct combination of a number of factors. Under certain culture conditions, it was found that cattle embryos developed readily to the blastocyst stage, with or without oviductal cell coculture, in SOF or M-199 and using a 5% or 20% oxygen gas phase.

In other studies in New Zealand, Pugh *et al.* (1991b) used SOF supplemented with 20% human serum under low (5%) oxygen tension to culture early sheep embryos to the 6-day stage after IVM/IVF. However, they found embryo viability at 6 days to be only 3% compared with 38% at 1 day, as tested by transfers to recipient ewes. They did, however, allude to other, unpublished results with the SOF medium which they claimed were much more promising.

In Arkansas, the development of IVM/IVF cattle embryos in SOF medium, with or without bovine serum albumin (BSA) or FCS supplementation and with or without oviductal cells, was reported by Rorie *et al.* (1991a, b); they found that the morphological quality of the embryos was better maintained when FCS and oviductal coculture were employed in conjunction with SOF medium.

In Australia, SOF was used in the culture of bovine nuclear transfer embryos by McLaughlin *et al.* (1992). Although the medium did support development, the low cell numbers of blastocysts developed *in vitro* and the low pregnancy rate observed after transfer to recipients indicated limitations in the capacity of this simple culture medium to maintain the full developmental potential of micromanipulated embryos.

In Belgium, Fontes *et al.* (1993) used frozen–thawed bovine oviductal cells with SOF to coculture early IVMFC embryos successfully to the blastocyst stage; it was felt that frozen BOEC may provide a more efficient and standardized system of coculture compared with the use of fresh oviductal cells.

In Japan, Matsuyama *et al.* (1993) found that the optimal levels of glucose for adding to their modified SOF medium (supplemented with 10% human serum) varied according to time of culture (days 1–3, 0.188 mM; days 4–8, 1.5 mM); the same authors showed that a glucose concentration above 3 mM was inhibitory to embryos at days 1 to 3.

In the routine production of bovine embryos described by Trounson *et al.* (1994) in Australia, four embryos in 30 μl droplets of SOF + MEM amino acids and 8 mg ml^{-1} BSA (under lightweight paraffin oil) were cultured from 24 hpi onwards in a gas atmosphere of 5% CO_2, 7% oxygen and 88% nitrogen. Data for more than 900 oocytes show a 39% blastocyst yield. It should be noted that this SOF medium needs to be replaced every 48 h to prevent toxic accumulation of ammonium as a result of amino acid degradation.

5.5.2 *Use of conventional culture media*

The culture of cattle embryos in the conventional media developed for cell culture led to inconsistent results. There were some exceptions, such as the SOF reports, but it was clear that for most embryos a development block occurred at the eight-cell stage which precluded *in vitro* development beyond this stage. In addition, only a minority of long-term cultured embryos showed regular cleavage and even that minority almost always failed to develop when transferred to recipients (Wright and Bondioli, 1981). The precise requirements of the early bovine embryo for various compounds have yet to be determined.

The composition of many formulations used in early embryo culture has been relatively simple, many being based on KRB salt solution, with the energy sources pyruvate, glucose and lactate. Such media are conventionally supplemented with a protein source such as FCS or BSA. The use of complex media such as M-199 and Ham's F-10, which contain mixtures of amino acids and other ingredients, may not always provide the optimal conditions for development. It was shown by Loutradis *et al.* (1987), for example, that hypoxanthine, at levels that are present in complex media such as Ham's F-10 (6–30 μM), is capable of blocking development in random-bred mouse embryos.

In studies with rabbit and hamster embryos, vitamin requirements have been examined (Kane, 1989; Kane and Bavister, 1988; Kane *et al.*, 1988, 1992). It is clear that inositol can stimulate protein synthesis and blastocyst expansion (Fahy and Kane, 1992); elsewhere, work by Bredbacka and Bredbacka (1993) in Finland also suggests that inositol at a concentration of 2 mM may be beneficial for short-term storage of cattle morulae and blastocysts in Ham's F-10 medium. In Australian Tsai and Gardner (1994) have shown that nicotinamide inhibits the development of mouse embryos and suggest that the role of vitamins in human embryo culture should be examined. In Japan, Takahashi *et al.* (1993a, b) found evidence that cysteamine, a low-molecular-weight thiol compound, can support *in vitro* development of bovine embryos in M-199 in the absence of somatic feeder cells; this effect is associated with intracellular glutathione synthesis.

In the view of some, consideration should be given to those compounds in media that may be inhibitory; this may be as important as seeking novel factors which may be required by the early embryo (Carney *et al.*, 1990a, b). A review by Bavister (1992a, b) dealt with the metabolism and nutrition of the early embryo. There is clearly a wide variety of epigenetic factors, including ions, energy substrates, amino acids, vitamins, growth factors, cytokines and hormones that play a part in early embryonic development. As well as that, it may be not only that optimal embryo development *in vitro*

Table 5.5. Effect of cleaning procedure on subsequent developmental competence of oocytes matured and fertilized *in vitro* (from Vergos, 1990).

		Presumptive zygotes cultured in TALP medium				
		48 h after IVF			72 h after IVF	
Treatment	Oocytes to IVF	CR (%)	CI (%)	4-cell (%)	< 8-cell (%)	> 8-cell (%)
Citrate-cleaned	73	55 (75.3)	40 (72.7)	28/40 (70.0)	16/28[a] (57.1)	12/28 (42.8)
Mechanically partially stripped	68	53 (77.9)	39 (73.5)	26/39 (66.6)	7/26[b] (26.9)	19/26 (73.1)

There was a significant difference in the percentage of embryos that proceeded to the greater than 8-cell stage at 72 h post-IVF when partial stripping of cumulus cells was carried out. It was also observed that embryo quality after IVC was poorer in the citrate-cleaned group.
a vs. b, $P < 0.05$

and *in vivo* requires culture media to contain many compounds, some at present unidentified, but also that successive stages of embryonic development may require changing metabolite levels (Gardner and Sakkas, 1993). It may be unrealistic to think of a single culture system as being adequate to cover all stages from one cell to blastocyst.

In Dublin, isolated instances of IVM/IVF bovine embryos cultured in M-199 passing through the eight-cell block have been recorded (Lu *et al.*, 1988a); such instances may have been the result of the activity of cumulus cells attached to such cells. Studies by Vergos (1990) showed evidence of a much higher percentage of zygotes progressing beyond the eight-cell stage when such zygotes were not fully denuded (see Table 5.5). When presumptive cattle zygotes are recovered from live cows (by flushing the oviducts after ovulation/fertilization), it is likely that they are invariably denuded and consequently incapable of receiving crucial support from cumulus cells. In Sweden, Shamsuddin *et al.* (1993a, b) were also able to develop IVM/IVF zygotes to the blastocyst stage in M-199 without somatic cells; the quality of such embryos, however, was significantly impaired in the absence of viable oviductal cells. Of the complex culture media, in the experience of some workers, M-199 has been reported to be preferable to Ham's F-10 (Lu *et al.*, 1988a; Kuwayama *et al.*, 1989). In Guelph, Semple *et al.* (1993) have shown bovine blastocyst yield to be comparable in M-199 and Menezo's B medium (41.1% vs. 35.4% of cleaved oocytes).

Effect of 2-mercaptoethanol

A low-molecular-weight thiol compound, 2-mercaptoethanol, is employed in certain tissue culture media. In one report, this compound has been found by Japanese workers to be as effective as coculturing when employed as an additive in M-199 medium (10% FCS) with bovine embryos (Takahashi *et al.*, 1992).

5.5.3 *Incubator problems*

The effectiveness of a cattle embryo production system is dependent on the quality of the different steps (IVM/IVF/IVC) involved. A study by Avery and Greve (1992a, b) has shown that consideration must even be given to the type of incubator employed in the work; they record a highly significant difference in blastocyst yield (51% vs. 6%) according to the incubator model employed. However, the Danish workers were unable to explain the reason for the observed differences. Problems have also been evident in work in Dublin which were believed to be the result of incubator usage. This can arise from numerous door openings when the incubator is used by several workers; such usage can disturb both the quality of the gas atmosphere and the temperature.

5.5.4 *Use of chemically defined media*

Bavister (1992c) has drawn attention to the need to investigate factors influencing early embryo development by way of well-defined, controllable culture conditions; this should be the means of effectively formulating new media capable of supporting embryo development and viability. There are already data in the literature showing that cattle embryos derived from IVM/IVF can develop *in vitro* to the morula stage in chemically defined, protein-free media, with no apparent advantage evident in using somatic feeder cells or serum (Pinyopummintr and Bavister, 1991). According to Bavister (1993), numerous laboratories using culture media without somatic cell support have obtained embryo development results equal to, and in some cases better than, those reported with coculture. In human embryo culture, Van Blerkom (1993) has also found no beneficial effect on development in the presence of Vero cells (epithelial cell line derived from green monkey kidneys used in human vaccine production). This author supports the view of Bavister (1993) that, with rigorous quality control of medium preparation and the elimination or modification of certain biochemical components, there may be no advantage in using a coculture system.

For effective blastocyst development, however, Pinyopummintr (1990) and Pinyopummintr and Bavister (1991) did find that serum factors were required. Serum was beneficial in stimulating morula compaction and blastocyst formation. It was also evident that the mean number of nuclei per blastocyst was significantly greater when serum was used. In Colorado, Seidel *et al.* (1991a, b) used a defined, protein-free IVC medium but found that embryos cultured with serum and/or coculture with somatic cells developed more readily to the blastocyst stage; in further work with that medium, lowering oxygen from 20% to 5% and the addition of certain chelators (5 μM EDTA) were found to be beneficial (Olson *et al.*, 1993).

Madison workers note that responses with cattle embryos are analogous to those obtained with rabbit and hamster embryos, in which early stages are extremely sensitive to culture conditions and require a relatively simple medium to undergo cleavage. As reported by Monis and Bavister (1990a, b), the development of four-cell hamster embryos *in vitro* is strongly influenced by specific components of the culture environment (i.e. amino acids, phosphate and carbon dioxide). On the other hand, components such as vitamins, fatty acids, growth factors and the like, present in serum, are believed to become more essential during the later stages of embryonic development.

There have been studies with mouse embryos which have examined the effect of FCS, recovered at different fetal stages (Olar and Potts, 1993); this was done in the knowledge that the glycosaminoglycan, hyaluronic acid (HA), plays a role in embryonic morphogenesis and that HA is secreted in response to a factor found in the serum. There may be qualitative differences in the HA content of FCS according to fetal age, which may contribute to variability in studies using such serum. Elsewhere, it has been suggested that cow serum recovered from superovulated donor cows 7 days after breeding may have merit as an ingredient in IVC medium (Suzuki, 1993).

In attempting to fully understand the needs of the early bovine embryo, it is clearly essential to know the precise composition of the medium in which the embryos are being cultured. It is essential to identify the important compounds in materials such as serum in order that these molecules may eventually be replaced with pure substances in the IVC medium. In view of the fact that conventional culture media contain many different ingredients, some authors have now developed simple, serum-free media.

In Madison, workers (Rosenkrans *et al.*, 1990; Rosenkrans and First, 1991) have concluded from studies with such a simple medium, Charles Rosenkrans 1 (CR1), that essential and non-essential amino acids are beneficial to bovine embryo development *in vitro* in the absence of feeder cells. A comparison between CR1 medium supplemented with amino acids and a buffalo rat liver (BRL) cell monolayer gave a similar outcome in terms of embryo yield (Moreno and Westhusin, 1993). A study by Takahashi and First (1992) had earlier shown that one-cell bovine embryos can be successfully cultured beyond the 8–16-cell block to the blastocyst stage, using a chemically semi-defined medium without glucose but containing pyruvate, lactate, amino acids and BSA.

A defined serum-free IVC medium was used by Moore and Bondioli (1993) to culture cattle embryos from the one-cell to the blastocyst stage. In an analysis of bovine oviductal fluid, these authors found glycine and alanine to be the most predominant amino acids; glycine and alanine supplementation of their IVC medium improved development. In Sweden, Shamsuddin *et al.* (1993a, b) used a cell-free, serum-free culture system (M-199 + BSA, insulin, transferrin and selenium); this semidefined medium supported development of embryos to the blastocyst stage as readily as a complex coculture system.

The effect of glucose on the development of bovine embryos, as already mentioned, is clearly dependent upon their developmental stages (Kim *et al.*, 1993a, b). These workers found that the presence of glucose in their semi-defined IVC medium (M-199 with BSA, lactate and pyruvate) inhibited early development, especially to the eight-cell stage; when added at 5 days after IVF, glucose improved development to the blastocyst stage. It may even be a question of the concentration of particular constituents of the medium. There is now ample evidence (Lawitts and Biggers, 1991, 1992) that most media used to culture early mouse embryos contain too much sodium chloride (NaCl). The same authors also note reports suggesting that the NaCl concentration in IVC media for pig embryos may be too high. Just why the NaCl concentration adversely affects embryonic development in the pig is not readily apparent (Beckmann and Day, 1993).

Chemically defined media used in many IVC systems usually contain a protein source, such as BSA. Yet the contribution of protein, or amino acids, to the early development of the embryo remains poorly defined. A study by Flood and Shirley (1991) examined the possibility that a principal function of protein (e.g. BSA) in an IVC medium is the chelation of embryotoxic substances. These workers showed that the rate of embryo development in BSA-supplemented media was greater than that in protein-free media only when media were prepared with the least purified water samples. Such results support the view that protein can protect embryos during culture by negating the effects of embryotoxic substances in the media. As to amino acids, it is evident that the presence of certain ones (glycine and alanine) in a simple IVC medium can markedly affect bovine embryo development (Moore and Bondioli, 1993).

In Japan, Matsuoka *et al.* (1992) used M-199 supplemented with Na-pyruvate and insulin to culture bovine embryos; they reported that blastocyst formation was greatest when superovulated cow serum (obtained on day 7 from superovulated cattle) was employed. The indication was that a growth factor was present in this serum.

Biphasic effect of serum supplementation

In serum-supplemented IVC media, the timing of the addition of serum, relative to embryo stage, may be important. A study reported by Pinyopummintr and Bavister (1991) demonstrated that serum-supplemented IVC medium had a biphasic effect, inhibitory at the first cleavage stage and stimulatory at the morula/blastocyst stage, on bovine embryo development *in vitro*. In a further report, Pinyopummintr and Bavister (1993) found that adding serum was most beneficial at 2 days post-insemination (when cattle embryos are at four- to eight-cell stages); the same workers found no difference between heated and non-heated serum in the IVC medium.

5.5.5 *Temperature and other environmental factors*

The critical importance of temperature regulation has already been mentioned in dealing with maturation and fertilization of the bovine oocyte (see Table 2.17 in Section 2.10.2). In this, it is evident that IVC at 40°C may lead to a significantly and substantially lower yield of blastocysts or hatched blastocysts than culture at 37–39°C (Wang, W.L., *et al.*, 1991). In the USA, on the other hand, Alfonso and Hunter (1992a, b) found evidence of significantly higher cleavage rates in early bovine embryos cultured at 37°C than at 39°C. In studies with cat embryos, a difference of 1°C was found to have a marked effect on development (Johnson *et al.*, 1990); significantly more embryos developed at 38°C than at 39°C.

Gas phase considerations

In terms of the most appropriate gas phase for embryo culture systems, the conventional 5% carbon dioxide in air environment is employed to provide a pH of about 7.4 with a 26 mM $NaHCO_3$ buffer; the same gas phase can be employed with higher levels of $NaHCO_3$ where higher pH values are required. Cambridge studies in the early 1970s suggested that an embryo culture medium (SOF) in combination with a particular gas phase (5%) may enable early cleavage cattle embryos to develop *in vitro* to the blastocyst stage (Tervit *et al.*, 1972). In the USA, Wright *et al.* (1976a, b) found some evidence suggesting that a 5% oxygen level may be preferable to the normal 20% level in air.

There are also a number of reports drawing attention to the oxygen tensions found in the oviduct; these are generally much below the 20% concentration. Some measurements have shown the oxygen concentration in the oviductal and uterine lumen of the rabbit to be about 40% or less of the atmospheric oxygen level (Bavister and Fischer, 1991). However, as noted in Section 4.8.3, there is no general agreement on the merit of using low oxygen tension in conjunction with IVC media.

In the USA, Voelkel and Hu (1992a) used two oxygen concentrations within two different IVC systems. When a bovine oviductal monolayer was used, then the low oxygen tension gave the highest embryo yields; when BRL cells were employed, then the yield of cattle embryos was higher with the normal atmospheric (20%) level of oxygen. The authors suggest that BRL cells are possibly adapted to the oxygen concentrations in which they were established and maintained. In Japan, Nagao *et al.* (1993) found that a low (5%) oxygen tension was most beneficial to bovine embryo development in a protein-free medium and in the absence of bovine oviductal cells.

The oxygen concentration most effective for culturing rabbit zygotes to hatching blastocysts was found to be about 5% in a study by Li and Foote (1993); these authors conclude, however, that certain types of coculture systems may require different oxygen concentrations for optimal results. The development of rabbit embryos cultured under 5% and 1% oxygen in protein-free medium was compared by Fischer and Lindenau (1993) with the commonly employed 5% CO_2 in air (i.e. about 19% oxygen); both reduced oxygen concentrations led to significantly better embryo development than 19%.

In Ireland, the effect of both oxygen and carbon dioxide levels on the culture of bovine embryos was examined (Wang, W.L., *et al.*, 1992); as evident in Table 5.6, it was found that 10% CO_2 in air, 5% CO_2, 5% O_2, 90% N_2 and 5% CO_2, 10% O_2, 85% N_2 had no advantage over the conventional 5%

Table 5.6. The effect of gas phase on the *in vitro* development of bovine embryos derived from *in vitro* maturation and fertilization of ovarian oocytes (from Wang, W.L., *et al.*, 1992).

Treatments		No. embryos (2–8 cells)	Blastocysts (% of cleaved)	Hatched blastocysts (% of blastocysts)
	1	349	159 (45.6)[b]	115 (72.3)
Exp. I	2	349	150 (42.9)[ab]	105 (70.0)
	3	349	126 (36.1)[a]	87 (69.1)
	4	349	141 (40.4)[ab]	95 (67.4)
	1	396	222 (56.1)	172 (77.5)
Exp. II	2	436	235 (53.9)	180 (76.6)
	3	463	237 (51.2)	166 (70.0)

Exp. I: (1) 5% CO_2 with 25 mM $NaHCO_3$; (2) 5% CO_2 with 50 mM $NaHCO_3$; (3) 10% CO_2 with 25 mM $NaHCO_3$; (4) 10% CO_2 with 50 mM $NaHCO_3$.
Exp. II: (1) 5% CO_2 in air; (2) 5% CO_2, 10% O_2, 85% N_2; (3) 5% CO_2, 5% O_2, 90% N_2.
a, b: $P < 0.05$.

carbon dioxide in air gas phase used as standard in the IVF laboratory. In the USA, working with rabbit embryos, concentrations of carbon dioxide higher than the 5% generally in use resulted in higher-quality blastocysts (Hallden *et al.*, 1992); the exact mechanism by which CO_2 exerts its action is not known. It may be an effect on intracellular pH, which is known to be a potent regulator of metabolism.

In pigs, Hagen *et al.* (1991) have shown that embryo development can be achieved in a simple medium with ambient air. Other investigators have examined the possibility of culturing mouse embryos in a chemically defined medium in an atmosphere of air rather than in the usual carbon dioxide gas mixture (Ali *et al.*, 1993); embryos were cultured in HEPES-buffered CZB medium. It was found that medium containing 10 mM $NaHCO_3$ supported embryo development as readily as medium incubated in 5% CO_2 in air.

Some workers have examined the effect of embryo culture systems on gas concentrations available to embryos. The standard technique for culturing cattle embryos is the 'microdrop' method, in which the embryos are placed in small droplets of culture medium under oil. Oxygen is supplied from the gas phase above the oil and must be transported through the oil and medium phases to the embryos for their consumption. Calculations by Baltze and Biggers (1991) suggest that droplets containing as many as 50 mouse embryos should have an adequate oxygen supply to meet their requirements.

Effect of light

Concern may be expressed about the possibility that the bovine zygote and early cleavage embryo may be adversely affected by light, particularly ultraviolet light. The inclusion of *para*-aminobenzoic acid (PABA), which is a naturally occurring compound in mammals, has been suggested as a means of providing protection (Robertson *et al.*, 1988). There are data indicating that rabbit morulae may be adversely affected by 24 h exposure to visible light (320–740 nm, 1600 lux) in reports by Schumacher and Fischer (1988) and Schumacher *et al.* (1991a, b). Clearly, the general principle should always be observed of keeping embryos in darkness rather than in light. In Japan, Umaoka *et al.* (1992a) compared the development of hamster one-cell embryos collected under dark and light conditions; severe light exposure adversely affected the development of embryos at all stages.

Preparation and quality control of IVC media

The preparation and quality control of IVC media requires time and expertise. It is normal for such media to be prepared at regular intervals and not to be used in IVC when more than 2 weeks old. In human IVF, there is evidence that Ham's F-10 medium can be stored in a refrigerator at 4°C for 6 months without compromising one-cell mouse embryo development *in vitro* (De Silva, 1993). In the operation of commercial cattle IVF facilities, there is need for similar information for the various media that are employed.

5.5.6 *Protection from oxygen toxicity*

Bovine early embryos cultured *in vitro* may be susceptible to oxidative stress. The development of cattle embryos *in vitro* was improved, in the studies of Rosenkrans *et al.* (1991) with the CR1 with amino acids (CR1aa) medium, by reducing the oxygen concentration to 5%. It is possible that the exposure to 20% oxygen and light during routine embryo manipulations may lead to the generation of superoxide radicals. According to Rieger (1992), an understanding of the mechanisms of production and elimination of oxygen radicals in embryos may result in significant improvements in the effectiveness of IVC systems. The same author suggests that oxygen radicals may be a common factor in the known detrimental effects of glucose and beneficial effects of glutamine on early embryonic development.

In Texas, Pabon *et al.* (1989) extended earlier findings with mouse embryos, in which they noted that oxygen toxicity occurred in protein-free Ham's F-10 culture medium, to demonstrate that a short exposure to atmospheric oxygen concentration (about 20%) had an adverse effect on early development in mouse embryos cultured in a 5% oxygen atmosphere. Results of work by Takeuchi *et al.* (1989) also suggested that oxygen has a toxic effect, mainly at about the two-cell stage, on mouse embryos *in vitro*. Subsequently, Umaoka *et al.* (1991, 1992b) attempted to investigate the effects of oxygen toxicity on the development of mouse embryos *in vitro*, using high (20%) and low (5%) oxygen tensions, with or without superoxide dismutase (SOD). Embryonic development was greatly affected by atmospheric oxygen throughout the early embryonic stages and this harmful effect was prevented by culturing embryos under low oxygen conditions and in the presence of the antioxidant enzyme SOD.

In Japan, Noda *et al.* (1991) demonstrated that SOD attenuates the two-cell block in mouse embryos cultured *in vitro* and had beneficial effects on subsequent embryo development. Other work in that country by Nonogaki *et al.* (1992) led them to conclude that protecting mouse embryos from oxygen radicals at all stage of IVF and embryo culture can improve culture efficiency. In rats, Kishi *et al.* (1991) record evidence that under low (5%) oxygen concentrations there was a permissive effect on embryo development in that species. Working with hamsters, Umaoka *et al.* (1993) also found evidence that the developmental block in that species may involve oxidative stress, which may result from exposure to high oxygen concentration and light during the manipulation of embryos. Some of the harmful effects of reactive oxygen species on cells are dealt with in a review by Peura (1993).

The inclusion of low-molecular-weight thiol compounds such as β-mercaptoethanol and cysteamine in IVC medium has been found to enhance development of six- to eight-cell cattle embryos to the blastocyst stage in the

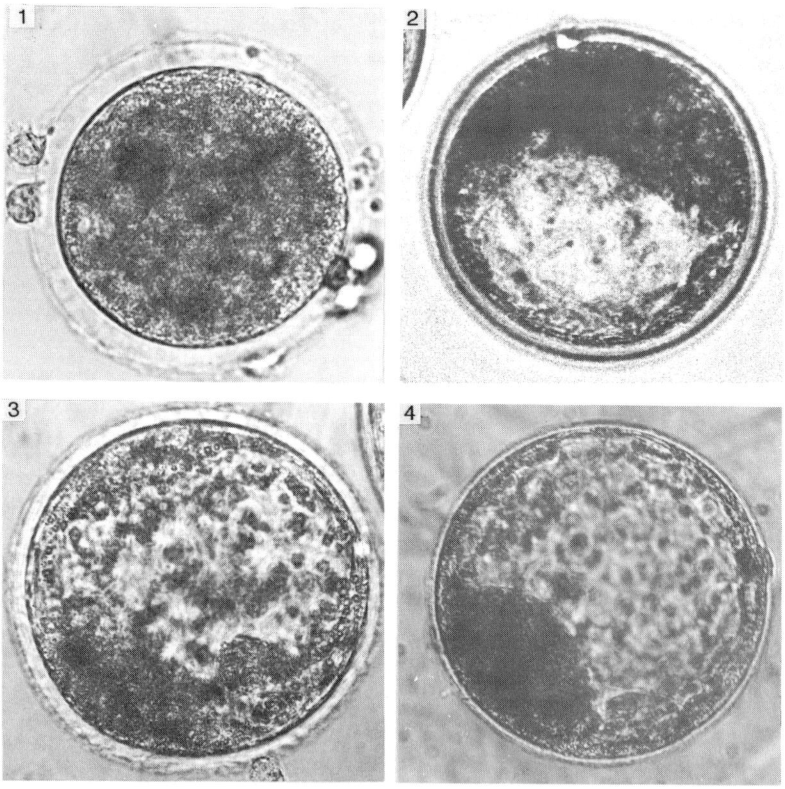

Fig. 5.7. Stages in the progress of the bovine blastocyst. A late morula/early blastocyst is shown in (1) with a blastocoele just starting to form; in (2) this day-7 blastocyst is showing a well-defined blastocoele; in (3) the blastocyst has reached the expanding stage, with a noticeable thinning of the zona pellucida; in (4) the blastocyst continues to expand with the zona becoming still thinner.

absence of feeder cells (Takahashi *et al.*, 1993a, b); this beneficial effect is apparently accompanied by an increase in intracellular glutathione. This would agree with evidence showing that glutathione plays an important role in protecting cells against the harmful effects of oxidative injuries. Rabbit zygotes cultured in a macromolecule-free medium have also developed more rapidly when several potential antioxidants were added to the medium (Li, J., *et al.*, 1993).

In human IVF, early embryos cultured *in vitro* may be susceptible to oxidative stress. A study by Taylor *et al.* (1992a) examined the addition of the antioxidants glutathione and SOD to the culture medium; results indicated that glutathione was without effect but SOD promoted blastocyst development. The effect of taurine and superoxide dismutase on rabbit zygotes *in vitro* is dealt with in a paper by Li and Foote (1992); both antioxidants facilitated more rapid cell growth and the development of expanding blastocysts.

Water quality

Among the factors of crucial importance in embryo culture is water quality in the preparation of IVC media; a number of reports may be instanced to illustrate this. A report by Keefer *et al.* (1991), for example, has shown that no development of IVMFC-derived bovine embryos (to the blastocyst stage) occurred when purchased deionized distilled water was used in comparison with water from an ultrafiltration system (Milli-Q). The various stages through which the bovine embryo passes in progressing from the early to the expanded blastocyst are shown in Fig. 5.7.

5.6 Coculture with Bovine Oviductal Cells

5.6.1 Historical

Studies in different countries during the 1960s and 1970s showed that *in vivo*-produced cattle and sheep embryos were generally unable to cleave in culture beyond the 8–12-cell stage. Despite many attempts to stimulate further cleavage by enriching the classical culture media with cattle, sheep or fetal calf serum, little or no success was obtained.

In the past 8 years, however, reports for sheep (Gandolfi *et al.*, 1986; Rexroad and Powell, 1986, 1987, 1988a, b, 1989, 1991a, b; Gandolfi and Moor, 1987; Fukui *et al.*, 1988b; Guillomot *et al.*, 1988), cattle (Eyestone and First, 1986, 1988, 1989; Eyestone *et al.*, 1987a, b; Lu *et al.*, 1987a, 1988a; Fukui and Ono, 1988; Le Guienne and Thibier, 1988; Le Guienne *et al.*, 1988; Fukui *et al.*, 1989; Marquant-Le Guienne *et al.*, 1989; Vergos *et al.*, 1989a, b; Ellington and Farrell, 1990; Ellington *et al.*, 1990c, d, e, f; Kim, C.I., *et al.*, 1990; McCaffrey *et al.*, 1990, 1991, 1992a, b, c; McCaffrey and Sreenan, 1991a, b; Fontes *et al.*, 1991, 1992b), pigs (White *et al.*, 1988, 1989), horses (Battut *et al.*, 1990, 1991), rabbits (Carney *et al.*, 1990a, b), mice (Ouhibi *et al.*, 1990; Sakkas and Trounson, 1990) and humans (Bongso *et al.*, 1990, 1991b, 1992) have all confirmed that it is possible to coculture the zygote or early embryo with oviductal cells and achieve development to the morula/blastocyst stage. In Belgium, Mermillod *et al.* (1991) have shown that rabbit oviduct cell monolayers can support bovine embryo growth; however, monolayers prepared from BOEC gave higher embryo yields. There are even studies showing that the bovine oviductal monolayer provides a suitable environment for the development of human embryos prior to transfer or freezing (Wiemer *et al.*, 1993).

Gandolfi *et al.* (1986) and Gandolfi and Moor (1987) were among the first to demonstrate that the eight-cell developmental block did not occur if sheep early embryos (*in vivo*-produced) were cultured on oviductal cell monolayers, or even on fibroblasts. These authors did much to establish the oviductal cell coculture system as equal or superior to alternative culture procedures available at that time.

5.6.2 Setting up the bovine oviductal monolayer

The bovine oviduct provides the microenvironment for the transport and final maturation of gametes, fertilization and early embryonic development. In setting up the bovine oviductal epithelial cell (BOEC) monolayer, researchers have attempted to provide a similar microenvironment in the laboratory. The methods employed in obtaining bovine oviductal cells have varied from collagenase digestion procedures (Eyestone *et al.*, 1987a, b; Lu *et al.*, 1987b, 1988a; Joshi, 1988) to simply flushing the oviducts of the live animal (McCaffrey *et al.*, 1990, 1991) or recently slaughtered animal with an appropriate medium (Fukui and Ono, 1988; Vergos, 1990). A procedure using trypsin for the isolation and culture of oviductal cells from several species, including cattle, is described by Ouhibi *et al.* (1988).

In early work, it was not evident how important it might be to obtain oviductal cells in the period corresponding to that when zygotes/embryos are normally in the bovine oviduct (early luteal phase). Studies by Vergos (1990) suggested that cycle stage may not be important. In studies using early bovine embryos (*in vivo*-produced) recovered by laparotomy 48 h after oestrus, McCaffrey *et al.* (1990) flushed oviductal cells from cattle, either on the day of embryo collection or 3 days previously (pro-oestrus phase of the cycle). No difference was evident in the yield of embryos between the two cycle stages. Results reported by Thibodeaux *et al.* (1991, 1992a) in the USA also showed that oviductal cells isolated at early and late luteal phases of the oestrous cycle were similarly capable of supporting early embryonic development.

However, the normality of epithelial cells in the BOEC monolayer system may be open to question (Gandolfi *et al.*, 1989b). Oviductal cells grown in a monolayer undergo rapid dedifferentiation; they flatten on the culture surface, losing the characteristic columnar shape and cilia. In culture, oviductal cells begin to divide regularly, in contrast to the *in vivo* state, where mitotic figures are rare. Such changes may result in the loss of steroidogenic effect on oviductal cells, possibly through inactivation or loss of receptors. For such reasons, the BOEC monolayer may not be the most appropriate substitute for the tubal environment. In Canada, Pollard *et al.* (1989), in the culture of

IVMF-derived bovine embryos, have shown an enhancement of BOEC function and morphological differentiation when cells were cultured on a Matrigel–Millicell substratum; this procedure was designed to maintain polarity, secretory character, morphology and steroid responsiveness of the BOEC cells.

5.6.3 *Culture of* in vitro- *and* in vivo-*produced embryos*

Some workers have failed to achieve the same results with monolayer culture systems using IVMF oocytes as with oocytes produced and fertilized in the live animal. Eyestone and First (1989b) found IVC of embryos, derived from ovulated oocytes, on their BOEC monolayer gave similar yields (42–46%) to those found using the sheep oviduct (37–53%). When IVM oocytes were used, the yield of embryos from the monolayer was 22% and from the sheep oviduct, it was 20–28%. Such evidence suggests a defect in the maturation or fertilization of the IVM oocytes.

In Belgium, Van Soom and Kruif (1992) cultured *in vivo*- and *in vitro*-produced bovine embryos under identical conditions (oviduct coculture system). They found that IVF embryos suffered from inadequacies in development compared with the *in vivo* embryos. There were obvious morphological differences between the two embryo categories.

Work in Ireland has also examined the development of cattle embryos from the one-cell to the blastocyst stage according to whether they have been produced *in vitro* (IVM/IVF) or *in vivo* (recovered from the oviducts). Clear evidence has been presented by McCaffrey and Sreenan (1991a, b) and McCaffrey *et al.* (1992c) showing that their BOEC monolayer culture system was highly effective in promoting development of *in vivo* embryos, but much less successful with those produced by *in vitro* methods (see Table 5.7). In fact, these workers found that, irrespective of culture system, medium or supplements employed, the final yield of viable IVF embryos was always below that of the *in vivo* embryos. It seems unlikely that the explanation for the difference lies in the possibility that the tubal embryos had access to oviductal

Table 5.7. Development of *in vitro* fertilized (IVF) and *in vivo* embryos in co-culture with bovine oviductal cells (from McCaffrey *et al.*, 1992a).

	In vivo	Day 1 IVF	Day 2 IVF
No. of ova	41	77	52
Initial ovum cell no. (M ± SEM)	2.47 ± 0.52	1.04 ± 0.03	4.9 ± 0.35
No. of morulae/blastocysts (%)	32 (78)[a]	28 (36)[b]	13 (25)[b]
No. of grade 1 and 2 mor./blast. (%)	20 (62)[a]	9 (32)[a]	3 (23)[b]

Values within rows with different superscripts are different, $P < 0.01$.

factors which may have enhanced their subsequent performance on the BOEC monolayer. No such evidence, for example, was found by Lu (1990) when presumptive cattle zygotes were transferred to the sheep oviduct 6 h after insemination. No such advantage was apparent in the studies of Sharif (1992) after fertilizing IVM cattle oocytes in the isolated mouse oviduct. The strong indication is that the IVM/IVF embryos used by these workers may have had less developmental potential and that this was likely to have stemmed from qualitative differences between oocytes matured *in vivo* and *in vitro*.

It should also be noted that the high yields of good-quality embryos from the BOEC monolayer system employed by McCaffrey *et al.* (1990, 1991, 1992a), when *in vivo* embryos were cultured, clearly demonstrates the effectiveness of this system. If zygotes and embryos are of the quality normally associated with the live cow, they appear to be quite capable of developing to the blastocyst stage in this particular culture system.

5.6.4 *Culture media employed with BOEC monolayers*

Early work in Ireland showed that M-199 was significantly more effective than Ham's F-10 in supporting embryo development when used in conjunction with a BOEC monolayer system (Lu *et al.*, 1987b, 1988a). The same work found one form of bovine serum (oestrous cow serum (OCS)) to be superior to others examined. Further studies in Ireland by Wang *et al.* (1989), using the BOEC monolayer, showed that M-199 was superior to Ham's F-10 and OCS was superior to FCS in the culture of early embryos (see Table 5.8).

In France, Marquant-Le Guienne *et al.* (1989) cultured bovine embryos on BOEC monolayers in B2 medium supplemented with OCS. The coculture of IVM/IVF embryos with bovine oviductal cells, as reported from Canada by Xu *et al.* (1992b), suggested that Menezo's B2 medium with 10% OCS was superior to other media tested. In the majority of reports on BOEC culture systems that have appeared during the past 5 years, however, M-199 has been the medium employed, usually supplemented with FCS.

Table 5.8. The effect of media and sera on the *in vitro* development of bovine embryos to the blastocyst stage (from Wang, 1991).

Treatments	No. embryos (2–8 cells)	Blastocysts (%)
M-199 + 20% OCS	146	40.4 (59/146)[c]
M-199 + 20% FCS	140	25 (35/140)[b]
HF_{10} + 20% OCS	131	20.6 (27/131)[b]
HF_{10} + 20% FCS	123	2.4 (3/123)[a]

Total of nine trials.
a b c: $P < 0.01$.

Serum supplementation

As well as the particular form of bovine serum employed in the culture media used in coculture systems, various workers have examined the question of serum concentration. In Ireland, Wang (1991) used varying concentrations of OCS in the coculture medium and concluded that 10% was appropriate (see Table 5.9).

In Arkansas, Gliedt *et al.* (1991) sought to determine if sequential supplementation of IVC medium (M-199 with BOEC) with serum collected from cows on days of the oestrous cycle corresponding to the developmental stages of the embryos *in vitro* was beneficial; no advantage was apparent.

In Japan, Nagao *et al.* (1991) used M-199, supplemented with 0.3% polyvinyl pyrrolidone (PVP) as the macromolecular supplement, in coculture with BOEC; the PVP proved to be as effective as FCS. In Belgium, Fontes *et al.* (1992a) found that OCS was superior to PVA as a supplement to their coculture BOEC system.

Table 5.9. The effect of oestrous cow serum (OCS) concentration on the *in vitro* development of bovine embryos to the blastocyst and hatched blastocyst stage (from Wang, 1991).

OCS conc. %	0	5	10	20
No. of embryos	359	331	366	444
Blastocysts	94	152	191	219
%	26.2^{a}	45.9^{b}	52.2^{b}	49.3^{b}
No. of blastocysts	92	141	177	202
Hatched blastocysts	69	109	157	165
%	75.0^{A}	77.3^{A}	88.7^{B}	$81.7^{A,B}$

Total of five trials; the basic medium was medium 199 with Earl's salts.
a b: $P < 0.01$, AB: $P < 0.01$.

5.6.5 *BOEC: monolayers and other culture systems*

Working with bovine oviductal cells, Xu *et al.* (1992b) describe a method whereby the BOEC remained in the form of a suspension and not a monolayer; the cells remained viable for at least 10 days, as judged by ciliary movement. The use of such a cell suspension, which is different from the conventional monolayer, may have contributed to the effectiveness of their coculture system. It may be relevant to note that Wiemer *et al.* (1991b), also working with BOEC, observed that the rate of IVF embryo development was closely related not only to the age of their BOEC monolayers but to the level of biological activity within the cells that made up the monolayers. They showed that granulosa and oviductal cell cultures were able to sustain a high rate of embryonic development for 72 h; the reduced rate of bovine embryonic development after this time may have been the result of a reduction in cell activity after the formation of the confluent monolayer. They speculated that the ability of the oviductal cells to maintain their differentiated state and to secrete embryotrophic factors may be lost after some time in culture.

5.6.6 *Additional factors employed with BOEC monolayers*

There are a number of reports setting out information on work with growth factors in conjunction with a BOEC monolayer culture system. One of the earliest was a study reported by Marquant-Le Guienne *et al.* (1989), in which they showed that TGF-β stimulated mitotic activity in ICM cells. Determination of the total cell number, as well as the number of ICM cells, may provide a useful standard against which IVC treatments can be compared. It was not possible to be certain whether the TGF-β effect was direct or indirect, via the stimulation of other growth factors from the cells used in coculture. The evidence does support the view, however, that growth factor receptors are present in the ICM cells.

In Ireland, Jiang (1991) used TGF-β ($3\,\mathrm{ng\,ml^{-1}}$) with various coculture systems, including the BOEC monolayer, and found evidence of increased cell numbers in the blastocyst (see Table 5.10); there was no significant effect,

Table 5.10. Effect of TGF-β on development of embryos cultured on different monolayers (from Jiang, 1991)).

			Eggs developed to		No. of cells in BL*	
Layer	TGF-β	Eggs cult.	Blastocysts [%]	Hatched blastocysts [%]	ICM (mean ± SE)	Trophoblasts (*n*)
G	–	270	99 [36.7]	54/73 [74.0][a]	68 ± 17.8[A]	89 ± 19.1 (26)
	+	311	107 [34.4]	63/73 [86.3][b]	109 ± 13.4[B]	104 ± 12.8 (31)
O	–	267	86 [32.2]	40/56 [71.4][a]	64 ± 11.7[A]	87 ± 13.3 (29)
	+	305	103 [33.8]	69/79 [87.3][b]	114 ± 15.8[B]	97 ± 16.4 (24)
U	–	245	70 [28.6]	31/49 [63.3][a]	57 ± 19.8[A]	77 ± 13.2 (21)
	+	278	83 [29.9]	40/52 [76.9][b]	92 ± 13.9[B]	97 ± 16.4 (31)
GO	–	277	95 [34.3]	51/70 [72.9][a]	67 ± 14.6[A]	87 ± 10.5 (23)
	+	324	116 [35.8]	69/78 [88.5][b]	112 ± 16.2[B]	94 ± 10.6 (38)
GU	–	281	91 [32.4]	44/62 [71.0][a]	66 ± 15.2[A]	81 ± 12.3 (29)
	+	309	105 [34.0]	45/53 [84.9][b]	98 ± 12.5[B]	106 ± 9.8 (36)
OU	–	266	79 [29.7]	40/58 [69.0][a]	59 ± 16.8[A]	77 ± 10.9 (18)
	+	294	91 [31.0]	60/69 [87.0][b]	101 ± 21.4[B]	91 ± 14.6 (22)
GOU	+	273	97 [35.5]	44/60 [73.3][a]	66 ± 13.2[A]	90 ± 11.6 (37)
	+	310	105 [33.9]	61/68 [89.7][b]	98 ± 16.9[B]	97 ± 18.1 (36)

a, b; A, B: $P < 0.05$ in pair.

* Expanding/expanded blastocysts.

G, granulosa cell monolayer; O, oviductal cell monolayer; U, uterine cell monolayer.

however, on the percentage of embryos developing to that stage, regardless of the form of monolayer used.

In a study involving several thousand bovine embryos produced by IVM/IVF/IVC methods, ranging from the one-cell stage to the hatched blastocyst, Watson *et al.* (1992a) found transcripts for TGF-α, TGF-β, insulin-like growth factor (IGF)-I, IGF-II and platelet-derived growth factor alpha (PDGF-α); receptors for PDGF-α, insulin, IGF-I and IGF-II were also evident during embryonic development. Such analyses should greatly aid an understanding of factors contributing towards early development of the bovine embryo.

In mice, the IGF family has already been implicated in early embryonic development (Harvey and Kaye, 1991, 1992a, b). An IVC system of bovine oviductal cells and bovine blood platelets was used by Thibodeaux *et al.* (1992b, 1993) to simulate events that may occur *in vivo*; the combination of BOEC and platelets resulted in a higher embryo yield than BOEC alone (29.4% vs. 35.9%). They concluded that oviductal cells may produce PDGF, which stimulates such embryonic development. PDGF was also found to enhance development of *in vitro*-derived bovine embryos in the studies of Eckert and Niemann (1993); their results suggested that PDGF is likely to be an important factor in a serum-free medium.

5.7 Coculture with Non-oviductal Cells

It has become apparent, from the many reports already published on coculture systems, that embryotrophic factors, if they exist, are not confined to the cells of a particular organ (e.g. oviduct), nor do they appear to be species-specific (Kim, H.-N., *et al.*, 1989, 1991).

5.7.1 Possible advantages of non-oviductal cells

There are several reasons why it may be preferable to employ non-oviductal cells in coculture systems. In human IVF, for example, certain cell lines (e.g. Vero epithelial cells derived from monkey kidneys) are highly controlled for viruses and other contaminants. In France, Menezo *et al.* (1990, 1992) have used Vero cells with B2 medium and 15% serum as a means of prolonging the IVC of human IVF embryos. The authors suggest that Vero cells would be safer than epithelial cells, obtained from the oviduct or uterus, in terms of incubation time of the acquired immune deficiency syndrome (AIDS) virus. Coculture of human embryos on a Vero cell monolayer and delayed transfer have been found to improve IVF results (Lai *et al.*, 1992a; Zetova *et al.*, 1992, 1993). In reporting studies dealing with coculture of mouse embryos on Vero cells, Lai *et al.* (1992b) conclude that these cells may function by metabolizing or absorbing inhibitory substances in the IVC medium rather than by secreting embryotrophic factors. The use of BRL cells has been reported in effective coculture systems; Hasler *et al.* (1994) record that different populations of BRL cells provide suitable coculture support for IVM/IVF bovine embryos.

In the early years of coculture systems, there were several papers suggesting that the beneficial effects of oviductal cells on the early embryo may be relatively specific. In sheep, for example, Cambridge workers concluded that, although cleavage of fertilized oocytes may proceed on different somatic cells, oviductal cells were apparently essential for the acquisition of full embryonic viability (Gandolfi and Moor, 1987). Elsewhere, trophoblastic vesicles, uterine cells and kidney cells (Rexroad and Powell, 1988a, b) apparently failed to stimulate *in vitro* cleavage of early-stage sheep embryos to the same extent as oviductal cells.

In Canada, on the other hand, Scodras *et al.* (1991) compared cell lines of various origins for their ability to support the development of *in vitro*-produced cattle zygotes to the morula/blastocyst stage. All the cell lines were cultured in 50 μl droplets as monolayers. Results showed that cell lines derived from a variety of tissues, not all epithelial, supported embryo development, including cell lines from a non-related (mouse) species. In The Netherlands, BOEC monolayers and BRL monolayers were found to support cattle embryo development with equal efficiency (Van Inzen *et al.*, 1992). Such findings made it clear that many types of cell can probably be used successfully in coculture. In Ireland, Jiang *et al.* (1991) arrived at that conclusion, having tested a variety of cell lines (see Table 5.11).

Further support for this conclusion came from work in Japan, reported by Goto *et al.* (1992c). They used various cell monolayers from different embryological origins and from different species (cow, chicken, mouse) without finding any significant difference in blastocyst yields. They did record, however, that embryos cultured on the BOEC monolayer developed faster than in the other coculture treatments. The Japanese workers used the granulosa cell monolayer as a control and found inexplicable differences over a 2-year period in blastocyst yield from that system (increasing from 26.9% in year 1 to 53.3% in year 2).

Other work in Japan has shown that IVMF-derived bovine embryos can be successfully developed in a coculture system using mouse embryonic fibroblast cells and CR1aa medium (Takahashi *et al.*, 1994). Working with mouse embryos and feeder-cell populations originating from the human reproductive tract, the specificity of feeder cell support was examined by Freeman *et al.* (1993); no evidence was found that such support was tissue-specific or dependent on cell type.

5.7.2 *Cumulus/granulosa cell monolayers*

Japanese workers were among the first to demonstrate clearly the potential of non-oviductal cells; they cocultured early IVM/IVF bovine embryos on a monolayer that was grown from the cumulus cells surrounding the oocyte (Fukuda *et al.*, 1988, 1990; Goto *et al.*, 1988a, b; Kajihara *et al.*, 1990; Nakao and Nakatsuji, 1990). In some of the earliest studies, about 15% of IVM/IVF embryos developed to the blastocyst stage after coculture for 7–9 days post-insemination (Goto *et al.*, 1988a); five of 19 embryos transferred in that study resulted in pregnancies. One obvious advantage in employing cumulus cells, rather than oviductal cells, is that they are readily obtainable as part of routine cattle IVM/IVF operations.

In studies reported by Nakao and Nakatsuji (1990), it was found that coculture with cumulus cells or trophoblastic vesicles promoted early cattle embryos to develop beyond the eight-cell block stage. Results were further improved by the addition of EDTA at a concentration of 0.1 mM and reducing oxygen concentration to 10%. It was also recorded that the absence of glucose and other energy substrates from the culture medium (M-199 + 2% FCS) during the earlier phases of embryo development improved subsequent embryonic development. Elsewhere in Japan, Iwasaki and Nakahara (1990a, b) reported that the average number of cells in rabbit-oviduct-cultured blastocysts was almost double that of blastocysts cultured on their cumulus cell monolayer. In Germany, on the other hand, Reintjes (1991) concluded that coculture with cumulus cells was an acceptable IVC system in comparison with the other alternatives examined. A cumulus monolayer was also successfully used in Italy by Dell'Aquita *et al.* (1992) in their studies in cattle embryo production.

As an alternative to using cumulus cells, it is possible to establish a monolayer from granulosa cells (recovered from follicles at the time of obtaining primary oocytes for IVM), using either aspiration or ovary dissection. A considerable amount of work has been carried out in Ireland using the granulosa cell monolayer, and the evidence suggests that this system is as effective as that based on cumulus cells (Chen-Lu and Lu, 1990; Jiang *et al.*, 1990; Ling and Lu, 1990; Shi *et al.*, 1990; Tan and Lu, 1990; Wang *et al.*, 1990; Yang, N.S. and Lu, 1990). In this, the granulosa cell monolayer is prepared in a 90 μl droplet under oil, using the same cells as are employed in supplementing the IVM medium. In human IVF, as well as with bovine embryos, granulosa cell coculture has been found to improve embryo development when compared with standard culture conditions (Plachot *et al.*, 1992).

As noted above (Section 5.6.6), studies in Ireland by Jiang (1991)

Table 5.11. Effect of different monolayers on *in vitro* development of IVF bovine embryos (from Jiang, H.S., *et al.*, 1991).

Cell monolayers	Embryos cultured	Embryos developed into: Blastocysts [%]	Embryos developed into: Hatched blastocysts [%]	No. of cells in blastocyst* (mean ± SE)
Without	470	62 [13.2][a]	31/45 [68.9][AB]	134 ± 14.3 ($n = 17$)
G	463	172 [37.2][c]	91/125 [72.8][B]	147 ± 17.8 ($n = 46$)
O	457	145 [31.7][bc]	80/115 [69.6][AB]	151 ± 11.7 ($n = 29$)
U	467	122 [26.1][b]	59/101 [58.2][A]	134 ± 19.8 ($n = 21$)
GO	477	155 [32.5][c]	81/112 [72.3][B]	164 ± 10.6 ($n = 43$)
GU	449	147 [32.7][c]	84/117 [71.8][B]	147 ± 15.2 ($n = 29$)
OU	466	141 [30.3][bc]	75/110 [68.2][AB]	138 ± 16.1 ($n = 31$)
GOU	482	161 [33.4][c]	88/124 [71.0][B]	146 ± 13.2 ($n = 37$)

a, bc; a, c: $P < 0.01$; b,c; A,B: $P < 0.05$, $^{*}P > 0.05$.

examined the effect of several different cell monolayers, including those prepared from granulosa cells, on *in vitro* development of early bovine embryos (two- to eight-cell). Embryo development was compared in further work with the seven different monolayers of bovine origin (granulosa, oviductal, uterine and combinations of these cells); with one exception, no difference was found in blastocyst yield (Jiang *et al.*, 1991a; see Table 5.11). The exception was between granulosa and uterine cells (37.2% vs. 26.1%), with the uterine cells giving a significantly lower blastocyst yield. In Japan, Kajihara *et al.* (1991b) found that with 3-day-old bovine embryos a combination of cumulus and uterine endometrial cells gave higher blastocyst yields than cumulus cells alone (39.4% vs. 25.9%). The coculture of bovine embryos with different cell monolayers has also been reported by Goto *et al.* (1992c).

In Germany, Berg and Brem (1989, 1990) reported that granulosa rather than oviductal cells may provide the more suitable coculture conditions (47% vs. 20%); they employed a modified Parker's medium with 20% OCS and a gas phase of 5% CO_2, 5% O_2 and 90% nitrogen. In Canada, on the other hand, Wiemer *et al.* (1991b) recorded a significantly higher blastocyst yield with oviductal cells than with granulosa cells (41.1% vs 24.6%).

5.7.3 *Use of cells in a particular sequence*

The general principle that culture with oviductal cells should be followed by culture with endometrial cells is based on what is known to occur *in vivo* in the cow, where the cow embryo spends the first 3 days in the oviduct and then enters the uterus and is sustained for some time by the histotrophe produced by endometrial cells. In Japan, Kajihara *et al.* (1990) reported on the use of a cumulus cell monolayer in combination with a uterine endometrial cell layer; they showed a 39.4% yield of bovine embryos with cumulus plus uterine and 25.9% with cumulus alone. This difference was significant and did suggest that this particular cell sequence may have merit.

In China, Xiaoxia *et al.* (1991), on the basis of their results, suggested culture to the four- to eight-cell stage should be by way of granulosa cells; from 8–16 cells onwards, it should be with oviductal cells. According to these workers, higher-quality embryos, which reached developmental stages at the same time as their *in vivo*-produced counterparts, were produced by way of this particular cell sequence. In Louisiana, Rodriguez, H.F., *et al.* (1990, 1991, 1992) examined various bilayer cell systems; it was found that, when BOEC and granulosa cell monolayers were used individually, they did not maintain development at the level achieved when both cell types were used as a single coculture unit. However, such reports have not always been supported by evidence recorded in Ireland (see Tables 5.8 and 5.9), where it was shown that no particular cell combination proved superior to the use of granulosa cells alone (Jiang, 1991; Jiang, H.S., *et al.*, 1991).

5.7.4 *Culture with trophoblastic vesicles*

Structures known as trophoblastic vesicles (TVs), which are blastocysts without the embryonic disc, can be formed by cutting the elongated blastocyst

(12-day-old) into several pieces and culturing these *in vitro* for 24 h or so, until they reform into vesicles. The procedure for preparing sheep TVs is well described in a paper by Flechon *et al.* (1986). French workers have been to the fore in reporting that the coculture of early-stage cattle embryos with TVs can promote development (Heyman and Menezo, 1987; Heyman *et al.*, 1987a); after 3–4 days of coculture, 46% of the one- to eight-cell cattle embryos developed to the morula stage as compared with 18% of the controls.

Elsewhere, in the USA, there are reports that the addition of TVs to IVC culture medium enhanced the *in vitro* viability of intact and demiembryos over that of controls (Rorie *et al.*, 1987a; Pool *et al.*, 1988a, b). There has also been work to show that TVs from 12–13-day-old cattle embryos can be successfully preserved by freezing in liquid nitrogen (Rorie *et al.*, 1987b). There have even been those who have produced a bovine trophoblastic cell line capable of supporting *in vivo*-produced two- to eight-cell cattle embryos (Machatkova *et al.*, 1988) and IVM/IVF zygotes to the morula/blastocyst stage (Scodras *et al.*, 1990).

5.7.5 *Action of feeder cells in monolayer systems*

The cells that have proved effective in coculture systems differ in their origin; they include cells taken from the oviduct, uterus, ovary, kidney, spleen and chick embryo (Imahie *et al.*, 1992). The feeder cells used have not always been from primary cultures, nor have they been limited to the same species as the embryos, nor have they always been derived from the reproductive tract. Although it has been speculated that feeder cells, such as those obtained from the oviductal epithelium, release embryotrophic factors into the embryo coculture medium (Gandolfi *et al.*, 1992a, b), it may be that such cells remove inhibitory or toxic factors present in the medium or derived from the embryo. It may also be a matter that embryo responses reflect the production and release of metabolites or growth factors by the feeder cells.

As noted above, EDTA, which chelates Ca^{2+} or Mg^{2+}, is known to play a beneficial role in embryonic development (Nakao and Nakatsuji, 1990) and chemically defined media containing low concentrations of Na^{+}, K^{+}, Cl^{-} or glucose can improve embryo development (Lawitts and Biggers, 1991). Such evidence supports the view that feeder cells may be beneficial by lowering the concentrations of these ions/glucose in the medium.

Then there is the evidence that SOD can play a valuable role in protecting embryos from oxygen toxicity. The consumption of oxygen around the embryos by feeder cells may detoxify oxygen. Differences in the efficiency of established cell lines in coculture systems may simply be due to characteristics such as metabolism or growth rate (Imahie *et al.*, 1992).

5.8 Use of Preconditioned Media

There have been several reports dealing with the culture of cattle embryos *in vitro*, using medium which has been preconditioned by exposure to oviductal/uterine or cumulus/granulosa cells. The use of such medium can have certain advantages over coculture procedures, such as eliminating the confounding presence of additional tissue and facilitating biochemical procedures employed in attempts to identify soluble embryotrophic factors (Pollard *et al.*, 1990).

5.8.1 *Preconditioning with oviductal cells*

Eyestone and First (1989b) were among the first to successfully employ preconditioned medium in the culture of cattle embryos. A further report by Eyestone *et al.* (1990) supported the view that medium conditioned by oviductal tissue was as effective as coculture in supporting bovine embryo development from the zygote to the blastocyst stage. These workers cultured embryos in 50 μl droplets of medium under oil, without changing the medium during the culture period. Optimum results were achieved after a conditioning period of 48 h and using the medium undiluted. No difference was evident in media prepared from oviductal cells collected at different stages of the oestrous cycle. The effectiveness of the preconditioned medium apparently remained stable through a single cycle of freezing and thawing (Eyestone *et al.*, 1991).

A pronounced influence of the oestrous cycle stage upon bovine oviducts used for producing conditioned medium for IVC has been reported by Harper *et al.* (1989); supernatants prepared from cells taken around oestrus were significantly more effective than those prepared at other times. In Belgium, Mermillod *et al.* (1992a, b, c) developed a serum-free M-199 medium conditioned on bovine oviductal cell monolayers and clearly demonstrated its effectiveness in cattle embryo culture. In a further report (Mermillod *et al.*, 1993a) it was again shown that this preconditioned medium improved blastocyst yield in comparison to serum-free M-199; it was apparent that the conditioned medium contained a low-molecular-weight factor(s) acting in early cleavage and a larger molecule(s) acting on blastocyst formation (Mermillod *et al.*, 1993b).

The advantages of serum-free conditioned medium over coculture, according to Massip *et al.* (1993b), include eliminating the confounding presence of cells and making the search for soluble embryotrophic factors easier; there is also the practical advantage that such media can be prepared in large quantities and used over a long period of time. A further point, arising from work reported by Vansteenbrugge *et al.* (1993), in which they found that good-quality cattle blastocysts can be produced using BRL-conditioned medium, is that the use of such medium can eliminate risks of contamination with bovine viruses.

Studies in the USA, reported by Sikes *et al.* (1993) and Hernandez-Ledezma *et al.* (1993a), have shown that BSA-supplemented CZB, a medium of low osmolarity lacking glucose but enriched in lactate and glutamine, was superior to M-199 (after conditioning by BOEC or RLC) in promoting early bovine embryo development. For optimal results, these authors suggest using CZB medium for the first 48 h and then M-199, with either BOEC or BRL conditioning throughout; their results show about 40% of oocytes developing to the blastocyst stage.

Table 5.12. Effect of preconditioned medium on the *in vitro* development of bovine embryos to the blastocyst stage (from Wang, 1991).

Treatments	Medium alone	Oviductal cell monolayers	Conditioned medium
No. of embryos	69	617	68
Blastocysts	11	174	22
%	15.9[a]	28.2[b]	31.3[b]

Total of four trials; the basic medium was M-199 + 20% OCS.
a b: $P < 0.05$.

In Ireland, studies reported by Wang (1991) showed a similar yield of embryos using preconditioned medium prepared from oviductal cells (32%) as with the standard BOEC monolayer system (28%). Details are in Table 5.12.

Vergos *et al.* (1991) reported that oviductal cells in suspension during medium preparation apparently provide factors that promote the development of higher-quality embryos than when the monolayer system is employed. Both Wiemer *et al.* (1993) in the USA and Xu *et al.* (1992a, b, c) in Canada have also provided some evidence indicating that actively dividing cells may produce substances not found with the monolayer.

In Belgium, Ectors *et al.* (1993a, b) evaluated the efficiency of different IVC methods, including the use of medium conditioned with BOEC; they found that cell number in embryos cultured in such medium was significantly lower than that for embryos cocultured with the BOEC monolayer or cultured *in vivo* in the rabbit oviduct. Other work in that country, reported by Massip *et al.* (1993a, b), included the use of BOEC-conditioned medium in the culture of IVMFC-derived blastocysts after freeze–thawing; in contrast to embryos cocultured with oviductal cells, the hatching rate of those cultured in conditioned medium was very low (6% vs. 41–54%). The authors, who employed a 5% carbon dioxide in air gas phase in their IVC system, speculate that the oviductal cells may have mitigated the adverse effects of high oxygen concentrations, probably by reducing oxygen concentration in the medium or by producing antioxidants. In Belgium, Van Soom *et al.* (1994) have reported an adverse effect on bovine embryo development using preconditioned medium overlaid with some forms of silicone oil; it was speculated that the oil may have altered the composition of the IVC medium by absorbing lipids essential for blastocyst formation. This is something which would not necessarily present a problem when cells are producing a constant supply of lipid in a coculture system.

In Thailand, Kitiyanant *et al.* (1993) have shown that both cattle and pig oviductal cells, and medium conditioned by these cells, supported bovine embryo development. It was found that protein secretion by the oviductal cells was similar in both species during different phases of their oestrous cycle. Attempts have been made to evaluate BOEC-conditioned medium for factors that promote embryo development (Long *et al.*, 1992).

5.8.2 *Preconditioning with non-oviductal cells*

In studies reported by Kobayashi, K., *et al.* (1991), conditioned media, prepared in a chemically defined medium (M-199 supplemented with insulin and aprotinin) from granulosa cells and oviductal cells, were used in cattle embryo culture; the authors concluded that both granulosa cells and BOEC produce soluble factors that promote early embryonic development. In The Netherlands, Inzen *et al.* (1993) found that conditioned medium prepared from BRL was as effective as coculture with BRL cells in developing cattle embryos from the two-cell to the blastocyst stage. Use of the preconditioned medium was seen by these authors as a further means of reducing variability in the IVC system.

5.8.3 *Embryo-conditioned medium*

In Ireland, studies by Shi (1991) produced evidence that factors produced by cattle embryos in culture (granulosa cell monolayer) could be employed to enhance development of early bovine embryos to the blastocyst stage. The effect of this 'blastocyst-conditioned medium' (BCM) is shown in Table 5.13. The proportion of early embryos developing to the blastocyst stage was significantly increased by the addition of BCM on either day 3 or day 0 of culture. The indications are that the beneficial effect became more pronounced as the period of exposure to BCM was extended.

As earlier noted, trophoblastic vesicles have been shown to favour the development of early cattle embryos (Heyman *et al.*, 1987a, b; Heyman and Menezo, 1987), but use of the intact embryo as a possible source of embryotrophic factors has not been explored to the same extent. In Germany, the rate of development of one-cell bovine embryos was significantly increased when they were cultured in groups of 40 rather than 20 per 0.5 ml medium (Khurana, 1992).

A paper by Keefer and Stice (1992) also showed that the number of IVM/IVF bovine embryos cultured per unit volume (10 μl droplet) of IVC medium can affect development. In Belgium, Ferry *et al.* (1992) presented evidence suggesting that greater embryonic development occurred because of embryo density rather than absolute number; a ratio of one embryo to 1 μl medium appeared to be an acceptable way of adapting culture conditions to

Table 5.13. Effect of blastocyst-conditioned medium (BCM) on the development of the early bovine embryo (from Shi, 1991).

Treatments	Embryos cult.	Blastocysts developed (%)
Control	349	168 (48.1)^{b}
BCM on day 3	349	194 (55.6)^{a}
BCM on day 0	349	207 (59.3)^{a}

Data shown in the table are from four replicates.
a, b: Values with different superscripts are different ($P < 0.05$).

the number of embryos. In Germany, Palma *et al.* (1992) reported evidence of a significant reduction in embryo yield with decreasing numbers of embryos per unit volume. In further work in Belgium, Dessy *et al.* (1993) cultured cattle embryos either singly or in groups in serum-free, oviduct-conditioned medium; when the number of embryos was increased from 4 to 40 in 40 μl culture drops, blastocyst yield increased from 2.5% to 22.0%.

In horses, there is evidence that the development of equine embryos may be enhanced *in vitro* by an increased embryo : medium ratio (Hinrichs *et al.*, 1990).

Embryo density can affect the development of mice embryos *in vitro*. Canseco *et al.* (1991) found that the percentage of live embryos when these were cultured singly was significantly lower than that when embryos were cultured in groups of 10 or 20 per droplet (38.7% vs. 65.6–61.8%). In a further report, Canseco *et al.* (1992) confirmed the importance of considering embryo numbers and medium volume as factors influencing mouse embryo development; the optimal arrangement appeared to be five embryos per 10 μl of medium. The clear indication was that embryo development increased when density was high.

A report by Quinn *et al.* (1991) also showed that mouse zygotes were able to develop more readily to fully expanded blastocysts in a protein-free medium if a sufficient density of embryos was employed. Lane and Gardner (1992) showed that culturing mouse embryos in groups significantly increased blastocyst cell numbers. In the USA, Li, J., *et al.* (1993) tested the effect of intact cocultured mouse embryos on the development of IVF-derived bovine embryos; results indicated that mouse embryos could enhance development of the cattle embryos.

The indications from such studies, both in cattle/horses and mice, are that autocrine factors secreted by embryos are beneficial rather than inhibitory. In the cattle IVF laboratory, it may be a matter of giving careful consideration to embryo density per unit volume of IVC medium and to avoid changing or otherwise replenishing the medium during the IVC period.

5.9 Hormones, Growth Factors and Cytokines in Embryo Culture

5.9.1 Hormones and their effects

There are various lines of evidence suggesting that hormones may be involved in the regulation of development in early embryos. Insulin has been shown to increase the rate of glucose transport in the blastocyst (Gardner and Kaye, 1991), blastocyst metabolism *in vitro* (Wales *et al.*, 1985), the rate of morphological development of mouse embryos during the preimplantation period *in vitro* (Gardner and Kaye, 1991), mouse embryonic protein synthesis during and after compaction (Harvey and Kaye, 1988, 1991, 1992a, b) and the fetal growth rate of embryos transferred after culture. There is also evidence of insulin acting synergistically with amino acids in improving development of rat embryos *in vitro*.

Table 5.14. Effect of PMSG, insulin, osmolarity and oestrous cow serum on the development of IVF embryos cultured on a granulosa cell monolayer (from Jiang *et al.*, 1990).

	OCS			PMSG			Insulin			Osmolarity		
	1	2	3	1	2	3	1	2	3	1	2	3
Embryos cult.	174	174	183	174	174	183	174	174	183	183	174	174
LM + BL	32	56	42	35	39	56	50	43	37	49	59	22
(%)	18.4[a]	32.2[b]	23.0	20.1[a]	22.4	30.6[b]	28.7	24.7	20.2	26.8	33.9[b]	12.6[a]

[a, b]: Each group of factors differs significantly ($P < 0.05$).
Three different levels of each of four factors were arranged by means of an orthogonal design (OCS, 0, 10, 20%; PMSG, 0, 10, 50 iu ml^{-1}; insulin, 0, 10, 50 μg ml^{-1}; and osm., 250–260, 290–300, 330–340 mM). Embryos were scored for development into late morulae (LM) or blastocysts (BL) at 5 to 7 days of the subsequent culture.

Separate receptors for insulin and IGF-I have been detected in early mouse embryos and it is known that insulin stimulates protein synthesis through its own receptor (Harvey and Kaye, 1992b); in this way, the hormone can promote blastocyst formation and mitogenesis in the ICM. Results reported by Dunglison and Kaye (1993) suggest that insulin acts to increase the endogenous protein reserves in the mouse embryo. According to Heyner *et al.* (1993a, b), insulin is the only hormone which has been shown to have a clear effect on early embryonic development. As noted in an earlier context (Section 5.2.3), there is some evidence that oestrogens are necessary in some species, including the pig, for the transformation of the compacted morula to the cavitated blastocyst (Niemann *et al.*, 1989).

In Ireland, various studies have used gonadotrophins (PMSG) and insulin in the culture *in vitro* of IVF-derived embryos (Jiang *et al.*, 1990). Data in Table 5.14 show that PMSG and insulin, as well as OCS and osmolarity, can influence the yield of embryos; however, results have not justified the use of these hormones in any routine supplementation of the IVC medium. In Colorado, Seidel *et al.* (1991b) were able to show that insulin had a beneficial effect in IVC, possibly due to binding to IGF-I receptors.

5.9.2 *Growth factors and their effects*

The ultimate aim of culturing cattle embryos *in vitro* is to simulate the microenvironment that these embryos experience in the bovine reproductive tract. Many growth factors are now recognized to exist in the maternal tract (Munson *et al.*, 1991, 1992; Simmen and Simmen, 1991; Heyner *et al.*, 1993a, b; Host-Hansen *et al.*, 1993); this accounts for the growing number of studies in this area of research (Menino *et al.*, 1989; Werb, 1990). Some of the growth factors that may be implicated in early embryonic growth and development are set out in Fig. 5.8. It is also apparent that growth factor

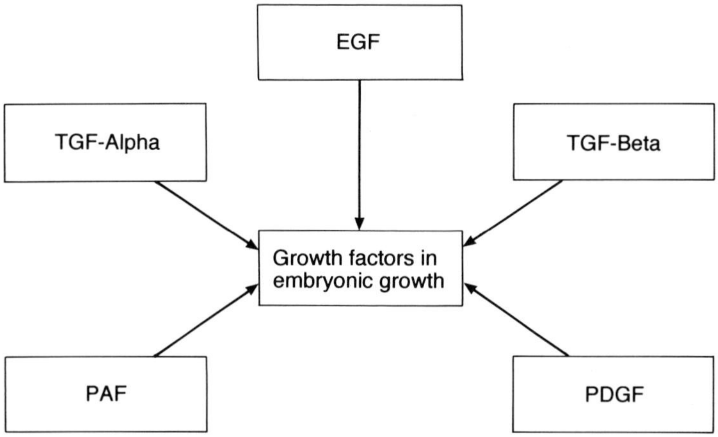

Fig. 5.8. Growth factors in embryo growth and development.

ligand and receptor genes are expressed before and after activation of the embryonic genome. Such findings show the finely regulated control of early embryonic development by growth factors.

The greatest research emphasis has been with three growth factor families: (i) insulin and the IGF family (IGF-I and II); (ii) TGF-β; and (iii) the epidermal growth factor (EGF) family, which comprises EGF and TGF-α. TGF-α bears 30% sequence homology to EGF and exerts its biological effects by binding to the EGF receptor (Derynck, 1990). TGF-α is a 50-amino acid polypeptide, while TGF-β is a homodimeric peptide consisting of two 112-amino acid monomers.

Insulin-like growth factor family

The hormone insulin is a protein, consisting of two polypeptide chains, A and B, linked by disulphide bonds; it is secreted by the β cells of the pancreas. IGF-I and II possess a high degree of amino acid sequence homology with insulin. Insulin and IGF-I receptors are structurally very similar; the IGF-II receptor has a very different structure. It is believed that oestrogen specifically induces the expression of IGF-I in uterine tissue, although the precise cells secreting the growth factor are unknown (Stevenson, K.R., *et al.*, 1992). In a study reported by Spicer and Zavy (1992), serum IGF-I decreased after oestrus in the ewe and increased as the next oestrus approached; this suggests that oestrogens may have a role in regulating IGF-I production.

Laboratory production of cattle embryos has been used to produce a developmental sequence of cattle embryos by which it is possible to study the expression of growth factor ligands and receptors (Watson *et al.*, 1992a). In contrast to the mouse, transcripts for IGF-II and the receptors for insulin, IGF-I and IGF-II are detectable in bovine embryos (Watson *et al.*, 1992a; Heyner *et al.*, 1993, a, b; Simmen *et al.*, 1993). In pigs, although the role of the growth factor is unclear, IGF-I expression in the period of rapid embryonic development has been recorded by Letcher *et al.* (1989).

Studies reported by Flood *et al.* (1993) and Palma *et al.* (1993b) showed no evidence that IGF-I supplementation of their IVC medium improved bovine embryo development. In mice, Harvey and Kaye (1992a, b) examined the effect of IGF-I on early embryos and found it did stimulate protein synthesis; they found that both insulin and IGF-I stimulates mitogenesis of the ICM and morphological development via insulin and IGF receptors.

The interactions of IGF ligands and receptors in early embryonic development may be modulated by IGF-binding proteins (Schultz, 1993). In pigs, IGF-I expression during the period of rapid embryonic development has been reported (Letcher *et al.*, 1989).

Transforming growth factor beta

Three TGF-β isoforms, termed TGF-β_1, TGF-β_2 and TGF-β_3, have been identified in mammals. It is a growth factor which can bring about enhanced rates of proliferation, depending on cell type and culture conditions. The production of the growth factor by rat granulosa cells has been demonstrated by Kim and Schomberg (1989). TGF-β_1 is a multifunctional growth factor having profound regulatory effects on many developmental and physiological processes. Evidence presented by Mulheron and Schomberg (1990) suggests that granulosa cells are a significant source of this growth factor in stilboestrol-treated rats. Evidence of the profound action of this growth factor is provided in the studies of Shull *et al.* (1992), who have shown that TGF-β_1-deficient mice succumb to a wasting syndrome about 20 days after birth. In France, Marquant-Le Guienne *et al.* (1989) and, in Ireland, Jiang (1991) have reported a significant effect after the addition of TGF-β to their IVC medium when embryos were at the early blastocyst stage; these authors found that TGF-β mainly stimulated mitotic activity in ICM cells. In the USA, TGF-β_1 in CR1aa medium (Rosenkrans and First, 1991) was also found to have a positive effect on improving bovine embryo development.

In Cornell, Larson, R.C., *et al.* (1992b, c), using a serum-free medium, tested the hypothesis that TGF-β and basic fibroblast growth factor (bFGF) synergistically promote the development of cattle embryos beyond the fourth cell cycle. TGF-β and bFGF permitted development to the blastocyst stage in only 9.6% of the embryos, which was markedly lower than the 22.1% development observed for coculture with preconditioned medium. They also found that such growth factor stimulated embryos were of inferior quality compared with embryos produced by conventional IVC procedures. However, it was evident that TGF-β and bFGF synergize to initiate embryonically controlled development, presumably by triggering embryonic gene expression.

Epidermal growth factor family

The EGF family is believed to be of particular interest in the period of early embryonic development. There is evidence that EGF/TGF-α may play an important role in the physiological regulation of follicular development (Bendell and Dorrington, 1990), but their interactions with other growth factors (e.g. TGF-β and IGF-I) are not well known or understood. EGF and TGF-α are known to be produced by the endometrium and their expression is believed to be under oestrogen control. According to Larson, R.C., *et al.* (1992c), EGF activity in small follicles may be enhanced by the presence of a binder in follicular fluid. Apart from their action on the endometrial cells (by autocine/paracrine stimulation), both ligands have been recorded bound to the EGF receptor in embryonic cells. Schultz and Dardik (1991) have suggested that either maternally derived EGF (the early embryo does not secrete EGF) or endogenously produced TGF-α may be involved in one of the first differentiated functions of the early embryo (blastocoele expansion).

It appears, from studies with cattle embryos (Watson *et al.*, 1992a), that EGF is not expressed in the early stages of cleavage. Studies by Wood and Kaye (1989) in mice suggest that EGF receptors appear at compaction. According to Fischer (1993), EGF may represent an important maternal signal sent to the embryo in early embryonic development which may have a bearing on subsequent maternal recognition of the embryo. In Utah, Flood and Bunch (1992) found evidence that EGF supplementation significantly improved bovine embryo development to the blastocyst stage. Studies reported by Keefer (1992) also showed some evidence of a response in cattle embryos to EGF (i.e. stimulated hatching).

The other factor in this family, TGF-α, is believed to be expressed throughout the early cleavage stages. According to Heyner *et al.* (1993a), the ICM of mouse embryos secretes TGF-α which binds to the EGF receptors in the apical trophectoderm. In the USA, studies reported by Flood *et al.* (1993) showed that supplementation of a chemically defined medium with TGF-α did not improve IVF-derived bovine embryo development, in agreement with an earlier report by Larson *et al.* (1991).

Work reported by Yang, B.K., *et al.* (1993a, b) has shown that EGF is capable of significantly improving the development of two- to eight-cell cattle embryos to the morula/blastocyst stage.

Other growth factors

Studies by Gandolfi *et al.* (1991) led to the detection of PDGF in bovine oviductal secretions. Larson *et al.* (1991) found that the addition of this growth factor induced cattle embryos to progress through the 8–16-cell block stage. A study by Thibodeaux *et al.* (1992b, 1993) demonstrated that PDGF stimulated bovine embryo development beyond the eight-cell stage; they concluded that at least part of the beneficial effect of BOEC coculture is due to PDGF. The PDGF was found to have the most beneficial effect of several growth factors examined by Yang, B.K., *et al.* (1993a, b); 70% of the two- to eight-cell embryos cultured reached the morula/blastocyst stage of development. In Germany, Eckert and Niemann (1993) record evidence that the addition of PDGF at various stages of the bovine embryo production system can enhance early embryo development.

Platelet-activating factor (PAF) is a potent metabolic activator of a variety of cell and tissue types. Studies reported by Ryan, J.P. *et al.*, (1990a, b) have demonstrated that exogenous PAF can act directly on the early mouse embryo to enhance the oxidative metabolism of energy substrates and to enhance the viability of the embryo itself. In a review, Ryan, J.P. *et al.* (1992) note that the production of PAF by mammalian embryos is highly variable and is adversely affected by IVC. A major site of action of embryo-derived PAF *in vivo* is believed to be the embryo itself. The growth factor has increased cell numbers in mice blastocysts and the mitotic index of mice embryos at the eight-cell and blastocyst stages. In human IVF, O'Neill (1991) reported that supplementation of culture media with PAF can result in a marked increase in the embryo's developmental potential.

5.9.3 Cytokines and their effects

The cytokines are hormone-like polypeptides that function as intracellular signals regulating cell–cell interactions in the uterus and elsewhere in the reproductive tract (Tabibzadeh and Sun, 1992). Cytokines differ from hormones because: (i) they are produced by a number of cells rather than by a particular gland; (ii) unlike hormones, which generally produce effects on distant target cells, cytokines usually act in a paracrine or autocrine manner; and (iii) they are not usually found in blood serum. A single cytokine may have multiple effects on the growth and differentiation of many cell types.

Leukaemia inhibitory factor (LIF)

Leukaemia inhibitory factor (LIF) is a glycoprotein, first isolated in mice, that induces differentiation and inhibits the proliferation of the M1 myeloid leukaemia cell line. In contrast to this action, LIF can also stabilize murine embryonic stem cells in culture, preventing them from undergoing spontaneous differentiation. There is now clear evidence that LIF plays a crucial role in enabling implantation to occur in mice (Stewart *et al.*, 1992). It is evident that LIF is an example of a cytokine with both embryonic and haemopoietic functions. Whether farm animals, such as the cow, share the need for LIF in early pregnancy remains to be seen. If they do, the cytokine may not only provide a means of controlling pregnancy but may also help to improve the establishment of successful pregnancies using IVF-derived embryos.

The addition of the cytokine LIF to SOF media has been shown to improve the viability of cultured sheep blastocysts (Fry *et al.*, 1991). Such findings may suggest that LIF is potentially an important embryotrophic agent. A further report by Fry *et al.* (1992) showed that the addition of human recombinant LIF to IVC medium led to a fourfold increase in the number of sheep blastocysts that hatched, with fewer embryos becoming degenerate. The Australian workers suggest that LIF may play a role in the live sheep in supporting the survival of the embryos when they enter the uterus (day 4 onwards). In France, Marquant-Le Guienne *et al.* (1993b) have shown that murine LIF has positive effects on embryo development, similar to those previously recorded in sheep. In a review, Fry (1992) proposes that LIF acts on the trophectoderm of the rapidly expanding sheep blastocyst and improves the attachment process. The signal for LIF expression in the live sheep appears to be of maternal origin, perhaps regulated by oestrogen.

In Japan, Fukui and Matsuyama (1994) found that human LIF (hLIF) improved the development of IVMFC-derived cattle embryos when SOF medium was supplemented with BSA or polyvinyl alcohol (PVA) but not when it was supplemented with human serum.

Interleukin 6

Interleukin 6 (IL-6) is a cytokine with multiple biological effects. It is known to influence cellular proliferation and differentiation and a range of immune and inflammatory responses. Its synthesis is not confined to cells of the immune system but has been observed in the cells of embryos. IL-6 expression during early pregnancy was studied by Mathialagan and Roberts (1991), who found evidence that the growth factor may be involved in evoking early maternal responses to the presence of the developing conceptus.

IL-6 was initially identified as a T-cell-derived cytokine which specifically promoted B cell differentiation and antibody production. There is also evidence to suggest that IL-6 may be an important regulator of certain endocrine phenomena. Studies reported by Gorospe *et al.* (1992) have shown that granulosa cells are capable of secreting IL-6 in response to follicle-stimulating hormone (FSH) stimulation. The same workers suggest that IL-6 produced by the granulosa cells may be involved in the local bidirectional communication between the granulosa cells and cellular elements of the immune system, such as macrophages and lymphocytes. There is a growing appreciation (see Seamark *et al.*, 1992) that all reproductive organs host a large population of white blood cells, whose composition changes in response to the cyclically changing hormonal environment. The immune system and cytokines are now clearly seen to be implicated in all aspects of reproductive function.

5.10 Methods Used in Evaluating IVMFC-derived Embryos

One difficulty in the field of laboratory embryo production in cattle is that adverse effects may not be evident in the mature oocytes, in the early cleaving embryo or even at the time of embryo transfer. Certainly, those who view the effectiveness of their IVM/IVF systems in terms of fertilization and early cleavage rates ignore at their peril the possibility that their procedures may be incapable of establishing viable pregnancies. It is important that due consideration be given to methods of accurately evaluating the normality and viability of embryos at the time of transfer or before cryopreservation.

Over the years, many techniques for evaluating the normality and viability of the early embryo have been reported. Many of these methods have centred around the morphological features of the embryo, such as uniformity of cell size, shape of embryo, its colour and overall size. Other methods have

Table 5.15. Measurements of *in vitro*-produced bovine embryos (μm) (from Lonergan, 1992).

	n	Outside	Inside	ZP	PB/blast.
Mature	21	150.47 ± 3.95	111.71 ± 3.12	13.56 ± 1.06	17.05 ± 0.98
2-Cell	13	148.42 ± 6.54	123.40 ± 9.12	13.73 ± 1.33	82.79 ± 5.18
4-Cell	31	149.98 ± 4.84	118.66 ± 5.76	14.12 ± 1.78	68.10 ± 4.03
8-Cell	15	149.46 ± 4.86	120.00 ± 6.42	13.39 ± 1.12	56.27 ± 2.87
D6 M	37	150.87 ± 4.67	115.84 ± 5.59	14.53 ± 1.44	–
D6 LM/EB	18	149.93 ± 6.69	123.60 ± 7.99	13.10 ± 1.71	–
D7 EB	18	158.31 ± 7.58	135.41 ± 9.77	11.75 ± 1.77	–
D7 B	30	167.73 ± 10.88	149.10 ± 15.54	9.63 ± 3.09	–
D8 B	46	182.99 ± 19.16	167.82 ± 23.66	7.92 ± 2.83	–
D9 B	56	200.60 ± 23.14	189.03 ± 25.15	5.95 ± 2.36	–
D9 HB	14	272.30 ± 52.80	–	5.77 ± 2.38	–

n, number of embryos measured; ZP, zona pellucida; PB, polar body; blast., blastomere diameter; D6, day 6 post-insemination; M, morula; B, blastocyst; LM/EB, late morula/early blastocyst; HB, hatched blastocyst.

aimed at detecting enzyme activity, measuring glucose uptake and other metabolic events. In human IVF, some studies have shown that oocytes with a high potential for development have a thicker ZP than those with a low potential (Chan, 1987). The development of reliable, rapid and practicable methods of assessing embryo normality and viability is a matter of considerable importance to the cattle ET industry.

Measurements of IVMFC-derived cattle embryos, produced by the methods employed in Dublin (see Appendix A), are provided in Table 5.15. These data are taken from Lonergan (1992).

It is evident, from studies reported by Suzuki *et al.* (1994a), that the diameter of the bovine oocyte changes significantly after maturation and fertilization; these workers concluded that the maturation process may involve surface modification of the vitelline membrane.

5.10.1 Morphological and morphometric parameters

In the USA, one system for the classification of cattle embryos, on a morphological basis, was that proposed by Lindner and Wright (1983). These authors classified embryos in four categories – excellent, good, fair and poor; pregnancy rates associated with each category after transfer to recipients were 45%, 44%, 27% and 20%, respectively. The report also provided a useful visual guide to the morphological evaluation of embryos, enabling them to be graded according to stage of development and their quality within that particular stage. In Dublin, a five-point scale for assessing the morphology of cattle embryos before and after cryopreservation was proposed by Kennedy *et al.* (1983). Details of this classification system are in Table 5.16.

The important feature of any embryo classification scheme is that it

Table 5.16. Grading bovine embryos by morphological features (from Kennedy *et al.*, 1983).

Grade 1. *Excellent.* Embryo perfectly symmetrical, showing even granulation and with a well-defined, distinct outline; no blastomere extrusion. The embryo should be at the expected stage of development for its age.

Grade 2. *Good.* Embryo showing even granulation with a well-defined distinct outline; some blastomere extrusion and some minor blastomere degeneration; occasionally somewhat asymmetric in shape.

Grade 3. *Fair.* Embryo intact but with a hazy outline in parts; obvious defects apparent such as extruded cells, vesiculation and some degenerate blastomeres.

Grade 4. *Poor.* Embryo showing uneven granulation and with a hazy outline; much blastomere extrusion and degeneration evident; sometimes shaped abnormally.

Grade 5. *Degenerate.* Degeneration so pronounced that it may not be possible to determine the exact developmental stage; sometimes shaped abnormally.

should be based on easily recognizable morphological features and should be backed by firm evidence on the pregnancy rates to be expected with each of the grades. Hasler *et al.* (1987), working with large numbers of cows, reported pregnancy rates of 83%, 75%, 63% and 46% after transfer of one fresh (*in vivo*-produced) embryo classified as grade 1, 2, 3 and 4, respectively.

It is of interest to note that, in mice embryos, the developmental potential of partially degenerate embryos can be restored by minimizing the exposure of the intact cells to degenerate material (Alikani *et al.*, 1993); this was done by appropriate micromanipulation. It is possible that exposure of intact blastomeres to perivitelline debris and degenerative material may influence their subsequent development. This may even have implications in explaining decreased embryo survival rates after freeze–thaw procedures, where it is expected that a number of cells inevitably degenerate.

In rabbits, Jung and Fischer (1988) compared morphometric and metabolic parameters in individual rabbit embryos to determine how closely such parameters were correlated; the hope was that a simple measurable morphometric criterion, such as blastocyst diameter, might allow assessment of the metabolic activity of the embryo. Although the correlation was useful with non-cultured rabbit blastocysts, it was of doubtful value with those that were cultured, due to substantially altered embryonic metabolism *in vitro*. Blastocyst size was examined as a determinant of fetal viability in mice by Harshbarger and Graves (1990); small-sized blastocysts were found to have lower survival rates due to embryonic mortality.

5.10.2 Age and developmental stage attained

Two factors are usually considered when visually evaluating the IVMFC-derived bovine embryo under the light microscope: (i) its general appearance, observing such features as the presence of irregular or degenerate cells; and (ii) the stage of development of the embryo in relation to the time since fertilization. The morphology and size of embryos recovered from superovulated cattle have been described by several authors (Nakagawa *et al.*, 1991).

The overall diameter of the bovine embryo is estimated to be between 150 and 190 μm, including a ZP thickness of about 12–15 μm (see Table 5.15). This diameter remains virtually unchanged from the zygote to the start of blastocyst expansion. The various developmental stages of the bovine embryo, after the early cleavage stages, are set out in Table 5.17, taken from the studies of Lonergan (1992). The development of the *in vivo*-produced bovine embryo during the first 7 days has been dealt with by several authors (Laing, 1945; Nakagawa *et al.*, 1991; Dorland, 1992); the development of the blastocyst beyond the hatched stage is well detailed and illustrated in the earlier paper of Chang (1952). As shown by Shea (1981), the majority of embryos recovered from superovulated donor cattle on day 7 are late morulae or early blastocysts.

Table 5.17. Morphological criteria employed in the evaluation of bovine embryonic development (from Lonergan, 1992).

Morula – Individual blastomeres are difficult to discern from one another. The cellular mass of the embryo occupies most of the perivitelline space.

Compact morula – Individual blastomeres have coalesced, forming a compact mass. The embryo mass occupies 60–70% of the perivitelline space.

Early blastocyst – This is an embryo that has formed a fluid-filled cavity or blastocoele and has the general appearance of a signet ring. The embryo occupies 70–80% of the perivitelline space. Visual differentiation between trophoblast and the inner cell mass may be possible at this stage of development.

Blastocyst or midblastocyst – Pronounced differentiation of the outer trophoblast layer and the darker, more compact inner cell mass is evident. The blastocoele is highly prominent with the embryo occupying most of the perivitelline space.

Expanded blastocyst – Overall diameter of the embryo dramatically increases (1.2 to 1.5 ×), with a concurrent thinning of the zona pellucida to approximately $\frac{1}{3}$ of its original thickness.

Hatched blastocyst – Embryos can be undergoing the process of hatching or may have completely shed the zona pellucida.

Hatched expanded blastocyst – A re-expanded embryo with a large blastocoele and round, very fragile appearance or, in later stages, an elongated shape.

The disadvantage of all morphological assessments of embryo quality and viability lies in the fact that they are subjective and require highly trained personnel, which makes comparisons between laboratories or even individuals difficult.

Several authors have reported on cell stage for age in cattle embryos produced *in vivo* (from superovulated donor cattle). In former Czechoslovakia, Holy *et al.* (1988) reported on 1761 7-day old *in vivo* cattle embryos and recorded that 18.7% were late morulae, 47.7% were very early blastocysts, 20.5% were early blastocysts, 9.5% were expanding blastocysts, 2.1% were hatching blastocysts and 0.5% were hatched blastocysts. Taking due account of the fact that ovulations in superovulated donors may extend over a period of 12 h and sex differences are now known to exist, such variation in developmental progress can only be regarded as normal.

Ellington *et al.* (1990b) also observed that the development of single embryos from unstimulated donor cattle appears to be ½–1 day behind the development of embryos from superovulated animals. There are strong grounds for believing that the incidence of abnormal embryos is likely to be higher in superovulated cattle than in spontaneously ovulating animals (King *et al.*, 1985a). According to Barnes and Eyestone (1990), the rate of cleavage of *in vivo*-cultured and IVC bovine embryos is similar up to the eight-cell stage; beyond the 8–16-cell stage, cleavage rate is very sensitive to environmental conditions. These authors note that synchronous and rapid cleavage is characteristic of maternally regulated early development.

In Japan, Goto *et al.* (1992e), examining IVMFC-derived embryos that formed blastocoeles 7, 8, 9 and 10 days after fertilization, noted that the mean cell number per embryo decreased and the incidence of chromosomal abnormalities increased over the 4-day period. In the USA, Monson *et al.* (1992) record work in which 2073 oocytes were matured, fertilized and cultured, resulting in 395 IVMFC blastocysts of transferable quality. These authors report that pregnancy rates were significantly lower for single day-8 blastocysts than for day-7 blastocysts (41% vs. 66%). Such findings would certainly agree with many data in Ireland showing that the later-developing blastocysts are of lower quality (Jiang, H.S., *et al.*, 1992).

A paper by McKiernan and Bavister (1993), recording studies with hamster embryos, warns against the practice of prolonging the culture time to maximize results, as expressed in terms of blastocyst yield; the same would certainly be true with cattle embryos. Time is an important parameter of embryo development. It should also be noted that retarded embryonic development has been shown to be the principal cause of repeat breeding in some cattle studies (Gustafsson, 1985).

In Belgium, Van Soom *et al.* (1993) examined the kinetics of compaction and blastocyst formation of IVMFC-derived bovine embryos. According to their data, *in vitro* embryos start blastulation at a very early stage and at a rather low cell number; they record blastocyst formation occurring over a prolonged period of time (from 114 until 162 hpi) during the sixth cleavage division (32- to 64-cell stage). The average cell number of embryos that showed compaction prior to blastulation was significantly greater than that in embryos in which there was no discernible compaction.

5.10.3 Fluorescent staining procedures

Schilling and Doepke (1978) and Schilling *et al.* (1979) were among the first to examine the possible value of fluorescent substrates to assess embryo viability in cattle, using 3′,6′-diacetylfluorescein (FDA); this dye is normally a non-fluorescent substance, which is converted to a fluorescing compound upon contact with hydrolases, which are only present in living cells. FDA fluorescence is green and limited to the cytoplasm of the blastomeres; the ICM of the blastocyst usually fluoresces more brightly than the surrounding trophoblast cells. An alternative test is one employing the dye 4′,6′-diamidino-2-phenylindole (DAPI), which has a high affinity for deoxyribonucleic acid (DNA). Only the nuclei of non-viable embryos or of dead blastomeres fluoresce after incubation with DAPI (Schilling *et al.*, 1979). DAPI fluorescence is characterized by the yellow coloration of the nuclei; the cytoplasm of the blastomeres does not fluoresce.

When the FDA and DAPI tests are combined, the living blastomeres fluoresce green while the nuclei of the dead blastomeres fluoresce yellow. According to Schilling *et al.* (1982), the selection of viable and transferable cattle embryos by these fluorescence tests may be 10% more accurate when compared with morphological evaluation by light microscopy.

In Japan, Iwasaki *et al.* (1990a, b) and Iwasaki and Nakahara (1990a, b) found a differential fluorochrome staining technique useful in counting cells of the ICM and of the trophectoderm in cattle blastocysts. The same authors described the morphology of ICM cells in cattle blastocysts produced by *in vivo* or *in vitro* (cumulus cell monolayer) procedures; *in vitro* embryos did not show the same clear compaction of blastomeres at the morula stage, although this was evident in IVM/IVF embryos cultured in the rabbit oviduct.

The use of fluorescent vital dyes calcein AM (which measures intracellular esterase activity) and ethidium homodimer (a DNA-binding dye impermeable to intact membranes) has also been suggested as possible methods of assessing cattle embryo viability (Duby *et al.*, 1993).

The use of fluorescein yellow dye in evaluating functional cell-to-cell communication in cattle embryos is reported by Prather and First (1993). These authors found evidence suggesting that less gap junction communication is present within bovine morula- and blastocyst-stage embryos in comparison with mouse embryos at similar stages. They also noted the absence of detectable cell-to-cell coupling in those embryos produced by *in vitro* methods; the implication is that such embryos may not be entirely normal.

For the commercial cattle IVF laboratory, it should be said, staining procedures for evaluating embryo viability are generally regarded as too difficult to perform on a routine basis or too inaccurate to warrant serious consideration.

5.10.4 Cell numbers

Results with IVF-derived cattle embryos in Ireland, in terms of cell numbers, have been in general agreement with reports that bovine blastocysts developed

Table 5.18. Examination of cell numbers in blastocysts derived from IVM, IVF and IVC of bovine follicular oocytes (from Jiang, H.S., *et al.*, 1992).

Stage	Blastocysts (BL) recov. at	Cell numbers (no. examined) (mean ± SD) G1	G2	G3	G4/5
E-BL	D7	110 ± 21.3 (29)[a]	78 ± 19.7 (20)[a]	72 ± 18.9 (31)[a]	52 ± 13.6 (46)[b]
	D8	95 ± 14.9 (12)[a]	78 ± 12.2 (12)[ab]	68 ± 15.3 (9)[b]	49 ± 12.2 (11)[c]
	D9	89 ± 13.8 (8)[a]	72 ± 9.7 (7)[b]	68 ± 14.9 (11)[b]	49 ± 19.3 (14)[c]
BL	D7	143 ± 21.6 (19)[a]	102 ± 20.8 (28)[b]	83 + 20.2 (23)[c]	64 ± 13.1 (22)[d]
	D8	134 ± 17.1 (27)[a]	103 ± 17.6 (17)[b]	75 ± 11.4 (15)[c]	57 ± 14.8 (17)[d]
	D9	123 + 16.6 (10)[a]	97 ± 13.2 (9)[b]	66 ± 16.8 (14)[c]	47 ± 13.3 (12)[d]
Exp-BL	D7	97 ± 24.8 (20)[a]	75 ± 24.6 (15)[b]	56 ± 16.4 (15)[b]	37 ± 12.0 (13)[c]
	D8	82 ± 13.5 (12)[a]	69 ± 8.9 (10)[a]	53 ± 10.1 (12)[b]	35 ± 11.5 (12)[b]
	D9	67 ± 9.9 (13)[a]	62 ± 8.6 (11)[a]	47 ± 10.2 (10)[b]	34 ± 11.0 (10)[b]

a, b, c, d: $P < 0.05$ within the row.

in vivo usually contain about 100, 120 and 160 cells at the early, expanding and expanded blastocyst stages, respectively (Mannaerts, 1986; Ushijima *et al.*, 1988). According to data summarized by Betteridge and Flechon (1988), early cattle blastocysts contain about 100 cells and blastocysts about to hatch contain about 160 cells. Cell number can certainly be regarded as a valuable indicator of the viability of cattle embryos. In Ireland, Jiang, H.S., *et al.* (1992) examined cell numbers in laboratory-produced cattle embryos; the results are in Table 5.18.

These data clearly show that, regardless of the developmental stage, the total cell number or the ICM cell number in grade 1 and grade 2 blastocysts was significantly higher than in grade 3 or grade 4/5 blastocysts. Day-7 and day-8 blastocysts had significantly greater cell numbers compared with day-9 embryos. Such data present clear evidence that later-developing blastocysts are of lower quality, as judged by cell number. The same authors also examined cell number in blastocysts developed on different monolayer culture systems; the mean number of nuclei per embryo did not differ significantly.

In Sweden, Shamsuddin *et al.* (1992c) have shown that the mean cell number in 7-day-old IVF embryos on a BOEC monolayer culture system is comparable to that of 7-day-old *in vivo*-produced embryos. On the other hand, Ectors *et al.* (1993a, b) in Belgium record significant differences in cell numbers according to whether IVM/IVF blastocysts had been cultured in a BOEC monolayer system or had used preconditioned medium (91 vs. 72 cells).

Cell numbers in IVMFC-derived cattle embryos, produced by Lonergan (1992) using the protocol currently employed in Dublin (see Appendix A), are provided in Table 5.19.

Some authors have reported on the relative numbers of cells in the ICM and the trophectoderm. In the report of Skrzyszowska and Smorag (1989), for example, using embryos from superovulated cattle, the average cell number

Table 5.19. Cell numbers in *in vitro*-produced bovine embryos (from Lonergan, 1992).

		n	Cell number
Day 6	Early morula	6	28.00 ± 5.21
	Morula	16	50.37 ± 17.65
	Late mor./early bl.	7	62.00 ± 22.80
Day 7	Late mor./early bl.	12	71.92 ± 28.01
	Early blastocyst	13	96.85 ± 47.30
	Blastocyst	26	120.35 ± 39.53
	Expanding bl.	16	130.56 ± 48.05
	Expanded bl.	3	191.00 ± 29.50
Day 8	Blastocyst	16	131.31 ± 41.87
	Expanded bl.	5	169.4 ± 18.95
	Hatched bl.	8	185.75 ± 51.64
Day 9	Blastocyst	59	117.15 ± 42.40
	Expanded bl.	8	105.00 ± 31.72
	Hatched bl.	12	204.40 ± 59.14

in late-stage cattle blastocysts is given as about 140, of which 93 cells were trophoblastic cells and 47 were ICM cells (1 ICM:2 trophoblastic). Laboratory-produced embryos should presumably match such standards. There is also the possibility that the ICM of the embryo is more susceptible to insult than the cells of the trophectoderm. This has been observed in studies with mouse embryos, where ICM cells were more sensitive to the effect of increased temperature than trophoblast cells (Capehart and White, 1990).

5.10.5 Computerized image analysis

For the future, new methods of assessing the normality of cattle embryos produced in the laboratory are likely to emerge. One technique worthy of mention is computerized image analysis (IA), which involves taking measurements of size and density of cells, which may identify objectively morphological features associated with the potential for normal embryonic development (Foote, 1987; Youngs *et al.*, 1988; Casey *et al.*, 1989). The use of videocinematography is described by Cohen *et al.* (1989), who employed this technique as a means of recording embryonic morphology in human embryos; they found that variations in the thickness of the ZP were related to the survival of the embryo after transfer. In Belgium, Grisart *et al.* (1992) developed a cinematographic technique which enabled them to analyse accurately the developmental sequences of early IVMFC-derived cattle embryos under different experimental conditions.

5.10.6 *Hatching as a measure of embryo viability*

Hatching of the bovine blastocyst can be expected to occur 8–10 days after ovulation and fertilization in the live cow (Chang, 1952; see Fig. 5.9). Hatching must be regarded as a key event in the development of the bovine embryo. If the blastocyst does not hatch, then pregnancy will not occur. There are reports in the literature showing that cattle embryos produced *in vivo* are well capable of hatching after IVC (Kuzan and Wright, 1982). Recent reports have shed some light on factors influencing the hatching process.

According to Gordon and Dapunt (1993a), dealing with mouse embryos, two factors have been identified as possible mediators of hatching in that

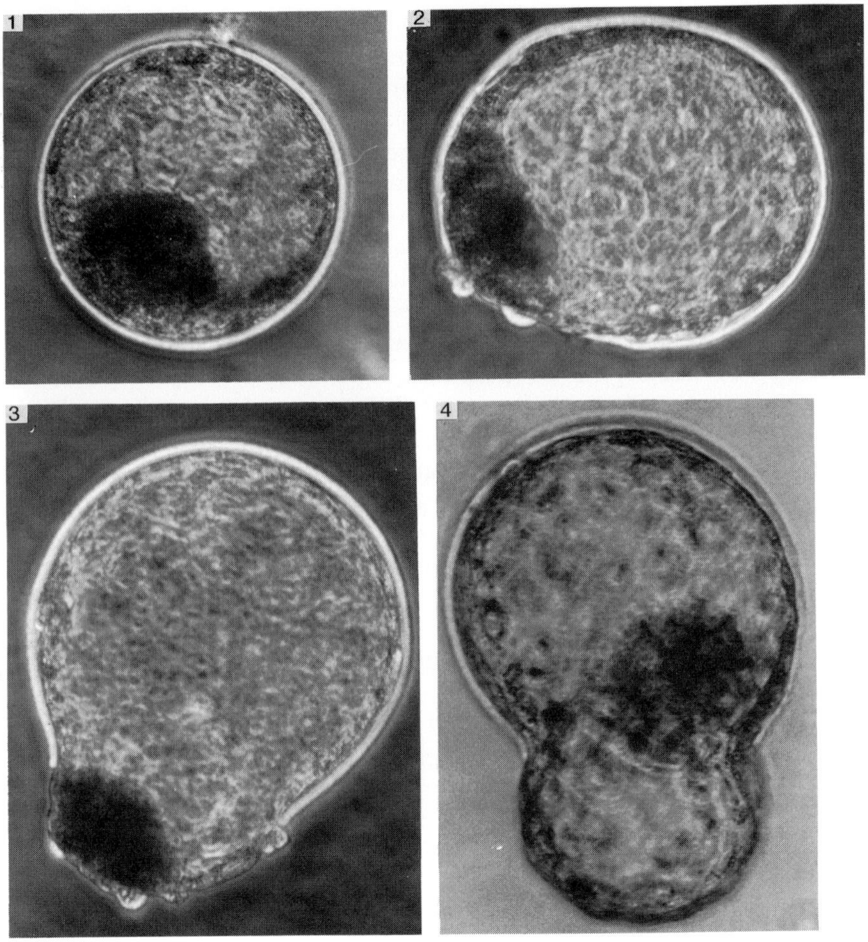

Fig. 5.9. Stages in the hatching of the bovine blastocyst. Hatching usually occurs when the bovine blastocyst has reached the 8–10-day stage of development. In this instance, the blastocyst has reached the expanded stage within the zona pellucida in (1); the subsequent photographs (2, 3 and 4) show stages in the hatching process. All photographs to the same scale.

species: (i) lysis of the ZP by substances secreted either by the embryo or the female reproductive tract; and (ii) pressure exerted on the zona by the expansion of the blastocyst. In regard to the latter, various studies have shown periodic, pulsatile expansion of the blastocyst before hatching and there is evidence of enzymatic mechanisms being involved in actively pumping fluid into the developing blastocoele. However, Gordon and Dapunt (1993a) have determined in the mouse that hatching is predominantly the result of zona lysis and that pressure by the expanding blastocyst plays little or no role in the emergence of the embryo from the zona. The same authors demonstrated that impaired hatching can result from a reduction in cell number and that normal hatching could be restored by circumferential zona thinning (Gordon and Dapunt, 1993b). It remains to be seen whether this explanation of embryo hatching in the mouse also applies to the cow.

In human IVF, there are those who believe that low pregnancy rates in IVF cycles, compared with natural cycles, may in part be due to impaired hatching caused by culture conditions. Compositional changes that occur in the hamster ZP, from the time the oocyte matures within the follicle until the embryo is released at hatching, were studied at Cambridge by Brown, C.R. *et al.* (1990); they suggest that some factors influencing changes in ZP composition may be due to the tubal environment. Deficiencies in reproducing such conditions *in vitro*, where fluid volumes could lead to considerable dilution effects, may be a factor in explaining differences that may arise between *in vivo*-cultured and IVC early embryos.

A number of workers have attempted to improve hatching rates by micromanipulation, often using the mouse embryo as the model. In the USA, Khalifa *et al.* (1992, 1993) have suggested that a combination of coculture and assisted hatching may be warranted in human IVF and embryo transfer, given the beneficial effects that they observed in the mouse. As shown by Gordon and Dapunt (1993a, b), circumferential thinning of the zona (using acid Tyrode solution) can be a highly effective assisted hatching technique in mice; these authors believe that the procedure may have important implications in certain instances in human IVF. In the USA, assisted hatching was performed in 330 human IVF patients with some evidence of a beneficial effect (Cohen *et al.*, 1992).

Hatching can be regarded as a significant event in the development of laboratory-produced embryos and as such has been employed as a measure of their ability to progress normally after various treatments or manipulations (e.g. after freeze–thawing). A paper by Delhaise *et al.* (1991) found no effect of culture media on the hatching rate of IVM/IVF embryos; however, hatching was observed from day 8 until day 15 after fertilization and no more than 64% hatched in any treatment group. In Belgium, studies by Massip *et al.* (1993a, b) with IVF-derived embryos cultured in BOEC-conditioned medium have shown a very low hatching rate after freeze–thawing (6%); hatching apparently required an epithelial cell coculture system. This was interpreted as evidence that the IVF embryos had reduced metabolic activity, which influenced their hatching ability.

5.10.7 *Ultrastructural features in evaluating embryos*

The ultrastructural analysis of cattle embryos has been reported by Chartrain and Picard (1988); this revealed the presence of dead cells or cellular debris, which could be related to embryo quality. In Denmark, Hyttel *et al.* (1989a) reported on the comparative ultrastructure of oocytes fertilized *in vivo* and *in vitro*; a high percentage of their IVF oocytes showed evidence of polyspermy, with the particular IVM/IVF system that was employed. Differences between *in vivo* and *in vitro* cattle embryos, evident at the ultrastructural level, were also reported by Kopecny *et al.* (1991). In eight-cell *in vivo* embryos examined by electron microscopy, chromatin blocks were distributed rather uniformly in the nuclei; in the IVM/IVF embryos, the degree of chromatin condensation was very high (supercondensation) and the chromatin blocks were localized on the periphery of the nuclei but not in contact with the nuclear envelope. As observed by Bergeron *et al.* (1992), the ultrastructure of the nucleolus, a prominent substructure of the nucleus, reflects morphologically changes that occur in the activity of the embryonic genome; the structure of the nucleolus corresponds to the level of transcriptional activity.

As noted earlier, there is evidence that cell-to-cell coupling in the IVMFC-derived bovine embryo may differ from that in embryos produced *in vivo* (Prather and First, 1993). In Japan, Iwasaki *et al.* (1990a, b) and Iwasaki (1992) have found that cell-to-cell contact of ICM cells in blastocysts derived from *in vivo* fertilization were tighter than those from IVF. The same authors concluded that low developmental ability of IVF-derived embryos may be the result of abnormal fertilization (resulting in polyploidy and other chromosomal abnormalities), reduced cell proliferation and loose cell-to-cell contact between ICM cells. In Sweden, Shamsuddin *et al.* (1992a) reported that IVMFC-derived cattle embryos showed a higher degree of vacuolation of the cytoplasm of the blastomeres and shorter and less numerous junctional complexes than embryos produced *in vivo*.

5.10.8 *Secretion of hormones and growth factors by embryos*

It is now clear that the bovine embryo, prior to the time of attachment, communicates its presence to the host cow via a series of signals. Quite apart from interferon (IFN) secretion, the studies of Wilson *et al.* (1992) have shown that the bovine embryo, as early as day 10, has the ability to synthesize and secrete both steroids and prostaglandins. In humans, there is evidence suggesting that the early embryo cultured *in vitro*, in the presence of exogenous androgen, is capable of secreting oestradiol (Edgar *et al.*, 1993); they may even be capable of progesterone secretion under certain conditions. A review by O'Neill (1991) notes that the production of embryo-derived PAF was first described and validated in rodents and subsequently confirmed in humans; however, it was not detected in media from embryos that failed to establish a pregnancy. Stock and Hansel (1992) sought to determine if the bovine embryo secretes PAF *in vitro* and if PAF measurements could be used to evaluate the embryo's potential to establish a pregnancy. They did find evidence suggesting that certain embryos secrete PAF or a PAF-like substance *in vitro*; however, it appears that PAF production is not a suitable marker for embryo viability.

In Canada, the ability of different grades of IVMFC-derived cattle embryos to interact with cultured endometrial cells has been evaluated (Fortier *et al.*, 1994); it is suggested that embryo quality may be related to differences in the prostaglandin E (PGE/PGF) ratio expressed by these cells.

Inhibin, with its paracrine and autocrine actions, is well recognized to be a regulator of gonadal function. There is even some evidence of a direct embryonic contribution to α-immunoreactive inhibin secretion in human early embryos (Phocas *et al.*, 1992); the peptide was secreted as early as 3–4 days post-fertilization. The Greek authors suggest that the hormone may play an important role in the endocrinology of early pregnancy.

5.11 Metabolic Tests

It has been recognized for many years that complex biochemical changes occur during the embryonic stages of development in farm mammals; evidence for this was reviewed some time ago by Biggers and Stern (1973). There are a number of tests of embryo viability which can be based on such biochemical knowledge (Rieger, 1984; Butler and Biggers, 1989; Meinecke and Meinecke-Tillman, 1990; Overstrom, 1992).

5.11.1 Enzyme assays

Bovine embryos cultured *in vitro* take up some metabolic substrates from the medium and release metabolic end-products. By measuring the amount of these substances, it may be possible to determine the viability of the embryo. According to Meinecke and Meinecke-Tillman (1990), the measurement of metabolic activity should meet three main conditions: (i) the assay should be applicable to individual embryos; (ii) the assay should not adversely affect subsequent embryonic survival; and (iii) the assay should be simple and effective.

The release of the enzyme L-lactate dehydrogenase (LDH) into IVC medium by dead and dying cells has been employed in measuring the viability of cattle embryos (Renard *et al.*, 1978). In subsequent studies reported by Jordan *et al.* (1989) and Johnson *et al.* (1991), a bioluminescence assay for measuring LDH was reported to be a more reliable indicator of bovine embryo viability than a morphological score. In Japan, Koike *et al.* (1991) recorded changes in 6-phosphogluconate dehydrogenase in rat embryos and found some evidence that this was linked to their viability.

5.11.2 Glucose, glutamine and pyruvate uptake

The bovine embryo is known to undergo a series of morphological changes associated with biochemical alterations in energy metabolism (Rieger and Guay, 1988). The pentose phosphate pathway (PPP) and Embden–Meyerhof pathway (EMP) each play an important role in embryo development. Data presented by Javed and Wright (1991) have indicated that, in good-quality cattle embryos, total glucose utilization is low until the 16-cell stage, with PPP the predominant pathway. The same workers showed that total glucose utilization increases significantly at the morula stage; EMP activity was observed to rise with increasing embryo development. There are some reports suggesting that glucose may be essential for the hatching of the bovine blastocyst (Renard *et al.*, 1980).

Several groups had previously examined the use of glucose and pyruvate uptake as a measure of embryo viability, in human embryos (Hardy *et al.*, 1989) as well as in cattle. Renard *et al.* (1978) found that day-10 cattle embryos which had a high rate of glucose utilization developed more readily *in vitro* and *in vivo* than did embryos using little or no glucose, regardless of their morphological appearance. Renard *et al.* (1980) reported a good correlation between glucose uptake by bovine blastocysts and their capacity to develop subsequently. Some 69% blastocysts consuming more than 5 μg glucose h^{-1} developed further, compared with 14% of those consuming less than 5 $\mu g\, h^{-1}$. Butler *et al.* (1988) measured glucose uptake in mouse embryos in two different media; the glucose uptake of degenerate embryos was significantly lower than that of normal embryos. At York, Brison and Leese (1991) developed a non-invasive assessment of energy metabolism in rat embryos using an ultramicrofluorometric technique.

Studies reported by Dorland *et al.* (1991, 1992) evaluated the rates of uptake of pyruvate and glucose by bovine embryos as a method of assessing bovine embryo viability. The ability of *in vivo*-produced embryos to develop during IVC was significantly related to their rate of uptake of both pyruvate and glucose, as measured by a microfluorescence assay. The authors concluded that nutrient uptake by day-7 cattle embryos could provide useful information about their viability. In a paper dealing with sheep embryos, glucose uptake was also found to be significantly higher in the expanded blastocyst than in earlier-stage embryos (Butler and Williams, 1989, 1992); the measurement of pyruvate uptake in sheep embryos has also been reported (Butler and Williams, 1990, 1991).

Little is known about the uptake of glutamine by the mammalian embryo and how this may be affected by the presence of other energy substrates in the culture medium. Gardner *et al.* (1989) have described a non-invasive, ultramicrofluorometric method for measuring the uptake of glutamine by single mouse embryos in the presence of various energy substrates (glucose, pyruvate and lactate); a highly significant reduction in glutamine uptake was observed in the presence of glucose at the one- to two-cell stage. At a later stage, glutamine uptake was only marginally reduced and no effect was observed at the blastocyst stage. The measurement of glucose and glutamine metabolism in bovine embryos (Rieger and Guay, 1988; Rieger *et al.*, 1989, 1991c, 1992) and in frozen–thawed equine embryos (Rieger *et al.*, 1990) has been reported from Canada.

A report by Gardner *et al.* (1993) proposes that the blastocysts of ruminants may utilize energy sources such as amino acids for oxidative metabolism, rather than carbohydrates. For such reasons, these authors suggest that ruminant embryo viability may be best assessed by the measurement of amino acid uptake and metabolism rather than glucose utilization.

5.12 *In Vitro* Culture of Embryos in Other Farm Species

5.12.1 Sheep and goats

As noted earlier (see Section 5.6.1), the first successful coculture studies with oviductal cells were those of Gandolfi and Moor (1987) in sheep. As observed at the time by Bavister (1988), this proved to be a landmark study that stimulated a great deal of research, in cattle rather than sheep, in the years that followed. Coculture studies using oviductal cells have been reported by Holm *et al.* (1991a) and Rexroad and Powell (1990). SOF has been commonly employed in the culture of sheep embryos; human serum-supplemented SOF has proved to be very effective (Thompson *et al.*, 1992a). A modified SOF medium, supplemented with 20% human serum and having a low glucose concentration (1.5 mM), was more effective, in terms of blastocyst development, than SOF with a higher (2.3 mM) level of glucose (Thompson *et al.*, 1992b). A report by McGinnis and Youngs (1991) draws attention to the danger of using morphology as the sole criterion to assess development of early-stage sheep embryos; the authors suggest this could be avoided by using blastocyst formation as the end point of culture.

In goats, results reported by Sakkas *et al.* (1989) demonstrated that coculture of goat embryos with oviductal cells (in various media) enabled a high percentage of the embryos to develop through the period of cleavage arrest; they observed, however, that the coculture system provided a less adequate environment for development to the blastocyst stage. In Clemson, in contrast to the work mentioned above, the studies of Richardson *et al.* (1992) indicated that goat embryos may not exhibit the stage-specific developmental block characteristic of cattle and sheep. It is evident that the nutrient uptake of the goat is very similar to that of the sheep embryo (Gardner *et al.*, 1994); low levels of glucose uptake prior to compaction are consistent with low levels of glucose utilization.

Pritchard *et al.* (1990, 1991, 1992) reported that coculture with oviductal cells resulted in a greater rate of goat embryo development than medium alone (0 vs. 70% blastocysts). They also showed that transfer of early-stage goat embryos (two- to four-cell) from oviductal to uterine cell monolayers during culture did not enhance development, in comparison with embryos maintained on oviductal cells alone. Such findings were in agreement with data recorded in studies with cattle, sheep and pig embryos. A study reported by Chartrain *et al.* (1987) has described nucleolar ultrastructure during early embryonic development in the goat.

5.12.2 Pigs and horses

More than 20 years ago, it was demonstrated that pig embryos continue developing when transferred to the ligated oviducts of oestrous and pseudopregnant rabbits (Polge *et al.*, 1972); a high embryo survival rate (90%) was achieved when embryos kept for 2 days in the rabbit were retransferred to recipients.

Embryo development during the early cleavage stages in the pig have been poorly documented, in comparison with the information available for cattle and sheep. Studies on the duration of the early cleavage stages are reported by Kronnie and Boer (1993). In the pig, activation of the embryonic genome occurs at the four-cell stage (Jarrell *et al.*, 1990b, 1991; Reed *et al.*, 1992; Kronnie and Boer, 1993). Improvements in the IVC of pig embryos in recent years is a result of increased understanding of the embryo's reqirements. Glucose metabolism by pig embryos, for example, was reported by Flood and Wiebold (1988); they concluded that glucose is important as an energy source for the later stage pig embryos. It is now possible to culture pig embryos from the one-cell to the blastocyst stage successfully. As already noted (Section 5.3.4), the isolated mouse oviduct can provide a suitable environment for pig embryo development.

Coculture of pig embryos with oviductal cells or supplementation of culture medium with oviductal fluid can result in improved embryonic development *in vitro*. Studies reported by Pipkin *et al.* (1991) suggested that pig embryos may be cocultured (with porcine oviductal cells) through the four-cell block with greater success using BMOC supplemented with FCS or BSA than in other media (Ham's F-10; Ham's F-12; Whitten's medium). In Denmark, Smith *et al.* (1993), although reporting a beneficial effect of oviductal coculture upon development of 4–16-cell embryos, failed to detect a similar benefit for earlier stages.

According to Galvin *et al.* (1993), the NaCl concentration of conventional culture media may be responsible for the developmental block observed *in vitro* in pig embryos; they showed that this NaCl-induced block may be avoided by the addition of glutamine. Previous reports by Petters and Reed (1991) and Illera *et al.* (1992c) had also shown that a medium developed in North Carolina (NCSU-23), containing glutamine, taurine and hypotaurine, would support one-cell pig embryos (*in vivo*-produced) to the blastocyst stage. In contrast to cattle embryos, glucose is apparently not inhibitory to the development of pig embryos. However, in Canada, Misener *et al.* (1991) have shown that glucose supplementation of the IVC medium did significantly decrease the mean cell number in blastocysts derived from the culture of zygotes. In Australia, Lewis *et al.* (1992) found evidence that the addition of insulin to the IVC medium increased the size of pig blastocysts and the rate of protein synthesis per unit size. In Canada, Xia and Armstrong (1994) report evidence that coculture of IVM/IVF-derived pig embryos with oestradiol-treated oviductal cells may facilitate their development *in vitro*.

At Madison, Hagen *et al.* (1991) reported the successful development of early pig embryos through the block stage in simple media with a carbon dioxide atmosphere. The medium was a modified Tyrode/albumin/sodium lactate/sodium pyruvate (TALP), using HEPES buffer, and the embryos were cultured in an air atmosphere; the authors note that the ability to culture embryos in a warm-air incubator may sometimes make embryo transport simpler.

In studies reported from Cornell by Ball *et al.* (1991) and Ball and Miller (1991, 1992), it was found that equine oviductal cells significantly improved the development of four- to eight-cell horse embryos (*in vivo*-produced) to the blastocyst stage and that such cocultured embryos were capable of establishing pregnancy in recipient mares. Subsequent reports from the same laboratory (Ball *et al.*, 1992a, b) were the first to record the development of one-cell horse embryos to blastocysts *in vitro*; it was recorded, however, that development of one- to two-cell embryos, even with coculture, was significantly less than that of four- to eight-cell embryos. In Idaho, in an earlier report, Freeman *et al.* (1991) cocultured day-5 to day-7 equine embryos with oviductal tissue (taken from the ampulla/isthmus of equine oviducts); cocultured embryos were more readily sustained than those cultured in the medium (Ham's F-12 and Dulbecco's Modified Eagle's Medium (F-12-DME)) alone.

According to Battut *et al.* (1991) in France, after reaching the early blastocyst stage *in vitro*, and whatever the culture conditions, equine embryo development can differ from that occurring in the live animal; there is, for example, no synthesis of a capsule. Although the studies of Weber *et al.* (1993) showed that the equine oviduct provides essential factors to the embryo that are missing in the uterus, this could not be effectively simulated *in vitro*; no difference was detected when oviductal and uterine cocultures were compared. In Louisiana, Wiemer *et al.* (1989) reported pregnancies after 24 h coculture of equine embryos on fetal bovine uterine monolayer cells; it was suggested that such coculture may be an alternative to freezing under some conditions.

5.12.3 Buffaloes

It is believed that the chronology of development to the early blastocyst stage in the buffalo is similar to that in cattle, although hatching would seem to occur earlier (Chantaraprateep *et al.*, 1989). In studies with IVF buffalo embryos, Suzuki *et al.* (1991b, 1992) reported on their cleavage capability when cultured in M-199 supplemented with 1% oestrous buffalo serum and using a 5% carbon dioxide in air gas phase. Late morulae were transferred non-surgically into 14 recipients, one of which was subsequently diagnosed pregnant. In India, Ahluwalia and Majumdar (1992) reported on the culture of IVM/IVF presumptive buffalo zygotes in the rabbit oviduct; of 200 transferred, only two of those recovered had progressed beyond the 16-cell stage.

6 Storage and Cryopreservation of Oocytes and Embryos

6.1 Introduction

The aim of embryo storage *in vitro* is to preserve the embryo in a state of suspended animation, from which it may be revived after a short or long period to continue its normal development. It is believed that sperm and embryonic cells can probably remain viable at a temperature of −196°C (liquid nitrogen) for perhaps a 1000 years or more, the only source of damage at such a temperature being direct ionization from background radiation. In normal commercial practice, however, there is little need to think in terms of effective storage for other than a limited number of months or years. As shown by Hruska (1991), the usual length of cryopreservation of cattle embryos in liquid nitrogen does not affect their viability after thawing. There may, however, be good practical reasons why relatively short-term storage (days) of cattle embryos, at refrigerator temperatures, can be useful in some commercial situations. The present chapter examines methods, both long- and short-term, that may be employed.

6.1.1 Historical

The first successful report on the freezing of mammalian embryos was that of Whittingham (1971). A useful review of storage methods available for mammalian embryos in the mid-1970s was provided by Maurer (1976). Some of the important milestones in farm animal embryo freezing are listed in Table 6.1.

In livestock farming, the application of artificial insemination in cattle was greatly facilitated by the development of effective sperm freezing procedures by Chris Polge and colleagues at Cambridge in the early 1950s (Polge and Rowson, 1952). Effective exploitation of the genetically valuable embryo also depends on the development of effective techniques for the low-temperature

Table 6.1. First young born after the freeze–thawing of mammalian embryos.

Species	Year	Authors
Mouse	1971	Whittingham (1971)
Cow	1973	Wilmut and Rowson (1973)
Rabbit	1974	Bank and Maurer (1974)
Sheep	1974	Willadsen *et al.* (1974)
Rat	1975	Whittingham (1975)
Goat	1976	Bilton and Moore (1976)
Horse	1982	Yamamoto *et al.* (1982)
Human	1983	Trounson and Mohr (1983)
Pig	1991	Kashiwazaki *et al.* (1991a, b)

storage of embryos. This is particularly important in developing trade in cattle embryos; only embryos frozen under carefully controlled and specified conditions are permitted to enter international trade.

As already noted, the first successful studies in the freeze–thawing of mammalian embryos were those reported by Whittingham (1971) in the mouse. Cambridge workers at the Animal Research Station (Wilmut and Rowson, 1973) were the first to show that cattle embryos would survive freezing; in this, only one embryo survived to term of the 21 transferred to recipients (< 5% survival). These early studies indicated that slow freezing of cattle embryos to low subzero temperatures (−80°C) required slow thawing. Subsequently, it became evident that slow freezing of cattle embryos to relatively high subzero temperatures (−25 to −35°C) required rapid thawing (Willadsen *et al.*, 1978).

Such findings, developed initially by Willadsen at Cambridge in the freezing of sheep embryos (Willadsen *et al.*, 1974), subsequently formed the basis of the cryopreservation technique most widely used in commercial practice. Data from freezing methods in current practice suggest that the viability of high-quality bovine embryos after freeze–thawing results in pregnancy rates little different from those achieved with fresh embryos. As observed by several authors (Schneider and Mazur, 1986; Wilmut, 1986; Munar and Hasler, 1989; Niemann, 1991; Lucas-Hahn, 1992), the survival rates of cryopreserved bovine embryos have improved steadily over the past 20 years; provided factors such as embryo quality, recipient selection and donor/recipient synchrony are held within acceptable limits, pregnancy rates of 60% or even higher may be achieved.

According to a review by Rall (1992), since the first report of normal offspring from cryopreserved mouse embryos by Whittingham in 1971, similar success has been recorded for 13 other mammalian species, including man (see Table 6.1). A review of freezing studies in farm and domestic animals is provided by Fahning and Garcia (1992). The advent of *in vitro* production of cattle embryos has been useful in permitting early stage embryos (two-cell to 16-cell) to become routinely available for cooling and freeze–thaw studies; prior to that, such embryos were only available to researchers after surgical recovery from the oviducts.

6.1.2 Advantages of storage

In view of the fact that all freeze–thaw procedures result in the loss of a proportion of cells, the option of chilling embryos to 0–5°C, or storing at ambient temperature under appropriate conditions, may occasionally be one to be considered in embryo storage.

The successful freezing of *in vitro* fertilization (IVF)-derived cattle embryos is essential for any widespread commercial application of this form of technology (see Fig. 6.1). In particular, the frozen embryo that can be thawed on the farm immediately prior to transfer is most desirable.

In view of the fact that the import and export of cattle embryos must be based on frozen rather than fresh ones, the international trade in embryos is very much dependent on effective freeze–thaw procedures. In this, it is a matter of careful washing and employing procedures which do not result in any fracture of the zona pellucida (ZP), which would make them unacceptable for health certification. In terms of international trade in cattle embryos between some countries, it may be necessary to think increasingly about replacing bovine sera or bovine serum albumin (BSA) in the freezing medium by defined macromolecules to minimize still further any risk of disease transmission.

Fig. 6.1. Cattle twins born after freeze–thawing of IVMF-derived embryos at University College, Dublin. Much research has focused on developing commercially acceptable freeze–thaw techniques for use with *in vitro*-produced embryos. This is a a major consideration in determining the widespread application of the technology.

6.2 Storing the Fresh Embryo

Over the period that cattle embryo transfer (ET) has been in operation commercially, relatively few studies have dealt with the short-term storage of embryos at temperatures of zero to +10°C. However, for several good practical reasons, the ability to store embryos at temperatures above freezing for short periods (1–2 days) may occasionally be a useful alternative to freezing. As observed by Singh and Hare (1986), the disease potential of frozen embryos may be regarded as somewhat greater than that of fresh embryos. An intact ZP affords protection against certain viruses and this membrane may occasionally be cracked during the freeze–thaw process.

6.2.1 Embryo storage at ambient temperature

As part of normal cattle ET operations, there is an obvious need to store temporarily embryos recovered from donor cattle until such time as they are either transferred to a recipient or frozen. Bovine embryo viability begins to decline after 12 h of storage in phosphate-buffered medium, supplemented with serum (Wright, 1986). In Japan, Aoyagi *et al.* (1990) found that the pregnancy rate using frozen–thawed cattle embryos kept less than 3 h from flushing to freezing was significantly higher than that for embryos kept more than 3 h (65% vs. 43%). The general principle to be followed is to either transfer or freeze the embryo as soon as possible after recovery from the donor animal (Wright, 1985). It has been shown, for example, that pregnancy rates after transfer of frozen–thawed cattle embryos may be inversely proportional to the time interval elapsing between embryo recovery and the onset of freezing (Pettit, 1985).

The choice of media for embryo collection and temporary storage has ranged from complex media such as M-199 and Ham's F-10 to simplex formulations such as Dulbecco's phosphate-buffered saline (PBS) supplemented with blood serum. After a comprehensive review of the literature at the time, Wright and Bondioli (1981) concluded that Ham's F-10 was probably the medium of choice. Several groups subsequently compared the progress of embryos in F-10 (with HEPES to provide buffering under normal air conditions) and PBS, and found F-10 to be superior in terms of growth, survival and hatching ability (Rajamahendran *et al.*, 1985; Smith *et al.*, 1986; Hasler *et al.*, 1987).

For those operating under farm conditions, the requirement may simply be for PBS containing antibiotics and 2% fetal calf serum (FCS) for embryo collection and PBS with 10–20% FCS for temporary storage prior to transfer. As to storage conditions, room temperature is usually found to be satisfactory; if embryos are held for any length of time, the culture medium should be changed occasionally.

Available evidence suggests that reduced embryo viability is the inevitable result of any prolonged (> 12 h) storage in conventional media, especially when this is at body temperature. It is, however, possible to store bovine blastocysts (*in vivo*-produced) at refrigerator temperature for 24–48 h.

6.2.2 Embryo sensitivity to cooling

As with sperm, mammalian embryos experience a suspension of development and metabolism when maintained *in vitro* below normal body temperatures; in this, freezing (long-term storage) is not the only option of interest in dealing with cattle embryos. Studies carried out in Dublin more than 20 years ago were among the first to show that it was possible to store cattle embryos on a short-term basis at 10°C for 24 h or longer and that the embryos were capable of resuming normal development on transfer to the rabbit oviduct (Sreenan *et al.*, 1970). Work at Cambridge in the mid-1970s showed a clear relationship between embryo quality and developmental stage and its ability to survive chilling and freezing (Trounson *et al.*, 1976a, b, 1978, 1980); early blastocysts were apparently resistant to cooling damage whereas preblastocyst stages were not. Similar evidence has been reported by Richards *et al.* (1988) in Missouri. In some instances, blastocysts are capable of retaining their viability after storage at refrigerator temperature for 48 h, as expressed in pregnancies established after their transfer to recipient cattle (Trounson *et al.*, 1976a, b).

In Australia, Mohr and Trounson (1981) showed that cattle embryos collected before compaction of the blastomeres were sensitive to cooling below 7.5°C; they concluded that the extent and nature of cryoinjury may be dependent on the developmental stage of the embryo. In more recent times, studies in Texas by Looney *et al.* (1989) have shown that 5-day-old precompacted cattle embryos cannot withstand cooling to 4°C; embryos appeared to be morphologically normal when evaluated immediately after cooling, but cellular membranes apparently lysed after 3–7 h in culture at 38°C.

6.2.3 Embryo storage at refrigerator temperature

Experience and experiments have clearly shown that bovine embryos maintained in non-nutrient media (e.g. PBS) for any extended interval at room temperature or higher exhibit a decrease in their ability to develop further. However, it is now evident that this decreased capacity can be prevented by chilling embryos (those beyond a certain stage of development) to 0°C to +4°C (Trounson *et al.*, 1976 a, b; BonDurant *et al.*, 1982; Lindner *et al.*, 1983; Lindner and Ellis, 1985; Leibo and Winniger, 1986; Nibart *et al.*, 1988a, b; Refsdal *et al.*, 1988); it is apparent that *in vivo*-produced cattle embryos can be safely stored at refrigerator temperature for 24 h and perhaps longer. In Norway, work in which cattle embryos were stored at 4°C or frozen was reported by Landsverk *et al.* (1992a); refrigerated embryos gave a pregnancy rate (56%) comparable to that with frozen (50%). Similar findings have been reported with mouse embryos (Herr and Wright, 1988a, b). The practical merit of the chilling technique lies in facilitating the movement of cattle embryos from one geographical region to another in a large country such as the USA.

There is also the possibility that embryo viability may be better retained by chilling rather than by freeze–thaw procedures, which inevitably result in a proportion of cells being lost. In passing, it may be noted that in horses there is ample evidence of successful embryo storage at 5°C for 24 h (Carnevale *et al.*, 1987; Sertich *et al.*, 1988; Carney *et al.*, 1991a, b; Martin *et al.*, 1991). A commercial device (Equitainer, Hamilton-Thorn) which was originally employed in fresh stallion semen storage has been employed successfully for storing both equine embryos (Carney *et al.*, 1991b) and sheep embryos (Driancourt *et al.*, 1988) at refrigerator temperature.

In terms of evidence of prolonging embryo storage beyond 1 day, there is that provided by Herr and Wright (1988a), who maintained mouse embryos for 6 days at +4°C; the embryos were revived by putting them through a temperature adjustment period.

6.2.4 *Storage of IVMFC-derived embryos*

It is clear that laboratory-produced cattle embryos can be sensitive to chilling in a way not found with those produced *in vivo*. Leibo and Loskutoff (1993) record that early *in vitro* maturation, fertilization and culture (IVMFC)-derived blastocysts, suspended in PBS, failed to survive rapid or slow cooling to 0°C in most instances; these apparently were embryos produced according to the protocol described by Xu *et al.* (1987a, 1992b). Results of work by Pollard and Leibo (1993) led them to conclude, after a comparison of *in vitro*- and *in vivo*-derived embryos, that fundamental cellular differences may exist between these two categories. Cooling compact morulae from 30°C to 0°C had little or no adverse effect on *in vivo* embryos whereas none of the IVMFC morulae cooled below 15°C survived.

Such data are reconcilable with results obtained in Ireland by Yang, N.S. (1991). This worker stored IVMFC blastocysts for periods of up to 48 h at different storage temperatures. Based on the evaluation of embryos by subsequent culture (up to hatching), storage at 20°C was preferable to storage at 4°C or 39°C. For practical purposes, it was concluded that embryo storage at 20°C for up to 24 h may be possible. The most striking feature of these studies was the fact that 20°C rather than refrigerator temperature was found to be optimal (see Table 6.2).

Table 6.2. Effect of storage temperature on subsequent hatching rate of bovine embryos (from Yang, N.S., 1991).

Temperature	Number of embryos stored	Hatching percentage
4°C	567	289 (51%)
20°C	551	369 (67%)
39°C	516	272 (52.7%)

6.3 Developments in Freeze–Thaw Procedures

Improvements in various areas of cattle ET technology, including cryopreservation, are being made continually by those commercial concerns that are carrying out the routine transfer of several thousand cattle embryos each year (Wilmut, 1986); in this, the quest is for simplicity and effectiveness of procedure. It is clearly of considerable economic concern for commercial companies to employ the most effective freeze–thaw procedures. Basic and applied research during recent decades has resulted in two approaches to cattle embryo cryopreservation. The first, slow-freezing method evolved from research on the effects of cooling and warming on the survival of mammalian cells. A second approach, involving vitrification, aimed at promoting glass rather than crystal formation of intracellular water and cryoprotectant, has also been applied with some success in recent years (Rall, 1992).

In human IVF, the success rate of *in vivo* hatching of IVF and *in vitro* cultured (IVC) human embryos from their ZPs is apparently rather low. It is also believed that hatching difficulties may arise from the thickness of the ZP or to alterations in this glycoprotein membrane due to IVC and/or freeze–thawing. For such reasons, techniques for assisted hatching have been of interest to those working in human IVF. Efforts to assist have generally involved opening the ZP mechanically or by 'drilling' with acid Tyrode's solution (pH = 2.8). However, according to Gordon and Dapunt (1993a), such zona opening may pose several risks to the embryo. The embryo may be lost from the zona during transfer or may 'escape' through the zona opening before blastulation. The procedure may result in herniation, with the embryo becoming trapped in the opening. For such reasons, some workers have employed a non-invasive 'zona rubbing' procedure, in which reduction of the thickness of the zona is achieved by gentle rubbing with a microneedle (Nijs *et al.*, 1993). Work with mouse embryos has also suggested that circumferential zona thinning with acid Tyrode's solution may be a possible procedure for use in certain instances (Gordon and Dapunt, 1993b).

6.3.1 Conventional freeze–thaw protocols

The principles of cryopreservation are believed to be the same for all living cells, the most important aspect of the process being the removal of most of the water from cells before they freeze intracellularly (Seidel, 1989). Research in freezing techniques has included numerous studies dealing with the type and concentration of cryoprotectant, cooling and freezing rates, seeding and plunging temperatures, thawing temperatures and rates and techniques used to ensure cryoprotectant removal (Lehn-Jensen, 1983, 1984; Lehn-Jensen and Ball, 1983a, b; Franks *et al.*, 1985, 1986; Wright, 1985; Prather *et al.*, 1987b; Takeda, 1987). It has even been demonstrated that excellent-quality cattle embryos can tolerate up to three freeze–thaw cycles and still maintain development *in vitro* (Vitale *et al.*, 1994).

The most commonly used freezing method involves placing the bovine embryo, at the blastocyst stage, in a concentrated glycerol solution (1.4 M in PBS supplemented with bovine serum or BSA) at room temperature for a 20-min equilibration period. Bovine serum is generally added to PBS at about the 15% level and BSA at 4 mg ml^{-1}. The straws are usually cooled abruptly from room temperature to 0°C and seeded at −4°C to −7°C. Seeding is the term used to describe the controlled initiation of ice formation at slightly supercooled temperatures, generally by touching the wall of the straw with very cold forceps. After seeding, ice forms quickly throughout the entire straw, and cooling is continued at a rate of 0.3°C min^{-1} to −35°C, when the embryo is plunged into liquid nitrogen.

Seeding is carried out because supercooled solutions do not form ice and cells would otherwise not dehydrate. When such undehydrated cells do freeze, they are likely to be damaged by large ice crystals.

One of the principal determinants of bovine embryo survival during freezing is the cooling rate. As a result of work in Cambridge (Willadsen, 1977), for many years it has been common practice to use a rate of about 0.3°C min^{-1} for cattle embryo cryopreservation. However, it was subsequently shown by Leibo (1988) that maximum survival of bovine embryos suspended in 1.5 M glycerol can be achieved with a cooling rate of 0.6°C min^{-1}.

As to thawing, various reports have dealt with procedures for the exposure of the frozen straw to air and water. A study by Rall and Meyer (1989), for example, used zona fracture as a guide to an appropriate procedure. These authors found a low incidence of zona damage when straws were warmed in air for 10 seconds prior to transfer into water; the short exposure to air allows the straw to warm to about −100°C prior to rapid warming in water.

6.3.2 Cryoprotectants in common use

The literature reveals many different methods for the cryopreservation of embryos that have been developed in various laboratories around the world. There are several crypoprotectants being employed in various concentrations and occasionally in combination with one another. Cryoprotectants generally fall into two categories: intracellular and extracellular. Of the intracellular cryprotectants, which are of rather low molecular weight, glycerol and dimethylsulphoxide (DMSO) have been those most commonly used in freezing cattle embryos. These agents permeate the cells of the embryo.

The extracellular cryoprotectants are larger molecules, such as sugars and proteins (e.g. BSA and hyaluronic acid (HA)). Little is known about the mechanisms involved in cryprotection by macromolecules. It has been suggested by Leibo (1988) that BSA protects cell membranes in the immediate post-thaw phase and its most valuable role may be in stabilizing cell membranes in some way at this time. Among the more novel forms of extracellular cryoprotectant to appear in recent years are the antifreeze glycoproteins extracted from Antarctic fish (Rall, 1992). As a general rule, embryos appear to survive freezing more readily when a combination of cryoprotective agents rather than when a single agent is used.

According to Gook *et al.* (1993), most human IVF clinics have altered their cryopreservation procedures to use the cryoprotectant 1,2-propanediol (PROH), following the method of Lassalle *et al.* (1985). This technique is known to give consistent results, maintain cellular integrity and achieve acceptable pregnancy rates with both embryos and pronuclear oocytes.

It is known that there are permeability differences between species and that cattle embryos are more permeable to glycerol than to DMSO, which was the first cryprotectant used in early experiments with cattle embryos (Wilmut and Rowson, 1973; Willadsen *et al.*, 1978). However, towards the end of the 1970s, glycerol emerged as a superior cryoprotectant to DMSO for the freezing of bovine compact morulae and blastocysts. Such differences in permeability can also be important when mammalian embryos are being cleared of cryoprotectant after thawing. In contrast to the thawing of frozen bull sperm, bovine embryos generally require specific steps to be taken to remove cryoprotectant after thawing.

Cattle embryos undergo rapid contraction and expansion as a result of marked changes in the osmolarity of the cryoprotectant medium in which they are held. Damage may result when embryonic cells are exposed to rapid changes in osmotic conditions and it is for that reason that cryoprotectants have usually been added or diluted out in a series of steps (the so-called multistep freezing and thawing procedures). Osmotic shock is an aspect of cell cryopreservation that is likely to be responsible for damage to the embryo. Oda *et al.* (1992) observed that mouse embryos could contract or swell to volumes of 30% to 290% of their isotonic volume without appreciable effects on their subsequent developmental capacity.

Accurate identification of those variables that influence osmotic shock in cattle embryos is required. It is also important to know that the survival of frozen cattle embryos after thawing may be as critically dependent on the warming rate as on the cooling rate (Leibo, 1988; Mazur, 1988). When dilution occurs at body temperature (37–39°C), glycerol can flow out of the bovine embryo sufficiently fast for it not to experience osmotic shock; however, dilution at 0°C causes an osmotic shock even with dilute glycerol solutions (Leibo, 1988). At the low temperature, water can flow into the embryo much more rapidly than glycerol can flow out and cells swell beyond their tolerable limits.

An alternative approach to the multistep thawing procedure is to place the cattle embryos in a medium containing a high concentration of a non-penetrating molecule such as sucrose. This enables the embryo to reach equilibrium, in terms of its isotonic volume, without undergoing the cell distension that otherwise occurs in a conventional medium. The use of sucrose as an osmotic buffer was first described by Leibo and Mazur (1978). In this, the cattle embryo is transferred directly from the concentrated cryoprotectant solution into a sucrose solution that is isotonic with the cryoprotectant. The non-permeating sucrose prevents swelling and the absence of cryoprotectant in the extracellular solution permits the rapid efflux of the intracellular solute. Sucrose concentrations between 0.25 M and 1.0 M have been used by various workers. When this range of sucrose concentrations is employed to dilute 1.5 M glycerol, the lower the sucrose concentration, the longer it takes to remove all glycerol from the cells.

It is believed that the protective effect of cryoprotectants is a result of factors such as: (i) reducing damage from solution effect and intracellular freezing; (ii) lowering the temperatures at which intracellular freezing occurs; and (iii) a stabilizing effect on cell membranes.

6.3.3 Simplification of freeze–thaw procedures

Simplifications of the conventional multistep freeze–thaw procedures have been reported by various workers. In France, studies suggested that cattle embryos could be equilibrated in 1.4 M glycerol in just the one step, of 10–30 min, rather than having to go through several steps (Chupin and Procureur, 1984; Chupin *et al.*, 1984). In Germany, it was also clearly demonstrated that the one-step addition of a final concentration of 1.4 M glycerol results in equally high survival rates as with the stepwise procedure (Niemann, 1985); cell membranes remained intact even though embryos shrank dramatically. It is apparent that membrane flexibility of the bovine embryo is sufficient to tolerate considerable changes in osmolarity (from 285 to 1700 mosm).

At the thawing end, workers such as Bielanski *et al.* (1986) found that removal of glycerol by sucrose yielded higher survival rates of embryos than did stepwise dilution of the cryoprotectant. It was also observed that the response to sucrose, i.e. shrinkage followed by re-expansion after exposure to isotonic medium, could be employed as an indicator of embryo viability. Embryos severely damaged during the freeze–thaw process would not shrink as visibly in sucrose or expand when placed in normal culture medium.

One of the more unusual observations in the first successful freezing of mammalian embryos was that survival apparently required rather slow warming at a rate of 25°C min^{-1} or less (Whittingham, 1971; Wilmut and Rowson, 1973). Slow cooling methods employed at that time led to a degree of cell dehydration which caused damage if embryos were thawed too rapidly. It is now clear that the optimum warming rate is dependent on the temperature at which slow cooling is terminated (and the degree of dehydration attained in cooling). It is now common practice to thaw cattle embryos by rapid warming in water baths at 37°C after having been cooled slowly to high subzero temperatures.

In evaluating the optimal methods to use in the thawing of straws, some workers have observed the effect of different thaw procedures on the ZP of the bovine embryo. Retention of the normal structure and function of the ZP is an important goal of cryopreservation; an intact zona is essential under certain conditions in preventing the transmission of viral, bacterial and fungal pathogens during cattle ET operations. Data presented by Rall and Meyer (1989) indicated that the use of plastic straws and careful control of the warming procedures can reduce or eliminate zona damage to bovine embryos. In their report, none of the ZPs of embryos frozen in 0.25 ml straws were damaged when the straws were warmed in air at 20°C (normal room temperature) or when a 10 second exposure to room temperature was provided before thawing in water at 37°C. In New Zealand, cattle studies reported by Subramaniam *et al.* (1990) also showed that rapid warming in water resulted in zona damage.

Freezing containers

Plastic straws of 0.5 or 0.25 ml capacity are favoured for embryo freezing by those working in commercial practice; the 0.25 ml ministraw is probably the preferred choice. Possible toxic factors associated with the straw, if they exist at all, can usually be eliminated by rinsing with sterile medium before introducing the embryo. The advantages of the straw over the earlier containers (e.g. 1 ml glass ampoules) include better exposure of the embryo and medium to the cooling effect of liquid nitrogen and the ease of storing large numbers of straws in nitrogen storage units. The effect of the embryo's position in the straw during freeze–thawing was an aspect investigated by Nakahara *et al.* (1990); the least damage to ZPs was observed with embryos located in the middle portion and held horizontally in the straw rather than vertically. Another Japanese report in the following year, this time with cattle oocytes, provided similar evidence on the merit of the horizontal position on the incidence of ZP damage (Suzuki and Hayashi, 1991).

Freezing apparatus

The freezing apparatus is the most expensive item in any bovine embryo cryopreservation programme. There are a number of commercial concerns that have marketed apparatus that provides for a variety of freezing and thawing programmes (e.g. Planar, IMV). In the research laboratory, several inexpensive machines have been described (e.g. Pye *et al.*, 1981; Staples and Page, 1987). The 'Pye apparatus' has been frequently used in the cattle IVF laboratory in Dublin (see Fig. 6.2). There are also those who have frozen

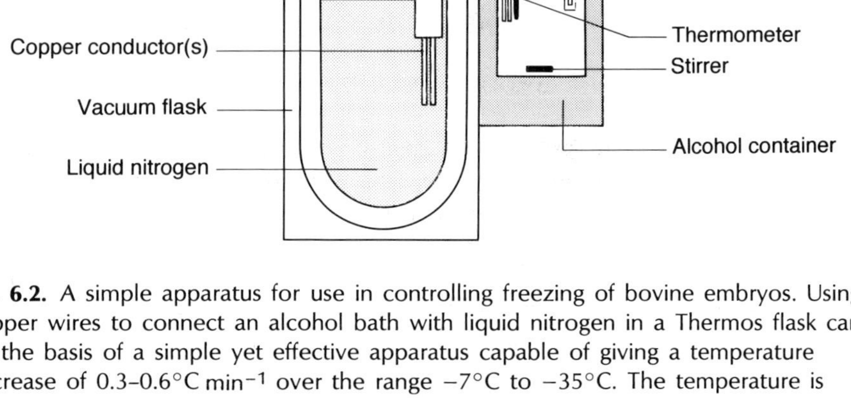

Fig. 6.2. A simple apparatus for use in controlling freezing of bovine embryos. Using copper wires to connect an alcohol bath with liquid nitrogen in a Thermos flask can be the basis of a simple yet effective apparatus capable of giving a temperature decrease of 0.3–0.6°C min^{-1} over the range −7°C to −35°C. The temperature is controlled by the number of copper wires used (from Pye *et al.*, 1981).

bovine embryos using computer-controlled thermoelectric cells (Voelkel *et al.*, 1984).

Seeding

Silver iodide has been used to induce crystal formation (automatic seeding) in the slow cooling of cattle embryos (Kojima *et al.*, 1988a, b). The development of an automatic and reliable ice nucleation method is further described by Kojima (1991), using silver iodide as the crystal-forming agent. The method was further improved for practical application by immobilizing the silver iodide in alginate gel, which made it possible to immerse the silver iodide with cattle embryos in any kind of freezing container; the author recorded a high post-thaw survival rate.

6.3.4 Substitutes for bovine serum/BSA

The cattle ET industry, as it currently operates, involves the movement of bovine embryos, both domestically and internationally (Joly *et al.*, 1992). Many thousands of embryos have been transported throughout the world, apparently without a single case of contamination by pathogens being reported. However, the replacement of bovine serum/BSA with chemically defined macromolecules for freezing would reduce the possibility of certain disease hazards in the application of cattle ET still further.

A report by Seidel *et al.* (1990) suggested that polyvinyl alcohol (PVA) may be an appropriate substitute where embryos have to be exported to countries that do not permit substances of animal origin in the freezing medium. It is also apparent that proteins used in conventional freezing media (bovine serum/BSA) may be replaced by HA, a macromolecule biosynthesized *de novo* commercially (Vetrapharm, Canada). It should be noted, however, that the handling of embryos in media containing HA may be more difficult than in media containing animal biological proteins, even though the outcome of freeze–thawing may otherwise be similar to that with normal protein supplementation.

6.4 One-step Thawing Procedures

6.4.1 Historical

Although many thousands of cattle embryos have been frozen since the first calf from a frozen embryo was born 21 years ago in Cambridge, a major obstacle to the more extensive use of freezing has been the tedious method required for the stepwise removal of cryoprotectant from the thawed embryo (see review by Niemann, 1991). In the conventional multistep dilution procedure, the embryo is exposed to decreasing concentrations of the protective agent, which usually requires a microscope and a minimum of 1–2 h to be carried out under laboratory-type conditions. Some workers have simplified multistep to one-step thaw procedures in the laboratory. In Norway, Landsverk *et al.* (1992a, b) used a one-step dilution of glycerol with 1 M sucrose and recorded higher pregnancy rates in recipient cattle (63% vs. 44%) than with a multistep procedure.

In the early to mid-1980s, however, an alternative to the multistep thaw procedure was described and tested by several groups, notably Stan Leibo and colleagues in North America, Renard and coworkers in France and Massip and associates in Belgium (Leibo, 1982a, b, 1984, 1985, 1986a, b, 1988; Renard *et al.*, 1983; Massip *et al.*, 1987). In this, the frozen–thawed bovine embryo was transferred from the cryoprotectant medium into one containing a hypertonic concentration of a non-permeating solute, such as sucrose. When this occurs, the intracellular cryoprotectant diffuses out of the embryo to maintain osmotic equilibrium. When the cattle embryo no longer contains intracellular protectant, it shrinks to a volume determined by the osmolarity of the impermeant sucrose solution and the embryo can then be rapidly diluted in an isotonic medium or directly transferred to a recipient animal.

In the development of the one-step thawing procedure, it may be noted that Leibo and Mazur (1978) were the first to demonstrate that frozen–thawed mouse embryos could have the cryoprotectant diluted out by washing the embryos in the sucrose solution in which they had been frozen; these findings were adapted by Leibo (1982a, b) to the freezing of cattle embryos. In this, cryoprotectant medium and sucrose diluent (0.25–1.0 M sucrose solution) were introduced successively into a straw in appropriate volumes and separated by air bubbles. After thawing, the straw was shaken (like a clinical thermometer) to dislodge the air bubbles; this exposed the embryo to the sucrose and resulted in the efflux of the glycerol cryoprotectant from the embryonic cells. The embryo was then incubated within the straw for an appropriate interval (2–20 min) at a controlled temperature (20–37°C) before being transferred to the waiting recipient animal (see Fig. 6.3).

It is known that one-step thawing, which involves shaking the straw, may suffer from variability, depending on the operator. In Belgium, therefore, a freeze–thaw procedure was devised to avoid the need for the dilution step after thawing. In this method, by incorporating sucrose in the freezing medium (1.36 M glycerol in PBS) at a concentration of 0.25 M, the embryos exposed to the glycerol/sucrose mixture are predehydrated prior to cooling; this means that slow cooling can be terminated at −25°C rather than at lower subzero temperatures (Massip *et al.*, 1987, 1993a). In Japan, workers have employed a somewhat similar technique (a mixture of 1.4 M glycerol and 0.2 M sucrose) with a plunge temperature of 20°C (Suzuki *et al.*, 1989b); although numbers were small, the 57% pregnancy rate established by the direct transfer of such embryos was regarded as encouraging.

From a commercial viewpoint, a one-step thawing procedure could clearly offer important advantages when cattle embryos are shipped to countries where there may be no laboratory facilities available for handling. As noted by Voelkel and Hu (1992b, c), direct transfer methods would eliminate the need for trained embryologists to be present during embryo transfer; trained inseminators could readily perform the transfers on the farm. It would bring

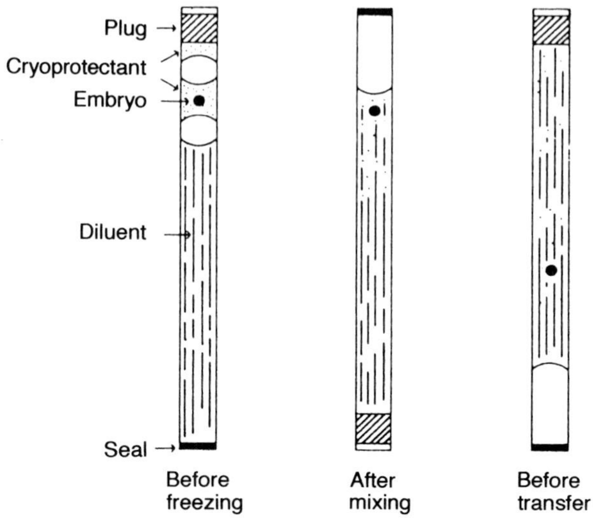

Fig. 6.3. The one-step *in-situ* thawing method described by Leibo (1984). Before freezing and after thawing, the embryo in its cryoprotectant solution is isolated from the diluent by an air bubble. The embryo is mixed with the diluent by shaking the straw upside down towards the plug, producing a single column of fluid. The embryo is then allowed to respond osmotically to the diluent before transfer.

embryo transfer on the farm to the same level of complexity as carrying out an insemination with frozen–thawed semen. This could lead on to the development of do-it-yourself (DIY) ET in cattle in the way that DIY artificial insemination has been applied for many years.

6.4.2 Use of bovine serum albumin

In the initial work, pregnancy rates after one-step thawing were often markedly below those obtained with multistep methods, a major reason for this being the fact that embryos could not be examined after thawing and a proportion of those transferred to recipients were presumably non-viable. Leibo (1984), for instance, reported the transfer of 1259 bovine embryos after one-step thawing with a subsequent pregnancy rate of 26%, much below the rate acceptable for most commercial purposes. In a later report, however, Leibo (1985) was able to demonstrate a substantial improvement in pregnancy rate. This was achieved by substituting BSA for FCS in the 1.5 M glycerol cryoprotectant solution used in freezing; pregnancy rates were 28% with FCS and 60% with BSA.

In the cryopreservation of mouse embryos, the optimal concentration of BSA in the DMSO/PBS freezing medium is recorded as $4\,g\,l^{-1}$ (Harrison *et al.*, 1987).

6.4.3 Use of ethylene glycol as a cryoprotectant

In recent years, ethylene glycol has been effectively employed as a cryoprotectant for cattle embryo preservation. The molecular weight of this agent is lower (62.07) than that of glycerol (92.10), propylene glycol (76.10) and DMSO (78.13) and it may be that its beneficial effect is partly due to its high permeability.

According to workers such as Suzuki *et al.* (1990a, b, c, 1993a, b) and Kocoski *et al.* (1991), one of the greatest disadvantages of the traditional method of cattle embryo freezing is the need to remove cryoprotectant after thawing. The latter authors presented data in former Yugoslavia suggesting that cattle embryos frozen with 10% ethylene glycol may not require removal of the agent before making the transfer. This view was based on embryo culture after thawing, rather than on evidence from transfers. However, it soon became clear from studies reported from the USA by Voelkel and Hu (1992a, b) that ethylene glycol is an effective cryoprotectant for bovine embryos, permitting direct rehydration of thawed embryos in a holding medium. It was established that a concentration of 1.5 M was optimal under the conditions of their work. It was considered by these authors that a direct transfer procedure based on ethylene glycol justified much further investigation.

In Japan, Suzuki *et al.* (1993a, b) showed that the pregnancy rate after using bovine embryos frozen in 1.8 M ethylene glycol was significantly higher than after using those frozen with other cryoprotectants; the authors suggested that one merit of ethylene glycol may be in reducing the toxic effects associated with cryopreservation. Higher pregnancy rates were achieved with IVMFC-derived cattle embryos, suspended in 1.8 M ethylene glycol, cooled at 0.3°C rather than at 0.5°C min^{-1} (Ooe *et al.*, 1993). In Canada, McIntosh and Hazeleger (1994) report adopting ethylene glycol as the cryoprotectant for routine operations in their commercial cattle ET operations on the basis of results achieved.

6.5 Freezing the IVMFC-derived Embryo

The viability of the *in vitro*-produced embryo after freeze–thawing is of crucial importance to the commercial exploitation of this technology. Commercial requirements are for a frozen IVMFC-derived embryo which can be thawed by a one-step procedure on the farm with a guaranteed survival rate approaching as closely as possible to 100%. To save time, it is also highly desirable that direct transfer to the recipient can be carried out immediately after thawing rather than after a delay. Such a technique has been employed commercially in Ireland and was used at one time to provide a beef cattle twinning service (Bourke, 1992). As noted by Leibo and Loskutoff (1993), it has been generally accepted that *in vitro*-produced cattle embryos are much more sensitive to freezing than those produced *in vivo*. However, as these authors point out, there is growing evidence to show that increasing the efficiency of cryopreservation techniques for use with IVF embryos may well require improvements in the techniques of maturation and culture.

A variety of methods have been employed in freezing IVMFC-derived embryos. Many workers have used glycerol as the cryoprotectant (Fukushima *et al.*, 1992a, b; Kajihara *et al.*, 1992, 1993), others have used 1,2-propanediol (Goto *et al.*, 1992b, d) or ethylene glycol (Leibo and Loskutoff, 1993; Pollard and Leibo, 1993). In Japan, Kajihara *et al.* (1993) found that the highest hatching rate was achieved when the pH of their freezing medium was 6.4. The toxicity of different cryoprotectants used for varying equilibration periods (10–120 min) was examined by Takagi *et al.* (1993); they found that ethylene glycol was relatively non-toxic, regardless of the length of the equilibration period. Ethylene glycol was also found to be a suitable cryoprotectant for IVMFC-derived embryos by Suzuki *et al.* (1993a, b); embryos survived direct rehydration by placing them into holding medium.

6.5.1 Differences between in vivo *and* in vitro *embryos*

As noted earlier (Section 6.3.1), it has been reported that the maximum survival rate of *in vivo*-produced embryos, suspended in conventional freezing medium (1.5 M glycerol), has been achieved using a cooling rate of 0.6°C min^{-1} (Leibo, 1988). In a recent study in Canada, Leibo and Loskutoff (1993) have shown that IVMFC embryos (produced by the method of Yang, N.S., *et al.*, 1990), when suspended in 1.5 M ethylene glycol and cooled to −35°C at a rate of 0.6°C min^{-1}, gave a virtually identical survival rate to that previously reported for the *in vivo* embryos cooled in glycerol. It should be noted that Yang, N.S., *et al.* (1990) in Ireland employed the *in vitro* maturation (IVM)/IVF protocol of Lu *et al.* (1987b) and an IVC system based on a granulosa cell monolayer supported by M-199. In Sweden, on the other hand, using a different embryo production system and glycerol as the cryoprotectant, Larsson *et al.* (1992) recorded poor survival of frozen–thawed IVMFC-derived embryos and concluded that this was indicative of inadequacies in their quality.

Elsewhere, the report of Pollard and Leibo (1993) has drawn attention to marked differences between *in vitro*- and *in vivo*-produced cattle embryos in their sensitivity to low temperatures. Similar differences have been reported in the freezing of *in vivo* and *in vitro* sheep embryos (Tervit *et al.*, 1994). When cattle embryos have been cooled in 1.5 M ethylene glycol, for instance, no IVMFC-derived morula cooled below +15°C survived; *in vivo* morulae suffered no adverse effects. After freeze–thawing, only 20% of *in vitro* embryos survived in contrast to the 80% of *in vivo* embryos. Such data, showing clear differences in favour of *in vivo*-produced embryos after cooling and freeze–thawing, are common in the literature (Waldmann, 1991; Larsson, R.C., *et al.*, 1992b).

The studies reported by Leibo and Loskutoff (1993) also found much evidence of adverse effects on *in vitro* embryos, when using those produced according to the protocol of Xu *et al.* (1992b). In this, the bovine oviductal epithelial cell (BOEC) coculture system with B2 medium was employed. When

1.5 M ethylene glycol was used as the cryoprotectant, higher hatching frequencies was shown by IVMFC-embryos cultured with granulosa cells and M-199 than with the other monolayer system involving BOEC and B2 medium. In Texas, Voelkel and Hu (1992a, b) had earlier reported that IVMFC-derived embryos produced by coculture with oviductal cells exhibited low tolerance to freezing; embryos cocultured with buffalo rat liver (BRL) cells, on the other hand, showed enhanced capacity to survive freezing. It is evident, from such findings, that certain culture systems may produce IVMFC embryos that can tolerate freeze–thawing as readily as *in vivo* embryos, provided the appropriate freezing agent (ethylene glycol rather than glycerol) is used.

IVM/IVF blastocysts developed *in vitro* have been shown in some studies to be more sensitive to cryoinjury than those developed in the rabbit oviduct (Fukushima *et al.*, 1992a, b). Studies have also shown that transfer of IVM/IVF bovine embryos to the rabbit oviduct can increase the proliferation of inner cell mass (ICM) cells to the level found in embryos fertilized *in vivo* (Iwasaki and Nakahara, 1990a). Greve *et al.* (1992) compared the viability of IVM/IVF embryos cultured either on BOEC or in the sheep oviduct and found that oviduct culture markedly improved the viability, as measured by freezing resistance.

It is also of interest to note that certain features associated with IVMFC embryos show similarities to features in pig embryos, which are known to be notoriously sensitive to cooling below 15°C and to freeze–thawing (Polge *et al.*, 1974). The presence of lipid droplets, for example, is much more in evidence with *in vitro* embryos, giving them their 'sunburnt' appearance (Tsuzuki *et al.*, 1992; Greve *et al.*, 1993). Pig embryos show a much greater lipid content than ruminant embryos. It is only in recent times that successful cryopreservation techniques for pig embryos have been reported and only then for the hatched blastocyst (Hayashi *et al.*, 1989; Nagashima *et al.*, 1989, 1992). In a review, Niemann (1991) notes information suggesting that the high lipid content of the pig embryo may lead to uneven intracellular ice formation and that this is the cause of the sensitivity to low temperatures; it may be that similar events occur with some of the IVMFC embryos that are produced.

There can be other differences between IVMFC embryos produced by some methods and *in vivo* embryos. It may, for example, take the enzyme pronase twice as long to digest the ZP of the *in vivo* embryo than that of the *in vitro* embryo (Pollard and Leibo, 1993). Such differences presumably are likely to influence the permeability of the zona to water and cryoprotectants.

Comparative studies of chilling injury in bovine embryos and those of other species have been reported by Pollard and Leibo (1994); it is likely that a close examination of factors influencing chilling sensitivity may lead to an increased understanding of embryo survival after freezing and other storage procedures.

6.5.2 *Embryo quality considerations*

There is ample evidence in conventional cattle ET literature to show that embryo quality and stage of development can strongly influence the success of freeze–thaw procedures, judged in terms of pregnancy rates established after transfer to recipients (Massip *et al.*, 1982; Kennedy *et al.*, 1983; Prather *et al.*, 1987a). In some reports, it has been suggested that expanded cattle blastocysts may freeze less readily than late morulae or blastocysts (Niemann *et al.*, 1982); this apparently may be related to ultrastructural damage to the tenofilaments and desmosomes within the trophoblastic layer, which adversely affects the hatching process. As to early-stage cattle embryos, there has been evidence for many years suggesting that, prior to day 5, *in vivo* embryos do not readily survive freeze–thawing (Willadsen *et al.*, 1978). In more recent times, a study by Vincent *et al.* (1985) examined the freezing of one- and two-cell cattle embryos; only 32% developed after freeze–thawing.

There has been some evidence that cattle embryos from certain donors freeze more readily than those from others (Niemann, 1991). In the mouse, it may be noted, significant differences in the tolerance of embryos from different genetic strains have been observed.

The grading of embryos on their morphological appearance is part of normal routine in cattle ET operations. The pregnancy rates associated with the different quality grades may show marked differences, as shown in data from transfers in a commercial multiple-ovulation embryo transfer (MOET) programme conducted in the UK (see Table 6.3) and reported by McGuirk (1990).

The general consensus with cattle embryos appears to be that the most suitable stages for freezing are from the late morula through to the expanded blastocyst, with the optimum stage being the early blastocyst. Based on experience in Ireland, the same would also hold true for IVMFC embryos, provided that the blastocyst in question was 7 days old and of good quality (grade 1).

Table 6.3. Summary of pregnancy rates, according to embryo grade, in the MMB-Premier MOET breeding scheme (from McGuirk, 1990).

Grade of embryo	Number transferred	Pregnancy rate (%)
First	256	74.2
Second	143	60.8
Third	36	33.3
Total	435	66.4

In vitro *embryos*

There is evidence to show that viability after cryopreservation of IVMFC embryos may be correlated with the rate of early embryonic development (Greve *et al.*, 1992); embryos cleaving between 26 and 32 h after fertilization reached the blastocyst stage sooner than those dividing later (36–48 h). It was found that embryos which reached the blastocyst stage on day 7 survived cooling and freezing more successfully than those in which blastocyst formation was delayed. The fact that the rate of compaction is higher in early-cleaving embryos (Van Soom *et al.*, 1992) may be a relevant quality consideration. In Belgium, Mermillod *et al.* (1992a) found evidence suggesting that early-appearing blastocysts are less sensitive than later ones to cryopreservation, using their freezing method (1.4 M glycerol and 0.25 M sucrose). In Canada, Campo *et al.* (1993) presented data suggesting that bovine blastocysts produced by IVF and cocultured with BOEC for 7 days after fertilization have the highest survival rate after traditional freezing procedures. The early blastocyst is given as the favoured stage of IVF-derived bovine embryo for freezing in the report of Pavasuthipaisit *et al.* (1993) in Thailand.

6.6 Evaluating the Embryo After Thawing

6.6.1 Morphological assessments

The frozen–thawed bovine embryo can be evaluated by putting it through a short-term period of IVC and assessing its progress by morphological examination. On the available evidence, it appears that post-thaw assessment of *in vivo*-produced cattle embryos can provide a reasonably accurate prediction of pregnancy rates after embryo transfer (Willadsen *et al.*, 1978; Elsden *et al.*, 1982; Seidel *et al.*, 1983; Takeda *et al.*, 1985; Rodriguez, J.H., *et al.*, 1991). In the early work in cattle embryo cryopreservation, Willadsen *et al.* (1978) followed the rule that no embryo was considered to have survived freeze–thawing unless it had expanded or re-expanded into a blastocyst with a visible embryonic disc by the end of the culture period. Attempts to assess embryo viability after thawing have included, among other procedures, the use of dye-exclusion tests (Dooley *et al.*, 1987), but these would have little to commend them for routine application.

6.6.2 Hatching as an indicator of embryo viability

The extended culture of the bovine blastocyst after thawing could probably take hatching as evidence of its viability/normality. In view of recent evidence in mice, showing clearly that hatching in that species is almost entirely the result of lysis of the ZP because of enzymes secreted by the embryo (Gordon and Dapunt, 1993a, b), it would be of interest to establish whether the same holds true in cattle. If it does, then hatching may be taken as an important quality expression, reflecting the ability of embryonic cells to secrete the lysin.

6.7 Freezing the Demiembryo

The use of identical cattle twins in research has a long history, dating back to work in Sweden in the 1930s and in many other countries (New Zealand,

UK, USA, France, Ireland, Germany) from the 1940s onwards. In this work, the identicals were usually collected from farmers a few days after birth. However, since the pioneering work of Willadsen (1979) with sheep, microsurgical techniques have been successfully applied to cattle embryos and normal monozygotic twin calves have been born after embryo bisection (Willadsen, 1986a; Sreenan and McEvoy, 1989a; Schmidt *et al.*, 1992a, b). Normal offspring have also been produced from cattle embryos containing only a quarter of the original cell number (Willadsen and Polge, 1981; Loskutoff *et al.*, 1991a, b, 1993; Tan *et al.*, 1991; Bredbacka *et al.*, 1992).

The relevance of demiembryos to the laboratory production of cattle embryos is likely to be through the use of live cattle as donors of oocytes (by way of transvaginal ultrasound-guided oocyte recovery). As mentioned elsewhere (see Section 1.5.4), recently developed follicular aspiration techniques are likely to enable high-yielding dairy cows (or other desirable cattle strains) to be used as oocyte donors. In this, micromanipulation techniques may be used to increase the number of embryos available for transfer, by splitting or even quartering IVMFC-derived embryos from genetically superior animals. Such procedures may be regarded, however, only in terms of acting as an interim measure before the advent of commercially acceptable large-scale cloning technology.

6.7.1 Options available

Several authors have reported on freezing cattle embryos either before or after they have been 'split' by microsurgery. Such work is of interest in showing the effect of freeze–thaw procedures on the subsequent viability of cattle embryos that have suboptimal cell numbers. Freeze–thawing of demiembryos results in lower pregnancy rates and viability than freezing intact embryos (Picard *et al.*, 1988). Freezing the embryo before splitting apparently resulted in lower survival rates of demiembryos than freezing after splitting in studies reported by Takeda *et al.* (1987) and Suzuki *et al.* (1991a). In the studies of Chesne *et al.* (1987), a short period of culture (4–6 h) prior to freezing apparently assisted survival of demiembryos as compared with immediate freezing after splitting. In contrast to this finding, Schmidt *et al.* (1992a, c) report that a 24 h period of coculture with BOEC prior to freezing reduced cell number in the demiembryo rather than increasing it.

In Canada, Picard *et al.* (1988) suggested that the combination of splitting the day-5/6 precompaction embryo, with subsequent agar-imbedding and intermediate transfer to a sheep oviduct to bring the demiembryo to the blastocyst stage before freezing, appeared to be a useful method of obtaining identical cattle twins. Other studies reported by King, W.A., *et al.* (1992) in Canada showed that splitting and culture of embryos before freezing resulted in a preferential loss of female embryos when they were thawed and transferred to recipients; it was not clear whether this effect was attributable to bisection or to culture.

The effect of microencapsulating bisected cattle embryos on their survival after freeze–thawing was examined by Hollingsworth and Page (1988); this appeared to have some beneficial effect. As to the effect of microsurgery on cell loss, Nibart *et al.* (1988b) split cattle embryos and recorded the loss of some 9% of cells. In Denmark, Schmidt *et al.* (1991) has reported on the culture of demiembryos with BOEC prior to freezing; there was a marked difference in survival rate in favour of whole embryos (75% vs. 41%).

In Germany, Lucas-Hahn (1992) concluded that current results after freezing bovine demiembryos are still unsatisfactory. According to this author, results can be improved by insertion of the bisected embryo into a ZP and a short IVC period between splitting and freezing; no beneficial effects were apparent when demiembryos were frozen by different procedures.

6.7.2 Studies with IVMFC-derived embryos

As observed by Loskutoff *et al.* (1993), for cattle embryos produced *in vitro*, micromanipulation at early cleavage stages may have advantages over splitting/quartering at the blastocyst stage. These workers showed that removing up to half the total number of blastomeres from two-, four-, eight- and 16-cell *in vitro* embryos had no adverse effect on subsequent development up to the hatching stage. However, two- and four-cell embryos have proved to be very sensitive to brief periods of cooling to 20–25°C and it was found necessary to maintain temperature during manipulation at about 38°C. Such extreme temperature sensitivity was not evident with early-cleavage-stage bovine embryos produced *in vivo* (Gatica, 1988). It seems likely that further studies with IVMFC-derived embryos may lead to efficient methods for producing blastocysts from early-cleavage embryos and that, after freeze–thawing, these may be successfully employed in transfers.

6.8 Freezing the Oocyte

There may be occasions when the freezing of cattle oocytes at the pronuclear stage would be of interest. One such occasion might be in genetic engineering research, where the pronuclear bovine oocyte may be used for the microinjection of gene constructs (McEvoy and Sreenan, 1990).

6.8.1 Freezing the pronuclear oocyte

The merits of cryopreserving pronuclear oocytes for future thaw and transfer are an important consideration in many human IVF programmes. Although embryo cryopreservation is both successful and widely used in human IVF, it has also created ethical and legal problems and is currently prohibited in some countries. It is several years since studies have shown that human pronuclear oocytes can be frozen successfully (Fugger *et al.*, 1988); the first successful pregnancy was reported by Abdalla and Leonard (1988). In more recent times, Veeck *et al.* (1993) reported results of a 5.5 year trial, in which it was shown that cryopreserved pronuclear oocytes that survived freezing and thawing had a similar developmental potential to fresh embryos. The freeze–thaw procedures already described for morulae and blastocysts (see Section 6.3.1) can also be successfully applied to pronuclear embryos. High survival rates were reported for the cryopreservation of the pronuclear oocyte by Testart *et al.* (1986), using the embryo protocol (PROH cryoprotectant) previously described by Lassalle *et al.* (1985).

In Belgium, Van der Auwera *et al.* (1992) showed that the age of the pronuclear mouse oocyte influenced its survival after thawing. They suggest that this is the result of the microtubular status of the cell before freezing begins. In human IVF, studies have recorded the importance of an accurate timing of zygote freezing to avoid periods of the cell cycle that seem to be particularly sensitive to cooling and freezing (Balakier *et al.*, 1993); it has been suggested that the optimal time for freezing human pronuclear oocytes may be about 20–22 h after insemination when the majority of oocytes have entered the G2 phase, before pronuclear dissolution and chromosome condensation (see Fig. 6.4).

It is clear that a detailed knowledge of the cell cycle is likely to be of considerable importance in the freezing of early-stage embryos. The presence of pronuclei permits oocytes in interphase to be accurately selected. According to Chedid *et al.* (1992), the freezing of embryos during mitosis (nuclear division) is detrimental for their survival, reflecting the effect of cryopreservation on the cytoskeleton, which is known to change during the cell cycle. It has been known for many years that the tolerance of cells to stress situations, such as dehydration and swelling, can be influenced by the cell cycle.

As noted by Chedid *et al.* (1992), a one-cell embryo in mitosis with a well-organized cytoskeleton may be more susceptible to cryopreservation damage

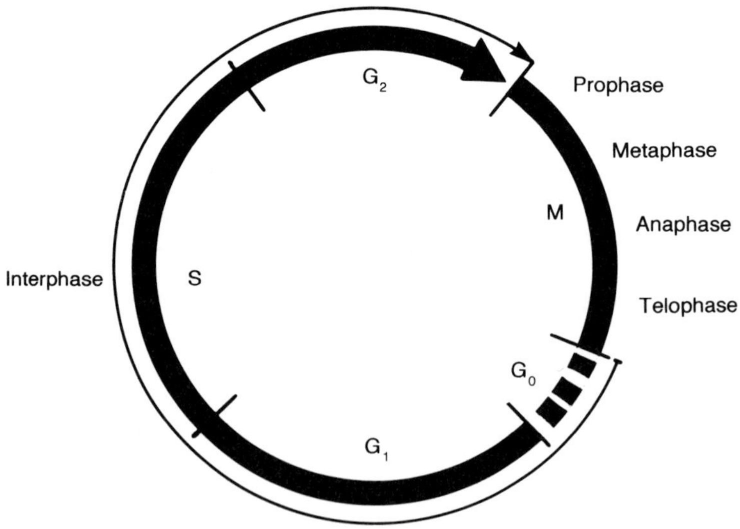

Fig. 6.4. The mammalian cell cycle, made up of phases of mitotic division (M) separated by interphase. During interphase, synthesis of DNA during the S phase is preceded and followed by gaps (G0, G1 and G2) of inactivity. During early cleavage of the embryo, there is no state of non-division (G_0) and the duration of the other phases differ as development progresses.

due to dehydration and swelling than a one-cell embryo during interphase; correct determination of the nuclear status of an embryo before cryopreservation may lead to more efficient techniques. Al-Hasani *et al.* (1992) reported on freezing rabbit pronuclear oocytes, using PROH as the cryoprotectant; these authors recorded a survival rate of 70% after thawing, but blastocyst formation was markedly below normal (10% vs 59%).

In cattle, data presented by Lim *et al.* (1991) have demonstrated that the post-thaw developmental capacity of oocytes frozen at the pronuclear stage was superior to that of those frozen as secondary oocytes; they also found that the concentration of glycerol was an important factor on post-thaw developmental capacity. One reason for this may be a change in permeability; there is evidence that fertilization can cause an increase in glycerol permeability. As noted above, a further reason may be related to the chromosomal status of oocytes. For various reasons, it may be preferable to freeze bovine oocytes at the pronuclear stage and early embryos at the interphase. Niemann *et al.* (1993) have also reported that IVM bovine oocytes are extremely sensitive to decreased temperatures; this sensitivity cannot be overcome by equilibration with the cryoprotectants DMSO or PROH.

6.8.2 *Using primary or secondary oocytes*

A review by Parks and Ruffing (1992) notes that the production of live young from the frozen oocytes of domestic animals had not been reported at that time. In cattle IVF programmes, there could be advantages in being able to freeze oocytes which could subsequently be thawed out and fertilized with sperm from an appropriate bull. Some MOET breeding programmes which have been proposed by animal breeders envisage collecting and freezing oocytes from all young females in the MOET herd and then, at a later date, thawing these out and using IVF to produce embryos for transfer (Woolliams, 1990).

As described elsewhere (see Section 4.14.2), zona-free hamster oocytes have been widely used in testing the fertilizing ability of sperm from various mammals. Tobback *et al.* (1991) showed that hamster oocytes survived freezing and were equivalent to fresh oocytes in sperm penetration tests. The use of frozen oocytes eliminated problems and costs involved in maintaining hamster colonies. There may be instances in cattle reproduction research when supplies of frozen oocytes may be of similar advantage.

In cattle, Heyman *et al.* (1986) were among the first to report the freezing of primary oocytes and zygotes; only 6% of oocytes survived freezing but 34% of the zygotes went on to cleave after thawing. Such studies emphasize the point that, although physically very similar, there are likely to be considerable differences in survival rates between primary oocytes and zygotes. Schellander *et al.* (1988) reported work in which they used PROH; although there was some evidence of pronucleus formation after IVM/IVF, none of the oocytes cleaved. In Dublin, a study with 481 secondary bovine oocytes showed a high incidence of polyspermy after freeze–thawing (Lu, 1990); only 4.5% of the oocytes cleaved after IVF.

Factors involved in this loss of viability in frozen bovine oocytes associated with specific steps in the freeze–thaw process were dealt with in a report by Glass and Voelkel (1990). They found that loss of viability occurred due to cooling to −7°C, seeding and slow-cooling to −35°C. No loss was evident due to equilibration in DMSO or plunging in liquid nitrogen after cooling to −35°C. It was concluded that significant losses in oocyte viability during cryopreservation are attributable to cooling and induction of ice crystallization. In Germany, Herrier *et al.* (1991a) examined the sensitivity of immature (primary) and IVM (secondary) bovine oocytes to the addition and removal of various cryoprotectants. Results suggested that DMSO and PROH might be suitable for IVM oocytes; a combination of DMSO and sucrose was superior to all the other treatments tested. In Denmark, Avery *et al.* (1992b) used three different cryopreservation protocols in studies with primary and secondary bovine oocytes; freezing with PROH gave the most promising results, although no embryo reached the morula/blastocyst stage after IVF.

In Japan, the ability of bovine oocytes at various stages of maturation to survive after freezing and thawing was examined by Lim *et al.* (1992); they concluded that the freezability of immature cattle oocytes was inferior to that of maturing or matured oocytes. These same authors did report, however, that some of their frozen–thawed bovine oocytes developed to the blastocyst stage after insemination. After considering the special features of mammalian oocytes which may contribute to their refractoriness to freezing and the points at which damage may occur, Parks and Ruffing (1992) concluded that additional studies are required to clarify mechanisms of cryoinjury and establish optimal freezing protocols.

The first report of an established pregnancy from embryos derived from mature bovine oocytes frozen with PROH was that of Otoi *et al.* (1992) in Japan; the blastocyst yield after IVF/IVC was markedly lower than that for the unfrozen controls (3.1% vs. 26.0%). Other work reported by Otoi *et al.* (1993a, b, c) has shown PROH to be superior to glycerol and DMSO in producing a higher number of morphologically normal oocytes after thawing and showing higher fertilization rates after IVF; however, despite such apparent advantages, the blastocyst yield showed no improvement with PROH.

Although embryo cryopreservation is both successful and widely employed in human IVF, it has also created many ethical and legal problems and is now prohibited in some countries. A reliable method for the cryopreservation of human oocytes could provide a valuable alternative. In studies related to human IVF, Trounson (1986a, b) was among the first to show that human secondary oocytes can survive freeze–thawing and remain capable of being fertilized. In cattle, Miller *et al.* (1988) found that the ZP of previously frozen bovine oocytes showed binding and penetration rates similar to those with fresh oocytes; the indications were that freeze–thawing did not alter zona penetrability. In the late 1980s, successful freezing of unfertilized eggs (secondary oocytes) was reported in species which include the mouse, rat, hamster, rabbit and monkey (DeMayo *et al.*, 1985; Al-Hasani *et al.*, 1986; Diedrich *et al.*, 1988; Ko and Threlfall, 1988; Sathananthan *et al.*, 1988; Carroll *et al.*, 1989; Vincent *et al.*, 1989; Wood *et al.*, 1989).

According to some authors at the time, oocyte freezing could possibly increase the efficiency of human IVF by presenting a real alternative to embryo storage (Chen, 1988; Hunter, J.E., *et al.*, 1991). However, according to a report by Gook *et al.* (1993), only five pregnancies by this method had been reported in 1993. Concern has been expressed at the possible effect of cold treatment on the human secondary oocyte (Mandelbaum *et al.*, 1988). It is known that cooling can cause depolymerization of microtubules and the disappearance of microtubule organizing centres in the oocyte. At the metaphase II stage, chromosomes of the second meiotic division are attached to cold-sensitive spindle microtubules and any disruption of this spindle may result in chromatid non-disjunction, leading to aneuploidy. Concern has also been expressed about the use of certain cryoprotectants. It is known that various alcohols and agents which have an −OH group are potent parthenogenetic activators of mammalian oocytes. Shaw and Trounson (1989) reported work showing that PROH can be a potent parthenogenetic activator of unfertilized mouse oocytes.

Intracellular and extracellular cryoprotectants have been employed in combination in attempts to protect mammalian oocytes from freezing damage. A report by Bergers *et al.* (1992) showed that dextran (with 1.5 M PROH as the intracellular cryoprotectant) protected mouse and human secondary ZP and oocytes from mechanical damage. Bernard *et al.* (1992), working with human secondary oocytes, found clear evidence that prefreeze cooling itself was not responsible for failures in fertilization after IVF.

A study by Overstrom *et al.* (1990) assessed changes in the organization of the filamentous actin (f-actin) cytoskeleton in ovulated mouse oocytes when exposed to cryoprotectant agents and thermal stress. They found evidence suggesting that glycerol (but not sucrose), by perturbing oocyte cytoskeletal organization, may be an inappropriate cryoprotectant for oocyte freezing protocols. In a report on ultrastructural features shown by frozen–thawed secondary bovine oocytes, Schmidt *et al.* (1993) record changes in the distribution of small membrane-bound vesicles; ultrastructural damage was incompatible with normal fertilization and embryonic development.

The freezing of primary and secondary oocytes

In an effort to avoid problems that may be associated with the cryopreservation of secondary oocytes, several workers have directed their attention at the fully grown but primary oocyte (germinal vesicle (GV) stage). At this stage, the oocyte contains a nucleus enveloping the prophase chromosomal complement and microtubular structures are not organized in a spindle. Van der Elst *et al.* (1992) evaluated the effects of different cryoprotectants and freezing protocols on mouse oocytes, using immunocytochemical examination of the spindle. Their results led them to assume that dehydration stress may destabilize the cytoskeleton and disrupt the anchorage of cytoskeletal elements. It has been shown elsewhere (Van Blerkom, 1991) that microtubules mediate not only nuclear maturation but also cytoplasmic maturation during the early stages of meiosis in mouse oocytes, in the sense that they organize mitochondrial redistribution.

According to Gook *et al.* (1993), their results showed that the mouse is not a suitable model for human oocyte cryopreservation. The same workers found that successful cryopreservation of human metaphase oocytes at the morphological and organelle level can be achieved in the presence of PROH; the high rate of preservation of the temperature-sensitive spindle and cortical granules was of particular interest. Further testing of this protocol, to determine the oocyte's ability to be fertilized and cleave, appears justified. Studies reported by Imoedemhe and Sigue (1992) indicated that, when PROH is precooled to 4°C prior to exposing the human oocyte to it, an acceptable level of cryopreservation can be obtained. The same workers found that survival was significantly improved when the oocyte had its cumulus cells attached. It should also be observed that, in a comparison of cryoprotectants, the viability of bovine oocytes frozen in PROH was found to be significantly higher that that of oocytes frozen in glycerol or DMSO (Otoi *et al.*, 1993b, c).

A paper by Carroll *et al.* (1993), on the other hand, suggests that their previous difficulties in IVF after thawing mouse oocytes (Carroll *et al.*, 1990b) could simply be resolved by freezing in the presence of a protective protein supplement such as FCS. They also note that many studies with human oocytes, in contrast to their protocol (dilution with DMSO at 0°C), have involved long equilibration periods (30–60 min) with DMSO at room temperatures or 37°C, which is known to cause severe cytoskeletal disturbances.

6.9 Rapid and Ultrarapid Freezing Procedures

Modifications of the conventional slow-cooling procedures, as originally described by Willadsen (1977) and others, have been attempted by various workers. In such protocols, there has usually been a considerable shortening in the time involved in cryopreservation, as well as a simplification of equipment requirements. When the technique involves some cooling period after equilibration below zero prior to plunging into liquid nitrogen, the term rapid freezing is employed; when embryos have been exposed to nitrogen vapour or plunged into liquid nitrogen after a brief period of equilibration, the term ultrarapid freezing is used.

There may be occasions, in using IVMFC-derived embryos, when a simple, inexpensive freezing procedure may have merit.

6.9.1 Rapid freezing of embryos

Lu and colleagues (Huang *et al.*, 1992) developed an effective means of freezing IVMFC-derived cattle embryos by a rapid freezing technique, which apparently eliminated the need for the conventional slow-cooling process requiring costly and time-consuming programmable biological cryopreservation units. The optimum cryoprotectant combination was 1.5 M glycerol with 0.25 M sucrose (in PBS + 15% FCS), evaluated in terms of embryo survival

Table 6.4. Survival of day-7 bovine embryos derived from IVMFC after rapid freezing (from Huang *et al.*, 1992).

Freezing medium	Thawing medium	No. of embryos recov./frozen	No. hatched (%)*
1.5 M G	0.5 M S	109/111	55 (50.5)[b]
	0.25 M S	100/103	49 (49.0)[b]
	0 M S	100/101	48 (48.0)[b]
1.5 M G + 0.25 M S	0.5 M S	111/113	76 (68.5)[a]
	0.25 M S	96/99	69 (71.9)[a]
	0 M S	100/101	79 (79.0)[a]
1.5 M S + 0.5 M S	0.5 M S	105/108	58 (55.2)[ab]
	0.25 M S	103/105	67 (65.0)[ab]
	0 M S	100/103	48 (48.0)[b]

*Values with different superscripts differ significantly ($P < 0.01$).
G, glycerol; S, sucrose.

and hatching rates. The embryos were loaded into 0.25 ml straws and directly placed in a domestic freezer (at −21°C) for 35 min before plunging into liquid nitrogen (see Table 6.4). It was concluded that 1.5 M glycerol with 0.25 M sucrose was optimal from the results of this study. The same protocol was followed subsequently in Ireland by Monaghan (1993) with similar success.

6.9.2 *Ultrarapid freezing of embryos*

A report by Barg *et al.* (1990) described ultrarapid freezing as a simple and low-cost method for human embryo preservation, using 4.5 M DMSO and 0.3 M sucrose; they reported pregnancies after transfer of embryos frozen by the method and concluded that the technique warranted further evaluation. Ultrarapid freezing of mouse embryos was described by Li and Trounson (1991), who concluded that blastocysts needed to be equilibrated at 0°C with a high concentration of cryoprotectant before freezing; exposure of blastocysts to 4.5 M cryoprotectant (DMSO/glycerol) and 0.25 M sucrose at ambient temperature was toxic or markedly reduced development after thawing. The replacement of biological proteins by chemically defined macromolecules (polyvinyl pyrrolidone (PVP); PVA; Ficoll) was examined in ultrarapid freezing studies with mouse embryos reported by Gutierrez *et al.* (1993); this was aimed at developing methods which would have particular merit for the international transport of embryos.

6.9.3 *Ultrarapid freezing of oocytes*

In the USA, Surrey and Quinn (1990) compared ultrarapid freezing of mouse secondary oocytes with the more conventional slow-cooling methods. They found that fertilization rates and embryonic development of oocytes undergoing ultrarapid freezing after exposure to 3.5 M DMSO and 0.5 M sucrose were similar or superior to those found after conventional slow cooling. They concluded that ultrarapid freezing may be a viable option to traditional methods. In Germany, Nowshari *et al.* (1993) attempted to develop a simple ultrarapid freezing protocol for the secondary oocytes of mice; freezing in a 4.5 M DMSO and 0.5 M sucrose combination and thawing in three or four steps of descending sucrose concentration were found to be successful.

6.10 Vitrification of Embryos and Oocytes

Cells can suffer cryoinjury either by the formation of intracellular crystals or by chemical toxicity/osmotic stress arising from high concentrations of salts. Vitrification is a process by which freezing of a solution is eliminated and the adverse effects of ice crystal formation are avoided.

6.10.1 Historical

Well over 40 years have passed since the original report by Chris Polge at Mill Hill that chicken sperm could be recovered alive after freezing and thawing if glycerol was incorporated in the holding medium. This discovery was followed by many reports of successful cryopreservation of various mammalian cells in the presence of glycerol or other cryoprotectants. More than 20 years ago, reliable techniques for the cryopreservation of mouse embryos were developed simultaneously by David Whittingham and associates and by Ian Wilmut.

Almost a decade ago, Rall and Fahy (1985) described an alternative form of cryopreservation; this involved a mixture of solutes (DMSO, acetamide, propylene glycol as permeating agents; polyethylene glycol as the macromolecule) that permitted vitrification of the embryo holding solution when cooled to very low temperatures. The whole procedure required about 35 min to allow stepwise equilibration before embryos were plunged into liquid nitrogen.

Vitrification is a process of solidification in which crystalline ice does not separate and there is no concentration of solutes, as in conventional freezing; there is an abrupt increase in the viscosity of the holding medium, producing a solid, glasslike condition. High cooling rates can be employed but initial exposure to the vitrifying solution has to be at refrigerator temperature and very brief to avoid adverse effects from cryoprotectant toxicity. Warming rate has to be rapid to avoid crystal formation as the temperature returns to normal.

The merit of vitrification may be seen, both in simplifying and speeding up freezing operations, in terms of avoiding the slow cooling period and the use of expensive equipment as required in conventional cryopreservation programmes. The first success in preserving mouse embryos by vitrification was followed by many other studies with that species (Scheffen *et al.*, 1986; Kasai et al., 1990; Valdez et al., 1990). In the original work of Rall and Fahy (1985), a mixture of four cryoprotectants was employed at different levels and using various temperatures and exposure times. A simpler and faster procedure was described by Scheffen et al. (1986), involving only two components, glycerol and propylene glycol, in differing proportions. A paper by Simmons and Graves (1990) suggested that vitrification may be the more desirable method of cryopreserving eight-cell- and morula-stage mouse embryos, while conventional controlled freezing may be more effective for two-cell embryos.

6.10.2 *Vitrification of* in vivo*-produced embryos*

Using the method previously described by Scheffen et al. (1986) for use with mouse embryos, Massip et al. (1986, 1989) successfully vitrified bovine morulae/blastocysts, which were subsequently used successfully in establishing pregnancies. The effectiveness of the method was also confirmed by Douchi et al. (1990) with cattle morulae, although not with blastocysts. In a comparative study of freezing methods, Leeuw et al. (1991, 1992) in The Netherlands, using a mixture of glycerol (10%) and PROH (20%) concluded that vitrification of bovine embryos, combined with one-step dilution within the straw, was an effective alternative to conventional slow-cooling procedures.

In former Czechoslovakia, Riha et al. (1991) preserved 7-day cattle embryos using a novel form of vitrification; this involved direct dropping of embryos in microdroplets of vitrification solution (30% glycerol, 20% FCS, 50% 2 M sucrose) into liquid nitrogen. The authors claimed that the method saves labour, material and cost and could be applied in difficult farm conditions.

A paper by Arav (1992), reviewing vitrification methods for use in embryos and oocytes, refers to a somewhat similar method, known as the minimum drop size (MDS) technique. This involved cooling and warming embryos and oocytes very rapidly in small drops (0.06 μl) containing a low concentration of cryoprotectant solution. The same author also noted that the addition of antifreeze glycoproteins, isolated from Antarctic fishes, resulted in a dramatic increase in the survival rate of embryos and oocytes vitrified with the MDS technique. This work stems from the discovery, in the late 1960s, that Antarctic fish possessed plasma which demonstrated a phenomenon known as 'thermal hysteresis'; it did not freeze until about $-2.5°C$ but, on rewarming, melted at about $-0.8°C$ (see Davenport, 1992).

These antifreeze glycoproteins, of which four major types have so far been characterized (Arav et al., 1993), are synthesized in the liver and are of crucial importance to the fish's survival in the cold Arctic/Antarctic waters during the winter months. The proteins reach peak concentrations in the body of fish during the periods of low water temperature. The mode of action is not well understood, but there is evidence that the glycoproteins bind to ice in such a way as to prevent ice crystal proliferation and growth. The antifreeze glycoproteins appear to possess the ability to interact with and stabilize the cell membranes of oocytes and embryos (see Arav et al., 1994).

Vitrification as used by Dobrinsky *et al.* (1991) for the routine cryopreservation of mouse and rabbit embryos was applied to bovine embryos recovered from superovulated cattle. They were first dehydrated in the equilibration medium (10% glycerol and 25% propylene glycol) for 7 min at 20°C. They were then loaded into vitrification medium (25% glycerol and 25% propylene glycol) and within 1 min lowered into liquid nitrogen. In thawing, straws were submerged in water at 20°C for 5–10 seconds. The authors concluded that the vitrification process may be a technically simple alternative for cattle embryo freezing. In Ireland, Yang, N.S., *et al.* (1992) examined the survival rate of IVMFC-derived bovine blastocysts after vitrification (see Table 6.5). They showed that survival was influenced by the type of cryoprotectant and exposure procedures. It was demonstrated that high hatching rates were possible after vitrification using glycerol and ethylene glycol (or proplylene glycol) as cryoprotectants. In Japan, Saito *et al.* (1994) have shown that the solution used by Yang, N.S., *et al.* (1992) was only effective for expanded blastocysts but not for early blastocysts or blastocysts. However, the addition of sugars to the solution significantly improved the survival rate of blastocysts, regardless of developmental stage.

A vitrification method successfully employed by Kasai *et al.* (1990, 1991) in mouse and rabbit embryos involved a mixture of ethylene glycol (40%, v/v), Ficoll (18%, v/v) and sucrose (0.3 M), all known to be compounds of low toxicity. The procedure only required exposure of embryos for 2 min at 20°C before they were plunged into liquid nitrogen. Normality of embryos after thawing was confirmed by the birth of young. Ishimori *et al.* (1992a, b), using ethylene glycol and DMSO as cryoprotectants, examined the influence of exposure temperature and other factors; it was found that mouse embryos could be handled at room temperature rather than low (4°C) temperature.

Studies reported by Valdez *et al.* (1990) examined the effects of equilibration time, precooling and embryo developmental stage on the survival rate of mouse embryos frozen by vitrification; results suggested that optimum vitrification conditions for high embryo survival rate may vary for each developmental stage. A subsequent paper by Valdez *et al.* (1992) described a

Table 6.5. Vitrification of IVMFC-derived bovine blastocysts (from Yang, N.S. *et al.*, 1992).

Group	No. of embryos frozen	No. recovered (%)	No. hatched (%)*
1	86	84 (97.7)	63 (75.0)^{bc}
2	93	91 (97.8)	55 (60.4)^{ab}
3	99	94 (94.9)	76 (80.8)^{c}
4	94	85 (90.4)	50 (58.8)^{ab}
5	94	87 (92.6)	64 (73.6)^{bc}
6	97	94 (96.9)	48 (51.1)^{a}

*Values with different superscripts differ significantly ($P < 0.01$).
Various treatment groups are shown in the table. The most effective treatment was (3), in which the embryos were in 10% glycerol for 5 min and then in 10% glycerol + ethylene glycol for 5 min.

simple vitrification solution (ethylene glycol; DMSO; 1,3-butanediol) which did not require precooling to achieve a high survival rate of mouse embryos. A vitrification process involving ethylene glycol, Ficoll and sucrose was found to be less toxic for *in vivo*-produced bovine compact morulae and early blastocysts than any of the other procedures tested by Mahmoudzadeh *et al.* (1993).

In Poland, factors affecting the survival of one- and two-cell rabbit embryos after vitrification were examined by Gajda and Smorag (1993); viability was reduced as a result of exposure to their vitrification solution rather than to the vitrification process itself. A method permitting direct transfer of vitrified *in vivo* cattle embryos has been reported by Ishimori *et al.* (1993a). The same group also reported on the use of a mixture of ethylene glycol and DMSO in vitrifying *in vivo*-produced morulae and blastocysts (Ishimori *et al.*, 1993b). In work with rabbit morulae, Kobayashi *et al.* (1990) have shown that their vitrification method was effective for the preservation of bisected embryos, which had not previously been demonstrated.

6.10.3 *Vitrification of IVMFC-derived embryos*

In Ireland, studies reported by Yang, N.S., *et al.* (1992) showed that the survival of IVMFC-derived embryos was influenced by the type of cryoprotectant and exposure procedures employed. Their results suggested that high hatching rates could be achieved after vitrification using glycerol (25%) and ethylene glycol (25%) as the cryoprotectants (see Table 6.5). In Japan, Kuwayama *et al.* (1992a, b) assessed the viability of IVMFC-derived cattle embryos after vitrification. The blastocysts were equilibrated in the vitrification solution (22.5% glycerol and 22.5% PROH in modified M-199) in 16 steps, a procedure which apparently avoided osmotic injury and resulted in high survival rates.

The vitrification procedure successfully employed by Kasai *et al.* (1990, 1991) in mice and rabbits was applied by Tachikawa *et al.* (1993) to IVMFC-derived cattle embryos. Exposure to the vitrification solution for longer than 2 min resulted in an abrupt decrease in embryo survival rate; this was attributed to ethylene glycol toxicity. These authors were among the first to report live births after transfer of IVMFC-derived embryos cryopreserved by vitrification.

The application of vitrification to IVMFC-derived embryos produced by different methods is dealt with in a report by Dinnyes *et al.* (1994); they concluded that further studies are required to improve survival rates and to examine the interaction of the age, quality and developmental stages on survival rates. One of the advantages of *in vitro* embryo production is that cattle embryos are now available in large numbers for detailed studies on the effect of freezing (see Fig. 6.5).

6.10.4 *Vitrification of oocytes*

Nakagata (1989) demonstrated that unfertilized mouse oocytes could be successfully frozen using a simple vitrification procedure. The method involved

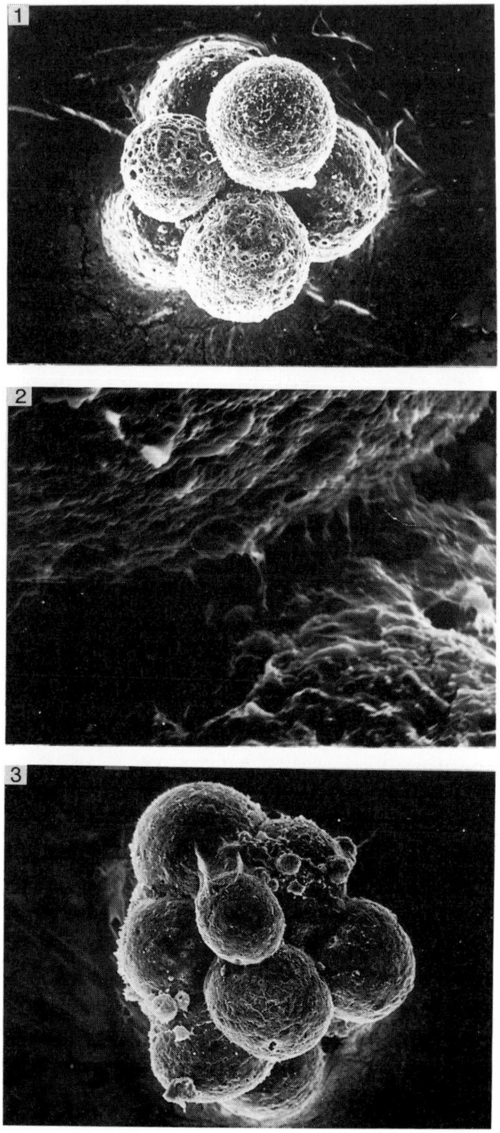

Fig. 6.5. Scanning electron micrographs of eight-cell cattle and sheep embryos. Differences are evident at the ultrastructural level between embryos produced *in vitro* and those produced in the living animal. Some differences are in the intercellular connections, which may be less in the *in vitro*-produced embryo. In these micrographs, (1) is a cattle embryo, with (2) an enlargement showing more detail of the intercellular bridges between blastomeres; a sheep embryo is shown in (3). It was found that cell union was greater in the eight-cell sheep embryo than in the cattle embryo at this stage (from Gatica, 1988).

plunging into liquid nitrogen immediately after exposure of the oocytes to the vitrification solution (DMSO, acetamide, propylene glycol and polyethylene glycol); 88% of oocytes appeared to be morphologically normal after thawing, and after fertilization 46% developed into normal young. Kono *et al.* (1991) also froze mouse oocytes in a similar vitrification mixture; they reported that fertilization occurred in 84–94% of the vitrified oocytes after thawing.

Shaw *et al.* (1991) used the same vitrification procedure and reported that 24% of vitrified mouse oocytes were fertilized subsequently; these authors concluded that vitrification may be an acceptable method of preserving mouse oocytes. Trounson and Shaw (1992) noted that several vitrification methods now permit high survival rates of secondary mouse oocytes. They report a freezing procedure with 4.5 M DMSO and 0.5 M sucrose in the medium; after 10–15 seconds of equilibration, straws are placed in liquid nitrogen vapour for 2–3 min before being plunged into liquid nitrogen. Bos-Mikich *et al.* (1992) also concluded, from their studies, that vitrification of mouse oocytes is a simple and safe procedure.

In Japan, Hamano and Kuwayama (1992) and Hamano *et al.* (1992) demonstrated that secondary bovine oocytes survive after vitrification using a solution of 2 M DMSO, 1 M acetamide and 3 M propylene glycol. The authors record that, after thawing and dilution in 12 steps of a sucrose solution (2 M to 0.1 M), most oocytes were deemed to be morphologically normal; some oocytes developed to the blastocyst stage after fertilization and two pregnancies were reported after transfers. The authors concluded that a gradual dilution of sucrose in 12 steps was sufficiently effective in mitigating the osmotic injury that otherwise would have occurred. Less successful results were reported by Watanabe and Oliveira Filho (1992), who attempted vitrification of bovine oocytes before and after IVM.

6.11 Storage of Embryos and Oocytes in Other Farm Animals

Although the majority of studies on embryo cryopreservation have been related to cattle ET technology, live young have been born after freeze–thaw procedures in all the other farm mammals. As yet, limited information is available on the cryopreservation of IVMFC-derived embryos in species other than cattle. Particular attention must be drawn to the problem of cryopreservation in the pig, a species which until recently has successfully resisted all efforts at embryo freezing. It is becoming increasingly apparent that problems of temperature sensitivity and inability to survive freeze–thaw procedures, common in pig embryos, are often found in studies with IVMFC cattle embryos. It seems likely that the role of lipids, in particular their effect on cell membranes, may be a key factor in both instances.

6.11.1 Sheep and goats

Although research and commercial interest in the cryopreservation of sheep embryos have been limited in comparison with those in cattle, from the research viewpoint the sheep has often been employed in testing methods. The first lamb born after transfer of a frozen–thawed embryo was born in Cambridge in the 1970s through the work of Steen Willadsen and associates, who used the sheep embryo as the model for developing the methods that were eventually applied widely in cattle (Willadsen *et al.*, 1974, 1976; Willadsen, 1977). Although early studies employed DMSO as the cryoprotectant, glycerol (1.4 M) was later to be regarded as the more effective agent.

In a recent review of embryo freezing methods in sheep and goats, Brebion *et al.* (1992) refer to the use of 1.5 M ethylene glycol as the cryoprotectant and 20% FCS in the medium; according to these authors, 85–90% of embryos survived freezing and 65% survived transfer. In the USA, Schiewe *et al.* (1991) evaluated several factors in the freezing of sheep embryos; they record that propylene glycol tended to be the optimum cryoprotectant. The use of ethylene glycol in the freezing of sheep embryos is reported by McGinnis *et al.* (1993). According to Sakul *et al.* (1992) in California, cryopreservation methods can contribute to genetic improvement in sheep by facilitating transfer of germplasm among countries while reducing the risk of disease transmission; they can also be used for conservation of endangered domestic breeds and wild species.

Sheep embryos have also been subjected to vitrification (mixture of glycerol and propanediol) and transfer of embryos has resulted in the birth of lambs (Gajda *et al.*, 1989). In Australia, Szell *et al.* (1990) used 25% glycerol and 25% propylene glycol in their vitrification attempts; embryo survival rates varied from 36% for late morulae to 70% for late blastocysts and lambs were born after transfers. Other work in that country reported results with day-6 embryos comparable to those found with conventional freeze–thaw procedures (Ali and Shelton, 1993); the use of ethylene glycol as the major permeating cryoprotectant in the vitrification medium was believed to be the main factor for the improved success rate. The efficacy of conventional freezing or vitrification of sheep embryos was examined by Schiewe *et al.* (1990b); the authors showed that sheep embryos were able to survive their simple and rapid vitrification procedure. As in cattle, vitrification is seen by some in sheep as a possible means of eliminating the need for expensive, programmable cryogenic units in freeze-storing embryos under field conditions. The difficulties of freezing sheep primary oocytes were apparent in a study reported by Moodie *et al.* (1992).

One of the first reports on the freeze–thawing of goat embryos was that of Bilton and Moore (1976) in Australia. In more recent times, the transfer of goat embryos frozen with glycerol (1.4 M) as the cryoprotectant has been reported by Baril *et al.* (1988); pregnancy rates were lower than with freshly transferred embryos. A comparison of cryoprotectants was made by Fieni *et al.* (1990), who concluded that ethylene glycol was the most suitable. As noted earlier, Brebion *et al.* (1992) reported satisfactory results with goat embryos using this cryoprotectant. In France, Le Gal *et al.* (1991, 1993) reported significant differences in the permeability of goat primary and secondary oocytes to cryoprotectants; they also showed that ethylene glycol was a more effective cryoprotectant than glycerol, in their studies with goat morulae and blastocysts. A report from Australia by Li *et al.* (1990) suggests that maximum survival of frozen goat embryos is attained at the expanded, hatching and hatched blastocyst stages. Studies in the vitrification of goat embryos, using a mixture of glycerol (25%) and propylene glycol (25%), was reported by Yuswiati and Holtz (1990) in Germany; two viable kids were born after transfer of thawed embryos to nine recipient goats.

6.11.2 Pigs

The international transportation of pig embryos, held at 36.5°C in 0.25 ml straws, has been recorded by Niemann (1989); after a 30–34 h journey from Germany in a thermos, most embryos were viable and many gave rise to piglets on transfer in Argentina.

As to freezing, modifications of methods originally used in 1972 by Whittingham and Wilmut with mice were found to be entirely unsuccessful with pigs. The fact that pig morulae are exceptionally sensitive to cooling below 15°C has been recognized for many years (Polge *et al.*, 1974). Even recent reports, which have shown that the lethal effect of chilling on pig morulae occurs abruptly just below 15°C, have been uncertain as to the precise cause of the problem (Plante *et al.*, 1993). There is a growing body of evidence suggesting that the sensitivity of pig embryos is the result of their high lipid content.

Recent work has shown that it is possible to store pig embryos at the expanded/hatched blastocyst stage at subzero temperatures (−20°C) and to produce young after thawing (Hayashi *et al.*, 1989; Nagashima *et al.*, 1989; Feng, S., *et al.*, 1991). Storage in liquid nitrogen (−196°C) rather than at subzero temperatures was reported by Kashiwazaki *et al.* (1991a, b) and Nagashima *et al.* (1992). It is known that the size of lipid droplets in pig embryos decreases as they develop to the hatched blastocyst stage. It has been suggested that such differences in lipid content are probably associated with changes in embryo freezability (Nagashima *et al.*, 1994); artificial removal of lipid from early pig embryos allows them to survive cooling. From the commercial viewpoint, working with hatched blastocysts has obvious disadvantages; the lack of a ZP, the internationally accepted barrier to disease pathogens, currently makes frozen hatched embryos unacceptable for importation into most countries.

In the USA, studies reported by Didion *et al.* (1990) demonstrated that pig oocytes at the GV stage did not survive cooling to 15°C or below in cryoprotectant medium (1.5 M glycerol/0.5 M sucrose). However, recent studies (Rubinsky *et al.*, 1991; Arav, 1992; Arav *et al.*, 1993) have now demonstrated that vitrification may be a successful procedure for the cryopreservation of pig oocytes; in this, survival rate was increased by the addition of antifreeze glycoproteins, derived from Antarctic fish, to the vitrification solution. It is believed that protection is provided to the cell membrane (oolemma) by the ability of the antifreeze proteins to interact with or perhaps bind to it. Preliminary results from work in the US Department of Agriculture (USDA) at Beltsville also indicate that day-6 pig embryos are capable of surviving vitrification (Dobrinsky, 1993; Dobrinsky and Johnson, 1993).

The development of an effective procedure for the cryopreservation of pig embryos would have many practical implications and could be expected to result in a significant impact on pig production worldwide (see review by Dobrinsky, 1993).

6.11.3. Horses

In the application of a freeze–thawing programme, the horse embryo presents certain problems not encountered in the farm ruminant species. According to a review of published results by Bruyas *et al.* (1993), of about 200 frozen–thawed equine embryos transferred, the mean pregnancy rates were lower than 20%. Pregnancy rates may be compromised when equine embryos above a certain size (250 μm in diameter) are frozen (Slade *et al.*, 1985; Squires *et al.*, 1989). A similar effect was noted by Rieger *et al.* (1991b) but within a relatively narrow range of embryo size (170–220 μm). However, using a slight modification of the freezing schedule employed in cattle embryo preservation, Allen *et al.* (1992) have been able to demonstrate that frozen horse embryos can be transported and transferred internationally. These authors also found that pregnancy rates after transfer was inversely related to embryo age and size before freezing. Using light and electron microscopy, Bruyas *et al.* (1993) found evidence that freeze–thawing with glycerol as the cryoprotectant induced considerable cellular damage; they suggest the need for a more suitable cryoprotectant for the horse embryo. An early attempt to preserve 12 equine embryos by the vitrification method resulted in no pregnancy after eight transfers to recipient mares (Kot *et al.*, 1991).

6.11.4 Buffaloes

A preliminary report on the cryopreservation of Thai swamp buffalo embryos was made by Techakumphu *et al.* (1989). Subsequently, a report from Kasiraj *et al.* (1993) in India showed that both buffalo morulae and blastocysts could be successfully frozen, thawed and transferred to establish pregnancies.

7 Transferring Embryos and Establishing Pregnancies

7.1 Introduction

7.1.1 Historical

In 1891, the first embryo transfer (ET) was recorded in Cambridge in the rabbit (see Table 7.1). A century later, Hahn and Aumuller (1992) were recording the fact that 169,651 embryos suitable for transfer had been recovered from 32,737 cows in 24 European countries; of these, 165,879 were transferred. As well as that, many thousands of cattle embryos were transferred in that same year in North America and elsewhere around the world.

Significant expansion of commercial cattle ET started with the introduction of non-surgical flushing in the early 1970, and grew further with the introduction of simple non-surgical transfer techniques towards the end of that decade (Kuzan and Seidel, 1986). Although the surgical transfer of embryos, in suitably experienced hands, probably remains the method capable of achieving the highest pregnancy rates (70%), any thought of low-cost exten-

Table 7.1. First records of successful embryo transfers.

Species	Year	Authors
Rabbit	1890	Heape (1890)
Goat	1932	Warwick and Berry (1949)
Rat	1933	Nicholas (1933)
Sheep	1933	Warwick *et al.* (1934)
Mouse	1942	Fekete and Little (1942)
Cow	1951	Willet *et al.* (1951)
Pig	1951	Kvasnickii (1951)
Horse	1974	Oguri and Tsutsumi (1974)
Human	1978	Steptoe and Edwards (1978)

sive use of cattle ET requires a non-surgical transfer technique essentially similar to that involved in routine artificial insemination (AI) (see Fig. 7.1). The animal welfare implications of using non-surgical procedures must also be kept very much in mind.

Work in Ireland and elsewhere showed that it is possible to establish pregnancies by a non-surgical procedure involving the use of the standard Cassou (IMV) inseminating instrument. The embryo is loaded, held in a small volume of medium (phosphate-buffered saline (PBS) supplemented with 15% bovine serum), into a 0.5 or 0.25 ml straw. At transfer, the straw is placed in the inseminating instrument ('gun') in the usual way and the same procedure adopted as for AI, the main difference being that the embryo is deposited around the midhorn position (ipsilateral horn). During the past 15 years, several variants of the standard inseminating instrument have been marketed. In the UK, for instance, Coultard (1991) has described his transfer procedure using a modified insemination gun with a special sheath having a metal tip and two side-outlets.

However, despite extensive research on follicular dynamics and the use of gonadotrophic hormones, there appears to be little evidence of significant

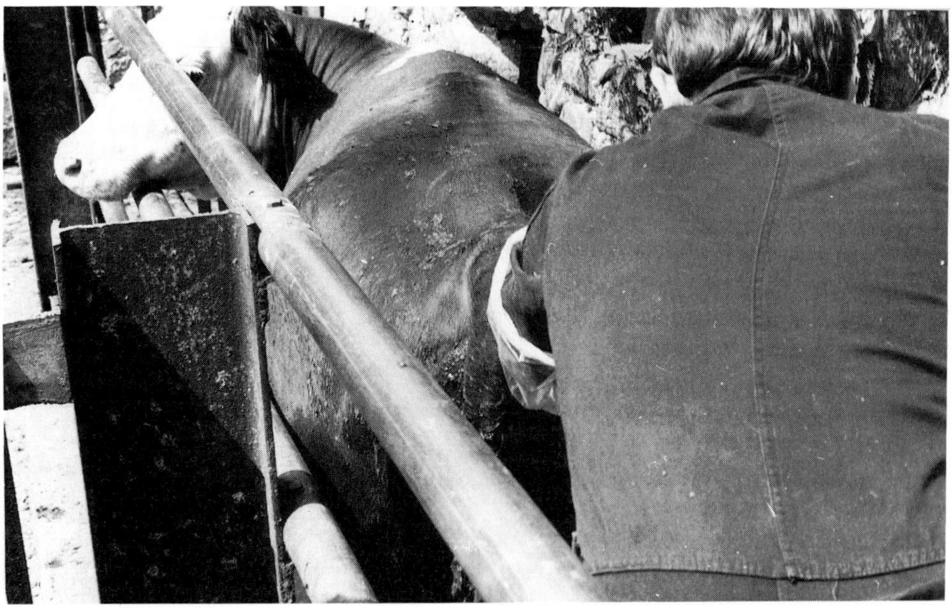

Fig. 7.1. Non-surgical embryo transfer in the cow. The development of non-surgical transfer methods in the mid-1970s was a major milestone in progress towards extensive application of embryo transfer technology in cattle. Those with skill and experience in artificial insemination could train to become competent in embryo transfer without difficulty. In Ireland, 130 inseminators were trained by Ovamass Ltd in 1991–92 to carry out the transfer of IVMFC-derived embryos through 56 sub-stations of the AI service.

improvements in the efficiency of superovulation protocols during the past two decades (Hasler, 1991); much the same can be said of the pregnancy rates after ET. Producing transferable embryos in greater numbers can follow embryo splitting, but together with sexing, such techniques have not been widely used in the cattle ET industry; it is largely a question of cost.

7.1.2 Technical requirements for easy on-farm application

Among the factors affecting the success and economics of a commercial cattle ET programme, the following would normally be mentioned: (i) skill and experience of the ET operator; (ii) selection and management of recipient animals, which must be cyclic, healthy and reproductively normal; (iii) synchrony of oestrus between donor and recipient; (iv) quality of embryos transferred; and (v) methods employed in embryo handling and transfer on the farm. In measuring success rate in cattle ET, the recipient pregnancy rate is usually the most crucial criterion. In general terms, using single embryos, the pregnancy rates which are achievable after non-surgical transfer should be broadly in line with those expected for the same category of recipient animal when bred normally (i.e. 50–70% pregnant to first service). Using frozen–thawed embryos is likely to result in pregnancy rates some 10 percentage points lower than those obtained for fresh embryos (Nelson and Nelson, 1988). In commercial practice, however, embryo transfers with frozen *in vivo*-produced embryos have given pregnancy rates below 40% (Veselinovic *et al.*, 1992).

7.2 Preparing Embryos for Transfer

7.2.1 Media employed

In the early Cambridge work on the recovery and transfer of cattle embryos, M-199 was commonly employed as the holding medium (Rowson *et al.*, 1969, 1971); the medium, however, requires a gas atmosphere of 5% carbon dioxide for its buffer system. During the second half of the 1970s, workers started using Dulbecco's PBS, either supplemented according to the modifications of Whittingham (1971) with glucose, sodium pyruvate and bovine serum albumin (BSA) or supplemented with fetal calf serum (FCS); the medium is now widely recognized as being particularly effective. Heat-inactivated FCS is usually added in volumes of 1–2% for flushing and 10–20% for storage; as a source of macromolecules, homologous or heterologous heat-inactivated blood serum or its albumin fraction has come to be an essential component of most culture media (Kane, 1978, 1987).

7.2.2 Measures in handling cattle embryos

In dealing with the laboratory handling of cattle embryos, where they inevitably come in contact with glassware, Petri dishes, plastic straws and other objects, their possible exposure to toxic factors must always be a consideration. It is, for example, essential to provide adequate aeration of straws after their sterilization with the powerful antimicrobial, ethylene oxide (Schiewe *et al.*, 1988). Other workers have reported on the safety of a range of other products employed in routine cattle ET operations (Lee *et al.*, 1988). Of various items examined by Lee *et al.* (1989), latex tubing, siliconized Foley catheters and syringe stoppers appeared to have the greatest potential for producing toxic effects.

7.2.3 *Protecting the embryo*

There are several problems associated with non-surgical embryo transfer in the cow which may influence the establishment of a pregnancy. As a result of its microscopic size, it is difficult to confirm that the embryo has been deposited in the uterine horn at a site suitable for its subsequent development. Although, in a research setting, a transfer straw can be inspected after removal to ensure that the embryo is not retained, this does not necessarily constitute proof of satisfactory transfer. The embryo may have moved away from the uterine wall or adhered to the exterior of the straw during withdrawal from the uterus and cervical canal.

For such reasons, there has been interest in various species, including the human, in encapsulating the embryo in a semipermeable and biodegradable material on the assumption that this might eliminate some of the transfer problems and improve protection of the embryo. Torner *et al.* (1986) in Germany reported on a method of encapsulating cattle embryos in microspheres measuring 3–4 mm in diameter; embryos were held in culture medium within the capsule without this apparently compromising their viability.

Working with mice, Cosby *et al.* (1989) showed that encapsulation in sodium alginate did not compromise embryo development. Studies with mice were also reported by Adaniya *et al.* (1987, 1991, 1993) involving encapsulating mouse embryos in biodegradable sodium alginate, which dissolved in 2 days; they also found that embryos were able to develop and implant normally. The authors concluded that the technique may be of future benefit for human *in vitro* fertilization (IVF) ET. They note that the technique may offer the possibility of incorporating long-chain molecules, growth factors and other beneficial substances into the matrix surrounding the embryo to nurture it in the early period after transfer.

Fixing the embryo's location after transfer

In human IVF ET, Feichtinger *et al.* (1992b) have drawn attention to the possibility that some transfers after IVF may fail because of expulsion of the embryos from the uterus. These authors used a fibrin sealant which also contains a fibrinolysis inhibitor (aprotinin) at various concentrations. After preliminary work with mouse embryos, the sealant was successfully employed in human ET. The possible merits of a biological adhesive, which may ensure that the embryo remains in the most appropriate location in the cow's uterus immediately after non-surgical transfer, have not been examined in cattle ET studies.

7.2.4 Number of embryos transferred

It was established some years ago in Cambridge and elsewhere that the surgical transfer of two embryos to recipient cattle can result in high pregnancy rates. A summary of such data is given in Table 7.2. There is also ample evidence in the literature to support the view that higher pregnancy rates can be achieved using two embryos rather than one (see Gordon, 1983). Work in France, using frozen–thawed embryos, showed further evidence of a significantly higher pregnancy rate after the transfer of two embryos rather than single or half-embryos (Heyman and Chesne, 1984).

Table 7.2. Pregnancy rates in cattle receiving two embryos (surgical transfer).

Number of recipients	Number of embryos and site of transfer: +CL	−CL	Pregnancy rate (%)	Authors
11	2	–	91	Rowson *et al.* (1969)
11	2	–	73	Rowson *et al.* (1971)
11	1	1	72	Rowson *et al.* (1971)
31	1	1	87	Sreenan and Beehan (1974)
55	1	1	71	Sreenan *et al.* (1975)
48	1	1	75	Anderson *et al.* (1978)

7.3 Surgical and Non-surgical Transfers

7.3.1 Surgical transfers

The work of Rowson *et al.* (1969) marked an important turning-point in cattle ET prospects by showing that an acceptable pregnancy rate could be achieved by a surgical technique. The midventral laparotomy technique employed in these studies involved general anaesthesia of the recipient, with surgical preparation of the midline just anterior to the mammary gland. Having brought the uterus to the incision site and confirmed the location of the corpus luteum (CL), a small puncture was made into the lumen of the uterine horn ipsilateral to the CL. The transfer pipette was introduced through the puncture and the embryo deposited. Although the midventral procedure was capable of being carried out quickly on heifer cattle, it was clearly not a procedure for use on the farm; as well as being both labour- and capital-intensive, the technique was not at all suitable for milking cows.

Where surgical transfers are now used, most commercial ET operators have adopted the flank surgical approach, with the recipient standing sedated and under local anaesthesia (paravertebral block). The site of incision is the sublumbar fossa and the embryo is transferred into the upper third of the uterine horn. Prior to surgery, the location of the CL is established to show the side on which the transfer should be made. In comparison with the midventral procedure, the flank approach is widely recognized as being a highly effective, practical and rapid means of performing surgical transfers with a minimum of equipment, facilities and labour.

As to the reasons for using the surgical approach, this is likely to be in an effort to give the recipient the greatest opportunity of becoming pregnant with the embryos that are available. According to Coultard (1991), if the embryo (*in vivo*-produced) has been frozen and thawed or is young or second-grade, an approximately 10–12% improvement in pregnancy rate can be achieved with the flank surgical approach, in comparison with non-surgical methods.

7.3.2 Non-surgical transfers

Although numerous instruments designed specifically for the non-surgical transfer of cattle embryos were described in the literature during the 1960s and early 1970s (see Gordon, 1983), it was the successful application of the Cassou AI gun which eventually proved to be the answer to this particular problem. Since the mid-1970s, several variants of the standard inseminating gun have been marketed for ET. Coultard (1991) describes his transfer procedure in the UK using a modified insemination gun with a special sheath having a metal tip and two side-outlets. In carrying out the transfer, the recipient is given an epidural anaesthetic and the sheathed, loaded gun is inserted into a sterile, loose, plastic outer sheath; this is passed up into the cervix and the gun forced through the outer sheath.

7.3.3 Factors influencing success with non-surgical transfers

There are a number of embryonic, maternal and environmental factors which may affect the pregnancy rates established in recipient cattle after non-surgical transfer (see Hasler *et al.*, 1987; Sreenan and Diskin, 1987; Sreenan, 1988; Coultard, 1991; Hasler, 1992). With heifer recipients, there can be about 10% that may be difficult, if not impossible, to transfer via the cervix (Coultard, 1991).

One important factor influencing the pregnancy rate is likely to be the skill and experience of the transfer operator, especially in the matter of avoiding trauma to the endometrium. A 6-year study by Park, R.L., *et al.* (1991) showed that an ET programme can be conducted by trained herdsmen, but only after an appropriate training period. In this regard, it must be remembered that transfer is being carried out at a time when the uterine environment is markedly different from that of a cow in oestrus, in terms of its susceptibility to infection and injury. In some species (e.g. hamsters) it is known that low pregnancy rates after ET are mainly due to trauma of the endometrium and prostaglandin (PG) release (Jarosz and Dukelow, 1990).

Avoiding trauma

For cattle ET, the importance of avoiding damage to the endometrium or otherwise traumatizing the uterus during the transfer process cannot be overemphasized. Introducing the transfer instrument for some distance up the uterine horn clearly increases the likelihood of such trauma unless appropriate care is taken by the operator. The manipulations involved in ET, if not conducted carefully, may create an endometrial inflammatory response, which would imply the migration of macrophages and immunocompetent cells to the inflammatory site; this could result in an environment hostile to the embryo. According to Thibier and Nibart (1992), high pregnancy rates occur when transfers take place quickly and smoothly; trainees may be too slow, as well as being inexperienced.

The importance of adequate restraint of the recipient animal during the transfer procedure, to avoid unexpected movements, must also be properly understood. The recipient is examined for the location of the CL and an epidural anaesthetic given to eliminate rectal contractions. According to Broadbent *et al.* (1991), the use of a sedative and epidural anaesthesia is not essential in recipient management but seems advisable and necessary for the welfare of the animal and for operator comfort and is conducive to the achievement of good pregnancy rates.

Maintaining adequate sterility

The success of non-surgical transfer depends on maintaining adequate sterility during the deposition of the embryo in the uterus, which is much more susceptible to low-grade infections 1 week after oestrus than at the time of heat. It was 35 years ago that Cambridge workers drew attention to the fact that uterine infections could be readily established during the luteal phase of the cow's oestrous cycle, but that this could be controlled using appropriate antibiotic cover (Rowson *et al.*, 1953); in practice, this means employing trace amounts of penicillin and streptomycin in the transfer medium.

Embryonic mortality

The viability of IVMFC-derived embryos *in utero* is an area that requires much further information. It is known that *in vitro* embryos can differ from those produced *in vivo* in several ways and it is clearly important to determine how this might influence the extent of embryonic and fetal mortality. In France, Guyader-Joly *et al.* (1993) examined the ability of IVMFC embryos to develop to term and observed some evidence that embryonic loss between day 35 and day 90 may be higher (20% vs. 12%) with *in vitro* embryos than with those produced *in vivo*.

7.4 Donor–Recipient Synchrony

A considerable body of evidence has been built up over the years (see review by Ashworth, 1992) on the importance of synchrony between donor (or IVMFC embryo) and recipient stage of cycle. Exact synchrony should be the aim, but recipients out of phase by ± 1 day can be regarded as acceptable, although some reduction in pregnancy rate is to be expected; cattle that are out of synchrony by as much as 2 days would not normally be used because of the reduced pregnancy rates. Attention is drawn by Pope (1988) to the fact that multiovulating sheep have additional latitude of asynchrony not found in single-ovulating animals; the same argument may well apply to multiovulating cattle. Pope (1988) noted evidence that unilaterally ovulating sheep consistently experienced higher embryonic losses than ewes with bilateral ovulations. One possibility was that transuterine-migrating sheep embryos may be particularly sensitive to uterine asynchrony upon entry to the contralateral horn.

7.4.1 Importance of synchronization

Rowson *et al.* (1969) at Cambridge were among the first to show that exact synchronization between embryo age and recipient cycle stage resulted in the optimum pregnancy rates and that a degree of variation between donor and recipient of ± 1 day could be tolerated. In the light of cattle ET practice over the years since that time, the aim would be to have exact synchrony (Hasler *et al.*, 1987; Hasler, 1992). Close synchrony must be regarded as essential because the embryo and the maternal uterine tissues form a complex communication system involving secretions from the embryo as well as from the uterine endometrium. Such secretions stimulate and mediate changes during early pregnancy in the cow.

7.4.2 Synchronization in the IVMFC embryo context

As a result of laboratory embryo production, day-7 blastocysts can be made available for transfer to recipients that are at day 7 of their oestrous cycle. This would be regarded as exact synchrony between donor (embryo age/stage) and recipient. It should also be noted that blastocysts can develop *in vitro* over a period of several days (from day 7 to day 10 and even beyond). However, it is generally recognized that IVMFC-derived embryos that reach the blastocyst stage at 7 days post-insemination are of higher quality than those forming blastocysts later. The aim with laboratory embryo production should be to produce blastocysts for transfer to day-7 recipients. However, it should be emphasized that retarded development of an IVMFC blastocyst may be symptomatic of defects which cannot be overcome simply by an adjustment in embryo/recipient synchrony. In other words, *in vitro* blastocysts should not be transferred at all if they are too far out of phase (stage for age) with what would be regarded as normal for *in vivo* embryos.

There have been those who have reported data in favour of asynchronous transfers using IVMFC embryos (Reichenbach *et al.*, 1990); these authors reported, although with limited numbers, a pregnancy rate that was significantly higher for −1 day than for +1 day recipient/embryo synchrony.

7.4.3 Hormones and embryo–recipient synchrony

In sheep, it has been recognized for many years that progesterone administration in the first few days of the oestrous cycle could enable recipient ewes to accept and maintain embryos older than the recipient stage. In the management of recipient cattle, there may also be opportunities for examining the use of asynchronous ETs, which may become possible by employing appropriate hormone treatments to adjust embryo–recipient synchrony. Studies in the USA and New Zealand have suggested that the administration of progesterone during the early part of the cow's oestrous cycle can enhance uterine receptivity and make such asynchronous transfers feasible using embryos that are older than the recipient stage (Fox *et al.*, 1988; Garrett *et al.*, 1988; Macmillan *et al.*, 1989; Geisert *et al.*, 1991). For cattle in the later stages of their oestrous cycles, it is possible to use synthetic gonadotrophin-releasing hormone (Gn-RH) to extend their cycles and to transfer embryos that are younger than the recipient stage (Drost *et al.*, 1989; Macmillan *et al.*, 1989).

The real practical interest in asynchronous transfers lies in devising systems whereby IVMFC blastocysts can be transferred to a group of recipient cattle which differ among themselves by several days in their stages of cycle. As one means of reducing the cost of cattle ET operations on the farm, it is an area worthy of examination.

7.5 Methods of Synchronizing Oestrus

Oestrus has been synchronized in cattle with progesterone or progestagens, contained in a variety of devices (see Fig. 7.2), $PGF_{2\alpha}$ and its analogues, progestagen–oestrogen combinations and progestagen–PG combinations (see review by Roche, 1989). As well as the synchronizing treatment, there is also the crucial need accurately to detect oestrus when it does occur. Of the various reproductive problems in dairy cattle, the single most important is often held to be poor detection of oestrus (Senger, 1993). As dairy herds continue to grow in size, the problem is likely to become more acute because of reduced manpower.

7.5.1 Devices for synchronizing oestrus

Synchronization of oestrus in recipient cattle may be achieved by several means, including the use of PGs (single or double injection regimes), progestagen-impregnated sponge pessaries, progestagen ear implants (Crestar), intravaginal progesterone-releasing devices such as the PRID (progesterone-releasing intravaginal device) or the more recently available CIDR (controlled

Fig. 7.2. Use of the PRID device for synchronizing oestrus in recipient cows. Control of the cow's oestrous cycle may involve shortening the lifespan of the corpus luteum by inducing luteolysis with prostaglandin or extending the cycle by use of progesterone or a progestagen.

internal drug release) devices. As noted in reviews by Odde (1990) and Larson and Ball (1992), despite many years of research, a protocol has not been found that will consistently synchronize oestrus in the cow with sufficient precision to permit high levels of success with fixed-time AI. Improvements have been achieved by combining progestagen treatments with the PG regime, but clearly this adds to the cost of synchronization and lessens the likelihood that the treatment will be cost-effective.

PRIDs are impregnated with progesterone and have an attached oestradiol benzoate (10 mg) capsule. They are inserted in the vagina, usually for a period of 12 days, and are effective at all cycle stages. The CIDR, which is also impregnated with progesterone, is a device which has been developed and tested widely in Australasia and elsewhere (Macmillan and Macmillan, 1989; Washburn *et al.*, 1989; Macmillan *et al.*, 1990a, 1991). According to the experiences of Broadbent *et al.* (1993), who examined the device in a comparison with the PRID in the UK, the CIDR was found to have superior retention rates in both heifers and parous cows. These authors concluded that it was a suitable device for synchronizing oestrus in recipient cattle.

Crestar implants are impregnated with a potent progestagen, norgestomet (3 mg), and are accompanied by a combined injection of norgestomet (3 mg) and oestradiol valerate (5 mg). The implants are inserted under the skin of the ear for a 9–10-day period and are also effective at all cycle stages (Broadbent *et al.*, 1992; Tregaskes *et al.*, 1994).

These short-term progestagen treatments (PRID, CIDR, Crestar) used in cyclic cattle can be expected to induce an oestrous response of 85–95%, with most animals (80%) coming in oestrus between 24 and 48 h after progestagen withdrawal. Of interest are results in a study reported by Wehrman *et al.* (1993) showing conception rates to be greater in cows receiving two PRIDs than in those receiving one for oestrus synchronization. This effect was believed to be due to the double PRID influencing the endocrine mechanisms involved in conception.

PGs (Dinolytic, Estrumate, Prosolvin) are given by intramuscular injection and are effective in the induction of oestrus in cattle that are between days 5 and 15 of their oestrous cycle. A double PG injection regimen (10–12-day intervals) used in cyclic heifers induces a high oestrous response (80–90%) with about 60% of animals responding between 48 and 72 h later and a further 20% between 72 and 96 h (Sreenan and Diskin, 1992; Diskin and Sreenan, 1993).

Combinations of progestagen and PGs have been widely employed in various research programmes (Tregaskes *et al.*, 1992; Sinclair *et al.*, 1994). As shown by Broadbent *et al.* (1991), a disadvantage of the PRID may be that a proportion of the devices are lost (5% in their report); using the CIDR, on the other hand, they recorded a loss rate close to zero and little trauma during application, with less discomfort and apparent vaginitis. The same workers preferred regimens in which the progestagen (PRID or CIDR) is withdrawn on day 8 to 10 and PG administered 2 days before progestagen withdrawal.

The effects of a norgestomet ear implant (Crestar) combined with a PG injection 2 days before implant removal in recipient cattle are reported by Sinclair *et al.* (1992); incorporation of PG into the synchrony regimen led to more recipients coming into oestrus simultaneously and a higher pregnancy rate.

7.5.2 Oestrus and its detection

The cow that is in oestrus will stand when mounted from the rear by a bull or companion cow. She indicates her willingness to stand by immobility when approached and a slight arching of the back (see Phillips, 1993). Sometimes a cow will mount another from the front (head-mounting) and in this situation it is the riding cow that is in oestrus and not the one underneath (Payas *et al.*, 1989; Broadbent *et al.*, 1991; Pullar and Collins, 1993). Cows that are in heat can be expected to display a variety of signs, including: discharge of clear mucus from the vulva; tail-raising and switching; licking, sniffing and rubbing against other cattle; swelling and reddening of the vulva; frequent bawling; general restlessness and attentiveness to the activities of other cattle and humans; ruffling of rump hair and mild abrasion of rump skin as a consequence of having been mounted; and a temporary decrease in milk yield (see Fig. 7.3).

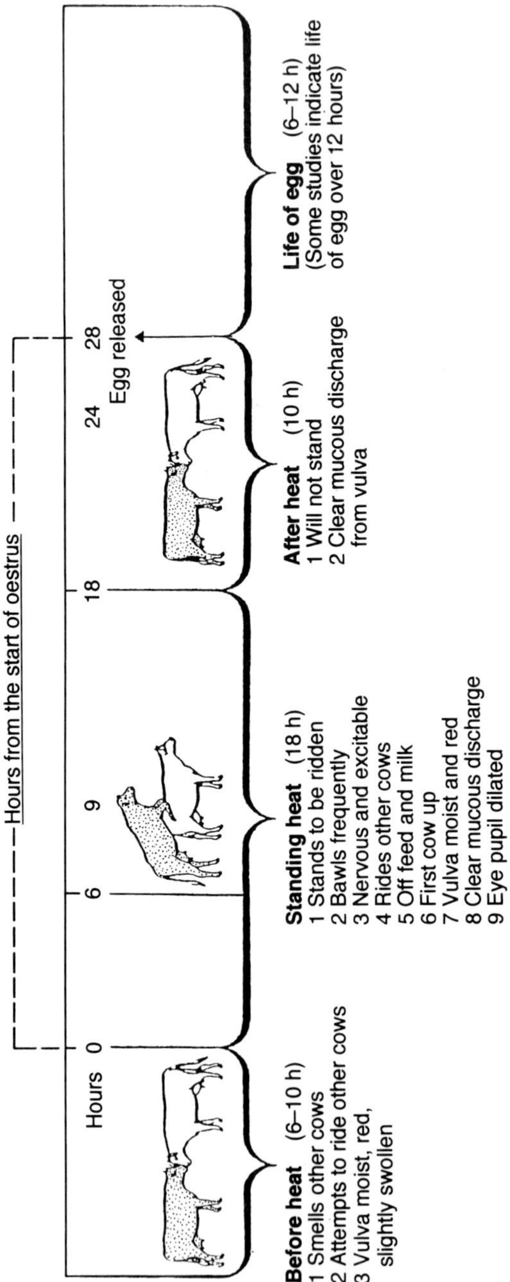

Fig. 7.3. Oestrus in the cow and associated features.

Oestrus in synchronized cattle

The observer has to be on the lookout for all or any of the signs mentioned above, if he/she is to be successful in oestrus detection. The most cost-effective method of detecting oestrus in a large group of synchronized recipients is by direct observation (Broadbent *et al.*, 1991). Observation periods should be spread over as much of the day as possible and should avoid feeding times, milkings or other routines which are likely to disturb or distract the cattle. It appears that the number of cattle in oestrus simultaneously may increase the frequency and duration of oestrous activity (Pullar and Collins, 1993); the same authors also record that the cow in oestrus is the best detector of oestrus in her herd mates.

The accurate identification of cows in heat after synchronizing treatments can occasionally be difficult because of the intense mounting activity among members of a group (Macmillan *et al.*, 1988); for that reason, it may be necessary to employ different methods of oestrus detection, with tail paints and other forms of animal identification.

Future developments in oestrus detection technology

The need to develop and apply new technology for oestrus detection in dairy cattle has been emphasized (Senger, 1993). Developments in oestrus detection in cattle in the past decade are reviewed by Lehrer *et al.* (1992); according to these authors, physical activity and, in the future, vulvar impedance, are the most promising variables for incorporation into an on-farm management system. There are many possible approaches towards automating the diagnosis of the behavioural and/or physiological signs of oestrus (Senger, 1993); these include the use of electronic technology such as activity monitors (Gauger *et al.*, 1991; Moore and Spahr, 1991; Pulvermacher, 1991), implantable impedance sensors, implantable and surface pressure sensors and in-line milking parlour hormone sensors.

Assessing accuracy of oestrus detection

Strategic use of on-farm progesterone-testing methods may enable farmers to improve their heat detection rates (Ruiz *et al.*, 1992). Under normal husbandry conditions in the UK, little more than 50% of cattle in oestrus are detected by farm staff in the course of a cycle interval (see Fig. 7.4); anything from 10 to 20% of non-oestrous animals may be recorded as being in heat (Esslemont, 1993). The cow in genuine oestrus, however, will normally have a low milk progesterone concentration; in some conditions this can be used to identify animals that are mistakenly assumed to be in heat (Foulkes, 1991; McLeod *et al.*, 1991; Williams, 1991; Williams and Esslemont, 1993). The use of a vaginal probe (to measure impedance) as an alternative means of confirming genuine oestrus detection is suggested in a paper by Kitwood *et al.* (1993).

Fig. 7.4. Standing oestrus in the recipient cow. For successful embryo transfer in cattle, it is essential that accurate data are available on the occurrence of oestrus in the recipients. Using the bred recipient for twinning means that AI is carried out a week before transfer of the embryo. This permits an additional check on oestrus.

7.6 Selection and Management of Recipients

7.6.1 Heifers or cows as recipients?

Maiden heifers are usually preferred as recipients in conventional cattle ET operations (Broadbent *et al.*, 1991); quite apart from being free of problems arising from previous pregnancies (e.g. endometritis), such animals are likely to cost less and to be easier to acquire than cows. However, in terms of ease of transfer, the parous cow is at an obvious advantage. Coultard (1991) in the UK recorded that about 10% of heifer recipients may be difficult, if not impossible, to use for cervical transfers. If twin pregnancy is to be attempted in dairy cows, then it should be with older parous cows that have a previous history of easy calving (Sreenan and Diskin, 1992).

The use of certain categories of maiden heifer cattle as recipients may pose very real welfare questions (Murray and Ward, 1993). In the UK, the use of beef-type heifers as recipients for embryos from large dairy breeds (Holstein) and double-muscled beef breeds may result in a proportion requiring surgery to deliver the fetus. It is clearly undesirable, for the long-term future of cattle ET, that inappropriate (e.g. Holstein) embryos should be transferred to unsuitable maiden heifers; this is only likely to be used as an example of modern farming practices compromising the quality of animal life. If farmers, and others in the cattle industry, expect to obtain the approval of society at large for certain forms of reproductive technology, they must exercise particular care in matching recipient with embryo.

7.6.2 *Recipient hormone levels*

Progesterone secreted by the CL is the major steroid influencing the physiological state of the recipient's uterus. However, the relationship between blood levels of progesterone and embryo survival after transfer in cattle has not been well resolved. There is some evidence that a low progesterone level can result in a stronger luteolytic mechanism and may predispose some cows to early embryo loss (Lamming and Mann, 1993). It may occasionally be useful to seek reassurance that recipients are at the anticipated stage of the cycle and that progesterone is actually being secreted in the amounts expected at that time.

Evaluation of CL function on the basis of examination per rectum has not been found to be sufficiently reliable for screening recipients before transfer (Humblot *et al.*, 1987). However, in a cattle ET programme conducted in Germany, Geim (1990) recorded that recipients with good CL development (assessed by rectal palpation) had a significantly higher pregnancy rate than those with a poor CL (58% vs. 14%). Screening recipients on the basis of their progesterone levels is probably regarded as more reliable than palpation of the CL (Britt and Holt, 1988). In Spain, Iriarte *et al.* (1991) concluded that a combination of plasma progesterone values for day zero and day 4 constituted an effective means of selecting their recipient heifers. In addition to the conventional kits that are now available for determining progesterone in milk and plasma samples (Nebel, 1988; Nebel *et al.*, 1989; Foulkes, 1991; McLeod *et al.*, 1991; Williams, 1991), developments include salivary determinations of the steroid (Gao *et al.*, 1988).

According to Scott (1993), real-time ultrasonics may occasionally have a role in the selection of recipients where the CL position and size cannot be reliably determined by palpation per rectum. Danish workers, on the other hand, incline to the view that careful gynaecological examination is probably as good as ultrasonic scanning and considerably cheaper (Greve and Purwantara, 1993).

7.6.3 *Using recipients on second and third occasions*

Although hard evidence appears to be limited, the experience of some commercial ET operators has led them to believe that recipients used for a second or third transfer show pregnancy rates below those recorded for first pregnancies (Cargill, 1992). There has apparently been some suggestion in former Czechoslovakian studies that this may be the result of an immunological reaction to cattle embryos (*in vivo*-produced from superovulated donors) due to proteins picked up in the oviduct prior to recovery. Such proteins subsequently act as an antigen, which results in an immune response in the recipient, which becomes increasingly sensitive in later transfers. Whatever the true explanation may be, it would be useful to have more information on this point.

7.6.4 Minimizing stress in recipients

The importance of the basic rule of avoiding stress in recipient cattle is rightly emphasized by Thibier and Nibart (1992). Any routine treatment (antiparasitic, etc.) should take place at least 3 weeks prior to transfer; changes in the feeding regimen should be prohibited for 3–4 weeks before and after transfer. Recipients should be located so that they can be handled easily and quietly on the day of transfer. Transfers should not be carried out by operators who are not well experienced in passing the transfer instrument through the cervix and manipulating it into the appropriate position in the uterus. The use of an epidural anaesthetic (lignocaine) and sedation (acetylepromazine), particularly for beef cattle recipients that are not well accustomed to handling, should always be considered (Stewart *et al.*, 1991). The need to have the recipient in a suitable restraining crush should require no emphasis.

7.6.5 Detecting early pregnancy in recipients

For research purposes, it may be useful to have early information on the establishment of pregnancy in the recipient animal after embryo transfer. Advances in the use of real-time ultrasonics now mean that, for research purposes, the bovine conceptus can be visualized as early as 9–10 days of age and with some accuracy by day 12 (Curran *et al.*, 1986b; Boyd *et al.*, 1988; Izaike *et al.*, 1988, 1991b; Wilson and Zalesky, 1988). In work reported by Jones *et al.* (1990), ultrasonic scanning was carried out at 20–24 days post-transfer (i.e. 27–31 days of gestation) with an overall accuracy of 94% (1379/1452). The scanning enabled non-pregnant recipients to be returned to breeding with the loss of only one oestrous cycle, allowing more effective animal utilization.

It is clearly possible, using such ultrasonic scanning, to follow the course of pregnancy with considerable accuracy through the early months of the cow's gestation period. This can add a new dimension to studies dealing with early pregnancy in the cow, by providing much more precise information on the progress of the embryo and fetus. Such information may be particularly valuable in comparative studies dealing with embryos produced by laboratory procedures (Guyader *et al.*, 1992, 1993).

7.7 Enhancing Pregnancy Rates in Recipients

As a result of increasing information and the development of new techniques, there are several ways in which it may be possible to enhance pregnancy rates in recipient cattle (see Fig. 7.5). However, it should be emphasized that much of the evidence in this field is so variable that it is difficult at this stage to formulate useful recommendations. Some of the methods, such as the use of Gn-RH preparations and interferons (IFNs), are also likely to be of general

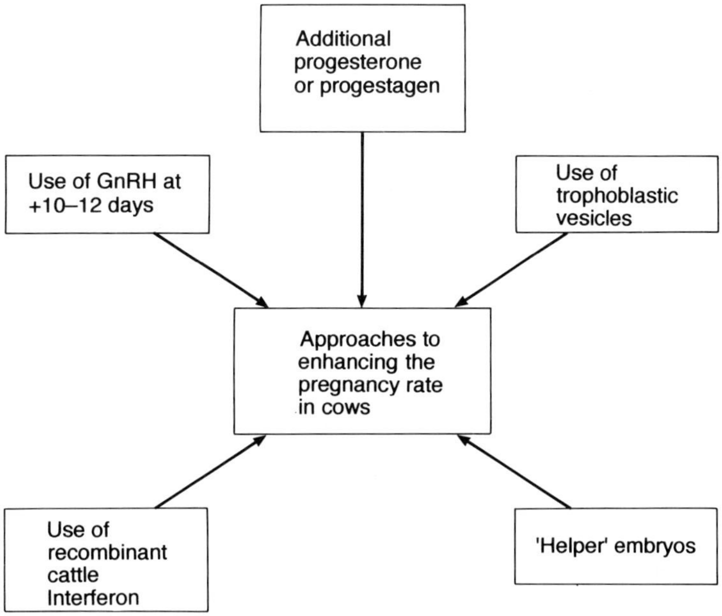

Fig. 7.5. Possible methods of enhancing pregnancy rates in cattle.

interest in enhancing pregnancy rates after AI (or even natural service) as well as after ET (Thatcher *et al.*, 1990). It is generally accepted that nearly one in five dairy cows pregnancy-tested at 24 days after insemination is not in calf and lost breeding days as a consequence may cost the farmer and the country substantial sums.

In human IVF ET, there is even some limited evidence suggesting that short-term immunosuppression over a 4-day period (using 16-β-methyl-prednisone) may improve pregnancy rate (Polak de Fried *et al.*, 1992, 1993). Whether a similar form of steroid treatment constitutes an option worthy of examination in cattle recipients remains unknown.

7.7.1 *Supplementary progesterone*

Progesterone levels in the cow increase rapidly from the fourth to the tenth day after oestrus in the cyclic animal. In some studies with Holstein cows in Canada, pregnancy rates have been increased by as much as 30% by using PRIDs inserted for a week on day 7 and day 10 after AI (Robinson, N.A. *et al.*, 1989). In Florida, Van Cleeff *et al.* (1991) found that treatment of dairy heifers with CIDR devices for days 7–13 after AI failed to increase conception rate (57.9% vs. 53.6% for controls). In New Zealand, Macmillan *et al.* (1990b) reported studies with 240 cows which received CIDRs on days 6, 7 or 9 of the cycle for a period of 6–12 days; these animals showed an overall pregnancy rate of 79.2% compared with 65.7% in the 245 contemporary control cattle, a highly significant difference.

In Australia, working with beef suckler cattle, Munro and Bertram (1990) used CIDRs without effect on the pregnancy rate; however, treatment significantly improved the proportion of non-pregnant cows observed in oestrus 3 weeks after AI. Similar findings were reported by Stevenson and Mee (1991a, b) in Kansas.

A report by Logue *et al.* (1992) describes the effect of using a progestagen (norgestomet) ear implant after breeding (day 10 to day 21); no improvement in pregnancy rate was recorded but there was a high degree of synchrony of oestrus in non-pregnant animals. Working in Aberdeen, Broadbent *et al.* (1992) recorded a significant increase in pregnancy rate among recipient cattle when Crestar ear implants (norgestomet) were given on day 7 (day of transfer).

An alternative option in attempting to increase progesterone levels is by using human chorionic gonadotrophin (HCG) to enhance the production of progesterone by the cow's own CL. In Canada, where Walton *et al.* (1991) had previously demonstrated a favourable effect of supplemental progesterone (between days 5 and 17) on pregnancy rate, it was found that treatment with 1500 iu HCG on day 5 resulted in the formation of accessory CLs and an increased progesterone level. A report by Rajamahendran and Sianangama (1992) also showed that HCG treatment on day 7 produced accessory CLs and reduced the incidence of embryonic mortality in cattle. The same authors refer to a large-scale field trial in which there was a significant increase in pregnancy rate after HCG treatment applied on day 7. Further results of HCG treatment at day 7 were provided by Sianangama and Rajamahendran (1992); a significant increase in pregnancy rate was recorded (62% vs. 47%).

7.7.2 *Use of Gn-RH*

In the past 22 years, it is estimated that more than 3000 analogues of Gn-RH, containing various structural modifications, have been synthesized (Schally, 1992); some of these analogues are 50–100 times more potent than the parent hormone and can be used to influence the reproductive process. In New Zealand, Macmillan *et al.* (1986) were among the first to demonstrate an enhancement of pregnancy rate in dairy cows when Gn-RH (10 μg buserelin) was injected on day 11 after breeding by AI. Similar findings were reported in the UK by Drew and Peters (1991, 1992, 1994) and Sheldon and Dobson (1993). Such ability to increase pregnancy rate by reducing embryo mortality is clear evidence that not all embryonic mortality in cattle is predetermined.

However, not all studies with Gn-RH have shown unequivocal evidence of increased pregnancy rates. In fact, reviewing some of the earlier literature, Drost and Thatcher (1992) concluded that there was no clear evidence that treatment with Gn-RH between days 11 and 14 after breeding would result in increased pregnancy rates. Ellington *et al.* (1990b, 1991a), in the USA, evaluated the use of a Gn-RH analogue (8 μg Receptal) in recipient cattle at the time of ET or 4–7 days after transfer; there was a non-significant tendency for pregnancy rates to increase. It appeared, however, that there was little benefit from the treatment, where control performance was judged adequate.

A trial by Jubb *et al.* (1990) in Australia revealed no increase in first-service pregnancy rate after administering Gn-RH on days 11–13.

In Colorado, Gn-RH treatment on day 12 in beef cattle produced no evidence of an enhanced pregnancy rate (LeFever *et al.*, 1991). In Ireland, Ryan *et al.* (1991, 1992a) used Receptal on day 12 without recording any increase in pregnancy rates in Friesian dairy cows. A large farm-to-farm variation in response was observed by Sheldon and Dobson (1993), but the cause of such variation remains unclear. It appears unlikely, however, that Gn-RH treatment leads to the formation of accessory CLs and increased progesterone levels. It may influence follicle development in such a way that there is a delay in the induction of uterine oxytocin receptors, with a consequential delay in luteolysis; this may provide the embryo with more time to produce bovine trophoblastic protein 1 (bTP-1).

In France, Lajili *et al.* (1991) found that treatment with Gn-RH (10 μg buserelin) 12–14 days after breeding only enhanced the pregnancy rate in cows which had previously been treated with $PGF_{2\alpha}$. In non-pregnant cows, Gn-RH had a beneficial effect on heat detection and improved the pregnancy rate at subsequent breeding. The French workers were not able to offer an explanation as to why only a selected group (PG-treated) of cattle responded favourably to treatment.

Reports by Rettmer *et al.* (1992a, b) showed that doses of a potent Gn-RH agonist (Fertirelin acetate; Upjohn) administered between day 11 and day 14 after breeding improved pregnancy rates in heifers and tended to do the same in suckler cows; the authors concluded that reduced follicular secretion of oestrogen was associated with enhanced, but not prolonged, luteal function in heifers resulting from the treatment. Treatment of Holstein cows in California led to variable results (Stevenson and Phatak, 1992; Stevenson, J.S., *et al.*, 1992); it was suggested that Gn-RH may increase the lifespan of the CL and the secretion rates of luteinizing hormone (LH) and progesterone, without necessarily altering the pregnancy rates.

It should be mentioned that Gn-RH treatment for enhanced pregnancy rates in cattle is now commercially available in some countries, based on the evidence produced in New Zealand and the UK. There are those who have used cost–benefit analysis to show a value from the treatment, based on fewer cull cows, a shorter calving interval and fewer semen straws.

7.7.3 Pregnancy-specific proteins

The maternal recognition of pregnancy in the cow involves physiological mechanisms that result in the protection of the CL from luteolysis by modifying or inhibiting the uterine production of luteolytic pulses of $PGF_{2\alpha}$ (Plante *et al.*, 1987, 1988, 1990; reviews by Parkinson, 1991; Roberts *et al.*, 1992). In the non-pregnant cow, luteal cells release oxytocin in a pulsatile manner during the late luteal phase; the peptide binds to its endometrial receptors and initiates luteolytic pulses of $PGF_{2\alpha}$. In the pregnant cow, early signals (day 16) from the elongating bovine blastocyst ensure maintenance of the functional and structural integrity of the CL and progesterone secretion ensures maintenance of an endometrium that is permissive to embryonic development. The signals are now known to be pregnancy-specific proteins secreted by cells of the trophectoderm for a limited period around the time of rapid blastocyst elongation. These proteins were initially known as bTP-1, but they were subsequently identified as functionally active IFNs in the IFN-α subfamily.

Interferons

At the time of their discovery in 1957, the term 'interferon' was used in referring to certain factors produced by cells in response to a viral infection. It is now known that the original concept of one IFN with one biological function was too simple and it is clear that IFNs are members of a large family of regulatory proteins (Balkwill, 1989), some of which are involved in the regulation of pregnancy. The IFNs are cytokines and, in addition to their direct antiviral action, stimulate antibody synthesis, the activity of natural killer cells and the antimicrobial activity of macrophages and neutrophils. The IFNs synthesized by the trophectoderm of the farm ruminant embryo are now particularly well characterized (Flint *et al.*, 1988; Imakawa *et al.* ,1988; Bazer *et al.*, 1990a; Hansen *et al.*, 1990; Thatcher *et al.*, 1990; Radford *et al.*, 1991). In the cow, it has been shown that bTP-1 messenger ribonucleic acid (mRNA) is present in the conceptus as early as day 12 of pregnancy (Farin *et al.*, 1990) and expression of the IFN is greatest around day 15 to 19. The fact that the protein is first expressed at the blastocyst stage may be of interest in terms of evaluating the viability of IVMFC-derived bovine embryos; one report has shown that hatched blastocysts graded good or excellent produced more bTP-1 than those of fair quality (Hernandez-Ledezma and Roberts, 1992). In other work in Missouri, evidence was found that continued viability and IFN production by IVMFC blastocysts may require some exposure to the uterine environment (Hernandez-Ledezma *et al.*, 1992a, b, 1993b). There is also evidence suggesting that the appropriate expression of maternal granulocyte-macrophage colony-stimulating factor (GM-CSF) may indirectly determine the fate of the ruminant conceptus by influencing the amount of IFN produced (Imakawa *et al.*, 1993). The secretion of progesterone by luteinized bovine granulosa cells *in vitro* has been stimulated by very low concentrations of bovine recombinant IFN (rboIFN) (Luck *et al.*, 1992); such results are consistent with a role for IFN in maintaining the CL of pregnancy.

rboIFN-α is available and has been employed in sheep (days 12–18) to enhance pregnancy rates in that species (Schalue-Francis *et al.*, 1989, 1990, 1991; Roberts *et al.*, 1990a, b; Martinod *et al.*, 1991). Studies by Lamming *et al.* (1988) and Parkinson *et al.* (1992) have confirmed that rboIFN-α can extend the luteal phase of cyclic ewes by inhibiting the secretion of $PGF_{2\alpha}$ and that it can stimulate luteal maintenance when administered by a systemic route. The studies of Wiltbank *et al.* (1992) led them to conclude that the early sheep embryo secretes a protein that blocks the actions of $PGF_{2\alpha}$ on large luteal cells.

In cattle, the commercial interest in rboIFN-α would presumably also be as a new means of enhancing the maintenance of pregnancy in that species.

It is a matter of devising a slow-release formulation of the IFN which would permit simple administration and determining those herd conditions in which the increase in fertility achieved can be in any way cost-effective.

7.7.4 *Other possibilities of enhancing pregnancy rates*

Trophoblastic vesicles

It is clear that early embryonic mortality is a major factor influencing the reproductive efficiency of the cow. As noted in the previous section, pregnancy-specific proteins are secreted by the elongating cattle embryo and these play a part in the early establishment of pregnancy in the cow. There may be possibilities in employing embryonic tissue capable of secreting such proteins to enhance pregnancy rates in the cow. Evidence was first provided in the report of Heyman *et al.* (1987a) showing that the cotransfer of trophoblastic vesicles (TVs) with frozen–thawed bovine embryos contributed to luteal maintenance in recipient cattle and an enhancement of the pregnancy rate. This was believed to be due to the vesicles augmenting the strength of the embryonic signals.

It had been earlier suggested that part of the losses experienced with cattle embryos after freeze–thawing may be due to some loss of functional activity by the embryonic cells (Heyman, 1984). In Louisiana, Pool *et al.* (1988a, b) also reported improved viability of cattle embryos cultured with TVs. In Arkansas, Lester *et al.* (1991) examined the toxicity of cryoprotectants to cattle TVs with the object of developing an effective cryopreservation procedure; it was concluded that ethylene glycol was less toxic than the other agents evaluated (glycerol, dimethylsulphoxide (DMSO), propylene glycol).

So far as recipient cattle are concerned, then, the laboratory production of TVs and their use alongside ETs as a means of enhancing pregnancy rate may be an area worthy of examination. Results reported by Loskutoff *et al.* (1993) in Canada supported the conclusions of Heyman *et al.* (1987a, b) that TVs can improve the survival of cattle embryos damaged by freeze–thawing; these same authors found fresh TVs to be more effective than frozen–thawed vesicles.

In Ireland, bovine TVs have been generated from day-13–14 embryos and transferred to the uterus of cows 5–7 days after breeding by AI; in postpartum Friesian dairy cows (those calved less than 55 days), pregnancy rate was increased by 13% (Ryan, 1993a, b; Ryan *et al.*, 1994).

Although horse embryos do not undergo the elongation characteristic of the farm ruminants and pigs, it has been shown that vesicles can be prepared (from the day-10 conceptus) and are capable of extending the cycle interval (Ball *et al.*, 1989; Hehnke *et al.*, 1990).

Producing 'helper embryos'

In former Czechoslovakia, Slavik *et al.* (1990) were among the earliest to demonstrate that bovine oocytes can be fertilized with ram sperm. In a subsequent report, Slavik and Fulka (1992) reported evidence that the zona pellucida of bovine and ovine oocytes does not necessarily present a barrier to penetration by goat sperm. In Canada, Kelk *et al.* (1991) reported that fertilization of bovine oocytes by ram sperm may be an *in vitro* phenomenon; attempts to produce hybrid embryos by inseminating superovulated cows with ram sperm failed.

An interesting possibility of using cattle × sheep hybrid embryos as 'helper embryos' is dealt with in a report by Lange *et al.* (1991), working in Germany. These workers speculate that such 'helper embryos' may be used to enhance pregnancy rates in recipient cattle by producing pregnancy-specific proteins but without developing further than the early embryonic stages. Using both *in vivo* and *in vitro* procedures, the German workers succeeded in producing a low percentage of morulae by fertilizing cattle oocytes with ram sperm.

Laser destruction of ICM

As long ago as the mid-1960s, the use of the laser for the selective destruction of blastomeres in the early embryo was demonstrated (Daniel and Takahashi, 1965). In view of the growing interest in laser beams as microsurgical tools, particularly in human IVF ET, there may be opportunities to examine the use of lasers to destroy the inner cell mass (ICM) of IVMFC-derived blastocysts. This might be the means of providing a TV capable of producing pregnancy-specific proteins, without this involving continued embryonic development. In the laboratory production of cattle embryos, there will inevitably be a proportion of blastocysts (grade 3 or poorer) unsuitable for transfer or freeze-storage; such lower-quality blastocysts could perhaps be employed for the production of TVs by laser treatment.

Production of parthenogenetic blastocysts

Parthenogenetic activation of cattle oocytes can be induced by various stimuli, such as exposure to ionophore (Ware *et al.*, 1989), ethanol (Lee, S.H., *et al.*, 1992; Nagai,1992; Monaghan, 1993; Wahid, 1993), ethanol and cycloheximide (Presicce and Yang, 1993, 1994) and electric current (Kono *et al.*, 1989; Landa and Hajkova, 1989; Kinis, 1990). An age dependency for activation of IVM bovine oocytes was reported by Ware *et al.* (1989). One of the earliest reports on *in vitro* development to the hatched blastocyst stage of parthenogenetically activated bovine oocytes after IVM was that of Lee, S.H., *et al.* (1992). Electrical activation was employed by Prochazka *et al.* (1993) to produce parthenogenomes capable of development to the blastocyst stage. The possible use of such parthenogenomes in a supportive role in the establishment of pregnancy in recipient cattle seems unlikely in the light of the report by Fukui *et al.* (1992) showing that IVM bovine oocytes treated with ethanol and cytochalasin B may not develop to the blastocyst stage *in vitro*, but are capable of continuing embryonic development for a further 1–2 months of early pregnancy.

In all of this, due consideration must be given to the ease with which any

Fig. 7.6. Avoid stress in the recipient cow at time of transfer. Welfare considerations are such as to require much care and attention in the non-surgical transfer of cattle embryos. It is a matter of appropriate training and an awareness that a humane approach must always be employed.

pregnancy-enhancing treatment can be applied. Those that can be employed simultaneously with ET would have an advantage in minimizing the handling of recipients. The welfare implications of the transfer process must be carefully considered at all times (see Fig. 7.6).

7.8 Embryo Transfer in Other Farm Animals

Whether for research or for commercial purposes, there are likely to be many reasons why non-surgical transfer approaches should be developed for the other farm ruminants (sheep/goats) and pigs. It is certain that effective procedures for producing transferable-quality embryos by *in vitro* means will be developed in due course for all the other farm species. This may be based on slaughterhouse ovaries as the source of primary oocytes or on recovering such oocytes from the live animals. Applications involving IVMFC-derived embryos would be greatly facilitated if non-surgical transfer methods were available. In the recent decade, workers in several countries have examined the possibility of developing suitable transfer techniques.

7.8.1 Sheep and goats

A review of studies in ET in sheep and goats is provided by Brebion *et al.* (1992) in France. An account of the transfer of fresh and imported sheep embryos in Denmark is given by Petersen and Holm (1993); they record pregnancy rates of 70–95% for fresh and 45–55% for frozen embryos. However, it is felt that the application of multiple-ovulation ET (MOET) programmes in sheep breeding improvement has been limited because of inadequacies in the technology of embryo transfer in sheep. According to Smith (1988), the rate of genetic improvement for different traits in sheep can be doubled by the use of MOET compared with normal reproduction and conventional selection schemes.

The use of fibre optics in the intrauterine insemination of superovulated sheep and in both embryo recovery and ET has been a major advance in the development of suitable technology in the recent decade (McKelvey *et al.*, 1985; Walker *et al.*, 1985; McKelvey and Robinson, 1986; Brebion *et al.*, 1989; Robinson, J., *et al.*, 1989). In the transfer of IVMFC-derived embryos to recipient ewes by laparoscopy, the procedure could be very similar to that of intrauterine insemination. The recipient uterine horn is gently grasped with forceps; this gives precision to the positioning of the transfer pipette and the expulsion of the embryo into the cavity of the uterine horn.

Unlike the cow, access to the uterus via its natural entry point, the cervix, has been regarded as extremely difficult in the ewe. There are, however, some workers who claim that the cervix of the dioestrous ewe can be penetrated successfully for the transfer of day-7 embryos (Buckrell *et al.*, 1993). Attempts to enhance pregnancy rates in sheep after ET may eventually employ uterine administration of ovine IFN (Ott *et al.*, 1991, 1993).

The first surgical goat ETs date back to the 1930s in the USA (Table 7.1), but the increasing importance of goat ET in breeding improvement programmes has led to the use of laparoscopy as an alternative transfer procedure. A report by Vallet *et al.* (1989) showed that comparable results could be achieved by laparoscopy in comparison with surgery. These authors list several advantages to laparoscopy under field conditions, including speed of transfer (5 min vs. 15 min) and freedom from genital tract adhesions.

7.8.2 Pigs

The use of ET in pigs makes it possible to introduce new genetic material into commercial specific pathogen-free herds in order to reduce the risk of disease transmission (Cameron *et al.*, 1989). As with sheep and goats, one factor limiting progress in the commercial application of pig ET has been the necessity to employ surgical procedures for the collection and transfer of embryos. The first non-surgical transfers of pig embryos (*in vivo*-produced) to result in the birth of living piglets are reported by Reichenbach *et al.* (1993b) in Germany. The embryos were transferred while the recipient pigs were under general anaesthesia to prevent stress; the animals were placed in dorsal recumbency on a modified laparotomy surgical table to limit movement.

However, the pregnancy rate of 10% was markedly below the 60–75% reported after surgical transfers (Cameron *et al.*, 1989) and the technique requires further refinement.

As with cattle, exposure of the pig embryo to an asynchronous uterus can adversely affect development. In Missouri, Jarrell *et al.* (1990a) found that 24 h exposure of day-4 pig embryos to an asynchronous uterus increased embryonic mortality rate.

7.8.3 Horses

It is possible that the production of horse embryos by *in vitro* procedures can find a useful place in the study of many aspects of reproduction in this species, as well as enabling progress to be made in areas such as embryo storage and freezing. Primary oocytes may be recovered from slaughterhouse ovaries (Choi *et al.*, 1993a, b) or by repeated transvaginal follicular aspiration in the live mare (Cook *et al.*, 1992, 1993; Parlevliet *et al.*, 1993). The commercial application of equine ET will continue to depend on a range of factors, including breed registration conditions and costs involved in transfers, as well as the level of results achieved.

In Ireland, attempts were made in the early 1980s to use ET commercially as a means of producing foals from valuable show-jumping mares (Wade *et al.*, 1987, 1989); non-surgical transfers in that work resulted in pregnancy rates of about 50%. Although non-surgical transfer of equine embryos via the cervix is a simple manipulative procedure in the horse, the maximum pregnancy rates achieved may be markedly lower than those following surgical transfer (Allen, 1991b). Methods suggested to improve non-surgical pregnancy rates include: (i) providing a protective sheath over the insemination catheter; (ii) high standards of cleanliness and aseptic technique; and (iii) deposition of the embryo in the uterine horn rather than in the body of the uterus (Squires, 1989; Squires and Tarr, 1991). It remains to be seen whether the non-surgical approach can become the technique of choice by refinements in procedure.

In the matter of recipient mares, there are commercial units who use ovariectomized mares (Biven *et al.*, 1989), a practice not found in conventional cattle ET. The use of ovariectomized mares eliminates the need for oestrus detection and ovary palpation, maintenance of a large number of recipient mares and control of ovulation in recipients. However, the practice obviously requires long-term progesterone or oral progestagen (altrenogest) administration (Squires and Tarr, 1991).

7.8.4 Buffaloes

In comparison with cattle ET, which is now being commercially applied in many countries, the use of ET technology in the buffalo is very limited. It is clear that the application of cattle ET technology directly to the buffalo has met with limited success and much remains to be done in developing technology specifically for use in this species (Drost, 1991). However, the possibility exists that advancements in oocyte recovery and maturation technology may eventually permit large numbers of buffalo embryos to be produced *in vitro*. The commercial application of such embryos would obviously be based on the employment of available freeze–thawing and non-surgical transfer methods.

8 Use of Embryos and Oocytes in Commercial Practice and Research

8.1 Introduction

The development of *in vitro* techniques for the laboratory production of farm animal embryos has excellent potential, both for basic research and for practical applications on the farm. The first offspring produced from *in vitro* fertilization (IVF) of a bovine oocyte matured *in vivo* was a bull calf born in the USA in 1981 (Brackett *et al.*, 1982). This was followed a few years later by work which led to the production of calves by totally *in vitro* procedures. Considerable interest now exists in this technology, which permits the low-cost production of large numbers of cattle embryos at specific stages of development. The technology also provides for the development of other techniques, such as cloning, which hitherto had only been possible using *in vivo*-produced embryos (see Fig. 8.1).

Papers reviewing progress in the *in vitro* production of cattle embryos have been published by workers in many countries, including Belgium (Mermillod *et al.*, 1992a, e, f), Ireland (Gordon, I., 1987, 1989a, b, c, 1990, 1991a, b; Gordon & Lu, 1990; Lu and Polge, 1992), Denmark (Greve, 1992; Avery and Greve, 1993), Russia (Sergejew and Baibekow, 1992), Japan (Kato and Iritani, 1993), China (Chang, Q.Z., *et al.*, 1992), France (Crozet, 1991), Greece (Vergos *et al.*, 1993), Canada (Stubbings *et al.*, 1990c), Germany (Hahn, 1993; Palma *et al.*, 1993c) and the USA (Fayrer-Hosken and Caudle, 1990; Foote, 1993; Takahashi and First, 1993; Yang, X., *et al.*, 1993a). Some of the potential research and commercial applications of the new *in vitro* maturation, fertilization and culture (IVMFC) technology relevant to cattle are provided in summary form in Table 8.1. In such applications, it should be remembered that it is cattle embryo production, linked to cattle embryo transfer (ET) technology that will allow advantage to be taken of the considerable advances that are currently taking place in developmental and molecular biology (see review by Seidel, 1991).

Fig. 8.1. Identical twins produced by embryo splitting in Dublin. In the early 1980s, twins were produced in sheep and cattle by micromanipulation procedures, using both early stage and morula/blastocyst embryos. The ability to split the blastocyst meant that identicals could be produced in cattle without this involving surgical intervention.

8.1.1 From research to commercial application

Only a few years ago, it would have appeared unlikely that large numbers of viable cattle embryos could be produced from oocytes collected indiscriminately at the local abattoir. Even at the present state of IVF technology, it is usually possible to match the yield of embryos achieved by the superovulation of donor cattle in conventional ET practice. At the same time, it must be recognized that considerable variability exists in the yields of transferable-quality embryos that have been reported by different laboratories. The most extensive studies must be those reported by Lu and Polge (1992) using the *in vitro* maturation (IVM)/IVF technology as initially described by Lu *et al.* (1987b). These data relate to almost 900,000 oocytes and show an average of 2.31 embryos per ovary graded as suitable for freezing.

However, it is certain that extremely careful attention to detail is required at all stages of bovine embryo production if consistently high yields are to be achieved. At the same time, existing technology is such as to permit commercial applications in which cattle embryos are transferred non-surgically by way of a technique little different from that used in artificial insemination (AI). The infrastructure of the commercial AI services that operate in Ireland and

Table 8.1. Applications of cattle IVF technology in commercial practice and research.

Commercial applications	Research applications
Embryos	
Twins in beef and dairy cattle	R & D in developing sperm separation by flow cytometry
Singleton beef calves from dairy cattle	Development of new embryo sexing methods
Producing sexed calves for beef and dairy herds	Improving IVF methods
Heifer calves for once-bred heifer beef systems	Developing a bull fertility test
Using IVF as part of normal cattle ET operations	Research in all aspects of freezing
Producing cattle hybrids for the tropics	Studies in embryo growth and development
	Research in the production of heat-tolerant embryos
Unfertilized oocytes	
Production of uniform calves and carcasses for the meat trade	Research in all aspects of cloning in cattle
Providing genetically superior and sexed calves for beef and dairy herds	Studies in oocyte freezing
	Testing effectiveness of embryonic stem cells
Pronucleate oocytes	
Transgenic cattle for the production of medically important proteins in milk	Development of gene injection procedures
Producing dairy cattle with novel milk proteins for the dairy industry	Studies in freezing
Trophoblastic vesicles	
Use to enhance pregnancy rates in recipient cattle	Research in all aspects of maternal recognition of pregnancy

many other countries should be capable of exploiting such developments. Of the population of over 2 million cows in the Irish Republic, at least 0.5 million could be available as potential recipients of high-quality beef embryos.

8.1.2 Cattle products and human health

Advances in reproductive technology as applied to cattle and the other farm species can be viewed as potentially valuable in increasing the overall efficiency of animal production and in improving the quality of the products available for human consumption (Bazer *et al.*, 1990b; Courot and Volland-Nail, 1991). As the pressures mount in the developed countries to move away from increased production using high inputs to greater efficiency with reduced inputs, there will be an ever-increasing need to find ways of utilizing the genetically superior farm animals, both male and female, to advantage. Although much of the current concern about food in the developed countries relates to health problems that arise from its abundance rather than its scarcity, there needs to be a constant awareness of those whose daily experience it is to be hungry. It is towards the developing countries, therefore, that much greater emphasis must be turned in getting new forms of animal and plant technology applied for the benefit of the human population.

Population growth and food resources

Advances in medical science, which now prevent and cure diseases which formerly regulated the human population in a natural way, have dramatically decreased infant mortality and increased life expectancy. The human population is currently increasing at a rate which may markedly affect the economic and political stability of the world, as a result of competition for resources and a steady deterioration in the quality of the environment. There are many who have written expressing the view that the most crucial problem facing the world today may be that of controlling population growth (Blaxter, 1987; Potts and Rosenfield, 1990a, b; Diczfalusy, 1992; Horiuchi, 1992; Leese, 1992; Johannisson, 1993). If the human population continues to grow at its present rate, science and technology may be quite unable to prevent widespread poverty and irreversible damage to the environment.

A thousand years ago, the world population was about 300 million. The first 1000 million was reached around 1800, the second (2000 million) about 1927. Currently, the population stands at 5500 million and is increasing by about 90 million a year, a fact that should influence every action that bears upon the balance between food supply and human demand. Were the world population to continue growing at the present rate up to AD 2100, it would reach 55,000 million at that time and double again in a further 35 years.

However, despite the religious, commercial and political pressures that currently restrict adequate fertility control measures, estimates by the World Health Organization (WHO) and other bodies suggest that different countries are likely to reach a net reproduction rate of 1.0 at various times between the years 2010 and 2050. Equilibrium population is forecast as about 10,400 million and this should be reached around AD 2100; to meet nutritional needs, as well as those arising from population growth, it has been estimated that food supplies would have to be trebled between now and AD 2100 (Blaxter, 1987). The required increase in food production would be greater in the initial period, however, in view of the slow rate of achieving zero population growth.

The appearance of the human immunodeficiency virus (HIV) and the fact that in some parts of the world death and morbidity from acquired immune deficiency syndrome (AIDS) have already reduced agricultural production are a further element that must be taken into account in looking realistically towards the future (Lincoln, 1992). For those reproductive biologists who are involved in fertility enhancement research, there is almost an obligation, as stated by Leese (1992), to make their knowledge available in the quest for the most acceptable forms of human fertility control. Zero population growth within the lifetime of our children is achievable, but only if action is taken without delay. It certainly cannot be achieved by science and technology alone; political will and appropriate policy formulation are equally important.

The crucial need is to provide for the needs of the malnourished in the developing countries, using the most appropriate forms of technology; in the developed countries, the need is to provide the quality of food that is most appropriate for the maintenance of good health. In regard to this latter consideration, there is the suggestion that new forms of reproductive technology may eventually prove to be the means of producing cattle that are capable of producing leaner meat and milk of more acceptable composition (Sreenan and McEvoy, 1989a); this may have significant health benefits for the consumer.

Red meat consumption and human health

In considering animal products and human health, it has to be said that the scientific validity of many of the official recommendations relating to fat consumption have been questioned; in particular, there are those who believe that the Committee on Medical Aspects of Food Policy (COMA) recommendations of 1984 to reduce fat and saturated fatty acids by the population as a whole cannot be sustained on scientific grounds. In reviewing dietary factors believed to be linked to the incidence of coronary heart disease (CHD), Ulbricht and Southgate (1991) note that there is ample evidence to support the view that meat consumption (total meat or red meat) is not related to the incidence of CHD in the European Union (EU) countries. The UK, for example, has the second lowest meat consumption per capita and the second highest CHD incidence; Greece, with the highest red meat consumption, has one of the lowest rates of CHD. Across the Atlantic, in the USA, where there has been a steady decline in the CHD incidence, there has been little change in red meat consumption. It might be mentioned, in reference to beef supplies in the EU, that if red meat consumption per capita was comparable to that in the USA, there would probably be a beef deficit, rather than a surplus.

8.1.3 Chapter contents

The ability to produce cattle embryos in the laboratory provides several opportunities for the commercial application of this new technology. Ireland was among the first countries in attempting the commercialization of the cattle embryo production process; this involved the formation of the Ovamass Company in 1987. At that time, this company was granted an exclusive licence by University College, Dublin, to exploit the new technology. The principal objective at that time was to employ good-quality beef embryos in cattle twin-

Fig. 8.2. IVF and the single-sexed, beef heifer production system. The most efficient beef production system is one in which recipient heifers are implanted with IVMFC-derived female embryos to give a self-replacing and self-perpetuating system. In this, all costs of production are carried by the slaughter animal, which is also capable of producing a high-quality carcass. Much work has been undertaken over the years at the Grange Research Station in Ireland on the once-calved heifer system.

ning in Ireland; a further aim was to explore markets abroad, in the UK and elsewhere, for low-cost embryo sales. Once the technology of low-cost embryo transfer was simplified to the point where it became possible to thaw embryos on the farm and carry out transfers in the same way as inseminations, it was envisaged by Ovamass that there would be a steady uptake of the technology.

In this final chapter, ways in which IVMFC-derived embryos may be employed are examined (see Fig. 8.2). The current *in vitro* production procedures can act as the foundation on which many other useful cattle breeding developments (e.g. sexing, large-scale cloning) can be based. There is also the rapidly expanding field of molecular biology, particularly that concerned with recombinant deoxyribonucleic acid (DNA) technology, which is capable of utilizing laboratory embryo production.

8.2 Potential Impact of Embryo Production Technology

8.2.1 Use of IVF/ET technology in cattle breeding improvement programmes

Breeding improvement in cattle in developed countries has made much progress over several decades due to the use of large-scale field data derived from many herds based on the progeny testing of males. At one time, breeding improvement programmes, as applied to dairy cattle, were based almost exclusively on field progeny testing methods, using AI. It has been estimated that about half the improvement in the efficiency of dairy cattle in developed countries since the Second World War can be attributed to the use of AI, equal to the combined contributions of improved animal health, husbandry and nutrition. Some of the breeding implications of new advances in reproductive technology were covered in a review by Nicholas (1985). It should be noted, however, that there are those in the UK who maintain, despite the increasing use of the technology, that cattle ET is not a cost-effective method of increasing genetic merit in commercial dairy herds (Rider, 1991); according to the costings given, the use of AI is still the commercial producer's best approach to genetic improvement.

It was Nicholas (1979) who first advanced the concept of using multiple-ovulation embryo transfer (MOET) in combination with AI as a new breeding method for use in dairy cattle. As noted by Shaw (1989), MOET is the use of ET technology to provide a more rapid rate of genetic improvement than that currently provided by progeny testing. The main feature of this scheme was that superovulation and ET would be carried out on all the heifers in the dairy herd before they had commenced their first lactations (juvenile method); a subsequent modification by Nicholas and Smith (1983) allowed heifers to complete one lactation before ET was attempted (adult method).

The MOET breeding schemes as currently used are characterized by the formation of a central breeding herd (nucleus herd) with extensive use of superovulation of donors selected among heifers and young cows (Woolliams and Wilmut, 1989; Smith, C., 1990). Such breeding schemes are likely to be superior to traditional AI breeding programmes in increasing the rate of genetic improvement, due to the higher selection intensity among females and a shorter generation interval (Ruane, 1988). According to Christensen (1991), the rate of genetic gain can be increased by 30–50% by using MOET on yearling dairy females to produce young bulls and potential bull dams. Dealing with the UK, McGuirk (1990) discusses how the Milk Marketing Board (MMB) Premier Breeders' nucleus MOET scheme may hopefully achieve much faster genetic improvement than using conventional progeny testing procedures. Experiences in operating a MOET scheme in Danish dairy cattle over a 6-year period are described by Callesen *et al.* (1992b); the use of MOET in proliferating breeding stock in Australia is covered by Evans (1991). Several authors note that the use of MOET in élite nucleus herds could also substantially increase the rate of genetic improvement in beef cattle (Gearheart *et al.*, 1989; King, 1989; Broadbent, 1990; Dekkers, 1991).

There are those who believe that MOET schemes are likely to be of considerable relevance and interest to developing countries, especially where progeny testing schemes are not feasible and where suitable bull semen cannot be imported (Gibson and Smith, 1989; McGuirk, 1989). Although the MOET concept initially found application among herds in the developed countries, the technique may yet prove to be even more appropriate in the developing world.

In the context of MOET breeding schemes, the use of IVF technology may be valuable, either as a means of permitting embryos to be produced from prepubertal cattle (Kinghorn *et al.*, 1991), or as a means of increasing the efficiency of embryo production in selected donors (i.e. via repeated transvaginal ultrasonic-guided oocyte recovery).

Open-nucleus breeding schemes

It has not generally been possible to introduce modern cattle breeding improvement schemes (e.g. contemporary comparison methods) for cattle and other ruminants in any of the developing countries where animals are kept in traditional environments. This is due to the lack of basic infrastructures required for production and pedigree recording of most of the animal population. According to Jasiorowski (1990), of the Food and Agriculture Organization (FAO), there is little hope of a change in this position in the foreseeable future.

In many developing countries, however, the government may own large cattle herds; it is possible that such herds may act as a nucleus for the use of ET programmes. In Mexico, for example, government agencies purchased 1000 dairy cattle of high genetic merit from the USA and used them to produce heifers for sale to the local farmers (Elsden, 1989).

Hodges (1990) has drawn attention to the breeding improvement method known as the 'open-nucleus breeding scheme' (ONBS), which, combined with MOET, may offer the potential for greater progress in genetic improvement in ruminants than the traditional or well-established animal selection methods. According to Jasiorowski (1990), the ONBS has the advantage of not requiring mass field recording and may be particularly applicable to livestock production in the developing countries. In the ONBS, a genetically superior nucleus herd or flock is established under controlled conditions where testing and genetic selection can be carried out. This test group is first established by screening the base population for outstanding females.

These outstanding animals in the herd or flock are then recorded individually and the élite females among them are used, by MOET with superior sires, to produce embryos that are carried by recipient animals from the base population. The resultant offspring are reared and recorded and the males among them are evaluated genetically, using the performance of their sibs and paternal half-sibs, and their own performance, where they exhibit the trait. From these animals, an élite group of males with high breeding value is selected to be used in the base population for genetic improvement by natural service or AI.

Table 8.2. Advantages and disadvantages of nucleus breeding units in comparison with field breeding schemes (from Hodges, 1990).

Advantages
Genetic lift in establishing the unit
Faster rates of genetic change
Control over husbandry and testing
More effective selection possible for economic merit
Use of traits difficult to record in field
Concentration of breeding resources
Possible use of expensive technologies
Economic benefits obtained sooner
Low costs on a national scale
Separate nucleus units for different sets of breeding objectives/environments
Disadvantages
Risk of disease and loss (avoid by dispersing age classes)
Risk from concentrating stock and resources in one unit
Possible genotype × environment interactions in commercial production
New funds needed to set up and operate
Re-education of producers to accept MOET-bred stocks

The advantages of ONBS combined with MOET, in comparison with traditional field breeding schemes, are summarized in Table 8.2. According to Hodges (1990), an important concern should always be the training of personnel so that they fully understand the theoretical basis of ONBS and MOET.

The ONBS/MOET schemes may prove to be the most appropriate route for channelling new developments in reproductive technology (e.g. *in vitro* embryo production, sexing and cloning) into livestock improvement programmes. In establishing the required support facilities for operating ONBS/MOET schemes, there may be particular merit in including those required for *in vitro* embryo production as well as for conventional ET purposes.

8.2.2 IVF in the preservation of genetic diversity

There is increasing interest in the establishment of programmes for the preservation of livestock breeds that may be in danger of extinction, in both developed and developing countries. According to the recently established FAO Global Data Bank for Domestic Livestock, which carries 2047 entries, there are some 221 breeds currently considered to be at risk, with most of these (60%) being in the developed countries (Ruane, 1993); intensification of agriculture in developed countries has led to a greater reliance on a small number of breeds, with the consequent neglect of the remainder.

In developing countries, the greatest threats to genetic diversity appear to be increased use of AI and indiscriminate crossbreeding of indigenous breeds (Ruane, 1993). The intensification of farming in these countries can mean that indigenous breeds are in danger of being pushed to extinction because native farmers, in aiming for greater production, have employed exotic breeds such as Holsteins or Friesians. In 1987, the FAO decided to establish seven Regional Animal Gene Banks to preserve the genetic diversity of breeds: two of these were to be in South America (Brazil and Argentina), two in Africa (Ethiopia and Senegal), two in Asia (China and India) and one in Mexico (to serve Central America and the Caribbean, a zone free of foot-and-mouth disease). The FAO initiative also included the compilation of a world inventory of native livestock breeds and proposals for conservation banks of frozen semen and embryo from threatened breeds.

The practical application of IVF technology in the preservation of an endangered cattle breed has been described by Solti *et al.* (1992), working with Hungarian Grey cattle. In Brazil, Mariante and Bem (1992) have emphasized the need to start preservation programmes before many of the native populations disappear due to systematic crossbreeding programmes with exotics. Future developments in IVF technology, such as methods for recovering much larger numbers of oocytes from ovaries and effective cloning techniques, are likely to be of great value in the preservation of embryos from the endangered species.

8.2.3 Crossbred embryos for tropical/subtropical regions

The application of IVF/ET technology as a means of upgrading the quality of cattle in developing countries may be one possibility worth exploring. As noted above, crossbreeding native cattle with exotic breeds can raise serious problems. From the time that crossbreeding programmes were first introduced in India in the 1960s with *Bos taurus* semen, it has been consistently observed that neither 100% nor less than 50% *Bos taurus* genotype proved to be really useful in that country's animal agriculture. There are a number of scientific studies (review by Cunningham, 1989) showing that *Bos taurus* × *Bos indicus* crossbreds may have particular merit in combining the qualities of productivity with the ability to withstand the rigours of the tropical and subtropical environment. Under conventional animal breeding schemes, however, such crossbreds cannot be sustained beyond the first generation.

The production of first crosses between Western breeds and native stock in a country such as India may be one area in which *in vitro* embryo production may prove valuable. Although it may be desirable to maintain the milking cow population at 50% *Bos taurus*:50% *Bos indicus* to exploit heterosis, such a population cannot be replaced under existing breeding programmes. The use of IVF technology to produce crossbred embryos derived from European dairy cattle (e.g. Holstein/Friesian, Jersey) and native Indian bulls (e.g. Red Sindhi, Sahiwal, Tharparkar) may be one solution to this problem. It would be possible to use ovaries from Irish or British dairy cattle and fertilize the oocytes with semen imported from the zebu bulls. The embryos could be used in India and elsewhere by way of direct non-surgical transfers by suitably trained personnel from cattle breeding stations.

The introduction of crossbred embryos could be the means of bringing new genetic material into a population of cows in which the use of AI would take too long to have a significant effect in upgrading the genetic base. Animals introduced as embryos are also likely to adapt more readily to their new environment than those introduced as grown animals; at birth, calves can receive passive immunity to local diseases from their surrogate mothers (native zebu cattle) and their immune systems will develop in the appropriate environment. In Venezuela, the non-surgical transfer of imported embryos into native Zebu recipients, with acceptable pregnancy rates, has been reported by Darrow (1989). The eventual availability of large-scale cloning and sexing procedures in embryo production laboratories would enable the genetic quality of dairy stock to be further upgraded.

8.2.4 Embryos for heat-stressed cattle

It is well recognized that high temperature and humidity are associated with marked seasonal declines in the reproductive efficiency of cattle. From the literature, it is clear that thermal conditions may be a serious constraint on the performance of farm animals, particularly in high-yielding dairy cattle originating in temperate regions (Holstein/Friesian). The dairy industry in Saudi Arabia, for example, is faced with problems associated with impaired reproductive performance under high environmental temperatures during the summer months (Gordon *et al.*, 1987; Boland and Gordon, 1989; Alfuraiji *et al.*, 1993). In Cuba, Fernandez Limia *et al.* (1990) record evidence of fertility problems in lactating Holstein cows in the summer months. In South Africa, Du Preez *et al.* (1991) recorded conception rate to first service in Friesian dairy cattle as being lowest (33%) when the temperature–humidity index (THI) was highest and highest (74%) when the THI was lowest.

Supplies of IVMFC-derived cattle embryos may have relevance in attempts to overcome certain forms of infertility arising from maternal heat stress (Gordon *et al.*, 1987; Putney *et al.*, 1987, 1989a, b; Ryan, D.P., *et al.*, 1992b). There is clear evidence now available to show that the bovine embryo is sensitive to maternal heat stress, particularly during the first 2 weeks after breeding (Ryan *et al.*, 1993a); ET may be a possible means of alleviating this problem, if the technique can be effectively employed to bypass this early period of embryonic sensitivity. It may even be possible to think in terms of preparing cattle embryos to show greater than usual heat tolerance, by stimulating the production of heat-shock proteins by way of controlled exposure to sublethal temperatures in the laboratory.

In Wyoming, Hunnicutt *et al.* (1990), for example, were able to demonstrate that day-15 sheep trophectoderm cells synthesized a protein, tentatively identified as heat-shock protein 70 (hsp70), in response to temperatures above 37°C. Similarly, in Louisiana, Ryan, D.P., *et al.* (1990b, 1992b) reported studies in which short-term heat stress was shown to trigger the development of heat-shock proteins in cultured bovine embryos. Although chronic elevated temperature (40°C) for 60 h resulted in embryo loss, when exposed to 43°C for 20 min of incubation (i.e. sublethal temperature) such bovine embryos were better able to survive the 40°C environment. The authors concluded that such findings might have useful implications for reducing embryonic mortality in heat-stressed cattle. In the University of Florida, workers have also reported studies in which it appears that cattle embryos may acquire mechanisms during early development to combat the otherwise deleterious effects of exposure to increased temperatures (Ealy *et al.*, 1991, 1992a, b). Other workers have shown how thermotolerance may be induced in mouse embryos if they are first exposed to mild heat shock (Malayer and Hansen, 1990; Arechiga *et al.*, 1992; Ealy and Hansen, 1992). A report by Gwazdauskas *et al.* (1992) showed that the first cell cycle in mice was the most sensitive to increased temperature; more advanced embryos could tolerate slight increases in temperature (39°C vs. 37°C) more readily than pronuclear embryos.

8.2.5 Producing high-quality beef IVMFC embryos

As noted earlier, a biotechnology company (Ovamass Ltd) was established in Dublin in 1987 with a view to commercializing the production of high-quality beef embryos for use in cattle twinning. In 1989, a 3-year CEC (ECLAIR) research programme was initiated, with Ovamass as the commercial partner, in an effort to continue refining the embryo production technology. It was felt, at that time, that twinning by IVMFC-derived embryo transfer may have a part to play in meeting beef calf requirements in a period of declining dairy cow populations in the EU.

In a review of beef cattle breeding programmes in Europe, attention has been drawn to the fact that the EU dairy herd not only produces milk but also 80% of the calves destined for rearing in the beef industry (Simm *et al.*, 1990). The dairy herd has been growing smaller with the imposition of milk quotas in 1984; in the twelve EU countries during the period 1984–94, dairy cow numbers fell by 7 million, resulting in fewer calves being available for beef production. The application of twinning by ET was seen to be one means of alleviating the calf scarcity that otherwise might arise.

In Ireland, about half a million heifers are currently available as donors of slaughterhouse ovaries (see Fig. 8.3). About a quarter of a million heifers have been sired by Continental bulls (Charollais, Limousin, Simmental, etc.); oocytes from such crossbreds are capable of giving rise to $\frac{3}{4}$-bred or even $\frac{7}{8}$-bred Continental beef embryos. In future times, the perfection of large-scale cloning technology should permit the production of high-quality bull calves of uniform type which go on to produce a consistent carcass type suitable for the needs of the large supermarkets.

There may be opportunities in Ireland, but more particularly in the UK, for using $\frac{7}{8}$-bred Continental beef embryos as a means of obtaining quality beef from extreme dairy-type cows (Holstein). The trend towards increased use of the more extreme dairy-type cattle in the UK has made the suitability of such calves for beef rearing very questionable (Mayne *et al.*, 1991; Reynolds and Street, 1991). The trend towards more efficient dairy production systems in Ireland and the UK is likely to accelerate the greater use of the more extreme dairy type sires, with a resultant adverse effect on beef carcass quality. This tendency could be reversed if bull embryos of $\frac{7}{8}$-bred Continental breeding are available at a cost which is not too far removed from that of a high-quality

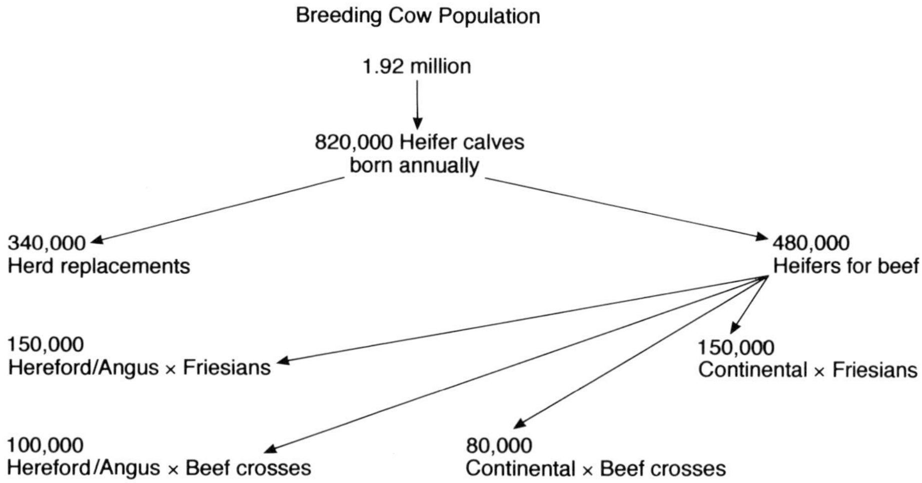

Fig. 8.3. Cattle available as source of oocytes in the Irish Republic (1990 data).

beef insemination. A few years ago, this would have seemed most unlikely, but increasing refinement of IVF technology should bring this low-cost possibility ever closer.

Another way in which sexing might assist in a beef production system, this time using heifer beef embryos, is mentioned in a paper by Rigby (1989); the author suggests using heifer embryos of the Continental crossbred type, as the means of permitting once-bred heifer production systems to become self-replacing and self-perpetuating. Calves would be weaned soon after birth and the recipient dam slaughtered; calves would be reared and made pregnant in the same way, and the cycle repeated. With this system, all costs of production are carried by the slaughter animal, which is also capable of producing a high-quality carcass.

Embryos for pure-bred beef (Hereford/Angus) herds in the USA

The USA is one of those countries that possesses many millions of beef cattle and there are likely to be a number of herds in that country in which nutritional and management conditions are suitable for twinning. The production of pure-bred Hereford or Angus embryos for use in such herds could mean that twinning by ET may be carried out on the farm by the stockperson, in much the same way as do-it-yourself (DIY) AI is currently practised by dairy farmers in the USA. As discussed later in the text, the most effective way of inducing cattle twins is by a one-embryo transfer to an already bred recipient, a week after the animal has taken the bull. It would not be unduly difficult to have the bull equipped with a marking harness (e.g. chinball device) and for cows marked a week earlier to receive an additional Hereford embryo by way of DIY ET from the farmer's stock of frozen IVMFC embryos.

8.2.6 Cattle embryos for research

In an age in which there is increasing resistance by animal welfare groups to the use of live animals in research, there are obvious advantages to techniques which enable farm animal embryos and oocytes to be made available for research without this involving live animals and superovulation intervention (see Fig. 8.4). It may be that the needs of many research programmes (e.g. development of suitable freezing techniques for use with pig embryos) can be

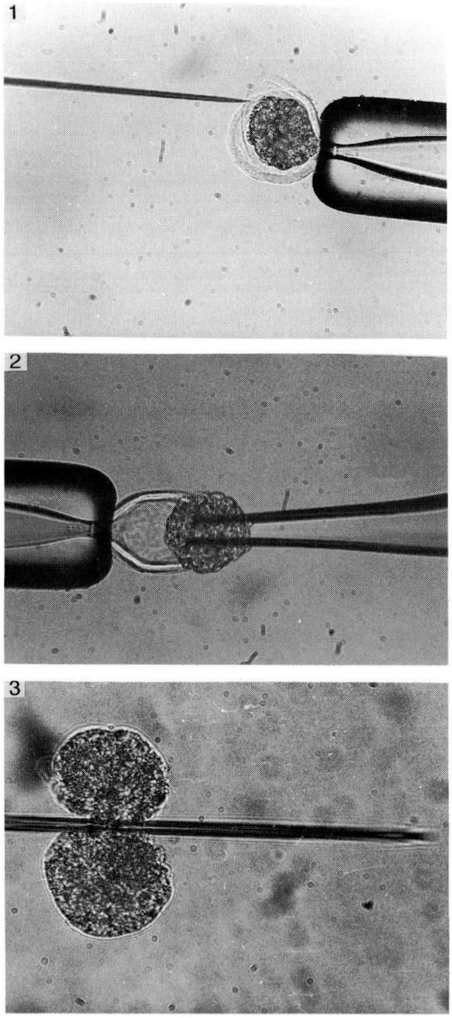

Fig. 8.4. Embryo splitting as seen under the microscope. Production of identical twins by splitting the sheep embryo at the late morula stage. (1) Opening the zona pellucida; (2) withdrawing the embryo; (3) cutting through the embryo with a glass needle (from Gatica, 1988).

satisfactorily met from slaughterhouse ovaries and a supply of semen, rather than this involving live animals. It is clear, from the many reports appearing in the literature in recent years, that a large number of cattle embryos are being produced by laboratories around the world for research purposes. By this means, many new and valuable developments may be expected over the next few years in sexing techniques, freeze–thaw technology, nuclear transfer (NT) and genetic engineering.

8.3 Animal Health and Welfare Considerations

8.3.1 *Transporting embryos rather than live animals*

There is increasing public concern about the welfare of farm animals during transportation, whether this is in course of their export from a country or during journeys within a country. Various measures have been proposed by bodies such as the European Commission to protect the welfare of livestock in transit. Coultard (1992), writing in regard to welfare problems associated with transportation of breeding animals, observes that such problems need not arise if the animals are transported as frozen embryos; transportation costs would obviously be much less. Although most instances of transportation of IVMFC-derived bovine embryos relate to those that are either chilled or frozen, some reports have mentioned shipping embryos in containers held at 38°C (Minoia *et al.*, 1991).

8.3.2 *International trade with cattle embryos*

There is ample evidence testifying to the value of employing ET as a means of avoiding the disease hazards normally associated with conventional methods of moving live cattle from one continent or country to another (Stringfellow, 1985; Stringfellow *et al.*, 1986, 1988, 1990a, b, 1991a, b; Campo *et al.*, 1987; Mahon and Rawle, 1987; Shelton, 1987; Singh, 1987, 1988; Barrios *et al.*, 1988; Echternkamp and Maurer, 1988; Rees, 1988; Shisong and Wrathall, 1989; Thibier, 1990; Joly *et al.*, 1991; Mebus and Singh, 1991; Wrathall, 1992). In one review, Stringfellow *et al.* (1991b) note that 20 years of commercial ET have not resulted in the transmission of a single infectious disease agent. As shown in Fig. 8.5, the transmission of an infectious agent through ET requires the uninterrupted occurrence of five events; each link in the 'chain of infection' must be intact for an infection to be transmitted, and transmission is prevented if any link in the chain is broken.

International trade in cattle embryos is always likely to be based on frozen rather than fresh embryos. The techniques employed to ensure that such frozen embryos are free of pathogens include the use of specific pathogen-free donors, the washing and trypsin treatment of the embryos, or a combination of these methods. Health certification procedures for international trade generally require embryos to be examined over all surfaces at a minimum of × 50 magnification to confirm that there is no material adhering to the zona

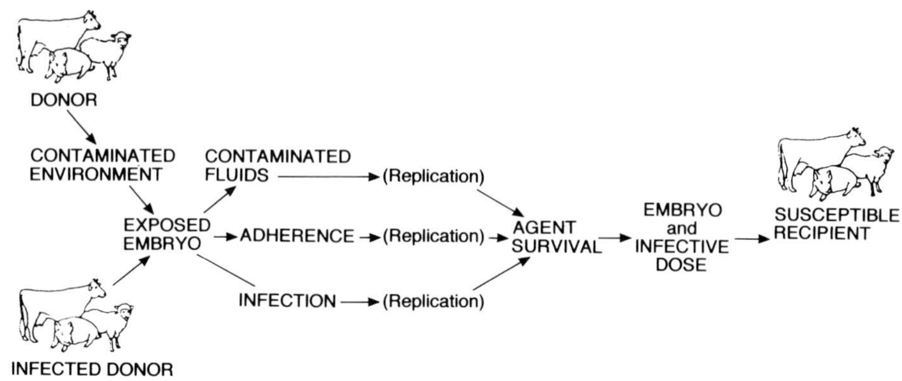

Fig. 8.5. Hypothetical chain of infection in embryo transfer. For an infectious agent to be transmitted through embryo transfer, every link of the chain must be intact. If a single link is broken, transmission is prevented (from Stringfellow *et al.*, 1991b).

pellucida (ZP) after washing. Cattle embryos appear to be less 'sticky' than those of sheep or pigs, and this eases the problem of materials and cells adhering to the ZP.

Trypsin treatment to ensure freedom from infectious bovine rhinotracheitis virus (IBRV) is required for the importation of bovine embryos into certain countries. Studies reported by Stringfellow *et al.* (1990a, b) confirmed that IBRV adheres to the ZP of the bovine embryo after exposure to the virus *in vitro* and that trypsin is effective in removing it; the same workers also showed that trypsin treatment is effective in removing bovine herpesvirus 4 (BHV-4). When trypsin treatment is employed, there are data indicating that this does not adversely affect embryo viability after freezing (Echternkamp and Maurer, 1988; Echternkamp and Kappes, 1989; Echternkamp *et al.*, 1989). Trypsin is a digestive enzyme whose role is to hydrolyse polypeptide chains. It is presumably this property of the enzyme that removes or inactivates pathogens adhering to the surface of the ZP of infected embryos.

All the indications are that cattle embryos, if appropriately washed according to the standards agreed by the International Embryo Transfer Association (IETA) (see Stringfellow and Seidel, 1990), should be free from bacterial and viral infections. In the preparation and prefreeze treatment of embryos, it is clearly essential that the integrity of the ZP is preserved (see Fig. 8.6).

Antimicrobial treatment of embryos

As noted above, there are some agents that are known to adhere to the ZP of ZP-intact embryos in such a way that they may not be entirely removed using a routine washing procedure. Methods of disinfecting embryos of such agents, by removing the risk of infectious disease transmission but without impairing the embryo's viability, are essential for the full development of disease control measures. Using mouse embryos, Bielanski and Hare (1991)

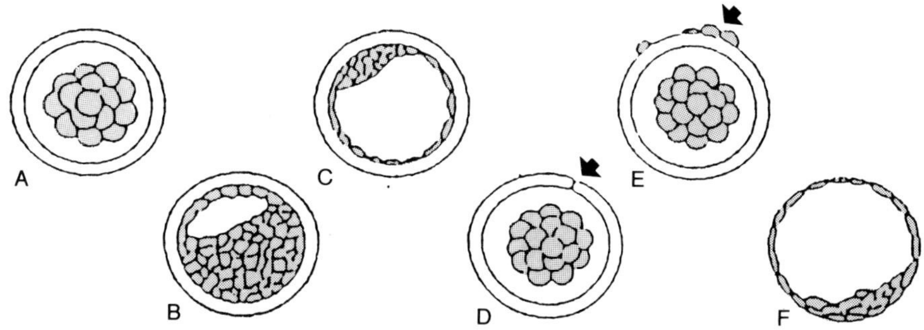

Fig. 8.6. Acceptable and unacceptable embryos in terms of disease control in cattle ET. The zona pellucida is an effective pathogen barrier; for that reason, it is essential that all cattle embryos possess an intact zona pellucida and are free from any adherent material. A, B and C are acceptable embryos. D, E and F are unacceptable embryos: D has a break in the zona pellucida; E has debris present on the zona pellucida; F has no zona (hatched embryo) (from Stringfellow and Seidel, 1990).

examined a number of substances and procedures known to have a germicidal effect on a number of pathogens. Elsewhere, Riddell *et al.* (1993a, b) tested selected antimicrobials for treatment of ZP-intact bovine embryos after *in vitro* exposure to *Mycoplasma bovis*; the antibiotic tylosin was found to be completely effective, at twice the dose recommended for cell culture, without any adverse effect being apparent on embryo development. In Alabama, Thomson *et al.* (1989) reported using the synthetic nucleic acid, polyriboinosinic acid–polyribocytidylic acid (poly I:C) to induce viral resistance in cultured cattle embryos.

Although many thousands of bovine embryos have been transported throughout the world without a single instance of disease being reported, continuous efforts are required to reduce any risk of disease transmission. In this regard, it is possible that proteins in embryo culture media (bovine serum/bovine serum albumin (BSA)) will be replaced by synthetic macromolecules, such as hyaluronic acid (HA). This possibility is dealt with in papers by Joly *et al.* (1991, 1992).

Use of IVMFC embryos in disease transmission research

The ability to produce cattle embryos by *in vitro* culture (IVC) will enable many studies to be conducted on the interaction between pathogens and embryo. There are still countries whose health regulations relating to embryos are so restrictive and costly that no embryo can be exported to them. Data derived from controlled studies and the general acceptance of internationally agreed health certification schemes by the various countries engaged in international trade will eventually overcome such difficulties. In Dublin, there has been a preliminary report, using IVMFC-derived bovine blastocysts, on infection with bovine viral diarrhoea virus (BVDV) and other viral agents (Mabruk *et al.*, 1990).

8.3.3 *Virus screening of laboratory-produced bovine embryos*

There is an obvious need to avoid any risk of transmitting infectious diseases by way of laboratory-produced cattle embryos. Where slaughterhouse ovaries are collected as the source of oocytes, then clearly there is the possibility of bacterial and viral agents being in contact with the oocytes. In Denmark, Avery *et al.* (1993) attempted to estimate the frequency of contamination with BVDV in a routine IVMFC embryo production system; they concluded that there was a significant risk of BVDV contamination and advised that those embryos with an intact ZP should be washed according to IETA recommendations (Stringfellow and Seidel, 1990).

In Canada, Bielanski *et al.* (1993) also found clear evidence that bovine oocytes recovered from ovaries collected at commercial abattoirs are not always free from pathogens. They showed some IVMFC embryos to be positive for herpesvirus 1; however, BVDV was not found in embryos which had previous contact with virus-positive follicular fluid or oviductal cells. Although Ireland has an excellent animal health status, with freedom from major viral diseases, there is clearly the need for constant vigilance in regard to the conditions under which IVMFC embryos are prepared for commercial use. It is of paramount importance that the integrity of the ZP is preserved in carrying out prefreezing manipulations of the IVMFC embryo. There is also need to examine the disease-resistant properties of the ZP at all stages from the oocyte to the day-7 embryo (Riddell *et al.*, 1993).

8.3.4 *Welfare implications of using IVMFC-derived embryos*

The technique of AI in cattle, which in many countries is employed to breed several million cattle annually, poses few welfare problems or ethical concerns. When carried out by trained inseminators or suitably experienced stock persons (DIY AI), AI can be regarded as an acceptable method of breeding cattle without imposing any unacceptable stress on the animal (Macaulay *et al.*, 1986). The use of IVMFC-derived embryos, prepared in straws and transferred a week after breeding by way of a technique similar to AI, should not pose welfare problems or ethical dilemmas that differ essentially from those involved in AI. However, there is the question of whether an epidural anaesthetic is routinely required for recipients. According to Lucke (1991), the risk of discomfiture in the cow is greater with non-surgical ET than with routine insemination; for that reason, recipients should routinely receive the benefit of epidural anaesthetic. There is some evidence that the use of a β-blocker (5 mg carazolol) a few moments before insemination can enhance conception rate in cattle (Krebs, 1988); such an effect could be of interest for recipients at time of ET.

Although the use of a sedative as well as epidural anaesthetic may not be essential, Broadbent *et al.* (1991) suggest that it is advisable and necessary for the welfare of the recipient and for operator comfort and to achieve optimal pregnancy rates. In the UK, appropriately trained ET technicians can now administer epidural anaesthetic under a new order, 'Veterinary Surgery

(Epidural Anaesthesia)', introduced in 1992. Also established in that country is a suitable course of instruction for cattle ET technicians at Cambridge University and the publication of a code of practice approved by the Royal College of Veterinary Surgeons.

It is in the matter of the suitability of recipients for the embryos transferred that most genuine concern for animal welfare must be directed. In the UK, the Ministry regulations currently under consideration envisage that recipient cattle are inspected within 30 days before ET to ensure that the embryo is appropriately matched to the recipient. As noted by Murray and Ward (1993), for economic reasons in conventional cattle ET, it is not uncommon to use beef crossbred heifers as recipients of expensive high-quality dairy embryos. It is not difficult to see that such a practice may give rise to welfare concerns, as indeed it would with AI, if dairy or beef maiden heifers were bred to certain Continental bulls. Careful selection of the bulls (for easy calving) used in producing IVMFC beef embryos and of the recipient cow (well-grown and with an easy calving history) must be exercised by those involved in the application of the new technology on the farm.

Cattle twins and cow welfare

Cattle twins occur naturally in a low proportion of calvings (2–3%) but it is certainly pertinent to ask questions about the welfare implications of substantially increasing the twinning percentage in this normally monotocous species. For beef cattle, the view has been expressed (see Gordon, 1983) that, with appropriate feeding and management, the cow can deal with twin pregnancy and with the birth of twins without undue concern. Indeed, in comparison with problems of dystocia raised by the large single calf, the birth of twins is seen to pose less of a welfare concern. However, there is a need to clearly determine that both cows and sheep are not subject to unreasonable stress in late pregnancy when fetal number is increased above the norm. A paper by Robinson (1990) examined some of the targets set for the reproductive performance of farm animals and asks whether they can always be regarded as reasonable. In view of those outside the livestock industry who now question the ethical reponsibilities of those engaged in modern farming practices, it is essential that every effort should be made to serve the best welfare interests of farm livestock at all times.

8.4 Approaches to Cattle Twinning

The percentage of twin births in cattle varies among breeds and according to factors such as age and environment. In the UK and Ireland, the incidence recorded in surveys usually falls somewhere between 2 and 3% (Gordon *et al.*, 1962; Scanlon, 1969). Elsewhere, in some reports, the twinning incidence is almost negligible; in others, an incidence as high as 10% has been recorded. It may be noted that the application of recombinant bovine somatotrophin (rBST) in dairy cows has usually been followed by marked increases in the twinning percentage (Aguilar *et al.*, 1991; Cole *et al.*, 1991; Wilkinson and Tarrant, 1991). rBST aside, there are a number of possible approaches to the induction of twin pregnancy in cattle, including the use of gonadotrophins, immunization procedures and selective breeding. The main approach used in Dublin during the past 20 years has been by way of ET.

8.4.1 *Advantages of twinning*

The possibility of increasing litter size in the beef cow as a means of increasing the animal's biological efficiency has attracted the attention of animal researchers for some time (see Gordon, 1983). There is general agreement that twinning is capable of increasing the biological and economic efficiency of beef production by 20–25%.

For Ireland, in which beef has been the single most important product of the farming industry, the availability of a commercially viable twinning technique may be of particular value in providing the required supply of suitable beef calves. Beef and dairying are based on a national herd consisting of some 2 million breeding cattle, about two thirds of which are dairy animals. With the introduction of milk quotas by the European Commission in 1984, the size of the dairy cattle population and the number of calves available for beef rearing have declined.

In the wider context of the EU cattle herd, drastic culling of dairy cows has taken place since 1984. On the other hand, suckler cattle herds in the EU have shown increases of 3–10% in most countries, including Ireland, in recent years. Twin calf production could be of interest in many of these herds as a means of improving the profitability of the beef cattle enterprise.

Dairy twins

In dairy cattle, it is well recorded that natural twinning may be associated with a higher incidence of calf mortality, a higher percentage of cows showing placental retention, shorter gestation lengths and longer calving-to-conception intervals (O'Farrell *et al.*, 1990, 1991; Eddy, 1991; Eddy *et al.*, 1991); for such reasons, there is a natural reluctance on the part of dairy farmers to increase the incidence of twins in their herds. However, controlled experiments carried out in Northern Ireland with Friesian dairy cattle have gone some way in showing that, given accurate identification of twin-bearing cows in early pregnancy, coupled with increased energy intake in late gestation, the adverse effects of twin-bearing on animal performance can be minimized (Mayne and McCaughey, 1990a, b; Mayne *et al.*, 1991; Unsworth and Mayne, 1991; Mayne and McEvoy, 1993a, b). In the Irish Republic, Sreenan and Diskin (1992) have found that many of the problems with dairy twins arise because the twin calvings are unexpected. Studies conducted by these authors in dairy cattle have shown that such problems can be largely avoided by planned management, particularly in regard to feeding and calving supervision.

Twins in Australia and North America

Further afield, it might be observed that Australian researchers in the Commonwealth Scientific and Industrial Research Organization (CSIRO) and the Department of Agriculture have conducted a well-coordinated project on twinning in beef cattle, with a view to having both the technology and management guidelines available to enable farmers in that country to produce twins from beef suckler cows successfully (Cummins *et al.*, 1992); the objective of the programme is to increase the net economic returns from beef farming. In the USA, several reports have examined the effect of twinning on the efficiency and profitability of beef production (Guerra-Martinez *et al.*, 1990; Swartz *et al.*, 1990; Rose and Wilton, 1991; Davis and Bishop, 1992); according to these studies, twins can contribute to beef cattle efficiency and research into suitable methods of producing and raising them are justified.

8.4.2 *Twins by gonadotrophin treatment*

The feasibility of inducing multiple births in beef cattle by gonadotrophin treatment was examined in a large-scale field trial undertaken by the MMB (England and Wales) in the 1959–61 period in the Welsh border counties. The gonadotrophin treatment at that time consisted of a single dose of pregnant mare serum gonadotrophin (PMSG) administered on day 16/17 of the cow's oestrous cycle; such treatment had previously been used with some success in boosting the twinning percentage in low-fertility breeds of sheep (Romney/Southdown) in the southern counties of England in the early 1950s (Gordon, 1955). The MMB trial dealt with upwards of 500 cattle which were treated with gonadotrophin and thereafter kept under observation through to calving (Gordon *et al.*, 1962); one useful feature of the trial was in providing some evidence, albeit limited, that the adverse effects of twin births in dairy cows might be alleviated by suitable modification of the feeding levels during late pregnancy. A major drawback of the PMSG approach, however, was the marked individual variation that occurred, together with the fact that the treatment not uncommonly resulted in the birth of triplets, quadruplets and even quintuplets; it was quickly apparent that litters in excess of twins had no merit.

Studies are still being reported today in which PMSG or follicle-stimulating hormone (FSH) preparations are employed in inducing twins in both beef (Davis and Bishop, 1992; Yang, L., *et al.*, 1992) and dairy (Isogai *et al.*, 1991) cattle; serious problems (e.g. high abortion rates) arising from cows carrying triplets and quadruplets are almost invariably mentioned in these reports. At Clay Center in Nebraska, Echternkamp (1992) used FSH-P in 379 cows and found that some cows had the capacity to gestate up to three fetuses per uterine horn, or a total of five fetuses; above this number, pregnancy failed.

8.4.3 *Twins by immunization techniques*

Ovarian physiology is a complex process involving both extragonadal (e.g. FSH and luteinizing hormone (LH)) and intragonadal regulators (steroids and peptides such as inhibin and activin). In regard to increasing the ovulation rate, it has been recognized for some years that ovarian activity in cattle, as in sheep, can be enhanced by the immunization of the animal against certain of the ovarian androgens/oestrogens or peptides (inhibin) that are involved in the negative feedback control of gonadotrophin release (see review by Reeves *et al.*, 1989). In Ireland, active immunization of heifer cattle against a number of steroids and steroid precursors has been reported (Sreenan *et al.*, 1987; Hanley *et al.*, 1988); such treatment in some experiments resulted in a marked increase (about 30%) in twin ovulations. In the UK, passive immunization of cows with ovine testosterone antiserum was employed by Morris, B.A., *et al.* (1988), with disappointing results. Attempts to influence the timing of puberty in heifers by removing steroid feedback effects have also been examined by some workers (D'Occhio *et al.*, 1988).

Recent years have seen substantial progress in the field of inhibin research. According to Knight (1990), the most significant developments occurred in 1984–86 when several groups independently reported the isolation and characterization of inhibin from cattle and pig follicular fluid, bringing to a close decades of uncertainty about the actual existence of this putative gonadal peptide, believed to exert a selective suppressive effect on pituitary FSH production. Understanding the chemistry of inhibin has been greatly advanced by the recombinant DNA studies that have shown the deduced amino acid sequences for pig, cattle, rat and human inhibin. Such studies have demonstrated that inhibin comprises an α and β subunit linked by disulphide bonds and that there are two forms of inhibin (A and B). The difference between these two forms is attributable to small sequence differences in the β-subunits.

A number of groups (Glencross *et al.*, 1990, 1992; Rhind *et al.*, 1991), including some in Ireland (Scanlon *et al.*, 1990; Morris *et al.*, 1992a, b, c, 1993; Taylor and Headon, 1993), have used synthetic peptide production to produce fragments of inhibin for immunization purposes. The work, in some instances, has been carried though to the point of cattle producing twins (Morris *et al.*, 1992a, 1993). However, both in Ireland and at Reading, it is evident that there is a low incidence of triplet and quadruplet ovulations. In Australia, O'Shea *et al.* (1994) also report variable multiple-ovulatory responses. The fact that it is not possible to limit ovulation rate strictly to two using the immunization approach would seem, however, to present that procedure with difficulties akin to those previously found in gonadotrophin-treated cows.

In Reading, studies have also demonstrated the effectiveness of inhibin immunization as a method of increasing ovulation rate in sheep (McLeod *et al.*, 1992; Wrathall *et al.*, 1992; Fray *et al.*, 1994); in this species, however, litters in excess of two do not present the problems experienced in cattle.

It should be mentioned, in reference to those studies involving immunization against inhibin, that in the mouse this peptide is now known to be a critical negative regulator of gonadal stromal cell proliferation and has been shown to exert a crucial tumour-suppressing role (Matzuk *et al.*, 1992). That aside, the fact remains that, for commercial acceptability, inhibin immunization treatments probably need to be capable of sustained action extending over several oestrous cycles; as yet, it is not clear whether any of the procedures reported in cattle satisfies that particular requirement. In Australia, it is noted that ovulation rate in the cow is much more difficult to manipulate by inhibin vaccination than in the ewe. However, a prototype vaccine against inhibin has been developed and is undergoing evaluation for cattle twinning in that country (O'Shea *et al.*, 1994).

8.4.4 Twins by selective breeding

Several groups in countries around the world have sought to explore the approach of selective breeding as the means of increasing the twinning rate in cattle. Reports by Morris, C.A., *et al.* (1988) and Morris (1990) describe four large-scale trials (established from foundation cows with a history of two or more twin sets) which have been in progress since the 1970s or early 1980s in the USA, Australia, France and New Zealand. Further reports by Morris (1991) and Morris, C.A., *et al.* (1992) deal with the New Zealand herd alone. The twin calving performance of the foundation cows in these herds averaged 15% (10–20% range).

Attention is drawn by the New Zealand authors to a review by Hanrahan (1983) showing that the probability of embryonic survival to term in cattle was 22 percentage points lower in unilateral than in bilateral twin pregnancies; a similar trend was evident in the New Zealand herd, with 11% twinning after unilateral twin ovulations and 38% after bilateral twin ovulations. The distribution of double ovulations in cattle may be a factor influencing embryo survival and twin-calving rates and, based on previous experiences with gonadotrophin-treated cows (Gordon *et al.*, 1962), may pose certain difficulties for any twinning method that involves an increase in ovulation rate.

Other studies on cattle twinning include those in the USA, where it is suggested that repeated measurements of ovulation rate in pubertal heifers may be an effective way of selecting cattle for the twinning trait (Leymaster and Bennett, 1990). Elsewhere in that country, Van Vleck *et al.* (1990) have also reported data supporting the concept of selecting for twinning by measuring ovulation rates in pubertal heifers over a period of several cycles. In Germany, there have been reports on selection studies for twinning in dairy cattle and in the selection of bulls for the twinning trait (Stolzenburg and Schonmuth, 1990a, b). In The Netherlands, on the other hand, a report on the economics of naturally occurring twins in dairy cattle by Beerepoot and Dykhuizen (1992) recommends that attempts to select for twin calves should be discouraged.

8.5 Twinning by Embryo Transfer

Reviewing a number of possibilities for induced twinning in beef cattle, in the light of a 3-year trial programme, Gordon *et al.* (1962) suggested that efforts might be profitably directed towards using ET to introduce a second embryo into the contralateral horn of the cow's uterus a few days after breeding. The beef cow might then carry twins in the form of her own calf and a calf that originated in a suitable donor animal. One unknown factor at that time was whether such a twin pregnancy could be successfully sustained by the recipient's single corpus luteum. The results of Rowson *et al.* (1971), in which they recorded a 73% incidence of twins when an embryo was placed in each uterine horn, clearly demonstrated that a high incidence of twinning could be induced in the absence of a second corpus luteum; such findings have been amply confirmed by other groups during the last 20 years. In advocating the ET approach to twins one important consideration is that the technique can ensure that there is little or no risk of cows carrying more than two calves, as well as ensuring that bilateral rather than unilateral twin pregnancy is established.

8.5.1 *Single embryos (*in vivo*) and bred recipients*

Although Rowson (1971) and Rowson *et al.* (1971), reporting the first successful production of twins by bilateral two-embryo transfer, did allude to the alternative possibility of twinning by one-embryo transfer to a previously bred recipient, it was work in Ireland that eventually showed that this approach might be the basis of a commercially feasible technique (review by Gordon, 1975). Dealing with Cambridge efforts at the time, Rowson (1971) mentioned various attempts to produce twins of differing breeds by transfer of an embryo to the uterus of the previously bred animal; apparently, results in both sheep and cattle at Cambridge suggested the possibility of some adverse physiological or immunological effect in the recipient uterus, which led to the unexpected loss of either the native or the transferred embryo. However, work in Ireland and elsewhere during the past two decades has made it clear that one-embryo transfers can be employed successfully in recipients that are already pregnant. A summary of results from several reports in Ireland (Boland *et al.*,

Table 8.3. Pregnancy and twinning rates in cattle after transfer of one embryo to the contralateral horn of the bred recipients's uterus.

		Other reports					
	Boland *et al.* (1975) Ir	Boland *et al.* (1979) Ir	Sreenan and McDonagh (1979) Ir	Renard *et al.* (1979) Fr	Sreenan *et al.* (1981) Ir	Holy *et al.* (1981) Cz	Total of 'other reports'
Cattle	24	52	25	63	84	95	319
Pregnant first service	15 (62%)	34 (65%)	15 (60%)	36 (57%)	49 (58%)	58 (62%)	192 (60%)
Pregnant with twins	6 (40%)	18 (53%)	9 (60%)	16 (44%)	20 (41%)	28 (48%)	91 (47%)

1975; Sreenan and McDonagh, 1979a, b; Sreenan *et al.*, 1981) and elsewhere (Renard *et al.*, 1979; Holy *et al.*, 1981) is provided in Table 8.3.

In more recent times, in the west of Ireland, Diskin *et al.* (1987a, b) have carried out a series of farm trials which have shown that a combined AI and ET technique is capable of consistently producing a twin-calving rate of 40–50% in pregnant cows. The economics of twinning by ET at that time was dealt with by Diskin and Hickey (1987).

It should be noted that the technical success of this approach to cattle twinning depends on accurate heat detection and the use of bulls of known high fertility in the initial breeding of recipient cows by AI. This is to ensure that the cow's own embryo will almost certainly be occupying the ipsilateral horn of the uterus. Otherwise, introducing the week-old donated embryo into the contralateral horn of a non-pregnant recipient might simply be followed by the loss of the conceptus. Studies of the relationship between the embryo and the corpus luteum in the cow have consistently shown that embryo survival is higher when transfer is made to the uterine horn associated with the ovulating ovary (see Sreenan, 1988); this may be due to the fact that embryonic signals (antiluteolytic and/or luteotrophic) arrive at the ovary ipsilateral to the conceptus by way of countercurrent exchange between the uterine venous and ovarian arterial vessels. On the other hand, Fuchs *et al.* (1991) have shown that the bovine conceptus does not suppress endometrial oxytocin secretion by any direct local effect; oxytocin receptors in the contralateral horn are similarly suppressed.

Studies in Scotland

At the North of Scotland Agricultural College, Broadbent and Dolman (1989) reported results from a research programme in which single beef embryos were transferred to the contralateral horns of 311 bred recipient cattle; a pregnancy rate of 64% to AI/ET was recorded and 45% of the pregnant cattle gave birth to twins. If cows that returned to oestrus after their first AI/ET exposure had been recycled and given further transfer treatment, the workers observed that the overall twinning rate would probably have been much higher.

8.5.2 Twinning studies in mainland Europe

Much work has been carried out in the past decade in former Czechoslovakia, using the approach of single ETs to bred recipient cows. Riha *et al.* (1984) employed surgical transfers (flank incision) of embryos recovered from slaughtered cull cows; they recorded a pregnancy rate of 58% and a twinning rate of 73%. The effect of using non-surgical or surgical transfers and other factors was examined in several other studies (Riha *et al.*, 1986, 1989; Riha and Polasek, 1987). The largest field trial reported to date, in which *in vivo*-produced cattle embryos have been employed in twinning, was that carried out in 1988 in former Czechoslovakia. In this, embryos were obtained from cull cows superovulated prior to slaughter; one-embryo transfers were made to the contralateral horn of 7500 cattle (Vachal and Sereda, 1990; Riha *et al.*,

Table 8.4. Twinning by one-embryo transfer in the Czech Republic (from Riha and Petelikova, 1990).

Superovulation details and conception rate in bred recipient cattle in Czech twinning trial		Yield of twins and single calves in Czech trial	
Mean no. eggs recovered per superovulated donor	12.08	No. calved recipients	3077
		Twins in total	1487
Average no. embryos recovered per donor	9.32	%	48.3%
		Single calves	1590
		Total calf production	4571
Embryos as % of all eggs recovered	70.4%	Per 1 parturition	1.49
Average no. transferable embryos per donor	4.81		
No. transfers carried out	7185		
Pregnant to first service	4124		
Conception rate in recipients	57.4%		

1992). In Table 8.4, data are provided from the report of Riha and Petelikova (1990) on embryo yield from cull cows, and on pregnancy and twinning rates achieved in the field trial. In view of the inherent difficulties in operating such a large programme with fresh embryos, the pregnancy and twinning rates as quoted can only be regarded as encouraging. A report on twinning in bred beef heifers, as opposed to cows, was dealt with in a report by Riha *et al.* (1990). Elsewhere in mainland Europe, small-scale studies on twinning by ET to bred recipient cattle have been reported from Hungary (Nagy *et al.*, 1986; Holdas *et al.*, 1987; Becze, 1990), Russia (Sergeev, 1986) and Germany (Bach and Muller, 1989; Huhn and Rommel, 1990; Papstein *et al.*, 1993); results appear to be in general agreement with data recorded in Ireland.

Twinning in Japan

In Japan, results reported by Suzuki, O., *et al.* (1994) suggested that transfer of bovine embryos, both *in vivo*- and *in vitro*-produced, to recipients previously inseminated was a more effective method of increasing calf numbers than transfers to non-bred recipients.

Using two-embryo transfers to non-bred recipients

The use of two embryos, one transferred to each horn, has been reported in some instances (Reid *et al.*, 1986); after such bilateral transfers, embryo survival rate has either not differed between horns (Heyman *et al.*, 1978) or has been lower in the contralateral horn (Izaike *et al.*, 1991b). There is some evidence that the developmental stage of the two embryos transferred bilaterally may influence subsequent pregnancy and twinning rates (Izaike *et al.*, 1991a).

8.5.3 Twins by IVMFC embryos in Ireland

A report by Lu and Polge (1991) demonstrated that a high pregnancy rate (77%) and twinning rate (65%) could be achieved when stored (20°C) IVMFC embryos were transferred non-surgically to the contralateral horn of bred recipient cattle. Information provided by Ovamass in 1992 to cattle AI stations in Ireland gave results of studies with bred recipients in which a pregnancy rate of 65% and a twinning rate of 49% was recorded using frozen–thawed embryos directly transferred to recipient cattle (see Fig. 8.7). For field application of the technology, transfer of the beef embryo was to be carried out without reference to the location of the corpus luteum. The beef embryos were produced by recovering follicular oocytes from Limousin, Simmental and Charollais-cross heifers which had graded with high quality scores at the abattoir. The oocytes were fertilized using sperm from bulls that had a record for ease of calving and good growth rates. Inseminators from many of the cattle AI stations in the country were trained by Ovamass personnel to transfer embryos. In the advisory information supplied by the company, it was suggested that farms already achieving good results with AI would be

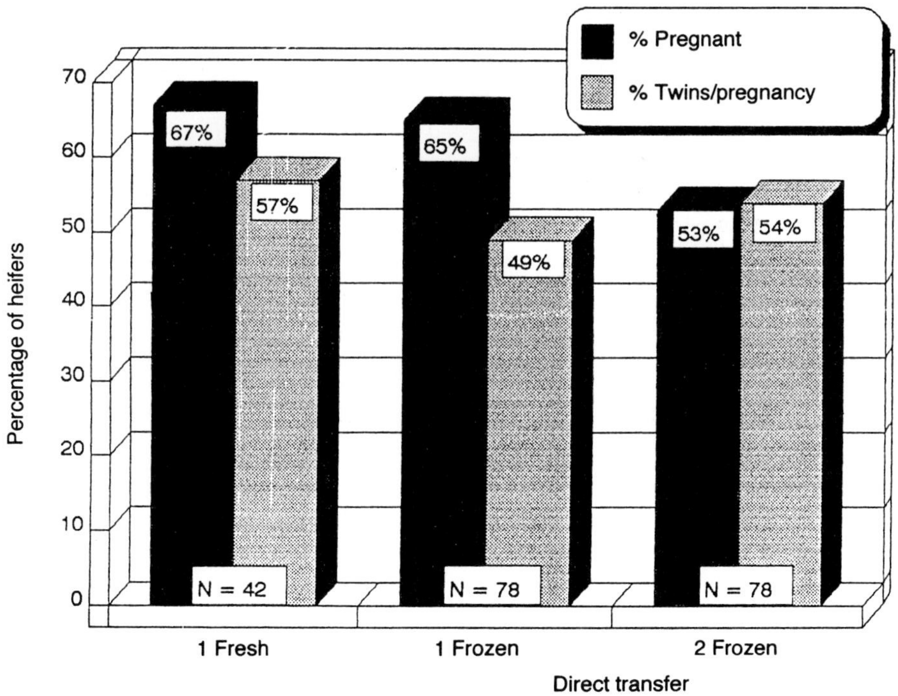

Fig. 8.7. Pregnancy and twinning rates in cattle following transfer of fresh or direct-thawed frozen embryos to bred recipients (from Ovamass data, 1991).

those most likely to benefit from the twinning application. The recommended procedure was to select cows that would normally be identified as suitable for breeding to a Continental bull, taking into account body size and condition of the cow (Bourke, 1992).

8.5.4 Factors affecting the efficiency of twinning by embryo transfer

There are a number of factors that are now known to influence the effectiveness of a twinning by ET technique. In conventional cattle ET, the main concern is with transferring a single embryo to the uterine horn ipsilateral to the ovulating ovary. With twinning transfers, however, there are several possible options, in terms of number of embryos employed, location of the transfers, age of recipient and whether bred or unbred cattle are used.

One- or two-embryo transfers?

Some of the earliest studies in twinning by ET were those of Rowson *et al.* (1971); they showed that surgical transfer of one embryo to each uterine horn resulted in higher twinning rates than transfer of two embryos to the ipsilateral horn. Such results agreed with earlier findings recorded by Gordon *et al.* (1962) in twin-ovulating cows; in that work, a much higher percentage of cows carried twins when twin ovulations involved both ovaries (see Table 8.5).

There have been few studies in which a direct comparison has been made between single-embryo transfers (to bred recipients) and two-embryo transfers (to unbred recipients). In Dublin, studies by Crosby (1976) involved the transfer of embryos to bred (one embryo) or unbred recipients (unbred); fewer transferred embryos survived when this involved unbred recipients. In Canada, Johnson, W.H., *et al.* (1989) found that their bred recipient cows (one embryo to the contralateral horn) had markedly higher pregnancy rates than their unbred (two-embryo transfers) recipients (67% vs. 42%). Bred recipients also performed better than unbred recipients in Japanese studies (Suzuki, O., *et al.*, 1994). It may be that the bred recipient's own embryo helps to provide conditions within the uterus that permit the transferred embryo to

Table 8.5. Egg loss in relation to site of ovulation in gonadotrophin-treated cattle shedding two eggs (from Gordon *et al.*, 1962).

	Fetuses present at 6 weeks		
	Both ovulations in left	One ovulation in each ovary	Both ovulations in right
Cows pregnant at 6 weeks	7	39	20
Cows with two fetuses	2	23	3
Percent to sustain twins	28.3%	59.0%	15.0%

survive. Such speculations are reconcilable with the most recent evidence on maternal recognition of pregnancy in the cow.

Unilateral or bilateral twin pregnancies

Acceptable pregnancy and twinning rates have been reported by several workers following transfer of two embryos to the ipsilateral uterine horn of unbred recipient cattle (Newcomb *et al.*, 1980; Williams and Evans, 1985; Sreenan and Diskin, 1988a, 1989). However, it seems apparent from the literature that bilateral twin pregnancies may provide a more favourable calving outcome. In Table 8.6, data are provided from studies in which direct comparisons were made between unilateral and bilateral twin pregnancies. Regardless of whether *in vivo* or *in vitro* embryos were employed, the results of studies in Ireland, UK, USA, France, Japan, Russia and Germany almost always showed that bilateral twin pregnancies resulted in higher embryo survival rates, lower abortion rates and a lower incidence of stillbirths.

Table 8.6. Summary of evidence provided in studies comparing bilateral and unilateral twin pregnancies in cattle.

Evidence	Reference
Higher embryo survival rate with bilateral twins	Rowson *et al.* (1971)
Higher embryo survival rate with bilateral twins	Renard *et al.* (1977)
Higher twinning rate with bilateral transfers	Ovamass data (1988)
Incidence of stillbirths higher with unilaterals	Suzuki *et al.* (1988)
Higher embryo survival rate with bilaterals	Davis *et al.* (1989)
Higher embryo survival with bilaterals	Suzuki *et al.* (1989a)
Abortion rate higher with unilaterals	Yamashima (1989)
Higher embryo survival with bilaterals	Svitoyus *et al.* (1990)
Higher abortion rate with unilaterals	Berg *et al.* (1991)
Higher twinning rate with bilateral transfers	Liebrich (1992)
Higher abortion rate with unilaterals	Reichenbach *et al.* (1992)
Stillbirth rate higher with unilaterals	Tachikawa *et al.* (1993)

Placental development in twin-bearing cattle

Much remains to be learnt about factors affecting the growth and demise of embryos and fetuses during twin pregnancy. Placental area, according to French data for single and twin-bearing cows (Testart and Du Mesnil du Buisson, 1966), is likely to be greater with bilateral twins (see Fig. 8.8). This is likely to be a factor influencing the size of twin calves and perhaps may even affect the incidence of retained fetal membranes (RFM) following calving. The French workers noted, for example, that, when twin pregnancy was unilateral, giant placentomes, twice the size of those observed in single or bilateral pregnancies, developed in the uterus. The significance of such observations, particularly in terms of RFM, is unclear.

Table 8.7. Number of cotyledons (placentomes) in slaughtered heifer cattle to which membranes were attached – in relation to pregnant or non-pregnant horns (from Rowson *et al.*, 1971).

Heifer no.	Pregnant horn (left/right)	Cotyledons/horn	Total cotyledons
Group 1			
149	+/+	55/59	114
179	+/+	53/56	109
180	+/d	45/2	47
199	+/+	66/62	128
53	+/+	57/36	93
207	+/–	71/44	115
260	+/+	56/48	104
Average			101.4
Group 2			
193	+/–	54/31	85
206	–/+	6/39	45
192	–/++	45/60	105
64	d+/–	42/22	64
209	+/+	45/46	91
203	dd/–	31/0	32
238	d+/–	54/4	58
Average			68.6

d, degenerating embryos; +, normal embryos.
Group 1 animals received two embryos by bilateral transfer; group 2 animals received two embryos in the ipsilateral horn.

In Cambridge, working with recipient heifers a few years later, Rowson *et al.* (1971) commented on the fact that when the pregnancy was unilateral, the number of cotyledons (placentomes) in the contralateral horn was often very small (see Table 8.7). Such evidence would be compatible with the view that bilateral twin pregnancies may be more 'normal' than those that occur unilaterally.

Transuterine migration of embryos in the cow

In contrast to what is known to occur in sheep and pigs, transuterine migration of embryos is rarely found in the cow (Gordon *et al.*, 1962; Scanlon, 1969). This observation is relevant to the site of embryo deposition in attempting twinning in cattle by ET. If two embryos are deposited in the one uterine horn, migration of an embryo into the opposite horn apparently does not occur. It is for that reason that in cattle bilateral transfer is required rather than unilateral transfer. In sheep, it is now evident that transuterine migration of embryos is associated with embryonic elongation and the synthesis of oestradiol by the embryo (Nephew *et al.*, 1989, 1990, 1992). In pigs, studies have shown oestradiol to be implicated in the migration and spacing of embryos in that species (Niemann and Elsaesser, 1986). It is not clear why the cow appears to differ from sheep and pigs in the migration of its embryos.

Heifers or cows as recipients?

Although many research groups have employed heifers as recipients in twinning research, it is doubtful whether this category of animal can be regarded as acceptable for the production of twin calves unless they are particularly well grown. There is no lack of reports in the literature showing that heifers may fail to sustain twin pregnancies to term. This may be related to a failure of such animals to establish a placental area sufficient to sustain the fetuses beyond a certain stage of gestation (Rowson *et al.*, 1971).

One large-scale study reported by Vandeplassche *et al.* (1979) for Red and White dual-purpose cattle in Belgium provides strong evidence that heifers of that breed may have a markedly reduced capacity for sustaining twins in comparison with cows (see Table 8.8). The Belgian authors also found evidence suggesting that the growth of the reproductive tract during pregnancy, resulting in a partly persisting hyperplasia, may play a more important role in determining the size of the placenta and uterine capacity than the growth that normally occurs with increasing age.

In Ireland, Sreenan and Diskin (1989), in some of their twinning studies, reported that twice as many heifers (15%) as cows (8%) lost pregnancies between day 50 and term.

Table 8.8. Incidence of double ovulations, twinning rates and embryo mortality in Red–White cattle in Belgium (from Vandeplassche *et al.*, 1979).

	Cattle with twin ovulations	Twins born at calving	Lost during pregnancy	Calves as % of ovulations
Heifer	58 (3.84%)	84 (1.04%)	73%	27%
Cow	143 (4.95%)	1445 (3.77%)	24%	76%

Effect of embryo genotype

In advocating twinning by ET to the bred recipient, note should be taken of the fact that where the two cattle fetuses differ markedly in their genetic constitution, the onset of parturition may be determined by the calf with the genotype for the shorter gestation period. This was amply demonstrated, for example, when Friesian calves, carried by Brahman-bred recipient cows, precipitated the early birth of their Brahman cotwin (see Shelton, 1988, for information on Friesian–Brahman twins). In Ireland, studies with Friesian dairy cows showed that recipients carrying twin crossbred calves had a gestation period of 282 days, whereas those with one crossbred and one Friesian calf had a duration of 276 days (O'Farrell *et al.*, 1991). For commercial application of twinning by one-embryo transfer, the aim should be to have both native and donated embryos of similar breed make-up.

8.6 Identifying Twin-bearing Cattle

For several reasons, it is highly desirable that farmers should be aware of cows carrying twins, in any commercial application of twinning by ET. The sooner this can be done in pregnancy, the better; in any event, it should be prior to the cow entering the final 2 months of the gestation period. It is known that the energy requirements for cows carrying twins are higher than for those with singles, both in beef (Koong *et al.*, 1982; Guerra-Martinez *et al.*, 1990) and dairy cattle (Mayne *et al.*, 1991). Oestrone sulphate, a metabolite formed by the placenta and fetus, occurs in the presence of a viable fetus from about 100 days after breeding; the assay of milk for oestrone sulphate has formed the basis of a confirmatory test for pregnancy which has been commercially available for some years (Fletcher and Worsfold, 1988).

8.6.1 Detection of fetal hormones

Work in Dublin in the recent decade sought to identify twin-bearing animals on the basis of an assay for total (conjugated and unconjugated) oestrogens (MacDonnell *et al.*, 1984, 1993); although significant differences in total oestrogens can be demonstrated to occur on a group basis at midpregnancy between twin-bearing and single-bearing cows (9.8 vs. 5.0 ng oestrogen ml^{-1}

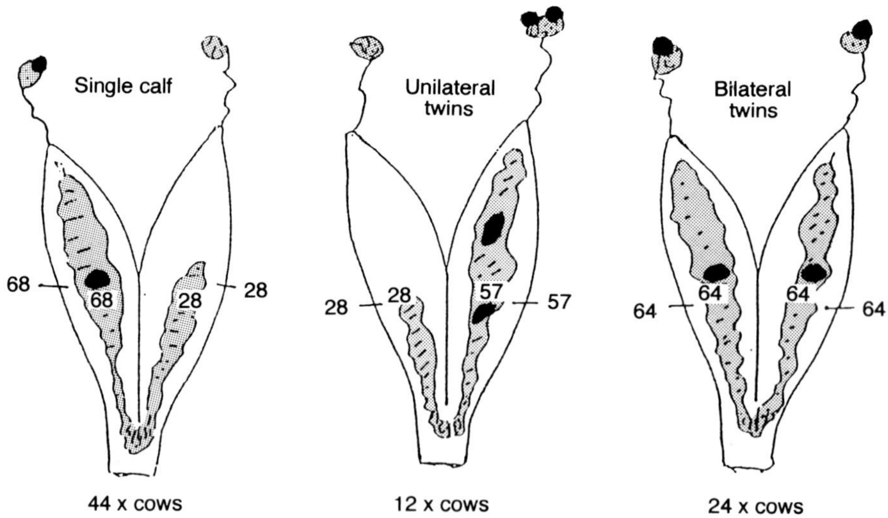

Fig. 8.8. Number of placentomes in single and twin pregnancies in cattle according to the location of the twins *in utero* (from Testart and Du Mesnil du Buisson, 1966).

plasma), individual variation proved too great for this assay to be usefully employed as a test on the farm. Reports by Worsfold *et al.* (1989) and Dobson *et al.* (1991, 1993) in the UK based, among other things, on the assay of oestrone sulphate, reached similar conclusions. It is noted by Dobson *et al.* (1993) that mammary gland tissue in the cow actively synthesizes oestrone sulphate and this may be a factor in masking the genuine differences arising from placental tissue.

The accuracy of pregnancy diagnosis by determining bovine pregnancy-specific protein B (bPSPB) by radio-immunoassay (RIA) has been reported (Humblot *et al.*, 1989); the assay was shown to have a positive accuracy of 90% at 30 days. Studies reported by Dobson *et al.* (1991, 1993) showed that from day 60 to the end of pregnancy samples from single and twin pregnancies had measurable bPSPB concentrations; between days 150 and 250, concentrations in twin-bearing animals were almost double those in single-bearing cattle. As with oestrone sulphate, however, individual values overlapped. Evidence suggests that measurement of bPSPB may give an assessment of fetal viability, whereas oestrone sulphate reflects placental well-being (Dobson *et al.*, 1993).

It might be mentioned that, in sheep, twin-bearers have been identified by an assay for ovine chorionic somatomammotrophin (Chene *et al.*, 1988). An earlier report by Bolander (1977) in the USA reported that serum levels of bovine placental lactogen were almost twice as high in twin pregnancies as in single.

8.6.2 *Use of real-time ultrasonics*

Real-time ultrasound scanning involves the use of a rectal probe which transmits harmless ultrasound waves through the body tissues. Waves are reflected to the transducer when they reach the fetus and are converted to produce an image on a display screen. Although real-time ultrasonics was first used in farm animals to depict pregnancy in the horse, the method has rapidly spread to the other species, including the cow (see review by Kahn, 1992). There have been a number of reports dealing with its use in identifying twin-bearing cows (Wright *et al.*, 1988; Hughes and Davies, 1989; Bach *et al.*, 1991; Davis and Haibel, 1992, 1993; Echternkamp and Gregory, 1991, 1993). The optimum time for scanning appears to be 50–60 days after the start of pregnancy; the large physical size of the uterus after about 80 days of gestation can make it difficult to scan the anterior tip of the uterine horns and to determine the number of young (Echternkamp and Gregory, 1991).

One problem requiring resolution, in using scanning as the diagnostic method, is the extent to which fetal deaths occur after scanning. Available evidence suggests that, for cattle, embryonic/fetal mortality should not exceed about 5% beyond day 30 (Sreenan and Beehan, 1976); scanning around day 50 by an experienced operator would probably seem to be the most suitable approach for attempting twin diagnosis.

Identifying the sex of the calf

With real-time ultrasound scanning, it is even possible to think in terms of identifying the sex of the calf, if the examination is made 60–70 days into pregnancy (Curran *et al.*, 1986a, b, 1989; Wideman *et al.*, 1989; Curran, 1992). The accuracy of gender diagnosis may be practically 100%, based on the location of the fetal genital tubercle. In a twinning herd context, if a single calf is identified, then knowledge that it is a bull calf may influence subsequent decisions about calving management (e.g. induction treatment may be more appropriate). In cows pregnant with singletons after conventional cattle ET, the sex of the unborn calf may greatly influence the monetary value of the pregnancy (e.g. pregnant heifers destined for export).

In the UK, a pregnancy testing service, operated by specially trained technicians, is now commercially available from the main cattle AI organization (Genus); cost of the service is currently £2.50 a cow plus a call-out fee of £10. For normal purposes, such a service can best be employed to determine non-pregnant cows as early as possible; according to Ryan (1993a) the scanning should be done between day 24 and day 30 after breeding.

8.6.3 Other methods

Pregnancy diagnosis by rectal palpation, usually at the 6–8-week stage of pregnancy, has been used routinely in cattle for many years. If rectal palpation is carried out by inexperienced operators, it is probably not without its hazards. There is some evidence from work in the USA that the incidence of embryonic death after palpation during the first 50 days of pregnancy may be significantly greater than that found thereafter (Vaillancourt *et al.*, 1979). In more recent times, Franco *et al.* (1987) concluded that operators need to be aware of the vulnerability of the bovine embryo to palpation per rectum around the end of the sixth week of pregnancy. In attempting to identify twin embryos by rectal palpation, it is inevitable that the examination involves a greater degree of manipulation of the uterus; for that reason alone, the technique cannot be recommended with any confidence.

8.7 Breeding, Feeding and Management Considerations in Twin-bearing Cattle

8.7.1 Embryos of appropriate genotype

In Ireland, high-quality beef embryos have been produced by recovering oocytes from Continental crossbred heifers (Limousin, Simmental and Charollais) and using Continental breed bulls in IVF (see Fig. 8.9). Such bulls, regardless of breed, must have a proven history of easy calving associated with them. Over the past 20 years, the use of the large, lean Continental bull breeds in Ireland has led to a corresponding decrease in the number of matings by the traditional Aberdeen Angus and Hereford beef bulls. If it were possible,

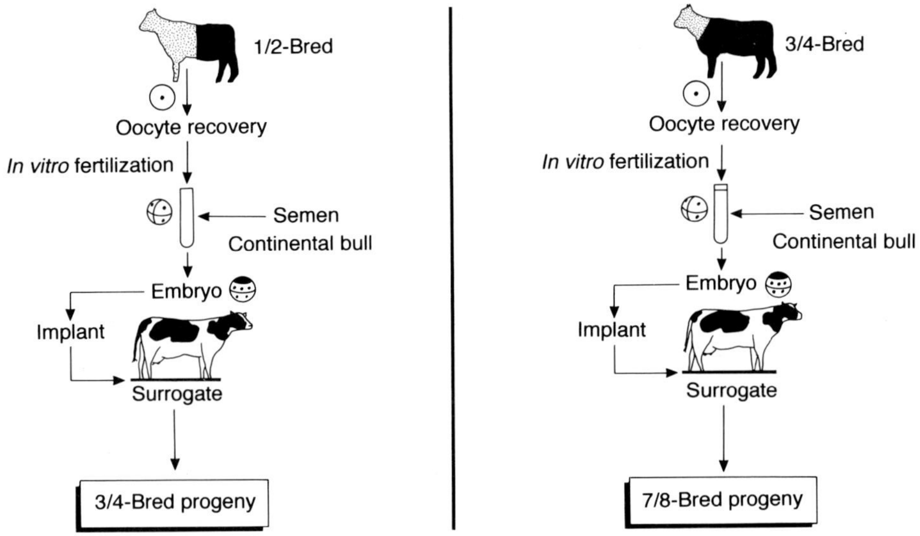

Fig. 8.9. Producing high-quality beef embryos for twinning and other applications.

as is occasionally claimed by breeders, to have Angus/Hereford bulls which give calves of low birth weight and yet with a high postnatal growth rate, there could be considerable commercial interest in such animals for beef rearing.

Elsewhere, the appropriate genotype of beef embryos may be quite different. In the pure-bred herds of beef cattle in countries such as the USA or Canada, it may well be a question of Angus or Hereford oocytes fertilized with Angus or Hereford semen.

In twinning by one-embryo transfer to bred recipient cattle, then it may well be possible to arrange with the AI station to use the same bull in breeding the recipient as was employed in siring the beef embryo. This facility has been provided in Ireland by some of the AI stations in cooperation with the embryo-producing company, Ovamass.

8.7.2 Nutritional considerations

It is essential that any introduction of cattle twinning by ET on the farm should be amply supported by back-up information on nutritional and management implications. The usual view is that nutritional considerations are especially important during the final trimester of pregnancy, but it would be wise to keep a watchful eye on what is happening at earlier stages. Studies reported by Lee *et al.* (1989) showed that embryos from cows subjected to nutritional stress (inadequate feeding levels) were of lower weight at days 29–34; the chemical composition of fetal fluids was also significantly altered. In Japan, conducting studies with cows that carried multiples, Izaike *et al.* (1991a) frequently recorded the loss of an embryo, especially at 38–57 days of gestation; such information is in contrast to that of Sreenan and Beehan (1976) mentioned above, which suggested that, once a twin pregnancy is established, it usually carries through to term.

In beef suckler cattle, work at Trawsgoed in Wales over a period of several years demonstrated that twinning by ET could have a favourable practical outcome (Williams and Evans, 1985). In Aberdeen, Topps *et al.* (1989, 1990) reported on trials which provided evidence on appropriate feeding patterns to apply in twin-bearing beef cows. In the same city, Williams *et al.* (1989) reported the increased energy metabolism that occurs in twin-bearing cows. In the Republic of Ireland, a considerable amount of data on the practical outcome of twinning in suckler cattle and some on Friesian dairy cows has been published by Sreenan and coworkers (Sreenan *et al.*, 1981; Diskin *et al.*, 1987a, b, 1990a, b; Sreenan and Diskin, 1988a, b); the problem was one of the cost of twinning by ET rather than any real difficulty with cow performance or the outcome of twin calvings. The usual routine employed by these workers was to use a Continental bull as the sire with beef suckler cows and a Hereford as the sire of twins in Friesian dairy cattle.

At Hillsborough in Northern Ireland, studies with spring-calving Friesian cows have shown that twinning by IVMFC embryos could be successfully applied, provided that twin-bearers are accurately identified and provided with additional food in the final trimester of pregnancy (Mayne and McCaughey, 1990a, b; Mayne *et al.*, 1991; Unsworth and Mayne, 1991; McEvoy, J.D., *et al.*, 1992; Mayne and McEvoy, 1993a, b).

8.7.3 *Management of cow and calf*

The selection of suitable cows to act as twin-bearers is an important practical consideration. The procedure recommended in Ireland is to select cows that the farmer would normally identify as suitable for breeding to a large Continental bull, taking into account body size and condition (Bourke, 1992); cows should be second or later calvers, reproductively normal, with a history of easy calving and having given birth at least 6 weeks earlier. As experience of twinning grows, the various criteria employed in cow selection may well be modified; in the early stages of commercial application, it is essential that they are rigorously applied. In the matter of breeding the beef cow by AI, it has been estimated by Diskin and Sreenan (1992) that, even in the best managed herds, oestrus detection rates rarely exceed 70% and quite often are much lower; as well as that, 10–20% of non-oestrous cows may be wrongly detected as in oestrus and inseminated when conception is not possible.

It is possible that factors known to influence the development, or even survival, of the singleton calf may be even more important in affecting the progress of twins. At Reading, for example, it was found that the calving outcome in cows suffering from leptospirosis was more severely affected where twins rather than single calves were involved (Reynolds and Street, 1991); this points up the need to ensure that induced twinning in cattle takes the disease status of recipients adequately into account.

Abortions and stillbirths

On the question of abortions and stillbirths with twin-bearing cattle, these problems may be associated with the form of twin pregnancy established (see Table 8.6, Section 8.5.4) and whether heifers rather than cows have been employed as recipients. In the normal course of events, the abortion rate in twin-bearing cows is apparently no different from that found with singles. A study of 10,000 calvings in dairy cattle reported by O'Farrell *et al.* (1990) presents evidence of this (see Table 8.9).

A report by Mee (1991) dealt with 10,082 calving records from nine research herds; the abortion rate was no higher for twin-bearing cows (1.2%) than for single bearers (1.8%). In some small-scale Japanese studies, in which IVMFC-derived embryos were used in bilateral transfers, a relatively high (25%) abortion rate in Holstein cow recipients has been recorded (Takada *et al.*, 1991); the cause of this is unknown.

Table 8.9. Effect of natural twinning on dairy cow and calf performance (data from more than 10,000 calvings in the Moorepark herds) (from O'Farrell *et al.*, 1990).

	Single	Twin
Abortion (%)	2	1
Perinatal mortality (%)	6	16
Calving difficulty (%)	8	7
Malpresentation (%)	3	11
Retained placenta (%)	3	19

Parturient behaviour and calf survival

Little information is available about the behaviour of twin-bearing cows at the time of parturition. In Australia, Owens *et al.* (1985) reported on parturient behaviour in a herd selected for twinning; twin-born calves were slower to begin sucking than those single-born. However, as shown in the report of O'Farrell *et al.* (1990), the higher calf mortality often associated with twin calves is largely due to entanglement of calves at birth rather than from any weakness after birth. If twin calvings are supervised, calf mortality may be much reduced. The immune status of single and twin beef calves was studied by Mackenzie *et al.* (1991); the immune response in twins was comparable to that in the single calves.

A routine adopted at Moorepark in Ireland for twin-bearing Friesians allows the cow to continue in labour until 1 hour after the feet of the calf first appear; if the cow has not calved by that time, it is essential to provide assistance and remove the two calves. There is a need for such information about factors affecting cow behaviour at parturition and calf survival to guide farmers in the management of twin-bearing beef suckler cattle. In sheep, for example, it is generally recognized that aberrant behaviour patterns are more common in twin births than in single births (Alexander *et al.*, 1974).

Growth of twin calves

Results of studies reported by Patterson (1993a, b) and Patterson *et al.* (1993) have shown that twin calves produced by IVF ET had equal beef-producing merit to that of single-born calves of similar genotype. This would be in agreement with previous reports (Gordon, 1983).

8.7.4 Birth weights and gestation length

In terms of size, the general rule in cattle, as in sheep, is that twins should be approaching 80% the weight of singles (see Table 8.10). Data from beef suckler and dairy cows in which twins were induced by ET (J.M. Sreenan, 1990, personal communication) have shown that the duration of gestation is shortened by rather less than a week in cows bearing twins (see Table 8.11).

In Northern Ireland, a 3-year research programme in Friesian dairy cattle induced to carry twins by ET produced evidence of fetal oversize in some of the singletons born (McEvoy, J.D., *et al.*, 1992). Although the authors conclude that the incidence of such fetal oversize may restrict the uptake of IVMFC embryo technology, it should be noted that the problem was confined to the first 2 years of their programme, when twin induction was by way of transfer of two IVMF embryos sired by Continental bulls not specifically selected for easy calving. A majority of cows carrying singles received meal-feeding in addition to *ad libitum* silage feeding in late gestation.

Table 8.10. Comparative ratios of twin to single birth weights in studies of cattle twinning by embryo transfer.

Country	Twin as % of single	Authors
Irish Republic	0.80	McCutcheon *et al.* (1991)
Irish Republic	0.77	Diskin *et al.* (1990a)
USA	0.77	Davis *et al.* (1989)
Irish Republic	0.78	Sreenan and Diskin (1988a)
Wales	0.78	Williams and Evans (1985)
USA	0.89	Anderson *et al.* (1978)

Table 8.11. Calving performance of single- and twin-calving beef and dairy cows at Belclare (from J.M. Sreenan, 1990, personal communication).

	Beef cows		Dairy cows	
	Singles	Twins	Singles	Twins
No. of cows	121	66	37	17
Gestation length (days)	285	280	282	279
Birth weight (kg)	43	33	42	36

In the third year, when fetal oversize was not a problem, the standard procedure employed was that of a one-embryo transfer to a bred recipient; single-bearing cows were identified by ultrasound scanning and provided with grass silage *ad libitum* without meal-feeding during the final trimester of pregnancy. Although the effects of maternal nutritional status on the birth weight of the single calf are known to be variable, it is possible that high-energy feeding in the final weeks of pregnancy may be a contributory factor to fetal oversize. For such reasons, it is again necessary to emphasize the practical importance of accurate identification of single-bearing cows in a twinning herd, to avoid such animals receiving high-energy feeding designed for twin-bearers.

Altered fetal development (at 7 months of gestation) of IVMFC-derived embryos, in comparison with those derived from *in vivo* embryos, has been noted in a report by Farin *et al.* (1994) in North Carolina; however, the implications of these observations, in terms of calf birth weights, is not clear. What is clear is that the phenomenon of fetal oversize observed in the Hillsborough herd has not been apparent in the studies of J.M. Sreenan (1990, personal communication) or in any of the studies reported elsewhere (see Sections 8.5.1 and 8.5.2). Although such twinning studies employed *in vivo* embryos, there is no reason to believe that the IVMFC procedure *per se* is likely to be a factor involved in fetal oversize. It is more likely to have occurred as a result of heterosis, which is known to be a factor that may profoundly influence birth weight in cattle (Holland and Odde, 1992; Bellows *et al.*, 1993). Although antigenic influences on placental development in cattle have not been reported, an immunological influence arising from the genetic dissimilarity of dam and embryo in the Hillsborough herd cannot be ruled out.

Induction of calving

There may be instances in which a safe and reliable method of inducing parturition in cattle, without the complication of placental retention, would be of practical advantage. Cows in calf to Continental bulls have longer gestation periods than those bred to traditional beef-breed bulls (Hereford/Angus); the growth of the fetus in the final week or so of pregnancy, particularly when it is a bull calf, can be substantial. Diskin *et al.* (1989) have described an induction technique that they employed as a means of controlling birth weights in cows carrying large single calves sired by a Continental bull.

These authors note that, although considerable information is available on the induction of calving using various corticosteroid preparations, the effect on calving difficulty has often been inconsistent. Where the response interval from treatment to calving is short, the incidence of calving difficulty has tended to increase, which is probably because of insufficient time being available to prepare the cow adequately for delivery. The protocol of Diskin *et al.* (1989) involved treatment with a long-acting corticosteroid on day 274 followed, 5 days later, with a short-acting preparation; calving occurred within 80 h of this second injection in about 90% of animals. Although the treatment was directed towards heifer cattle, it may also be relevant to cows in a twinning herd that are likely, for various reasons, to be carrying a large single calf.

It might be mentioned that the potent progesterone antagonist, RU 486, which is a 19-norsteroid with a high affinity for progesterone receptors, has been used alone, or in combination with relaxin, in small-scale studies in beef cattle as an induction agent (Li *et al.*, 1991a, b). Injection of RU 486 on days 277 and 278 was effective in inducing parturition some 55 h later; no detrimental effects (dystocia, retained membranes, postpartum infertility) were evident.

In the pregnant bovine uterus, progesterone binds with progesterone receptors to maintain pregnancy to normal term. As well as that, progesterone is believed to inhibit myometrial contractility and ensure the quiescent state of the uterus throughout pregnancy; the antagonist RU 486 renders progesterone biologically inactive and precipitates parturition. There may be a place for using this agent to achieve a highly predictable calving time and a low incidence of placental retention without compromising calf viability.

8.7.5 Postpartum events in twin-bearing cows

Retained fetal membranes (RFM)

The lack of dehiscence and expulsion of the fetal membranes by 12–24 h postpartum is normally what constitutes retention. It is well documented that twin calvings, particularly those occurring naturally in dairy herds, are associated with a higher incidence of RFM (Gordon *et al.*, 1962; O'Farrell *et al.*, 1990; Eddy *et al.*, 1991). The underlying mechanism of RFM in cattle is not fully understood (Paisley *et al.*, 1986; Eiler and Hopkins, 1991; Joosten and Hensen, 1992). In one sense, all cows that calve show some evidence of RFM because there is always an interval between parturition and expulsion of the placenta. There is an increasing risk of RFM with parity, even when single births are involved (Van Werven *et al.*, 1992). According to these authors, heifers not only expel their membranes earlier than cows, but, even when expulsion is delayed, it causes almost no effect on reproductive parameters. They suggest that better conditions of the endometrium, faster involution of the uterus or a more effective immune system may be responsible for the minimal effect of RFM on the performance of first-calving heifers. In Ireland, there is evidence that as the severity of calving difficulty increases, there is a clear decrease in conception rate to first and later services (Sreenan and Diskin, 1994).

It seems possible that one contributory factor in the aetiology of RFM may be the nutritional status of the cow in late pregnancy. The cow would normally be fed and managed in expectation of a singleton and this is unlikely to be adequate for a twin-bearing animal. When the cow is prepared for the twin calving, evidence supports the view that this may reduce the problem of RFM. In France, Chassagne and Barnouin (1992), on the basis of their studies on dairy cows, suggest that there is a need for an adequate energy status at parturition to minimize the incidence of RFM.

Hormonal and antibiotic treatments

Whether oxytocin or prostaglandin $F_{2\alpha}$ ($PGF_{2\alpha}$) administered immediately after a twin calving in an effort to facilitate the expulsion of the fetal membranes has any real merit appears unlikely (see Gordon, 1983). Fetal membranes not expelled within 36–48 h undergo necrosis and the uterus is at risk of infection; the efficacy of intrauterine antibiotic treatment (e.g. oxytetracycline), with or without a non-steroidal anti-inflammatory agent (benzydamine), in the prevention of endometritis in dairy cattle has been examined and reported on (Cairoli *et al.*, 1993).

Freemartinism

Any discussion of cattle twinning inevitably raises the questions posed by the occurrence of freemartins, the sterile heifer calf born cotwin to a bull calf (see Gordon, 1983). For beef rearing, freemartins present no difficulty; they grow as readily as normal heifers. In using ET for twin induction, a proportion of cows that give birth to a single calf may well have started pregnancy with twin embryos. The occurrence of freemartinism in single female calves produced by multiple ET has been studied in Japan (Kadokawa *et al.*, 1993); evidence supports the view that heifer calves born in such instances have normal reproductive function.

Rebreeding the cow

Although Eddy *et al.* (1991) found that an increased culling rate was a major problem with twin-bearing cows in their study (35% vs. 23% in single-bearing cattle), controlled experimentation in Northern Ireland with Friesian cows showed that similar proportions of twin and single bearers were rebred successfully in each of the 3 years of that study (McEvoy, J.D., *et al.*, 1992).

8.7.6 The postpartum suckler cow

The reproductive wastage which occurs in suckler cattle as a result of the long postpartum anoestrus is dealt with in papers by Sreenan *et al.* (1991) and Diskin *et al.* (1992a, b). They record that, whereas more than 90% of dairy cows resume oestrous cycles by 42 days after calving, only 20% or less of suckler cattle do so. In first-calvers and in any cow that is in poor body condition at calving, the problem of delayed return to cyclicity is likely to be even more acute. The frequency of suckling is a predominant factor affecting the postpartum interval (Williams, 1990); it is possible, however, to shorten the postpartum interval by up to 3 weeks by restricting suckling to once or twice daily, starting at 35 days after calving (Diskin *et al.*, 1992a, b). In twin-rearing cattle, where the frequency of suckling is likely to be an even greater factor influencing the postpartum interval, such management measures should be borne in mind.

In Scotland, Wright *et al.* (1992a) found evidence that pheromonal influences could alter the duration of the postpartum period in suckling beef cows. They showed that cervical mucus from cows in oestrus contains agents capable of reducing the postpartum interval, particularly in cows with extended anoestrous periods. The authors suggest that this finding may have implications for the design of housing for postpartum cows; lack of opportunity to mix with other cows or keeping cows in small herds may exacerbate the problem of extended postpartum intervals. In the USA, Burns and Spitzer (1992) have reported that beef cows exposed to biostimulation (either bulls or testosterone-treated cows) immediately post partum returned to oestrus earlier than animals isolated from such stimulation.

It must be emphasized constantly that suckler cattle should not be used for twin production unless the nutritional environment is suitable. The importance of good body condition at calving (condition score 2.25) has been demonstrated in several studies (Wright and Peters, 1991; Wright *et al.*, 1992b); level and pattern of feeding post partum may have little effect if cows are in reasonable body condition at calving. It is worth mentioning that reproductive failure (e.g. lack of postpartum ovarian activity) in response to energy balance is a phenomenon of widespread occurrence. In humans, ovarian function is sensitive to weight loss in late pregnancy (e.g. negative energy balance), which may be independent of current nutritional status (Panter-Brick *et al.*, 1993).

In terms of the endocrinology of the postpartum dairy cow, it appears that high oestrogen concentrations during late gestation induce a refractory state such that the brain cannot respond, in the absence of previous progesterone action, to the oestrus-inducing action of ovarian oestradiol; this results in the first postpartum ovulation often being 'silent' (Allrich, 1993). The corpus luteum formed after the first ovulation provides the progesterone necessary to sensitize receptors in the brain to respond to follicular oestrogen. Experimentally, McDougall *et al.* (1992) have shown that a short period (5 days) of progesterone pretreatment can sensitize the postpartum cow to the behavioural effects of oestradiol.

In the beef suckler cow, where resumption of ovarian activity usually occurs much later, the first ovulation post partum, although accompanied by oestrus, is usually associated with development of a corpus luteum with a short lifespan and low fertility. Such low fertility may be due, not only to premature regression of the corpus luteum but also to defects in oocyte quality or, more probably, adverse conditions in the reproductive tract (Breuel *et al.*, 1993; Schrick *et al.*, 1993).

Stimulating rebreeding

Various workers have reported on attempts to induce the resumption of ovarian activity in beef cattle. In Canada, Twagiramungu *et al.* (1992) used a treatment involving two gonadotrophin-releasing hormone (Gn-RH) doses

(8 and 4 μg) within a 3-day period and showed that this increased the precision of oestrus after PGF-induced luteolysis in postpartum cows and induced the resumption of cyclicity. In Ireland, beef suckler cows that have not resumed cyclical breeding activity after calving have been stimulated into doing so by short-term progesterone/progestagen treatments (Diskin and Sreenan, 1993) with or without exogenous gonadotrophin, in the form of a low dose (700 iu) of PMSG at the time of terminating the steroid treatment.

Reducing gestation length

It is worth noting the observations of Marshall *et al.* (1993), who found that gestation length is a highly heritable trait in Limousin cattle; they suggest that selection for reduced gestation length could be of some economic benefit, especially when choosing bulls to increase the postpartum period.

8.8 Using IVF in Embryo Production from Valuable Donor Cattle

The establishment of commercial IVF facilities by those engaged in cattle ET is likely to increase markedly in the years ahead. Not only may this be one means of overcoming the problem of low and variable embryo yields associated with conventional ET, but it may also provide an opportunity to cash in on current advances in cattle IVF, such as semen sexing (Cran *et al.*, 1993), as well as future advances, such as cloning and genetic engineering. Some authors have argued that conventional cattle ET has limited application and recommend IVM/IVF as an alternative (Stubbings, 1990).

8.8.1 Embryo production after ultrasound-guided retrieval of bovine oocytes

As already mentioned (see Section 2.6.6), there has been an increasing number of research reports in recent years dealing with oocyte recovery from the live animal (review by Rath, 1993). The procedure is illustrated in diagrammatic form in Fig. 8.10. In the USA, commercial IVF facilities are already being employed as a means of obtaining embryos from clinically infertile but valuable cattle (Stroud and Myers, 1993; Looney *et al.*, 1994). Infertile animals apart, there is a case for using IVF technology in normal cattle as an alternative to the conventional technology. For more than 20 years, superovulation has been employed in commercial cattle ET operations as the principal means of increasing the number of embryos that may be recovered from genetically superior beef and dairy donor cattle.

There are, however, well-recognized difficulties associated with such technology, including marked variability in superovulatory response, embryo recovery rate and embryo quality among donor cattle. In dairy cattle, for example, superovulation and subsequent embryo recovery almost inevitably result in a marked increase in the calving interval (CI) of the donor cow (Bak

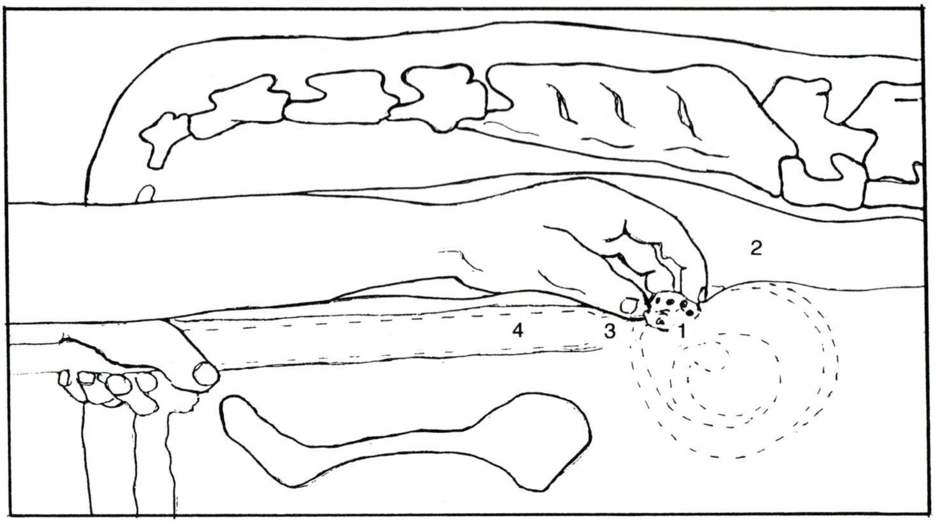

Fig. 8.10. Oocyte recovery by transvaginal ultrasonic-guided aspiration of vesicular follicles in the ovaries of the live cow. Diagrammatic representation of the position of the ovary (1) in the cow during transvaginal puncturing of vesicular follicles. With one hand in the rectum (2) the ovary is manipulated against the anterior part of the vagina, just beside the cervix (3). With the other hand, the transducer (4) is advanced into the vagina and, as the needle is pushed forward, it perforates the vaginal wall and punctures the ovary (from Kruip *et al.*, 1991).

et al., 1989). Ryan, D.P., *et al.* (1990a, 1993a, b), working in Louisiana, have shown that oocytes recovered from gonadotrophin-stimulated pregnant dairy cows could be employed to produce embryos for transfer to recipients after IVF and IVC. By using genetically superior pregnant cows for embryo production, it may well be possible to maintain a normal CI while improving the genetic quality of the herd.

It would appear from studies in several European countries that twice-weekly collection of primary oocytes can be attempted for periods of several months without this having adverse effects on the ovaries or genital tract (Vander Schans *et al.*, 1991, 1992; Boni *et al.*, 1993a, b; Kruip *et al.*, 1993; Roelofsen *et al.*, 1993; Simon *et al.*, 1993; Walton *et al.*, 1993; Wiebke *et al.*, 1993). There is no requirement for gonadotrophin stimulation in donor animals and, with twice-weekly collections, the animals remain acyclic. It is essential to have prior knowledge of bulls that are used in the IVF to avoid low fertilization rates; the necessary screening can be done beforehand, using oocytes obtained at the abattoir.

However, it must be remembered that the success of an IVF laboratory depends on meticulous attention to detail and it would be essential to have a proved capacity to deal with all aspects of IVM/IVF/IVC, using abattoir ovaries, before embarking on any commercial application of the technology. According to Stroud and Myers (1993), transvaginal aspiration of follicles in the donor cow is simple, fast, relatively easy to learn and very efficient. Weekly oocyte recovery, 3 days after FSH stimulation, resulted in an average of 8.6 oocytes per aspiration in work reported by Looney *et al.* (1994).

In vitro production of embryos, however, although clearly not too complicated to be established in a veterinary clinic, remains a multistep procedure requiring considerable attention to detail if it is to be successful. Initially, such an IVF service could provide for donors referred by ET practitioners as being clinically infertile; subsequently, in the light of growing knowledge of long-term effects of follicle aspiration, it could possibly be widened to include other categories of cattle, particularly prepubertal heifers, postpartum anoestrous cows and animals in their first trimester of pregnancy.

8.8.2 Developing a genetic salvage service

Workers in Ireland were among the first to report the birth of calves after casualty slaughter of a valuable donor cow. In this, 20 oocytes were recovered from an 8-year-old Simmental cow; after IVMFC embryo production and ET, three calves were eventually born (Keating, 1992). In Aberdeen, workers reported an average of three transferable-quality embryos per animal from a group of 14 aged, infertile and even chronically diseased cows (Mylne *et al.*, 1992). In Canada, Xu *et al.* (1992a) recovered 222 oocytes from two 11-year-old Holstein cows; 27 embryos were obtained for freezing or immediate transfer to recipients. In the USA, Stringfellow *et al.* (1992, 1993) report producing an average of 5.3 transferable embryos per animal from cows culled from dairy herds (for mastitis, injury, reproductive failure); their IVF procedure is apparently available to cattle producers in Alabama. In Italy, it is estimated that, on average, one to three live calves per slaughtered donor could be a reasonable expectation for a farmer who decides to send the ovaries of his cow to an IVF laboratory (Lazzari and Galli, 1992, 1993a, b; Galli *et al.*, 1994). Workers in Belgium have also shown that blastocysts can be produced from the ovaries of single cows, although individual results were highly variable (Mermillod *et al.*, 1992b, d).

In terms of embryo yield from slaughtered cattle, studies by Carolan *et al.* (1993, 1994), using an ovary dissection procedure, resulted in an average of 15.4 embryos per animal. It is clear that a considerable number of oocytes can be recovered by such methods, and subsequently employed to salvage much useful genetic material. There is also scope, as shown by Loskutoff *et al.* (1991a, 1993), for the possible use of micromanipulation techniques to increase the supply of embryos available for transfer.

8.9 Sex Control by Sperm Separation

Manipulation of the sperm cell as a means of selecting the sex of choice has been of great interest to animal scientists since the time when it was first appreciated that gender is determined in farm mammals by the male gamete. The literature contains an abundance of accounts claiming varying degrees of success in separating the two types of sperm in human and domestic animals; convincing evidence of success has been much less easy to find until recent years (see reviews by Miller, 1991; McEvoy, 1992). It is still true that sexed semen for AI may be some years down the road, but, for the IVF laboratory, certain forms of sperm separation technology may have immediate application. Although possible differences in physical, biochemical and immunological properties of X- and Y-bearing sperm have been suggested by many workers, the only certain difference appears to be that contributed by the sex chromosomes themselves (Gledhill, 1988). It is on this basis that there is ample scope for refinement of sperm separation technology, which should prove capable of making sexed bull sperm available for IVF, although not in the first instance for AI.

The primary sex ratio

In the testes of the bull, sperm carrying the X and Y sex chromosome are produced in equal numbers. It should also be mentioned that studies in Canada have shown that the male : female ratio of bovine embryos produced either by fertilization and culture *in vitro* or fertilization and development *in vivo* does not differ from 1:1 (Yadav *et al.*, 1990; King, W.A., *et al.*, 1991); the same workers have also shown that male and female embryos at equivalent developmental stages have similar cell numbers and show a similar rate of mitosis.

8.9.1 Advantages of gender preselection

A summary of possible advantages of sex ratio control in cattle is provided in Table 8.12, which is taken from an article by Foote and Miller (1971).

Table 8.12. Possible practical value of sex ratio control in commercial cattle: advantages of one sex (from Foote and Miller, 1971).

Dairy cattle
More heifer progeny from good cows as herd replacements for milk production, or getting heifers from better cows
More bull progeny for meat, especially from 'cull' cows
Ensuring birth of bulls as potential sires from top cow × sire
Ensuring birth of heifers when progeny-testing young bulls
Avoiding freemartins in multiple births
Beef cattle
More bull calves for meat via beef rearing
Ensuring bull progeny as potential sires from top cow × sire
Ensuring heifers from next best cows × top sires as future brood cows
Avoiding birth of freemartins in multiple births

Gender preselection in cattle could be of immense value in dairying because the female is the unit of production in the dairy industry. At present, the dairy farmer must breed about half the herd with semen from dairy bulls to obtain the necessary number of herd replacements, assuming a natural sex ratio of 50:50 and an annual replacement rate of 25%. As observed by Morrell (1991), if sexed semen were available for AI in the UK, then only the top-quality cows would be inseminated with X-bearing sperm and the remainder could be inseminated with Y-bearing sperm to produce bull calves for beef rearing. In beef production, a change-over to bull calves could be the means of improving growth performance and allowing increases in slaughter weights (Spedding, 1989). It is estimated that sexing embryos for beef production could increase saleable meat production by about 20% (Reynolds and Street, 1991). An exception to the rule of 'bulls for beef' might be the single-sex bred-heifer (SSBH) system of beef production envisaged by Taylor *et al.* (1985) and Rigby (1989); in this, the birth of heifer calves could be the means of maintaining the system on a continuous basis. When a dam is slaughtered shortly after her first calf is weaned, there is a marked increase in efficiency because the dam herself assumes the role of slaughter offspring and most of the conventional maternal overhead cost of producing a calf disappears by becoming part of productive growth. Such an SSBH system requires that each heifer must in turn wean a heifer calf, which will serve as her replacement. As to the meat quality of these heifers, provided slaughter occurs before 30 months of age, available evidence in Ireland suggests that there is essentially no difference in quality between once-calved heifers and maiden heifers (Keane and Harte, 1990).

8.9.2 Flow-cytometric separation of X and Y sperm

The technique which currently appears most likely to be of immediate value in the separation of bull sperm is that involving the use of flow cytometry (review by Mackenzie and Pinder, 1987). As in many other mammals, including humans, the Y sex chromosome in cattle is smaller than the X chromosome. It is for that reason that the separation of X-bearing sperm and Y-bearing sperm on the basis of their relative total DNA content is possible. Flow cytometry permits such separation after measurement of the relative DNA content, based on the fluorescence emission intensity after staining with a DNA-specific fluorochrome, such as Hoechst 33342 (Keeler *et al.*, 1983). Work on separating sperm by this approach has been reported from several centres in recent years (see Fig. 8.11 for a schematic diagram of a typical flow cytometer system).

Mill Hill (Medical Research Council (MRC) Institute) studies

In the UK, workers at Mill Hill were the first to use sperm sorted into X- and Y-enriched populations to inseminate small numbers of cattle (Morrell *et al.*, 1988); the offspring produced showed a statistically significant bias from the normal sex ratio. No evidence of teratogenic effects arising from the use

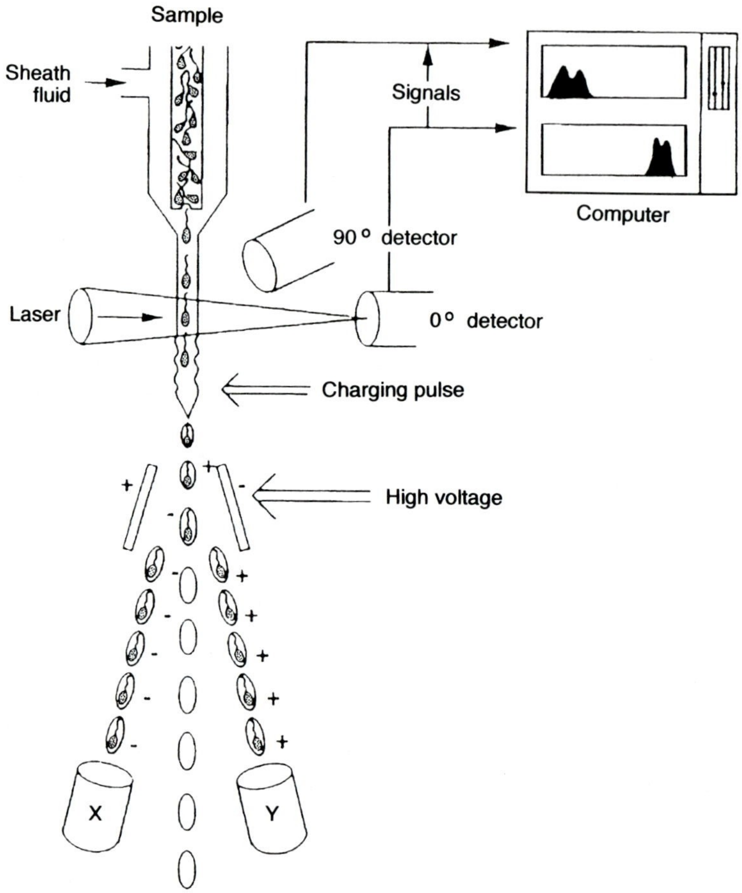

Fig. 8.11. Schematic diagram showing the principal components of a typical flow cytometer system.

of the H.33342 DNA-staining dye was found in studies with bull, sheep, pig and rabbit sperm, even where rabbits were carried through several generations (Morrell and Dresser, 1989, 1992).

Early attempts to apply sorted bull sperm in IVF, in collaboration with Animal Biotechnology Cambridge Ltd, are described by Morrell (1991); one difficulty mentioned at that time was the fact that fresh bull sperm required a longer capacitation time than frozen–thawed cells. The method of bull sperm separation developed at Mill Hill and described by Dresser *et al.* (1993) appears capable of giving a useful throughput of accurately sorted bull sperm.

Beltsville studies

In the USA, the flow cytometric determination of the proportions of X- and Y-chromosomes bearing sperm in bull semen samples obtained from several commercial and academic sources after attempted sexing was reported in a paper by Pinkel *et al.* (1985). A paper by Johnson and Pinkel (1986) described the modification of a laser-based flow cytometer to enable it to separate mammalian sperm more effectively. This instrument was further employed by the Beltsville workers, using stained dead sperm, as a means of verifying claims made for various sexing procedures (Johnson, 1987, 1988, 1989, 1990; Johnson and Clarke, 1988a, b). Continued refinement of the cell-sorting equipment enabled the Beltsville workers to separate viable rabbit sperm (86% purity for X-bearing sperm; 81% purity for Y-bearing sperm) and produce young after the insemination of does (Johnson, L.A., *et al.*, 1989a, b). A later report showed similar success in pigs (Johnson, 1992); although it was evident that porcine sperm survived the staining and sorting procedure, the pregnancy rates in sows were markedly below normal. Reduced sperm motility, attributable to increased sensitivity of pig sperm after flow cytometric treatment, is also reported by Rath *et al.* (1994).

Cambridge studies

Results from using bull sperm separated by the Beltsville procedure have been reported in several papers from Cambridge (Cran, 1992; Cran *et al.*, 1993, 1994; Miller *et al.*, 1993). In this procedure, fresh semen is stained with H.33342 for 1 h at 35°C before being put through the flow cytometer; however, the small number of sorted sperm (about 1 million 5 h^{-1}) means that the application of this method is limited to IVF. Under ideal conditions, sufficient sperm for 100 oocytes can be sorted in about 60 minutes; embryo yield has been about half (17% vs. 35%) that found using standard procedures (Cran *et al.*, 1994). When the sorted sperm were employed in routine IVF (Lu *et al.*, 1987b), subsequent sexing of the resultant embryos showed values in close agreement with those recorded in the analysis of the sperm. Transfers of embryos to nine recipient heifers in initial trials resulted in four pregnancies and the birth of three male and three female calves of the predicted sex; a second trial on commercial farms using frozen embryos and direct transfer has resulted in a pregnancy rate of 36% and 34 pregnancies.

French studies

A flow-cytometric method of separating X- and Y-chromosome-bearing bull sperm has been described by Metezeau *et al.* (1991); semen fractions were tested by molecular hybridization using X- and Y-chromosome-specific DNA probes.

Factors affecting efficiency of sorted sperm in IVF

It is clear, however, from the studies reported by Dresser *et al.* (1993) that there may be considerable variation in the fluorescence of H.33342-stained sperm, even those collected from an individual bull; the same authors note that dilution media, concentration of the DNA stain, temperature and duration of staining are all factors that may influence the staining pattern of bull sperm. There has been the need to employ fresh rather than frozen–thawed semen (Johnson *et al.*, 1994); this obviously places limitations on the application of the separation technology. There is also the problem of having a flow cytometer close to the IVF laboratory. In New Zealand, Gurnsey *et al.* (1994) showed that bull sperm can be transported for considerable distances before and after separation and give acceptable results when used in IVF.

According to McNutt and Johnson (1992), stained and sorted bull sperm are less effective in fertilization than normal, and embryonic development may also be retarded; they showed that certain proteins present on the sperm surface apparently change during processing, and concluded that such changes may influence sexed sperm function and sperm/oocyte interactions. There is an obvious need for staining and processing to be refined to overcome such problems. The advent of more efficient flow cytometers in the future may well enable much higher sperm separation rates to be achieved.

A review article by Johnson (1992) notes that, given the progress made in the field of gender preselection over the past decade, it is probable that 10 years hence will see a practical sexing procedure for livestock semen. In Babraham, now that it is possible to use flow cytometry to achieve a 90% enrichment of either X- or Y-bearing bovine sperm populations, studies of the surface characteristics of such sorted sperm are being pursued (Howes *et al.*, 1993). If specific surface antigens should be present on the two sperm types, it may be possible to use an appropriate immunological technique to separate sperm on a large scale (Coulter, 1992). In the embryo production laboratory, the use of intracytoplasmic sperm injection rather than IVF may eventually be one means of making the more effective use of limited numbers of accurately sorted sperm (see Section 4.11.4).

Gender preselection in humans

Although it has been recognized for some time that the efficiency of sperm-sorting techniques varies with species, it was believed that bullet-shaped sperm such as human sperm may be much less easily separated by flow cytometry than the paddle-shaped sperm of farm ruminants and pigs. However, despite the fact that the difference in the amount of DNA in human X and Y chromosomes is only 2.9% (in ruminants and pigs it ranges from 3 to 4.5%), Johnson *et al.* (1993) have now reported the successful separation of human sperm into populations of more than 80% purity for X-bearing sperm and 75% purity for Y-bearing sperm. Any practical application of such sperm sorting is likely to be in preventing X-linked diseases in families at risk. Sperm sorting by flow cytometry now joins a growing list of innovations in human reproductive technology which pose new ethical issues (Schulman, 1993; Shushan and Schenker, 1993).

8.9.3 Electrophoresis, thin-layer countercurrent distribution (TLCCD) and immunological methods

Other methods under examination in laboratories are outlined in Fig. 8.12. They include free-flow electrophoresis, which is based on sex chromosome-dependent charge differences on the outer surface of the sperm cells, as reported in Germany by Engelmann *et al.* (1988) for human sperm; when this procedure was used with bull sperm, some evidence of separation was also obtained (Blottner *et al.*, 1992). In Japan, an attempt to use free-flow electrophoresis to separate bull sperm was apparently unsuccessful (Masuda, 1989). In mice, using sperm from the cauda epididymidis, Ishijima *et al.* (1992) in the same country employed free-flow electrophoresis to separate gametes into cathodic and anodic fractions, which were confirmed to be rich and poor, respectively, in Y-bearing sperm.

By using thin-layer countercurrent distribution (TLCCD), it is possible to detect small differences in cell surface characteristics. Cartwright *et al.* (1993) in Manchester were apparently able to show that ejaculated bull sperm could be fractionated into two discrete populations on the basis of non-charge-associated surface molecules. Using a Y-chromosome-specific DNA marker, it was found that one of these populations was enriched in Y-bearing sperm (about 80%). The majority of sperm appeared to be viable, but were not used in IVF in this study.

The H-Y antigen and Y-bearing sperm

The existence of a male-specific antigen, the H-Y antigen, was described in the mid-1950s by Eichwald and Silmser. Studies to determine whether or not anti-H-Y antibodies bind preferentially to Y-chromosome-bearing sperm have,

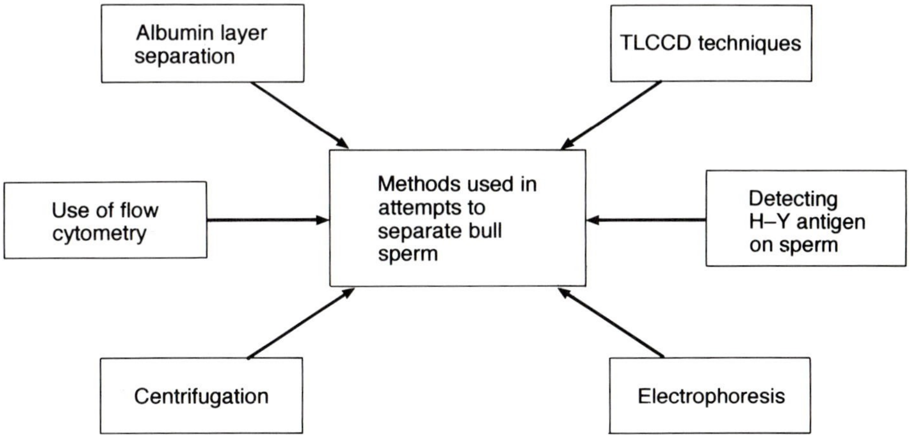

Fig. 8.12. Sexing by methods to separate X- and Y-chromosome-bearing sperm in cattle.

until recently, lacked a reliable method of estimating the percentages of X and Y sperm in the fractions examined. In the USA, Ali (1987) attempted to separate bull semen into X- and Y-bearing sperm using monoclonal H-Y antibodies. New Zealand and Australian researchers have also attempted to isolate H-Y-positive and H-Y-negative sperm populations (Bradley and Heslop, 1988; Bradley, 1989). With the advent of cell sorting by flow cytometry, however, it has now been suggested by Hendriksen *et al.* (1992, 1993) that the presence of the H-Y antigen on sperm may not be restricted to Y-bearing-sperm, as earlier reports had indicated.

In the USA, on the assumption that only Y-chromosome-bearing sperm possess cell-surface H-Y antigens, a novel, rapid immunomagnetic method has been reported for identifying sperm that contain these antigens (Peter *et al.*, 1993); it is claimed that a 98% purity of X-chromosome-bearing sperm was obtained.

8.9.4 Albumin separation of sperm

Work with human sperm cells (Ericsson *et al.*, 1973; Corson *et al.*, 1984; Ericsson and Ericsson, 1992) suggested that the male-producing sperm fraction could be enriched by up to 85% by allowing sperm into a modified Tyrode solution containing a high concentration of BSA; the mechanism of sperm separation was believed to be one in which the Y-bearing gametes swim faster than the X-bearing sperm into a fluid with high density and viscosity. The possibility that such density-separated human sperm might be employed in gender preselection led to the introduction of the albumin separation method into clinical practice. Beernink *et al.* (1993) mention 65 clinical practices (designated as sperm centres) in which gender preselection has been offered; these authors also provided evidence that separation of human sperm on columns of liquid albumin can affect the sex ratio at birth. A modified swim-up sperm preparation method as used by Check and Katsoff (1993) resulted in a high male birth-rate, although the mechanism involved, if genuine, remained unclear.

There have been those who believe that binding of BSA to bull sperm under suitable ionic conditions may amplify subtle differences in bull sperm thereby allowing X- and Y-bearing sperm to be separated (Upreti *et al.*, 1988). However, attempts to use Percoll density gradients as a means of influencing the sex ratio have met with no convincing success (Iwasaki *et al.*, 1988b; Masuda *et al.*, 1988; Hamano and Uchiyama, 1989; Suzuki, 1989).

However, it has been clearly demonstrated, by accurately identifying sex chromosome content, that techniques such as Sephadex filtration and albumin gradients do not separate human sperm on a chromosomal basis (Vidal *et al.*, 1993); the authors question the validity of similar protocols previously described and note the risks incurred when they are used to prevent the birth of male children in families affected by X-linked diseases. If the outcome of insemination with separated sperm does result in a high male birth rate, it may be that the procedure in some way reduces the viability of X-bearing sperm rather than enriching the Y-bearing population.

8.9.5 Checking validity of sperm-sorting procedures

For more than 20 years, quinacrine staining has been employed to detect the fluorescent Y body in human sperm; this fluorescent body has also been detected in the head of the Y-bearing sperm in cattle and other farm species (Ogawa *et al.*, 1987; Saito *et al.*, 1992). A more accurate estimation of sex chromosome content in human sperm is by way of double fluorescence *in situ* hybridization (FISH), as shown by Han *et al.* (1993); they found that double FISH allowed a more accurate assessment of sex chromosomal content than single FISH and quinacrine staining for Y bodies.

Chromosomal analysis was employed by Hamano and Uchiyama (1989) to identify X- and Y-bearing bull sperm processed by a Percoll density gradient method; metaphase plates prepared from hamster oocytes penetrated by the bull sperm showed the expected 1:1 ratio. A method of examining the chromosomes of bull sperm, using zona-free hamster oocytes, is also reported by Kovacs and Foote (1989). Monitoring sperm separation procedures by the use of Y-chromosome-specific probes is commonly reported (Schwerin *et al.*, 1991).

8.10 Embryo Sexing Procedures

The long-term prospects of gender preselection by sperm separation must certainly include the use of AI as a means of bringing sexing to the farm; embryo sexing, on the other hand, must inevitably involve the use of cattle ET technology. Some of the embryo sexing methods are outlined in Fig. 8.13.

The first success with gender preselection in cattle was a calf born on Christmas Day in 1975 in Canada after embryo sexing. Many thousands of sexed calves have been born since that time, almost all of them involving embryo sexing (Hare *et al.*, 1976; Hare, 1986; Seidel, 1988). In the laboratory,

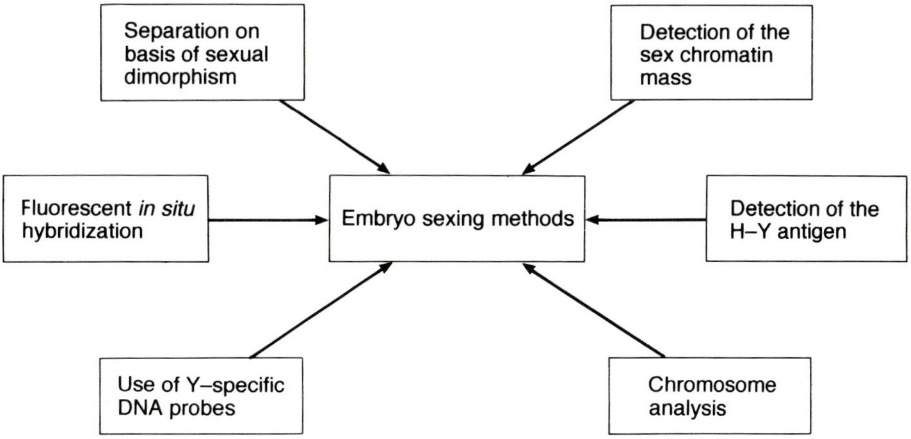

Fig. 8.13. Methods employed in the sexing of bovine embryos.

it is even possible at this stage to employ genetic engineering as a means of inducing embryos to develop into male young (Koopman *et al.*, 1991); when female mouse zygotes carrying the normal pair of XX chromosomes were injected with a small fragment of Y chromosome DNA containing the Sry gene, most developed as males with testes and male behaviour.

8.10.1 Early studies

Control of the sex ratio in the offspring of rabbits by the transfer of sexed blastocysts was first reported by Gardner and Edwards (1968); sexing was based on the presence or absence of the sex chromatin mass in a sample of trophoblastic cells removed from the embryos. In farm animals, where sexing on that basis is not feasible because of the granular nature of the nucleus, the first sexing was apparently that recorded with sheep in Cambridge after karyotypic analysis in 12-day embryos by Dane and Rowson in 1973 (quoted in Polge and Rowson, 1975). As mentioned above, Canadian workers were later to report success in cattle (Hare *et al.*, 1976); they found it possible to sex, also by karyotypic analysis, most 2-week-old cattle embryos before transfer to recipients. A serious disadvantage of the Cambridge and Canadian procedures was the fact that 2-week-old sexed embryos were totally unsuitable for commercial purposes; attempts to sex embryos at 7 rather than 14 days is not feasible because of the low mitotic rate making chromosomal analysis difficult. The number of metaphases per embryo is a critical factor in cytogenetic analysis of cattle embryos at an early stage of development (Iwasaki *et al.*, 1988a, b, 1989a; Kajihara *et al.*, 1988; Iwasaki, 1989; Filho *et al.*, 1992).

In embryo sexing, there are several possibilities that may be explored. According to White (1989), among the techniques employed at that time by commercial embryo transfer companies were the following: (i) cytological analysis of demiembryos; (ii) immunological detection of embryonic H-Y antigen; and (iii) the use of Y-specific DNA probes. The first of these involved embryo splitting, with one demiembryo sexed by karyotypic analysis and the other one retained for transfer, once its gender was known. The feasibility of this procedure is obviously dependent on the availability of facilities for carrying out the cytogenetic work. A review of non-invasive and invasive sexing procedures available to workers in the late 1980s was provided by Van Vliet *et al.* (1989).

8.10.2 Sexing by male-specific antigen

In the immunological detection of a male-specific antigen on the bovine embryo, it is believed that the histocompatibility antigen (H-Y antigen) is expressed in some species as early as the eight-cell stage; in cattle, the antigen is apparently expressed at the late morula stage. Several workers have recorded their attempts at sexing by identifying this male-specific antigen (White *et al.*, 1987a, b; Wachtel *et al.*, 1988; Booman *et al.*, 1989; White, 1989). In Dublin, working with sheep, Ware (1986) employed the H-Y sexing procedure with limited success; one problem was in the apparently reduced viability of the sheep embryos after sexing. For sexing using the H-Y approach, antibodies to H-Y are usually raised in rodents. The bovine embryos are exposed to H-Y antibodies, which bind to those that are male; a second antibody, usually with a fluorescent label, is then employed to permit identification under the microscope. A major disadvantage of this procedure is that the accuracy of sexing does not appear to exceed 80–90% (Van Vliet *et al.*, 1989).

Attempts to sex bovine embryos with such antibodies, including monoclonal and polyclonal antibodies (Booman *et al.*, 1989; Hossepian de Lima *et al.*, 1993) or to use them to inhibit blastocoele formation (Avery and Schmidt, 1989) have been described in the literature. The use of a male-specific antibody, raised in the rat, to suppress blastocoele formation in bovine embryos, is described by Utsumi *et al.* (1993) and Utsumi and Iritani (1993). When morula-stage embryos are cultured in the presence of this antibody, the development of male embryos is arrested whereas female embryos continue to develop to the blastocyst stage. The authors conclude that the technique is capable of further refinement. Even the possibility of using bovine freemartin serum as an agent for sexing cattle embryos has been explored (Goto *et al.*, 1993); in this, serum samples from certain freemartins inhibited development of morulae to the blastocyst stage and chromosomal analysis showed most of these embryos to be male.

8.10.3 Use of Y-specific DNA probes

The sexual differentiation of the bovine embryo is determined by the presence or absence of elements normally located on the Y chromosome. Detection of DNA sequences present on the Y chromosome may be the basis for a sexing procedure, if the sequences are repetitive and male-specific. Since the mid-1980s, the cloning of repeated sequences specific for the bovine Y chromosome has been possible and these DNA fragments have been employed in sexing attempts (Ellis *et al.*, 1988; Marquant-Le Guienne *et al.*, 1992a). The use of a biotinylated Y-specific probe enabled the sexing of bovine embryos within 30 h in the report of Leonard *et al.* (1987). This technique was subsequently used, in combination with polymerase chain reaction (PCR) technology, in field trials in France (Nibart, 1992; Nibart *et al.*, 1992, 1993); pregnancy rates in recipient cattle in excess of 50% were recorded by these authors. In the USA, Bondioli *et al.* (1989a) described the use of DNA probe technology in determining the sex of cattle embryos from a biopsy sample; although accurate, their method involved an 8-day delay and the use of radioactive isotopes for the detection of DNA hybridization.

Dual fluorescent in situ *hybridization (FISH)*

The availability of DNA probes for chromosome-specific repeated DNA sequences has made DNA–DNA *in situ* hybridization a feasible approach for detecting human cells with a Y chromosome (West *et al.*, 1987, 1988); sexing was possible using either the tritiated or the biotinylated Y probe. *In situ* hybridization can be employed to identify sex chromosomes in interphase and metaphase nuclei. A paper by Munne *et al.* (1993) describes a short FISH procedure using fluorochrome and digoxigenin-labelled DNA probes for application in human embryos; successful FISH analysis was achieved in most of the available blastomeres.

8.10.4 Sex determination using a DNA probe in combination with polymerase chain reaction

Although some functional genes are known to be present on particular regions of the mammalian Y chromosome, most of the chromosome is composed of non-functional repeated sequences. The PCR method is capable of identifying a particular repeated DNA sequence in a minute amount of sample material by specific amplification of the sequence. The advent of the PCR technique (Saiki *et al.*, 1988) led to a marked improvement in cattle embryo sexing methodology and it has been successfully applied to sexing in other species, including humans (Handyside, 1990; Handyside *et al.*, 1989, 1990).

An Australian company (AB Technology) was one of the first to offer a commercial service based on a rapid Y-chromosome-detecting test. According to Cole (1989), the method was highly accurate and could be carried out on the farm within 3 h. The technique was described as the rapid Y-chromosome-detecting (YCD) assay and details were presented in several papers (Herr *et al.*, 1990a, b; Herr and Reed, 1991). The YCD assay involves removal of a small number of blastomeres (two to ten), requiring, according to the authors, less than 2 min, and the subsequent amplification of the male-specific DNA sequences. A study by Hilgers and Herr (1993) examined reagents used in the culture and storage of bovine embryos for sexing; they concluded that it is advisable to avoid using products derived from animal blood and tissue (e.g. serum preparations).

Nivot and Liegeois (1990, 1991) reported work in which the YCD assay was employed over an 18-month period in France, Belgium, The Netherlands, Switzerland, Italy and Spain; 1554 bovine embryos were biopsied for sexing and acceptable pregnancy rates in recipients achieved. Other reports on PCR-based methods for sexing bovine embryos have come from several countries, including Germany (Schroder *et al.*, 1990; Agrawala *et al.*, 1992), Sweden (Setiabudi and Gustavsson, 1991), Finland (Peura *et al.*, 1991a, b), Japan (Higashi *et al.*, 1991; Utsumi *et al.*, 1992; Kameyama *et al.*, 1993), China (Zeng *et al.*, 1992), Belgium (Grisart *et al.*, 1993) and the USA (Kirkpatrick and Monson, 1993).

Bovine embryo sexing is now routinely used on farms in France, using the Institut National de la Recherche Agronomique (INRA) DNA probe (Leonard *et al.*, 1987) in combination with the PCR technique (Nibart, 1992; Nibart *et al.*, 1992, 1993). On average, five to eight cells are taken for sampling; it is recorded that 95% of biopsied embryos can be sexed with a 95% accuracy. A marked reduction in the percentage of recipients required to produce heifer replacements in the dairy herd is one of several advantages claimed for this sexing procedure (Nibart *et al.*, 1993). A mean pregnancy rate of 58% after the transfer of biopsied and sexed embryos is given by Esposito *et al.* (1993); the same authors show that only high-grade embryos can be successfully frozen after sexing.

Near-normal pregnancy rates after cervical transfer of biopsied bovine embryos was reported by Nibart *et al.* (1989) from their studies in France. In Germany, Roschlau *et al.* (1992) describe their experiences in sexing more than 2500 cattle embryos as part of ET activities; no adverse effect on pregnancy rate was recorded with fresh embryos but a 10% reduction occurred when embryos were frozen.

A report by Machaty *et al.* (1992) in Hungary showed that removing one cell from the 16–32-cell stage IVMFC-derived bovine embryo did not adversely affect its subsequent development *in vitro*. A further report from the same group showed that the sexing procedure had no significant effect on pregnancy rate (Machaty *et al.*, 1993).

8.10.5 Sex and early embryonic development

It has long been believed that phenotypic sexual differentiation in mammals is secondary to gonadal development. In the past decade, however, it has become evident in the bovine embryo that sexual dimorphism may occur even before activation of the embryonic genome (Yadav *et al.*, 1993); there may even be a relationship between embryonic growth rate and lifespan (review by Mittwoch, 1993). In mice, although male and female embryos commence cleavage at about the same time, males are more advanced by the blastocyst stage (Tsunoda *et al.*, 1985; Seller and Perkins-Cole, 1987). These early reports dealt with mouse embryos fertilized *in vivo*. In Finland, Peippo and Bredbacka (1992) showed that female mouse embryos grew faster up to the eight-cell stage, after which, at least in those cultured *in vitro*, the males grew faster. Valdivia *et al.* (1993) used IVC to synchronize the start of embryonic development accurately and concluded that the male mouse embryos developed faster than the female.

Some early evidence of sex-dependent developmental changes in cattle embryos produced *in vivo* was provided by workers in Denmark (Avery and Schmidt, 1987; Avery *et al.* 1989a, b, c, 1992a). However, when cattle embryos are recovered from superovulated donor cattle, the impact of the sex effect differences may be masked by the fact that ovulations probably occur over a period of several hours (Avery, 1989; Callesen *et al.*, 1992a, b). An early-ovulated oocyte fertilized by an X-bearing sperm (female embryo) might therefore be further developed than a later-ovulated oocyte fertilized by a Y-bearing sperm (male embryo). With IVF-produced embryos, such sources of variation would not be expected.

In Canada, studies reported by Tiffin *et al.* (1991) showed that total glucose metabolism in male cattle embryos was twice that in female embryos and increased between the morula and expanded-blastocyst stages. Although it is possible that such differences are related to sex-dependent developmental changes, the authors note that it is uncertain which is cause and which is effect.

Table 8.13. Sex-related differences in the development of cattle embryos produced by *in vitro* techniques (from Xu *et al.*, 1991b).

Sex	Morula/ early blastocyst	Blastocyst	Expanded blastocyst	Hatching blastocyst	Hatched blastocyst
Male	4	15	49	19	20
Female	18	21	36	12	3
Ratio (M/F)	0.22	0.68	1.36	1.58	6.67
χ^2	$P < 0.01$	$P > 0.05$	$P > 0.05$	$P > 0.05$	$P < 0.01$

Two groups, one in Denmark (Avery *et al.*, 1991, 1992b), the other in Canada (Xu *et al.*, 1991b, 1992d), duly reported that the apparent relationship between sex and developmental rate may be employed as a method for non-invasive sexing of IVMFC-derived bovine embryos. Data from the Canadian studies are shown in Table 8.13. Such results demonstrate that male embryos generally develop to a more advanced stage than females in the first 8 days after IVF. A further report by Xu *et al.* (1992b) recorded a pregnancy rate of 63% after ET and provided evidence of the normality of their sexual dimorphism system.

Elsewhere, Marquant-Le Guienne *et al.* (1991, 1992a) in France showed that in fast-developing IVMFC-derived cattle embryos there may be a preponderance of males and in slow developers a preponderance of females. The same authors showed that, as early as 3 days after IVF, the fastest-developing embryos were males. In the USA, there are even data showing that the male predominance of bovine IVMFC-derived embryos is evident as early as the two-cell stage (Dominko and First, 1993).

8.10.6 Gender preselection by X-chromosome-linked enzymes

The availability of effective procedures for the IVC of IVF-derived cattle embryos may provide novel opportunities for sexing the early cattle embryo. The early cleavage period, for example, is the only time in the life of the embryo when both X sex chromosomes are active. The products coded for by the active genes on the X chromosome are secreted in double the usual amounts and such differences may be detected by suitably sensitive assay procedures. The process of X chromosome inactivation in the bovine embryo has been identified by Fuente *et al.* (1993) to occur at the early blastocyst stage and to increase progressively as the blastocyst elongates.

A non-invasive method of sexing embryos, based on a colorimetric assay for glucose-6-phosphate dehydrogenase (G6PD) activity has been reported in mice (Womack, 1987; Williams, 1988; Miyano *et al.*, 1990a) and cattle (Iwasaki *et al.*, 1989b, 1991; Tiffin *et al.*, 1990, 1991); enzymatic activity was not sufficiently different to permit accurate separation of the sexes. An alternative approach to sexing mice embryos, using the technique of Monk (1987) for the measurement of the activity of the X-chromosome-linked enzyme, hypoxanthine phosphoribosyl transferase (HPRT), was reported by

Monk and Handyside (1988); of 15 embryos examined, 14 were correctly predicted on the basis of HPRT activity in a single blastomere.

8.10.7 Production of chimeric phenotypic males

Mammalian chimeras have been valuable in the study of physiology, behaviour and developmental biology (Papaioannou and Dieterlen-Lievre, 1984; Anderson, 1987; Picard *et al.*, 1990; Anderson *et al.*, 1991). There may, however, be ways of using bovine chimeras to influence the sex ratio in calves. At the time in the early 1960s when the first experimental chimeras were formed in mammals, little was known of cell mingling during development and even less about the cellular basis of sex determination (review by McLaren, 1984a). From the earliest reports of the birth of chimeric mice, it was evident that the sex ratio was heavily biased in favour of the male and intersexes were relatively uncommon. It is believed that this may be due to the gene for H-Y antigen production being located on the Y chromosome and influencing the development of the undifferentiated gonad into a testis.

Reports summarized by McLaren (1984a) showed about 15% of XX–XY chimeras in mice developed as females, 6% as intersexes and 80% as males. In sheep, Tucker *et al.* (1974) reported that two XX–XY chimeras were male at birth. In a subsequent report of Fehilly *et al.* (1984), 26/36 of chimeric lambs were male at birth. Various attempts to produce chimeric cattle embryos in Dublin have been reported (Kinis, 1990; Kinis *et al.*, 1990a, e).

In former Czechoslovakia, studies reported by Landa and Riha (1992) involved the disaggregation of 16-cell bovine embryos and the construction of 14-cell aggregates; two male and one female calves were born after transfer of such chimeras. The aggregation of eight-cell IVF cattle embryos (Holstein–Japanese Red) has been reported by Boediono *et al.* (1993a, b) in Japan, with the majority of aggregated embryos progressing to the blastocyst stage. It may be possible to use such technology to produce chimeric beef embryos, in the expectation that the sex ratio would favour the male.

8.11 Cloning and Its Potential Impact in Cattle Production

For farmers and animal scientists, there has always been a certain fascination with the possibility of producing copies, or 'clones', of outstanding beef or dairy animals in some number (see Fig. 8.14). At the time McLaren (1984b) reviewed work on NT in mammals, there was some doubt, based on studies in mice (McGrath and Solter, 1983a, b, 1984a, b, 1986; Robl *et al.*, 1986a), as to whether that particular approach to cloning was likely to prove successful. Fortunately, research efforts during the past decade have shown that NT can be achieved in cattle, albeit with nuclei from embryos rather than adult animals (reviews and papers by Robl *et al.*, 1986b; Marx, 1988; Wilmut and Smith, 1988; Robl and Stice, 1989; Prather and First, 1990a, b; Renard, 1990; Smith, L.C., 1990; Yang, X., 1991; Wilmut and Campbell, 1992; Wilmut *et al.*, 1992a, b; Wolfe and Kraemer, 1992). However, it must be emphasized

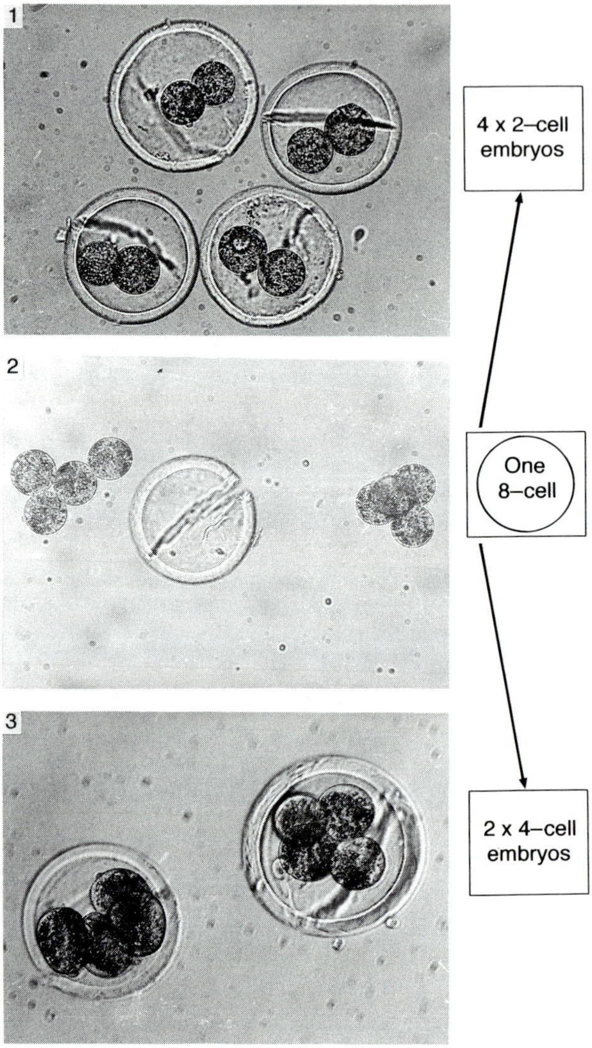

Fig. 8.14. Producing two or four embryos from one 8-cell embyro. There may be opportunities for employing micromanipulation to produce extra embryos for transfer when IVF is the means of producing embryos from cattle after ultrasonic-guided aspiration of follicles. Tubal embryos (8 cell, etc.) are not normally available after superovulation, but IVF provides a ready source of such early embryos.

that present cloning methods in cattle have a low success rate (1–6% overall efficiency) and the application of the technology in commercial practice is likely to be a few years distant.

8.11.1 Historical

The first to report experiments in which the nuclei of embryonic rabbit cells were introduced into unfertilized rabbit oocytes was Bromhall (1975), working in Oxford; although some development of embryos to the morula stage occurred, the data were limited. A decade was to pass before the first unequivocal evidence of successful mammalian cloning by NT appeared; this was a report of work in sheep (Willadsen, 1986b). This report was followed by considerable research activity worldwide in attempts to refine techniques and improve the success rate; in some countries, an attempt was made to commercialize the cloning procedure in cattle.

Cloning could be useful in dairy cattle by allowing genotypes of the highest genetic merit to be introduced directly into commercial herds, thus circumventing the traditional breeding process of upgrading. At the same time, appropriate progeny testing programmes are required to ensure that the stocks that are proliferated are genetically outstanding (Nicholas and Smith, 1983). In beef cattle, cloning could be the means of providing a slaughter generation with the most appropriate genotype for producing uniform carcasses (Smith, 1989). On a more limited scale, cloning in cattle could be employed to produce genetically similar multiple calves for research and testing. An obvious advantage with large-scale cloning is the fact that gender preselection is simultaneously achieved.

One of the earliest reports of an NT calf was that of Prather *et al.* (1987a) in the USA. In Ireland, the work of Kinis (1990) and Kinis *et al.* (1989, 1990b, d, 1991b) resulted in the birth of the first calf after NT in which both donor embryo and recipient cytoplast were produced by totally *in vitro* procedures. The Irish work was part of efforts directed towards the eventual production of genetically uniform and sexed beef embryos for cattle twinning applications in that country. In Japan, Mitani *et al.* (1993) have also reported the development of embryos to the blastocyst stage after NT, using *in vitro*-produced donor and recipient cells.

However, according to recent reports, cloning investigations in several countries have largely been suspended because of unexpected problems, including calf abnormalities and fetal oversize at birth (Willadsen *et al.*, 1991; Bondioli, 1992, 1993; Seidel, 1992; Hahn, 1993). The biological explanation for the 'large-calf' syndrome is not apparent. As observed by Ozil (1992), such problems may reveal an important feature of mammalian development; certain crucial regulatory events that normally occur at fertilization may be absent or otherwise inoperative after cloning procedures. What happens in the very early embryo may well influence developmental processes occurring much later in pregnancy. The cost involved in cattle NT research, particularly in carrying clones through pregnancy to full term, makes it necessary to consider using other, less costly, species. The rabbit has been suggested as an inexpensive biologically relevant model for such work (Yang, X., 1991; Yang, X., *et al.*, 1992d).

8.11.2 *Cloning by embryo splitting and quartering*

Work at Cambridge with two-cell sheep embryos was the first to show that each blastomere has the potential to develop into a normally organized blastocyst (Willadsen, 1979); the more limited developmental capacity of single blastomeres from four- and eight-cell sheep embryos was reported in a later paper (Willadsen, 1981; Willadsen and Fehilly, 1983). A technique was developed which resulted in the production of identical twins in both cattle

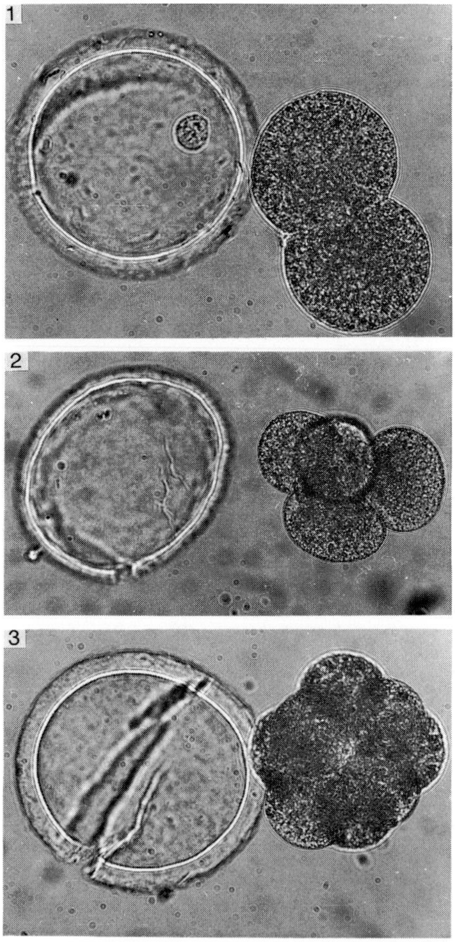

Fig. 8.15. Early-stage cattle embryos (two to eight cells) may be employed in micromanipulation research. The use of IVMFC-derived cattle embryos will enable many aspects of early embryo development to be studied. Such embryos cannot be recovered from live cattle without surgical intervention and considerable cost. The photographs show (1) blastomeres from a two-cell embyro; (2) blastomeres from a four-cell embyro; (3) blastomeres from an eight-cell embyro.

and the other farm mammals (see Fig. 8.15). Willadsen and Polge (1981) were the first to report the production of calves from embryos containing only a quarter of their original cell mass. However, the technique employed to produce demi- and quarter-embryos at that time, although of obvious interest to those in developmental biology, was too tedious and time-consuming to warrant consideration for possible commercial application.

By the early 1980s, the technology had moved to splitting sheep and cattle embryos at later stages of development. For cattle, this meant that embryos could now be flushed from a superovulated donor cow and split at the late morula/early blastocyst stage, without the surgical intervention previously required with early tubal embryos. This permitted the number of pregnancies obtained from a given collection of embryos to be markedly increased.

In more recent times, the need for rapid biopsy techniques to provide cells for embryo sexing in cattle has helped to improve the effectiveness of the splitting process (reviews and papers by Williams and Moore, 1988; Suzuki *et al.*, 1991a; King, W.A., *et al.*, 1992; Nibart, 1992; Godke and Rorie, 1993). Australian and other workers have described the 'splitting medium' which they employ (Herr and Reed, 1991; Lopes *et al.*, 1992). This is a modified Dulbecco's phosphate-buffered saline (PBS) medium, containing no protein and 200 mM sucrose, which results in the embryo being attracted to and held to the bottom of the Petri dish. After splitting, a protein solution is introduced to allow the demiembryos to float free.

The pregnancy rates with demiembryos produced from excellent/good-quality embryos can be expected to be 50% and above (Nibart *et al.*, 1988a; Seike *et al.*, 1989; Schwiderski *et al.*, 1990a; Gray, B.W., *et al.*, 1991; Kippax *et al.*, 1991; Nibart, 1992). Where two demiembryos are transferred, then pregnancy rates are usually comparable to those in recipients receiving intact embryos (Schwiderski *et al.*, 1990a; Holm *et al.*, 1991b).

In Japan, Utsumi and Iritani (1990) developed a special metal microblade for splitting blastocysts and obtained a 40% pregnancy rate after transfer of pairs of demiembryos into the ipsilateral horns of recipient Holstein–Friesians. In Brazil, however, Larocca *et al.* (1992) recorded the incidence of embryonic death to be significantly higher when both demiembryos were transferred to the ipsilateral horn than when transferred separately.

Attempts to produce calves from embryos containing only a quarter of their original cell mass have been reported from several countries, including China (Tan *et al.*, 1991, 1992; Tao *et al.*, 1991) and Denmark (Bredbacka *et al.*, 1992); embryos at the late morula/blastocyst stage have been used in these studies. The general experience is that the survival rate of such quarter-embryos is only half that found with demiembryos. There has been no report of calves being born from embryos containing less than one quarter of the original cell number. In sheep, Willadsen (1989) has described work in which five lambs were produced from single blastomeres taken from an eight-cell embryo, when the blastomeres were aggregated with blastomeres taken from earlier-stage embryos.

Use of IVMFC-derived embryos in splitting and quartering

For those who may be attempting to produce the maximum number of embryos for transfer, the linking of oocyte recovery from the live cow, IVF and embryo micromanipulation may have certain attractions. For example, when IVMFC embryos are available, there may be some advantages in micromanipulating these at the early cleavage stage (8–16 cells), rather than applying the conventional splitting technique at the later, blastocyst stage. There is, it has been shown, an inevitable loss of a proportion of cells in the conventional splitting process; such loss can be avoided by using early-stage embryos. Loskutoff *et al.* (1991a, 1993) showed that removal of half the total number of blastomeres from 2–16-cell IVMF embryos did not adversely affect development to the blastocyst stage. Particular care was necessary, however, in handling two-to four-cell embryos, which proved to be metabolically sensitive to cooling below body temperature.

The Canadian workers also showed that multiple calves may be produced from single blastomeres isolated from four-cell IVMF embryos; there was some evidence that pregnancy rates could be increased if these quarter-embryos were cotransferred with fresh trophoblastic vesicles.

At this stage, however, it is wise to stress the need for much further data upon which to base commercial usage of such micromanipulation techniques.

8.11.3 Developments in large-scale cloning technology

Although the technique of embryo splitting and quartering has been available for some years, as an effective means of increasing the number of calves of high genetic value, as noted above, it has obvious limitations. Despite much work in nuclear transplantation in amphibians, dating back to the early 1950s, it is only in recent decades that similar cloning approaches have been successfully applied in farm mammals. In Oxford, nuclei from morula-stage rabbit embryos were injected into unfertilized oocytes by Bromhall (1975); although subsequent development of embryos was limited, the work used many of the micromanipulation methods later to prove effective in NT in rabbits and farm species.

The work at Cambridge by Willadsen (1986a, b), in which he showed that it was possible to produce multiples (clones) by fusing a whole nucleated blastomere from a donor sheep embryo into an enucleated recipient oocyte, opened up an important new area of exploration in embryology and developmental biology. In the Cambridge technique, a simple slit was first made in the ZP of the sheep secondary oocyte at a point close to the first polar body. Enucleation of the oocyte was undertaken by aspirating the polar body and part of the ooplasm; using the same pipette, a nucleated blastomere was then introduced into the perivitelline space of the oocyte and fused into the ooplasm. Fusion of the blastomere and cytoplast was achieved by various means, including electrofusion. In this, alignment of the cell and cytoplast was achieved by alternating current (AC) and pulses of direct current (DC) used to cause fusion. The micromanipulation was carried out in the presence of a cytoskeletal inhibitor (cytochalasin B); using such an inhibitor, it was possible to remove the metaphase II chromosomes without piercing the ooplasm membrane.

According to Smith and Wilmut (1989) in Edinburgh, evidence was found of an increased rate of development in sheep NT embryos treated with cytochalasin B post fusion; in cattle, however, studies by Levanduski and Westhusin (1990) have apparently not shown this effect. In Madison, Sims *et al.* (1991) used calcium- and magnesium-free Tyrode/albumin/sodium lactate/sodium pyruvate (TALP)–HEPES (without cytochalasin B) during manipulation of donor and recipient cells. The same authors record a beneficial effect from using delayed cell fusion (18 h delay) after NT on subsequent embryo development.

The enucleated recipient oocyte (a cytoplast) appears to have the ability to reprogramme the donor nucleus and the reconstituted embryo commences development as though it is a recently fertilized oocyte (Stice and Robl, 1988). In sheep, and presumably cattle, donor nuclei appear to be reprogrammed more easily when the recipient ooplasm is derived from oocytes rather than from cleavage-stage embryos (Smith and Wilmut, 1988, 1989, 1990); the reason for this is still uncertain.

Nuclear reprogramming may involve the uptake of regulatory proteins by the transferred nucleus and it is believed that such proteins are released into the oocyte's cytoplasm when the metaphase II chromosomes are being removed in the enucleation process. Prather *et al.* (1989a, b, 1990, 1992) conclude that the somatic nucleus is remodelled in such a way as to resemble and behave like a pronucleus in a zygote. It is believed that this is accomplished by the exchange of proteins between the nucleus and cytoplasm.

The early Cambridge work suffered from the need to develop the cloned embryo (embedded in agar) in the sheep oviduct, due to the absence at that time of a suitable culture system (Willadsen, 1986a, b). Subsequently, however, it became possible to grow the cloned embryos in coculture systems and these in due course replaced the sheep oviduct. In Ireland, Kinis (1990) used a bovine oviductal cell coculture system in his work.

8.11.4 The use of in vitro*-matured oocytes*

Early in cattle cloning research, it became evident that any widespread commercial application of the technology would be very dependent on the cost of recipient oocytes. All the early cloning reports used, as recipient cells, metaphase II bovine oocytes matured *in vivo* after surgical collection from superovulated cattle (Prather *et al.*, 1986, 1987a; Bondioli *et al.*, 1989b, 1990a, b; Hackett *et al.*, 1989, 1990; Westhusin *et al.*, 1991). In some studies, when *in vitro* oocytes were employed, results were poorer than with those matured *in vivo* (Prather *et al.*, 1987a; Chesne *et al.*, 1991a, b). However, subsequent reports of *in vitro–in vivo* comparisons were to demonstrate convincingly the efficacy of IVM bovine oocytes for cloning when selected on the basis of certain parameters (Bondioli *et al.*, 1990a, b; Barnes *et al.*, 1991, 1993a, b). These include the duration of maturation and subsequent culture, the selection of oocytes on the basis of follicle size (> 3 mm) and the presence of a polar body.

8.11.5 Factors known to influence efficiency of cloning procedures

It is clear that a number of factors may influence the efficiency of NTs; some of these are outlined in Fig. 8.16.

Quantity of cytoplasm in recipient oocyte

In cattle, as in sheep, NT requires removal of a substantial amount of ooplasm from recipient oocytes to ensure that the metaphase chromosomes are taken out. Northey *et al.* (1991) showed that removing half the cytoplasm in bovine oocytes prior to fertilization led to a corresponding decrease in the total number of cells in the blastocyst. The same process would presumably occur in cloned embryos, where one third or more of the ooplasm is frequently removed. This reduced cell number may contribute to decreased viability in cloned embryos. According to Stice and Keefer (1993), oocytes only having a minimal amount of cytoplasm removed had 30% more cells than the NT embryos obtained with half of the normal amount of cytoplasm. In Dublin, Greising *et al.* (1994) found evidence of a higher developmental capacity in cloned embryos derived from recipients consisting of two cytoplasts rather than one (38% blastocysts vs. 14%); the larger available cytoplasmic mass may have been a contributory factor in this.

It should also be mentioned that the transfer of embryonic nuclei (with some embryonic cytoplasm) into recipient cells containing a large amount of foreign cytoplasm means that members of a given clone may not share a common cytoplasm; there is the potential for mitochondrial genes to differ. A report by Plante *et al.* (1992) describes a unique cytoplasmic marker which can be used to identify each individual animal within a clone.

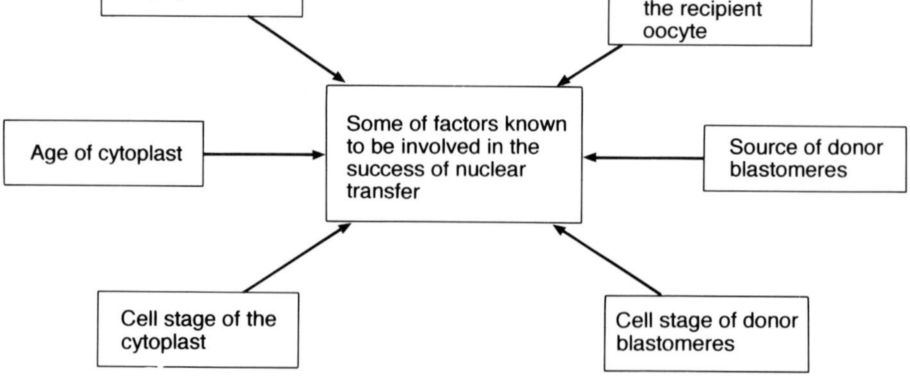

Fig. 8.16. Factors influencing the effectiveness of cloning procedures in cattle.

Selection criteria for IVM oocytes used in cloning

The general consensus of reports suggests that oocyte ageing beyond the time required for nuclear maturation (i.e. 24–26 h) is a general requirement for activation competence of IVM cattle oocytes (Ware *et al.*, 1989; Yang, X., *et al.*, 1991, 1993b; Chesne *et al.*, 1992; Heyman *et al.*, 1992b; Todorov *et al.*, 1992; Van Stekelenburg-Hamers *et al.*, 1992, 1993a, b; Takano *et al.*, 1993; Smith *et al.*, 1994). In France, Rao *et al.* (1993) concluded that there was an optimum stage between 32 and 40 h of age for recipient oocytes used for NT.

In Dublin, Kinis *et al.* (1989a) reported the least success with 36–48 h matured/cultured oocytes in comparison with those normally matured (24–26 h). The selection of oocytes showing the first polar body at 24 h is described in the report of Barnes *et al.* (1993a, b); such selection may effectively remove oocytes that have not shown a normal rate of meiosis. The same authors note that the size of ovarian follicle from which the oocyte is derived influences its subsequent developmental potential when used for cloning.

Workers at Cornell attempted to capitalize on the better enucleation rate of young oocytes (22–24 h) and the better activation of the aged oocytes (30–40 h); they developed a technique involving enucleation at 22–24 h and performed NT after incubating the cytoplast to a maturation age of 30–40 h (Yang, X., *et al.*, 1993a, b). According to the protocol described by Bondioli (1993), oocytes are matured for 20–22 h, cumulus cells removed (by vortexing), polar body identified and the oocytes returned to culture for an additional 18–20 h to gain activation competence.

Source of donor blastomeres

All blastomeres of a single bovine embryo are genetically identical. The genome of each blastomere is derived from replication through mitosis of the original single diploid genome constructed by the fusion of the male and female pronuclei. The use of donor embryos at advanced rather than at early cleavage stages has the advantage of larger numbers of blastomeres to work with. However, it is clear that other considerations apply. The problems of obtaining *in vivo* embryos at the 16–32-cell stage are dealt with in a paper by Hackett *et al.* (1993). Barnes *et al.* (1987) reported development of NT embryos to the blastocyst stage after fusion of nucleated blastomeres from four-cell to 32-cell donor embryos.

Subsequently, Bondioli *et al.* (1990a, b) and Marek *et al.* (1990) used 16–64-cell cattle embryos as nucleus donors without observing differences either in the fusion frequency or in the developmental potential of the transplanted embryos. In Japan, Ushijima *et al.* (1991) used 16-cell-stage bovine embryos and showed that their nuclei could be reprogrammed within the cytoplasm of IVM oocytes. Van Stekelenburg-Hamers *et al.* (1994) have shown that advanced nuclei from bovine embryos up to the 50-cell stage can support normal development after use in NTs; they found that nuclei from such embryos reverted to the equivalent of an earlier developmental stage when transferred to ooplasm.

Attempts have been made to preserve bovine blastomeres by freezing (Lucas-Hahn and Niemann, 1991a); single blastomeres from eight-cell embryos were frozen using 1,2-propanediol (PROH), either in an empty ZP or in the denuded state. On the basis of morphologically intact blastomeres after thawing, the presence of a ZP was found to be necessary.

A report by Navara *et al.* (1992), however, has shown that bovine embryos undergo a process of polarization at about the 32-cell stage, which is just prior to compaction. They found that, after NT, polarized blastomeres gave a much lower rate of development to the morula/blastocyst stage than did non-polarized blastomeres (27% vs. 3%). Prior to polarization, microvilli are evenly distributed over the entire surface of the blastomere and organelles are distributed throughout the cytoplasm; after polarization, microvilli are only to be found at the apical surface of the blastomere.

In view of such evidence on cell polarization, Keefer *et al.* (1993) sought to determine whether bovine inner cell mass (ICM) cells could be used as donor nuclei. Their study showed that ICM cells are pluripotent and after NT can give rise to embryos capable of normal embryonic and fetal development.

Effect of donor and recipient cell stage

During the recent decade, there have been considerable advances in the understanding of the cell cycle (see Murray, 1992). In cattle cloning research, a number of reports have now drawn attention to the effect of donor cell cycle stage on the development of embryos after NT. For high rates of development, the donor cell should be in the G1 phase (Collas *et al.*, 1992a, b; Campbell *et al.*, 1993); the short duration of that phase led Collas *et al.* (1992a, b) to develop a method for synchronizing sheep blastomeres in G1. These workers found that the morphology of prematurely condensed G1 and early S chromatin has a minor influence on the chromosome constitution of manipulated embryos. The morphology of late S chromatin, however, can effect the chromosome constitution in embryos and may account for reduced development of nuclear transplant embryos when late-S-phase donor nuclei are employed. There are certain agents that can be employed to bring blastomeres to the most appropriate stage of the cell cycle.

Bovine blastomeres were synchronized using demecolcine or nocodazole in studies reported by Stice *et al.* (1993); those synchronized to the midportion of the cell cycle and fused to aged (42 h) oocytes developed to the blastocyst stage at higher rates than non-synchronized blastomeres. Nocodazole was also employed by Techakumphu *et al.* (1993) to induce temporary metaphase arrest without this impairing the viability of the blastomeres in NT studies. Demecolcine and nocodazole are agents that can induce cell arrest at metaphase by depolymerizing microtubules.

The development of bovine NT embryos was improved by fusion of the donor blastomere into activated oocyte cytoplasm in the studies of Stice and Keefer (1992). It was apparent that, when donor blastomeres of unknown cell cycle stage were used, activation of enucleated oocytes prior to fusion resulted in enhanced development of NT embryos. In the studies of Barnes *et al.* (1993b), further evidence was presented to show that cell cycle stage synchrony between donor nucleus and recipient oocyte is critical for the development of the NT embryos. These workers found that S-phase nuclei transferred into S-phase cytoplasm resulted in embryos with normal chromosome content and in high rates of development to the blastocyst stage.

Identifying enucleated recipient oocytes

Chromatin in oocytes and embryos can be readily visualized under the fluorescent microscope when they are stained with Hoechst 33342, a dye which specifically binds to the adenine and thymine bases of DNA. The employment of such fluorescent staining to identify enucleated demioocytes used in NT in cattle is dealt with in several reports (Westhusin *et al.*, 1990, 1991, 1992; Barnes *et al.*, 1991; First and Prather, 1991; Bondioli, 1993); results suggest that there should be no difference in fusion rates or embryo development when stained or unstained demioocytes are used as recipients. It is important that oocytes should be viewed under ultraviolet (UV) light just long enough (less than 10 seconds) to determine whether metaphase chromatin is present (Bondioli, 1993).

The H.33342 stain can also be employed to check the cytoplasm withdrawn from the oocyte for the presence of the metaphase II chromosomes (Van Stekelenburg-Hamers *et al.*, 1993a, b).

According to Westhusin *et al.* (1992), significantly fewer embryos produced by NT to nucleated demioocytes develop to the morula/blastocyst stage (10% vs. 23% in enucleated oocytes). In Dublin, Kinis (1990) recorded that 15% of enucleated oocytes developed to the morula/blastocyst stage after NT whereas none of the nucleated oocytes reached that stage.

However, oocyte enucleation requires considerable manipulative skill and it is likely that the procedure also removes important cytoplasmic components which may reduce cytoplast viability. For such reasons, workers in Cambridge and Quebec have examined non-invasive methods of enucleation (Fulka and Moor, 1993; Fulka *et al.*, 1993; Smith, 1993). They found that the use of UV irradiation induced undesirable cytoplasmic changes in bovine oocytes; a chemical enucleation method, however, was found to be highly effective in mouse oocytes but less so in cattle. There is scope for much further work in this area.

8.11.6 Blastomere–cytoplast fusion and oocyte activation

Fusion of donor nuclei to recipient oocytes can be undertaken using viral methods (usually the inactivated Sendai virus) or electrofusion. In both sheep (Willadsen, 1986a) and cattle (Robl *et al.*, 1987) electrofusion appears to be much more efficient than using the Sendai virus. In the rabbit, Fissore and Robl (1992) report that electrical pulse parameters and extracellular Ca^{2+} concentrations can be used to influence intracellular Ca^{2+} levels and optimize oocyte activation rates. A report by Long *et al.* (1991) deals with the effect of DC voltage and pulse duration on electrofusion and cell lysis in NT in cattle.

The karyoplast–cytoplast fusion in cattle work is usually carried out in a chamber attached to the electrofusion apparatus and in a non-electrolyte fusion solution (e.g. Zimmermann cell fusion medium; see Table 8.14). The fusion chamber is a microscope slide with electrodes on its surface set at a gap varying from 0.2 to 1.0 mm. Prior to fusion, oocytes are either aligned mechanically or electrically using AC and fusion is induced by DC.

Table 8.14. Formula of the Zimmermann mammalian cell fusion medium.

Compound	Mol. wt	$g\ l^{-1}$	Molarity
Sucrose	342.3	95.84	0.28 M
$Mg(C_2H_3O_2)_2 \cdot 4H_2O$	214.5	0.107	0.5 mM
$Ca(C_2H_3O_2)_2$	158.2	0.016	0.1 mM
K_2HPO_4 (anh)	174.2	0.174	1.0 mM
Glutathione	307.3	0.031	0.1 mM
Bovine serum albumin pH 7.0	25,000	0.01	0.01 mg ml^{-1}

Both fusion of the donor blastomere to the recipient oocyte and activation of the oocyte are crucial steps in NT. Both events are usually initiated by applying an electrical pulse to the cells; this pulse is believed to cause the formation of pores, or at least a destabilization of the cell membranes which mediate cell fusion, and, by an influx of Ca^{2+} into the cell, oocyte activation. Robl *et al.* (1992) suggest that the most effective electrical activation stimulus would be one which mimics the Ca^{2+} response after sperm–oocyte fusion. In this, transient increases in Ca^{2+} results in cortical granule exocytosis and initiates resumption of meiosis, with consequent extrusion of the second polar body.

It is evident that electrostimulation may induce a form of partial activation but is not always capable of eliciting the full range of biological changes associated with fertilization. As recorded by Moor *et al.* (1992a) and Sun and Moor (1992), sperm penetration causes calcium oscillations to persist for several hours, whilst an individual electrical stimulation induces only a single transient increase. It is a matter of developing approaches which can reproduce the pattern of the calcium transients associated with normal fertilization in reconstituted bovine oocytes.

Activation of bovine oocytes by way of other approaches has been reported. The synergistic effect of a calcium ionophore (A23187) and a protein inhibitor (cycloheximide) was found to be highly effective, regardless of the age of the oocytes (Presicce and Yang, 1993; Shi *et al.*, 1993). A comparison of ethanol and electrical activation was made by Clement-Sengewald *et al.* (1993); they concluded that electrical pulses alone are sufficient to activate aged oocytes. Activation of the recipient cytoplast several hours prior to receiving the donor nucleus has been a method employed by several workers (Kono *et al.*, 1993; Ushijima and Eto, 1993).

The absence of a sperm-derived aster in the NT oocyte may be a factor contributing to decreased viability of cloned embryos; although the exact function of the aster is not known, it is believed to be involved in organizing the early events of the first cell cycle (Pinto-Correia *et al.*, 1993).

Effect of temperature

Experiments reported by Powell and Barnes (1992) have demonstrated that IVM and aged bovine oocytes are activated by room-temperature conditions (24–26°C) preceding NT. The authors note that their data established for the first time an additional variable to be controlled when NT is attempted. These authors subsequently conducted all manipulations at 37–39°C, with enucleation at 19–21 h of maturation and fusion occurring at 44–45 h of maturation. The costs involved in equipping a micromanipulation laboratory may be considerable (Fig. 8.17).

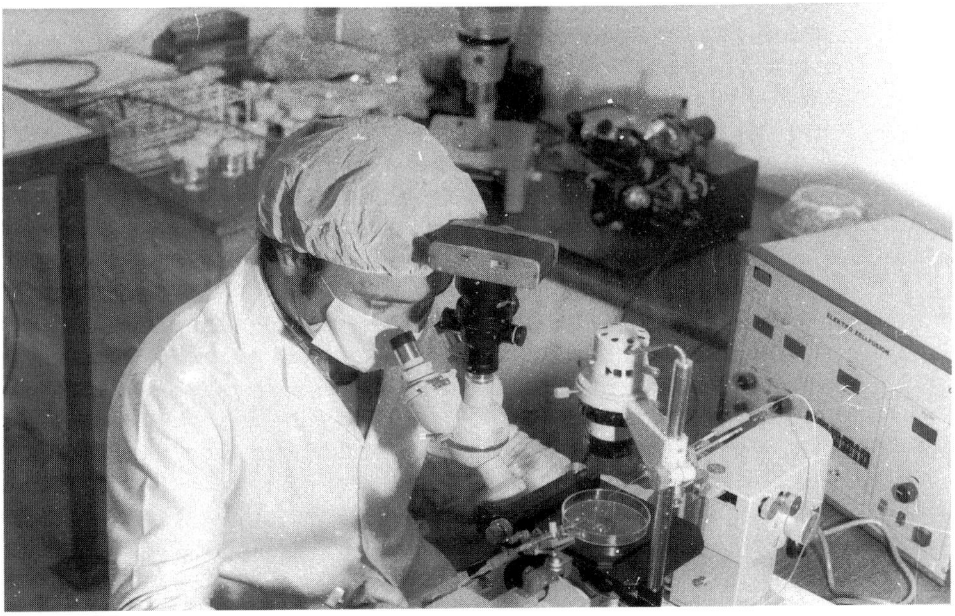

Fig. 8.17. Carrying out nuclear transfer in the laboratory with the Leitz Micromanipulator.

8.11.7 *Pregnancy rates and calf birth weights with cloned embryos*

It is now clear that a major limiting factor in the commercial application of NT procedures is the pregnancy rate achieved with cloned embryos. Using ovulated oocytes as recipient cells and blastomeres from 16–64-cell embryos as donor cells, Bondioli *et al.* (1990a) reported a pregnancy rate of 22.5% for 463 cloned embryos transferred to recipient cattle. A total of 302 transfers was involved in studies reported by Willadsen *et al.* (1991); 42.4% of recipients were pregnant at 35 days. Of 70 NT embryos transferred to 35 recipients in work reported by Yang, X., *et al.* (1993b), only one calf was eventually born.

The importance of obtaining accurate information about the occurrence of embryonic loss in cloned embryos was stressed by Heyman *et al.* (1992a, 1993); of 30 recipients in their study, only nine animals (30%) remained pregnant at 3 months. A pregnancy rate of 15% was reported by Van Stekelenburg-Hamers *et al.* (1993a, b) but involved only 13 recipients. Small-scale studies in Dublin with NT embryos resulted in a pregnancy rate at 60 days of 44% (Kinis, 1990). As observed by Bondioli (1992), it is highly unlikely that commercial marketing of cloned embryos will become a reality until current pregnancy rates are markedly improved.

Fetal oversize

Several exceptionally large calves (up to 155 lb) occurred among the 100 or so births reported by Willadsen *et al.* (1991). There is, however, no mention of instances of fetal oversize in the calving reports of Bondioli *et al.* (1990a, b). A review by Seidel (1992), on the other hand, states that 20–30% of cloned calves are larger than normal (up to twice normal size). Such information has been used by some to indicate that current cloning procedures may result in adverse effects in later embryonic life (Ozil, 1992). However, there have been several small-scale studies in which calves have been born without evidence of fetal oversize or congenital anomalies (Kinis, 1990; Heyman *et al.*, 1993; Stice *et al.*, 1993). Much further evidence is required before an accurate assessment of the seriousness of the large-calf syndrome can be made.

In sheep, data have been reported showing that the IVC of embryos from the zygote stage for either 3 or 5 days can result in extended gestation length and increases in birth weight and lamb mortality (Walker *et al.*, 1992a, b); it is suggested that such anomalies may be due to inappropriate activation of the embryonic genome induced by deficiencies in the culture medium.

8.11.8 Producing unlimited numbers of identical young

As noted by Stice (1992), the ultimate goal for a commercial embryo cloning project is likely to be the production of an unlimited number of young from an original high-genetic-quality embryo. One way would be multiple-generation cloning. In this, the NT embryos are themselves used as nuclear donors to produce the next generation of identical embryos. Data reported by Stice *et al.* (1991) indicated that multiple-generation embryos had a lower fusion rate and in some cases a lower developmental rate than NT embryos from original embryos. Westhusin *et al.* (1991) have also produced multiple identical calves after frozen storage of embryos and serial recloning. In France, Rao *et al.* (1992), working with rabbits, showed that it was possible to combine freezing of cloned embryos with subsequent recloning to produce the required number of identical young from selected parents.

Use of embryonic stem cells

The establishment of embryonic stem (ES) cell lines in mice has already been reported (Evans and Kaufmann, 1981; Wells *et al.*, 1991) and it would be of great value, both for large-scale cloning and for genetic engineering, if the same could be done in cattle. It would be possible to use such stem cells as donors of nuclei for transfer to enucleated oocytes; the number of clones would then become practically limitless. Attempts to isolate pluripotential bovine embryonic cells have been reported from several countries, including Germany (Schellander *et al.*, 1989a; Strelchenko *et al.*, 1991), the UK (Evans *et al.*, 1990), The Netherlands (Weima *et al.*, 1991, 1992) and the USA (Ware and First, 1988). Attempts have also been made to isolate stem cells from rabbit, hamster, pig, goat and sheep embryos (Handyside *et al.*, 1987; Anderson *et al.*, 1990; Flechon *et al.*, 1990; Galli *et al.*, 1990; Notarianni *et al.*, 1990; Piedrahita *et al.*, 1990; Giles *et al.*, 1991; Hochereau-de Reviers and Perreau, 1991, 1992; Meinecke-Tillman and Meinecke, 1991; Talbot *et al.*, 1991; Niemann and Strelchenko, 1994).

The first real success in the use of stem cells, including several pregnancies, appears to be that reported by Sims and First (1993). In this, cell lines were derived from immunologically isolated ICM cells and became established; NT clones were produced by fusing the ES-like cells into enucleated oocytes using polyethylene glycol as the fusion agent. Other studies in Wisconsin have succeeded in establishing pluripotent cell lines from both morula-stage and blastocyst cattle embryos and in producing NT embryos from them (Stice *et al.*, 1994; Strelchenko and Stice, 1994); cell lines did not change in morphological appearance over 9–12 months of IVC.

8.11.9 Future developments in cloning research

According to Yang, X. (1991), the multistep nature of cloning technology, with current levels of efficiency, means that no more than 1–6% of recipient oocytes are likely to give rise to a calf; this assumes an efficiency of oocyte enucleation of 70–80%, fusion 70–80%, activation 70–80%, development to morula/blastocyst 20–30% and survival to term 20–30%. As with the *in vitro* production of cattle embryos, the ongoing research in many laboratories around the world in cloning is likely to refine the technology to the stage at which it is commercially relevant (Fig. 8.18). NT studies will also aid in elucidating parameters involved in the transition from maternal to embryonic control in the early embryo (Telford *et al.*, 1990). Much useful cloning work has been conducted with the rabbit as the model (Collas and Robl, 1989, 1990; Collas *et al.*, 1991; Yang, X., *et al.*, 1992). It appears that the developmental abilities of reconstituted rabbit embryos after NT differ from those observed in the mouse and rat and follow the developmental patterns of sheep and bovine NT embryos (Modlinski and Smorag, 1991).

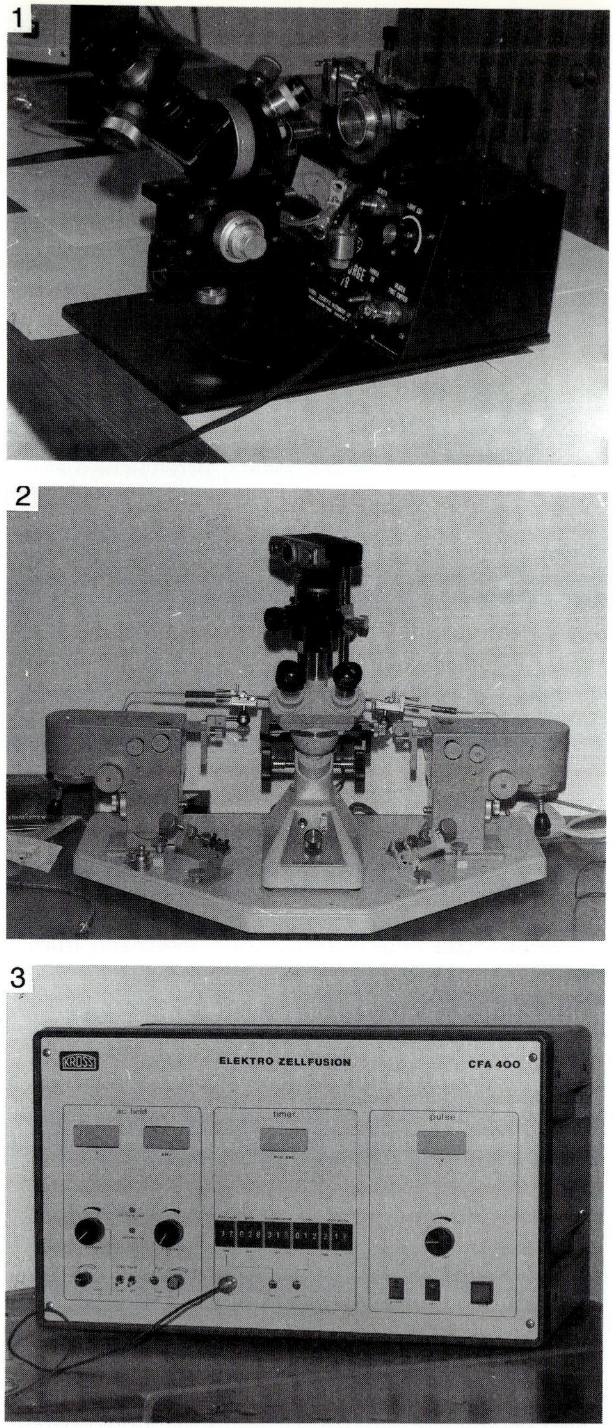

Fig. 8.18. Some of the equipment used in nuclear transfer studies in cattle. (1) An MF-79 Micro-Forge, used in the preparation of micropipettes; (2) Leitz Micromanipulator for the manipulation of oocytes and embryos; (3) the electropulse generator used for effecting cell fusion (Elektro-Zellfusion CFA, Kruss, Germany).

8.12 Production of Transgenic Cattle

The elucidation of the structure of DNA, the genetic material, by Watson and Crick in Cambridge in 1953 heralded the start of an era of biochemical research in cellular biology that has steadily gathered momentum during the past four decades. The application of genetic engineering in animals and plants is certainly likely to be a factor in expanding the range of options available in the twenty-first century to meet global food supplies (see McConnell, 1986; Petters, 1986; Seidel, 1986; Heap *et al.*, 1992).

The technology that permits stable and transmissible integration of cloned DNA segments into the mammalian genome was first employed in the mouse (Gordon *et al.*, 1980). Although farmers have for centuries manipulated the genetic constitution of cattle, usually in attempts to make them more productive, it is only in the last decade that developments in recombinant DNA technology have enabled the molecular biologist to isolate and modify genes in such a way as to allow novel constructs to be introduced into the bovine genome. Cattle that integrate this recombinant DNA into their genome are known as transgenes. Economically important traits in cattle, such as disease resistance, growth rate, feed conversion efficiency and milk yield, are controlled by numerous genes, and it will be some time before these are identified and become amenable to control. In the meantime, however, considerable progress can be made in the use of those genes that are available (reviews by Lathe, 1985; Lovell-Badge, 1985; Ebert, 1989; Clark, 1990; Clark *et al.*, 1990; Muller, 1990; Ebert and Schindler, 1993; Kelley *et al.*, 1993).

8.12.1 Historical

In the early 1980s, only a few dozen scientific papers concerning transgenic procedures in mammals had been published; a decade later, the number of such publications runs to several thousand. The first reported transgenic mammals were mice, produced by microinjecting a recombinant DNA construct into the pronuclei of zygotes and (see Fig. 8.19) subsequently transferring the injected embryos to recipients. It can be expected that 1–5% of microinjected mouse zygotes will develop into transgenic animals; the frequency of success has usually been markedly lower in farm animals (Hammer *et al.*, 1986; Pursel *et al.*, 1989; Brem, 1992). In one study in the USA involving 3398 DNA-injected bovine embryos, the overall efficiency of DNA transfer was 0.2% (Bondioli *et al.*, 1989b). Gene constructs have, however, been introduced with varying degrees of success into most species of food animal at this time, including cattle, sheep, goats, pigs, rabbits, chickens and fish.

However, before transgenic farm animals become a common feature of the livestock industry, a number of formidable obstacles must be overcome (reviews and papers by Wagner *et al.*, 1984; Miller *et al.*, 1986; Simons and Land, 1987; Smith *et al.*, 1987; Jaenisch, 1988; Pursel *et al.*, 1989; Roschlau *et al.*, 1989; Ward *et al.*, 1989; McWhir *et al.*, 1990; Wilmut *et al.*, 1990a, b; Ward and Nancarrow, 1991; Wilmut, 1991; McEvoy, T.G., *et al.*, 1992; Peura *et al.*, 1992a, b; Wall and Seidel, 1992; Wall *et al.*, 1992). In pigs,

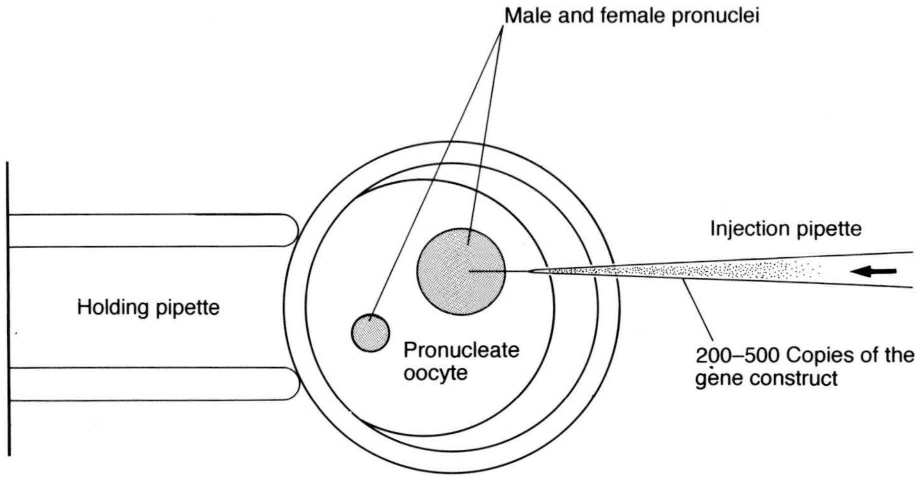

Fig. 8.19. Method for microinjection of DNA into cattle egg. Gene transfer by direct transfer of gene construct into the male pronucleus of the bovine pronucleate oocyte.

for example, although it has been possible to stimulate growth and enhance food conversion to protein, this has sometimes been at the expense of the general health of the animal (Pursel *et al.*, 1990). In transgenic sheep, health problems have also been recorded in Australia (Nancarrow *et al.*, 1990) and the USA (Rexroad *et al.*, 1991). The need is clearly to develop tissue-specific genes while ensuring that any undesirable side-effects are eliminated.

In biomedical research, on the other hand, transgenic laboratory animals (mice/rats) are now routinely used in large numbers in basic studies aimed at exploring the function of specific genes. The genetic modification of animals is, however, a controversial subject and there is opposition to work in this area of biotechnology (Vines, 1992); a major concern of workers must be in ensuring that studies in recombinant DNA technology do not unnecessarily compromise animal welfare (Campbell, 1990).

8.12.2 Transgenic modification of milk quality in cattle

Milk from the dairy cow is a unique source of certain proteins, such as those in whey (α-lactalbumin and β-lactoglobulin (BLG)) and in casein. The options and prospects for altering milk composition in dairy cattle have been discussed by Gibson (1988). In New Zealand, Thompson *et al.* (1989b) have reported the initiation of a research programme at Ruakura to develop transgenic cattle with modified milk composition. The impact of such technology on animal production in New Zealand is discussed by Fennessey (1990). A review by McEvoy, T.G., *et al.* (1992) draws attention to genetic engineering studies designed to enhance the suitability of milk as a food for human consumption. In countries where the dairy industry plays a major role in the agricultural economy, modifications of milk protein so as to provide novel forms of cheese may eventually be one commercial possibility, using gene transfer. Milk contributes significantly to the protein intake of humans in the developing countries and it may be possible to improve its nutritional value as a food by reducing its lactose and BLG content where the population is lactose-intolerant. In the developed countries, modifications of milk content to give a reduced saturated fat content may be a further way in which the new technology can be employed to the advantage of human health and welfare.

Expression of pharmaceutical proteins in milk

The use of transgenic animals for the production of human proteins of biomedical importance is one area in which considerable research emphasis is currently concentrated (Moffat, 1991; Wilmut *et al.*, 1991; Wright *et al.*, 1991). The strategy is to target expression of the appropriate gene to the mammary gland and then harvest the product from the milk. Mice have been widely used as model systems to evaluate expression of foreign genes in the mammary glands. It has been shown by Edinburgh workers (Simons *et al.*, 1987, 1988, 1990; Harris *et al.*, 1989) that the composition of mouse milk can be dramatically changed by the expression of the sheep protein gene for BLG. In further work, a gene construct (designated AATB) was evaluated in mice and then introduced into sheep. Among the female sheep produced was one (Tracy) capable of producing more than $30\,g\,l^{-1}$ of human plasma α_1-antitrypsin in her milk (Clark *et al.*, 1992); obtaining therapeutic proteins in this way has the potential to markedly reduce costs of production.

Although there may be those who see rats (Hochi *et al.*, 1992) and rabbits (Brem, 1993) as eminently suitable mammals for the mass production of pharmaceutical proteins by transgenic technology, the very considerable efficiency of the bovine mammary gland for protein synthesis should not be lightly discounted. A dairy cow is capable of producing 1 kg or more of endogenous protein daily and the milk can obviously be taken from the animal in the conventional way without compromising welfare. Clearly, dairy cattle offer the advantage of producing markedly higher yields of milk than can be obtained from sheep or goats; animal welfare considerations in mammalian species that are quite unaccustomed to routine milking should not be overlooked. Meanwhile, in The Netherlands, Krimpenfort *et al.* (1991) have already produced transgenic dairy cattle containing the bovine casein–human lactoferrin coding sequences. In the USA, studies on the production of transgenic goats have been reported by Ebert *et al.* (1991); one of these secreted human tissue-type plasminogen activator in her milk at a concentration of $3\,\mu g\,ml^{-1}$ and transmitted the gene to some of her progeny.

8.12.3 Gene transfer techniques

The transfer of cloned foreign DNA in such a way that all cells of the animal contain the foreign gene was first reported for the mouse by Gordon *et al.*

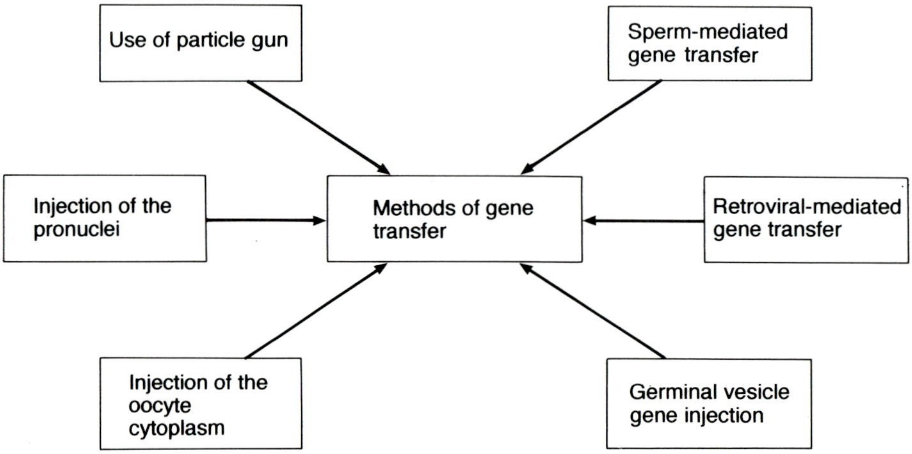

Fig. 8.20. Possible gene transfer methods for use in cattle.

(1980); this was by way of pronuclear injection. It became clear, in the following years, that foreign DNA can be stably integrated into the mammalian genome and can be inherited according to Mendelian principles. Some of the methods employed to incorporate foreign genes into the bovine genome are outlined in Fig. 8.20.

Gene transfer by pronuclear microinjection

Microinjection of several hundred copies of a gene construct into the pronuclei of the recently fertilized oocyte was the method successfully used in mice and a similar technique has been employed in dealing with cattle. The microinjection method takes advantage of the unique DNA processing events that apparently occur in the pronuclei, especially the male structure; the pronucleus provides the specialized nuclear environment for the incorporation of DNA sequences and for their inclusion in a functional chromosomal region. Work in Ireland resulted in a total of 66 microinjected bovine zygotes being transferred to the oviducts of recipient cattle (McEvoy *et al.*, 1987, 1990; Sreenan and McEvoy, 1989b; McEvoy, 1990); 11 recipients calved to produce 21 young, although no calf proved to be transgenic.

The ease with which nuclear microinjection can be achieved varies with species; in mice, it is relatively easy, whereas in cattle, because of the lipid content and opacity of the oocyte, it is difficult. For that reason, it is essential to employ procedures which will allow visualization of the pronuclei, such as the use of differential interference-contrast and phase-contrast microscopy (Kay *et al.*, 1991a, b), staining (Minhas *et al.*, 1984) or centrifugation (McEvoy *et al.*, 1990; Thomas and Seidel, 1991). Once injected, the development of the embryo to the transferable stage is usually required. In Poland, Jura *et al.* (1992) found that coculture with bovine oviductal epithelial cells (BOEC) was an effective method.

In mice, results of work with fresh and frozen pronuclear oocytes have suggested that the use of cryopreserved oocytes may reduce logistic problems in producing transgenic animals (Leibo *et al.*, 1991a, b).

A reduced developmental potential of zygotes after microinjection of DNA has been recorded in cattle (Gagne *et al.*, 1990, 1991a, b; Hyttinen *et al.*, 1992), sheep (Rexroad *et al.*, 1990) and goats (Selgrath *et al.*, 1990). Studies in Finland by Peura *et al.* (1993a, b) showed that handling of oocytes, mimicking gene transfer manipulation conditions, was not in itself responsible for such decreased developmental potential; microinjection of DNA conducted at the mid to late pronuclear stage did, however, decrease embryo development.

Germinal vesicle gene injection

Although the visualization of pronuclei in bovine oocytes can present difficulties, in most morphologically normal oocytes the germinal vesicle (GV) should be evident. With that in mind, Jura *et al.* (1990) in Poland attempted to microinject the GV of such oocytes with a buffer solution and then mature the cells. In Canada, Gagne *et al.* (1991b) reported some evidence to suggest that this GV injection approach may have some merit as a method of gene transfer.

Cytoplasmic gene injection

Several authors have examined the possibility of using direct cytoplasmic injection of DNA. Galli, C., *et al.* (1991), working with cattle, sheep and pigs, administered a DNA repetitive sequence at different stages of the cell cycle of oocytes and zygotes and analysed its metabolic fate; it is apparent that DNA metabolism varies among the farm ungulates and at different cell stages. The frequency of integrating DNA into the sheep genome by cytoplasmic injection is also dealt with in a report by Powell *et al.* (1992).

Sperm cell-mediated gene transfer

Although the first evidence of sperm-mediated gene transfer goes back more than 20 years to the pioneering studies of Ben Brackett and associates in the USA (Brackett *et al.*, 1971), it was the recent work of Italian researchers that brought this possibility to wider attention (Gandolfi *et al.*, 1989c, d; Lavitrano *et al.*, 1989, 1992a, b). They were able to demonstrate that mouse epididymal sperm cells behave as vectors transferring exogenous DNA into the oocytes during the fertilization process. It is apparent that exogenous DNA, as well as other negatively charged molecules, specifically bind to the equatorial and postacrosomal region of the sperm head and that such binding is largely reversible and may occur by the interaction of the nucleic acid with specific protein components. Optimal binding of the foreign gene occurs when both sperm motility and DNA concentrations are high.

Although such results were of great interest, they were also controversial, particularly when molecular biologists in the USA and elsewhere were initially unable to confirm them. Brinster *et al.* (1989) reported that studies in sperm-mediated gene transfer in their own and in seven other laboratories failed to produce a single transgenic mouse in more than 1300 births. At one time, the American scientific press appeared to view sperm-mediated gene transfer as in the same league as cold fusion in physics.

However, since that time, reports have appeared in several countries confirming the Italian findings and shedding light on likely causes of variation in results among laboratories. In Cuba, Castro *et al.* (1990) showed that the sperm of the mouse, bull, boar, buffalo, ram, goat and chicken were all able to bind foreign DNA; the authors suggested that such a capacity may be a general feature of all living sperm cells. In New Zealand, Peterson *et al.* (1990) reported on the adsorption of DNA to the sperm of bulls, rams and fallow deer; capacitated ram sperm were found to be more effective in binding DNA than uncapacitated sperm. In Canada, Gagne *et al.* (1991b) used electroporation to test the ability of bull sperm to carry foreign DNA into IVM bovine oocytes; they showed that foreign DNA was stably captured by the sperm and carried into the oocytes at fertilization. The authors concluded that sperm-mediated DNA transfer may be the means of providing transgenic cattle embryos by a less traumatic procedure than microinjection.

In Australia, Atkinson *et al.* (1991) were also able to demonstrate that bull sperm could bind and internalize foreign DNA; they concluded that the inability of many laboratories to validate this method of gene transfer was due to the relatively low frequency with which sperm took up the DNA. Other studies in the same country showed that ram sperm were able to take up linear DNA fragments (Imai *et al.*, 1992). In Ireland, Horan *et al.* (1991) have demonstrated that DNA molecules interact with pig sperm during incubation; the DNA remained closely associated, despite extensive washing.

In the mean time, Italian researchers continue their efforts in cattle and pigs to better define the conditions required for effective sperm-mediated gene transfer using an AI procedure (Lavitrano *et al.*, 1992a, b; Lauria and Gandolfi, 1993); these workers believe that further studies may lead to a reliable technique, capable of being employed both in basic and applied research. Some of the possible consequences of sperm acting as carriers of DNA are dealt with by Cohen (1991b).

Use of the particle gun

The particle gun was developed to deliver gene constructs into plant cells by firing DNA-coated tungsten microparticles into them. At Cornell, Hough and Foote (1990) have attempted to determine the effects of the particle gun on bull sperm viability and to separate sperm containing microparticles.

Retroviral-mediated gene transfer

Retroviruses are a subgroup of viruses with unusual properties; they have a genome of ribonucleic acid (RNA) rather than DNA, and are distinctive in being capable, by way of a process termed reverse transcription, of making DNA from RNA. Once a cell is infected by a retrovirus, the resultant viral DNA, after reverse transcription and integration, becomes a part of the host cell genome. Unlike DNA microinjection, integration of a viral gene apparently does not induce rearrangements of the host genome, other than duplication of several bases at the site of integration. Genes have been effectively transferred into mouse embryos by way of retroviral vectors. As yet, however, attempts to develop retroviral vector systems specific for farm animals have been limited to chickens. One difficulty in applying the retroviral approach in cattle has been the fact that retroviruses of high infectivity have not been identified in ruminants.

The first successful use of retrovirus-mediated gene transfer in the bovine embryo appears to be that reported by Kim, T., *et al.* (1993) in the USA, using murine leukaemia virus (MLV)-based replication-defective retroviral vectors encapsidated with gibbon ape leukaemia virus (GaLV) envelope proteins; these workers succeeded in infecting not only trophectoderm but also ICM cells of IVMFC-derived bovine embryos. The possibility of producing calves from cultured ICM cells by nuclear transfer or chimerization is noted by the authors; in comparison with conventional DNA microinjection, retrovirus-mediated transfer of DNA appears to justify some consideration.

In pigs, it has been shown that early pig embryos can develop normally after culture with cells producing a retrovirus (Jin *et al.*, 1991); the implications were that retrovirus-mediated gene transfer in that species was possible.

8.12.4 Gene transfer using embryonic stem cells

ES cells from mouse embryos are now routinely used in gene transfer work in that species. The main advantage in using such cells is that their use allows the directed modification of endogenous genes, known as 'gene targeting'. It is possible, by way of such gene targeting and subsequent germ-line transmission from chimeras, to make almost any desired change to the genome of the mouse (Robertson, 1991). As noted by Wilmut *et al.* (1992a), application of similar technology to farm livestock can only follow a clear demonstration of germ-line competence of stem cells from these animals.

In the stem cell approach, cells are initially isolated from the ICM of the blastocyst and grown in culture. This can be done in mice and the stem cells multiply without differentiating (Evans and Kaufmann, 1981). This provides large numbers of cells for use in gene transfer. The stem cells are subsequently introduced into blastocysts to produce chimeric embryos. Some of the cells may get incorporated into the gonads, resulting in germ cells carrying the new DNA. In attempting to derive ES cells from cattle, problems arise because exactly analogous stages do not exist in the embryos of mice and ungulates owing to differences in their embryonic development. The ICM of the bovine embryo undergoes little development, relative to the trophectoderm, for several days in the elongating blastocyst; this differs markedly from the way development of the ICM occurs in the mouse.

The greatest disadvantage of using ES cells in the conventional manner for the transfer of foreign DNA in cattle is the fact that the calves born after the injection of stem cells will be chimeras. In this species, it could, as noted by Wilmut *et al.* (1992a) be a matter of 6 years before calves are born carrying the transfer genes.

However, it has already been noted in reference to cloning by NT (see Section 8.11.8) that considerable progress has been made in recent years in establishing apparently stable stem cell lines from cattle embryos (Sims and First, 1993). There are many who believe that the ES cell route is likely to become the method of choice for genetic manipulation in the farm animals (review by Notarianni and Laurie, 1992). It may well be possible to combine NT technology with gene transfer into ES cells to produce non-chimeric transgenic cattle.

Germ cells as an alternative to embryonic stem cells

It should be noted that reports have appeared showing the long-term survival and proliferation of mouse germ cells in culture (Resnick *et al.*, 1992). If this can be similarly done with bovine germ cells and such cells prove capable of being an alternative to ES cells, this may open up a new route for transgenic technology in cattle. There is already some evidence from Australian studies of the pluripotency of bovine primordial germ cell (PGC)-derived cell lines maintained in long-term culture (Cherny and Merei, 1994). In Canada, Lavoir *et al.* (1993, 1994) have succeeded in isolating and identifying female germ cells from bovine gonadal cell suspensions between days 35 and 130 of gestation; the optimum period to isolate such cells appears to be between days 50 and 70.

8.12.5 IVMF-derived embryos for transgenic cattle production

The production of transgenic cattle by way of superovulating and breeding donors, with subsequent recovery of recently fertilized oocytes for DNA microinjection, involves procedures that are both labour-intensive and extremely costly (Ellington *et al.*, 1990g; Tervit, 1990). A caudal flank approach for the collection of oviductal-stage bovine embryos has been described by Wolfe *et al.* (1990). The obvious alternative to oviductal embryos is to employ zygotes produced *in vitro*; as well as being inexpensive, such zygotes can be obtained at a predictable cell stage. It is also possible, using the appropriate PCR techniques in the laboratory, to identify transgenic cattle embryos before the stage is reached for their transfer to recipient cattle (Sparks *et al.*, 1992a, b; Bowen *et al.*, 1993).

In view of the very substantial costs involved in acquiring and maintaining recipient cattle over the 9-month gestation period, the elimination of calves other than those likely to express the foreign gene is an important consideration. According to Thomas *et al.* (1993), the prohibitive expense is a factor contributing to the lack of literature dealing with the production of transgenic cattle and for the limited success reported in the few studies that have been conducted. The same authors deal with studies in which an *in vitro* system was used exclusively to produce transgenic bovine embryos of transferable quality.

In Texas, Leibo and Rall (1990) developed a practical method to diagnose the sex of bovine pregnancies by amniocentesis at the 8-week stage of gestation; they suggest that amniocentesis, with subsequent use of DNA probes specific for a gene construct, could be useful in identifying transgenic calves at 2 months rather than 9.

8.13 Use of Embryo Production Technology in Other Farm Mammals

The technology described for the cow (see Fig. 8.21) can, with certain modifications, be employed in each of the other farm mammals. Much further work within each species is required to bring the technology to the stage which it is currently at in cattle. However, much of the equipment in the IVF laboratory can be used for sheep and the other species as it is for cattle. In the matter of transferring the IVMFC-derived embryo to recipients on the farm, simple non-surgical procedures are available for the horse and buffalo and in the small farm ruminants (sheep, goat, deer) transfers are possible using a laparoscopic technique.

8.13.1 Sheep and goats

There are many countries in which sheep and goats feature prominently in the livestock industry; in these, new forms of reproductive technology, some of it based on *in vitro* production of oocytes and embryos, may possibly be employed to economic advantage (reviews by Naitana *et al.*, 1992; Thompson *et al.*, 1992b, c, d). The transgenic modification of sheep was first reported by Hammer *et al.* (1985) in the USA; at this point in time, Edinburgh in Scotland has become a well-recognized world centre of transgenic research in this species (Cherfas, 1990); among the successes recorded is the first sheep to carry human genes and the first demonstration that genetic defects can be corrected in an animal and the correction passed on to the offspring. Progress in gene transfer research in sheep is dealt with in recent papers (Rexroad and Wall, 1987; Walker *et al.*, 1990a, b) and reviews (Ebert and Schindler, 1993; Pursel and Rexroad, 1993).

The recovery of sheep oocytes and the production of embryos is likely to be of interest to those involved in conventional sheep ET applications for breeding improvement purposes. An evaluation of MOET in Suffolk sheep by Dingwall *et al.* (1993) notes that current ET rates in this species fall some way short of what is required for an effective MOET programme (Smith, 1988). As with cattle (see Sections 8.8.1 and 8.8.2), the repeated recovery of oocytes from live sheep may be of good commercial interest in the development of effective breeding improvement programmes. In this, it is possible that sexed sperm can eventually be used in sheep IVF, as for cattle. For a valuable pedigree sheep, at the end of its productive life, the recovery of the ovaries at slaughter, with subsequent IVF of oocytes, may be an additional way of obtaining high-quality lambs from her.

Fig. 8.21. Nuclear transfer calf born in 1990 at University College, Dublin. Several pregnancies were established using blastomeres from 32-cell embyros (produced *in vitro*) and fusing them with *in vitro*-produced oocytes. This birth marked the end-point of work by Kinis and coworkers at that time.

The first sets of identical twins were produced in sheep by Willadsen (1979) in Cambridge using a blastomere separation procedure. Subsequently, an alternative and less tedious method of producing such twins by embryo splitting was used (Gatica *et al.*, 1984; Wells *et al.*, 1990; Szell and Hudson, 1991; Vivanco *et al.*, 1991; Shelton, 1992). The technology of cloning by NT was first employed in sheep at Cambridge (Willadsen, 1986a, b) and further developed in that species by Smith and Wilmut (1988, 1989) at Edinburgh. In Australia, McLaughlin *et al.* (1990) have reported the birth of identical twin and quadruplet Merino lambs after NT.

Those who have reported using IVM oocytes as cytoplasts, however, have not found them to be as effective as those matured *in vivo* (McLaughlin *et al.*, 1991; Loi *et al.*, 1992). However, most of the NT procedures are similar to those employed in cattle; some progress towards the use of ES cells as the donors of nuclei has been reported (Rust *et al.*, 1993).

The amount of information on goats is much less than for sheep, but there is little reason to believe that this species should present problems that cannot be solved using procedures based on cattle and sheep work. Goat embryos have been split and young produced, after transfer of the demiembryos to recipients (Nowshari and Holtz, 1991). Normal kids have been born in China from NT (Yong *et al.*, 1991; Zhang, Y., *et al.*, 1991); they employed *in vivo*-produced oocytes and blastomeres from 4–32-cell embryos in their work. A transgenic goat programme has been described by Ebert and Schindler (1993).

8.13.2 Pigs

The ability to mature, fertilize and develop pig oocytes *in vitro* has been demonstrated by several workers (Mattioli *et al.*, 1989; Yoshida *et al.*, 1993a). This is of value in providing a supply of porcine oocytes and embryos for use in basic research as well as work aimed at commercial application. Although the first transgenic modification of pigs was reported almost a decade ago (Hammer *et al.*, 1985; Miller *et al.*, 1986), research in this area would obviously be facilitated by an abundant supply of zygotes produced by *in vitro* techniques. For those interested, the collection of *in vivo* one-cell porcine embryos from gilts has been detailed by Williams *et al.* (1990). Microinjection of DNA, for example, is known to be detrimental to early zygote development and may be responsible for the low pregnancy rates with transgenic embryos (Williams *et al.*, 1992); *in vitro*-produced zygotes might assist in resolving this problem.

A method of sexing the pig embryo without adversely affecting its subsequent development was described by Kato *et al.* (1987); this involved chromosomal analysis of trophectoderm cells taken from expanding blastocysts. The possibility of using sexed semen in pig IVF has received preliminary examination in the studies of Rath *et al.* (1993) in Germany. Although further work is required to optimize sperm sorting and IVF protocols, this is a further example of utilizing a system which has been pioneered in cattle IVF. Elsewhere, studies relevant to cloning in pigs by NT have been reported (Prather *et al.*, 1989a, b, 1990, 1991a, b; Sun *et al.*, 1990; Onishi and Youngs, 1993; Saito *et al.*, 1993; Stumpf *et al.*, 1993).

For research, particularly that dealing with the development of commercially acceptable techniques for embryo cryopreservation, the production of embryos in the laboratory would have obvious advantages. In the production of transgenics, there may well be areas relevant to human health and welfare, such as the germ-line modification of pig organs for human transplantation, which can be facilitated by a ready supply of *in vitro* embryos.

8.13.3 Horses

The first equine pregnancy to reach full term after IVF (oocyte collected prior to ovulation) was that reported by Palmer *et al.* (1990, 1991). The possible advantage of IVF technology to produce equine embryos would be as an adjunct to conventional ET activities, where the technique is employed to obtain foals from subfertile mares, competition mares and horses that would otherwise have difficulty in either achieving or maintaining a pregnancy of their own. There would be the obvious possibility of providing a genetic salvage service in instances where the death of a valuable mare occurs unexpectedly. The slicing of equine ovaries has been found useful in increasing the number of oocytes for IVF (Choi *et al.*, 1993a, b).

8.13.4 Buffaloes

Buffaloes are valuable draught animals and sources of milk and meat in a number of countries; the buffalo population has been steadily increasing throughout the 1980s and is estimated at 130–150 million (one eighth of the cattle population). About half the buffalo population is in India, where about 2.1 million are slaughtered annually for meat, providing 19.5% of the total meat production in that country (Agnihotri, 1992).

In some Western countries, such as the USA, Italy and Bulgaria, buffaloes have been introduced for the production of products such as cheese and low-fat buffalo beef. There is interest in developing IVF technology in this species, particularly in view of the low superovulatory responses that occur with conventional ET practice. However, sufficient work has been done to make it clear that *in vitro* technology must proceed to the point where relatively large numbers of oocytes can be routinely recovered from ovaries before IVF technology is likely to be commercially relevant in this species.

The sexing of *in vitro*-produced buffalo embryos has been reported from India by Appa Rao *et al.* (1993a, b), using the PCR to amplify bovine Y-chromosome-specific primers. Clearly, many procedures that were developed for use in cattle can be usefully employed in developing reproductive technology in the buffalo.

8.13.5 Deer

In a number of countries, there are farmers who are turning to alternative farming practices, including deer farming, as a means of increasing their income. ET procedures in deer are based on those used in sheep and cattle.

The application of MOET breeding schemes has been described by Fennessy *et al.* (1994). Some preliminary attempts have already been made in New Zealand to mature and fertilize oocytes recovered from slaughterhouse ovaries. It is possible to collect semen from stags with an artificial vagina (extended arm technique) in a stress-free manner (Krzywinski and Wierzchos, 1992), so the production of embryos after slaughter of high-quality females is one of the possibilities that can be developed in this species.

Appendix A
Embryo Production Protocol Used in University College, Dublin

Numerous protocols now exist for the laboratory production of cattle embryos. The details provided in this appendix are those that have been used in the Dublin *in vitro* fertilization (IVF) laboratory. They differ in minor detail from those first used in the work of Lu *et al.* (1987b). The cattle IVF laboratory is arranged in three areas: (i) an ovary reception area, which must be regarded as non-sterile; (ii) an area for the handling of gametes and embryos; and (iii) an area for ancillary activities, such as media preparation and embryo freezing. Both (ii) and (iii) must be regarded as tissue culture areas and fully aseptic laboratory precautions should be followed when using them.

1 Ovary Collection and Oocyte Recovery

Ovaries are collected on a daily basis from sexually mature heifers or cows slaughtered in a local abattoir and returned to the laboratory within 1 hour of slaughter. Ovaries are removed within 30 min of animal slaughter without reference to stage of oestrous cycle or breed type. Appropriate records are kept of the source of the ovaries (breed type/age).

Ovaries are transported to the laboratory in vacuum flasks containing presterilized phosphate-buffered saline (PBS) held at 30°C; this medium has antibiotic cover (kanamycin at $0.05\ \mathrm{mg\ ml^{-1}}$).

An alternative is to use sterile NaCl (0.9%) as the ovary storage medium. It is desirable that antibiotic cover (kanamycin) is provided; a brief (5 seconds) rinse of the ovary in 70% ethyl alcohol can be used to eliminate surface organisms before processing begins.

Oocyte recovery

In commencing processing, ovaries are washed twice in fresh, sterile PBS and dried lightly on sterile filter paper. Primary oocytes are recovered from

2–8 mm vesicular follicles by aspiration of the follicle containing the oocyte with an 18-gauge needle attached to a 5 ml syringe.

The medium used for oocyte recovery and washing (washing medium: see Appendix B) is Medium 199 (M-199) (with Earle's salts, L-glutamine, 25 mM HEPES, supplemented with 10% (v/v) steer serum (SS) (heat-treated at 56°C for 30 min) and trace amounts of antibiotic (50 μg gentamicin sulphate)). The pH is adjusted to 7.3–7.4, using either HCl or NaOH, and the medium is sterilized by filtration through a 0.22 μm Millipore filter and equilibrated in an atmosphere of 5% CO_2 in air with maximum humidity at a temperature of 38.5–39.0°C for at least 2 h prior to use.

Before commencing aspiration, the needle and syringe are first 'primed' with approximately 0.25–0.5 ml of the aspiration medium. After aspiration, the contents of the syringe are slowly dispelled into a sterile test-tube (Vacutainer; Becton Dickinson) with minimum disruption of the cumulus–oocyte complex (COC). Once the last ovary of a particular batch is processed, the oocytes are allowed a few minutes to settle to the bottom of the test-tube. The supernatant is then decanted and the remaining medium, containing the COCs and other follicular debris, is poured into a series of 60 mm sterile Petri dishes (Sterilin Ltd) for subsequent searching. The recovery work is undertaken in a sterile environment in a laminar flow cabinet with the room temperature held at 25°C.

Once located in the Petri dish, oocytes are lifted in a sterile glass pipette (bore diameter 400 μm) and transferred into a dish of fresh medium. Oocyte quality is assessed on the basis of morphological appearance and only oocytes exhibiting a compact cumulus investment and uniform ooplasm are selected for culture.

Harvesting of granulosa cells

After removal of all oocytes from the Petri dishes, the remaining aspiration medium containing granulosa cells and follicular debris is washed twice by centrifugation for 5 min at 500 *g*. The final pellet of granulosa cells is suspended in the *in vitro* maturation (IVM) medium (see below) and the resulting suspension is passed three times through an 18-gauge needle attached to a 1 ml syringe, to disperse the cells.

2 *In Vitro* Maturation of the Primary Oocyte

Selected oocytes (20–50 in 2 ml volume) are matured for 24 h in M-199 (with Earle's salts, L-glutamine and 25 mM HEPES) supplemented with 20% heat-inactivated SS, additional granulosa cells (3–5 million ml^{-1}) and trace amounts of antibiotic (50 iu penicillin and 50 μg streptomycin per ml). The pH is adjusted to 7.3–7.4, the medium sterilized by filtration through a 0.22 μm Millipore filter and equilibrated with 5% CO_2 in air for 2 h before use. The medium is contained in 35 mm Petri dishes (Nunclon), which are gently 'rocked' throughout maturation (flux system); the temperature is held at

38.5–39.0°C and the medium exposed to the 5% CO_2 atmosphere at maximum humidity.

3 *In Vitro* Capacitation of Bovine Sperm

Frozen bull semen is obtained from an appropriate source (cattle artificial insemination (AI) centre); to avoid variation due to individual bulls, three straws from each of three bulls is used for each 'swim-up'.

Straws are thawed in a water-bath at 35–37°C for 30–40 seconds; approximately 0.25 ml of thawed, pooled semen is layered under 1 ml of capacitation medium in each of four conical test-tubes. Each straw is wiped with 70% alcohol before being opened. The capacitation medium consists of modified Ca^{2+}-free Tyrode's medium (see Appendix B), the pH of which is adjusted to 7.3–7.4 prior to sterilization by filtration through a 0.22 μm Millipore filter. The medium is equilibrated overnight in an atmosphere of 5% CO_2 in air at a temperature of 38.5°C with maximum humidity.

After incubation for 1 h (tubes angled at 45°), the uppermost 0.5–0.8 ml of medium containing the more motile sperm is removed from each tube and pooled in a sterile conical centrifuge tube. This sample is washed twice by centrifugation at 500 *g* for 7–10 min. The resultant pellet is measured using an adjustable micropipette. A 50–80 μl aliquot of the swim-up-separated sperm is then diluted with an equal volume of capacitation medium containing 200 μg heparin (Sigma, H-3125 sodium salt) ml^{-1}.

After incubation for 15 min at 38.5°C in 5% CO_2 in air, the suspension is further diluted with two volumes of fresh capacitation medium to give a final concentration of 50 million sperm ml^{-1}. The sperm suspension is passed three times through a 25-gauge needle to separate the sperm before the insemination of the fertilization microdroplets is carried out.

4 *In Vitro* Fertilization Procedures

Fertilization is carried out in 46 μl microdroplets of modified Tyrode's (see Appendix B) under sterile mineral oil (Sigma M3516). The pH of the fertilization medium is adjusted to 7.7–7.8 prior to sterilization through a 0.22 μm Millipore filter. The sperm motility-stimulating mixture, containing penicillamine, hypotaurine and epinephrine (PHE), is added at a concentration of 40 μl PHE ml^{-1} of fertilization medium.

After preparation of the fertilization microdroplets, the Petri dishes are placed in the incubator and allowed to equilibrate for at least 1 h at 38.5–39.0°C in a CO_2 in air gas atmosphere at maximum humidity.

After IVM, oocytes are partially denuded of surrounding expanded cumulus cells (mechanical stripping in a micropipette). Before being transferred to the fertilization microdroplets, oocytes are washed by transfer through two successive dishes of fresh wash medium. Groups of 5–10 oocytes are then arranged in each of the microdroplets before a volume of 2–5 μl of sperm suspension is added so as to achieve a final concentration of sperm in the microdroplet of 1–1.5 million ml^{-1}.

The Petri dish containing the microdroplets is then placed in the incubator at 38.5–39.0°C for 44–48 h, maintaining a gas phase of 5% CO_2 in air and at high relative humidity. In carrying out the IVF, strict aseptic precautions are observed throughout to avoid any contamination of the medium, by performing all operations in a sterile atmosphere at 30°C.

Assessing fertilization

At 44–48 h after post-insemination (hpi), oocytes are evaluated under a stereoscope for evidence of cleavage. Results are recorded in terms of cleavage rate (percentage of embryos that have cleaved to the two-cell stage or beyond at 44–48 hpi) and cleavage index (percentage of embryos that have cleaved beyond the two-cell stage at 44–48 hpi).

5 *In Vitro* Culture of the Early Embryo

After cleavage assessment, those embryos deemed to be morphologically normal (on the basis of shape, symmetry, colour and texture of the blastomeres) are selected for further *in vitro* culture. They are removed from the fertilization microdroplets and washed by transferring them through two to three successive dishes of fresh M-199 supplemented with 20% SS before being placed on a granulosa cell monolayer for further development.

Preparing the granulosa cell monolayer

After removal of oocytes from the searching Petri dishes (see above under 1), the remaining aspiration medium, containing granulosa cells and cellular debris, is washed twice by centrifugation for 5 min at 500 *g*. The final pellet of granulosa cells is then suspended in M-199 supplemented with 20% SS and trace amounts of gentamicin sulphate until the appropriate cell concentration (1.5 million ml^{-1}) is achieved; the resultant suspension is passed several times through an 18-gauge needle to disperse the granulosa cells.

Droplets of 90 μl volume of this dilute suspension are then placed in a 60 mm Petri dish and covered with sterile mineral oil. After 72 h of culture, a confluent monolayer of cells should have formed at the base of the droplet. Before transferring embryos on to the layer at 44–48 hpi, the media is replenished by removing 45 μl and replacing with 45 μl of fresh culture medium. This 'cleaning' procedure is carried out every 48 h subsequently until termination of culture. Early embryos (two to four or more cells) showing regular blastomeres are placed in the droplets of medium in groups of 10–20 and cultured at 38.5–39.0°C in a gas phase of 5% CO_2 in air at maximum humidity.

Appendix B
Preparation of Culture Media

Table B1. Preparation of phosphate-buffered saline (PBS).

Solution A		
Ingredient	mM	$g\,l^{-1}$ (10×)
NaCl	136.89	80.0
KCl	2.68	2.0
$Na_2HPO_4 \cdot 12H_2O$	8.09	29.0
or		
Anhydrous Na_2HPO_4	8.09	11.5
KH_2PO_4	1.47	2.0
Solution B		
Ingredient	mM	$g\ 100\ ml^{-1}$ (10×)
$CaCl_2 \cdot 2H_2O$	0.457	0.67
or		
$CaCl_2 \cdot 6H_2O$	0.457	1.0
$MgCl_2 \cdot 6H_2O$	0.492	1.0

Sterilize solutions A and B separately and then take 100 ml of A and 10 ml of B and 890 ml of H_2O to make 1000 ml of PBS.

Table B2. Preparation of oocyte washing medium (modified HEPES-buffered Tyrode's medium).

Ingredient	mM	mg 100 ml $^{-1}$
NaCl	114.0	666.0
KCl	3.2	23.8
$NaHCO_3$	2.0	16.8
$NaH_2PO_4 \cdot 2H_2O$	0.4	6.2
Sodium lactate	10.0	112.1
$MgCl_2 \cdot 6H_2O$	0.5	10.0
$CaCl_2 \cdot 2H_2O$	2.0	29.4
HEPES	10.0	240.0
Phenol red		1.0
Sodium pyruvate	0.5	5.5
Bovine serum albumin		300.0
Glucose	5.6	100.0
pH		7.4
mosm		270–290

Table B3. Preparation of sperm capacitation medium (modified Ca^{2+}-free Tyrode's medium).

Ingredient	mM	mg 100 ml $^{-1}$
NaCl	112.0	654.5
KCl	2.7	20.0
$NaHCO_3$	25.1	210.0
$NaH_2PO_4 \cdot 2H_2O$	0.4	6.2
Na lactate	10.0	112.1
$MgCl_2 \cdot 6H_2O$	0.5	10.0
HEPES	5.0	120.0
Sodium pyruvate	1.0	11.0
Bovine serum albumin		600.0
Glucose	13.9	250.0
pH		7.4
mosm		290–310

Table B4. Preparation of fertilization medium (modified Tyrode's medium).

Ingredient	mM	mg 100 ml^{-1}
NaCl	114.0	666.0
KCl	3.2	73.8
$NaHCO_3$	25.0	209.0
$NaH_2PO_4 \cdot 2H_2O$	0.4	6.2
Sodium lactate	10.0	112.1
$MgCl_2 \cdot 6H_2O$	0.5	10.0
$CaCl_2 \cdot 2H_2O$	2.0	29.4
HEPES	10.0	240.0
Phenol red		1.0
Sodium pyruvate	0.5	5.5
Bovine serum albumin		600.0
Glucose	5.6	100.0
pH		7.8
mosm		290–310

Table B5. Preparation of motility-stimulating mixture: penicillamine–hypotaurine–epinephrine (PHE) solution.

MUST be prepared in low density of light.

Solution A

Step 1: Measure 200 ml of saline in beaker.
Step 2: (a) 0.2003 g sodium metabisulphite ($Na_2S_2O_5$);
(b) 0.6562 g sodium lactate (CH_3.CHOH.COONa, 70% w/w).
Step 3: Add them into 200 ml of saline and measure pH 4.004 at room temperature 21.3°C.

Solution B

Separately prepared before mixing with solution A.

Step 4: Add 0.011 g hypotaurine (Sigma, H-1384, lot 69F7800) in 100 ml saline.
Step 5: Add 0.0301 g penicillamine (Sigma, N-55F-0634) in 100 ml saline.
Step 6: Add 0.0185 g epinephrine (Sigma, E-4125, lot 105F7702) into 10 ml of solution A (low density of light).
Step 7: Put 1 ml of step 6 solution into 39 ml of solution A.
Step 8: Put in one container and add:
32 ml saline;
16 ml of epinephrine stock (step 7 solution);
20 ml of penicillamine stock;
20 ml of hypotaurine stock.
Step 9: Make aliquots in small vials and freeze at −20°C

Table B6. Procedures for fixation and staining of oocytes and embryos.

Fixative: acetoethanol

1. Mix one part of acetic acid into three parts of ethanol (1 : 3 v/v mixture).
2. Store mixture in sealed container at 5°C.

Stain

0.5% Acetolacmoid stain

Acetic acid	45.00 ml
Lacmoid	0.50 g
Distilled water	55.00 ml

Aceto-orcein stain

Purchased from BDH company.
Filter both stains before use (0.22 μm).

Acetoglycerol decolorizer

Acetic acid	20.00 ml (1 part)
Glycerol	20.00 ml (1 part)
Distilled water	60.00 ml (3 parts)

Preparation of vaseline : paraffin wax mixture

1. Weigh out 2 g paraffin wax and 40 g vaseline (1 : 10, w/w) into a 100 ml beaker.
2. Heat the beaker to melt contents and stir gently with a glass rod to mix the mixture.
3. When the mixture had melted and mixed completely, put into a 5 or 10 ml syringe.
4. Cool the syringe at 4°C.
5. Attach a blunt 18-gauge needle before use.

APPENDIX C FREEZING AND TRANSFER OF IVMFC-DERIVED EMBRYOS

1 Acceptability of Embryos

For acceptance, the cattle blastocysts must be day-7 and of highest grade (Grade 1).

The blastocyst is suspended in 1.8 M ethylene glycol in phosphate-buffered saline (PBS) supplemented with 10% steer serum at 25°C and held for 10 min. It is then loaded into the midportion of a 0.25 ml ministraw (IMV) in a small volume of the freezing medium, separated by air bubbles on each side from larger volumes of the same medium, and cooled to $-7°C$ at $-1°C\ min^{-1}$.

At $-7°C$, the freezing medium is seeded, held for 10 min at that temperature, and cooled at $0.3°C\ min^{-1}$ to $-30°C$. The straw is then plunged directly into liquid nitrogen for storage.

For thawing, the straw is held in air for 6 seconds and plunged into a water-bath held at 35°C. It is then loaded into a standard Cassou cattle embryo transfer instrument (IMV) and immediately transferred into a bred synchronized recipient cow.

Before transfer for twinning, the bred recipient cow receives an epidural anaesthesia (lignocaine) and an appropriate tranquillizer (Buscopan; Boehringer Ingelheim) for its comfort during the transfer procedure. The ovary containing the corpus luteum is identified and the embryo deposited in the midhorn position of the opposite (contralateral) uterine horn. In the transfer procedure, care is taken at all points to maintain sterility in the handling and manipulation of the straw.

2 Pregnancy Monitoring

In the absence of information about a 'repeat' oestrus, the cow's uterus is checked by suitably experienced operatives using appropriate real-time ultrasonics at 50–60 days post-breeding for the presence of twins. More than one such examination should be avoided to minimize the risk of disturbing the pregnancy.

Appendix D
Books and Journals Dealing with *In Vitro* Fertilization and Embryo Transfer in Farm Animals

1 Books

Adams, C.E. (ed.) (1982) *Mammalian Egg Transfer*. CRC Press, Boca Raton, Florida, USA.

Allen, W.R., Antczak, D.F. and Wade, J.F. (eds) (1989) Equine *Embryo Transfer II*. Supplement No. 8, Equine Veterinary Journal Ltd, Newmarket.

Amann, R.P. and Seidel, G.E., Jr (eds) (1982) *Prospects for Sexing Mammalian Semen*. Colorado Associated University Press, Boulder, Colorado.

Babiuk, L.E. and Phillips, J.P. (eds) (1989) *Animal Biotechnology*. Pergamon Press, Oxford.

Bavister, B.D. (ed.) (1993) *Preimplantation Embryo Development*. Serono Symposium, Springer-Verlag, New York.

Bavister, B.D., Cummins, J. and Roldan, E.R.S. (eds) (1990) *Fertilization in Mammals*. Serono Symposia Norwell, Massachusetts, USA.

Betteridge, K.J. (ed.) (1977) *Embryo Transfer in Farm Animals: a Review of Techniques and Applications*. Monograph 16, Canada Department of Agriculture, Ontario, 92 pp.

Biggers, J.D. and Schuetz, A.W. (eds) (1972) *Oogenesis*. Butterworths, London.

Brackett, B.G., Seidel, G.E., Jr and Seidel, S.M. (eds) (1982) *New Technologies in Animal Breeding*. Academic Press, New York.

Brooks, N., Challis, J., McNeilly, A. and Doberska, C. (eds) (1992) *Frontiers in Reproductive Biology*. Supplement No. 45, Journal of Reproduction and Fertility Ltd, Cambridge.

Cole, D.J.A., Foxcroft, G.R. and Weir, B.J. (eds) (1990) *Control of Pig Reproduction, III*. Supplement No. 40, Journal of Reproduction and Fertility Ltd, Cambridge.

Curtis, J.L. (1991) *Cattle Embryo Transfer Procedure*. Academic Press, New York.

Dale, B. (ed.) (1989) *Mechanisms of Fertilization: Plants to Humans*. Springer-Verlag, Berlin.

Daniel, J.C. Jr (ed.) (1978) *Methods in Mammalian Reproduction*. Academic Press, New York.

Dunbar, B.S. and O'Rand, M.G. (eds) (1991) *A Comparative Overview of Mammalian Fertilization*. Plenum Press, New York.

First, N.L. and Haseltine, F.P. (eds) (1988). *Transgenic Animals*. Butterworth–Heinemann, London.

Fishel, S. and Symonds, M. (eds) (1993) *Gamete and Embryo Micromanipulation in Human Reproduction*. Edward Arnold, London.

Hafez, E.S.E. (ed.) (1993) *Reproduction in Farm Animals*, 6th edn. Lea and Febiger, Philadelphia.

Hartmann, J.F. (ed.) (1983) *Mechanism and Control of Animal Fertilization*. Academic Press, New York.

Hunter, R.H.F. (1988) *The Fallopian Tubes: Their Role in Fertility and Infertility*. Springer-Verlag, Berlin.

Knobil, E. and Neill, J.D. (eds) (1988) *The Physiology of Reproduction*. Raven Press, New York.

Lamming, G.E., Flint, A.P.F. and Weir, B.J. (eds) (1991) *Reproduction in Domestic Ruminants II*. Supplement No. 43, Journal of Reproduction and Fertility Ltd, Cambridge.

Lauria, A. and Gandolfi, F. (eds) (1992) *Embryonic Development and Manipulation in Animal Production*. Portland Press, London.

Mastroianni, L. and Biggers, J.D. (eds) (1981) *Fertilization and Early Embryonic Development In Vitro*. Plenum Press, New York.

Pedersen, R.A., McLaren, A. and First, N.L. (eds) (1991). *Animal Applications of Research in Mammalian Development*. CSHL Press, Plainview, New York.

Phillips, C.J.C. (1993) *Cattle Behaviour*. Farming Press, Ipswich.

Sreenan, J.M. (ed.) (1978) *Control of Reproduction in the Cow*. Current Topics in Veterinary Medicine, Vol. 1, Martinus Nijhoff, The Hague.

Sreenan, J. and Diskin, M. (1992) *Breeding the Dairy Herd*. Teagasc Publication, Dublin.

Thibault, C., Levasseur, M.-C. and Hunter, R.H.F. (eds) (1993) *Reproduction in Mammals and Man*. Ellipses, Paris.

Trounson, A. and Gardner, D.K. (eds) (1993) *Handbook of In Vitro Fertilization*. CRC Press, Boca Raton, Florida, USA.

Wassarman, P.M. (ed.) (1991). *Elements of Mammalian Fertilization*. Vol. II. *Practical Application*. CRC Press, Boca Raton, Florida, USA.

2 Journals

AgBiotech News and Information. CAB INTERNATIONAL, Wallingford, Oxon.

American Journal of Veterinary Research. American Veterinary Medical Association, 600 S. Michigan Avenue, Chicago, Illinois.

Animal Breeding Abstracts. CAB INTERNATIONAL, Wallingford, Oxon.

Animal Production. Longmans, Harlow, Essex.

Animal Reproduction Science. Elsevier Science Publishers BV, Amsterdam.

Biology of Reproduction. Society for the Study of Reproduction, 309 West Clark Street, Champaign, Illinois 61820.

British Veterinary Journal. Baillière Tindall, 33 Red Lion Square, London WC1R 4SG.

Canadian Journal of Animal Science. Canadian Society of Animal Science, Slater Street, Ottawa.

Embryo Transfer Newsletter. International Embryo Transfer Society, 309 West Clark Street, Champaign, Illinois 61820.

Fertility and Sterility. American Fertility Society, 2140 11th Avenue South, Suite 200, Birmingham, Alabama 35205.

Human Reproduction. European Society of Human Reproduction and Embryology, Oxford University Press, Oxford.

International Journal of Fertility. Ben Franklin Press, Pittsfield, Massachusetts.

Journal of Agricultural Science (Cambridge). Cambridge University Press, Cambridge.

Journal of the American Veterinary Medical Association. 600 S. Michigan Avenue, Chicago, Illinois.

Journal of Animal Science. American Society of Animal Science, 309 West Clark Street, Champaign, Illinois.

Journal of Assisted Reproduction and Genetics. Plenum Publishing Corporation, 233 Spring Street, New York.

Journal of Dairy Science. American Dairy Science Association, 309 West Clark Street, Champaign, Illinois 61820.

Journal of Reproduction and Fertility. Journal of Reproduction and Fertility Ltd, 22 Newmarket Road, Cambridge CB5 8DT.

Molecular Reproduction and Development. John Wiley and Sons, Inc., 1 Wiley Drive, Somerset, New Jersey 08875.

Reproduction, Nutrition, Development. Institut National de la Recherche Agronomique, Paris.

Theriogenology. Butterworths, 80 Montvale Avenue, Stoneham MA 02180.

Veterinary Record. British Veterinary Association, 7 Mansfield Street, London.

References

Aalseth, E.P. and Saacke, R.G. (1986) Vital staining and acrosomal evaluation of bovine sperm. *Gamete Research* 15, 72–81.

Abate, V., Stinchi, A., Abate, M. and Abate, V., Jr (1992) Successful aspiration of multiple oocytes from recruiting follicles: is ovarian stimulation necessary? *Human Reproduction* 7 (Suppl. 2), 164.

Abdalla, H.I. and Leonard, T. (1988) Cryopreserved zygote intrafallopian transfer for anonymous oocyte donation. *Lancet* i, 835.

Abeydeera, L.R. and Niwa, K. (1992) Ability of *in vitro* maturing bovine oocytes to transform sperm nuclei to metaphase chromosomes. *Journal of Reproduction and Fertility* 96, 565–572.

Adams, G.P. (1994) Control of ovarian follicular wave dynamics in cattle: implications for synchronization and superstimulation. *Theriogenology* 41, 19–24.

Adams, C.E., Moor, R.M. and Rowson, L.E.A. (1968) Survival of cow and sheep eggs in the rabbit oviduct. *Proceedings of the Sixth International Congress on Animal Reproduction and Artificial Insemination* (Paris) 1, 573–574.

Adams, G.P., Matteri, R.L., Kastelic, J.P., Ko, J.C.H. and Ginther, O.J. (1992a) Association between surges of follicle-stimulating hormone and the emergence of follicular waves in heifers. *Journal of Reproduction and Fertility* 94, 177–188.

Adams, G.P., Kot, K, Smith, C.A. and Ginther, O.J. (1992b) Role of FSH in the mechanism of selection of the dominant follicle in heifers. *Proceedings of the Twelfth International Congress on Animal Reproduction* (The Hague) 1, 179–181.

Adams, G.P., Matteri, R.L. and Ginther, O.J. (1992c) Effect of progesterone on ovarian follicles, emergence of follicular waves and circulating follicle-stimulating hormone in heifers. *Journal of Reproduction and Fertility* 95, 627–640.

Adaniya, G.K., Rawlins, R.G., Miller, I.F. and Zaneveld, L.J.D. (1987) Effect of sodium alginate encapsulation on the development of preimplantation mouse embryos. *Journal of In Vitro Fertilization and Embryo Transfer* 4, 343–345.

Adaniya, G.K., Rawlins, R.G., Quigg, J.M., Roblero, L. and Miller, I.F. (1991) First pregnancies and livebirths from transfer of sodium alginate encapsulated embryos in the mouse. *Biology Reproduction* 44 (Suppl. 1), 58 (Abs. 21).

Adaniya, G.K., Rawlins, R.G., Quigg, J.M., Roblero, L., Miller, I.F. and Zaneveld, L.J.D.

(1993) First pregnancies and livebirths from transfer of sodium alginate encapsulated embryos in a rodent model. *Fertility and Sterility* 59, 652–656.

Adashi, E.Y. (1990) Growth factors and the oocyte: is there a connection? In: Asch, R.H. Balmaceda, J.P. and Johnston, I. (eds) *Gamete Physiology*. Serona Symposia, Norwell, Massachusetts, USA, pp. 187–195.

Adashi, E.Y., Resnick, C.E., Hurwitz, A., Ricciarelli, E., Hernandez, R., Roberts, C.T., Leroith, D. and Rosenfeld, R. (1991) Insulin-like growth factors: the ovarian connection. *Human Reproduction* 6, 1213–1219.

Agnihotri, M.K. (1992) Current status and future prospects of buffalo meat production in India. *Buffalo Journal* 8(2), 95–102.

Agrawal, K.P. and Polge, C. (1988) *In vitro* maturation of sheep oocytes. *Indian Journal of Animal Reproduction* 9(1), 8–9.

Agrawala, P.L., Wagner, V.A. and Geldermann, H. (1992) Sex determination and milk protein genotyping of preimplantation stage bovine embryos using multiplex PCR. *Theriogenology* 38, 969–978.

Aguilar, A.A., Green, H.B., Basson, R.P. and Overpeck-Alvey, M.J. (1991) Effect of somidobove on reproductive performance in lactating dairy cows. *Journal of Dairy Science* 74 (Suppl. 1), 192 (Abs. P150).

Ahluwalia, R.S. and Majumdar, A.C. (1992) *In vivo* culture in rabbit oviduct of *in vitro* fertilized bubaline oocytes. *Indian Journal of Animal Sciences* 62, 528–531.

Aktas, H., Leibfried-Rutledge, M.L. and Wheeler, M.B. (1991a) Bovine oocytes arrested in meiosis *in vitro* retain developmental capacity. *Biology of Reproduction* 44 (Suppl. 1), 75 (Abs. 90).

Aktas, H., Leibfried-Rutledge, M.L. and First, N.L. (1991b) Bovine oocyte maturation is regulated by interactions between cAMP, cumulus and granulosa. *Serono Symposium on Preimplantation Embryo Development*, Abstract Vol., 25 (Abs. I–2).

Alexander, G., Signoret, J.P. and Hafez, E.S.E. (1974) Sexual and maternal behaviour. In: Hafez , E.S.E. (ed.) *Reproduction in Farm Animals*, 3rd edn. Lea and Febiger, Philadelphia, pp. 222–254.

Alfonso, M.F. and Hunter, A.G. (1991) Effect of concentration of bovine seminal plasma on the reversal of bovine sperm capacitation. *Journal of Dairy Science* 74 (Suppl. 1), 229 (Abs. P261).

Alfonso, N.F. and Hunter, A.G. (1992a) Effects of temperature (37°C or 39°C) on bovine *in vitro* fertilization and post-fertilization culture media on subsequent cleavage. *Journal of Dairy Science* (Suppl. 1), 242 (Abs. P293).

Alfonso, N.F. and Hunter, A.G. (1992b) Effect of different temperature combinations during *in vitro* bovine culture on subsequent cleavage. *Journal of Animal Science* 70 (Suppl. 1), 252 (Abs. 458).

Alfuraiji, M.M., Alibrahim, R.M. and Elnouty, F.D. (1993) Seasonal variations in superovulatory responses of Holstein cows treated with pregnant mare serum gonadotrophin in Saudi Arabia. *Journal of Reproduction and Fertility*, Abstract Series No. 11, 75 (Abs. 140).

Al-Hasani, S., Toldsdorf, A., Diedrich, K., Van der Ven, H. and Krebs, D. (1986) Successful *in vitro* fertilization of frozen–thawed rabbit oocytes. *Human Reproduction* 1, 309–312.

Al-Hasani, S., Hepnar, C., Diedrich, K., van der Ven, H. and Krebs, D. (1992) Cryopreservation of rabbit zygotes. *Human Reproduction* 7 (Suppl. 1), 81–83.

Ali, J.I. (1987) Separation of bovine X- from Y-chromosome bearing spermatozoa using monoclonal H-Y antibodies. *Dissertation Abstracts International B, Sciences and Engineering* 47(7), 2689.

Ali, J. and Shelton, J.N. (1993) Successful vitrification of day-6 sheep embryos. *Journal of Reproduction and Fertility* 99, 65–70.

Ali, J., Whitten, W.K. and Shelton, J.N. (1993) Effect of culture systems on mouse early embryo development. *Human Reproduction* 8, 1110–1114.

Alikani, M., Adler, A., Reing, A.M., Malter, H.E. and Cohen, J. (1992) Subzonal sperm insertion and the frequency of gamete fusion. *Journal of Assisted Reproduction and Genetics* 9, 97–101.

Alikani, M., Olivennes, F. and Cohen, J. (1993) Microsurgical correction of partially degenerate mouse embryos promotes hatching and restores their viability. *Human Reproduction* 8, 1723–1728.

Allen, W.R. (1991a) Extraspecies embryo transfer in equids. *Proceedings of the Seventh Congress of the European Embryo Transfer Association* (Lyon), 87–100.

Allen, J. (1991b) Biotechnology is just an extension of evolution. *Dairy Farmer* September, 114–115.

Allen, W.R., Meadows, F., Lascombres, F. and Pashen, R.L. (1992) International transport of frozen equine embryos. *Proceedings of the Third International Symposium on Equine Embryo Transfer* (Buenos Aires), 10.

Allrich, R.D. (1993) Endocrine and neural control of estrus expression in dairy cattle. *Journal of Dairy Science* 76 (Suppl. 1), 310.

Allworth, A.E. and Albertini, D.F. (1992) Cytoskeletal reorganization in bovine cumulus oocyte complexes is maturation associated. *Biology of Reproduction* 46 (Suppl. 1), 139 (Abs. 353).

Allworth, A.E. and Albertini, D.F. (1993) Meiotic maturation in cultured bovine oocytes is accompanied by remodeling of the cumulus cell cytoskeleton. *Developmental Biology* 158, 101–112.

Alm, H., Kauffold, P., Makarowa, S.N. and Sirotkin, A.V. (1990) Effect of granulosa cells on *in vitro* maturation, fertilization and cleavage. *Arch. exper. Vet. med. Leipzig* 44, 83–91.

Alvarez, J.G., Lasso, J.L., Blasco, L., Nunez, R.C., Heyner, S., Caballero, P.P. and Storey, B.T. (1993) Centrifugation of human spermatozoa induces sublethal damage; separation of human spermatozoa from seminal plasma by a dextran swim-up procedure without centrifugation extends their motile lifetime. *Human Reproduction* 8, 1087–1092.

Anderson, G.B. (1987) Use of chimaeras to study development. *Journal of Reproduction and Fertility* Suppl. 34, 251–259.

Anderson, G.B., Cupps, P.T., Drost, M., Horton, M.B. and Wright, R.W. (1978) Induction of twinning in beef heifers by bilateral embryo transfer. *Journal of Animal Science* 46, 449–452.

Anderson, G.B., BonDurant, R.H. and Cupps, P.T. (1982) Induction of twins in different breeds of cattle. *Journal of Animal Science* 54, 485–490.

Anderson, G.B., Piedrahita, J.A. and BonDurant, R.H. (1990) On the isolation of embryonic stem (ES) cells: comparative behaviour of murine, porcine and ovine embryos. *Biology of Reproduction* 42 (Suppl. 1), 175.

Anderson, G.B., Ruffing, N.A., BonDurant, R.H. and Pashen, R.L. (1991) Preliminary observations on reproduction in a female sheep–goat chimaera. *Veterinary Record* 129, 467–469.

Anderson, S.H. and Killian, G. (1992) Bovine sperm alter the electrophoretic profile of proteins produced by ampullar oviductal explants. *Biology of Reproduction* 46 (Suppl. 1), 75 (Abs. 97).

Andersson, M., Aalto, J. and Gustavsson, I. (1992) Embryo quality and andrological study of two subfertile bulls versus five control bulls with normal fertility. *Theriogenology* 38, 623–631.

Andrews, J.C. and Bavister, B.D. (1988) Chemical capacitation of hamster spermatozoa in a protein free medium with divalent cation chelators and the subsequent requirements for albumin for zona pellucida penetration. *Biology of Reproduction* 38 (Suppl. 1), Abs. 32.

Andrews, J.C. and Bavister, B.D. (1989) Capacitation of hamster spermatozoa with the divalent cation chelators D-penicillamine, L-histidine, and L-cysteine in a protein free culture medium. *Gamete Research* 23, 159–170.

Anon. (1993) Statistics describing the international embryo transfer industry. *Embryo Transfer Newsletter* 11 (4), 10–13.

Aoyagi, Y., Fujii, K., Irazumi, Y., Furudate, M., Fukui, Y. and Ono, H. (1988) Effects of two treatments on semen from different bulls on *in vitro* fertilization results of bovine oocytes. *Theriogenology* 30, 973–985.

Aoyagi, Y., Iwazumi, Y., Kosaka, Y. and Furudate, M. (1990) Effects of interval to seeding and temperature to removing glycerol on conception of frozen embryo of Holstein cows. *Japanese Journal of Animal Reproduction* 36, 245–248.

Appa Rao, K.B.C., Totey, S.M., Pawshe, C.H. and Singh, G.P. (1993a) Sex determination by DNA amplification of buffalo (*Bubalis bubalis*) embryos produced *in vitro*. *Theriogenology* 39, 179.

Appa Rao, K.B.C., Pawshe, C.H. and Totey, S.M. (1993b) Sex determination of *in vitro* developed buffalo (*Bubalus bubalis*) embryos by DNA amplification. *Molecular Reproduction and Development* 36, 291–296.

Arav, A. (1992) Vitrification of oocytes and embryos. In: Lauria, A. and Gandolfi, A. (eds) *Embryonic Development and Manipulation in Animal Production*. Portland Press, London, pp. 255–264.

Arav, A., Rubinsky, B., Fletcher, G. and Seren, E. (1993) Cryogenic protection of oocytes with antifreeze proteins. *Molecular Reproduction and Development* 36, 488–493.

Arav, A., Rubinsky, B., Seren, E., Roche, J.F. and Boland, M.P. (1994) The role of thermal hysteresis proteins during cryopreservation of oocytes and embryos. *Theriogenology* 41, 107–112.

Archibong, A.E., Stephens, D.T., Vijayaraghavan, S., Wolf, D.P. and Hoskins, D.D. (1992) 2-Chloro-adenosine stimulates human sperm motility and elevates cAMP levels. *Biology of Reproduction* 46 (Suppl. 1), 53 (Abs. 11).

Arechiga, C.F., Ealy, A.D. and Hansen, P.J. (1992) Induction of thermotolerance in mouse embryos is blocked by the glutathione synthesis inhibitor, DL-buthionine-{S, R}-sulfoximine (BSO). *Journal of Animal Science* 70 (Suppl. 1), 208 (Abs. 505).

Arellano, J., Illera, M.J., Lorenzo, P., Sanchez, J., Silvan, G., Illera, M. and Petters, R.M. (1993) Epidermal growth factor enhances porcine oocyte maturation *in vitro* in the absence of follicular fluid or hormones. *Theriogenology* 39, 180.

Arkins, S. and Kelley, K.W. (1992) Neuroendocrine control of immune competence. *Journal of Dairy Science* 75 (Suppl. 1), 218 (Abs. P219).

Arlotto, T.M., Leibfried-Rutledge, M.L. and First, N.L. (1990) Size distribution and meiotic competence of bovine primary oocytes from two locations in the ovary. *Theriogenology* 33, 188.

Arlotto, T., Leibfried-Rutledge, M.L. and First, N.L. (1992) Correlation of oocyte diameter with time of maturation in bovine oocytes. *Biology of Reproduction* 46 (Suppl. 1), 66 (Abs. 63).

Armstrong, D.T. (1991) Factors influencing superovulation success. *Embryo Transfer Newsletter* 9 (1), 11–17.

Armstrong, D.T. (1993) Recent advances in suprovulation of cattle. *Theriogenology* 39, 7–24.

Armstrong, D.T. and Xia, P. (1993) Differential mitogenic actions of insulin-like growth factor-I and FSH on bovine cumulus cells and granulosa cells. *Theriogenology* 39, 181.

Armstrong, D.T., Holm, P., Irvine, B., Petersen, B.A., Stubbings, R.B., McLean, D., Stevens, G.F. and Seamark, R.F. (1991a) Laparoscopic aspiration and *in vitro* maturation of oocytes from calves. *Theriogenology* 35, 182.

Armstrong, D.T., Zhang, X., Vanderhyden, B.C. and Khamsi, F. (1991b) Hormonal actions during oocyte maturation influence fertilization and early embryonic development. *Annals of the New York Academy of Sciences* 626, 137–158.

Armstrong, D.T., Holm, P., Irvine, B., Petersen, B.A., Stubbings, R.B., McLean, D., Stevens, G. and Seamark, R.F. (1992) Pregnancies and live birth from *in vitro* fertilization of calf oocytes collected by laparoscopic follicular aspiration. *Theriogenology* 38, 667–678.

Artiga, C.G., Gutierrez, A., Garde, J., Pintado, B. and Vazquez, I. (1993) Physiological changes of fresh or frozen ram spermatozoa during capacitation '*in vitro*'. *Theriogenology* 39, 182.

Asdell, S.A. (1969) Historical introduction. In: Cole, H.H. and Cupps, P.T. (eds) *Reproduction in Domestic Animals*, 2nd edn. Academic Press, New York, pp. 1–16.

Ashworth, C.J. (1992) Synchrony embryo–uterus. *Animal Reproduction Science* 28, 259–267.

Assey, R.J., Hyttel, P., Purwantara, B., Greve, T. and Schmidt, M. (1992) Oocyte structure in dominant versus subordinate follicles during the first follicular wave in cattle. *Proceedings of the Twelfth International Congress on Animal Reproduction* (The Hague) 1, 303–305.

Assey, R.J., Hyttel, P., Greve, T. and Purwantara, B. (1994). Oocyte morphology in dominant and subordinate follicles. *Molecular Reproduction and Development* 37, 335–344.

Atkinson, P.W., Hines, E.R., Beaton, S., Matthaei, K.I., Reed, K.C. and Bradley, M.P. (1991) Association of exogenous DNA with cattle and insect spermatozoa *in vitro*. *Molecular Reproduction and Development* 29, 1–5.

Aurich, C. and Hahn, J. (1993) Male effects on *in vivo* and *in vitro* production of bovine embryos. *Proceedings of the Ninth Congress of the European Embryo Transfer Association* (Lyon); 117.

Austin, C.R. (1951) Observations on the penetration of the sperm into the mammalian egg. *Australian Journal of Scientific Research* 4, 581–596.

Averill, R.L.W., Adams, C.E. and Rowson, L.E.A. (1955) Transfer of mammalian ova between species. *Nature* 176, 167–170.

Avery, B. (1989) Impact of asynchronous ovulations on the expression of sex-dependent growth rate in bovine preimplantation embryos. *Journal of Reproduction and Fertility* 87, 627–631.

Avery, B. and Greve, T. (1992a) The effect of incubator type on the development of *in vitro* produced bovine embryos. *Proceedings of the Nordic A.I. Virtanen Institute Symposium on In Vitro Culture of Domestic Animal Embryos*, 19–20.

Avery, B. and Greve, T. (1992b) Impact of incubator type on the yield of *in vitro* produced bovine blastocysts. *Acta Veterinaria Scandinavica* 33, 341–348.

Avery, B. and Greve, T. (1993) Some aspects of bovine IVP. *Ars Veterinaria* 9(2), 35–52.

Avery, B. and Schmidt, M. (1987) Embryo sexing in cattle. *Proceedings of the Third Scientific Meeting, European Embryo Transfer Association* (Lyon), 80–81.

Avery, B. and Schmidt, M. (1989) Sex determination of bovine embryos using H-Y antibodies. *Acta Veterinaria Scandinavica* 30, 155–164.

Avery, B., Bak, A. and Schmidt, M. (1989a) Sex dependent developmental cleavage rates in bovine embryos. *Journal of Reproduction and Fertility*, Abstract Series No. 3, 49.

Avery, B., Bak, A. and Schmidt, M. (1989b) Differential cleavage rates and sex determination in bovine embryos. *Theriogenology* 32, 139–147.

Avery, B., Schmidt, M. and Greve, T. (1989c) Sex determination of bovine embryos based on embryonic cleavage rates. *Acta Veterinaria Scandinavica* 30, 147–153.

Avery, B., Madison, V. and Greve, T. (1991) Sex and development in bovine *in vitro* fertilized embryos. *Theriogenology* 35, 953–963.

Avery, B., Jorgensen, C.B., Madison, V. and Greve, T. (1992a) Morphological development and sex of bovine *in vitro*-fertilized embryos. *Molecular Reproduction and Development* 32, 265–270.

Avery, B., Schmidt, M. and Greve, T. (1992b) Cryopreservation of *in vitro* produced bovine oocytes. *Proceedings of the Twelfth International Congress on Animal Reproduction* (The Hague) 3, 1390–1392.

Avery, B., Greve, T., Ronsholt, L. and Botner, A. (1993) Virus screening of a bovine *in vitro* embryo production system. *Veterinary Record* 132, 660.

Ayoub, M. and Hunter, A.G. (1990) Characteristics of a bovine oocyte maturation inhibitor in bovine follicular fluid. *Biology of Reproduction* 42 (Suppl. 1), 458.

Ayoub, M.A. and Hunter, A.G. (1991) Effect of sperm washing, sperm concentration and protein source in capacitating media on bovine sperm motility and capacitation. *Journal of Dairy Science* 74 (Suppl. 1), 229 (Abs. P262).

Ayoub, M.A. and Hunter, A.G. (1993a) Inhibitory effect of bovine follicular fluid on *in vitro* maturation of bovine oocytes. *Journal Dairy Science* 76, 95–100.

Ayoub, M.A. and Hunter, A.G. (1993b) Parthenogenetic activation of *in vitro* matured bovine oocytes. *Journal Dairy Science* 76, 421–429.

Azambuja, R.M. De, Moreno, J.F., Kraemer, D. and Westhusin, M. (1993) Effect of gas atmosphere on *in vitro* maturation of bovine oocytes. *Theriogenology* 39, 184.

Bacci, M.L., Galeati, G., Mattioli, M., Boni, R. and Seren, E. (1991) *In vitro* maturation and *in vitro* fertilization of buffalo oocytes. *Proceedings of the Third World Buffalo Congress* (Varna Bulgaria) 3, 599–603.

Bach, S. and Muller, F. (1989) The induction of twinning in beef cattle by means of embryo transfer. *Tierhygiene-Information, Eberswalde-Finow* 21(80), 83–87.

Bach, S., Kahn, W., Muller, F., Schulz, J. and Geigenmuller, S. (1991) Early ultrasonic diagnosis of twin pregnancies in cattle following artificial insemination and unilateral embryo transfer and calving results. *Tierarztliche Praxis* 19(4), 365–368.

Badinga, L., Driancourt, M.A., Savio, J.D., Wolfenson, D. and Thatcher, W.W. (1991) Changes in follicular development, aromatase activity and follicular steroids in dominant and subordinate follicles at days 5, 8 and 12 of the estrous cycle in cattle. *Biology of Reproduction* 44 (Suppl. 1), 70 (Abs. 72).

Bailey, J.L., Robertson, L. and Buhr, M.M. (1992) Relationship between the Ca^{2+} control, motility and fertility of cryopreserved bull sperm. *Biology of Reproduction,* 46 (Suppl. 1), 52 (Abs. 8).

Baird, D.T. (1993) Isoforms of FSH: implications for therapy. In: Jacob, H.S. (ed.) *New Frontiers in Ovulation Induction*. Parthenon Publishing Group, pp. 23–26.

Bak, A., Greve, T. and Schmidt, M. (1989) Effect of superovulation on reproduction. *Theriogenology,* 31 169.

Baker, H.W.G., Liu, D.Y., Bourne, H. and Lopata, A. (1993) Diagnosis of sperm defects in selecting patients for assisted fertilization. *Human Reproduction* 8, 1779–1780.

Baker, T.G. (1971) Electron microscopy of the primary and secondary oocyte. *Advances in the Biosciences* 6, 7–23.

Baker, T.G. (1982) Oogenesis and ovulation. In: Austin, C.R. and Short, R.V. (eds) *Reproduction in Mammals,* Book 1, 2nd edn. Cambridge University Press, Cambridge, pp. 17–45.

Baker, T.G. and Franchi, L.L. (1967) The fine structure of chromosomes in bovine primordial oocytes. *Journal of Reproduction and Fertility* 14, 511–513.

Baker, T.G. and Hunter, R.H.F. (1978) Oogenesis and follicular growth in the cow: implications for superovulation. In: Sreenan, J.M. (ed) *Control of Reproduction in the Cow*. C.E.C. Publications, Luxembourg, pp. 34–49.

Balakier, H., MacLusky, N.J. and Casper, R.F. (1993) Characterization of the first cell cycle in human zygotes: implications for cryopreservation. *Fertility and Sterility* 59, 359–365.

Baldassarre, H., Castro, T.E., Furnus, C.C. and de Matos, D.G. (1994) *In vitro* maturation and fertilization of sheep oocytes collected by laparoscopic folliculocentesis. *Theriogenology* 41, 159.

Balkwill, F.R. (1989) Peptide regulatory factors: interferons. *Lancet* i, 13 May, 1060–1063.

Ball, B.A. and Miller, P.G. (1991) Viability of equine embryos co-cultured with equine oviductal epithelium from the 4–8-cell to blastocyst stage. *Theriogenology* 35, 183.

Ball, B.A. and Miller, P.G. (1992) Survival of equine embryos cocultured with equine oviductal epithelium from the four- to eight-cell to the blastocyst stage after transfer to synchronous recipient mares. *Theriogenology*, 37, 979–991.

Ball, B.A., Altschul, M. and Hillman, R.B. (1989) Luteal maintenance in mares after transfer of equine trophoblastic vesicles. *Equine Veterinary Journal*, Suppl. 8, 21–24.

Ball, B.A., Altschul, M. and Ellington, J.E. (1991) *In vitro* development of day-2 equine embryos co-cultured with oviductal explants or trophoblastic vesicles. *Theriogenology* 35, 669–682.

Ball, B.A., Thomas, P.G.A., Brinsko, S.P., Miller, P.G. and Ellington, J.E. (1992a) Development of 1- to 2-cell equine embryos cocultured with oviductal epithelial cells. *Theriogenology* 37, 189.

Ball, B.A., Brinsko, S.P., Thomas, P.G.A., Miller, P.G. and Ellington, J.E. (1992b) Development of 1–8 cell equine embryos co-cultured with equine oviductal epithelial cells. *Proceedings of the Third International Symposium on Equine Embryo Transfer* (Buenos Aires), 15.

Ball, G.D. and First, N.L. (1983) Fertility of ejaculated bull sperm *in vitro* after exposure to caffeine and cAMP. *Journal of Animal Science* 57 (Suppl. 1), 317 (Abs.).

Ball, G.D., Lenz, R.W., Ax, R.L. and First, N.L. (1983a) Effects of cAMP on bovine oocyte maturation and cumulus expansion. In: Greenwald, G.S. and Terracova, P.F. (eds) *Factors Regulating Ovarian Function*. Raven Press, New York, pp. 61–64.

Ball, G.D., Leibfried, M.L. Lenz, R.W., Ax, R.L., Bavister, B.D. and First, N.L. (1983b) Factors affecting successful *in vitro* fertilization of bovine follicular oocytes. *Biology of Reproduction* 28, 717–725.

Baltussen, R.M.W.J., Vos, P.L.A.M., Pieterse, M.C., de Loos, F.A.M., Bevers, M.M. and Dieleman, S.J. (1992) Transvaginal ultrasound-guided follicle puncture in PMSG/PG-treated cows: a comparison between three puncture needles. *Proceedings of the Twelfth International Congress on Animal Reproduction* (The Hague) 1, 129–131.

Baltze, J.M. and Biggers, J.D. (1991) Oxygen transport to embryos in microdrop cultures. *Molecular Reproduction and Development* 28, 351–355.

Bank, H. and Maurer, R.R. (1974) Survival of frozen rabbit embryos. *Experimental Cell Research* 89, 188–196.

Barandi, Z., Solti, L., Cseh, S., Varga, Z.S., Machaty, Z. and Vajta, G. (1993) Comparison of the *in vitro* fertilizing ability of sperm from endangered Hungarian Grey bulls. *Animal Reproduction Science* 31, 13–19.

Barboni, B., Mattioli, M. and Seren, E. (1992) Meiotic competence of growing oocytes fused with enucleated fully grown oocytes. *Proceedings of the Twelfth International Congress on Animal Reproduction* (The Hague) 1, 306–308.

Barg, P.E., Barad, D.H. and Feichtinger, W. (1990) Ultrarapid freezing (URF) of mouse and human preembryos: a modified approach. *Journal of In Vitro Fertilization and Embryo Transfer* 7, 355–357.

Baril, G., Casamitjana, P., Perrin, J. and Vallet, J.C. (1988) Embryo production, freezing and transfer in Angora, Alpine and Saanen goats. *Proceedings of the Fourth Meeting, European Embryo Transfer Association* (Lyon), 67–93.

Barlow, P., VanDer Zwalmen, P., Schoysman, R. and Leroy, F. (1991) Improvement of fertilizing performance by normal and abnormal mouse semen after zona opening of mature oocytes. *Molecular Reproduction and Development* 30, 385–389.

Barlow, P., Puissant, F., Van Der Zwalmen, P., Vandromme, J., Trigaux, P. and Leroy, F. (1992) *In vitro* fertilization, development, and implantation after exposure of mature mouse oocytes to visible light. *Molecular Reproduction and Development* 33, 297–302.

Barnes, F.L. and Eyestone, W.H. (1990) Early cleavage and the maternal transition in bovine embryos. *Theriogenology* 33, 141–152.

Barnes, F.L., Parrish, J.J., Susko-Parrish, J.L. and First, N.L. (1987) Morphological and molecular aspects of early development in the bovine. *Theriogenology* 27, 210.

Barnes, F.L., Westhusin, M.E. and Looney, C.R. (1990) Embryo cloning: principles and progress. *Proceedings of the Fourth World Congress on Genetics Applied to Livestock Production* 16, 323–333.

Barnes, F.L., Looney, C.R. and Westhusin, M.E. (1991) Embryo cloning in cattle: the current state of technology. *Embryo Transfer Newsletter* 6(1), 1–5.

Barnes, F., Endebrock, M., Looney, C., Powell, R., Westhusin, M. and Bondioli, K. (1993a) Embryo cloning in cattle: the use of *in vitro* matured oocytes. *Journal of Reproduction and Fertility* 97, 317–320.

Barnes, F.L., Collas, P., Powell, R., King, W.A., Westhusin, M. and Shepherd, D. (1993b) Influence of recipient oocyte cell cycle stage on DNA synthesis, nuclear envelope breakdown, chromosome constitution, and development in nuclear transplant bovine embryos. *Molecular Reproduction and Development* 36, 33–41.

Barrios, D.R., Kraemer, D.C., Bessoudo, E. and Adams, L.G. (1988) Failure to isolate *Brucella abortus* from embryos or ova from culture-positive superovulated cows. *Theriogenology* 29, 353–361.

Barth, A.D. and Bowman, P.A. (1988) Determination of the best practical method of thawing bovine semen. *Canadian Veterinary Journal* 29, 366–369.

Battut, I., Bezard, J. and Palmer, E. (1990) Establishment of equine oviduct cell monolayer for coculture with early equine embryos. *Proceedings of the Fifth International Symposium on Equine Reproduction* (Deauville), 128–129 (Abs.).

Battut, I., Bezard, J. and Palmer, E. (1991) Establishment of equine oviduct cell monolayers for co-culture with early equine embryos. *Journal of Reproduction and Fertility*, Suppl. 44, 393–403.

Bavister, B.D. (1986) Animal *in vitro* fertilization and embryo development. In: Gwatkin, R.L. (ed.) *Developmental Biology*, Vol. 4, *Manipulation of Mammalian Development*. Plenum Press, London, pp. 81–148.

Bavister, B.D. (1988) Role of oviductal secretions in embryonic growth *in vivo* and *in vitro*. *Theriogenology* 29, 143–154.

Bavister, B.D. (1989) A consistently successful procedure for *in vitro* fertilization of Golden Hamster eggs. *Gamete Research* 23, 139–158.

Bavister, B.D. (1992a) Metabolism and nutrition of the preimplantation embryo. *Proceedings of the Nordic A.I. Virtanen Institute Symposium on the In Vitro Culture of Domestic Animal Embryos*, 5–12.

Bavister, B.D. (1992b) Problems of *in vitro* embryo development in domestic animals. *Proceedings of the Nordic A.I. Virtanen Institute Symposium on In Vitro Culture of Domestic Animal Embryos*, 21–27.

Bavister, B.D. (1992c) Co-culture for embryo development: is it really necessary? *Human Reproduction* 7, 1339–1341.

Bavister, B.D. (1993) Response to the use of co-culture for embryo development. *Human Reproduction* 8(8), 1160–1162.

Bavister, B.D. and Fischer, B. (1991) What is the physiological oxygen tension for mammalian pre-and early post-implantation embryos. *Journal of Reproduction and Fertility*, Abstract Series No. 7, Abs. 4.

Bavister, B.D. and McKiernan, S.H. (1991) Different lots of bovine serum albumin stimulate or inhibit hamster embryo development *in vitro*. *Serono Symposium on Preimplantation Embryo Development*, Abstract Vol., 52 (Abs. II–3).

Bavister, B.D. and Yanagimachi, R. (1977) The effects of sperm extracts and energy sources on the motility and acrosome reactions of hamster spermatozoa *in vitro*. *Biology of Reproduction* 16, 228–237.

Bavister, B.D., Rose-Hellekant, T.A. and Pinyopummintr, T. (1992) Development of *in vitro* matured/*in vitro* fertilized bovine embryos into morulae and blastocysts in defined culture media. *Theriogenology* 37, 127–146.

Bazer, F.W., Thatcher, W.W., Hansen, P.J., Mirando, M.A., Ott, T.L. and Plante, C. (1990a) Physiological mechanisms of pregnancy recognition in ruminants. *Proceedings of the Third International Ruminant Reproduction Symposium* (Nice), 1 Abstract.

Bazer, F.W., Terqui, M. and Martinat-Botte, F. (1990b) Physiological limits to reproduction. *Proceedings of the Fourth World Congress on Genetics Applied to Livestock Production* 16, 292–298.

Bazer, F.W., Ott, T.L. and Spencer, T.E. (1994) Pregnancy recognition in ruminants, pigs and horses: signals from the trophoblast. *Theriogenology* 41, 79–94.

Beard, A.J., Groome, N. and Knight, P.G. (1991) Estimation of immunoreactive (ir) inhibin concentrations in bovine ovarian follicles using a novel 2-site immunoradiometric assay (IRMA) specific for dimeric inhibin. *Journal of Reproduction and Fertility*, Abstract Series No. 7, Abs. 40.

Beckmann, L.S. and Day, B.N. (1993) Effects of media NaCl concentration and osmolarity on the culture of early-stage porcine embryos and the viability of embryos cultured in a selected superior medium. *Theriogenology* 39, 611–622.

Becze, J. (1990) Breeding and biotechnological (economic) aspects of twinning in Simmental cows. *Tierarztliche Umschau* 44, 719–722.

Bedford, J.M. and Dobrenis, A. (1989) Light exposure of oocytes and pregnancy rates after their transfer in rabbits. *Journal of Reproduction and Fertility* 85, 477–481.

Bedirian, K.N. and Baker, R.D. (1975) Follicular development, oocyte maturation and ovulation in gonadotropin-treated prepuberal calves. *Canadian Journal of Animal Science* 55, 193–199.

Bedirian, K.N., Shea, B.F. and Baker, R.D. (1975) Fertilization of bovine follicular oocytes in bovine and porcine oviducts. *Canadian Journal of Animal Science* 55, 251–256.

Beerepoot, G.M.M. and Dykhuizen, A.A. (1992) The economics of naturally occurring twinning in dairy cattle. *Journal of Dairy Science* 75 1044–1051.

Beernink, F.J., Dmowski, W.P. and Ericsson, R.J. (1993) Sex preselection through albumin separation of sperm. *Fertility and Sterility* 59, 382–386.

Behalova, E. and Greve, T. (1993) Penetration rate of cumulus enclosed versus denuded bovine eggs fertilized *in vitro*. *Theriogenology* 39, 186.

Behboodi, E., Anderson, G.B. and BonDurant, R.H. (1991) Development of *in vitro* fertilized oocytes from pregnant and nonpregnant cows in oviductal epithelial and cumulus cell co-culture systems. *Biology of Reproduction*, 44 (Suppl. 1), 148 (Abs. 384).

Behboodi, E., Anderson, G.B. and BonDurant, R.H. (1992) Development of *in vitro* fertilized oocytes from pregnant and nonpregnant cows in oviductal epithelial and cumulus cell co-culture systems. *Theriogenology* 38, 1077–1084.

Behrens, C. (1992) Investigations on *in vitro* maturation, fertilization and culture of bovine oocytes in modified Menezo-B2 medium. Thesis, Tierarztliche Hochschule Hannover, Germany, 117 pp.

Bellin, M.E., Hawkins, H.E., Somoza, J.N. and Ax, R.L. (1993) Sperm from high fertility bulls contain high affinity heparin binding proteins. *Biology of Reproduction* 48 (Suppl. 1), 108.

Bellows, R.A., Staigmiller, R.B., Wilson, J.M., Phelps, D.A. and Darling, A. (1991) Use of bovine FSH for superovulation and embryo production in beef heifers. *Theriogenology* 35, 1069–1082.

Bellows, R.A., Staigmiller, R.B., Orme, L.E., Short, R.E. and Knapp, B.W. (1993) Effect of sire and dam on late-pregnancy conceptus and hormone traits in beef cattle. *Journal of Animal Science*, 71, 714–723.

Bendell, J.J. and Dorrington, J.H. (1990) Epidermal growth factor influences growth and differentiation of rat granulosa cells. *Endocrinology* 127, 533–540.

Bennett, W.A., Watts, T.L., Blair, W.D., Waldhalm, S.J. and Fuquay, J.W. (1988) Patterns of oviductal motility in the cow during the oestrous cycle. *Journal of Reproduction and Fertility* 83, 537–543.

Ben-Rafael, Z. and Orvieto, R. (1993) Cytokines – involvement in reproduction. *Fertility and Sterility* 58, 1093–1099.

Berg, U. and Brem, G. (1989) *In vitro* production of bovine blastocysts by *in vitro* maturation and fertilization of oocytes and subsequent *in vitro* culture. *Zuchthygiene* 24, 134–139.

Berg, U. and Brem, G. (1990) Developmental rates of *in vitro* produced IVM/IVF bovine oocytes in different cell co-culture systems. *Theriogenology* 33, 195.

Berg, U. and Brem, G. (1991) *In vitro* embryo production from oocytes from ovaries of a single slaughtered cow. *J. Deutsche Tierarztliche Wochenschrift* 98(3), 89–91.

Berg, U., Reichenbach, H.D., Liebrich, J. and Brem, G. (1991) Twin pregnancies after transfer of *in vitro* produced bovine embryos, *Proceedings of the 7th Meeting European Embryo Transfer Association (Cambridge)*, 122.

Bergeron, H., King, W.A. and Goff, A.K. (1992) Ultrastructural study of nucleolar changes in rat embryos during diapause and reactivation. *Reproduction, Nutrition, Development* 32, 415–427.

Bergers, J.M., Dumoulin, J.C.M., Ignoul, R.C.M., Michiels, A.H.J.C., Pieters, M.H.E.C., Land, J.A., Geraedts, J.P.M. and Evers, J.L.H. (1992) Protection of mammalian oocytes against damage from the growth of ice crystals during freezing. *Human Reproduction* 7 (Suppl. 2), 204.

Bergfelt, D.R., Kastelic, J.P. and Ginther, O.J. (1991) Continued periodic emergence of follicular waves in non-bred progesterone-treated heifers. *Animal Reproduction Science* 24, 193–204.

Bernard, A., Hunter, J.E., Fuller, B.J., Imoedemhe, D., Curtis, P. and Jackson, A. (1992) Fertilization and embryonic development of human oocytes after cooling. *Human Reproduction* 7, 1447–1450.

Bernard, C., Valet, J.P., Beland, R. and Lambert, R.D. (1983) Prediction of bovine ovulation by a rapid radioimmunoassay for plasma LH. *Journal of Reproduction and Fertility* 69, 425–430.

Betancourt, M., Fierro, R. and Ambriz, D. (1993) *In vitro* fertilization of pig oocytes matured *in vitro*. *Theriogenology* 40, 1155–1160.

Betteridge, K.J. and Flechon, J.E. (1988) The anatomy and physiology of pre-attachment bovine embryos. *Theriogenology* 29, 155–187.

Betteridge, K.J. and Loskutoff, N.M. (1993) Prospects for improving the survival rate of transferred embryos. *Molecular Reproduction and Development* 36, 262–265.

Betteridge, K.J. and Rieger, D. (1993) Embryo transfer and related techniques in domestic animals, and their implications for human medicine. *Human Reproduction* 8, 147–167.

Bevers, M.M., van Tol, H.T.M. and de Loos, F.A.M. (1992) Protein patterns of immature bovine oocytes incubated with *S*-methionine in either medium or medium supplemented with dibutyryl-cyclic AMP or follicular fluid. *Proceedings of the Twelfth International Congress of Animal Reproduction* (The Hague) 1, 312–314.

Bezard, J., Magistrini, M., Duchamp, G. and Palmer, E. (1989) Chronology of equine fertilization and embryonic development *in vivo* and *in vitro*. *Equine Veterinary Journal*, Suppl. 8, 105–110.

Bhattacharyya, A., Roldan, E.R.S. and Yanagimachi, R. (1986) Requirements of monovalent cations in the acrosome reaction of guinea pig spermatozoa. *Gamete Research* 15, 285–294.

Bielanski, A. and Hare, W.C.D. (1991) Investigation of some antimicrobial procedures on the *in vitro* development of early murine embryos aimed toward developing methods for the disinfection of mammalian embryos prior to transfer. *Journal of In Vitro Fertilization and Embryo Transfer* 8, 24–32.

Bielanski, A.S., Schneider, U., Pawlyshyn, V.P. and Mapletoft, R.J. (1986) Factors affecting survival of deep-frozen bovine embryos *in vitro*: the effect of freezing container and method of removing cryoprotectant. *Theriogenology* 25, 429–437.

Bielanski, A., Lowen, K.S. Del Campo, M.R., Sirard, M.A. and Willadsen, S. (1993) Isolation of bovine herpesvirus-1 (BHV-1) and bovine viral diarrhea virus (BVDV) in association with the *in vitro* production of bovine embryos. *Theriogenology* 40, 531–538.

Biery, K.A., Bondioli, K.R. and De Mayo, F.J. (1988) Gene transfer by pronuclear injection in the bovine. *Theriogenology* 29, 224.

Biggers, J.D. (1991) Walter Heape, FRS: a pioneer in reproductive biology. Centenary of his embryo transfer experiments. *Journal of Reproduction and Fertility* 93, 173–186.

Biggers, J.D. and Stern, S. (1973) Metabolism of the preimplantation mammalian embryo. In: Bishop, M.W.H. (ed.) *Advances in Reproductive Physiology*, Vol. 6. Paul Elek, London, pp. 1–59.

Bilodeau, S., Fortier, M.A. and Sirard, M.A. (1993) Effect of adenylate cyclase stimulation on meiotic resumption and cyclic AMP content of zona-free and cumulus enclosed bovine oocytes *in vitro*. *Journal of Reproduction and Fertility* 97, 5–11.

Bilton, R.J. and Moore, N.W. (1976) *In vitro* culture, storage and transfer of goat embryos. *Australian Journal Biological Science* 29, 125–129.

Bird, J.M. and Houghton, J.A. (1988) Heterologous *in vitro* fertilization of zona-free hamster oocytes as a method for the visualization of bovine sperm chromosomes. *Proceedings of the Eleventh International Congress of Animal Reproduction and Artificial Insemination* (Dublin) 3, 320 (3 pp.).

Bird, J.M., Carey, S. and Houghton, J.A. (1989) Motility and acrosomal changes in ionophore-treated bovine spermatozoa and their relationship with *in vitro* penetration of zona-free hamster oocytes. *Theriogenology* 32, 227–242.

Biven, J., Vogelsang, S., Palmer, E. and Cook, V. (1989) Commercial equine embryo transfer. *Equine Veterinary Journal*, Suppl. 8, 75.

Blakewood, E.G. and Zhang, L. (1993) The use of chick embryo amniotic fluids for the *in vitro* culture of early stage mammalian embryos. *Theriogenology* 39, 189.

Blakewood, E.G., Jaynes, J.M. and Godke, R.A. (1988) Culture of pronuclear mammalian embryos using domestic chicken eggs. *Theriogenology* 29, 226.

Blakewood, E.G., Pool, S.H., Wiemer, K.E. and Godke, R.A. (1989a) Culturing early-stage bovine morulae using domestic chicken eggs. *Theriogenology* 31, 176.

Blakewood, E.G., Wiemer, K.E. and Godke, R.A. (1989b) Post-thaw viability of bovine embryos cultured in domestic chicken eggs or on epithelial monolayers prior to freezing in liquid nitrogen. *Theriogenology* 31, 177.

Blakewood, E.G., Jaynes, J.M., Johnson, W.A., Godke, R.A. (1990) Using the amniotic cavity of the developing chick embryo for the *in vivo* culture of early-stage mammalian embryos. *Poultry Science* 68, 1695–1702.

Blanchet, G.B., Russell, J.B., Fincher, C.R. and Portmann, M. (1992) Laser micromanipulation in the mouse embryo: a novel approach to zona drilling. *Fertility and Sterility* 57, 1337–1341.

Blaxter, K. (1987) Future hunger. *Lancet* i, 7 February, 309–313.

Blondin, P. (1993) The influence of oocyte and follicular morphology on developmental competence following *in vitro* maturation in cows. *Biology Reproduction* 48 (Suppl. 1), 125.

Blottner, S., Nehring, H. and Torner, H. (1990) Individual differences in capacitation of bull spermatozoa by heparin *in vitro*: relationship to fertility. *Theriogenology* 34, 619–628.

Blottner, S., Bostedt, H., Mewes, K. and Schill, W.B. (1992) Electrophoretic enrichment of bovine X- and Y-spermatozoa quantified by use of Y-specific DNA and the F-body assay. *Proceedings of the Twelfth International Congress on Animal Reproduction* (The Hague) 1, 411–413.

Blue, B.J., McKinnon, A.O., Squires, E.L., Seidel, G.E. and Muscari, K.T. (1989) Capacitation of stallion spermatozoa and fertilization of equine oocytes *in vitro*. *Equine Veterinary Journal*, Suppl. 8, 111–116.

Bo, G.A., Hockley, D., Tribulo, H., Jofre, F., Tribulo, R., Busso, N., Barth, A.D. and Mapletoft, R.J. (1991) The effect of dose schedule and route of administration on superovulatory response to Folltropin in the cow. *Theriogenology* 35, 186.

Bo, G.A., Rabella, R., Caccia, M., Gomez, A., Adams, G.P., Pierson, R.A. and Mapletoft, R.J. (1993) The effect of estradiol-17B on follicular growth and wave emergence in heifers. *Theriogenology* 39, 190.

Boatman, D.E., Andrews, J.C. and Bavister, B.D. (1988) A quantitative assay for capacitation: evaluation of multiple sperm penetration through the zona pellucida of salt-stored Hamster eggs. *Gamete Research* 19, 19–29.

Boediono, A., Takagi, M., Saha, S. and Suzuki, T. (1993a) Chimeric blastocyst production using IVF bovine embryos (Holstein-Japanese Red) without zona pellucida. *Theriogenology* 39, 191.

Boediono, A., Ooe, M., Yamamoto, M., Takagi, M., Saha, S. and Suzuki, T. (1993b) Production of chimeric calves by aggregation of *in vitro*-fertilized bovine embryos without zonae pellucidae. *Theriogenology* 40, 1221–1230.

Boice, M.L., Geisert, R.D., Blair, R.M. and Verhage, H.G. (1990a) Characterization of secretory glycoproteins synthesized by the bovine oviduct at oestrus. *Biology of Reproduction* 42 (Suppl. 1), 128–129.

Boice, M.L., Geisert, R.D., Blair, R.M. and Verhage, H.G. (1990b) Identification and characterization of bovine oviductal glyoproteins synthesized at estrus. *Biology of Reproduction* 43, 457–465.

Boland, M.P. (1984) Use of the rabbit oviduct as a screening tool for the viability of mammalian eggs. *Theriogenology* 21, 127–137.

Boland, M.P. and Gordon, I. (1989) Embryo transfer in developing countries – Saudi Arabia. *Theriogenology* 31, 41–42.

Boland, M.P., Crosby, T.F. and Gordon, I. (1975) Twin pregnancy in cattle established by non-surgical egg transfer. *British Veterinary Journal* 131, 738–740.

Boland, M.P., Crosby, T.F. and Gordon, I. (1979) Twinning in beef and dairy cows. *Faculty of General Agriculture, Research Report (1978–79)*, 80.

Boland, N.I., Humpherson, P.G., Leese, H.J. and Gosden, R.G. (1991) Carbohydrate metabolism by mouse ovarian follicles. *Journal of Reproduction and Fertility*, Abstract Series No. 8, 8 (Abs. 5).

Boland, N.I., Humpherson, P.G., Leese, H.J. and Gosden, R.G. (1993) Pattern of lactate production and steroidogenesis during growth and maturation of mouse ovarian follicles *in vitro*. *Biology of Reproduction* 48, 798–806.

Bolander, F.F. (1977) The purification and characterization of bovine and rabbit placental lactogen. *Dissertation Abstracts International* 37B, 5990–5991.

Bolton, V.N. (1992) Controversies and opinions in embryo culture: two to four cell transfer vs blastocyst. *Journal of Assisted Reproduction and Genetics* 9, 506–508.

Bondioli, K.R. (1992) Commercial cloning of cattle by nuclear transfer. *Proceedings of a Symposium on Cloning Mammals by Nuclear Transplantation* (Fort Collins), 35–38.

Bondioli, K.R. (1993) Nuclear transfer in cattle. *Molecular Reproduction and Development* 36, 274–275.

Bondioli, K.R. and Wright, R.W. (1980) Influence of culture media on *in vitro* fertilization of ovine tubal oocytes. *Journal of Animal Science* 51, 660–667.

Bondioli, K.R. and Wright, R.W. (1983a) *In vitro* fertilization of bovine oocytes by sperm capacitated *in vitro*. *Journal of Animal Science* 57, 1001–1005.

Bondioli, K.R. and Wright, R.W. (1983b) *In vitro* fertilization of pig and sheep oocytes matured *in vitro*. *Journal of Animal Science* 57, 1006–1012.

Bondioli, K.R., Ellis, S.B., Pryor, J.H., Williams, M.W. and Harpold, M.M. (1989a) The use of male-specific chromosomal DNA fragments to determine the sex of bovine implantation embryos. *Theriogenology* 31, 95–104.

Bondioli, K.R., Biery, K.A., Hill, K.G., Jones, K.B. and Mayo, F.J.De (1989b) Production of transgenic cattle by pronuclear injection. *Proceedings of the Agbiotch '89 Conference*, 292–299.

Bondioli, K.R., Westhusin, M.E. and Looney, C.R. (1990a) Production of identical bovine offspring by nuclear transfer. *Theriogenology* 33, 165–174.

Bondioli, K.R., Westhusin, M.E. and Looney, C.R. (1990b) Production of identical bovine offspring by nuclear transfer. *Proceedings of the Third Symposium on Advanced Topics in Animal Reproduction*, FCAV-UNESP, Jaboticabal, 143–151.

BonDurant, R.H., Anderson, G.B., Boland, M.P., Cupps, P.T. and Hughes, M.A. (1982) Preliminary studies on bovine embryo survival following short-term storage at 4°C. *Theriogenology* 17, 223–230.

Bongso, A. and Fong, C.Y. (1991) Improving embryo quality via co-cultures. *Contributions to Obstetrics and Gynaecology* 1, 53–60.

Bongso, T.A., Sathananthan, A.H., Wong, P.C., Ratnam, S.S., Ng, S.C., Anandakumar, C. and Ganatra, S. (1989) Human fertilization by micro-injection of immotile spermatozoa. *Human Reproduction* 4, 175–179.

Bongso, A., Ng, S.-C. and Ratnam, S. (1990) Co-cultures: their relevance to assisted reproduction. *Human Reproduction* 5, 893–900.

Bongso, A., Ng, S.-C., Fong, C.-Y. and Ratnam, S. (1991a) Improved fertilization rates of human oocytes in coculture. *Journal of In Vitro Fertilization and Embryo Transfer* 8(4), 216–221.

Bongso, A., Ng, S.-C., Fong, C.-Y. and Ratnam, S. (1991b) Cocultures: a new lead in embryo quality improvement for assisted reproduction. *Fertility and Sterility* 56, 179–191.

Bongso, A., Ng, S.-C., Fong, C.Y., Anandakumar, C., Marshall, B., Edirisinghe, R. and Ratnam, S. (1992) Improved pregnancy rate after transfer of embryos grown in human fallopian tubal cell co-culture. *Fertility and Sterility* 58, 569–574.

Boni, R., Santella, L., Dale, B., Barbieri, V. and Zicarelli, L. (1991) An ultrastructural study of maturation in buffalo oocytes. *Proceedings of the Seventh Meeting of the European Embryo Transfer Association* (Cambridge), 128.

Boni, R., Roelofsen, M.W.M., Pieterse, M.C., Wurth, Y.A. and Kruip, Th.A.M. (1992a) Ovum pick-up and embryo production *in vitro*: an established procedure. *Proceedings of the Eighth Conference of the European Embryo Transfer Association* (Lyon), 130.

Boni, R., Campanile, G., Zicarelli, L., Taccone, W. and Kruip, Th.A.M. (1993a) Health and some related blood parameters in cows, punctured twice-weekly during three months to collect immature oocytes. *Proceedings of the Ninth Congress of the European Embryo Transfer Association* (Lyon), 160.

Boni, R., Roelofsen-Vendrig, M., Pieterse, M. and Wurth, Y. (1993b) Follicular recruitment after repeated removal of all follicles >2 mm in the bovine ovary. *Journal of Reproduction and Fertility*, Abstract Series No. 12, 40.

Booman, P., Kruijt, L., Veerhuis, R., Hengst, A.M., Tieman, M. and Ruch, F.E. (1989) Sexing bovine embryos with monoclonal antibodies against the H-Y antigen. *Livestock Production Science* 23(1–2), 1–16.

Boone, W.R. and Shapiro, S.S. (1990) Quality control in the *in vitro* fertilization laboratory. *Theriogenology* 33, 23–50.

Bos-Mikich, A., Candy, C., Wood, M.J. and Whittingham, D.G. (1992) Cytogenetic analysis and developmental potential of vitrified mouse oocytes. *Journal of Reproduction and Fertility*, Abstract Series No. 9, 26 (Abs. 38).

Bottcher, M., Alm, H., Lange, W. and Kauffold, P. (1989) Current methods of collecting ova for *in vitro* cell techniques. *Wissenschaftliche Zeitschrift der Wilhem-Pieck-Universitat Rostock, Naturwissenschraftliche Reihe* 38(4), 21–24.

Bottcher, M., Blottner, S., Kruger, S., Rettig, B. and Lange, W. (1990) Effect of cumulus oophorus on the fertilization of cattle oocytes matured *in vitro*. *Archiv für Experimentelle Veterinarmedizin* 44, 53–57.

Bottcher, M., Kluge, B., Tomalak, K., Meyer, E. and Lange, W. (1992) Influence of follicular fluid on fertilization of *in vitro* matured bovine oocytes. *Reproduction in Domestic Animals* 27(2), 101–111.

Bourke, S. (1992) Beef embryo transfer twinning at commercial level. *Irish Farmers' Journal* 44(13), 43.

Bou Shorgan, Zhang, S.L., Xue, X.X., Pang, Y.F., Chan, H.W., Li, X., Hai, Q.L. and Si, Qu. (1990) *In vitro* development of ovine oocytes matured and fertilized *in vitro* and lambing after embryo transfer. *Japanese Journal of Animal Reproduction* 36, 225–230.

Bousquet, D., Goff, D., King, A.K., Greve, T. and Betteridge, K.J. (1984) Fertilization of bovine oocytes *in vitro*: analysis of some factors affecting the fertilization rates. *Proceedings of the Tenth Congress on Animal Reproduction and Artificial Insemination* (Illinois) 3, 386 (3 pp).

Bowen, R.A., Reed, M., Schnieke, A. and Seidel, G.E. (1993) Production of transgenic cattle from PCR-screened embryos. *Theriogenology* 39, 194.

Boyd, J.S., Omran, S.N. and Ayliffe, T.R. (1988) Use of high frequency transducer with realtime B-mode ultrasound scanning to identify early pregnancy in cows. *Veterinary Record* 123, 8–11.

Brackett, B.G. (1981) Applications of *in vitro* fertilization. *New Technologies in Animal Breeding*. Academic Press, New York, pp. 141–161.

Brackett, B.G. (1983) A review of bovine fertilization *in vitro*. *Theriogenology* 19, 1–15.

Brackett, B.G. (1992) *In vitro* fertilization in farm animals. In: Lauria, A. and Gandolfi, F. (eds) *Embryonic Development and Manipulation in Animal Production*. Portland Press, London, pp. 59–76.

Brackett, B.G. and Bousquet, D. (1984) Comparative aspects of fertilization *in vitro*. In: Trounson, A. and Wood, C. (eds) *In Vitro Fertilization and Embryo Transfer*. Churchill Livingstone, London, pp. 32–43.

Brackett, B.G. and Oliphant, G. (1975) Capacitation of rabbit spermatozoa *in vitro*. *Biology of Reproduction* 12, 260–274.

Brackett, B.G. and Zuelke, K.A. (1993) Analysis of factors involved in the *in vitro* production of bovine embryos. *Theriogenology* 39, 43–64.

Brackett, B.G., Baranska, W., Sawicki, W. and Koprowski, H. (1971) Uptake of heterologous genome by mammalian spermatozoa and its transfer to ova through fertilization. *Proceedings of the National Academy of Sciences (USA)* 68, 353–357.

Brackett, B.G., Evans, J.F., Donawick, W.J., Boice, M.L. and Cofone, M.A. (1980a) *In vitro* penetration of cow oocytes by bull sperm. *Archives of Andrology* 5, 69–71.

Brackett, B.G., Oh, Y.K., Evans, J.F. and Donawick, W.J. (1980b) Fertilization and early development of cow ova. *Biology of Reproduction* 23, 189–205.

Brackett, B.G., Schellander, K., Fayrer-Hosken, R.A. and Keefer, C.L. (1987) *In vitro* fertilization of *in vivo* matured oocytes. *Proceedings of a Symposium on the Application of Egg and Embryotechnologies to Domestic Animals,* Copenhagen, 32a–32c.

Brackett, B.G., Younis, A.I. and Fayrer-Hosken, R.A. (1989) Enhanced viability after *in vitro* fertilization of bovine oocytes matured *in vitro* with high concentrations of luteinizing hormone. *Fertility and Sterility* 52, 319–324.

Brackett, R.G., Bousquet, D., Boice, M.L., Donawick, W.J., Evans, J.F. and Dressel, M.A. (1982) Normal development following *in vitro* fertilization in the cow. *Biology of Reproduction* 27, 147–158.

Brackett, R.G., Keefer, C.L., Troop, C.G., Donawick, W.J. and Bennett, K.A. (1984) Bovine twins resulting from *in vitro* fertilization. *Theriogenology* 21, 224.

Bradley, M.P. (1989) Immunological sexing of mammalian semen: current status and future options. *Journal of Dairy Science* 72, 3372–3380.

Bradley, M.P. and Heslop, B.F. (1988) The development of strategies for the immunological sexing of mammalian semen. *Journal of Dairy Science* 71 (Suppl. 1), 318 (Abs.).

Braun, J. (1988) Influence of protein supplement and culture conditions on cumulus cell expansion and nuclear maturation of sheep follicular oocytes. *Theriogenology* 29, 189–202.

Brebion, P., Vallet, J.C., Cognie, Y., Lajous, D. and Baril, G. (1989) Ovine embryo transfer in the Lacaune breed: efficiency of a new laparoscopic technique. *Proceedings of the Fifth Congress of the European Embryo Transfer Association* (Lyon), 134.

Brebion, P., Baril, G., Gognie, Y. and Vallet, J.C. (1992) Embryo transfer in sheep and goats. *Annales de Zootechnie* 41, 331–339.

Bredbacka, P. and Bredbacka, K. (1993) Effect of inositol on *in vitro* development of murine and bovine morulae and blastocysts. *Theriogenology* 39, 195.

Bredbacka, P., Huhtinen, M., Aalto, J. and Rainio, V. (1992) Viability of bovine demi- and quarter-embryos after transfer. *Theriogenology* 38, 107–113.

Brem, G. (1992) Gene transfer in farm animals. In: Lauria, A. and Gandolfi, F. (eds) *Embryonic Development and Manipulation in Animal Production.* Portland Press, London, 147–164.

Brem, G. (1993) Inheritance and tissue-specific expression of transgenes in rabbits and pigs. *Molecular Reproduction and Development* 36, 242–244.

Breuel, K.F., Spitzer, J.C., Gimenez, T., Henricks, D.M. and Gray, S.L. (1988) Effect of holding time and temperature of whole blood on concentration of progesterone, estradiol-17B and estrone in plasma and serum samples. *Theriogenology* 30, 613–627.

Breuel, K.F., Lewis, P.E., Schrick, F.N., Lishman, A.W., Inskeep, E.K. and Butcher, R.J.L. (1993) Factors affecting fertility in the post-partum cow: role of the oocyte and follicle in conception rate. *Biology of Reproduction* 48, 655–661.

Brinster, R.L. (1968) *In vitro* culture of mammalian embryos. *Journal of Animal Science,* Suppl. 1 1–14.

Brinster, R.L., Sandgren, E.P., Behringer, R.R. and Palmiter, R.D. (1989) No simple solution to making transgenic mice. *Cell* 59, 239–241.

Brison, D.R. and Leese, H.J. (1991) Energy metabolism in late preimplantation rat embryos. *Journal of Reproduction and Fertility* 93, 245–251.

Brison, D.R., Hewitson, L.J., Hinde, P.S., Tomos, A.D. and Leese, H.J. (1992) The microenvironment of the blastocoel cavity in mouse and rat. *Journal of Reproduction and Fertility*, Abstract Series No. 9, 57 (Abs. 100).

Brison, D.R., Hewitson, L.C. and Leese, H.J. (1993) Glucose, pyruvate and lactate concentrations in the blastocoel cavity of rat and mouse embryos. *Molecular Reproduction and Development* 35, 227–232.

Britt, J.H. and Holt, L.C. (1988) Endocrinological screening of embryo donors and embryo transfer recipients: a review of research with cattle. *Theriogenology* 29, 189–202.

Broadbent, P.J. (1990) The NOSCA Simmental MOET Project. *Proceedings of the British Cattle Breeders Club*, Digest No. 45, 44–48.

Broadbent, P.J. and Dolman, D.F. (1989) Twinning cattle by embryo transfer. In: Phillips, C.J.C. (ed.) *New Techniques in Cattle Production.* Butterworths, London, pp. 238–239.

Broadbent, P.J., Stewart, M. and Dolman, D.F. (1991) Recipient management and embryo transfer. *Theriogenology* 35, 125.

Broadbent, P.J., Sinclair, K.D., Dolman, D.F., Mullan, J.S. and McNally, J.R. (1992) The effect of a Norgestomet ear implant (Crestar) on pregnancy rate in embryo transfer recipients. *Proceedings of the Twelfth International Congress of Animal Reproduction* (The Hague) 2, 782–784.

Broadbent, P.J., Tregaskes, L.D., Dolman, D.F., Franklin, M.F. and Jones, R.L. (1993) Synchronization of estrus in embryo transfer recipients after using a combination of PRID or CIDR-B plus PGF2 alpha. *Theriogenology* 39, 1055–1065.

Bromhall, J.D. (1975) Nuclear transplantation in the rabbit egg. *Nature* 258, 719–722.

Brown, C.R., Clarke, N., Aiken, M. and Bavister, B.D. (1990) Changes in the composition of the hamster zona pellucida after fertilization *in vivo* but not *in vitro*. *Journal of Reproduction and Fertility* 90, 447–454.

Brown, J.J.G. and Whittingham, D.G. (1990) Glucose requirement for morula to blastocyst transition during culture of 1-cell F2 hybrid mouse embryos. *Journal of Reproduction and Fertility*, Abstract Series No. 5, Abs. 21.

Brucker, C., Alexander, N.J., Hodgen, G.D. and Sandow, B.A. (1991) Transforming growth factor-alpha augments meiotic maturation of cumulus cell-enclosed mouse oocytes. *Molecular Reproduction and Development* 28, 94–98.

Bruyas, J.-F., Bezard, J., Lagneaux, D. and Palmer, E. (1993) Quantitative analysis of morphological modifications of day 6.5 horse embryos after cryopreservation: differential effects on inner cell mass and trophoblast cells. *Journal of Reproduction and Fertility* 99, 15–23.

Buccione, R., Schroeder, A.C. and Eppig, J.J. (1990) Interactions between somatic cells and germ cells throughout mammalian oogenesis. *Biology of Reproduction* 43, 543–547.

Buckrell, B.C., Gartley, C.J., Buschbeck, C., Jordan, P. and Walton, J.W. (1993) Evaluation of a transcervical AI technique for transferring embryos in sheep. *Theriogenology* 39, 197.

Buhi, W.C., Alvarez, I.M., Sudhipong, V. and Dones-Smith, M.M. (1990) Identification and characterization of *de novo*-synthesized porcine oviductal secretory proteins. *Biology of Reproduction* 43, 929–938.

Bungartz, L. and Niemann, H. (1993) Effects of a dominant follicle on ovarian responses of dairy cows following various superovulatory treatment schedules. *Theriogenology* 39, 198.

Burkle, K., Bispink, L., Rader, C. and Gehring, W. (1993) Artificial insemination with frozen–thawed and Percoll treated donor semen. *Journal of Reproduction and Fertility*, Abstract Series No. 12, 57.

Burns, P.D. and Spitzer, J.C. (1992) Influence of biostimulation on reproduction in postpartum beef cows. *Journal of Animal Science* 70, 358–362.

Butler, J.E. and Biggars, J.D. (1989) Assessing the viability of preimplantation embryos *in vitro*. *Theriogenology* 31, 115–126.

Butler, J.E. and Williams, J.E. (1989) Non-invasive measurement of glucose uptake by individual ovine preimplantation embryos. *Journal of Animal Science* 67 (Suppl. 1), 368–369 (Abs. 894).

Butler, J.E. and Williams, J.E. (1990) Preliminary characterization of pyruvate uptake by one-cell ovine embryos. *Theriogenology* 33, 423–432.

Butler, J.E. and Williams, J.E. (1991) Non-invasive measurement of pyruvate uptake by ovine preimplantation embryos and unfertilized ova. *Theriogenology* 36, 1043–1048.

Butler, J.E. and Williams, J.E. (1992) Noninvasive measurement of glucose uptake by preimplantation sheep embryos and unfertilized ova. *Small Ruminant Research* 7, 347–352.

Butler, J.E., Lechene, C. and Biggars, J.D. (1988) Non-invasive measurement of glucose uptake by two populations of murine embryos. *Biology of Reproduction* 39, 779–786.

Byrd, W. (1981) *In vitro* capacitation and the chemically induced acrosome reaction in bovine spermatozoa. *Journal of Experimental Zoology* 215, 35–46.

Byskov, A.G. (1982) Primordial germ cells and regulation of meiosis. In: Austin, C.R. and Short, R.V. (eds) *Reproduction in Mammals*, Book 1, 2nd edn. Cambridge University Press, Cambridge, pp. 1–16.

Byskov, A.G., Hoyer, P.E. and Gualiang, X. (1992) An *in vitro* model for studying oocyte follicular growth initiation. *Journal of Reproduction and Fertility*, Abstract Series No. 9, 16 (Abs. 18).

Byun, T.H., Lee, S.H. and Song, H.B. (1991) Development of a rapid staining method for nucleus of the oocyte from domestic animals. *Korean Journal of Animal Sciences* 33(1), 25–31.

Cairoli, F., Ferrario, L., Carli, S. and Soldano, F. (1993) Efficacy of oxytetracycline and tetracycline-benzydamine in the prevention of infection after placental retention in cattle. *Veterinary Record* 133, 394–395.

Calder, M. and Rajamahendran, R. (1992) Follicular growth, ovulation and embryo recovery in dairy cows given FSH at the beginning of middle of the estrous cycle. *Theriogenology* 38, 1163–1174.

Calder, M.C., Taylor, C., Sianangama, C. and Rajamahendran, R. (1991) The effect of superovulatory treatment at the beginning of the cycle on follicular number, ovulation rate and embryo recovery in dairy cows. *Journal of Dairy Science* 74 (Suppl. 1), 227 (Abs. P256).

Callesen, H., Greve, T. and Christensen, F. (1987) Ultrasonically guided aspiration of bovine follicular oocytes. *Theriogenology* 27, 217.

Callesen, H., Bak, A. and Greve, T. (1992a) Prediction of sex of embryos from superovulated cattle based on developmental stages at recovery. *Proceedings of the Eighth Conference of the European Embryo Transfer Association* (Lyon), 134.

Callesen, H., Greve, T. and Bak, A. (1992b) Embryotechnology in dairy cattle breeding. In: Lauria, A. and Gandolfi, F.(eds) *Embryonic Development and Manipulation in Animal Production*. Portland Press, London, pp. 207–214.

Cameron, R.D.A., Durack, M., Fogarty, R., Putra, D.K.H. and McVeigh, J. (1989) Practical experience with commercial embryo transfer in pigs. *Australian Veterinary Journal* 66, 314–318.

Campbell, A.V. (1990) Ethics and animal production. *Proceedings of the New Zealand Society for Animal Production* 50, 211–213.

Campbell, B.K. and Scaramuzzi, R.J. (1991) Induction by FSH of oestradiol and inhibin production by ovine granulosa cells cultured in serum free media containing insulin. *Journal of Reproduction and Fertility*, Abstract Series No. 7, 13 (Abs. 15).

Campbell, K.H.S., Ritchie, W.A. and Wilmut, I. (1993) Nuclear-cytoplasmic interactions during the first cell cycle of nuclear transfer reconstructed bovine embryos: implications for deoxyribonucleic acid replication and development. *Biology of Reproduction* 49, 933–942.

Campo, M.R., Tamayo, R. and Campo, C.H. (1987) Embryo transfer from brucellosis-positive donors: a field trial. *Theriogenology* 27, 221.

Campo, M.R. Del., Donoso, M.X., Parrish, J.J. and Ginther, O.J. (1990) *In vitro* fertilization of *in vitro*-matured equine oocytes. *Journal of Equine Veterinary Science* 10(1), 18–22.

Campo, M.R. Del, Donoso, M.X., Palasz, A.T., Garcia, A. and Mapletoft, R.J. (1993) The effect of days in co-culture on survival of deep frozen bovine IVF blastocysts. *Theriogenology* 39, 208.

Canseco, R.S., Sparks, A.E.T., Pearson, R.E. and Gwazdauskas, F.C. (1991) Effect of microdrop size and number of embryos per microdrop on early murine embryo development. *Serono Symposium on Preimplantation Embryo Development*, Abstract Vol., 29 (Abs. I-10).

Canseco, R.S., Sparks, A.E.T., Pearson, R.E. and Gwazdaukas, F.C. (1992) Embryo density and medium volume effects on early murine embryo development. *Journal of Assisted Reproduction and Genetics* 9, 454–457.

Capehart, J.S. and White, K.L. (1990) The differential effects of thermal stress on subsequent development of preimplantation murine embryo cell types. *Journal of Animal Science* 68 (Suppl. 1), 429.

Cargill, L.J. (1992) *Beef Prospects and Embryo Technology in an Expanded Europe*. Nuffield Farming Scholarships Trust Publication, Uckfield, East Sussex, 60 pp.

Carlsson, B., Bergh, C., Bentham, J., Olsson, J.-H., Norman, M.R., Billig, H., Roos, P. and Hillensjo, T. (1992) Expression of functional growth hormone receptors in human granulosa cells. *Human Reproduction* 7 (Suppl. 2), 1205–1209.

Carnevale, E.M., Squires, E.L. and McKinnon, A.O. (1987) Comparison of Ham's F-10 with CO or Hepes buffer for storage of equine embryos at 5°C for 24 h. *Journal of Animal Science* 65, 1775–1781.

Carney, E.W., Tobback, C., Ellington, J.E. and Foote, R.H. (1990a) Co-culture of rabbit 2-cell embryos with rabbit oviduct epithelial cells and other somatic cells. *Molecular Reproduction and Development* 27, 209–215.

Carney, E.W., Tobback, C. and Foote, R.H. (1990b) Co-culture of rabbit one-cell embryos with rabbit oviduct epithelial cells. *In Vitro Cellular and Developmental Biology* 26, 629–635.

Carney, N.J., Squires, E.L., Cook, V.M. and Jasko, D.J. (1991a) Three year study of pregnancy rates resulting from transfer of fresh versus cooled, transported equine embryos. *Theriogenology* 35, 189.

Carney, N.J., Squires, E.L., Cook, V.M., Seidel, G.E. and Jasko, D.J. (1991b) Comparison of pregnancy rates from transfer of fresh versus cooled, transported equine embryos. *Theriogenology* 36, 23–32.

Carolan, C., Monaghan, P., Mehmood, A., Lonergan, P., Gallagher, M. and Gordon, I. (1992a) Slicing of bovine ovaries as a means of oocyte recovery. *Journal of Reproduction and Fertility*, Abstract Series No. 9, 51 (Abs. 88).

Carolan, C., Monaghan, P., Lonergan, P., Gallagher, M. and Gordon, I. (1992b) Effect of granulosa cell source and serum concentration on the cleavage and blastocyst yield of bovine oocytes recovered by dissection. *Proceedings of the Eighth Conference of the European Embryo Transfer Association* (Lyon), 136.

Carolan, C., Monaghan, P., Gallagher, M. and Gordon, I. (1993) Developmental capacity of unselected bovine oocytes recovered from individual pairs of ovaries by surface dissection. *Proceedings of the Ninth Congress of the European Embryo Transfer Association* (Lyon), 180.

Carolan, C., Monaghan, P., Gallagher, M. and Gordon, I. (1994) Effect of recovery method on yield of bovine oocytes per ovary and their developmental competence after maturation, fertilization and culture *in vitro*. *Theriogenology* 41, 1061–1068.

Carroll, J. and Whittingham, D.G. (1993) The generation of Ca^{2+} oscillations during growth and maturation of mouse oocytes. *Journal of Reproduction and Fertility,* Abstract Series No. 11, 7 (Abs. 3).

Carroll, J., Warnes, G.M. and Matthews, C.D. (1989) Increase in digyny explains polyploidy after *in vitro* fertilization of frozen–thawed mouse oocytes. *Journal of Reproduction and Fertility* 85, 489–494.

Carroll, J., Whittingham, D.G., Wood, M.J., Telfer, E. and Gosden, R.G. (1990a) Extra-ovarian production of mature viable mouse oocytes from frozen primary follicles. *Journal of Reproduction and Fertility* 90, 321–327.

Carroll, J., Depypere, H. and Matthews, C.D. (1990b) Freeze–thaw-induced changes of the zona pellucida explains decreased rates of fertilization in frozen–thawed mouse oocytes. *Journal of Reproduction and Fertility* 90, 547–553.

Carroll, J., Whittingham, D.G. and Wood, M.J. (1991a) Effect of dibutyryl cyclic adenosine monophosphate on granulosa cell proliferation, oocyte growth and meiotic maturation in isolated mouse primary ovarian follicles cultured in collagen gels. *Journal of Reproduction and Fertility* 92, 197–207.

Carroll, J., Whittingham, D.G. and Wood, M.J. (1991b) Growth *in vitro* and acquisition of meiotic competence after the cryopreservation of isolated mouse primary ovarian follicles. *Reproduction, Fertility and Development* 3(5), 593–599.

Carroll, J., Wood, M.J. and Whittingham, D.G. (1993) Normal fertilization and development of frozen–thawed mouse oocytes: protective action of certain macromolecules. *Biology of Reproduction* 48, 606–612.

Cartwright, E.J., Harrington, P.M., Cowin, A. and Sharpe, P.T. (1993) Separation of bovine X and Y sperm based on surface differences. *Molecular Reproduction and Development* 34, 323–328.

Casey, P.L., Looney, C.R., Casey, D.C., Youngs, C.R. and Godke, R.A. (1989) Evaluating bovine embryo quality by computerized image analysis. *Theriogenology* 31, 181.

Castro, F.O., Hernandez, O., Uliver, C., Solano, R., Milanes, C., Aguilar, A., Perez, A., de Armas, R., Herrera, L. and de la Fuente, J. (1990) Introduction of foreign DNA into the spermatozoa of farm animals. *Theriogenology* 34, 1099–1110.

Catt, J.W. (1990) Operation of a bovine IVF centre. *Proceedings of the Third Symposium on Advanced Topics in Animal Reproduction* FCAV-UNESP, Jaboticabal, 89–100.

Catt, J.W., Ryan, J.P. and O'Neill, C. (1994) Do sperm differ in their ability to fertilize human oocytes after intracytoplasmic injection? *Theriogenology* 41, 173.

Cha, K.Y., Koo, J.J., Ko, J.J., Choi, D.H., Han, S.Y. and Yoon, T.K. (1991) Pregnancy after *in vitro* fertilization of human follicular oocytes collected from nonstimulated cycles, their culture *in vitro* and their transfer in a donor oocyte program. *Fertility and Sterility* 55, 109–113.

Chan, P.J. (1987) Developmental potential of human oocytes according to zona pellucida thickness. *Journal of In Vitro Fertilization and Embryo Transfer* 4, 237–241.

Chang, M.C. (1951) Fertilizing capacity of spermatozoa deposited into the fallopian tube. *Nature* 168, 697–698.

Chang, M.C. (1952) Development of bovine blastocyst with a note on implantation. *Anatomical Record* 113, 143–161.

Chang, M.C. (1955) The maturation of rabbit oocytes in culture and their maturation, activation, fertilization and subsequent development in the fallopian tubes. *Journal of Experimental Zoology* 123, 379–386.

Chang, M.C. (1968) *In vitro* fertilization of mammalian eggs. *Journal of Animal Science* 27 (Suppl. 1), 15–22.

Chang, M.C. and Marden, W.G.R. (1954) The aerial transport of fertilized mammalian ova. *Journal of Heredity* 45, 75–78.

Chang, M.C., Austin, C.R., Bedford, J.M., Brackett, B.G., Hunter, R.H.F. and Yanagimachi, R. (1977) Capacitation of spermatozoa and fertilization in mammals. In: Greep, R.O. and Kobrinsky, M.A. (eds) *Frontiers in Reproduction and Fertility Control,* Part 2. Ford Foundation Publications, Cambridge, Massachusetts, pp. 434–451.

Chantaraprateep, P., Lohachit, C., Techakumphu, M., Kobayashi, G., Virakul, P., Kunayongkrit, A., Prateep, P. and Limskul, A. (1989) Early embryonic development in Thai swamp buffalo (*Bubalis bubalis*). *Theriogenology* 31, 1131–1139.

Chartrain, I. and Picard, L. (1988) Ultrastructural analysis of bovine embryos at days 6–8: correlation with embryo quality. *Theriogenology* 29, 236.

Chartrain, I., Niar, A., King, W.A., Picard, L. and St. Pierre, H. (1987) Development of the nucleus in early goat embryos. *Gamete Research* 18, 201–213.

Chassagne, M. and Barnouin, J. (1992) Circulating PGF2-alpha and nutritional parameters at parturition in dairy cows with and without retained placenta: relation to prepartum diet. *Theriogenology* 38, 407–418.

Chatot, C.L., Ziomek, C.A., Bavister, B.D., Lewis, J.L. and Torres, I. (1989) An improved culture medium supports development of random-bred 1-cell mouse embryos *in vitro*. *Journal of Reproduction and Fertility* 86, 679–688.

Chatot, C.L., Lewis, J.L., Torres, I. and Ziomek, C.A. (1990) Development of 1-cell embryos from different strains of mice in CZB medium. *Biology of Reproduction* 42, 432–440.

Chauhan, M.S. and Anand, S.R. (1991) *In vitro* maturation and fertilization of goat oocytes. *Indian Journal of Experimental Biology* 29, 105–110.

Check, J.H. and Katsoff, D. (1993) A prospective study to evaluate the efficacy of modified swim-up preparation for male sex selection. *Human Reproduction* 8, 211–214.

Chedid, S., Van den Abbeel E. and Van Steirteghem, A. (1992) Effects of cryopreservation on survival and development of interphase- and mitotic-stage 1-cell mouse embryos. *Human Reproduction* 7, 1451–1456.

Chen, C. (1986) Pregnancy after human oocyte cryopreservation. *Lancet* i, 19 April, 884–886.

Chen, C. (1988) Pregnancies after human oocyte cryopreservation. *Annals of the New York Academy of Sciences* 541–549.

Chen, H.B., Lu, K.H and Polge, C. (1992) *In vitro* fertilization (IVF) of bovine oocytes with frozen–thawed sperm after storage at different temperatures and periods of time. *Proceedings of the Twelfth International Congress of Animal Reproduction* (The Hague) 2, 629–631.

Chen, L., Wert, S.E., Russell, P., Cannon, M., Hendrix, M. and Larsen, W.J. (1989) The addition of substrates of hyaluronic acid synthesis to the maturation medium increases the degree of expansion of mouse and sheep cumulus-oocyte complexes and significantly improves fertilization rates of mouse oocytes matured *in vitro*. *Biology of Reproduction* 40 (Suppl. 1), 54 Abs. 19.

Chen, L., Russell, P.T. and Larsen, W.J. (1990a) Hormonal regulation and serum factors in mouse cumulus expansion *in vitro*. *Biology of Reproduction* 42 (Suppl. 1), 92.

Chen, L., Wert, S.E., Hendrix, E.M., Russell, P.T., Cannon, M. and Larsen, W.J. (1990b) Hyaluronic acid synthesis and gap junction endocytosis are necessary for normal expansion of the cumulus mass. *Molecular Reproduction and Development* 26, 236–247.

Chene, N., Martal, J. and Charrier, J. (1988) Ovine chorion somatomammotrophin and foetal growth. *Reproduction, Nutrition, Development* 28, 1707–1730.

Cheng, W.T.K. (1985) *In vitro* fertilization of farm animal oocytes. PhD Thesis, Council of National Academic Awards Cambridge, UK.

Cheng, W.T.K., Moor, R.M. and Polge, C. (1986) *In vitro* fertilization of pig and sheep oocytes matured *in vivo* and *in vitro*. *Theriogenology* 25, 146.

Chen-Lu, H.B. and Lu, K.H. (1990) Effect of protein supplements on *in vitro* maturation, fertilization and culture to blastocyst and hatching stage of bovine oocytes. *Theriogenology* 33, 205.

Chen-Lu, H.B., Huang, F.L., Lu, K.H. and Polge, C. (1991) Effect of oviductal cell suspensions on *in vitro* fertilization and development of bovine oocytes. *Proceedings of the Seventh Conference of the European Embryo Transfer Association*. (Cambridge), 134.

Cherfas, J. (1990) Molecular biology lies down with the lamb. *Science* 249, 124–126.

Cherny, R.A. and Merei, J. (1994) Evidence for pluripotency of bovine primordial germ cell derived cell lines maintained in long-term culture. *Theriogenology* 41, 175.

Chesne, P., Heyman, Y., Chupin, D., Procureur, R. and Menezo, Y. (1987) Freezing the cattle demi-embryo: influence of a period of culture between splitting and freezing on survival. *Theriogenology* 27, 218.

Chesne, P., Heyman, Y., Peynot, N., Desmedt, V., Rao, V.H. and Renard, J.P. (1991a) Effect of *in vitro* or *in vivo* maturation of cattle oocytes on cleavage rate following nuclear transfer. *Proceedings of the Seventh Meeting of the European Embryo Transfer Association* (Cambridge), 136.

Chesne, P., Heyman, Y., Peynot, N., Desmedt, V., Rao, V.H. and Renard, J.P. (1991b) Nuclear transfer in cattle: comparison of *in vivo* or *in vitro* matured oocytes as recipient cytoplasts. *Journal of Reproduction and Fertility*, Abstract Series No. 8, 41 (Abs. 72).

Chesne, P., Le Bourhis, D., Peynot, N., Rao, V.H., Renard, J.P. and Heyman, Y. (1992) *In vitro* development to the blastocyst stage of cattle nuclear transfer embryos: effect of the age of recipient oocytes. *Proceedings of the Eighth Conference of the European Embryo Transfer Association*. (Lyon), 138.

Chew, B.P., Holpunch, D.M. and O'Fallon, J.V. (1984) Vitamin A and B-carotene in bovine and porcine plasma, liver, corpora lutea and follicular fluid. *Journal of Dairy Science* 67, 1316–1322.

Chian, R.C. and Niwa, K. (1994) Effect of cumulus cells present during different periods of culture on maturation *in vitro* of bovine oocytes. *Theriogenology* 41, 176.

Chian, R.C., Niwa, K. and Okuda, K. (1991) *In vitro* penetration of zona pellucida of salt-stored bovine oocytes before and after maturation by frozen–thawed spermatozoa. *Theriogenology* 36, 209–219.

Chian, R.C., Nakahara, H., Niwa, K. and Funahashi, H. (1992a) Fertilization and early cleavage *in vitro* of aging bovine oocytes after maturation in culture. *Theriogenology* 37, 666–672.

Chian, R.C., Niwa, K. and Nakahara, H. (1992b) Effect of sperm penetration *in vitro* on completion of first meiosis by bovine oocytes arrested at various stages in culture. *Journal of Reproduction and Fertility* 96, 73–78.

Chikamatsu, N., Urakawa, M., Fukui, Y., Aoyagi, Y. and Ono, H. (1989) *In vitro* fertilization and early development of bovine follicular oocytes matured in different cultured systems and inseminated with spermatozoa treated by different methods. *Japanese Journal of Animal Reproduction* 35, 154–158.

Choi, Y.H., Fukui, Y. and Ono, H. (1991) Effects of media and the presence of bovine oviduct epithelial cells during *in vitro* fertilization on fertilizability and developmental capacity of bovine oocytes. *Theriogenology* 36, 863–873.

Choi, Y.H., Hochi, S., Braun, J. and Oguri, N. (1993a) *In vitro* maturation of equine oocytes collected by aspiration and additional slicing of ovaries. *Theriogenology* 39, 200.

Choi, Y.H., Hochi, S., Braun, J., Sato, K. and Oguri, N. (1993b) *In vitro* maturation of equine oocytes collected by follicle aspiration and by the slicing of ovaries. *Theriogenology* 40, 959–966.

Choudary, J.B., Gier, H.T. and Marion, G.B. (1968) Cyclic changes in bovine vesicular follicles. *Journal of Animal Science* 27, 468–471.

Christensen, L.G. (1991) Use of embryo transfer in future cattle breeding schemes. *Theriogenology* 35, 141–149.

Christman, G.M. and Halme, J.K. (1992) Growth hormone revisited. *Fertility and Sterility* 57, 12–14.

Christmann, L., Jung, T. and Moor, R.M. (1993) MPF components and meiotic competence in growing pig oocytes. *Journal of Reproduction and Fertility*, Abstract Series No. 11, 6 Abs. 2).

Chuangsoongneon, U. and Kamonpatana, M. (1991) Oocyte maturation, *in vitro* fertilization and culture system for developing preimplantation swamp buffalo embryos using frozen–thawed semen. *Buffalo Journal* 7, 189–198.

Chung, Y.C., Kim, C.K., Kim, H.R., Lee, J.H. and Chung, J.T. (1991) Effect of individual variation in acrosome reaction of bull sperm on *in vitro* fertilization and embryonic development. *Journal of Animal Science* 69 (Suppl. 1), 427.

Chung, Y.G. and Seidel, G.E. (1993) Development of *in vitro*-derived bovine embryos after maturation of oocytes in TCM-199 supplemented with EDTA and pyruvate. *Theriogenology* 39, 202.

Chupin, D. (1989) Selection of frozen–thawed bovine spermatozoa for IVF by swim-up or Percoll gradient. *Proceedings of the Fifth Conference of the European Embryo Transfer Association* (Lyon), 140.

Chupin, D. and Procureur, R. (1984) Glycerol equilibration for deep freezing of cattle blastocysts: effect of number of steps and of total duration. *Theriogenology* 21, 230.

Chupin, D., Florin, B. and Procureur, R. (1984) Comparison of two methods for one-step in-straw thawing and direct transfer of cattle blastocysts. *Theriogenology* 21, 455–459.

Chupin, D., Hamidi, J., Vallet, J.C. and Menezo, Y. (1988) Capacity of prepubertal oviducts to support early embryonic development in mouse and cow. *Proceedings of the Fourth Conference of the European Embryo Transfer Association* (Lyon), 39.

Chupin, D., Perez, E. and Saumande, J. (1992) Cryopreservation of bovine granulosa cells. *Proceedings of the Eighth Conference of the European Embryo Transfer Association* (Lyon), 140.

Clark, A.J. (1990) Transgenic animals applied to livestock production. *Proceedings of the Fourth World Congress on Genetics Applied to Livestock Production* 13, 37–40.

Clark, A.J., Archibald, A.L., McClenaghan, M., Simons, J.P., Whitelaw, C.B.A. and Wilmut, I. (1990) The germline manipulation of livestock: progress during the past five years. *Proceedings of the New Zealand Society of Animal Production* 50, 167–179.

Clark, A.J., Archibald, A.L., McClenaghan, M., Simons, J.P., Whitelaw, C.B.A. and Wilmut, I. (1992) High level expression of biomedical proteins in the milk of transgenic animals. *IAPGR Babraham and Roslin Biennial Report (1990–1991)*, 53.

Clarke, R.N. and Johnson, L.A. (1987) Effect of liquid storage and cryopreservation of boar spermatozoa on acrosomal integrity and the penetration of zona-free hamster ova *in vitro*. *Gamete Research* 16, 193–204.

Clarke, R.N. and Johnson, L.A. (1988) Factors related to successful sperm microinjection of hamster eggs: the effect of sperm species, technical experience, needle dimensions and incubation medium on egg viability and sperm decondensation following microinjection. *Theriogenology* 30, 447–460.

Clement-Sengewald, A., Palma, G.A. and Winkler, B. (1993) Ethanol versus electrical activation of oocytes in cattle cloning experiments. *Theriogenology* 39, 203.

Coddington, C.C., Veeck, L.L., Swanson, R.J., Kaufmann, R.A., Lin, J., Simonetti, S. and Bocca, S. (1992) The YAG laser used in micromanipulation to transect the zona pellucida of hamster oocytes. *Journal of Assisted Reproduction and Genetics* 9, 557–563.

Cognie, Y., Guerin, Y., Guyader, C., Poulin, N. and Crozet, N. (1989) *In vitro* fertilization of sheep oocytes matured *in vivo*. *Proceedings of the Fifth Congress of the European Embryo Transfer Association* (Lyon), 142.

Cognie, Y., Guerin, Y., Guyader, C., Poulin, N. and Crozet, N. (1991) *In vitro* fertilization of sheep oocytes matured *in vivo*. *Theriogenology* 35, 393–400.

Cohen, J. (1991a) Assisted hatching of human embryos. *Journal of In Vitro Fertilization and Embryo Transfer* 8(4), 179–190.

Cohen, J. (1991b) Spermatozoa as carriers of extrinsic DNA: consequences for intentional and unintentional gene transfer procedures. *Human Reproduction* 6, 1181.

Cohen, J. and Feldberg, D. (1991) Effects of the size and number of zona pellucida openings on hatching and trophoblast outgrowth in the mouse embryo. *Molecular Reproduction and Development* 30, 70–78.

Cohen, J., Malter, H., Fehilly, C., Wright, G., Elsner, C., Kort, H. and Massey, J. (1988) Implantation of embryos after partial opening of oocyte zona pellucida to facilitate sperm penetration. *Lancet* 8603, 162.

Cohen, J., Inge, K.L., Suzman, M., Wiker, S.R. and Wright, G. (1989) Videocinematography of fresh and cryopreserved embryos: a retrospective analysis of embryonic morphology and implantation. *Fertility and Sterility* 51, 820–827.

Cohen, J., Alikani, M., Trowbridge, J. and Rosenwaks, Z. (1992) Implantation enhancement by selective assisted hatching using zona drilling of human embryos with poor prognosis. *Human Reproduction* 7, 685–691.

Cole, A.R. (1989) Embryo sexing is here-now. *Farmers Weekly* 111(7), 13.

Cole, W.J., Madsen, K.S., Hintz, R.L. and Collier, R.J. (1991) Effect of recombinantly-derived bovine somatotrophin on reproductive performance of dairy cattle. *Theriogenology* 36, 573–595.

Collas, P. and Barnes, F.L. (1992) Parthenogenetic development of bovine oocytes. *Biology of Reproduction* 46 (Suppl. 1), 65 (Abs. 59).

Collas, P. and Robl, J.M. (1989) Age of recipient oocyte for nuclear transplantation in the rabbit. *Biology of Reproduction* 40 (Suppl. 1), 110 (Abs. 188).

Collas, P. and Robl, J.M. (1990) Factors affecting the efficiency of nuclear transplantation in the rabbit embryo. *Biology of Reproduction* 43, 877–884.

Collas, P., Duby, R.T. and Robl, J.M. (1991) *In vitro* development of rabbit pronuclear embryos in rabbit peritoneal fluid. *Biology of Reproduction* 44, 1100–1107.

Collas, P., Balise, J.J. and Robl, J.M. (1992a) Influence of cell cycle stage of the donor nucleus on development of nuclear transplant rabbit embryos. *Biology of Reproduction* 46, 492–500.

Collas, P., Pinto-Correia, C., Abel Ponce De Leon, F. and Robl, J.M. (1992b) Effect of donor cell cycle stage on chromatin and spindle morphology in nuclear transplant rabbit embryos. *Biology of Reproduction* 46, 501–511.

Conaghan, J., Hardy, K., Winston, R.M.L. and Leese, H.J. (1992a) Pyruvate uptake in relation to pregnancy outcome after human IVF. *Journal of Reproduction and Fertility*, Abstract Series No. 9, 27 (Abs. 39).

Conaghan, J., Handyside, A.H., Winston, R.M.L. and Leese, H.J. (1992b) Glucose inhibits cleavage of human preimplantation embryos *in vitro*. *Journal of Reproduction and Fertility*, Abstract Series No. 10, 4 (Abs. 4).

Conaghan, J., Handyside, A.H., Winston, R.M.L. and Leese, H.J. (1993) Effects of pyruvate and glucose on the development of human preimplantation embryos *in vitro*. *Journal of Reproduction and Fertility* 99, 87–95.

Cook, N.L., Squires, E.L., Ray, B.S., Cook, V.M. and Jasko, D.J. (1992) Transvaginal ultrasonically guided follicular aspiration of equine oocytes: preliminary results. *Journal of Equine Veterinary Science* 12, 204–207.

Cook, N.L., Squires, E.L. and Jasko, D.J. (1993) Repeated transvaginal follicular aspiration in cyclic mares. *Theriogenology* 39, 204.

Corson, S.L., Batzer, F.R., Alexander, N.J., Schlaff, S. and Otis, C. (1984) Sex selection by sperm separation and insemination. *Fertility and Sterility* 42, 756–760.

Cosby, N.C., Ye, L. and Dukelow, W.R. (1989) Development of multiple encapsulated mouse embryos in sodium alginate. *Biology of Reproduction* 40 (Suppl. 1), 110 (Abs. 187).

Coskun, S. and Lin, Y.C. (1992) Site of action of epidermal growth factor (EGF) on *in vitro* porcine oocyte maturation in chemically defined medium. *Biology of Reproduction* 46 (Suppl. 1), 138 (Abs. 350).

Coskun, S., Sanbuissho, A., Lin, Y.C. and Rikihisa, Y. (1991) Fertilizability and subsequent developmental ability of bovine oocytes matured in medium containing epidermal growth factor (EGF). *Theriogenology* 36, 485–494.

Coultard, H. (1991) On farm embryo transfer in cattle. *In Practice* 13, 16–22.

Coultard, H. (1992) Embryo transfer. *Veterinary Record* 130, 211.

Coulter, G.H. (1992) Bovine spermatozoa *in vitro*: a review of storage, fertility estimation and manipulation. *Theriogenology* 38, 197–207.

Courot, M. and Volland-Nail, P. (1991) Management of reproduction in farm animals: present and future. In: *On the Eve of the Third Millenium: the European Challenge for Animal Production*. Pudoc, Wageningen, Netherlands, pp. 23–37.

Cox, J.F. (1991) Effect of the cumulus on *in vitro* fertilization of *in vitro* matured cow and sheep oocytes. *Theriogenology*, 35, 191.

Cox, J.F., Hormazabal, J. and Santa Maria, A. (1993) Effect of the cumulus on *in vitro* fertilization of bovine matured oocytes. *Theriogenology* 40, 1259–1267.

Coy, P., Martinez, E., Ruiz, S., Vazquez, J.M., Roca, J., Matas, C. and Pellicer, M.T. (1993a) *In vitro* fertilization of pig oocytes after different coincubation intervals. *Theriogenology* 39, 1201–1208.

Coy, P., Martinez, E., Ruiz, S., Vazquez, J.M., Roca, J. and Matas, C. (1993b) Sperm concentration influences fertilizaion and male pronuclear formation *in vitro* in pigs. *Theriogenology* 40, 539–546.

Cran, D.G. (1987) The distribution of organelles in mammalian oocytes following centrifugation prior to injection of foreign DNA. *Gamete Research* 18, 67–76.

Cran, D.G. (1989) Cortical granules during oocyte maturation and fertilization. *Journal of Reproduction and Fertility*, Suppl. 38, 49–62.

Cran, D.G. (1992) Gender preselection in mammals. In: Lauria, A. and Gandolfi, F. (eds) *Embryonic Development and Manipulation in Animal Production*. Portland Press, London, pp. 125–134.

Cran, D.G. and Esper, C.R. (1990) Structural and molecular changes during mammalian fertilization. *Proceedings of the Third Symposium on Advanced Topics in Animal Reproduction* FCAV-UNESP, Jaboticabal, 1–12.

Cran, D.G., Johnson, L.A., Miller, N.G., Cochrane, D. and Polge, C. (1993) Production of bovine calves following separation of X- and Y-chromosome bearing sperm and *in vitro* fertilisation. *Veterinary Record* 132, 40–41.

Cran, D.G., Cochrane, D.J., Johnson, L.A., Wei, H., Lu, K.H. and Polge, C. (1994) Separation of X- and Y-chromosome bearing bovine sperm by flow cytometry for use in IVF. *Theriogenology* 41, 183.

Critser, E.S., Leibfried-Rutledge, M.L. and First, N.L. (1984) The effect of semen extension, cAMP and caffeine on *in vitro* fertilization of bovine oocytes. *Theriogenology* 21, 625–631.

Critser, E.S., Leibfried-Rutledge, M.L. and First, N.L. (1986a) Influence of cumulus cell association during *in vitro* maturation of bovine oocytes on embryonic development. *Biology of Reproduction* 34 (Suppl. 1), 286 (Abs.).

Critser, E.S., Leibfried-Rutledge, M.L. Eyestone, W.H., Northey, D.L. and First, N.L. (1986b)

Acquisition of developmental competence during maturation *in vitro*. *Theriogenology* 25, 150.

Crosby, I.M. and Moor, R.M. (1984) Oocyte maturation. In: Trounson, A. and Wood, C. (eds) *In Vitro Fertilization and Embryo Transfer.* Churchill Livingstone, London, pp. 19–31.

Crosby, I.M. and Moor, R.M. (1985) Oocyte maturation *in vitro*: manipulation of cellular signals. In: Rolland, R. *et al.* (eds) *Gamete Quality and Fertility Regulation.* Elsevier Press, Amsterdam, pp. 85–96.

Crosby, I.M., Osborn, J.C. and Moor, R.M. (1981) Follicle cell regulation of protein synthesis and developmental competence in sheep oocytes. *Journal of Reproduction and Fertility* 62, 575–582.

Crosby, I.M., Gandolfi, F. and Moor, R.M. (1988) Control of protein synthesis during early cleavage of sheep embryos. *Journal of Reproduction and Fertility* 82, 769–775.

Crosby, T.F. (1969) Studies in the culture and fertilization of sheep oocytes. M. Agr. Sc. Thesis, University College, Dublin.

Crosby, T.F. (1976) Studies in the induction of twin-pregnancy in cattle. PhD Thesis, National University of Ireland, Dublin.

Cross, B.A. (1988) Jumping the queue: new methodologies for genetic improvement. *British Cattle Breeders Club Digest* 43, 18–19.

Cross, N.L. and Meizel, S. (1989) Methods for evaluating the acrosomal status of mammalian sperm. *Biology of Reproduction* 41, 635–641.

Crozet, N. (1984) Ultrastructural aspects of *in vivo* fertilization in the cow. *Gamete Research* 10, 241–251.

Crozet, N. (1986) Fine structure of sheep fertilization *in vitro*. *Gamete Research* 19, 291–303.

Crozet, N. (l989) Nucleolar structure and RNA synthesis in mammalian oocytes. *Journal of Reproduction and Fertility*, Suppl. 38, 9–16.

Crozet, N. (1991) Manipulation of oocytes and *in vitro* fertilization. *Journal of Reproduction and Fertility*, Suppl. 43, 235–243.

Crozet, N. and Dumont, M. (1984) The site of the acrosome reaction during *in vivo* penetration of the sheep oocyte. *Gamete Research* 10, 97–105.

Crozet, N., Kanka, J., Motlik, J. and Fulka, J. (1986) Nucleolar fine structure and RNA synthesis in bovine oocytes from antral follicles. *Gamete Research* 14, 65–73.

Crozet, N., Theron, M.-C. and Chemineau, P. (1987a) Ultrastructure of *in vivo* fertilization in the goat. *Gamete Research* 14, 65–73.

Crozet, N., Huneau, D., De Smedt, V., Theron, M.-C., Szollosi, D., Torres, S. and Sevellec, C. (1987b) *In vitro* fertilization with normal development in sheep. *Gamete Research* 16, 159–170.

Crozet, N., Smedt, V. De, Ahmed-Ali, M. and Sevellec, C. (1993) Normal development following *in vitro* oocyte maturation and fertilization in the goat. *Theriogenology* 39, 206.

Cummins, L.J., Bindon, B.M., Hillard, M.A., Clark, A.J., McLeod, I.K., Wilkins, J.F., Williamson, P.J., Hennessy, D.W., Pashen, R.L., Darrow, M.D., Andrews, C.M., Graham, J.F., Makings, B.J., Farquarson, R.J. and McLaren, C.E. (1992) Twinning in cattle. *Proceedings of the Australian Society of Animal Production* 19, 438–447.

Cunningham, E.P. (1989) The genetic improvement of cattle in developing countries. *Theriogenology* 31, 17–28.

Curran, S. (1992) Fetal sex determination in cattle and horses by ultrasonography. *Theriogenology* 37, 17–21.

Curran, S., Pierson, R.A. and Ginther, O.J. (1986a) Embryonic loss and ultrasonic anatomy of the bovine conceptus. *Journal of Animal Science* 63 (Suppl. 1), 448 (Abs.).

Curran, S., Pierson, R.A. and Ginther, O.J. (1986b) Embryonic loss and ultrasonic anatomy of the bovine conceptus from days 10 through 20. *Journal of the American Veterinary Medicine Association* 189, 1289–1290.

Curran, S., Kastelic, J.P. and Ginther, O.J. (1989) Determining sex of the bovine fetus by ultrasonic assessment of the relative location of the genital tubercle. *Animal Reproduction Science* 19, 217–227.

Czlonkowska, M., Eysymont, U., Guszkiewicz, A., Kossakowski, M. and Dziak, J. (1991) Birth of lambs after *in vitro* maturation, fertilization, and co-culture with oviductal cells. *Molecular Reproduction and Development* 30, 34–38.

Dahlhausen, R.D., Bonham, J.B., Meyers, G. and Ludwick, T.M. (1981) Characterization and maturation of prepubertal calf follicular oocytes *in vitro*. *Theriogenology* 15, 111.

Dandekar, P.V., Martin, M.C. and Glass, R.H. (1991) Maturation of immature oocytes by coculture with granulosa cells. *Fertility and Sterility* 55, 95–99.

Daniel, J.C., Jr and Takahashi, K. (1965) Selective laser destruction of rabbit blastomeres and continued cleavage of survivors *in vitro*. *Experimental Cell Research* 39, 475–482.

Daniel, S.A.J., Armstrong, D.T. and Gore-Langton, R.E. (1989) Growth and development of rat oocytes *in vitro*. *Gamete Research* 24, 109–121.

Darrow, M.D. (1989) Transport of imported embryos to zebu recipients in tropical Venezuela. *Theriogenology* 31, 182.

Das, K., Stout, L.E., Hensleigh, H.C., Tagatz, G.E., Phipps, W.R. and Leung, B.S. (1991a) Direct positive effect of epidermal growth factor on the cytoplasmic maturation of mouse and human oocytes. *Fertility and Sterility* 55, 1000–1004.

Das, K., Phipps, W.R., Hensleigh, H.C. and Tagatz, G.E. (1991b) The epidermal growth factor component of human follicular fluid is a stimulatory factor for mouse oocyte maturation *in vitro*. *Serono Symposium on Preimplantation Embryo Development*, Abstract Vol., 30 (Abs. I-12).

Dasgupta, S., O'Toole, C., Mills, C.L. and Fraser, L.R. (1993) Pentoxifylline and progesterone promote capacitation and acrosomal exocytosis in human spermatozoa. *Journal of Reproduction and Fertility*, Abstract Series No. 11, 53 (Abs. 96).

Davenport, J. (1992) The precarious life of high-latitude marine fish. *Biologist* 39(5), 218–221.

Davis, A.P., Graham, J.K. and Foote, R.H. (1987a) *In vitro* hetereospermic insemination of zona free hamster eggs by fluorochrome labelled bull sperm: a fertility test. *Journal of Dairy Science* 69 (Suppl. 1), 147 (Abs.).

Davis, A.P., Graham, J.K. and Foote, R.H. (1987b) Homospermic versus heterospermic insemination of zona-free hamster eggs to assess fertility of fluorochrome-labeled acrosome-reacted bull spermatozoa. *Gamete Research* 17, 343–354.

Davis, M.E. and Bishop, M.D. (1992) Induction of multiple births in beef cows with FSH: calving rate and subsequent performance. *Livestock Production Science* 32, 41–62.

Davis, M.E. and Haibel, G.K. (1992) Use of real-time ultrasound to identify multiple fetuses in beef cattle. *Animal Science Departmental Series – Ohio Agricultural Research and Development Center* 92–1, 10–18.

Davis, M.E. and Haibel, G.K. (1993) Use of real-time ultrasound to identify multiple fetuses in beef cattle. *Theriogenology* 40, 373–382.

Davis, M.E., Harvey, W.R., Bishop, M.D. and Gearheart, W.W. (1989) Use of embryo transfer to induce twinning in beef cattle: embryo survival rate, gestation length, birthweight and weaning weight of calves. *Journal of Animal Science* 67, 301–310.

Daya, S., Gwatkin, R.B.L. and Bissessar, H. (1987) Separation of motile human sperm by means of a glass bead column. *Gamete Research* 17, 375–387.

Deb, S.M. and Goswami, S.L. (1990) Cytogenetic analysis of goat oocytes cultured *in vitro*. *International Journal of Animal Sciences* 5, 137–143.

DeJarnette, J.M., Saacke, R.G. and Vogler, C.J. (1990) Effect of sperm dose and retrograde sperm blockage on accessory sperm number and embryo quality in artificially inseminated cows. *Journal of Animal Science* 68 (Suppl. 1), 434.

DeJarnette, J.M., Saacke, R.G., Bame, J. and Vogler, C.J. (1992) Accessory sperm: their importance to fertility and embryo quality and attempts to alter their numbers in artificially inseminated cattle. *Journal of Animal Science* 70, 484–491.

Dekkers, J.C.M. (1991) Structure of breeding programs to capitalize on reproductive technology for genetic improvement. *Journal of Dairy Science* 74 (Suppl. 1), 197 (Abs. P167).

Delhaise, F., Mermillod, P., Boccart, C., Bralion, V. and Dessy, F. (1991) Influence of culture media on the hatching and implantation of *in vitro* produced bovine blastocysts. *Serono Symposium on Preimplantation Embryo Development*, Abstract Vol., 57 (Abs. II–13).

Dell'Aquita, M.E., Fusco, S., Maritato, F. and Lacalandra, G.M. (1992) Development and cytogenetic evaluation of embryos produced *in vitro* in cattle. *Atti della Societa Italiana di Buiatria* 24, 425–432.

DeMayo, F.J., Rawlins, R.G. and Dukelow, W.R. (1985) Xenogenous and *in vitro* fertilization of frozen–thawed primate oocytes and blastomere separation of embryos. *Fertility and Sterility* 43, 295–300.

DeMott, R.P. and Suarez, S.S. (1991) Evidence for an advantage of hyperactivated mouse sperm in remaining free of the oviductal wall. *Biology of Reproduction* 44 (Suppl. 1), 117 (Abs. 258).

DeMott, R.P. and Suarez, S.S. (1992) Hyperactivated sperm progress in the mouse oviduct. *Biology of Reproduction* 46, 779–785.

Den Daas, J.H.G., de Leeuw, A.M. and Woelders, H. (1992) Use of the fix-vital stain method to monitor capacitation of frozen–thawed bovine spermatozoa *in vitro*. *Proceedings of the Twelfth International Congress on Animal Reproduction* (The Hague) 1, 432–434.

Depypere, H.T., McLaughlin, K.J., Seamark, R.F., Warnes, G.M. and Matthews, C.D. (1988) Comparison of zona cutting and zona drilling as techniques for assisted fertilization in the mouse. *Journal of Reproduction and Fertility* 84, 205–211.

Depypere, H.T., Carroll, J.C., Vandekerckhove, D. and Matthews, C.D. (1991) Normal survival and *in vitro* development after cryopreservation of zona-drilled embryos in mice. *Human Reproduction* 6, 432–435.

Derynck, R. (1990) Transforming growth factor-alpha. *Molecular Reproduction and Development* 27, 3–9.

De Silva, M. (1993) Effects of storage of Ham's F-10 medium on one-cell mouse embryo development *in vitro*. *Journal of Assisted Reproduction and Genetics* 10, 238–241.

De Silva, M. and Pearl, A.W. (1990) Thyroid stimulating hormone (TSH) causes cumulus expansion in mouse oocytes. *Biology of Reproduction* 42 (Suppl. 1), 78.

De Smedt, V., Crozet, N., Ahmed-Ali, M., Martino, A. and Cognie, Y. (1991) *In vitro* oocyte maturation and fertilization in the goat. *Journal of Reproduction and Fertility*, Abstract Series No. 8, 13 (Abs. 15).

De Smedt, V., Crozet, N., Ahmed-Ali, M., Martino, A. and Cognie, Y. (1992) *In vitro* maturation and fertilization of goat oocytes. *Theriogenology* 37, 1049–1060.

Dessy, F., Ferry, L., Mermillod, P. and Massip, A. (1993) Bovine embryos cultured in serum-free conditioned-medium need cooperation to reach the blastocyst stage. *Biology of Reproduction* 48 (Suppl. 1), 172.

De Sutter, P., Dozortsev, D., Cieslak, J., Wolf, G., Verlinsky, Y. and Dyban, A. (1992) Parthenogenetic activation of human oocytes by Puromycin. *Journal of Assisted Reproduction and Genetics* 9, 328–337.

Devroey, P., Mannaerts, B., Smitz, J., Coelingh Bennink, H. and Van Steirteghem, A. (1993) First established pregnancy and birth after ovarian stimulation with recombinant human follicle stimulating hormone (Org 32489). *Human Reproduction* 8, 863–865.

Diczfalusy, E. (1992) Human reproduction: lessons learned and new challenges. *Proceedings of the Fourteenth World Congress on Fertility and Sterility* (Caracas), Abstract Vol., 3.

Didion, B.A. and Graves, C.N. (1986a) The influence of preincubation on the occurrence of the true acrosome reaction in bovine spermatozoa induced by heparin. *Biology of Reproduction* 34 (Suppl. 1), 356 (Abs.).

Didion, B.A. and Graves, C.N. (1986b) *In vivo* capacitation and acrosome reaction of bovine sperm in estrous and diestrous cows. *Journal of Animal Science* 62, 1029–1033.

Didion, B.A., Dobrinsky, J.R., Gilles, J.R. and Graves, C.N. (1987) A direct staining procedure to determine viability and the acrosome reaction in sperm of various species. *Journal of Animal Science* 65 (Suppl. 1), 356 (Abs.).

Didion, B.A., Pomp, D., Martin, M.J. and Markert, C.L. (1990) Observations on the cooling and cryopreservation of pig oocytes at the germinal vesicle stage. *Journal of Animal Science* 68, 2803–2810.

Diedrich, K., Al-Hasani, S., Van Der Ven, H. and Krebs, D. (1988) Successful *in vitro* fertilization of frozen-thawed rabbit and human oocytes. *Annals of the New York Academy of Sciences* 541, 562–570.

Dieleman, S.J. and Bevers, M.M. (1987) Development of preovulatory follicles in the cow from luteolysis until ovulation. In: Roche, J.F.and O'Callaghan, D. (eds) *Follicular Growth and Ovulation Rate in Farm Animals*. Martinus Nijhoff, Dordrecht, The Netherlands, pp. 31–44.

Dieleman, S.J. and Bevers, M.M. (1993) Folliculogenesis and oocyte maturation in superovulated cattle. *Molecular Reproduction and Development* 36, 271–273.

Dieleman, S.J. and Blankenstein, D.M. (1987) Changes in the oestrogen-synthesizing ability of preovulatory bovine follicles relative to the peak of LH. *Journal of Reproduction and Fertility* 72, 487–494.

Ding, J. and Foxcroft, G.R. (1992) Follicular heterogeneity and oocyte maturation *in vitro* in pigs. *Biology of Reproduction* 47, 648–655.

Ding, J.C., Nagai, T. and Moor, R.M. (1988) Effects of follicle cells and hormones on maturation and fertilization *in vitro* of oocytes from immature gilts. *Journal of Reproduction and Fertility*, Abstract Series No. 2, Abs. 42.

Ding, J., Clarke, N., Nagai, T. and Moor, R.M. (1991) Reprogramming of protein synthesis during fertilization in pig oocytes. *Serono Symposium on Preimplantation Embryo Development*, Abstract Vol., 31 (Abs. I-13).

Ding, J., Clarke, N., Nagai, T. and Moor, R.M. (1992a) Protein and nuclear changes in pig eggs at fertilization. *Molecular Reproduction and Development* 31, 287–296.

Ding, J., Moor, R.M. and Foxcroft, G.R. (1992b) Effects of protein synthesis on maturation, sperm penetration, and pronuclear development in porcine oocytes. *Molecular Reproduction and Development* 33, 59–66.

Dingwall, W.S., McKelvey, W.A.C., Mylne, J. and Simms, G. (1993) An evaluation of MOET in Suffolk sheep. *British Society of Animal Production* (Winter Meeting), Paper 103 (2 pp.).

Dinnyes, A., Keefer, C.L., Stice, S.L., Solti, L., Vajta, G., Machaty, Z. and Rall, W.F. (1994) Vitrification of IVMFC bovine embryos in VS3a and EFS solutions: a preliminary report. *Theriogenology* 41, 189.

Dippert, K.D., Ray, B.S. and Squires, E.L. (1994) Maximizing ultrasound-guided retrieval of equine oocytes. *Theriogenology* 41, 190.

Diskin, M.G. (1987) Studies related to embryonic mortality in the cow. PhD Thesis, National University of Ireland, Dublin.

Diskin, M.G. and Hickey, B.C. (1987) The impact of twin-calving on beef output and financial returns. *Proceedings of the Thirteenth Annual Meeting of the Irish Grassland and Animal Production Association*, 33–34.

Diskin, M.G. and Sreenan, J.M. (1980) Fertilization and embryonic mortality rates in beef heifers after artificial insemination. *Journal of Reproduction and Fertility* 59, 463–468.

Diskin, M.G. and Sreenan, J.M. (1992) Reproductive management of sucklers. *Irish Farmers' Journal* 44(27), 20–21.

Diskin, M. and Sreenan, J. (1993) Heat synchronization in suckler cows. *Irish Farmers' Journal* 45(24), 22.

Diskin, M.G., McEvoy, T.G., Hickey, B.C. and Sreenan, J.M. (1987a) More twins in the beef herd: impact on beef output and financial returns. *Farm and Food Research* 18(5), 4–6.

Diskin, M.G., McDonagh, T. and Sreenan, J.M. (1987b) The experimental induction of twin calving in beef cows by embryo transfer. *Theriogenology* 27, 224.

Diskin, M.G., McEvoy, T.G. and Sreenan, J.M. (1989) Induction of parturition in heifers. *Farm and Food Research* 20(6), 4–6.

Diskin, M.G., McEvoy, T.G. and Sreenan, J.M. (1990a) Calving performance and calf survival in a twin-calving beef herd. *British Society of Animal Production* (Winter Meeting), Paper 1.

Diskin, M.G., McEvoy, T.G. and Sreenan, J.M. (1990b) A comparison of the growth rate of single and twin born beef calves. *British Society of Animal Production* (Winter Meeting), Paper 5.

Diskin, M.G., Grealy, M. and Sreenan, J.M. (1992a) Effect of body condition score at calving and suckling frequency on post-partum interval in suckler cows. *British Society of Animal Production* (Winter Meeting), Paper 67.

Diskin, M.G., Grealy, M. and Sreenan, J.M. (1992b) Reducing the suckling frequency shortens the post-partum and calving to conception intervals in suckler cows. *Proceedings of the Irish Grassland and Animal Production Association* (Eighteenth Annual Research Meeting), 31–32.

Dobrinsky, J.R. (1993) Cryopreservation of porcine embryos: current status and perspectives. *Embryo Transfer Newsletter* 11(4), 13–18.

Dobrinsky, J.R. and Johnson, L.A. (1993) Effect of vitrification media on the *in vitro* development of porcine embryos. *Theriogenology* 39, 209.

Dobrinsky, J.R., Hess, F.F., Duby, R.T. and Robl, J.M. (1991) Cryopreservation of bovine embryos by vitrification. *Theriogenology* 35, 194.

Dobson, H., Hopkinson, C.R.N. and Ward, W.R. (1973) Progesterone, oestradiol and LH in relation to ovulation in cows. *Veterinary Record* 93, 76 (Abs.).

Dobson, H., Rowan, T.G., Kippax, I.S. and Humblot, P. (1991) Assessment of fetal number and placental viability throughout pregnancy in cattle. *Journal of Reproduction and Fertility*, Abstract Series 3, 23.

Dobson, H., Rowan, T.G., Kippax, I.S. and Humblot, P. (1993) Assessment of fetal number, and fetal and placental viability throughout pregnancy in cattle. *Theriogenology* 40, 411–425.

D'Occhio, M.J., Gifford, D.R., Hoskinson, R.M., Weatherly, T. and Setchell, B.P. (1988) Gonadotrophin secretion and ovarian responses in prepubertal heifers actively immunized against androstenedione and oestradiol-17B. *Journal of Reproduction and Fertility* 83, 159–168.

Dokras, A., Sargent, I.L. and Barlow, D.H. (1993) Human blastocyst grading: an indicator of developmental potential? *Human Reproduction* 8, 2119–2127.

Dominko, T. and First, N.L. (1991) Maturation of denuded bovine oocytes results in normal fertilization and development. *Biology of Reproduction* 44 (Suppl. 1), 141 (Abs. 353).

Dominko, T. and First, N.L. (1993) Male predominance of bovine embryos can be observed at the 2-cell stage. *Biology of Reproduction* 48 (Suppl. 1), 168.

Donaldson, L.E. and Ward, D.N. (1985) Superovulation in cattle: dose response to FSH-P with and without LH contamination. *Theriogenology* 23, 189.

Donaldson, L.E. and Ward, D.N. (1987) LH effects on superovulation and fertilization rates. *Theriogenology* 27, 225.

Donaldson, L.E., Ward, D.N. and Glenn, S.D. (1986) Use of porcine follicle stimulating hormone after chromatographic purification in superovulation of cattle. *Theriogenology* 25, 747–757.

Dooley, M.P., Wichtel, J.J., Plante, C. and Hopkins, S.M. (1987) The influence of Eosin-B exposure on the *in vitro* development of bovine embryos. *Biology of Reproduction* 36 (Suppl. 1), 78 (Abs.).

Dooley, V.D., Yorozu, Y. and Fan Biqin (1983) Bovine oocyte *in vitro* maturation and xenogenous fertilization. *Biology of Reproduction* 28 (Suppl. 1), 166 (Abs.).

Dorland, M. (1992) *Day 7 Bovine Embryo Quality: a Cell Biological Study.* Rijksuniversiteit Utrecht, Utrecht, the Netherlands, 127 pp.

Dorland, M., Kruip, T.A.M. and Van der Donk, J.A. (1991) Pyruvate uptake by day-7 bovine embryos determined by a non-invasive microfluorescence technique. *Theriogenology* 35, 196.

Dorland, M., Kruip, Th.A.M. and Van der Donk, J.A. (1992) Assessing day 7 bovine embryo viability by measuring the rate of nutrient uptake. *Journal of Reproduction and Fertility,* Abstract Series No. 9, 27 (Abs. 40).

Douchi, O., Takakura, H. and Imai, K. (1990) Transfer of bovine embryos cryopreserved by vitrification. *Japanese Journal of Animal Reproduction* 36, 69–72.

Dow, M.P.D. and Bavister, B.D. (1989) Direct contact is required between serum albumin and hamster spermatozoa for capacitation *in vitro. Gamete Research* 23, 171–180.

Dowling, D.F. (1949) Problems of transplantation of fertilized ova. *Journal of Agricultural Science (Cambridge)* 39, 374–390.

Downs, S.M. (1989) Specificity of epidermal growth factor action on maturation of the murine oocyte and cumulus oophorus *in vitro. Biology of Reproduction* 41, 371–379.

Downs, S.M. (1992) The participation of carbohydrates in the meiotic maturation of mouse cumulus cell-enclosed oocytes. *Biology of Reproduction* 46 (Suppl. 1), 139 (Abs. 355).

Downs, S.M., Schroeder, A.C. and Eppig, J.J. (1986) Serum maintains the fertilizability of mouse oocytes *in vitro* by preventing hardening of the zona pellucida. *Gamete Research* 15, 115–122.

Dresser, D.W., Atkins, C.J., Pinder, A. and Morrell, J.M. (1993) Analyses of DNA content of living spermatozoa using flow cytometry techniques. *Journal of Reproduction and Fertility* 98, 357–365.

Drew, S.B. and Peters, A.R. (1991) The effect of treatment with a gonadotrophin releasing hormone on the fertility of dairy cows. *Proceedings of the British Society of Animal Production* (Winter Meeting), Paper 171 (2 pp.).

Drew, S.B. and Peters, A.R. (1992) The effect of treatment with a gonadotrophin-releasing hormone analogue on the fertility of dairy cows. *Proceedings of the Twelfth International Congress on Animal Reproduction* (The Hague) 3, 1121–1123.

Drew, S.B. and Peters, A.R. (1994) Effect of buserelin on pregnancy rates in dairy cows, *Veterinary Record* 134, 267–269.

Driancourt, M.A. (1991) Follicular dynamics in sheep and cattle. *Theriogenology* 35, 55–79.

Driancourt, M.A., Lorentz, R., Chupin, D., Webb, R. and Wilmut, I. (1988) Survival of ovine embryos stored at 4°C for 24 h. *Theriogenology* 30, 441–446.

Drost, M. (1991) Training manual for embryo transfer in water-buffaloes. *FAO Animal Production and Health Paper* 84, 58 pp.

Drost, M. and Thatcher, W.W. (1992) Application of gonadotrophin releasing hormone as therapeutic agent in animal reproduction. *Animal Reproduction Science* 28, 11–19.

Drost, M., Tan, H.S., Macmillan, K.L. and Thatcher, W.W. (1989) Successful asynchronous embryo transfer in cattle. *Theriogenology* 31, 136.

Duby, R.T. and Robl, J.M. (1987) Oocyte collection from gonadotropin-stimulated calves and their subsequent fertilization *in vitro. Journal of Animal Science* 65 (Suppl. 1), 387 (Abs.).

Duby, R.T., Dobrinsky, J.R., Robl, J.M., Overstrom, E.W., Boland, M.P., Bagusi, A., Lonergan, P. and Roche, J.F. (1993) Development rate, esterase activity and death of cells in *in vitro* and *in vivo* produced bovine embryos. *Theriogenology* 39, 210.

Dufour, J.J. and Roy, G.L. (1985) Distribution of ovarian follicular populations in the dairy cow within 35 days after parturition. *Journal of Reproduction and Fertility* 73, 229–235.

Dufrasnes, E., Vanderheyden, I., Robin, D., Delcourt, J., Pampfer, S. and De Hertogh, R. (1993) Glucose and pyruvate metabolism in preimplantation blastocysts from normal and diabetic rats. *Journal of Reproduction and Fertility* 98, 169–177.

Dumoulin, J.C.M., Menheere, P.P.C.A., Evers, J.L.H., Kleukers, A.P.G., Pieters, M.H.E.C., Bras, M. and Geraedts, J.P.M. (1991) The effects of endotoxins on gametes and preimplantation embryos cultured *in vitro*. *Human Reproduction* 6, 730–734.

Dunglison, G.F. and Kaye, P.L. (1993) Insulin regulates protein metabolism in mouse blastocysts. *Molecular Reproduction and Development* 36, 42–48.

Du Preez, J.H., Terblanche, S.J., Giesecke, W.H., Maree, C. and Welding, M.C. (1991) Effect of heat stress on conception in a dairy herd model under South African conditions. *Theriogenology* 35, 1039–1049.

Eaglesome, M.D. and Miller, S.A. (1989) Prediction of fertility of bovine semen: preliminary studies with the hamster egg penetration test. *Theriogenology* 31, 643–651.

Ealy, A.D. and Hansen, P.J. (1992) Ontogeny of thermotolerance during murine embryonic development. *Biology of Reproduction* 46 (Suppl. 1), 278 (Abs. 278).

Ealy, A.D., Drost, M., Barros, C.M. and Hansen, P.J. (1991) Protection of murine and bovine preimplantation embryos from heat shock. *Biology of Reproduction* 44 (Suppl. 1), 154 (Abs. 405).

Ealy, A.D., Drost, M., Barros, C.M. and Hansen, P.J. (1992a) Thermoprotection of preimplantation bovine embryos from heat shock by glutathione and taurine. *Cell Biology International Reports* 16(2), 125–131.

Ealy, A.D., Drost, M. and Hansen, P.J. (1992b) Developmental changes in resistance of bovine embryos to heat stress. *Journal of Animal Science* 70 (Suppl. 1), 272 (Abs. 530).

Ebert, K.M. (1989) Gene transfer through embryo microinjection. In: Babiuk, L.A. and Phillips, J.P.(eds) *Animal Biotechnology*. Pergamon Press, Oxford, pp. 233–250.

Ebert, K.M. and Papaioannou, V.E. (1989) *In vivo* culture of embryos in the immature mouse oviduct. *Theriogenology* 31, 299–308.

Ebert, K.M. and Schindler, J.E.S. (1993) Transgenic farm animals: progress report. *Theriogenology* 39, 121–135.

Ebert, K.M., Selgrath, J.P., DiTullio, P., Denman, J., Smith, T.E., Memon, M.A., Schindler, J.E., Monastersky, G.M., Vitale, J.A. and Gordon, K. (1991) Transgenic production of a variant of human tissue-type plasminogen activator in goat milk: generation of transgenic goats and analysis of expression. *Bio/Technology* 9, 835–843.

Echternkamp, S.E. (1992) Fetal development in cattle with multiple ovulations. *Journal of Animal Science* 70, 2309–2321.

Echternkamp, S.E. and Gregory, K.E. (1991) Identification of twin pregnancies in cattle by ultrasonography. *Journal of Animal Science*, 69 (Suppl. 1), 220.

Echternkamp, S.E. and Gregory, K.E. (1993) Identification of twin pregnancies in cattle by ultrasonography to improve neonatal calf survival. *Journal of Animal Science* 71 (Suppl. 1), 73.

Echternkamp, S.E. and Kappes, S. (1989) Effect of trypsin washing on survivability of bovine embryos. *Theriogenology* 31, 187.

Echternkamp, S.E. and Maurer, R.R. (1988) Capability of bovine embryos to develop *in vitro* after trypsin treatment and cryopreservation. *Theriogenology* 29, 241.

Echternkamp, S.E., Kappes, S.M. and Maurer, R.R. (1989) Exposure of bovine embryos to trypsin during washing does not decrease embryonic survival. *Theriogenology* 32, 131–137.

Eckert, J. and Niemann, H. (1993) PDGF enhances development of *in vitro* derived bovine embryos. *Journal of Reproduction and Fertility*, Abstract Series No. 12, 7.

Ectors, F.J., Vander Zwalmen, P., Touati, K., Beckers, J.F. and Ectors, F. (1989a) *In vitro* maturation and fertilization of bovine oocytes. *Theriogenology* 31, 188.

Ectors, F.J., Delval, A., Bormans, M., Toutai, K., Beckers, J.-F. and Ectors, F. (1989b) *In vitro* maturation and fertilization of bovine oocytes. *Proceedings of the Fifth Conference of the European Embryo Transfer Association* (Lyon), 148.

Ectors, F.J., Thonon, F., Delval, A., Fontes, R.S., Touati, K. and Beckers, J.F. (1993a) Comparison between *in vitro* culture of bovine embryos in conditioned medium and in presence of bovine oviductal cells versus development *in vivo* and in rabbit oviducts. *Theriogenology* 39, 211.

Ectors, F.J., Thonon, F., Delval, A., Fontes, R.S., Touati, K., Beckers, J.-F. and Ectors, F. (1993b) Comparison between culture of bovine embryos *in vitro* versus development in rabbit oviducts and *in vivo*. *Livestock Production Science* 36, 29–34.

Eddy, R. (1991) Twinning survey. *Stock* 4(3), 12.

Eddy, R.G., Davies, O. and David, C. (1991) An economic assessment of twin births in British dairy herds. *Veterinary Record* 129, 526–529.

Edgar, D.H., James, G.B. and Mills, J.A. (1993) Steroid secretion by human early embryos in culture. *Human Reproduction* 8, 277–278.

Edirisinghe, W.R., Wales, R.G., Chapman, H.M. and Yovich, J.L. (1991) Assisted fertilization of mouse oocytes and preliminary results for human oocytes using zona drilling. *Journal of In Vitro Fertilization and Embryo Transfer* 8, 48–55.

Edwards, R.G. (1962) Meiosis in ovarian oocytes of adult mammals. *Nature (London)* 196, 446–450.

Edwards, R.G. (1965) Maturation *in vitro* of mouse, sheep, cow, pig, rhesus monkey and human ovarian oocytes. *Nature* 196, 349–351.

Edwards, R.G. (1966) Mammalian eggs in the laboratory. *Scientific American* 215, 72–81.

Edwards, R.G. and Van Steirteghem, A.C. (1993) Intracytoplasmic sperm injections (ICSI) and human fertilization: does calcium hold the key to success? *Human Reproduction* 8, 988–989.

Ehrenwald, E., Foote, R.H. and Parks, J.E. (1990) Bovine oviductal fluid components and their potential role in sperm cholesterol efflux. *Molecular Reproduction and Development* 25, 195–204.

Eiler, H. and Hopkins, F.M. (1991) Effects of collagenase and hyaluronidase on detachment of bovine retained placenta. *Biology of Reproduction* 44 (Suppl. 1), 166 (Abs. 456).

El-Danasouri, I., Westphal, L.M., Neev, Y., Gebhardt, Louie, D. and Berns, M.W. (1993) Zona opening with 308 nm XeCl excimer laser improves fertilization by spermatozoa from long-term vasectomized mice. *Human Reproduction* 8, 464–466.

El-Gaafary, M.N., Daader, A.H. and Ziedan, A. (1990) Effects of caffeine on bull semen quality and sperm penetration into cervical mucus. *Animal Reproduction Science* 23, 13–19.

Ellington, J.E. (1991) The role of the bovine oviduct in reproduction. *Cornell Veterinarian* 81, 313–320.

Ellington, J.E. and Farrell, P.B. (1990) Early bovine embryo development in oviduct cell coculture versus in the cow for 6 days. *Biology of Reproduction* 42 (Suppl. 1), 176.

Ellington, J.E., Carney, E.W., Simkin, M.E. and Foote, R.H. (1989a) Comparison of media in an early bovine embryo and oviduct epithelial cell co-culture system. *Theriogenology* 31, 189.

Ellington, J.E., Vredenburgh, W.L., Padilla, A.W. and Foote, R.H. (1989b) Acrosome reaction of sperm in bovine oviduct cell coculture. *Journal of Reproduction and Fertility*, Abstract Series No. 3, 46.

Ellington, J.E., Farrell, P.B., Carney, E.W., Simkin, M.E. and Foote, R.H. (1990a) *In vitro* developmental potential of bovine zygotes in oviduct epithelial cell co-culture systems. *Theriogenology* 33, 223.

Ellington, J.E., Farrell, P.B., Simkin, M.E., Goldman, E.E. and Foote, R.H. (1990b) Bovine zygote development on oviduct cell coculture versus rabbit oviducts. *Theriogenology* 33, 224.

Ellington, J., Foote, R.H., Webb, J., Hasler, J. and McGrath, A. (1990c) The use of a GnRH analog in an embryo transfer field trial. *Theriogenology* 33, 225.

Ellington, J.E., Farrell, P.B., Simkin, M.E., Foote, R.H., Goldman, E.E. and McGrath, A.B. (1990d) Development and survival after transfer of cow embryos cultured from 1–2-cells to morulae or blastocysts in rabbit oviducts or in a simple medium with bovine oviduct epithelial cells. *Journal of Reproduction and Fertility* 89, 293–299.

Ellington, J.E., Carney, E.W., Farrell, P.B., Simkin, M.E. and Foote, R.H. (1990e) Bovine 1–2 cell embryo development using a simple medium in three oviduct epithelial cell coculture systems. *Biology of Reproduction* 43, 97–104.

Ellington, J.E., Farrell, P.B. and Foote, R.H. (1990f) Comparison of six-day bovine embryo development in uterine tube (oviduct) epithelial cell co-culture versus *in vivo* development in the cow. *Theriogenology* 134, 837–844.

Ellington, J.E., Farrell, P.B., Simkin, M.E. and Foote, R.H. (1990g) Method for obtaining bovine zygotes produced *in vivo*. *American Journal of Veterinary Research* 51, 1708–1710.

Ellington, J.E., Padilla, A.W., Vredenburgh, W.L., Dougherty, E.P. and Foote, R.H. (1991a) Behaviour of bull spermatozoa in bovine uterine tube epithelial cell co-culture: an *in vitro* model for studying the cell interactions of reproduction. *Theriogenology* 35, 977–989.

Ellington, J.E., Foote, R.H., Farrell, P.B., Hasler, J.F., Webb, J., Henderson, W.B. and McGrath, A.B. (1991b) Pregnancy rates after the use of a gonadotropin releasing hormone agonist in bovine embryo transfer recipients. *Theriogenology* 36, 1035–1042.

Ellington, J.E., Wilker, C.W., Hillman, R.B. and Ball, B.A. (1993a) Ability of epididymal or ejaculated bull spermatozoa to bind to cow zonae before and after oviduct epithelial cell co-culture. *Theriogenology* 39, 213.

Ellington, J.E., Ignotz, G.G., Ball, B.A., Meyers-Wallen, V.N. and Currie, W.B. (1993b) *De novo* protein synthesis by bovine uterine tube (oviduct) epithelial cell changes during co-culture with bull spermatozoa. *Biology of Reproduction* 48, 851–856.

Ellington, J.E., Ball, B.A. and Yang, X. (1993c) Binding of stallion spermatozoa to the equine zona pellucida after coculture with oviductal epithelial cells. *Journal of Reproduction and Fertility* 98, 203–208.

Ellis, S.B., Bondioli, K.R., Williams, M.E., Pryor, J.H. and Harpold, M.M. (1988) Sex determination of bovine embryos using male-specific DNA probes. *Theriogenology* 29, 242.

Elsden, R.P. (1989) Mexican government uses embryo transfer to increase production of national dairy herd. *Theriogenology* 31, 47–48.

Elsden, R.P., Seidel, G.E., Takeda, T. and Farrand, G.D. (1982) Field experiments with frozen–thawed bovine embryos transferred non-surgically. *Theriogenology* 17, 1–10.

Eng, L.A., Kornegay, E.T., Huntighton, J. and Wellman, T. (1986) Effects of incubation temperature and bicarbonate on maturation of pig oocytes *in vitro*. *Journal of Reproduction and Fertility* 76, 657–662.

Engelmann, U., Krassnigg, F., Schatz, H. and Schill, W.B. (1988) Separation of human X and Y spermatozoa by free-flow electrophoresis. *Gamete Research* 19, 151–159.

Eppig, J.J. (1991a) Development of murine preantral follicles *in vitro*. *Biology of Reproduction* 44 (Suppl. 1), 145 (Abs. 369).

Eppig, J.J. (1991b) Intercommunication between mammalian oocytes and companion somatic cells. *Bioessays* 13, 569–574.

Eppig, J.J. (1992) Development-related gene expression in oocytes. *Development and Function of the Reproductive Organs: Serona Symposium* (Peebles), 11.

Eppig, J.J. and Schroeder, A.C. (1986) Culture systems for mammalian oocyte development: progress and prospects. *Theriogenology* 25, 97–106.

Eppig, J.J. and Schroeder, A.C. (1989) Capacity of mouse oocytes from preantral follicles to undergo embryogenesis and development to live young after growth, maturation and fertilization *in vitro*. *Biology of Reproduction* 41, 268–276.

Eppig, J.J., Ward-Baily, P.F. and Coleman, D.L. (1985) Hypoxanthine and adenosine in murine ovarian follicular fluid: concentration and activity in maintaining oocyte meiotic arrest. *Biology of Reproduction* 33, 1041–1049.

Eppig, J.J., Daniel, S.A.J. and Schroeder, A.C. (1988) Growth and development of mouse oocytes *in vitro*. *Annals of the New York Academy of Sciences* 541, 205–210.

Eppig, J.J., Schroeder, A.C., van de Sandt, J.J.M., Ziomek, C.A. and Bavister, B.D. (1990) Developmental capacity of mouse oocytes that grow and mature in culture: the effect of modification of the protocol. *Theriogenology* 33, 89–100.

Eppig, J.J., Schroeder, A.C. and O'Brien, M.J. (1992a) Developmental capacity of mouse oocytes matured *in vitro*: effects of gonadotrophin stimulation, follicular origin and oocyte size. *Journal of Reproduction and Fertility* 95, 119–127.

Eppig, J.J., Wigglesworth, K. and O'Brien, M.J. (1992b) Comparison of embryonic developmental competence of mouse oocytes grown with and without serum. *Molecular Reproduction and Development* 32, 33–40.

Eppig, J.J., Peters, A.H.F.M., Telfer, E.E. and Wigglesworth, K. (1993) Production of cumulus expansion enabling factor by mouse oocytes grown *in vitro*: preliminary characterization of the factor. *Molecular Reproduction and Development* 34, 450–456.

Erickson, B.H. (1966a) Development and radio-response of the prenatal bovine ovary. *Journal of Reproduction and Fertility* 10, 97–105.

Erickson, B.H. (1966b) Development and senescence of the postnatal bovine ovary. *Journal of Animal Science* 25, 800–805.

Erickson-Lawrence, M.F., Turner, T.T., Thomas, T.S. and Oliphant, G. (1989) Effect of steroid hormones on sulfated oviductal glyoprotein secretion by oviductal explants *in vitro*. *Biology of Reproduction* 40, 1311–1319.

Ericsson, R.J., Langevin, C.N. and Nishino, M. (1973) Isolation of fractions rich in human Y sperm. *Nature* 246, 421–424.

Ericsson, S.A. and Ericsson, R.J. (1992) Couples with exclusively female offspring have an increased probability of a male child after using male sex preselection. *Human Reproduction* 7, 372–373.

Ericsson, S.A., Garner, D.L., Redelman, D. and Ahmad, K. (1989) Assessment of the viability and fertilizing potential of cryopreserved bovine spermatozoa using dual fluorescent staining and two-flow cytometric systems. *Gamete Research* 22, 355–368.

Ericsson, S.A., Boice, M.L., Funahashi, H. and Day, B.N. (1993) Assessment of porcine oocytes using brilliant cresyl blue. *Theriogenology* 39, 214.

Eriksen, T., Terkelsen, O., Hyttel, P., Mollgard, K. and Greve, T. (1993) Regional morphology and localization of PDGF in bovine oviduct epithelium. *Theriogenology* 39, 215.

Eroglu, A. and Meinecke, B. (1990) Polypeptide synthesis of *in vivo* and *in vitro* matured porcine oocytes. *Reproduction in Domestic Animals* 25, 261–268.

Esposito, L., Thuard, J.M., Nibart, M. and Thibier, M. (1993) Sexed and frozen bovine embryos: *in vitro* and *in vivo* survival rate. *Proceedings of the Ninth Congress of the European Embryo Transfer Association* (Lyon), 194.

Esslemont, R.J. (1993) Relationship between herd calving to conception interval and culling rate for failure to conceive. *Veterinary Record* 133, 163–164.

Estienne, M.J., Knight, J.W. and Beal, W.E. (1988) Isolation of a population of intact highly motile porcine spermatozoa using a discontinuous bovine serum albumin gradient. *Theriogenology* 29, 771–778.

Evans, A.C.O., Adams, G.P. and Rawlings, N.C. (1992) Follicular waves and gonadotrophins in 36 week old heifers. *Biology of Reproduction* 46 (Suppl. 1), 323 (Abs. 323).

Evans, A.C.O., Adams, G.P. and Rawlings, N.C. (1993) Prepubertal ovarian and endocrine changes preceding first ovulation in heifers. *Journal of Reproduction and Fertility,* Abstract Series No. 11, 76 (Abs. 141).

Evans, G. (1991) Application of reproductive technology to the Australian livestock industries. *Reproduction Fertility and Development* 3(6), 627–650.

Evans, G., Brooks, J., Crow, W. and McNeilly, A.S. (1992) Oocyte maturation and embryo development in GnRH agonist treated ewes given FSH. *Journal of Reproduction and Fertility,* Abstract Series No. 10, 14 (Abs. 24).

Evans, M. and Kaufmann, M.H. (1981) Establishment in culture of pluripotential cells from mouse embryos. *Nature* 292, 154–156.

Evans, M.J., Notarianni, E., Laurie, S. and Moor, R.M. (1990) Derivation and preliminary characterization of pluripotent cell lines from porcine and bovine blastocysts. *Theriogenology* 33, 125–128.

Eyestone, W.H. and Boer, H.A. de (1991) Inhibition of bovine oocyte maturation *in vitro* with a phosphodiesterase inhibitor, iso-butyl-methyl-xanthine (IBMX): reversibility, kinetics of meiotic resumption after IBMX inhibition and timing of an IBMX-sensitive event. *Serono Symposium on Preimplantation Embryo Development,* Abstract Vol., 32 (Abs. I-16).

Eyestone, W.H. and Boer, H.A. de (1993) FSH enhances developmental potential of bovine oocytes matured in chemically defined medium. *Theriogenology* 39, 216.

Eyestone, W.H. and First, N.L. (1986) A study of the 8-to-16 cell developmental block in bovine embryos cultured *in vitro*. *Theriogenology* 25, 152.

Eyestone, W.H. and First, N.L. (1988) Co-culture of bovine embryos with oviductal tissue. *Proceedings of the Eleventh International Congress on Animal Reproduction and Artificial Insemination* (Dublin) 3, 431 (3 pp.).

Eyestone, W.H. and First, N.L. (1989a) Variation in bovine embryo development *in vitro* due to bulls. *Theriogenology* 31, 191.

Eyestone, W.H. and First, N.L. (1989b) Co-culture of early cattle embryos to the blastocyst stage with oviductal tissue or in conditioned medium. *Journal of Reproduction and Fertility* 85, 715–720.

Eyestone, W.H. and First, N.L. (1991) Characterization of developmental arrest in early bovine embryos cultured *in vitro*. *Theriogenology* 35, 613–624.

Eyestone, W.H., Northey, D.L. and Leibfried-Rutledge, M.L. (1985) Culture of 1-cell bovine embryos in the sheep oviduct. *Biology of Reproduction* 32 (Suppl. 1), 100 (Abs.).

Eyestone, W.H., Vignieri, J. and First, N.L. (1987a) Co-culture of early bovine embryos with oviductal epithelium. *Theriogenology* 27, 228.

Eyestone, W.H., Leibfried-Rutledge, M.L., Northey, D.L., Gilligan, B.G. and First, N.L. (1987b) Culture of one- and two-cell bovine embryos to the blastocyst stage in the ovine oviduct. *Theriogenology* 28, 1–7.

Eyestone, W.H., Jones, J.M. and First, N.L. (1990) The use of oviduct-conditioned medium for culture of bovine oocytes to the blastocyst stage. *Theriogenology* 33, 226.

Eyestone, W.H., Jones, J.M. and First, N.L. (1991) Some factors affecting the efficacy of oviduct tissue-conditioned medium for the culture of early bovine embryos. *Journal of Reproduction and Fertility* 92, 59–64.

Faddy, M.J., Gosden, R.G., Gougeon, A., Richardson, S.J. and Nelson, J.F. (1992) Accelerated disappearance of ovarian follicles in mid-life: implications for forecasting menopause. *Human Reproduction* 7, 1342–1346.

Fahning, M.L. and Garcia, M.A. (1992) Status of cryopreservation of embryos of domestic animals. *Cryobiology* 29(1), 1–18.

Fahy, M.M. and Kane, M.T. (1992) Inositol stimulates DNA and protein synthesis, and expansion by rabbit blastocysts *in vitro*. *Human Reproduction* 7, 550–552.

Fair, T. (1992) Factors affecting *in vitro* fertilization of the bovine follicular oocyte. M. Agr. Sc. Dissertation, University College, Dublin.

Falcone, L., Gianni, S., Piffarettii-Yanez, A., Marchini, M., Eppenberger, U. and Balerna, M. (1991) Follicular fluid enhances sperm motility and velocity *in vitro*. *Fertility and Sterility* 55, 619–623.

Fallon, M.N., Rammell, C.G. and Hoogenboom, J.J.L. (1988) Amino acids in bovine sera. *New Zealand Veterinary Journal* 36, 96–98.

Fanning, A. and Edwards, S. (1991) *Mycobacterium bovis* infection in human beings in contact with elk (*Cervus elaphus*) in Alberta, Canada. *Lancet* 338, 1253–1255.

Farin, C.E., Imakawa, K., Hansen, T.R., McDonnell, J.J., Murphy, C.N., Farin, P.W.and Roberts, R.M. (1990) Expression of trophoblastic interferon genes in sheep and cattle. *Biology of Reproduction* 43, 210–218.

Farin, P.W., Farin, C.E. and Yang, L. (1994) *In vitro* production of bovine embryos is associated with altered fetal development. *Theriogenology* 41, 193.

Farlin, M.E., Jasko, D.J., Graham, J.K. and Squires, E.L. (1992) Heparin-induced capacitation: a comparison between the stallion and bull. *Proceedings of the Third International Symposium on Equine Embryo Transfer* (Buenos Aires), 31.

Fassi-Fihri, N., Chupin, D., Marquant-Le Guienne, B. and Thibier, M. (1991) Critical effect of granulosa cells from preovulatory follicles on IVM, IVF and embryonic development in cattle. *Theriogenology* 35, 198.

Fattouh, El-S.M. and Abdou, M.S.S. (1991) Effect of caffeine on the post-thaw motility of buffalo spermatozoa. *Theriogenology* 36, 149–154.

Faundez, R., Spohr, I. and Boryczko, Z. (1988) Effect of follicle cells on maturation and *in vitro* fertilization of cattle oocytes. *Proceedings of the Eleventh International Congress on Animal Reproduction and Artificial Insemination* (Dublin) 3, 325 (3 pp.).

Fayed, A.H. and Hattab, S.A. (1991) Effect of theophylline on motility of bull spermatozoa *in vitro*. *Assiut Veterinary Medical Journal* 25, 225–229.

Fayrer-Hosken, R.A. and Caudle, A.B. (1990) Bovine *in vitro* fertilization: will the technique be practical? *Embryo Transfer* 5(2), 1–5.

Fayrer-Hosken, R.A. and Caudle, A.B. (1991) The laparoscope in follicular oocyte collection and gamete intrafallopian transfer and fertilization (GIFT). *Theriogenology* 36, 709–725.

Fayrer-Hosken, R.A. and Stroud, B. (1989) Multiple embryo production in the cow after laparoscopic oocyte collection and oviductal transfer. *Theriogenology* 31, 192.

Fayrer-Hosken, R.A., Younis, A.I., Brackett, B.G., McBride, C.E., Harper, K.M., Keefer, C.L. and Cabaniss, D.C. (1988) Pregnancy after laparoscopic oviductal transfer of *in vitro* matured and *in vitro* fertilized bovine oocytes. *Biology of Reproduction* 38 (Suppl. 1), 70 (Abs.).

Fayrer-Hosken, R.A., Younis, A.I., Brackett, B.G., McBride, C.E., Harper, K.M., Keefer, C.L. and Cabaniss, D.C. (1989) Laparoscopic oviductal transfer of *in vitro* matured and *in vitro* fertilized bovine oocytes. *Theriogenology* 32, 413–420.

Fayrer-Hosken, P.M., Brooks, A., Miller, C.C., Willis P. and Caudle, A.B. (1992). *In vitro* definition and characterization of equine gametes. *Proceedings of the 3rd International Symposium of Equine Embryo Transfer (Buenos Aires)* 22.

Fazeli, A.R., Steenweg, W., Bevers, M., de Loos, F. and Colenbrander, B. (1991) Sperm-zona binding: an additional parameter to assess quality of bull semen? *Journal of Reproduction and Fertility*, Abstract Series No. 8, 30 (Abs. 49).

Fazeli, A.R., Steenweg, W., Bevers, M., Colenbrander, B. (1992) Sperm–zona binding in the bovine is not influenced by the day to day variations in ejaculates. *Journal of Reproduction and Fertility*, Abstract Series No. 10, 19 (Abs. 34).

Fehilly, C.B., Willadsen, S.M. and Tucker, E.M. (1984) Experimental chimaerism in sheep. *Journal of Reproduction and Fertility* 70, 347–351.

Feichtinger, W., Strohmer, H., Fuhrberg, P., Antinori, S., Pepe, G. and Versaci, C. (1992a) Photoablation of oocyte zona pellucida by erbium-yag laser for *in vitro* fertilization in severe male infertility. *Lancet* 339, 811.

Feichtinger, W., Strohmer, H., Radner, K.M. and Goldin, M. (1992b) The use of fibrin sealant for embryo transfer: development and clinical studies. *Human Reproduction* 7, 890–893.

Fekete, E. and Little, C.C. (1942) Observations on the mammary tumour incidence of mice born from transferred ova. *Cancer Research* 2, 525–530.

Feng, P., Catt, K.J. and Knecht, M. (1991) Transforming growth factor-beta stimulates meiotic maturation of the rat oocyte. *Endocrinology* 122, 181–186.

Feng, S., Zhang, Y., Li, S., Ma, Z, Wanmg, R. and Lu, D. (1991) Piglets from frozen (−20°C) embryos were born in China. *Theriogenology* 35, 199.

Fennessey, P.F. (1990) Animal Production in the future – a New Zealand response. *Proceedings of the New Zealand Society of Animal Production* 50, 361–371.

Fennessy, P.F. and Taylor, P.G. (1989) Deer farming in Oceania. In: Hudson, R.J., Drew, K.R. and Baskin, L.M. (eds) *Wildlife Production Systems: Economic Utilisation of Wild Ungulates*. Press Syndicate of the University of Cambridge, Cambridge, pp. 309–322.

Fennessy, P.F., Asher, G.W., Beatson, N.S., Dixon, T.E., Hunter, J.W. and Bringans, M.J. (1994) Embryo transfer in deer. *Theriogenology* 41, 133–138.

Fenton, S.E. and Ax, R.L. (1990) Glycoaminoglycans enhance cumulus expansion and hyaluronic acid production, but not fertilization success of bovine oocytes *in vitro*. *Biology of Reproduction* 42 (Suppl. 1), 89.

Fenton, S.E., Dentine, M.R. and Ax, R.L. (1993) Modulation of bovine oocyte–cumulus cell complex maturation and fertilization *in vitro* by glycosaminoglycans. *Journal of Dairy Science* 76, 701–712.

Fernandez Limia, O., Alonso, J.C., Barbier, R. and Faure, R. (1990) Effect of summer on the corpus luteum and oocyte fertilization in Holstein cows in the tropics. *Revista de Salud Animal* 12, 50–54.

Ferrandi, B., Lange Consiglio, A., Cremonesi, F., Carnevali, A. and Porcelli, F. (1988) Effects of deep freezing on energy metabolism of bovine spermatozoa during *in vitro* capacitation: a cytochemical approach. *Theriogenology* 30, 563–573.

Ferry, L., Mermillod, P., Massip, A. and Dessy, F. (1992) Adaptation of culture conditions to small numbers of embryos obtained *in vitro* from individual cows. *Proceedings of the Eighth Conference of the European Embryo Transfer Association* (Lyon), 150.

Fieni, F., Buggin, M., Tainturier, D., Bruyas, J.F., Perrin, J., Dumont, P., Beckers, J.P., Chupin, D. and Daubie, M. (1990) Comparison of the efficiency of three cryoprotectants for freezing goat embryos. *Proceedings of the Sixth Congress of the European Embryo Transfer Association* (Lyon), 144.

Figueiredo, J.R., Hulshof, S.C.J., Ectors, F.J., Fontes, R.S., Nusgens, B., Le Point, A. and Beckers, J.F. (1992) Mechanical and enzymatic method for the isolation of intact preantral follicles from the fetal calf, calf and cow ovaries. *Proceedings of the Eighth Conference of the European Embryo Transfer Association* (Lyon), 152.

Figueiredo, J.R., Hulshof, S.C.J., Van den Hurk, R., Ectors, F.J., Fontes, R.S., Nusgens, B., Bevers, M.M. and Beckers, J.F. (1993) Development of a combined new mechanical and enzymatic method for the isolation of intact preantral follicles from fetal, calf and adult bovine ovaries. *Theriogenology* 40, 789–799.

Filho, I.M.B., Pinheiro, L.E.L., Rodrigues, C.F.M. and Carvalho, C. (1992) Preparation of embryo chromosomes under farm conditions. *Theriogenology* 37, 521–525.

Findlay, J.K. (1990) Peripheral and intragonadal actions of inhibin-related peptides in the female. *Proceedings of the Third International Ruminant Reproduction Symposium* (Nice), Abstract.

Findlay, J.K. (1993) An update on the roles of inhibin, activin and follistatin as local regulators of folliculogenesis. *Biology of Reproduction* 48, 15–23.

First, N.L. (1987) Gene transfer. *Proceedings of a Symposium on the Application of Egg and Embryo Technologies to Domestic Animals* (Copenhagen), 30–32.

First, N.L. (1989) Technology of the 90s. *British Cattle Breeders Conference,* Digest No. 44, 2–7.

First, N.L. (1990a) Multiplication of embryos by nuclear transfer. *Proceedings of the Third International Ruminant Reproduction Symposium* (Nice), Abstract.

First, N.L. (1990b) The application of new reproductive and genetic technologies to animal breeding. *Proceedings of the Fourth World Congress on Genetics Applied to Livestock Production* 16, 321–322.

First, N.L. and Parrish, J.J. (1987a) Preparation of sperm for *in vitro* fertilization. *Proceedings of a Symposium on the Application of Egg and Embryo Technologies to Domestic Animals* (Copenhagen), 33–35.

First, N.L. and Parrish, J.J. (1987b) *In vitro* fertilization of ruminants. *Journal of Reproduction and Fertility,* Suppl. 34, 151–165.

First, N.L. and Parrish, J.J. (1988) Sperm maturation and *in vitro* fertilization. *Proceedings of the Eleventh International Congress on Animal Reproduction and Artificial Insemination* (Dublin) 5, 161–168.

First, N.L. and Prather, R.S. (1991) Production of embryos by oocyte cytoplast–blastomere fusion in domestic animals. *Journal of Reproduction and Fertility,* Suppl. 43, 245–254.

First, N.L., Critser, E.S. and Robl, J.M. (1986) Bovine embryo: development, cloning, sexing and transfer of genes. In: Turner, G.P. (ed.) *Immunological Approaches to Contraception and Promotion of Fertility* Plenum Press, New York, pp. 375–392.

Fischer, B. (1993) Effects of EGF and TGF-alpha on preimplantation embryo development. *Proceedings of the Pre-European Society of Human Reproduction and Embryology Congress,* Post-Graduate Course No. 2, 47–54.

Fischer, B. and Lindenau, A. (1993) Effect of oxygen concentration on the development of cultured preimplantation rabbit embryos. *Journal of Reproduction and Fertility,* Abstract Series No. 11, 69 (Abs. 127).

Fishel, S., Dowell, K., Timson, J., Green, S., Hall, J. and Klentzeris, L. (1993a) Micro-assisted fertilization with human gametes. *Human Reproduction* 8, 1780–1784.

Fishel, S., Timson, J., Green, S., Hall, J., Dowell, K. and Klentzeris, L. (1993b) Micromanipulation. *Reproductive Medicine Review* 2, 199–222.

Fissore, R.A. and Robl, J.M. (1992) Intracellular Ca^{2+} response of rabbit oocytes to electrical stimulation. *Molecular Reproduction and Development* 32, 9–16.

Flechon, J.-E., Guillomot, M., Charlier, M., Flechon, B. and Martal, J. (1986) Experimental studies on the elongation of the ewe blastocyst. *Reproduction, Nutrition, Development* 26, 1017–1024.

Flechon, J.-E., Notarianni, E., Laurie, S., Galli, C., Evans, M. and Moor, R.M. (1990) Characterization of ovine and porcine embryonic stem cells. *Journal of Reproduction and Fertility,* Abstract Series No. 6, 20, Abs. 40.

Fleming, W.N. and Saacke, R.G. (1972) Fine structure of the bovine oocyte from the mature Graafian follicle. *Journal of Reproduction and Fertility* 30, 493–497.

Fletcher, N.A. and Worsfold, A.I. (1988) A direct radioimmunoassay for oestrone sulphate in serum. *British Veterinary Journal* 144, 269–272.

Fletcher, T.J. (1989) Deer farming in Europe. In: Hudson, R.J., Drew, K.R. and Baskin, L.M.

(eds) *Wildlife Production Systems: Economic Utilisation of Wild Ungulates*. Press Syndicate of the University of Cambridge, Cambridge, pp. 322–333.

Flint, A.P.F., Lamming, G.E., Stewart, H.J. and McCann, S.H.E. (1988) Evidence that the ovine trophoblast antiluteolysin is an interferon confirmed by receptor binding studies. *Journal of Reproduction and Fertility* Abstract Series No. 1, 18, Abs. 32.

Flood, L.P. and Shirley, B. (1991) Reduction of embryotoxicity by protein in embryo culture media. *Molecular Reproduction and Development* 30, 226–231.

Flood, M.R. and Bunch, T.D. (1992) Effect of growth factors on preimplantation bovine embryo development.*Journal of Animal Science* 70 (Suppl. 1), 265 (Abs. 506).

Flood, M.R. and Wiebold, J.L. (1988) Glucose metabolism by preimplantation pig embryos. *Journal of Reproduction and Fertility* 84, 7–12.

Flood, M.R., Kern, D., Spendlove, R. and Bunch, T.D. (1992) Fetal clone versus fetal bovine serum for *in vitro* maturation, fertilization and culture. *Theriogenology* 37, 212.

Flood, M.R., Gage, T.L. and Bunch, T.D. (1993) The effect of various growth-promoting factors on preimplantation bovine embryo development *in vitro*. *Theriogenology* 39, 823–833.

Fontes, R.S., Ectors, F.J., Delval, A., Touati, K., Beckers, J.-F.and Ectors, F. (1991) Rabbit and bovine oviductal epithelial cell culture media are able to support early development from zygote to the blastocyst stage. *Theriogenology* 35, 201.

Fontes, R.S., Ectors, F.J., Thonon, F., Figueiredo, J.R., Beckers, J.-F. and Ectors, F. (1992a) Development of bovine oocytes to blastocyst and hatching in co-culture supplemented with estrous cow serum or polyvinyl alcohol. *Theriogenology* 37, 214.

Fontes, R.S., Figueiredo, J.R., Ectors, F.J., Beckers, J.F. and Ectors, F. (1992b) Effect of fetuin on *in vitro* maturation of bovine oocytes in defined media. *Proceedings of the Eighth Conference European Embryo Transfer Association* (Lyon), 156.

Fontes, R.S., Noel, A., Figueredo, J.R., Beckers, J.-F., Foidart, J.M. and Ectors, F. (1993) Use of frozen–thawed bovine oviduct cells for *in vitro* culture of bovine embryos: effect of matrigel. *Proceedings of the Ninth Congress of the European Embryo Transfer Association* (Lyon), 202.

Foote, R.H. (1987) *In vitro* fertilization and embryo transfer in domestic animals: applications in animals and implications for humans. *Journal of In Vitro Fertilization and Embryo Transfer* 4, 73–88.

Foote, R.H. (1993) Current state of the art and future of embryo manipulation and engineering to enhance breeding programs in cattle. *Ars Veterinaria* 9(2), 1–21.

Foote, R.H. and Ellington, J.E. (1988) Is a superovulated oocyte normal? *Theriogenology* 29, 111–123.

Foote, R.H. and Miller, P. (1971) What might sex ratio control mean in the animal world? In: Kiddy, C.A. and Hafs, H.D. (eds) *Proceedings of the American Animal Science Symposium*. Boyd Printing Company, Inc., Albany, NY, pp. 1–9.

Fortier, M.A., Nadeau, Y. and Blouin, L. (1994) Cultured endometrial cells as a detection system for quantitative evaluation of bovine embryo quality. *Theriogenology* 41, 196.

Fortune, J.E., Sirios, J. and Quirk, S.M. (1988) The growth and differentiation of ovarian follicles during the bovine estrous cycle. *Theriogenology* 29, 95–109.

Fortune, J.E., Sirois, J., Turzillo, A.M. and Lavoir, M. (1990) Follicle selection in domestic ruminants. *Proceedings of the Third International Ruminant Reproduction Symposium* (Nice), Abstract.

Foulkes, J.A. (1991) Milk progesterone: a changing application. *Proceedings of the British Cattle Veterinary Association* (Reading), 112–121.

Fox, T.C., Geisert, R.D., Morgan, G.L., Wells, M.E. and Zevy, M.T. (1988) Survival of bovine embryos transferred synchronously to progesterone treated recipients. *Journal of Animal Science* 66 (Suppl. 1), 414 (Abs.).

Franco, O.J., Drost, M., Thatcher, M.-J., Shille, V.M. and Thatcher, W.W. (1987) Fetal survival in the cow after pregnancy diagnosis by palpation per rectum. *Theriogenology* 27, 631–644.

Franks, G.C., Coley, S.L., Betterbed, B. and Page, R.D. (1985) The effects of cryoprotective agents, dilution rates, freezing rates and freezing units on the survival of bovine embryos. *Theriogenology* 23, 194.

Franks, G.C., Coley, S.L., Betterbed, B. and Page, R.D. (1986) The effect of freezer type, cryoprotectant and processing methods on viability of frozen embryos. *Theriogenology* 26, 135–144.

Franks, S. (1992) Superovulation strategy. *Development and Function of the Reproductive Organs: Serona Symposium* (Peebles), 21.

Fraser, L.R. (1984) Mechanisms controlling mammalian fertilization. *Oxford Review of Reproductive Biology* 6, 174–225.

Fray, M.D., Wrathall, J.H.M. and Knight, P.G. (1994) Active immunization against inhibin promotes a recurrent increase in litter size in sheep. *Veterinary Record* 134, 19–20.

Freeman, D.A., Butler, J.E., Weber, J.A., Geary, R.T. and Woods, G.L. (1991) Co-culture of day-5 to day-7 equine embryos in medium with oviductal tissue. *Theriogenology* 36, 815–822.

Freeman, M.R., Bastias, M.C., Hill, G.A. and Osteen, K.G. (1993) Coculture of mouse embryos with cells isolated from the human ovarian follicle, oviduct, and uterine endometrium. *Fertility and Sterility* 59, 138–142.

Frei, R.E., Schultz, G.A. and Church, R.B. (1989) Qualitative and quantitative changes in protein synthesis occur at the 8–16 cell stage of embryogenesis in the cow. *Journal of Reproduction and Fertility* 86, 637–641.

Fricke, P.M., Jablonka, A. and Redmer, D.A. (1992) Fibroblast growth factors in the ovine follicle and effects on proliferation of thecal cells. *Journal of Reproduction and Fertility*. Abstract Series No. 9, 62 (Abs. 109).

Fry, R.C. (1992) The effect of leukaemia inhibitory factor (LIF) on embryogenesis. *Reproduction, Fertility and Development* 4, 449–458.

Fry, R.C., Batt, P.A., Fairclough, R.J., Cameron, A.W.N. and Parr, R.A. (1991) Leukemia inhibitory factor (LIF) improves the health of cultured ovine embryos. *Journal of Reproduction and Fertility*, Abstract Series No. 7, 34 (Abs. 58).

Fry, R.C., Batt, P.A., Fairclough, R.J. and Parr, R.A. (1992) Human leukemia inhibitory factor improves the viability of cultured ovine embryos. *Biology of Reproduction* 46, 470–474.

Fry, R.C., Simpson, T.L., Squires, T.J., Parr, R.A. and Damanik, R.M. (1994) Factors affecting transvaginal oocyte pick-up in heifers. *Theriogenology* 41, 197.

Fuchs, A.-R., Liu, H.C., Nevarro, D., Chang, S.-M., Stallings-Mann, M. and Fields, M.J. (1991) Preimplantation bovine conceptus suppresses endometrial oxytocin receptors in both ipsi- and contralateral horn. *Biology of Reproduction* 44 (Suppl. 1), 159 (Abs. 425).

Fuente, R. de la, Plante, L. and King, W.A. (1993) X chromosome inactivation in the preattachment bovine embryo. *Theriogenology* 39, 220.

Fugger, E.P., Bustillo, M., Katz, P., Dorfmann, A.D., Bender, S.A. and Schulman, J.D. (1988) Embryonic development and pregnancy from fresh and cryopreserved sibling pronucleate human zygotes. *Fertility and Sterility* 50, 273–278.

Fuhrer, F., Mayr, B., Schellander, K., Kalat, M. and Schleger, W. (1989) Maturation competence and chromatin behaviour in growing and fully grown cattle oocytes. *Journal of Veterinary Medicine* 36, 285–291.

Fukuda, A., Noda, Y., Tsukui, S., Matsumoto, H., Yano, J. and Mori, T. (1987) Influence of water quality on *in vitro* fertilization and embryo development for the mouse. *Journal of In Vitro Fertilization and Embryo Transfer* 4, 40–45.

Fukuda, Y., Ichikawa, M., Naito, K. and Toyoda, Y. (1988) Normal development of bovine oocytes matured, fertilized and cultured with cumulus cells *in vitro*. *Proceedings of the Eleventh International Congress on Animal Reproduction and Artificial Insemination* (Dublin) 3, 327 (3 pp.).

Fukuda, Y., Ichikawa, M., Naito, K. and Toyoda, Y. (1990) Birth of normal calves resulting from bovine oocytes matured, fertilized, and cultured with cumulus cells *in vitro* up to the blastocyst stage. *Biology of Reproduction* 42, 114–119.

Fukui, Y. (1990) Effect of follicle cells on the acrosome reaction, fertilization and developmental competence of bovine oocytes matured *in vitro*. *Molecular Reproduction and Development* 26, 40–46.

Fukui, Y. and Matsuyama, K. (1994) Development of bovine embryos matured and fertilized *in vitro* in media containing human leukemia inhibitory factor. *Theriogenology* 41, 199.

Fukui, Y. and Ono, H. (1988) *In vitro* development to blastocyst of *in vitro* matured and fertilized bovine oocytes. *Veterinary Record* 122, 282.

Fukui, Y. and Ono, H. (1989) Effects of sera, hormones and granulosa cells added to culture medium for *in vitro* maturation, fertilization, cleavage and development of bovine oocytes. *Journal of Reproduction and Fertility* 86, 501–506.

Fukui, Y. and Sakuma, Y. (1980) Maturation of bovine oocytes cultured *in vitro* in relation to ovarian activity, follicular size and the presence or absence of cumulus cells. *Biology of Reproduction* 22, 669–673.

Fukui, Y., Fukushima, M. and Ono, H. (1983a) Fertilization and cleavage of bovine follicular oocytes in rabbit reproductive tracts after maturation *in vitro*. *Journal of Experimental Zoology* 226, 137–142.

Fukui, Y., Fukushima, M. and Ono, H. (1983b) Fertilization *in vitro* of bovine oocytes after various sperm procedures. *Theriogenology* 20, 651–660.

Fukui, Y., Glew, A.M. and Moor, R.M. (1987a) Method of sperm preparation for *in vitro* fertilization of sheep. *Japanese Journal of Animal Reproduction* 33, 44–47.

Fukui, Y., Imai, K., Alfonso, N.F. and Ono, H. (1987b) Fertilizability and cleavage of bovine oocytes. *Journal of Animal Science* 64, 935–941.

Fukui, Y., Glew, A.M., Gandolfi, F. and Moor, R.M. (1988a) Ram-specific effects on *in vitro* fertilization and cleavage of sheep oocytes matured *in vitro*. *Journal of Reproduction and Fertility* 82, 337–340.

Fukui, Y., Glew, A.M., Gandolfi, F. and Moor, R.M. (1988b) *In vitro* culture of sheep oocytes matured and fertilized *in vitro*. *Theriogenology* 29, 883–891.

Fukui, Y., Urakawa, M., Sasaki, C., Chikamatsu, N. and Ono, H. (1989) Development to the late morula or blastocyst stage following *in vitro* maturation and fertilization of bovine oocytes. *Animal Reproduction Science* 18, 139–148.

Fukui, Y., Sonoyama, T., Mochizuki, H. and Ono, H. (1990) Effects of heparin dosage and sperm capacitation time on *in vitro* fertilization and cleavage of bovine oocytes matured *in vitro*. *Theriogenology* 34, 575–591.

Fukui, Y., McGowan, L.T., James, R.W., Asher, G.W. and Tervit, H.R. (1991a) Effects of culture duration and time of gonadotrophin addition on *in vitro* maturation and fertilization of red deer (*Cervus elaphus*) oocytes. *Theriogenology* 35, 499–512.

Fukui, Y., McGowan, L.T., James, R.W., Pugh, P.A. and Tervit, H.R. (1991b) Factors affecting the *in vitro* development to blastocysts of bovine oocytes matured and fertilized *in vitro*. *Journal of Reproduction and Fertility* 92, 125–131.

Fukui, Y., Sawai, K., Furudate, M., Sato, N., Iwazumi, Y. and Ohsaki, K. (1992) Parthenogenetic development of bovine oocytes treated with ethanol and Cytochalasin B after *in vitro* maturation. *Molecular Reproduction and Development* 33, 357–362.

Fukunari, A., Sasabe, M., Maeda, T., Terada, T. and Tsutsumi, Y. (1989) A mechanical and enzymatic approach to recovery of ovarian follicular oocytes in rabbits. *Japanese Journal of Animal Reproduction* 35(1), 26–33.

Fukushima, M. and Fukui, Y. (1985) Effects of gonadotropins and steroids on the subsequent fertilizability of extra follicular bovine oocytes cultured *in vitro*. *Animal Reproduction Science* 9, 323–332.

Fukushima, M., Shioya, Y., Kuwayama, M., Iwasaki, S. and Hanada, A. (1991) Effects of gonadotrophins and estradiol added into maturation medium of bovine oocytes *in vitro* on the subsequent capacity for fertilization and embryonic development. *Japanese Journal of Animal Reproduction* 37, 127–132.

Fukushima, M., Tominaga, K., Utsumi, K. and Iritani, A. (1992a) Freezing bovine blastocyst stage embryos derived from oocytes matured and fertilized *in vitro*. *Proceedings of the Twelfth International Congress on Animal Reproduction* (The Hague) 3, 1412–1414.

Fukushima, M., Tominaga, K., Hataya, Y. and Utsumi, K. (1992b) Freezing sensitivity of bovine blastocysts derived from oocytes matured, fertilized and cultured *in vitro*. *Journal of Reproductive Development* 38, 49–54.

Fulka, J. and Moor, R.M. (1993) Noninvasive chemical enucleation of mouse oocytes. *Molecular Reproduction and Development* 34, 427–430.

Fulka, J. and Motlik, J. (1980) *In vitro* maturation. *Proceedings of the Ninth International Congress of Animal Reproduction and Artificial Insemination* (Madrid) 2, 55–62.

Fulka, J., Bradshaw, J., Jung, T., Antalikova, L. and Moor, R.M. (1993) UV irradiation of metaphase II bovine oocytes. *Journal of Reproduction and Fertility*, Abstract Series No. 11, 29 (Abs. 48).

Funahashi, H. and Day, B.N. (1993a) Effects of different serum supplements in maturation medium on meiotic and cytoplasmic maturation of pig oocytes. *Theriogenology* 39, 965–973.

Funahashi, H. and Day, B.N. (1993b) Effects of the duration of exposure to hormone supplements on cytoplasmic maturation of pig oocytes *in vitro*. *Journal of Reproduction and Fertility* 98, 179–185.

Funahashi, H. and Day, B.N. (1993c) Effects of follicular fluid at fertilization *in vitro* on sperm penetration in pig oocytes. *Journal of Reproduction and Fertility* 99, 97–103.

Funahashi, H., Aoyagi, Y., Takeda, T. and Onihara, T. (1991) Developmental capacity of bovine oocytes collected from ovaries of individual heifers and fertilized *in vitro*. *Therioegenology* 36, 427–434.

Furuya, S., Endo, Y., Oba, M., Matsui, Y., Nozawa, S. and Suzuki, S. (1992) Protein phosphorylation regulates the mouse sperm acrosome reaction induced by the zona pellucida. *Journal of Assisted Reproduction and Genetics* 9, 384–390.

Gagne, M., Pothier, F. and Sirard, M.A. (1990) Developmental potential of early bovine zygotes submitted to centrifugation and microinjection following *in vitro* maturation of oocytes. *Theriogenology* 34, 417–425.

Gagne, M.B., Pothier, F. and Sirard, M.A. (1991a) Effect of microinjection in *in vitro* matured bovine oocytes on *in vitro* development of embryos. *Biology of Reproduction* 44 (Suppl. 1), 76 (Abs. 93).

Gagne, M.B., Pothier, F. and Sirard, M.-A. (1991b) Electroporation of bovine spermatozoa to carry foreign DNA in oocytes. *Molecular Reproduction and Development* 29, 6–15.

Gajda, B. and Smorag, Z. (1993) Factors affecting the survival of one-and two-cell rabbit embryos cryopreserved by vitrification. *Theriogenology* 39, 499–506.

Gajda, B., Smorag, Z., Wierzbowski, S., Jura, J. and Wieczorek, B. (1989) Transfer of vitrified sheep morulae. *Zuchthygiene* 24, 97–100.

Galeati, G., Mattioli, M., Bacci, M.L. and Seren, E. (1990) Effect of LH and FSH on pig oocyte maturation in vitro. *Journal of Reproduction and Fertility*, Abstract Series No. 5, 40, Abs. 70.

Galli, A., Basetti, M., Balduzzi, D., Martignoni, M., Bornaghi, V. and Maffii, M. (1991) Frozen bovine semen quality and bovine cervical mucus penetration test. *Theriogenology* 35, 837–844.

Galli, C. and Moor, R.M. (1991) Gonadotrophin requirements for the *in vitro* maturation of sheep oocytes and their subsequent embryonic development. *Theriogenology* 35, 1083–1093.

Galli, C., Notarianni, E., Laurie, S., Moor, R.M. and Evans, M. (1990) Derivation of pluripotent, embryonic cell lines from porcine and ovine blastocysts. *Journal of Reproduction and Fertility,* Abstract Series No. 6, 24, Abs. 39.

Galli, C., Powell, D.J. and Moor, R.M. (1991) Stability of DNA injected in oocytes and embryos of domestic animals. *Proceedings of the Seventh Meeting of the European Embryo Transfer Association* (Cambridge), 146.

Galli, C., Duchi, R. and Lazzari, G. (1994) Production of purebred embryos from beef cattle by *in vitro* embryo technology. *Theriogenology* 41, 201.

Galvin, J.M., Stewart, A.N.V. and Meredith, S. (1993) Higher sodium chloride concentration can induce a four-cell block in porcine embryos. *Theriogenology* 39, 224.

Gandolfi, F. (1994) Autocrine, paracrine and environmental factors influencing embryonic development from zygote to blastocyst. *Theriogenology* 41, 95–100.

Gandolfi, F. and Moor, R.M. (1987) Stimulation of early embryonic development in the sheep by co-culture with oviduct epithelial cells. *Journal of Reproduction and Fertility* 81, 23–28.

Gandolfi, F. and Moor, R.M. (1988) Interactions between somatic and germinal cells during early development. *Proceedings of the Eleventh International Congress on Animal Reproduction and Artificial Insemination* (Dublin) 5, 170–176.

Gandolfi, F., Brevini, T.A.L. and Moor, R.M. (1986) Role of somatic cells in the long-term culture of ovine embryos. *Proceedings of the Society of the Study of Fertility* (Bristol), Abs. 111.

Gandolfi, F., Galli, C., Modina, S., Artini, P. and Lauria, A. (1988a) An *in vitro* approach to the growth of ovarian follicles from primary to antral stage. *Proceedings of the Eleventh International Congress on Animal Reproduction and Artificial Insemination* (Dublin) 3, 328 (3 pp.).

Gandolfi, F., Brevini, T.A.L. and Blakeley, D.M. (1988b) Detection of an oviduct mitogenic factor in the sheep. *Journal of Reproduction and Fertility,* Abstract Series No. 1, 63 (Abs.).

Gandolfi, F., Brevini, T.A.L. and Moor, R.M. (1989a) Effect of oviduct environment on embryonic development. *Journal of Reproduction and Fertility* Suppl. 38, 107–115.

Gandolfi, F., Brevini, T.A., Richardson, L., Brown, C.R. and Moor, R.M. (1989b) Characterization of proteins secreted by sheep oviduct epithelial cells and their function in embryonic development. *Development* 106, 303–312.

Gandolfi, F., Lavitrano, M., Camaioni, A., Spadafora, C., Siracusa, G. and Lauria, A. (1989c) The use of sperm-mediated gene transfer for the generation of transgenic pigs. *Journal of Reproduction and Fertility,* Abstract Series No. 4, Abs. 16.

Gandolfi, F., Brevini, T.A.L., Modina, S., Cran, D.G. and Moor, R.M. (1989d) Oviduct environment and embryonic development: *in vivo* and *in vitro* relationship. *Proceedings of a Symposium on in vitro Approaches to Mammalian Gamete Maturation and Embryonic Development* (Milan), 39–55.

Gandolfi, F., Brevini, T.A.L., Modina, S. and Lauria, A. (1990) Embryogenesis *in vitro*: role of culture media and co-culture. *Proceedings of the Third Symposium on Advanced Topics in Animal Reproduction* FCAV-UNESP, Jaboticabal, 13–25.

Gandolfi, F., Brevini, T.A.L., Modina, S .and Lauria, A. (1991) Detection and characterization of a growth factor in bovine oviduct secretions. *Journal of Reproduction and Fertility,* Abstract Series No. 7, 6 (Abs. 2).

Gandolfi, F., Brevini, T.A.L., Modina, S. and Passoni, L. (1992a) Early embryonic signals:

embryo-maternal interactions before implantation. *Animal Reproduction Science* 28, 269–276.

Gandolfi, F., Brevini, T.A.L., Modina, S., Passoni, L. and Lauria, A. (1992b) Maternal control of early embryonic development. In: Lauria, A. and Gandolfi, F. (eds) *Embryonic Development and Manipulation in Animal Production*. Portland Press, London, pp. 93–101.

Gandolfi, F., Brevini, T.A.L., Modina, S., Bianchi, R. and Passoni, L. (1993) Role of the oviduct during early embryogenesis. *Reproduction in Domestic Animals* 28, 189–192.

Gao, Y., Short, R.V. and Fletcher, T.P. (1988) Progesterone concentration in plasma, saliva and milk of cows in different reproductive states. *British Veterinary Journal* 144, 262–268.

Garbers, D.L., First, N.L., Sullivan, J.J. and Lardy, H.A. (1971) Stimulation and maintenance of ejaculated bovine spermatozoan respiration and motility by caffeine. *Biology of Reproduction* 5, 336–339.

Garcia, J.M., Pinheiro, L.E.L., Almeida, I.L., Lima, V.F.R. and Mikich, A.B. (1988) Studies on *in vitro* maturation of bovine oocytes in chemically defined media. *Ars Veterinaria* 4, 69–75.

Garcia-Bojalil, C.M., Staples, C.R., Savio, J.D., Drost, M. and Thatcher, W.W. (1991) Effect of dietary protein on follicle growth and embryo development of superovulated non-lactating Holstein cows. *Journal of Dairy Science* 74 (Suppl. 1), 195.

Garde, J., Gutierrez, A., Artiga, C.G. and Vazquez, I. (1993) Influence of freezing process on '*in vitro*' capacitation of ram semen. *Theriogenology* 39, 225.

Gardner, D.K. and Sakkas, D. (1993) Mouse embryo cleavage, metabolism and viability: role of medium composition. *Human Reproduction* 8, 288–295.

Gardner, D.K., Clarke, R.N., Lechene, C.P. and Biggers, J.D. (1989) Development of a noninvasive ultramicrofluorometric method for measuring net uptake of glutamine by single preimplantation mouse embryos. *Gamete Research* 24, 427–438.

Gardner, D.K., Lane, M. and Batt, P. (1993) Uptake and metabolism of pyruvate and glucose by individual sheep preattachment embryos developed *in vivo*. *Molecular Reproduction and Development* 36, 313–319.

Gardner, D.K., Lane, M. and Batt, P.A. (1994) Nutrient uptake and enzyme activity of the preattachment goat embryo developed *in vivo*. *Theriogenology* 41, 204.

Gardner, H.G. and Kaye, P.L. (1991) Insulin increases cell numbers and morphological development in mouse pre-implantation embryos *in vitro*. *Reproduction, Fertility and Development* 3, 79–91.

Gardner, R.I. and Edwards, R.G. (1968) Control of the sex ratio at full term in the rabbit by transferring sexed blastocysts. *Nature (London)* 218, 346–348.

Garrett, J.E., Geisert, R.D., Zavy, M.T. and Morgan, G.L. (1988) Evidence for maternal regulation of early conceptus growth and development in beef cattle. *Journal of Reproduction and Fertility* 84, 437–446.

Garrisi, G.J., Chin, A.J., Dolan, P.M. and Nagler, H.M. (1993) Analysis of factors contributing to success in a program of micromanipulation-assisted fertilization. *Fertility and Sterility* 59, 366–374.

Gatica, R. (1988) Studies in the micromanipulation of farm animal embryos. PhD Thesis, National University of Ireland, Dublin.

Gatica, R., Boland, M.P., Crosby, T.F. and Gordon, I. (1984) Micromanipulation of sheep morulae to produce monozygotic twins. *Theriogenology* 21, 555–560.

Gauger, P., Zachary, E., Timms, L.L. and Eness, P. (1991) Evaluation of activity monitors as a tool for heat detection in dairy cattle in a tie stall barn. *Journal of Dairy Science* 74 (Suppl. 1), 228 (Abs. P259).

Gayerie de Abreu, F., Lamming, G.E. and Shaw, R.C. (1984) A cytogenetic investigation of early stage bovine embryos: relation with embryonic mortality. *Proceedings of the Tenth International Congress on Animal Reproduction and Artificial Insemination* (Illinois) 2, 82.

Gearheart, W.W., Smith, C. and Teepker, G. (1989) Multiple ovulation and embryo manipulation in the improvement of beef cattle: relative theoretical rates of genetic change. *Journal of Animal Science* 67, 2863–2871.

Gehlaut, B.S. and Srivastava, R.K. (1987) Effect of caffeine citrate on post-thaw motility of buffalo bull spermatozoa. *Theriogenology* 28, 767–772.

Geim, A. (1990) Trials on the synchronization of oestrus in donor and recipient cows in connection with embryo transfer. Thesis, Tierarztliche Hochschule, Hannover, Germany, 109 pp.

Geisert, R.D., Buhi, W.C. and Fields, M. (1987) Synthesis and release of polypeptides by bovine oviductal tissue in culture. *Biology of Reproduction* 36 (Suppl. 1), 84 (Abs.).

Geisert, R.D., Fox, T.C., Morgan, G.L., Wells, M.E., Wettemann, R.P. and Zavy, M.T. (1991) Survival of bovine embryos transferred to progesterone-treated asynchronous recipients. *Journal of Reproduction and Fertility* 92, 475–482.

George, M.A., Johnson, M.H. and Vincent, C. (1992) Use of fetal bovine serum to protect against zona hardening during preparation of mouse oocytes for cryopreservation. *Human Reproduction* 7, 408–412.

Georges, M. and Massey, J.M. (1991) Velogenetics, or the synergistic use of marker assisted selection and germ line manipulation. *Theriogenology* 35, 151–159.

Gerena, R.L. and Killian, G.J. (1988) Estrus-associated 98 k Dalton protein in bovine oviductal fluid (ODF). *Biology of Reproduction*, 35 (Suppl. 1), 246 (Abs.).

Gibbons, J.R., Beal, W.E., Krisher, R.L., Faber, E.G., Pearson, R.E. and Gwazdauskas, F.C. (1994) Effects of once versus twice weekly transvaginal follicular aspiration on bovine oocyte recovery and embryo development. *Theriogenology* 41, 206.

Gibson, J.P. (1988) The options and prospects for genetically altering milk composition in dairy cattle. *Animal Breeding Abstract* 55, 231–234.

Gibson, J.P. and Smith, C. (1989) The incorporation of biotechnologies into animal breeding strategies. In: Babiuk, L.A. and Phillips, J.P. (eds) *Animal Biotechnology*. Pergamon Press, Oxford, pp. 203–231.

Giles, J.R., Foote, R.H. and Mark, W. (1991) Development of embryo stem cell technology for production of chimeric rabbits. *Biology of Reproduction* 44 (Suppl. 1), 57 (Abs. 19).

Giudice, L.C., Dsupin, B.A., Irwin, J.C. and Eckbert, R.L. (1992) Identification of insulin-like growth factor binding proteins in human oviduct. *Fertility and Sterility* 57, 294–301.

Glass, K.W. and Voelkel, S.A. (1990) Loss of viability in frozen bovine oocytes associated with specific steps in the cryopreservation process. *Biology of Reproduction* 42 (Suppl. 1), 52.

Gledhill, B.L. (1988) Selection and separation of X- and Y-chromosome bearing mammalian sperm. *Gamete Research* 20, 377–395.

Glencross, R.G., Bleach, E.C., McLeod, B.J., Beard, A.J. and Knight, P.G. (1990) Increased ovulation rate in heifers immunized against a synthetic peptide sequence of bovine inhibin. *Journal of Reproduction and Fertility*, Abstract Series No. 6, Abs. 30.

Glencross, R.G., Bleach, E.C.L., McLeod, B.J., Beard, A.J. and Knight, P.G. (1992) Effect of active immunization of heifers against inhibin on plasma FSH concentrations, ovarian follicular development and ovulation rate. *Journal of Endocrinology* 134, 11–18.

Gliedt, D.W., Miller, G.F., Lester, T.D., Rakes, J.M. and Rorie, R.W. (1991) Development of *in vitro* produced bovine embryos co-cultured with oviductal cells in M-199 medium supplemented with either fetal calf serum or sequentially supplemented with day 2, 4 and 6 cow serum. *Journal of Animal Science* 69 (Suppl. 1), 48–49.

Godke, R.A. and Rorie, R.W. (1993) Embryo microsurgery for large animals. In: Fishel, S. and

Symonds, M. (eds) *Gamete and Embryo Micromanipulation in Human Reproduction.* Arnold, London, pp. 155–164.

Gong, J.G., Bramley, T.A. and Webb, R. (1990) The effect of bovine somatotropin (BST) on ovarian folliculogenesis and ovulation rate in heifers. *Journal of Reproduction and Fertility,* Abstract Series No. 6, Abs. 9.

Gong, J.G., Bramley, T.A. and Webb, R. (1991a) The effect of recombinant bovine somatotropin (BST) on ovarian follicular dynamics in heifers. *Journal of Reproduction and Fertility,* Abstract Series No. 8, 6 (Abs. 2).

Gong, J.G., Bramley, T. and Webb, R. (1991b) The effect of recombinant bovine somatotrophin on ovarian function in heifers: follicular population and peripheral hormones. *Biology of Reproduction* 49, 941–949.

Gong, J.G., McBride, D., Bramley, T.A. and Webb, R. (1992) Effects of bovine somatotropin (BST), IGF-I and insulin on the proliferation of bovine granulosa cells *in vitro. Journal of Reproduction and Fertility,* Abstract Series No. 10, 5 (Abs. 6).

Gong, J.G., Bramley, T.A. and Webb, R. (1993) The effect of recombinant bovine somatotrophin on ovarian follicular growth and development in heifers. *Journal of Reproduction and Fertility* 97, 247–254.

Goode, S. and Donaldson, L.E. (1990) Clinical field trials in England and U.S.A. with Super-Ov. *Proceedings of the Sixth Conference of the European Embryo Transfer Association* (Lyon), 148.

Goodeaux, L.L. (1979) Biochemical effects of bovine serum albumin on *in vitro* capacitation, acrosome reaction and oocyte fertilization of beef cattle. *Dissertation Abstracts International* B40, 2448.

Gook, D.A., Osborn, S.M. and Johnston, W.I.H. (1993) Cryopreservation of mouse and human oocytes using 1, 2-propanediol and the configuration of the meiotic spindle. *Human Reproduction* 8, 1101–1109.

Gordon, A.C., Lee, S.H., Ko, T.H., Kondaveeti, U., Harrison, R.F. and Gordon, I. (1992) Nuclear configuration of human oocytes following fertilization failure *in vitro. Proceedings of the Fourteenth World Congress on Fertility and Sterility* (Caracas), 151 (Abs. 428).

Gordon, I. (1955) The hormonal augmentation of fertility in sheep. *Proceedings of the British Society of Animal Production,* 55–63.

Gordon, I. (1975) Problems and prospects in cattle embryo transfer. *Irish Veterinary Journal* 39, 39–62.

Gordon, I. (1976) Cattle twinning by the egg transfer approach. *Proceedings of an EEC Seminar on Egg Transfer in Cattle* (Cambridge), 305–319.

Gordon, I. (1983) *Controlled Breeding in Farm Animals* Pergamon Press, Oxford.

Gordon, I. (1987) Embryo transfer breakthrough. *British Friesian Journal* 69, 663.

Gordon, I. (1988) Use of embryo transfer for twinning in cattle. *Journal of the Irish Grassland and Animal Production Association* 22, 60–66.

Gordon, I. (1989a) Cattle *in vitro* fertilization: research and practical application. *British Cattle Breeders Conference Digest* 44, 29–31.

Gordon, I. (1989b) Modern techniques for enhancing reproduction in suckler cattle systems. *British Society of Animal Production* (Winter Meeting), Paper 6 (2 pp.).

Gordon, I. (1989c) Large-scale production of cattle embryos by *in vitro* culture methods. *AgBiotech News and Information* 1(3), 345–348.

Gordon, I. (1990) *In vitro* maturation (IVM) and fertilization (IVF) of cattle ova. *Embryo Transfer Newsletter* 8(3), 6–11.

Gordon, I. (1991a) Potential for *in vitro* fertilization in cattle breeding. *Proceedings of the British Cattle Veterinary Association* (Reading), 16–24.

Gordon, I. (1991b) Potential application of cattle *in vitro* fertilization in commercial practice and research. *Embryo Transfer Newsletter* 9(2), 4–9.

Gordon, I. and Lu, K.H. (1990) Production of embryos *in vitro* and its impact on livestock production. *Theriogenology* 33, 77–87.

Gordon, I., Williams, G. and Edwards, J. (1962) The use of serum gonadotrophin (PMS) in the induction of twin-pregnancy in the cow. *Journal of Agricultural Science (Cambridge)* 59, 143–198.

Gordon, I., Boland, M.P., McGovern, H. and Lynn, G. (1987) Effect of season on superovulatory responses and embryo quality in Holstein cattle in Saudi Arabia. *Theriogenology* 27, 231.

Gordon, I., Lu, K.H., Gallagher, M. and McGovern, H. (1988) Production of cattle embryos by *in vitro* and *in vivo* culture. *Animal Production* 46, 498 (Abs.).

Gordon, J.W. (1988) Use of micromanipulation for increasing the efficiency of mammalian fertilization *in vitro*. *Annals of the New York Academy of Sciences* 541, 601–613.

Gordon, J.W. (1990) Zona drilling: a new approach to male infertility. *Journal of In Vitro Fertilization and Embryo Transfer* 7, 223–228.

Gordon, J.W. (1992) Current unresolved controversies in micromanipulation-assisted fertilization. *Journal of Assisted Reproduction and Genetics* 9, 184–186.

Gordon, J.W. and Dapunt, U. (1993a) A new mouse model for embryos with a hatching deficiency and its use to elucidate the mechanism of blastocyst hatching. *Fertility and Sterility* 59, 1296–1301.

Gordon, J.W. and Dapunt, U. (1993b) Restoration of normal implantation rates in mouse embryos with a hatching impairment by use of a new method of assisted hatching. *Fertility and Sterility* 59, 1302–1307.

Gordon, J.W.and Laufer, N. (1988) Applications of micromanipulation to human *in vitro* fertilization. *Journal of In Vitro Fertilization and Embryo Transfer* 5, 57–60.

Gordon, J.W., Scangos, G.A., Plotkin, D.J., Barbosa, J.A. and Ruddle, F.H. (1980) Genetic transformation of mouse embryos by microinjection of purified DNA. *Proceedings of the National Academy of Sciences (USA)* 77, 7380–7384.

Gordon, J.W., Grunfeld, L., Garrisi, G.J., Talansky, B.E., Richards, C. and Laufer, N. (1988) Fertilization of human oocytes by sperm from infertile males after zona pellucida drilling. *Fertility and Sterility* 50, 68–73.

Gorospe, W.C., Hughes, F.M. and Spangelo, B.L. (1992) Interleukin-6: effects on and production by rat granulosa cells *in vitro*. *Endocrinology* 130, 1750–1752.

Gosden, R.G. (1992a) Transplantation of fetal germ cells. *Journal of Assisted Reproduction and Genetics* 9, 118–123.

Gosden, R.G. (1992b) Ovarian germ cell manipulation and transplantation. *AgBiotech News and Information* 4, 337N–340N.

Gosden, R.G. and Telfer, E.E. (1992) Development of germ cells and their supporting cell lineages. *Journal of Reproduction and Fertility*, Abstract Series No. 9, 4 (Abs. 54.).

Gosden, R.G., Boland, N.I., Spears, N., Murray, A.A., Chapman, M., Wade, J.C., Zohdy, N.I. and Brown, N. (1993) The biology and technology of follicular oocyte development *in vitro*. *Reproductive Medicine Review* 2, 129–152.

Goto, K. (1993) Bovine microfertilization and embryo transfer. *Molecular Reproduction and Development* 36, 288–290.

Goto, K. and Iritani, A. (1992) Oocyte maturation and fertilization. *Animal Reproduction Science* 28, 407–413.

Goto, K., Kajihara, Y., Kosaka, S., Koba, M., Nakanishi, Y. and Ogawa, K. (1988a) Pregnancies after *in vitro* fertilization of cow follicular oocytes, their incubation *in vitro* and their transfer to the cow uterus. *Theriogenology* 29, 251.

Goto, K., Kajihara, Y., Kosaka, S., Koba, M., Nakanishi, Y. and Ogawa, K. (1988b) Pregnancies after co-culture of cumulus cells with bovine embryos derived from *in vitro* fertilization of *in vitro* matured follicular oocytes. *Journal of Reproduction and Fertility* 83, 753–758.

Goto, K., Kajihara, Y., Koba, M., Kosaka, S., Nakanishi, Y. and Ogawa, K. (1989) *In vitro* fertilization and development of *in vitro* matured bovine follicular oocytes. *Journal of Animal Science* 67, 2181–2185.

Goto, K., Kinoshita, A., Takuma, Y. and Ogawa, K. (1990a) Fertilization by sperm injection in cattle. *Theriogenology* 33, 238.

Goto, K., Takuma, Y., Ooe, N. and Ogawa, K. (1990b) *In vitro* development of bovine oocytes collected from ovaries of individual cows after *in vitro* fertilization. *Japanese Journal of Animal Reproduction* 36, 110–113.

Goto, K., Kinoshita, A., Takuma, Y. and Ogawa, K. (1990c) Fertilization of bovine oocytes by the injection of immobilised, killed spermatozoa. *Veterinary Record* 127, 517–520.

Goto, K., Kinoshita, A., Takuma, Y. and Ogawa, K. (1991) Birth of calves after the transfers of oocytes fertilized by sperm injection. *Theriogenology* 35, 205.

Goto, K., Matsumoto, T., Takuma, Y. and Nakanishi, Y. (1992a) Microfertilization of bovine oocyte by the injection of sperm head. *Japanese Journal of Animal Reproduction* 37, 277–280.

Goto, K., Matsumoto, T., Takuma, Y. and Nakanishi, Y. (1992b) *In vitro* development of bovine oocytes fertilized by sperm-head injection. *Theriogenology* 37, 216.

Goto, K., Iwai, N., Takuma, Y. and Nakanishi, Y. (1992c) Co-culture of *in vitro* fertilized bovine embryos with different cell monolayers. *Journal of Animal Science* 70, 1449–1453.

Goto, K., Takuma, Y., Matsuura, S., Nakanishi, Y., Yanagida, K., Katahira, K., Toguchi, M. and Nonaka, K. (1992d) A study on the viability of bovine blastocysts derived from *in vitro* fertilization after freezing and thawing by two cryopreservation methods adapted to practical use in the field. *Journal of Reproduction and Development* 38(5), 15–19.

Goto, K., Iwai, N., Ichikawa, K., Ishihara, A., Takuma, Y., Motoishi, M., Tokumaru, M. and Nakanishi, Y. (1992e) *In vitro* culture of early bovine embryos produced by *in vitro* technique. *Journal of Reproduction and Development* 38, 165–171.

Goto, K., Tokumaru, M., Motoishi, M. and Nakanishi, Y. (1993) Sexing of mouse and cow embryos with freemartin serum. *Theriogenology* 39, 226.

Gotze, M., Kauffold, P., Schuffenhauer, A., Torner, H. and Spitchak, M. (1990) Investigations on the inhibition of meiosis in cattle oocytes by means of biological or synthetic inhibitors. *Archiv für Experimentalle Veterinarmedizin* 44, 19–27.

Gradl, E., Lutterbach, A. and Brem, G. (1988) Influence of follicle cell co-culture and 17-B estradiol on *in vitro* maturation and fertilizaation of cattle oocytes. *Proceedings of the Eleventh International Congress of Animal Reproduction and Artificial Insemination* (Dublin), 3, 339 (3 pp.).

Graham, C.F. (1974) The production of parthenogenetic mammalian embryos and their use in biological research. *Biological Review* 49, 399–422.

Graham, J.K. and Foote, R.H. (1987a) Dilauroylphosphatidylcholine liposome effects on the acrosome reaction and *in vitro* penetration of zona-free hamster eggs by bull sperm: I. A fertility assay for fresh semen. *Gamete Research* 16, 133–145.

Graham, J.K. and Foote, R.H. (1987b) Dilauroylphosphatidylcholine liposome effects on the acrosome reaction and *in vitro* penetration of zona-free hamster eggs by bull sperm: II. A fertility assay for frozen–thawed semen. *Gamete Research* 16, 147–158.

Graham, J.K., Foote, R.H. and Hough, S.R. (1987) Penetration of zona-free hamster eggs by liposome-treated sperm from the bull, ram, stallion and boar. *Biology of Reproduction* 37, 181–188.

Graham, J.K., Foote, R.H. and Parrish, J.J. (1986) Effect of dilauroylphosphatidyl-choline on the acrosome reaction and subsequent penetration of bull spermatozoa into zona- free hamster eggs. *Biology of Reproduction* 35, 413–424.

Gray, B.W., Cartee, R.E., Stringfellow, D.A., Riddell, M.G., Riddell, K.P. and Wright, J.C. (1991) The effects of dominant follicular regression and FSH priming on the superovulatory response of cattle. *Theriogenology* 35, 207.

Gray, B.W., Cartee, R.E., Stringfellow, D.A., Riddell, M.G., Riddell, K.P. and Wright, J.C. (1992) The effects of FSH-priming and dominant follicular regression on the superovulatory response of cattle. *Theriogenology* 37, 631–639.

Gray, B.W., Stringfellow, D., Riddell, M., Riddell, K., Daenport, G. and Wright, J. (1993) The effect of treatment with bovine somatotropin (BST) on the superovulatory response of cattle. *Theriogenology* 39, 227.

Gray, C.W., Morgan, P.M. and Kane, M.T. (1992) Purification of an embryotrophic factor from commercial bovine serum albumin and its identification as citrate. *Journal of Reproduction and Fertility* 94, 471–480.

Gray, K.R., Bondioli, K.R. and Betts, C.L. (1991) The commercial application of embryo splitting in beef cattle. *Theriogenology* 35, 37–44.

Green, D.P.L. (1993) Mammalian fertilization as a biological machine: a working model for adhesion and fusion of sperm and oocyte. *Human Reproduction* 8, 91–96.

Greenstein, J.S. and Foley, R.C. (1958) The early embryology of the cow with notes on comparable human development. *International Journal of Fertility* 3, 67–79.

Greenwald, G.S. and Moor, R.M. (1989) Isolation and preliminary characterization of pig primordial follicles. *Journal of Reproduction and Fertility* 87, 561–571.

Greising, T., Monaghan, P., Healy, A., Gallagher, M., Wahid, H. and Gordon, I. (1994) The influence of cytoplasmic composition and quantity of cytoplasm in recipient oocytes on the developmental capacity of bovine nuclear transfer embryos. *Theriogenology* 41, 208.

Greve, T. (1992) *In vitro* embryo technologies in cattle with particular reference to their use in cattle breeding. *Reproduction in Domestic Animals* 27(1), 22–28.

Greve, T. and Madison, V. (1991) *In vitro* fertilization in cattle: a review. *Reproduction, Nutrition and Development* 31, 147–157.

Greve, T. and Purwantara, B. (1992) Ultrasonography in embryo transfer practice. *Proceedings of the Eighth Conference of the European Embryo Transfer Association* (Lyon), 93–104.

Greve, T. and Purwantara, B. (1993) Ultrasonography in embryo transfer practice. *Proceedings of the Ninth Congress of the European Embryo Transfer Association* (Lyon), 137–147.

Greve, T., Bousquet, D., King, W.A. and Betteridge, K.J. (1984) *In vitro* fertilization and cleavage of *in vivo* matured bovine oocytes. *Theriogenology* 22, 151–165.

Greve, T., Goff, A., King, W.A. and Bousquet, D. (1986) LH surge, oocyte maturation and *in vitro* fertilization in superovulated heifers. *Proceedings of the Fourth World Conference on In Vitro Fertilization* (Melbourne). *Journal of In Vitro Fertilization and Embryo Transfer* 3, 136 (Abs.).

Greve, T., Xu, K.P., Callesen, H. and Hyttel, P. (1987) *In vivo* development of *in vitro* fertilized bovine oocytes matured *in vivo* versus *in vitro*. *Journal of In Vitro Fertilization and Embryo Transfer* 4, 281–285.

Greve, T., Callesen, H. and Hyttel, P. (1989a) Follicular correlates with *in vitro* fertilization in cattle. *Journal of Reproduction and Fertility* Suppl. 38, 117–126.

Greve, T., Xu, K.P., Callesen, H. and Hyttel, P. (1989b) Calves resulting from *in vitro* fertilization of oocytes. *Zuchthygiene* 24, 79–83.

Greve, T., Loskutoff, N.M., Buckrell, B.C., Christian, C.R., Leibo, S.P. and Betteridge, K.J. (1992) Morphology and freezing tolerance of *in vitro* derived bovine embryos after culture *in vivo* or *in vitro*. *Cinquième colloque Franco-Tchecoslovaque sur la réproduction des animaux domestiques*. (Jouy-en-Josas, France) 36.

Greve, T., Avery, B. and Callesen, H. (1993) Viability of *in vivo* and *in vitro* produced bovine embryos. *Reproduction in Domestic Animals* 28, 164–169.

Grimek, H.J., Bellin, M.E. and Ax, R.L. (1984) Characteristics of proteoglycans isolated from small and large bovine ovarian follicles. *Biology of Reproduction* 30, 397–409.

Grimes, R.W. and Ireland, J.J. (1985) Can the appearance of the surface of an ovarian follicle be used to predict whether an oocyte will mature *in vitro*? *Journal of Animal Science* 59 (Suppl. 1), 318 (Abs.).

Grimes, R.W. and Ireland, J.J. (1987) Relationship of macroscopic appearance of the surface of bovine ovarian follicles, concentration of steroids in follicular fluid and maturation of oocytes *in vitro*. *Biology of Reproduction* 35, 725–732.

Grimes, R.W., Matton, P. and Ireland, J.J. (1987) A comparison of histological and non-histological indices of atresia and follicular function. *Biology of Reproduction* 37, 82–88.

Grippo, A.A., Henault, M.A., Anderson, S.H. and Killian, G.J. (1992) Cation concentrations in fluid from the oviduct ampulla and isthmus of cows during the estrous cycle. *Journal of Dairy Science* 75, 58–65.

Grisart, B., Massip, A. and Dessy, F. (1992) Cinematographic study of early bovine embryo development in a serum-free medium. *Proceedings of the Eighth Conference of the European Embryo Transfer Association* (Lyon), 160.

Grisart, B., Perret, J., Georges, M., Vassart, G. and Dessy, F. (1993) A new quick and reliable sexing method: application to the sex ratio of *in vitro* produced bovine embryos. *Proceedings of the Ninth Congress of the European Embryo Transfer Association* (Lyon), 204.

Grizzle, J., Roberts, A. and Wise, T. (1991) Human recombinant interleukin 1 beta stimulates bovine thecal cell steroidogenesis *in vitro*. *Biology of Reproduction* 44 (Suppl. 1), 177 (Abs. 497).

Gualtieri, R., Santella, L. and Dale, B. (1992) Tight junctions and cavitation in the human pre-embryo. *Molecular Reproduction and Development* 32, 81–87.

Guerra-Martinez, P., Dickerson, G.E., Anderson, G.B. and Green, R.D. (1990) Embryo transfer twinning and performance efficiency in beef production. *Journal of Animal Science* 68, 4039–4050.

Guilbault, L.A. (1991) The influence of the presence of a dominant follicle on superovulatory response in Holstein heifers. *Bulletin-Agriculture Canada, Research Branch* 14, 25–27.

Guilbault, L.A., Grasso, F., Roy, G.L. and Lussier, J.G. (1990) Effect of ovarian condition on the ovulation rate and embryo production of superovulated heifers. *Bulletin-Agriculture Canada, Research Branch* 13, 54–56.

Guilbault, L.A., Matton, P., Rouillier, P., Glencross, R.G., Beard, A.J. and Knight, P.G. (1991) Level of atresia and inhibin contents (alpha-subunit and alpha-beta dimer) in dominant follicles during their growing and regressing phases of development in cattle. *Biology of Reproduction* 44 (Suppl. 1), 67 (Abs. 60).

Guilbault, L.A., Lussier, J.G. and Grasso, F. (1992) Interrelationships of hormonal and ovarian responses in superovulated heifers pretreated with FSH-P at the beginning of the estrous cycle. *Theriogenology* 37, 1030–1040.

Guilbault, L.A., Bolamba, D., Desaulniers, D.M. and Lussier, J.G. (1993a) Follicular and hormonal events associated with the transient increase in estradiol concentrations during the first wave of follicular development in cattle. *Theriogenology* 39, 228.

Guilbault, L.A., Rouillier, P., Matton, P., Glencross, R.G., Beard, A.J. and Knight, P.G. (1993b) Relationships between the level of atresia and inhibin contents (α subunit and

α–β dimer) in morphologically dominant follicles during their growing and regressing phases of development. *Biology of Reproduction* 48, 268–276.

Guillomot, M., Reinaud, P., Charpigny, G. and Martal, J. (1988) Embryo–uterine interactions during early stages of pregnancy in domestic mammals. *Reproduction, Nutrition and Development* 28, 1629–1648.

Gurnsey, M.P., Hagemann, L.J., Xu, D. and Welch, R.A.S. (1994) Preliminary results from the use of FACS-sorted bull sperm in an *in vitro* embryo production system. *Theriogenology* 41, 210.

Gustafsson, H. (1985) Characteristics of embryos from repeat breeder and virgin heifers. *Theriogenology* 23, 487–498.

Gutierrez, A., Garde, J., Artiga, C.G., Munoz, I. and Pintado, B. (1993) *In vitro* survival of murine morulae after quick freezing in the presence of chemically defined macromolecules and different cryoprotectants. *Theriogenology* 39, 1111–1120.

Guyader, C. and Chupin, D. (1991) Capacitation of fresh bovine spermatozoa on bovine epithelial oviduct cell monolayers. *Theriogenology* 36, 505–512.

Guyader, C., Procureur, R. and Chupin, D. (1989) Capacitation of fresh bovine sperm on oviductal cell monolayer. *Proceedings of the Fifth Conference of the European Embryo Transfer Association* (Lyon), 156.

Guyader, C., Marquant-Le Guienne, B., Humblot, P. and Thibier, M. (1990) Effect of heparin concentration on *in vitro* development of bovine eggs. *Proceedings of the Sixth Conference of the European Embryo Transfer Association* (Lyon), 150.

Guyader, C., Charbonnier, G., Durand, M., Marquant-Leguienne, B. and Thibier, M. (1992) Pregnancy rates following transfer of *in vitro* produced and fresh bovine embryos to asynchronous heifers. *Proceedings of the Eighth Conference of the European Embryo Transfer Association* (Lyon), 162.

Guyader-Joly, C., Charbonnier, G., Durand, M., Marquant-Le Guienne, B., Humblot, P., Jeanguyot, N. and Thibier, M. (1993) Ability of *in vitro* produced bovine embryos to develop to term. *Proceedings of the Ninth Congress of the European Embryo Transfer Association* (Lyon), 208.

Gwatkin, R.B.L. (1993) Micromanipulation-assisted fertilization. *Molecular Reproduction and Development* 36, 285–287.

Gwazdauskas, F.C., McCaffrey, C., McEvoy, T.G. and Sreenan, J.M. (1992) *In vitro* preimplantation mouse embryo development with incubation temperatures of 37 and 39°C. *Journal of Assisted Reproduction and Genetics* 9, 149–154.

Hackett, A.J., Durnford, R., Marcus, G.J., McAllister, A.J. and Nagai, J. (1989) Critical aspects of hormonal treatment of cows for nuclear transfer cloning. *Journal of Animal Science* 67 (Suppl. 1), 354 (Abs. 859).

Hackett, A.J., Durnford, R., Sasada, H., Lee, A.J., Marcus, G.J. and Nagai, J. (1990) The use of advanced bovine embryos to produce clones by nuclear transfer. *Journal of Animal Science* 68 (Suppl. 1), 444.

Hackett, A.J., Durnford, R., Mapletoft, R.J. and Marcus, G.J. (1993) Location and status of embryos in the genital tract of superovulated cows 4 to 6 days after insemination. *Theriogenology* 40, 1147–1153.

Hafez, E.S.E. and Gordon, I. (1962) Female reproductive organs of farm animals. In: Hafez, E.S.E. (ed.) *Reproduction in Farm Animals*. Baillière, Tindall and Cox, London, pp. 79–96.

Hafez, E.S.E. and Ishibashi, I. (1964) Maturation division in bovine oocytes following gonadotropin injections. *Cytogenetics* 3, 167–183.

Hafez, E.S.E. and Sugie, T. (1963) Reciprocal transfer of cattle and rabbit embryos. *Journal of Animal Science* 22, 30.

Hagen, D.R. and Graboski, R.A. (1990) Effects of porcine pituitary growth hormone (pGH) on cytoplasmic maturation of porcine oocytes *in vitro*. *Journal of Animal Science* 68 (Suppl. 1), 446.

Hagen, D.R., Prather, R.S., Sims, M.M. and First, N.L. (1991) Development of one-cell porcine embryos to the blastocyst stage in simple media. *Journal of Animal Science* 69, 1147–1150.

Hahn, J. (1992) Attempts to explain and reduce variability of superovulation. *Theriogenology* 38, 269–275.

Hahn, J. (1993) The development of ET in Germany over the last 20 years – a retrospective and personal view. *Proceedings of the Ninth Congress of the European Embryo Transfer Association.* (Lyon), 21.

Hahn, R. and Aumuller, R. (1992) 100 years of embryo transfer. *Tierzuchter* 44(3), 14–15.

Hallden, K., Li, J., Carney, E.W. and Foote, R.H. (1992) Increasing carbon dioxide from five percent to ten percent improves rabbit blastocyst development from cultured zygotes. *Molecular Reproduction and Development* 33, 276–280.

Hamano, K. and Uchiyama, K. (1989) Identification of X- and Y-bearing bull spermatozoa using chromosomal analysis. *Japanese Journal of Animal Reproduction* 35, 46–52.

Hamano, S. and Kuwayama, M. (1992) Development of *in vitro* matured bovine oocytes after cryopreservation by vitrification and *in vitro* fertilization. *Proceedings of the Twelfth International Congress on Animal Reproduction* (The Hague) 3, 1421–1423.

Hamano, S. and Kuwayama, M. (1993) *In vitro* fertilization and development of bovine oocytes recovered from the ovaries of individual donors: a comparison between the cutting and aspiration method. *Theriogenology* 39, 703–712.

Hamano, S., Toyoda, Y., Naito, K. and Fukuda, Y. (1990) *In vitro* fertilization of porcine oocytes with spermatozoa preserved in Kiev extender at 18°C. *Japanese Journal of Animal Reproduction* 36, 231–234.

Hamano, S., Koikeda, A., Kuwayama, M. and Nagai, T. (1992) Full-term development of *in vitro*-matured, vitrified and fertilized bovine oocytes. *Theriogenology* 38 1085–1090.

Hamilton, S.A., Xu, Z.Z., Kieborz, K.R., Youngquist, R.S. and Garverick, H.A. (1992) Relationship between ovarian follicular dynamics and follicle stimulating hormone levels during the bovine estrous cycle. *Journal of Animal Science* 70 (Suppl. 1), 261 (Abs. 492).

Hamilton, W.J. and Laing, J.A. (1946) Development of the egg of the cow up to the stage of blastocyst formation. *Journal of Anatomy* 80, 194–204.

Hammer, R.E., Pursel, V.G., Rexroad, C.E., Wall, R.J., Bolt, D.J., Ebert, K.M., Palmiter, R.D. and Brinster, R.L. (1985) Production of transgenic rabbits, sheep and pigs by microinjection. *Nature* 315, 680–685.

Hammer, R.E., Pursel, V.G., Rexroad, C.E., Wall, R.J., Bolt, D.J., Palmiter, R.D. and Brinster, R.L. (1986) Genetic engineering of mammalian embryos. *Journal of Animal Science* 63, 269–278.

Han, T.L., Ford, J.H., Webb, G.C., Flaherty, S.P., Correll, A. and Matthews, C.D. (1993) Simultaneous detection of X- and Y-bearing human sperm by double fluorescence *in situ* hybridization. *Molecular Reproduction and Development* 34, 308–313.

Hanada, A. (1985a) *In vitro* fertilization in goats. *Japanese Journal of Animal Reproduction* 31, 21–26.

Hanada, A. (1985b) *In vitro* fertilization in cattle, with particular reference to sperm capacitation by ionophore A23187. *Japanese Journal of Animal Production* 31, 56P–61P.

Hanada, A., Enya, Y. and Suzuki, T. (1986) Birth of calves by non-surgical transfer of *in vitro* fertilized embryos obtained from oocytes matured *in vitro*. *Japanese Journal of Animal Reproduction* 32, 208 (Abs.).

Handrow, R.R., Lenz, R.W. and Ax, R.L. (1982) Structural comparisons among glycosaminoglycans to promote an acrosome reaction in bovine spermatozoa. *Biochemical and Biophysical Research Communications* 107, 1326–1332.

Handrow, R.R., Parrish, J.J. and Susko-Parrish, J.L. (1986) Effects of glycosaminoglycans on capacitation and the acrosome reaction of bovine and hamster sperm. *Biology of Reproduction* 34 (Suppl. 1), 93 (Abs.).

Handrow, R.R., First, N.L. and Parrish, J.J. (1989) Calcium requirement and increased association with bovine sperm during capacitation by heparin. *Journal of Experimental Zoology* 252, 174–182.

Handyside, A. (1990) Sex and the single cell. *New Scientist* 126(1713), 34–35.

Handyside, A.H., Hooper, M.L., Kaufmann, M.H. and Wilmut, I. (1987) Towards the isolation of embryonic stem cell lines from the sheep. *Roux' Archives of Development Biology* 196, 185–190.

Handyside, A.H., Penketh, R.J.A., Winston, R.M.L., Pattinson, J.K., Delhanty, J.D.A. and Tuddenham, E.G.D. (1989) Biopsy of human preimplantation embryos and sexing by DNA amplification. *Lancet* i, 347–349.

Handyside, A.H., Kontogianni, E.H., Hardy, K. and Winston, R.M.L. (1990) Pregnancies from biopsied human preimplantation embryos sexed by Y-specific DNA amplification. *Nature* 344, 768–770.

Hanley, T., McDermott, M. and Sreenan, J.M. (1988) Active immunisation against dehydroepiandrosterone (DHEA). *Proceedings of the Eleventh International Congress on Animal Reproduction and Artificial Insemination* (Dublin) 4, 494 (3 pp.).

Hanrahan, J.P. (1983) The inter-ovarian distribution of twin ovulations and embryo survival in the bovine. *Theriogenology* 20, 3–11.

Hansel, W., Alila, H.W., Dowd, J.P. and Milvae, R.A. (1990) Differential origin and control mechanisms in small and large luteal cells. *Proceedings of the Third International Ruminant Reproduction Symposium* (Nice), Abstract.

Hansen, C., Srikandakumar, A. and Downey, B.R. (1991) Presence of follicular fluid in the porcine oviduct and its contribution to the acrosome reaction. *Molecular Reproduction and Development* 30, 148–153.

Hansen, T.R., Bixby, J.A., Leaman, D.W., Cross, J.C., Mathialagan, N. and Roberts, R.M. (1990) Isolation and sequence of genes for bovine trophoblast protein-1. *Biology of Reproduction* 42 (Suppl. 1), 168.

Hardy, K., Hooper, M.A.K., Handyside, A.H., Rutherford, A.J., Winston, R.M.L. and Leese, H.J. (1989) Non-invasive measurement of glucose and pyruvate uptake by individual human oocytes and preimplantation embryos. *Human Reproduction* 4, 188–191.

Hare, W.C.D. (1986) Embryo sexing of farm animals. In: Gwatkin, R.B.L. (ed.) *Developmental Biology*, Vol. 4, *Manipulation of Mammalian Development*. Plenum Press, London, pp. 195–216.

Hare, W.C.D., Mitchell, D., Betteridge, K.H., Eaglesome, M.D. and Randall, G.C.B. (1976) Sexing 2-week old bovine embryos by chromosomal analysis prior to surgical transfer: preliminary methods and results. *Theriogenology* 5, 243–253.

Harper, K.M. and Brackett, B.J. (1991) Effect of gonadotrophins with epidermal growth factor (EGF) during maturation on embryo viability *in vitro*. *Biology of Reproduction* 44 (Suppl. 1), 85 (Abs. 132).

Harper, K.M. and Brackett, B.G. (1992) Enhanced bovine oocyte quality after *in vitro* maturation (IVM) with insulin-like growth factor-I (IGF-I) and gonadotropins. *Biology of Reproduction* 46 (Suppl. 1), 67 (Abs.67).

Harper, K.M. and Brackett, B.G. (1993) Bovine blastocyst development after *in vitro* maturation in a defined medium with epidermal growth factor and low concentrations of gonadotropins. *Biology of Reproduction* 48, 409–416.

Harper, K.M., Younis, A.I., Zuelke, K.A. and Brackett, B.G. (1989) Early bovine embryo development during culture in different estrous stage oviductal cell suspension supernatants. *Biology of Reproduction* 40 (Suppl. 1), 121 (Abs. 222).

Harris, S., McClenaghan, M., Simons, J.P., Ali, S. and Clark, A.J. (1989) Gene expression in the mammary gland. *Journal of Reproduction and Fertility*, Abstract Series No. 3, 4.

Harrison, K.L., Pope, A.K., Wilson, L.M., Breen, T.M. and Cummins, J.M. (1987) The optimum concentration of albumin as an embryo cryoprotectant. *Journal of In Vitro Fertilization and Embryo Transfer* 4, 288–291.

Harrison, K.L., Sherrin, D.A., Hawthorne, T.A., Breen, T.M., West, G.A. and Wilson, L.M. (1990) Embryotoxicity of Micropore filters used in liquid sterilization. *Journal of In Vitro Fertilization and Embryo Transfer* 7, 347–350.

Harshbarger, C.H. and Graves, C.N. (1990) Blastocyst size as a determinant of fetal survivability. *Journal of Animal Science* 68 (Suppl. 1), 444–445.

Hartshorne, G.M. and Barlow, D.H. (1992) Factors affecting the growth of mouse ovarian follicles *in vitro*. *Journal of Reproduction and Fertility*, Abstract Series No. 10, 14 (Abs. 23).

Harvey, M.B. and Kaye, L. (1988) Insulin stimulates protein synthesis in compacted mouse embryos. *Endocrinology* 122, 1182–1184.

Harvey, M.B. and Kaye, P.L. (1991) Physiological levels of IGF-I stimulate mouse embryos. *Serono Symposium on Preimplantation Embryo Development* Abstract Vol., 61 (Abs. II–21).

Harvey, M.B.and Kaye, P.L. (1992a) Insulin-like growth factor-I stimulates growth of mouse preimplantation embryos *in vitro*. *Molecular Reproduction and Development* 31, 195–199.

Harvey, M.B. and Kaye, P.L. (1992b) Mediation of the actions of insulin and insulin-like growth factor-I on preimplantation mouse embryos *in vitro*. *Molecular Reproduction and Development* 33, 270–275.

Hashizume, T., Tanimura, I. and Kanematsu, S. (1990) Ultrastructures of the acrosome in frozen–thawed boar spermatozoa by the pellet freezing method. *Japanese Journal of Animal Reproduction* 36, 195–202.

Hasler, J.F. (1989) Observations on bovine embryo transfer in the People's Republic of China. *Theriogenology* 31, 43–44.

Hasler, J.F. (1991) Current status and potential of reproductive technology. *Journal of Dairy Science* 74 (Suppl. 1), 197 (Abs. P166).

Hasler, J.F. (1992) Current status and potential of embryo transfer and reproductive technology in dairy cattle. *Journal of Dairy Science* 75, 2857–2879.

Hasler, J.F., McCauley, A.D., Lanthrop, W.F. and Foote, R.H. (1987) Effect of donor–embryo–recipient interactions on pregnancy rate in a large-scale bovine embryo transfer program. *Theriogenology* 27, 139–168.

Hasler, J.F., Stokes, J.E. and Merton, J.S. (1994) Comparison of two different populations of BRL cells in a bovine *in vitro* culture system. *Theriogenology* 41, 214.

Hatanaka, Y., Nagai, T., Tobita, T. and Nakano, M. (1992) Changes in the properties and composition of zona pellucida of pigs during fertilization *in vitro*. *Journal of Reproduction and Fertility* 95, 431–440.

Hattingh, H. and Kay, G.W. (1992) The sperm penetration assay: a useful technique for the evaluation of sheep freezing methodology. *Animal Technology* 43(2), 121–126.

Hawk, H.W. and Wall, R.J. (1993) Experiments to improve the yield of bovine blastocysts from *in vitro* produced oocytes. *Journal of Animal Science* 71 (Suppl. 1), 220.

Hawk, H.W., Wall, R.J. and Conley, H.H. (1988) *In vivo* culture of pronuclear cow embryos injected with foreign DNA. *Journal of Animal Science* 66 (Suppl. 1), 407 (Abs.).

Hawk, H.W., Wall, R.J. and Conley, H.H. (1989) Survival of DNA-injected cow embryos temporarily cultured in rabbit oviducts. *Theriogenology* 32, 243–253.

Hawk, H.W., Nel, N.D., Waterman, R.A. and Wall, R.J. (1992) Investigation of means to improve rates of fertilization in *in vitro* matured/*in vitro* fertilized bovine oocytes. *Theriogenology* 38, 989–998.

Hayashi, S., Kobayashi, K., Mizuno, J., Saitoh, K. and Hirano, S. (1989) Development of pig fetuses from embryos frozen at −35°C. *Japanese Journal of Animal Reproduction* 35, 135–139.

Hazeleger, N.L. and Stubbings, R.B. (1992) Developmental potential of selected bovine oocyte cumulus complexes. *Theriogenology* 37, 219.

Hazeleger, N.L., Hill, D.J., Walton, J.S. and Stubbings, R.B. (1993) The interrelationship between the development of bovine oocytes *in vitro* and their follicular fluid environment. *Theriogenology* 39, 231.

Heap, R.B. (1992) Mr Walter Heape, MA, FRS. (1855–1929). *Proceedings of the Eighth Conference of the European Embryo Transfer Association* (Lyon), 109–124.

Heap, R.B., Ingram, D.L.and Wathes, C.M. (1992) Scientific and technical options for sustainable livestock production. *CAS Paper* (Centre for Agricultural Strategy, University of Reading) 25, 13–30.

Heape, W. (1890) Preliminary note on the transplantation and growth of mammalian ova within a uterine foster mother. *Proceedings of the Royal Society* (London) 48, 457–458.

Hedges, A.K. and Saunders, D.M. (1993) Clinical outcomes of pregnancies achieved by microinsemination by sperm transfer. *Fertility and Sterility* 60, 720–723.

Hehnke, K.E., Thompson, D.L., Barry, B.E., White, K.L. and Wood, T.C. (1990) Formation and characterization of vesicles from day-10 horse conceptuses. *Theriogenology* 34, 709–719.

Henault, M.A. and Killian, G.J. (1991) Synthesis and secretion of lipids by bovine oviductal explants. *Biology of Reproduction* 44 (Suppl. 1), 161 (Abs. 434).

Henault, M.A. and Killian, G.J. (1993a) Synthesis and secretion of lipids by bovine oviduct mucosal explants. *Journal of Reproduction and Fertility* 98, 431–438.

Henault, M.A. and Killian, G.J. (1993b) Neutral lipid droplets in bovine oviductal epithelium and lipid composition of epithelial cell homogenates. *Journal of Dairy Science* 76, 691–700.

Henault, M.A., Way, A.L., Grippo, A.A. and Killian, G.J. (1992a) A heterospermic, competitive penetration assay utilizing zona-free bovine oocytes. *Biology of Reproduction* 46 (Suppl. 1), 69 (Abs. 74).

Henault, M.A., Way, A.L. and Killian, G.J. (1992b) A sperm penetration assay utilizing zona-free bovine oocytes. *Journal of Animal Science* 70 (Suppl. 1), 257 (Abs. 476).

Hendriksen, P.J.M., Johnson, L.A., Themmen, A.P.N., Hoogerbrugge, J.W., Grootegoed, J.A. and Vander Lende, T. (1992) X and Y spermatozoa do not differ in H-Y antigen expression. *Development and Function of the Reproductive Organs: Serona Symposium* (Peebles), 55.

Hendriksen, P.J.M., Tieman, M., Van der Lende, T. and Johnson, L.A. (1993) Binding of anti-H-Y monoclonal antibodies to separated X and Y chromosome bearing porcine and bovine sperm. *Molecular Reproduction and Development* 35, 189–196.

Henery, C.C. and Kaufman, M.H. (1992) Cleavage rate of haploid and diploid parthenogenetic mouse embryos during the pre-implantation period. *Molecular Reproduction and Development* 31, 258–263.

Hensleigh, H.C. and Hunter, A.G. (1983a) Influence of FSH and HCG upon the *in vitro* maturation of bovine oocytes. *Theriogenology* 19, 133.

Hensleigh, H.C. and Hunter, A.G. (1983b) Fertilization of bovine oocytes matured *in vitro*. *Proceedings of the Eighteenth Annual Meeting of the American Dairy Association*, 237 (Abs.).

Hensleigh, H.C. and Hunter, A.G. (1983c) *In vitro* maturation of bovine cumulus enclosed primary oocytes and their subsequent *in vitro* fertilization and cleavage. *Journal of Dairy Science* 68, 1456–1462.

Hernandez-Ledezma, J.J. and Roberts, R.M. (1992) Factors affecting bTP-1 production by bovine blastocysts produced *in vitro*. *In Vitro Fertilization and Embryo Transfer in Primates: Serona Symposium* (Beaverton), 30.

Hernandez-Ledezma, J.J., Sikes, J.D. and Murphy, C.N. (1992a) Expression of bovine trophoblast interferon (bTP-1) in embryos cultured entirely *in vitro*. *Biology of Reproduction* 46 (Suppl. 1), 72 (Abs.85).

Hernandez-Ledezma, J.J., Sikes, J.D., Murphy, C.N., Watson, A.J., Schultz, G.A. and Roberts, R.M. (1992b) Expression of bovine trophoblast interferon in conceptuses derived by *in vitro* techniques. *Biology of Reproduction* 47, 374–380.

Hernandez-Ledezma, J.J., Villanueva, C., Sikes, J.D. and Roberts, R.M. (1993a) Effects of CZB versus medium 199 and of conditioning culture media with either bovine oviductal epithelial cells or buffalo rat liver cells on the development of bovine zygotes derived by *in vitro* maturation and *in vitro* fertilization procedures. *Theriogenology* 39, 1267–1277.

Hernandez-Ledezma, J.J., Mathialagan, N., Villanueva, C., Sikes, J.D. and Roberts, R.M. (1993b) Expression of bovine trophoblast interferons by *in vitro*-derived blastocysts is correlated with their morphological quality and stage of development. *Molecular Reproduction and Development* 36, 1–6.

Herr, C.M. and Reed, K.C. (1991) Micromanipulation of bovine embryos for sex determination. *Theriogenology* 35, 45–54.

Herr, C.M. and Wright, R.W. (1988a) Cold storage of mouse embryos of different stages of development. *Theriogenology* 29, 765–770.

Herr, C.M. and Wright, R.W. (1988b) Cold culture of different stage mouse embryos in bicarbonated and bicarbonate free media. *Theriogenology* 30, 159–168.

Herr, C.M., Holt, N.A., Matthaei, K.I. and Reed, K. (1990a) Sex of progeny from bovine embryos sexed with a rapid Y-chromosome detection assay. *Theriogenology* 33, 247.

Herr, C.M., Matthaei, K.I., Bradley, M.P. and Reed, K.C. (1990b) Rapid, accurate sexing of livestock embryos. *Proceedings of the Fourth World Congress on Genetics Applied to Livestock Production* 16, 334–343.

Herrier, A., Rath, D. and Niemann, H. (1991a) Effects of cryoprotectants on fertilization and cleavage of bovine oocytes *in vitro*. *Theriogenology* 35, 212.

Herrier, A., Elsaesser, F., Parvizi, N. and Niemann, H. (1991b) Superovulation of dairy cows with purified FSH supplemented with defined amounts of LH. *Theriogenology* 35, 633–643.

Herrier, A., Lucas-Hahn, A. and Niemann, H. (1992) Effects of insulin-like growth factor-I on *in vitro* production of bovine embryos. *Theriogenology* 37, 1213–1224.

Herz, Z., Northey, D., Lawyer, M. and First, N.L. (1985) Acrosome reaction of bovine spermatozoa *in vivo*: sites and effects of stages of the estrous cycle. *Biology of Reproduction* 32, 1163–1168.

Heuwieser, W., Yang, X., Jiang, S. and Foote, R.H. (1991) Fertilization of bovine oocytes by microinjection of spermatozoa into the perivitelline space. *Theriogenology* 35, 213.

Heuwieser, W., Yang, X., Jiang, S. and Foote, R.H. (1992a) Fertilization of bovine oocytes after microsurgical injection of spermatozoa. *Theriogenology* 38, 1–9.

Heuwieser, W., Yang, X., Jiang, S. and Foote, R.H. (1992b) A comparison between *in vitro* fertilization and microinjection of immobilized spermatozoa with defects. *Molecular Reproduction and Development* 33, 489–491.

Hewitson, L.C. and Leese, H.J. (1992) Energy metabolism of the trophectoderm and inner cell mass of the mouse blastocyst. *Journal of Reproduction and Fertility*, Abstract Series No. 9, 57.

Heyman, Y. (1984) Embryonic loss in cattle after transfer of fresh, frozen or cultured embryos. *Proceedings of the Tenth International Congress on Animal Reproduction and Artificial Insemination* (Illinois) 2, 228.

Heyman, Y. and Chesne, P. (1984) Freezing bovine embryos: survival after cervical transfer of one half, one or two blastocysts frozen in straws. *Theriogenology* 21, 240.

Heyman, Y. and Menezo, Y. (1987) Interaction of trophoblastic vesicles with bovine embryos developing *in vitro*. In: Bavister, B.D. (ed.) *The Mammalian Preimplantation Embryo*. Plenum Press, New York, pp. 175–191.

Heyman, Y., Renard, J.P., Ozil, J.P. and Mesnil du Buisson, F. (1978) Cervical embryo transfer at different stages in cattle. In: Sreenan, J.M. (ed.) *Control of Reproduction in the Cow* Vol. 1. Current Topics in Veterinary Medicine, Martinus Nijhoff, The Hague, pp. 330–335.

Heyman, Y., Smorag, Z., Katska, L. and Vincent, C. (1986) Influence of carbohydrates, cooling and rapid freezing on viability of bovine non-matured oocytes or 1-cell fertilized eggs. *Cryo-Letters* 7, 170–183.

Heyman, Y., Menezo, Y., Chesne, P., Camous, S. and Garnier, V. (1987a) *In vitro* cleavage of bovine and ovine early embryos: improved development using co-culture with trophoblastic vesicles. *Theriogenology* 27, 59–68.

Heyman, Y., Chesne, P., Chupin, D. and Menezo, Y. (1987b) Improvement of survival rate of frozen cattle blastocysts after transfer with trophoblastic vesicles. *Theriogenology* 27, 477–484.

Heyman, Y., Chesne, P., Rao, V.H., Marchal, J., Camous, S. and Renard, J.P. (1992a) Gestational profile of recipient heifers following transfer of *in vitro* produced cloned blastocysts. *Proceedings of the Eighth Conference of the European Embryo Transfer Association* (Lyon), 164.

Heyman, Y., Rao, V.H., Chesne, P. and Renard, J.P. (1992b) Bovine nuclear transfer: embryo development is influenced by age of the recipient oocyte. *Proceedings of the Twelfth International Congress on Animal Reproduction* (The Hague) 2, 688–690.

Heyman, Y., Chesne, P., Lebourhis, D. and Renard, J.P. (1993) Survival of bovine nuclear transfer embryos. *Proceedings of the Ninth Congress of the European Embryo Transfer Association* (Lyon), 210.

Heyner, S., Shah, N., Smith, R.M., Watson, A.J. and Schultz, G.A. (1993a) The role of growth factors in embryo production. *Theriogenology* 39, 151–161.

Heyner, S., Shi, C.-Z., Garside, W.T. and Smith, R.M. (1993b) Functions of the IGFs in early mammalian development. *Molecular Reproduction and Development* 35, 421–426.

Higashi, Y., Iwaya, M. and Ogawa, S. (1991) Newly found bovine male specific DNA fragments amplified by polymerase chain reaction. *Japan Journal of Animal Reproduction* 37, 115–120.

Hilgers, L.J. and Herr, C. (1993) DNA contamination of reagents used in embryo transfer and culture. *Theriogenology* 40, 923–932.

Hillery, F.L., Parrish, J.J. and First, N.L. (1990) Bull specific effect on fertilization and embryo development *in vitro*. *Theriogenology* 33, 249.

Hillier, S.G. (1989) Gonadotrophic control of granulosa cell function. *Serona Symposia Publications* 63, 105–109.

Hillier, S.G. (1991) Regulatory functions for inhibin and activin in human ovaries. *Journal of Endocrinology* 131, 171–175.

Hillier, S.G. (1993) Folliculogenesis in the human ovary. *Proceedings of the ESHR Post-Graduate Course* 3, 7–12.

Hinkovski, T. (1990) Buffalo breeding in Bulgaria. *Animal Science Papers and Reports* (Warsaw) 6, 95–98.

Hinrichs, K. (1991) The relationship of follicle atresia to follicle size, oocyte recovery rate on aspiration, and oocyte morphology in the mare. *Theriogenology* 36, 157–168.

Hinrichs, K. and DiGiorgio, L.M. (1990) Embryonic development after intrafollicular transfer of equine oocytes. *Proceedings of the Fifth International Symposium on Equine Reproduction* (Deauville), Abstracts, 122–123.

Hinrichs, K. and DiGiorgio, L.M. (1991) Embryonic development after intra-follicular transfer of horse oocytes. *Journal of Reproduction and Fertility* Suppl. 44, 369–374.

Hinrichs, K. and Schmidt, A.L. (1993) Reversible meiotic arrest of horse oocytes by coculture with follicular components. *Theriogenology* 39, 232.

Hinrichs, K., Schmidt, A.L., Memon, M.A., Selgrath, J.P. and Ebert, K.M. (1990) Culture of 5-day horse embryos in microdroplets for 10 to 20 days. *Theriogenology* 34, 643–653.

Hinrichs, K., Schmidt, A.L., Friedman, P.P., Selgrath, J.P. and Martin, M.G. (1993) *In vitro* maturation of horse oocytes: characterization of chromatin configuration using fluorescence microscopy. *Biology of Reproduction* 48, 363–370.

Hirao, Y., Miyano, T. and Kato, S. (1990) Fertilization of *in vitro* grown mouse oocytes. *Theriogenology* 34, 1071–1077.

Hirao, Y., Nagai, T., Kubo, M., Miyano, T., Miyake, M. and Kato, S. (1993) *In vitro* growth and maturation of pig oocytes. *Journal of Reproduction and Fertility* (in press).

Hirshfield, A.N. (1991) Theca cells may be present at the outset of follicular growth. *Biology of Reproduction* 44, 1157–1162.

Hirst, P.J., DeMayo, F.J. and Dukelow, W.R. (1981) Xenogenous fertilization of laboratory and domestic animals in the oviduct of the pseudopregnant rabbit. *Theriogenology* 15, 65–67.

Hochereau-de Reviers, M. T. and Perreau, C. (1991) *In vitro* culture of porcine ICM cells: influence of genetical origin and role of piglet serum. *Journal of Reproduction and Fertility*, Abstract Series No. 8, 42 (Abs. 73).

Hochereau-de Reviers, M.T. and Perreau, C. (1992) Isolation and *in vitro* culture of inner cell mass cells from porcine blastocysts. *Proceedings of the Eighth Conference of the European Embryo Transfer Association* (Lyon), 168.

Hochi, S-I., Ninomiya, T., Wagga-Homma, M., Sagara, J. and Yuki, A. (1992) Secretion of bovine alpha-lactalbumin into the milk of transgenic rats. *Molecular Reproduction and Development* 33, 160–164.

Hockley, D.K., Bo, G.A., Palasz, A.T., Del Campo, M.R. and Mapletoft, R.J. (1992) Superovulation with a single subcutaneous injection of Folltropin in the cow: effect of dose and site of injection. *Theriogenology* 37, 224.

Hodgen, G.D. (1990) Follicle selection. In: Asch, R.H., Balmaceda, J.P. and Johnston, I. (eds) *Gamete Physiology*. Serona Symposia, Norwell, Massachusetts, USA, pp. 153–161.

Hodges, J. (1990) Genetic improvement of livestock in developing countries using the open nucleus breeding system. *Animal Science Papers and Reports (Warsaw)* 6, 13–22.

Hofig, A., Simmen, F.A. and Simmen, R.C.M. (1990) Effects of dihydroepiandrosterone (DHEA), insulin-like growth factor (IGF-1), insulin (I) on estradiol production by porcine conceptuses *in vitro*. *Biology of Reproduction* 42 (Suppl. 1), 135.

Holdas, S., Nagy, Z., Bardny, I., Papp, D., Koppany, A., Meszaros, J. and Becze, J. (1987) Induction of twinning in beef cattle. *Allattenyesztes es Takarmanyozaz* 36, 227–230.

Holland, M.D. and Odde, K.G. (1992) Factors affecting calf birth-weight: a review. *Theriogenology* 38, 769–798.

Hollingsworth, T.S. and Page, R.D. (1988) The effects of microencapsulation of freezing bisected bovine embryos. *Theriogenology* 28, 262.

Holm, P., Irvine, B., Armstrong, D.T. and Seamark, R.F. (1991a) *In vitro* production of sheep blastocysts from IVM-oocytes using frozen semen and oviduct epithelial cell co-culture for IVF. *Theriogenology* 35, 214.

Holm, P., Greve, T., Bak, A. and Schmidt, M. (1991b) Bisection of bovine morulae and blastocysts from superovulated Danish dairy cows. *Acta Veterinaria Scandanavica* 32, 47–53.

Holmgren, W.J. and Jeyendran, R.S. (1993) Synergistic effect of TEST-yolk buffer treatment and glass wool filtration of spermatozoa on the outcome of the hamster oocyte penetration assay. *Human Reproduction* 8, 425–427.

Holy, L., Jiricek, A., Vanatka, F., Vrtel, M. and Fernandez, V. (1981) Artificial induction of twinning in cattle by means of supplemental embryo transfer. *Theriogenology* 16, 483–488.

Holy, L., Lopatarova, M. and Krontorad, P. (1988) The quality and stage of development of cattle embryos in relation to their survival *in vivo* following an ipsilateral non-surgical transfer. *Veterinarni Medicina* 33, 577–588.

Homa, S.T. (1991) Neomycin, an inhibitor of phosphoinoditide hydroysis, inhibits the resumption of bovine oocyte spontaneous meiotic maturation. *Journal of Experimental Zoology* 258, 95–103.

Homa, S.T. and Brown, C.A. (1992) Changes in linoleic acid during follicular development and inhibition of spontaneous breakdown of germinal vesicles in cumulus-free bovine oocytes. *Journal of Reproduction and Fertility* 94, 153–160.

Homa, S.T., Racowsky, C. and McCaughey, R.W. (1986) Lipid analysis of immature pig oocytes. *Journal of Reproduction and Fertility* 77, 425–434.

Homa, S.T., McCaughey, R.W. and Racowsky, C. (1988) Isolated subpopulation of mass-harvested pig oocytes and their characterization by size, incidence of atresia and competence to mature in culture. *Journal of Reproduction and Fertility* 82, 81–86.

Homa, S.T., Carroll, J. and Swann, K. (1993) The role of calcium in mammalian oocyte maturation and egg activation. *Human Reproduction* 8(8), 1274–1281.

Hopper, H.W., Silcox, R.W., Byerley, D.J. and Kiser, T.E. (1993) Follicular development in prepubertal heifers. *Animal Reproduction Science* 31, 7–12.

Horan, R., Powell, R., McQuaid, S., Gannon, F. and Houghton, J.A. (1991) Association of foreign DNA with porcine spermatozoa. *Archives of Andrology* 26, 83–92.

Horiuchi, S. (1992) Stagnation in the decline of the world population growth during the 1980s. *Science* 257, 761–765.

Hoshi, K., Yanagida, K. and Sato, A. (1992) Pretreatment of hamster oocytes with Ca^{2+} ionophore to facilitate fertilization by ooplasmic micro-injection. *Human Reproduction* 7, 871–875.

Hoskins, D.D., Hoskins, M.K. and Vijayaraghavan, S. (1990) 2-Chloroadenosine is transported across the bovine sperm flagellar membrane by passive diffusion prior to motility stimulation. *Biology of Reproduction* 42 (Suppl. 1), 65.

Hosoi, Y. and Iritani, A. (1993) Rabbit microfertilization. *Molecular Reproduction and Development* 36, 282–284.

Hossepian de Lima, V.F.M., Moreira-Filho, C.A., De Bem, A.R. and Jorge, W. (1993) Sex determination of murine and bovine embryos using cytotoxicity and immunofluorescence assays. *Theriogenology* 39, 1343–1352.

Host-Hansen, Th., Avery, B., Hyttel, P. and Greve, T. (1993) Embryonic growth factor receptors and growth factor secretion by cultured oviduct epithelial cells in cattle (*Bos taurus*). *Reproduction in Domestic Animals* 28, 193–194.

Hough, S. and Foote, R.H. (1990) The effect of the Cornell particle gun on bull and rabbit spermatozoa. *Biology of Reproduction* 42 (Suppl. 1), 65.

Howes, E.A., Miller, N.G.A., Richardson, L. and Jones, R. (1993) Monoclonal antibodies to FACS sorted X and Y-bearing bovine spermatozoa. *Journal of Reproduction and Fertility*, Abstract Series No. 11, 53 (Abs. 95).

Hruska, K. (1991) The effect of length of cryopreservation on the viability of bovine embryos in a commercial operation. *Theriogenology* 36, 477–484.

Huang, F.L., Lu, K.L. and Polge, C. (1992) Effect of sucrose concentration on viability of bovine IVF embryos frozen by a rapid method. *Proceedings of the Twelfth International Congress on Animal Reproduction* (The Hague) 3, 1430–1432.

Huang, J.C., Lin, J.H. and Cheng, W.T.K. (1989) Studies on *in vitro* maturation and fertilization of bovine oocytes. II. Effects of some reproductive hormones on the maturation, fertilization and the subsequent development potential of bovine oocytes *in vitro*. *Memoirs of the College of Agriculture, National Taiwan University* 29(1), 150–160.

Hughes, E.A. and Davies, D.A.R. (1989) Practical uses of ultrasound in early pregnancy in cattle. *Veterinary Record* 124, 456–458.

Hugues, J.N., Torresani, T., Herve, F., Martin- Pont, B., Tamboise, A. and Santarelli, J. (1991) Interest of growth hormone-releasing hormone administration for improvement of ovarian responsiveness to gonadotropins in poor responder women. *Fertility and Sterility* 55, 945–951.

Huhn, R. and Rommel, P. (1990) Basic studies on the induction of twin pregnancies in cows by means of embryo transfer. *Monatshefte für Veterinarmedizin* 45, 529– 532.

Huhtinen, M., Rainio, V., Aalto, J., Bredbacka, P. and Maki-Tanila, A. (1992) Increased ovarian responses in the absence of a dominant follicle in superovulated cows. *Theriogenology* 37, 457–463.

Hulshof, S.C.J., Bevers, M.M., van der Donk, H.A. and van den Hurk, R. (1992) The isolation and characterization of preantral follicles from foetal bovine ovaries. *Proceedings of the Twelfth International Congress on Animal Reproduction* (The Hague), 1, 336–338.

Hulshof, S.C.J., Figueiredo, J.R., Beckers, J.F., Bevers, J.F., Bevers, M.M., Dijkstra, G. and Van Den Hurk, R. (1993) Bovine preantral follicle viability *in vitro*: effects of pyruvate. *Journal of Reproduction and Fertility*, Abstract Series No. 11, 80 (Abs. 149).

Humblot, P., Perrin, J., Jeanguyot, N., Nibart, M. and Thibier, M. (1987) Effects of age and quality of thawed embryos, synchronization and corpus luteum function on pregnancy rates of bovine embryo recipients. *Theriogenology* 27, 240.

Humblot, P., Mechekour, F., Jeanguyot, N., Payen, B., Nibart, M., Sasser, G. and Thibier, M. (1989) Accuracy of pregnancy diagnosis by PSPB and P4 RIA after embryo transfer in dairy heifers. *Proceedings of the Fifth Conference of the European Embryo Transfer Association* (Lyon), 160.

Huneau, D. and Crozet, N. (1988) Effect of elevated calcium concentration on *in vitro* fertilization in the sheep. *Journal of Reproduction and Fertility*, Abstract Series No. 2, Abs. 85.

Hunnicutt, L.K., Petersen, N.S., Dunn, T.G. and Moss, G.E. (1990) Synthesis of stress proteins (hsp70) by ovine embryonic trophectoderm. *Journal of Animal Science* 68 (Suppl. 1), 419.

Hunter, A.G. and Moor, R.M. (1986) *In vitro* blockage of germinal vesicle breakdown (GVBD) in bovine oocytes by RNA inhibitors. *Journal of Dairy Science* 69 (Suppl. 1), 145 (Abs.).

Hunter, A.G. and Moor, R.M. (1987) Stage-dependent effects of inhibiting ribonucleic acids and protein synthesis on meiotic maturation of bovine oocytes *in vitro*. *Journal of Dairy Science* 70, 1646–1651.

Hunter, J.E., Bernard, A., Fuller, B., Amso, N. and Shaw, R.W. (1991) Fertilization and development of the human oocyte following exposure to cryoprotectants, low temperatures and cryopreservation: a comparison of two techniques. *Human Reproduction* 6, 1460–1465.

Hunter, M.G., Biggs, C., Faillace, L.S. and Picton, H.M. (1992) Current concepts of folliculogenesis in monovular and polyovular farm species. *Journal of Reproduction and Fertility*, Suppl. 45, 21–38.

Hunter, R.H.F. (1985) Fertility in cattle: basic reasons why late insemination should be avoided. *Animal Breeding Abstracts* 53, 83–87.

Hunter, R.H.F. (1988) *The Fallopian Tubes: Their Role in Fertility and Infertility*. Springer-Verlag, Berlin, pp. 109–126.

Hunter, R.H.F. (1989) Ageing of the unfertilized cow egg *in vivo*: how soon is fertility compromised? *Veterinary Record* 124, 489–490.

Hunter, R.H.F. (1990) Fertilization of pig eggs *in vivo* and *in vitro*. *Journal of Reproduction and Fertility* Suppl. 40, 211- 226.

Hunter, R.H.F. (1991) Oviduct function in pigs, with particular reference to the pathological condition of polyspermy. *Molecular Reproduction and Development* 29, 385–391.

Hunter, R.H.F. (1993) Sperm : egg ratios and putative molecular signals to modulate gamete interactions in polytocous mammals. *Molecular Reproduction and Development* 35, 324–327.

Hunter, R.H.F., Lawson, R.A.S. and Rowson, L.E.A. (1972) Maturation, transplantation and fertilization of ovarian oocytes in cattle. *Journal of Reproduction and Fertility* 30, 325–328.

Hunter, R.H.F., Flechon, B. and Flechon, J.E. (1991) Distribution, morphology and epithelial interactions of bovine spermatozoa in the oviduct before and after ovulation: a scanning electron microscope study. *Tissue and Cell* 23, 641–656.

Hurk, van den R. and Dijkstra, G. (1992) Development of antral bovine follicles after prostaglandin induced luteolysis. *Proceedings of the Twelfth International Congress of Animal Reproduction* (The Hague) 1, 283–285.

Hurwitz, A., Payne, D.W., Packman, J.N., Andreani, C.L., Resnick, C.E., Hernandez, E.R. and Adashi, E.Y. (1991) Cytokine-mediated regulation of ovarian function: interleukin-1 inhibits gonadotrophin-induced androgen biosynthesis. *Endocrinology* 129, 1250–1256.

Hyttel, P., Callesen, H. and Greve, T. (1986a) Ultrastructural features of preovulatory oocyte maturation in superovulated cattle. *Journal of Reproduction and Fertility* 76, 645–656.

Hyttel, P., Callesen, H., Greve, T. and Smith, S. (1986b) Ultrastructure of fertilised and unfertilised tubal ova in superovulated cattle. *Proceedings of the Annual Meeting of the Society for the Study of Fertility* (Bristol), Abs. 110.

Hyttel, P., Xu, K.P., Smith, S. and Greve, T. (1986c) Ultrastructure of *in vitro* maturation in cattle. *Journal of Reproduction and Fertility* 78, 615–625.

Hyttel, P., Xu, K.P., Smith, S., Callesen, H. and Greve, T. (1987a) Ultrastructure of the final nuclear maturation of bovine oocytes *in vitro*. *Anatomy and Embryology* 176, 35–40.

Hyttel, P., Callesen, H. and Greve, T. (1987b) Pronucleus formation in ova in superovulated cattle. *Theriogenology* 27, 251.

Hyttel, P., Greve, T., Xu, K.P. and Callesen, H. (1987c) Ultrastructural aspects of bovine fertilization *in vivo* and *in vitro*. *Proceedings of a Symposium on the Application of Egg and Embryo Technologies to Domestic Animals* (Copenhagen), 41–44.

Hyttel, P., Callesen, H. and Greve, T. (1988a) Fertilization *in vivo* versus *in vitro* of bovine oocytes matured *in vivo*: an ultrastructural study. *Journal of Reproduction and Fertility*, Abstract Series No. 1, Abs. 17.

Hyttel, P., Greve, T. and Callesen, H. (1988b) Ultrastructure of *in vivo* fertilization in superovulated cattle. *Journal of Reproduction and Fertility* 82, 1–13.

Hyttel, P., Greve, T. and Callesen, H. (1989a) Cortical granule dynamics in cattle ova. *Theriogenology* 31, 205.

Hyttel, P., Greve, T. and Callesen, H. (1989b) Ultrastructural aspects of oocyte maturation and fertilization in cattle. *Journal of Reproduction and Fertility*, Suppl. 38, 35–47.

Hyttel, P., Callesen, H. and Greve, T. (1989c) A comparative ultrastructural study of *in vivo* versus *in vitro* fertilization of bovine oocytes. *Anatomy and Embryology* 179, 435–442.

Hyttel, P., Callesen, H., Greve, T. and Schmidt, M. (1991a) Oocyte maturation and sperm transport in superovulated cattle. *Theriogenology* 35, 91–108.

Hyttel, P., Greve, T. and Callesen, H. (1991b) Is deviant fertilization a significant problem in superovulated cattle? *Embryo Transfer Newsletter* 9(3), 9–13.

Hyttel, P., Assey, R.J., Viuff, D. and Grondahl, C. (1993) Transcriptional regulation of oocyte development, fertilization and initial embryonic development in cattle. *Reproduction in Domestic Animals* 28, 154–156.

Hyttinen, J.M., Peura, T., Utriainen, M., Kosonen, R. and Janne, J. (1992) Transgene integration delays the development of pronucleus injected bovine embryos *in vitro*. *Proceedings of the Nordic A.I. Virtanen Institute Symposium on the In Vitro Culture of Domestic Animal Embryos*, 34–35.

Ibrahim, Z.H., Matson, P.L., Buck, P. and Lieberman, B.A. (1991) The use of biosynthetic human growth hormone to augment ovulation induction with buserelin acetate/human menopausal gonadotropin in women with a poor ovarian response. *Fertility and Sterility* 55, 202–204.

Ijaz, A. and Hunter, A.G. (1989a) Evaluation of TEST-yolk buffer capacitated sperm for use in bovine *in vitro* fertilization. *Biology of Reproduction* 40 (Suppl. 1), 147 (Abs. 298).

Ijaz, A. and Hunter, A.G. (1989b) Induction of bovine sperm capacitation by TEST-yolk semen extender. *Journal of Dairy Science* 72, 2683–2690.

Ijaz, A. and Hunter, A.G. (1989c) Effect of washing and capacitating media pH on bull sperm motility, acrosome integrity, and ability to penetrate zona-free hamster oocytes. *Journal of Dairy Science* 72, 2691–2699.

Ijaz, A. and Hunter, A.G. (1989d) Evaluation of calcium-free Tyrode's sperm capacitation medium for use in bovine *in vitro* fertilization. *Journal of Dairy Science* 72, 3280–3285.

Ijaz, A. and Hunter, A.G. (1992) Evaluation of a TEST-yolk sperm capacitation system for use in bovine *in vitro* fertilization. *Journal of Dairy Science* 75, 394–398.

Ijaz, A. and Sirard, M.A. (1993) Bovine granulosa and oviductal cell secretions support sperm motility. *Theriogenology* 39, 234.

Ijaz, A., Hunter, A.G. and Graham, E.F. (1989) Identification of the capacitating agent for bovine sperm in egg yolk-TEST semen extender. *Journal of Dairy Science* 72, 2700–2706.

Ijaz, A., Lambert, R.D. and Sirard, M.-A. (1994) *In vitro* cultured bovine granulosa and oviductal cells secrete sperm motility-maintaining factor(s). *Molecular Reproduction and Development* 37, 54–60.

Illera, M.J. and Petters, R.M. (1993) Effects of EGF and IGF-I on porcine oocyte fertilization *in vitro* in the absence of follicular fluid or hormones. *Theriogenology* 39, 235.

Illera, M.J., Arellano, J., Lorenzo, P., Sanchez, J., Silvan, G. and Illera, J.C. (1992a) Porcine oocyte maturation with EGF (epidermal growth factor). *Proceedings of the Eighth Conference of the European Embryo Transfer Association* (The Hague), 170.

Illera, M.J., Lorenzo, P., Illera, J.C., Sanchez, J., Silvan, G. and Illera, M. (1992b) Epidermal growth factor increase the bovine oocyte maturation *in vitro*. *Proceedings of the Twelfth International Congress on Animal Reproduction*. (The Hague) 1, 339–341.

Illera, M.J., Lorenzo, P., Illera, J.C., Portela, A. and Petters, R.M. (1992c) A simple medium supports *in vitro* development of one-cell embryos from miniature pigs to the blastocyst stage. *Theriogenology* 37, 225.

Imahie, H., Nito, S., Kobayashi, T. and Ariyuki, F. (1992) Overcoming the mouse 2-cell block by co-culture with established cell lines. *Journal of Reproduction and Development* 38, 267–270.

Imai, H., Walton, C.E., Marshall, J.T. and Nancarrow, C.D. (1992) Characteristics of DNA binding to sheep spermatozoa. *Journal of Reproduction and Development* 38, 255–262.

Imakawa, K., Hansen, T.R., Anthony, R.V., Polites, H.G., Marotti, K.R. and Roberts, R.M. (1988) An embryonic-derived interferon is formed in the bovine and ovine species. *Biology of Reproduction* 38 (Suppl. 1), Abs. 425.

Imakawa, K., Helmer, S.D., Nephew, K.P., Meka, C.S.R. and Christenson, R.K. (1993) A novel role for GM-CSF: enhancement of pregnancy specific interferon production, ovine trophoblast protein-1. *Endocrinology* 132, 1869–1871.

Imoedemhe, D.G. and Sigue, A.B. (1992) Survival of human oocytes cryopreserved with or without the cumulus in 1, 2-propanediol. *Journal of Assisted Reproduction and Genetics* 9, 323–327.

Imoedemhe, D.A.G. and Sigue, A.B. (1993) Subzonal multiple sperm injection in the treatment of previous failed human *in vitro* fertilization. *Fertility and Sterility* 59, 172–176.

Imoedemhe, D.A.G., Sigue, A.B., Pacpaco, E.A. and Olazo, A.B. (1992a) Successful use of the sperm motility enhancer 2-deoxyadenosine in previously failed human *in vitro* fertilization. *Journal of Assisted Reproduction and Genetics* 9, 53–56.

Imoedemhe, D.A.G., Sigue, A.B., Pacpaco, A. and Olazo, A.B. (1992b) The effect of caffeine on the ability of spermatozoa to fertilize mature human oocytes. *Journal of Assisted Reproduction and Genetics* 9, 155–160.

Imoedemhe, D.A.G., Sigue, A.B. and Luciano, E.C. (1992c) Deactivation of serum prior to medium supplementation: a necessity or a myth? *Human Reproduction* 7 (Suppl. 2), 159.

Imoedemhe, D.A.G., Sigue, A.B. and Luciano, E.C. (1993) Successful twin pregnancy and delivery after microinseminated oocyte fallopian transfer for male factor infertility. *Fertility and Sterility* 59, 662–663.

Ing, R.M.Y., Li, D.Q., Harding, A.M. and Jones, W.R. (1991) A comparison of swim-down and swim-up methods for the extraction of high motility sperm. *Fertility and Sterility* 55, 817–819.

Inzen, W.G. van, Kruip, Th.A.M. and Weima, S.M. (1993) Use of conditioned medium for IVM–IVF bovine embryos *in vitro* culture systems. *Theriogenology* 39, 236.

Iqbal, N. and Hunter, A.G. (1991) Comparison of bovine sperm capacitation systems. *Journal of Dairy Science* 74 (Suppl. 1), 228 (Abs. P260).

Iqbal, N. and Hunter, A.G. (1992a) Comparison of various capacitating systems as to their ability to alter bovine sperm membrane proteins during capacitation. *Biology of Reproduction* 46 (Suppl. 1), 94 (Abs. 173).

Iqbal, N. and Hunter, A.G. (1992b) Comparison of bovine sperm capacitation systems and effect of temperature on their efficiency in fertilizing bovine oocytes. *Journal of Dairy Science* 75 (Suppl. 1), 243 (Abs. P294).

Ireland, J.J., Coulson, P.B. and Murphee, R.L. (1979) Follicular development during four stages of the estrous cycle of beef cattle. *Journal of Animal Science* 49, 1261–1269.

Ireland, J.J., Murphee, R.L. and Coulson, P.B. (1980) Accuracy of predicting stages of bovine estrous cycle by gross appearance of the corpus luteum. *Journal of Dairy Science* 63, 155–160.

Ireland, J.L.H., Maloney, T.E., Dickerson Weber, P., Knight, P.G. and Ireland, J.J. (1992) Evidence for a role of inhibin in growth and atresia of bovine follicles. *Biology of Reproduction* 46 (Suppl. 1), 85 (Abs. 139).

Iriarte, J., Garcia, F. and Mendizabal, J. (1991) Selection of bovine recipients according to the plasma progesterone concentrations. *Proceedings of the Seventh Meeting of the European Embryo Transfer Association* (Cambridge), 148.

Iritani, A. (1980) Fertilization *in vitro* of follicular oocytes matured in culture in cattle, pig and human. *Archives of Andrology* 5, 77–78.

Iritani, A. (1985) *In vitro* fertilization in cattle. *Japanese Journal of Animal Production* 31, 52P–55P.

Iritani, A. (1991) Micromanipulation of gametes for *in vitro* assisted fertilization. *Molecular Reproduction and Development* 28, 199–207.

Iritani, A. and Niwa, K. (1977) Capacitation of bull spermatozoa and fertilization *in vitro* of cattle follicular oocytes matured in culture. *Journal of Reproduction and Fertility* 50, 119–121.

Iritani, A., Kasai, M., Niwa, K. and Song, H.B. (1984) Fertilization *in vitro* of cattle follicular oocytes with ejaculated spermatozoa capacitated in a chemically defined medium. *Journal of Reproduction and Fertility* 70, 487–492.

Iritani, A., Utsumi, K., Miyake, M. and Yamaguchi, Y. (1986) Individual variation in the *in vitro* fertilizing ability of bull spermatozoa. *Development, Growth and Differentiation,* Suppl. 28, 45.

Iritani, A., Utsumi, K., Miyake, M., Hosoi, Y. and Saeki, K. (1988) *In vitro* fertilization by a routine method and by micromanipulation. *Annals of the New York Academy of Sciences* 541, 583–590.

Iritani, A., Cheng, Z.S. and Utsumi, K. (1992a) Effects of follicular factors on the IVM–IVF bovine oocytes. *Proceedings of the Twelfth International Congress on Animal Reproduction (The Hague)* 1, 342–344.

Iritani, A., Utsumi, K. and Hosoi, Y. (1992b) Fertilization by assisted micromanipulation of gametes. In: Lauria, A. and Gandolfi, F. (eds) *Embryonic Development and Manipulation in Animal Production*. Portland Press, London, pp. 51–57.

Irvine, B., Armstrong, D.T., Earl, C., McLean, D. and Seamark, R.F. (1993) Follicle development and oocyte recovery from calves with repeated gonadotropin stimulation and follicular aspiration. *Theriogenology* 39, 237.

Ishijima, S.A., Okuno, M., Odagiri, H., Mohri, T. and Mohri, H. (1992) Separation of X- and Y-chromosome-bearing murine sperm by free-flow electrophoresis: evaluation of separation using PCR. *Zoological Science* 9, 601–606.

Ishimori, H., Takahashi, Y. and Kanagawa, H. (1992a) Viability of vitrified mouse embryos using various cryoprotectant mixtures. *Theriogenology* 37, 481–487.

Ishimori, H., Takahashi, Y. and Kanagawa, H. (1992b) Factors affecting survival of mouse blastocysts vitrified by a mixture of ethylene glycol and dimethyl sulfoxide. *Theriogenology* 38, 1175–1185.

Ishimori, H., Saeki, K., Inai, M., Itasaka, J., Miki, Y., Nozaki, N., Seike, N. and Kainuma, H. (1993a) Direct transfer of vitrified bovine embryos. *Theriogenology* 39, 238.

Ishimori, H., Saeki, K., Inai, M., Nagao, Y., Itasaka, J., Miki, Y., Seike, N. and Kainuma, H. (1993b) Vitrification of bovine embryos in a mixture of ethylene glycol and dimethyl sulfoxide. *Theriogenology* 40, 427–433.

Isogai, T., Nakanishi, T., Sasaki, K. and Tagami, M. (1991) Twin pregnancy and births in cattle induced by FSH and PGF or PGF-analogue treatment. *Japanse Journal of Animal Reproduction* 37, 105–113.

Iwasaki, S. (1989) Sexing of bovine sperm by application of chromosome analysis of embryos fertilized *in vitro. Japanese Journal of Animal Reproduction* 35, 53–56.

Iwasaki, S. (1992) Studies on cytogenetic and morphological assessment of bovine embryos fertilized *in vitro. Journal of Reproduction and Development* 38, 109–119.

Iwasaki, S. and Hamano, S. (1991) Occurrence of haploid cells bearing Y chromosomes in bovine embryos fertilized *in vitro. Journal of Heredity* 82, 420–421.

Iwasaki, S. and Nakahara, T. (1990a) Cell number and incidence of chromosomal anomalies in bovine blastocysts fertilized *in vitro* followed by culture *in vitro* or *in vivo* in rabbit oviducts. *Theriogenology* 33, 669–675.

Iwasaki, S. and Nakahara, T. (1990b) Incidence of embryos with chromosomal anomalies in the inner cell mass among bovine blastocysts fertilized *in vitro. Theriogenology* 34, 683–690.

Iwasaki, S., Kono, T., Nakahara, T., Shioya, Y., Fukushima, M. and Hanada, A. (1987) New methods for the recovery of oocytes from bovine ovarian tissue in relation to *in vitro* maturation and fertilization. *Japanese Journal of Animal Reproduction* 33, 188–192.

Iwasaki, S., Shioya, Y., Hanada, A. and Nakahara, T. (1988a) Chromosome preparation from 2-cell bovine embryos derived from follicular oocytes fertilized *in vitro. Japanese Journal of Animal Reproduction* 34, 79–83.

Iwasaki, S., Shioya, Y., Masuda, H., Hanada, A. and Nakahara, T. (1988b) Sex ratio of early embryos fertilized *in vitro* with spermatozoa separated by Percoll. *Theriogenology* 30, 1191–1198.

Iwasaki, S., Kono, T. and Nakahara, T. (1989a) Sexing rate and developmental ability of bovine two- to eight-cell stage embryos split by treatment with pronase. *Japanese Journal of Animal Reproduction* 35, 147–153.

Iwasaki, S., Shioya, Y., Monji, Y. and Nakahara, T. (1989b) Developmental changes and sex difference in glucose-6-phosphate dehydrogenase activity of bovine oocytes and embryos fertilized *in vitro*. *Japanese Journal of Animal Reproduction* 35, 198–203.

Iwasaki, S., Yoshida, Y., Watanabe, S. and Nakahara, T. (1990a) Differential nuclear staining and cell count of trophectoderm and inner cell mass of bovine blastocysts fertilized *in vitro* by double fluorochrome dye technique. *Japanese Journal of Animal Reproduction* 36, 60–65.

Iwasaki, S., Yoshiba, N., Ushijima, H., Watanabe, S. and Nakahara, T. (1990b) Morphology and proportion of inner cell mass of bovine blastocysts fertilized *in vitro* and *in vivo*. *Journal of Reproduction and Fertility* 90, 279–284.

Iwasaki, S., Shioya, Y. and Nakahara, T. (1991) Measurement of glucose 6-phosphate dehydrogenase activity in bovine oocytes and embryos using quantitative fluorescence microscopy. *Japanese Journal of Animal Reproduction* 37, 273–276.

Iwasaki, S., Hamano, S., Kuwayama, M., Yamashita, M., Ushijima, H., Nagaoka, S. and Nakahara, T. (1992) Developmental changes in the incidence of chromosome anomalies of bovine embryos fertilized *in vitro*. *Journal of Experimental Zoology* 261, 77–85.

Izaike, Y., Suzuki, O., Shimada, K., Fujita, K. and Kosugiyama, M. (1988) Twin pregnancy diagnosis and early embryonic loss after bilateral egg transfer in beef cattle. *Japanese Journal of Animal Reproduction* 34, 236–242.

Izaike, Y., Suzuki, O., Shimada, K. and Takahashi, M. (1991a) Association of corpus luteum size, body condition and combination of developmental stage of two embryos with pregnancy rate in beef cows after bilateral transfer of embryos. *Bulletin of the Chugoku National Agriculture Experimental Station* 8, 79–89.

Izaike, Y., Suzuki, O., Shimada, K., Takenouchi, N. and Takahashi, M. (1991b) Observation by ultrasonography of embryonic loss following the transfer of two or three embryos in beef cows. *Theriogenology* 36, 939–947.

Jablonka, A., Fricke, P.M., Redmer, D.A., and Reynolds, L.P. (1991) Effects of FSH on size, number and labeling index of follicles in ewes. *Journal of Animal Science* 69 (Suppl. 1), 438.

Jaenisch, R. (1988) Transgenic animals. *Science* 240, 1468–1474.

Jagiello, G.M., Miller, W.A., Ducayen, M.B. and Lin, J.S. (1974) Chiasma frequency and disjunctional behaviour of ewe and cow oocytes matured *in vitro*. *Biology of Reproduction* 10, 354–363.

Jagiello, G., Graffeo, J., Ducayen, M. and Prosser, R. (1977) Further studies of inhibitors of *in vitro* mammalian oocyte maturation. *Fertility and Sterility* 28, 476–481.

Jainudeen, M.R., Takahashi, Y., Nihayah, M. and Kanagawa, H. (1993) *In vitro* maturation and fertilization of swamp buffalo (*Bubalis bubalis*) oocytes. *Animal Reproduction Science* 31, 205–212.

Jarosz, S.J. and Dukelow, W.R. (1990) Embryo transfer and pregnancy rate in the golden hamster (*Mesocricetus auratus*). *Zoological Science* 7, 85–91.

Jarrell, V.L., Beckmann, L.S., Cantley, T.C., Rieke, A.R. and Day, B.N. (1990a) The effect of exposure to an asynchronous uterus on development of day 4 swine embryos. *Journal of Animal Science* 68 (Suppl. 1), 429.

Jarrell, V.L., Day, B.N. and Prather, R.S. (1990b) Synthesis of protein in pig oocytes and embryos. *Biology of Reproduction* 42 (Suppl. 1), 43.

Jarrell, V.L., Day, B.N. and Prather, R.S. (1991) The transition from maternal to zygotic control of development occurs during the 4-cell stage in the domestic pig, *Sus scrofa*: quantitative and qualitative aspects of protein synthesis. *Biology of Reproduction* 44, 62–68.

Jasiorowski, H.A. (1990) Open nucleus breeding schemes – new challenge for the developing countries. *Animal Science Papers and Reports* (*Warsaw*) 6, 7–12.

Javed, M.H. and Wright, R.W. (1991) Determination of pentose phosphate and Embden-Meyerhof pathway activities in bovine embryos. *Theriogenology* 35, 1029–1037.

Jiang, H.S. (1991) Studies in the culture of bovine embryos derived from oocytes matured and fertilized *in vitro*. PhD Thesis, National University of Ireland, Dublin.

Jiang, H.S., Wang, W.L., Lu, K.H., McCarthy, D. and Gordon, I. (1989) *In vitro* culture of early bovine embryos derived from *in vitro* fertilization, comparison with *in vivo* culture. *Journal of Reproduction and Fertility*, Abstract Series No. 3, 50.

Jiang, H.S., Wang, W.L., Lu, K.H. and Gordon, I. (1990) Effects of PMSG, insulin, osmolality and oestrous cow serum on development of IVF early embryos on granulosa cell monolayers. *Theriogenology* 33, 258.

Jiang, H.S., Wang, W.L., Lu, K.H., Gordon, I. and Polge, C. (1991) Roles of different cell monolayers in the co-culture of IVF bovine embryos. *Theriogenology* 35, 216.

Jiang, H.S., Wang, W.L., Lu, K.H., Gordon, I. and Polge, C. (1992) Examination of cell numbers of blastocysts derived from IVM, IVF and IVC of bovine follicular oocytes. *Theriogenology* 37, 229.

Jiang, J.Y., Zhong, S. and Fan, B.Q. (1991) Calf born after the transfer of frozen–thawed embryos from follicular oocytes matured, fertilized and developed *in vitro*. *Theriogenology* 35, 217.

Jiang, S., Yang, X., Chang, S., Heuwieser, W. and Foote, R.H. (1991) Effect of sperm capacitation and oocyte maturation procedures on fertilization and development of bovine oocytes *in vitro*. *Theriogenology* 35, 218.

Jiang, S., Yang, X. and Foote, R.H. (1992) Sperm capacitation procedures for *in vitro* fertilization and development of bovine embryos. *Theriogenology* 37, 230.

Jin, D.I., Petters, R.M., Johnson, B.H. and Shuman, R.M. (1991) Survival of early preimplantation porcine embryos after co-culture with cells producing an avian retrovirus. *Theriogenology* 35, 521–526.

Jinghe, T., Yang, Z.M. and Qin, P.C. (1989) Studies on the fine structure and maturation potential of follicular oocytes of different types from the Chinese yellow cows prior to *in vitro* culture. *Scientia Agricultura Sinica* 22, 83–88.

Jinyi, J., Sheng, Z., Huiqing, X. and Biqin, F. (1990) Studies on *in vitro* fertilization in pigs. *Jiangsu Journal of Agricultural Science* 6(3), 66–68.

Johannisson, E. (1993) A poisoned inheritance to our children. *Human Reproduction* 8, 1346–1348.

Johnson, L.A. (1987) Separation of X and Y chromosome bearing mammalian sperm by DNA content using cytometric analysis and sorting. *Biology of Reproduction* 36 (Suppl. 1), 80 (Abs.).

Johnson, L.A. (1988) Flow cytometric determination of sperm sex ratio in semen purportedly enriched for X- or Y-bearing sperm. *Theriogenology* 29, 265.

Johnson, L.A. (1989) Progress towards achieving sex preselection in farm animals. *Journal of Animal Science* 67 (Suppl. 1), 342 (Abs. 831).

Johnson, L. (1990) Spermatogenesis (animal species and humans). In: Asch, R.H, Balmaceda, J.P. and Johnson, I. (eds) *Gamete Physiology* Serona Symposia, Norwell, Massachusetts, USA, pp. 3–18.

Johnson, L.A. (1992) Advances in sex-predermination using flow cytometrically sorted X- and Y-chromosome bearing sperm. *Embryo Transfer Newsletter* 10(2), 5–11.

Johnson, L.A. and Clarke, R.N. (1988a) Flow sorting of X and Y chromosome bearing mammalian sperm: activation and pronuclear development of sorted bull, boar and ram sperm microinjected into hamster oocytes. *Gamete Research* 21, 335–343.

Johnson, L.A. and Clarke, R.N. (1988b) Flow cytometric isolation of X and Y chromosome-bearing mammalian sperm and subsequent development of eggs microinjected with sperm nuclei. *Proceedings of the Eleventh International Congress on Animal Reproduction and Artificial Insemination* (Dublin), 4, Paper 472 (3 pp.).

Johnson, L.A. and Pinkel, D. (1986) Modification of a laser-based flow cytometer for high-resolution DNA analysis of mammalian spermatozoa. *Cytometry* 7, 268–273.

Johnson, L.A., Flook, J.P. and Look, M.V. (1987) Flow cytometry of X and Y chromosome bearing sperm for DNA using an improved preparation method and staining with Hoechst 33342. *Gamete Research* 17, 203–212.

Johnson, L.A., Flook, J.P. and Hawk, H.W. (1989a) Altered sex ratios in offspring after surgical insemination of flow-sorted populations of X and Y chromosome-bearing sperm. *Biology of Reproduction* 40 (Suppl. 1), 162 (Abs. 344).

Johnson, L.A., Flook, J.P. and Hawk, H.W. (1989b) Sex preselection in rabbits: live births from X and Y sperm separated by DNA and cell sorting. *Biology of Reproduction* 41, 199–203.

Johnson, L.A., Welch, G.R., Keyvanfar, K., Dorfmann, A., Fugger, E.F. and Schulman, J.D. (1993) Gender preselection in humans? Flow cytometric separation of X and Y spermatozoa for the prevention of X-linked diseases. *Human Reproduction* 8, 1733–1739.

Johnson, L.A., Cran, D.G. and Polge, C. (1994) Recent advances in sex preselection of cattle: flow cytometric sorting of X- and Y-chromosome bearing sperm based on DNA to produce progeny. *Theriogenology* 41, 51–56.

Johnson, S.K., Jordan, J.E., Dean, R.G. and Page, R.D. (1991) The quantitation of bovine embryo viability using a bioluminescent assay for lactate dehydrogenase. *Theriogenology* 35, 425–433.

Johnson, W.H., Etherington, W.G., de Rose, E.P., Wilton, J.W. and Savage, N.C. (1989) The production of twins in beef cattle utilizing embryo transfer technology. *Theriogenology* 31, 206.

Johnson, L.A., O'Brien, S.J. and Wildt, D.E. (1989) *In vitro* maturation and fertilization of domestic cat follicular oocytes. *Gamete Research* 24, 343–356.

Joly, T., Nibart, M. and Thibier, M. (1991) A synthetic freezing medium for cattle embryos. *Proceedings of the Seventh Meeting of the European Embryo Transfer Association* (Cambridge), 150.

Joly, Th., Nibart, M. and Thibier, M. (1992) Hyaluronic acid as a substitute for proteins in the deep-freezing of embryos from mice and sheep: an *in vitro* investigation. *Theriogenology* 37, 473–480.

Jones, A.L., Marek, D.E., Wilson, J.M. and Looney, C.R. (1990) The use of ultrasonography to increase recipient efficiency through early pregnancy diagnosis. *Theriogenology* 33, 259.

Jones, J.M. and First, N.L. (1990) Effect of transcriptional inhibition on bovine embryo development. *Biology of Reproduction* 42 (Suppl. 1), 57.

Jones, J.M. and First, N.L. (1991) Blocking cleavage of 2-cell bovine embryos induces the synthesis of an 8-cell stage specific 52 kD protein. *Serono Symposium on Preimplantation Embryo Development*, Abstract Vol., 35 (Abs. I-22).

Joosten, I. and Hensen, E.J. (1992) Retained placenta: an immunological approach. *Animal Reproduction Science* 28, 451–461.

Jordan, J.E., Johnson, S.K.and Page, R.D. (1989) A rapid non-subjective enzyme assay to measure bovine embryo viability. *Theriogenology* 31, 209.

Jordan, R.M. (1992) Deer farming education – is it an extension concern? *Journal of Animal Science* 70 (Suppl. 1), 53 (Abs. 63).

Joris, H., Liu, J., Nagy, Z., Bocken, G., Van Ranst, H., Camus, M., Devroey, P. and Van Steirteghem, A.C. (1993) Controlled comparison of subzonal insemination and intracytoplasmic sperm injection. *Human Reproduction* 8 (Suppl. 1), 21–22.

Joshi, M.S. (1988) Isolation, cell culture and immunocytochemical characterization of oviduct epithelial cells of the cow. *Journal of Reproduction and Fertility* 83, 249–261.

Jubb, T.F., Abhayaratne, D., Malmo, J. and Anderson, G.A. (1990) Failure of an intramuscular injection of an analogue of gonadotrophin-releasing hormone 11 to 13 days after insemination to increase pregnancy rates in dairy cattle. *Australian Veterinary Journal* 67, 359–361.

Juengel, J.L., Youngquist, R.S. and Garverick, H.A. (1991) Pattern of expression of inhibin/activin mRNA in the preovulatory follicle during the follicular phase. *Biology of Reproduction* 44 (Suppl. 1), 65 (Abs. 51).

Jung, T. and Fischer, B. (1988) Correlation between diameter and DNA or protein synthetic activity in rabbit blastocysts. *Biology of Reproduction* 39, 1111–1116.

Jung, T., Lee, C. and Moor, R.M. (1991) Effects of protein kinase inhibitors on pig oocyte maturation *in vitro*. *Journal of Reproduction and Fertility*, Abstract Series No. 7, 33 (Abs. 56).

Jung, T., Lee, C. and Moor, R.M. (1992) Effects of protein kinase inhibitors on pig oocyte maturation *in vitro*. *Reproduction, Nutrition and Development* 32, 461–473.

Jura, J., Smorag, Z., Katska, L. and Bak, M. (1990) *In vitro* maturation of bovine oocytes following buffer microinjection into germinal vesicle or cytoplasm. *Theriogenology* 33, 263.

Jura, J., Kopchick, J.J., Chen, W.Y., Modlinski, J.A., Reed, M.A. and Smorag, Z. (1992) Production of transferable bovine embryos from exogenous DNA microinjected zygotes and 2-cell embryos. *Proceedings of the Twelfth International Congress on Animal Production* (The Hague), 2, 697–699.

Kadokawa, H., Minezawa, M., Takahashi, H., Sasaki, O. and Kariya, T. (1993) Analysis of the occurrence of freemartinism among single female calves born from multiple embryo transfer. *Theriogenology* 39, 239.

Kahn, W. (1992) Ultrasonography as a diagnostic tool in female animal reproduction. *Animal Reproduction Science* 28, 1–10.

Kajihara, Y., Goto, K., Tokumaru, M., Nakanishi, Y. and Ogawa, K. (1988) Number of blastomeres and chromosome abnormalities of bovine blastocysts obtained from *in vitro* matured, fertilized and cultured follicular oocytes. *Japanese Journal of Animal Reproduction* 34, 191–198.

Kajihara, Y., Kometani, N., Kobayashi, S., Shitanaka, Y., Koshiba, Y., Hishiyama, K., Shiraiwa, K. and Goto, K. (1990) Pregnancy rates and births after co-culture of cumulus cells with bovine embryos derived from *in vitro* fertilization of *in vitro* matured follicular oocytes. *Theriogenology* 33, 264.

Kajihara, Y., Blakewood, E.G., Myers, M.W., Kometani, N., Goto, K. and Godke, R.A. (1991a) *In vitro* maturation and fertilization of follicular oocytes obtained from calves. *Theriogenology* 35, 220.

Kajihara, Y., Kometani, N., Kobayashi, S., Shitanaka, Y. and Goto, K. (1991b) Pregnancy by bovine blastocysts developed in co-culture with cumulus/uterine endometrial cells after *in vitro* fertilization. *Japanese Journal of Animal Reproduction* 37, 177–184.

Kajihara, Y., Kometani, N., Shitanaka, Y., Saito, S., Yamaguchi, Y., Hishiyama, K. and Endo, M. (1992) Pregnancy rates and births after the direct transfers of frozen–thawed bovine IVF embryos. *Theriogenology* 37, 233.

Kajihara, Y., Hira, A., Fukudome, H., Hishiyama, K., Endo, M. and Goto, K. (1993) Effect of the pH of freezing medium on *in vitro* survival rate of bovine blastocysts after freezing and thawing. *Theriogenology* 39, 240.

Kalchreuter, S. and Horneber, F. (1992) Is litter size in sheep affected by the sire? *Kleinviehzuchter* 40(14), 854–855.

Kalous, J., Sutovsky, P., Rimkevicova, Z., Shioya, Y., Lie, B.-L. and Motlik, J. (1993) Pig membrana granulosa cells prevent resumption of meiosis in cattle oocytes. *Molecular Reproduction and Development* 34, 58–64.

Kameyama, K., Aoyagi, Y., Takedomi, T., Konishi, M. and Itakura, H. (1993) Transfer of bovine embryos sexed with rapid Y-chromosome-detection assay after cryopreservation. *Theriogenology* 39, 241.

Kane, M.T. (1978) Culture of mammalian ova. In: Sreenan, J.M. (ed.) *Control of Reproduction in the Cow.* Martinus Nijhoff, The Hague, pp. 381–397.

Kane, M.T. (1987) Culture media and culture of early embryos. *Theriogenology* 27, 49–57.

Kane, M.T. (1989) Effects of the putative phospholipid precursors, inositol, choline, serine and ethanolamine, on formation and expansion of rabbit blastocysts *in vitro. Journal of Reproduction and Fertility* 87, 275–279.

Kane, M.T. (1990) Control of growth in preimplantation embryos. *Irish Journal of Medical Science* 160(1), 17–22.

Kane, M.T. and Bavister, B.D. (1988) Vitamin requirements for development of eight-cell hamster embryos to hatching blastocysts *in vitro. Biology of Reproduction* 39, 1137–1143.

Kane, M.T. and Fahy, M.M. (1993) Blastocyst development and growth: role of inositol and citrate. In: Bavister, B.D. (ed.) *Preimplantation Embryo Development.* Serona Symposium, Springer-Verlag, New York, pp. 183–195.

Kane, M.T., Bavister, B.D. and Fahy, M. (1988) Water-soluble vitamins stimulate *in vitro* development of rabbit and hamster blastocysts. *Proceedings of the Eleventh International Congress on Animal Reproduction and Artificial Insemination* (Dublin), 4, 473 (3 pp.).

Kane, M.T., Carney, E.W. and Ellington, J.E. (1992) The role of nutrients, peptide growth factors and co-culture cells in development of preimplantation embryos *in vitro. Theriogenology* 38, 297–313.

Karabinus, D.S., Evenson, D.P. and Kaproth, M.T. (1991) Effects of egg yolk-citrate and milk extenders on chromatin structure and viability of cryopreserved bull sperm. *Journal of Dairy Science* 74, 3836–3848.

Kasai, M., Komi, J.H., Takakamo, A., Tsudera, H., Sakurai, T. and Machida, T. (1990) A simple method for mouse embryo cryopreservation in a low toxicity vitrification solution, without appreciable loss of viability. *Journal of Reproduction and Fertility* 89, 91–97.

Kasai, M., Hamaguchi, Y., Zhu, S.E., Sakurai, T. and Machida, T. (1991) High survival of rabbit morulae after vitrification in an ethylene glycol-based solution by a simple method. *Biology of Reproduction* 44 (Suppl. 1), 158 (Abs. 421).

Kashiwazaki, N., Ohtani, S., Nagashima, H., Yamakawa, H., Cheng, W.T.K., Lin, A.-C., Ma, R.C.-S. and Ogawa, S. (1991a) Production of normal piglets from hatched blastocysts frozen at −196°C. *Theriogenology* 35, 221.

Kashiwazaki, N., Ohtani, S., Miyamoto, K. and Ogawa, S. (1991b) Production of normal piglets from hatched blastocysts frozen at −196°C. *Veterinary Record* 128, 256–257.

Kasiraj, R., Misra, A.K., Mutha Rao, M., Jaiswal, R.S. and Rangareddi, N.S. (1993) Successful culmination of pregnancy and live birth following the transfer of frozen–thawed buffalo embryos. *Theriogenology* 39, 1187–1192.

Kastrop, P.M.M. (1990) Bovine oocyte maturation: biochemical aspects. Thesis, Rijksuniversiteit, Utrecht, 114 pp.

Kastrop, P.M.M., Bevers, M.M., Destree, O.H.J.and Kruip, Th.A.M. (1988) Protein synthesis of cultured bovine oocytes. *Journal of Reproduction and Fertility,* Abstract Series No. 2, Abs. 69.

Kastrop, P.M.M., Bevers, M.M., Destree, O.H.J. and Kruip, T.A.M. (1990a) Analysis of protein synthesis and phosphorylation during bovine oocyte maturation *in vivo. Proceedings of the Sixth Meeting of the European Embryo Transfer Association* (Lyon), 160.

Kastrop, P.M.M., Bevers, M.M., Destree, O.H.J. and Kruip, Th.A.M. (1990b) Changes in protein synthesis and phosphorylation patterns during bovine oocyte maturation *in vitro. Journal of Reproduction and Fertility* 90, 305–310.

Kastrop, P.M.M., Hulshof, S.C.J., Bevers, M.M., Destree, O.H.J. and Kruip, Th.A.M. (1991a)

The effects of α-Amanitin and cycloheximide on nuclear progression, protein synthesis, and phosphorylation during bovine oocyte maturation *in vitro*. *Molecular Reproduction and Development* 28, 249–254.

Kastrop, P.M.M., Bevers, M.M., Destree, O.H.J. and Kruip, Th.A.M. (1991b) Protein synthesis and phosphorylation patterns of bovine oocytes maturing *in vivo*. *Molecular Reproduction and Development* 29, 271–275.

Katayose, H., Matsuda, J. and Yanagimachi, R. (1992) The ability of dehydrated hamster and human sperm nuclei to develop into pronuclei. *Biology of Reproduction* 47, 277–284.

Kato, H. and Iritani, A. (1993) *In vitro* fertilization in cattle. *Molecular Reproduction and Development* 36, 229– 231.

Kato, Y., Nagashima, H., Shibata, K., Ishiwata, M. and Ogawa, S. (1987) Microsurgical bisection of expanding blastocysts for the sexing of pig embryos. *Japanese Journal of Animal Reproduction* 33, 151–154.

Katska, L. (1984) Comparison of two methods for recovery of ovarian oocytes from slaughter cattle. *Animal Reproduction Science* 7, 461–463.

Katska, L. and Smorag, Z. (1984) Number and quality of oocytes in relation to age of cattle. *Animal Reproduction Science* 7, 451–460.

Katska, L. and Smorag, Z. (1985) The influence of culture temperature on the *in vitro* maturation of bovine oocytes. *Animal Reproduction Science* 9, 205–212.

Katz, D.F., Drobnis, E.Z. and Overstreet, J.W. (1989) Factors regulating mammalian sperm migration through the female reproductive tract and oocyte vestments. *Gamete Research* 22, 443–469.

Katz, E., Ricciarelli, E. and Adashi, E.Y. (1993) The potential relevance of growth hormone to female reproductive physiology and pathophysiology. *Fertility and Sterility* 59, 8–34.

Kauffold, P., Torner, H., Schmidt, D. and Jacobi, K. (1986) Results of *in vitro* fertilization of bovine oviductal ova. *Tagungsbericht, Berlin* 252, 161–167.

Kauffold, P., Torner, H., Schmidt, D., Blottner, S. and Rommel, P. (1988) Fertilization *in vitro* of bovine oocytes matured *in vivo*. *Theriogenology* 29, 266.

Kauffold, P., Torner, H., Konig, I., Alm, H., Gotze, M. and Ebersbach, T. (1990) The way to *in vitro* embryo production. *Tagungsbericht, Berlin* 287, 91–97.

Kay, G.W. and Frylinck, S. (1992) Recovery of ovine follicular oocytes: effect of different methods on yield and quality. *Proceedings of the Twelfth International Congress on Animal Reproduction* (The Hague) 1, 345–347.

Kay, G.W., Hawk, H.W., Waterman, R.A. and Wall, R.J. (1991a) Identification of pronuclei in *in vitro* fertilized cow embryos. *Biology of Reproduction* 44 (Suppl. 1), 74 (Abs. 87).

Kay, G.W., Hawk, H.W., Waterman, R.A. and Wall, R.J. (1991b) Identification of pronuclei in *in vitro* fertilized cow embryos. *Animal Biotechnology* 2(1), 45–59.

Kay, V.J., Coutts, J.R.T. and Robertson, L. (1993) Pentoxifylline stimulates hyperactivation in human spermatozoa. *Human Reproduction* 8, 727–731.

Keane, M.G. and Harte, F.J. (1990) Beef production from once-calved heifers. *British Society of Animal Production* (Winter Meeting), Paper No. 173.

Keating, B. (1992) Production of calves from the ovaries of cows after casualty slaughter. *Irish Veterinary News* 14(9), 27–28.

Keefer, C.L. (1990) New techniques for assisted fertilization. *Theriogenology* 33, 101–112.

Keefer, C.L. (1992) Development of *in vitro* produced bovine embryos cultured individually in a simple medium: effects of EGF and TGF-beta 1. *Theriogenology* 37, 236.

Keefer, C.L. and Brackett, B.G. (1987) Microinjection of sperm nuclei into bovine and rabbit oocytes. *Biology of Reproduction* 36 (Suppl. 1), 52 (Abs.).

Keefer, C.L. and Stice, S. (1992) *In vitro* culture of bovine nucleus transfer embryos. *Biology of Reproduction*, 46 (Suppl. 1), 166 (Abs. 461).

Keefer, C.L., Brackett, B.G. and Troop, C.G. (1985) Bovine fertilization after *in vitro* insemination with motile sperm fractions. *Theriogenology* 23, 198.

Keefer, C.L., Younis, A.I. and Brackett, B.G. (1990) Cleavage development of bovine oocytes fertilized by sperm injection. *Molecular Reproduction and Development* 25, 281–285.

Keefer, C.L., Stice, S.L. and Maki-Laurila, M. (1991) Bovine embryo development *in vitro*: effect of *in vitro* maturation conditions on fertilization and blastocyst development. *Theriogenology* 35, 223.

Keefer, C.L., Koppang, R., Paprocki, A.M., Golueke, P., Stice, S., Maki-Laurila, M. and Matthews, L. (1993) Bovine inner cell mass (ICM) cells as donor nuclei in the production of nuclear transfer embryos. *Theriogenology* 39, 242.

Keeler, K.D., Mackenzie, N.M. and Dresser, D.W. (1983) Flow microfluorometric analysis of living spermatozoa stained with Hoechst 33342. *Journal of Reproduction and Fertility* 68, 205–212.

Kelk, D.A., Plante, L., Gartley, C.J., Buckrell, B.C. and King, W.A. (1991) Production of cattle × sheep hybrid embryos. *Theriogenology* 35, 224.

Kelk, D.A., Gartley, C.J. and King, W.A. (1992) Birth of lambs from embryos produced by *in vitro* maturation, fertilization and culture of oocytes. *Theriogenology* 37, 237.

Keller, M.L., Olson, S.E. and Seidel, G.E. (1993) Storage of bovine oocytes in calcium-free medium for 1 hour markedly lowers *in vitro* fertilization and embryonic development. *Theriogenology* 39, 243.

Kellerman, S.A. and Hunter, A.G. (1992) Identification of a receptor on the vitelline membrane of bovine oocytes that recognizes bovine immunoglobin G. *Journal of Dairy Science* 75, 1443–1447.

Kelley, K.M., Johnson, T.R., Gwatkin, R.B.L., Ilan, J. and Ilan, J. (1993) Transgenic strategies in reproductive endocrinology. *Molecular Reproduction and Development* 34, 337–347.

Kennedy, L.G., Boland, M.P. and Gordon, I. (1983) The effect of embryo quality at freezing on subsequent development of thawed cow embryos. *Theriogenology* 19, 823–832.

Khalifa, E.A.M., Tucker, M.J. and Hunt, P. (1992) Cruciate thinning of the zona pellucida for more successful enhancement of blastocyst hatching in the mouse. *Human Reproduction* 7, 532–536.

Khalifa, E.A.M., Tucker, M.J., Hunt, P.P. and Hamidi, J. (1993) Improved hatching in mouse embryos brought about by combined partial zona dissection and co-culture. *Human Reproduction* 8, 599–603.

Khurana, N. (1992) Comparative aspects of the energy metabolism of bovine embryos from oocyte to the hatched blastocyst derived *in vitro* and *in vivo*. Thesis, Tierarztliche Hochschule, Hannover, Germany, 211 pp.

Kiessling, A.A., Loutradis, D., McShane, P.M. and Jackson, K.V. (1988) Fertilization in trypsin-treated oocytes. *Annals of the New York Academy of Sciences* 541, 614–620.

Kikuchi, K. and Nagai, T. (1993) Pig follicular oocytes penetrated at metaphase-I can form both female and male pronuclei. *Theriogenology* 39, 244.

Kikuchi, K., Nagai, T., Motlik, J., Shioya, Y. and Izaike, Y. (1993) Effect of follicle cells on *in vitro* fertilization of pig follicular oocytes. *Theriogenology* 39, 593–599.

Killian, G.J. and Grippo, A.A. (1992) Incubation in ampullary oviductal fluid improves the ability of sperm to fertilize bovine oocytes *in vitro*. *Proceedings of the Twelfth International Congress of Animal Reproduction* (The Hague) 2, 650–652.

Killian, G.J., Chapman, D.A., Kavanaught, J. and Deaver, D.R. (1987) Changes in the composition of oviductal fluid during the cow estrous cycle. *Biology of Reproduction* 36 (Suppl. 1), 84 (Abs.).

Killian, G.J., Chapman, D.A., Kavanaugh, J.F., Deaver, D.R. and Wiggin, H.B. (1989) Changes in phospholipids, cholesterol and protein content of oviduct fluid of cows during the oestrous cycle. *Journal of Reproduction and Fertility* 86, 419–426.

Kim, C.I., Ellington, J.E. and Foote, R.H. (1990) Maturation, fertilization and development of bovine oocytes *in vitro* using TCM 199 and a simple defined medium with co-culture. *Theriogenology* 33, 433–440.

Kim, C.K., Im, K.S., Zheng, X. and Foote, R.H. (1989) *In vitro* capacitation and fertilizing ability of ejaculated rabbit sperm treated with lysophosphatidylcholine. *Gamete Research* 22, 131–141.

Kim, C.K., Chung, Y.C., Yoon, J.T., Choi, S.H., Ryu, B.Y., Chung, K. J., Kim, H.R. and Song, H.B. (1990) Studies on *in vitro* fertilization and development of bovine follicular oocytes matured *in vitro*. *Korean Journal of Animal Sciences* 32, 27–36.

Kim, H.-N., Roussel, J.D., Amborski, G.F., Hu, Y.X. and Godke, R.A. (1989) Monolayers of bovine fetal spleen cells and chick embryo fibroblasts for co-culture of bovine embryos. *Theriogenology* 31, 212.

Kim, H.-N., Zhang, L., Roussel, J.D. and Godke, R.A. (1991) Development of *in vitro* fertilized (IVF)-bovine oocytes using fetal bovine spleen cell and chick embryo fibroblast monolayers. *Biology of Reproduction* 44 (Suppl. 1), 73 (Abs. 83).

Kim, I.C. and Schomberg, D.W. (1989) The production of transforming growth factor-β activity by rat granulosa cell culture. *Endocrinology* 124, 1345–1351.

Kim, J.-H., Funahashi, H., Niwa, K. and Okuda, K. (1993a) Glucose requirement at different developmental stages of *in vitro* fertilized bovine embryos cultured in semi-defined medium. *Theriogenology* 39, 875–886.

Kim, J.-H., Niwa, K., Lim, J.-M. and Okuda, K. (1993b) Effects of phosphate, energy substrates, and amino-acids on the development of *in vitro*-matured, *in vitro*-fertilized bovine oocytes in a chemically defined, protein-free culture medium. *Biology of Reproduction* 48, 1320–1325.

Kim, S.K. and Park, H.K. (1990) Studies on the *in vitro* maturation and fertilization rate of bovine follicular oocytes. *Theriogenology* 33, 267.

Kim, T., Leibfried-Rutledge, M.L. and First, N.L. (1993) Gene transfer in bovine blastocysts using replication-defective retroviral vectors packaged with gibbon ape leukemia virus envelopes. *Molecular Reproduction and Development* 35, 105–113.

Kim, T.D.H. and Babcock, D.F. (1991) Determinants of bull sperm membrane potential: possible activation of K^+ channels during capacitation. *Biology of Reproduction* 44 (Suppl. 1), 118 (Abs. 261).

Kincy, V.L., Williams, J.E. and Butler, J.E. (1992) Expression of the insulin-like growth factor II (IGF-II) gene by ovine oviductal cells. *Biology of Reproduction* 46 (Suppl. 1), 67 (Abs. 68).

King, G., Chapeau, C. and Ackerly, C. (1991) Structure function relationships during preovulatory development of eCG stimulated porcine follicles. *Biology of Reproduction* 44 (Suppl. 1), 69 (Abs. 68).

King, J.W.B. (1989) New techniques in cattle breeding. In: Phillips, C.J.C. (ed.) *New Techniques in Cattle Production.* Butterworths, London, pp. 61–66.

King, R.S., McNutt, T.L., Henault, M.A. and Killian, G. (1990) Effect of cumulus oophorus and follicular cells on meiotic maturation in bovine oocytes. *Biology of Reproduction* 42 (Suppl. 1), 91.

King, R.S., Grippo, A.A. and Killian, G.J. (1991) Bovine estrus-associated protein stimulates capacitation of bull sperm. *Biology of Reproduction* 44 (Suppl. 1), 135 (Abs. 331).

King, R.S., Anderson, S.H. and Killian, G.J. (1992) Bovine estrus-associated protein facilitates sperm capacitation and fertilizing ability. *Proceedings of the Twelfth International Congress on Animal Reproduction* (The Hague) 1, 487–489.

King, W.A. (1985) Intrinsic embryonic factors that may affect survival after transfer. *Theriogenology* 23, 161–174.

King, W.A., Linares, T., Gustavsson, I. and Bane, A. (1979) A method for preparation of chromosomes from bovine zygotes and blastocysts. *Veterinary Science Communications* 3, 51–56.

King, W.A., Betteridge, K.J., Bousquet, D. and Greve, T. (1985a) Cytologically detectable abnormalities in bovine zygotes after *in vivo* and *in vitro* fertilization. *Journal of Dairy Science* 68 (Suppl. 1), 148–149.

King, W.A., Niar, A. and Betteridge, K.J. (1985b) The nucleolus organizer regions of early bovine embryos. *Journal of Dairy Science* 68 (Suppl. 1), 149.

King, W.A., Bousquet, D., Greve, T. and Goff, A.K. (1986) Meiosis in bovine oocytes matured *in vitro* and *in vivo*. *Acta Veterinaria Scandinavica* 27, 267–279.

King, W.A., Niar, A., Chartrain, I., Betteridge, K.J. and Guay, P. (1988a) Nucleolus organizer regions and nucleoli in preattachment bovine embryos. *Journal of Reproduction and Fertility* 82, 87–95.

King, W.A., Xu, K.P., Sirard, M.A., Greve, T., Leclare, P., Lambert, R.D. and Jacques, P. (1988b) Cytogenetic study of parthenogenetically activated bovine oocytes matured *in vivo* and *in vitro*. *Gamete Research* 20, 265–274.

King, W.A., Chartrain, I., Kopecny, V., Betteridge, K.J. and Bergeron, H. (1989) Nucleolus organizer regions and nucleoli in mammalian embryos. *Journal of Reproduction and Fertility* Suppl. 38, 63–71.

King, W.A., Yadav, B.R., Xu, K.P., Picard, L., Sirard, M.-A., Verini-Supplizi, A. and Betteridge, K.J. (1991) The sex ratios of bovine embryos produced *in vivo* and *in vitro*. *Theriogenology* 36, 779–788.

King, W.A., Picard, L., Bousquet, D. and Goff, A.K. (1992) Sex dependent loss of bisected bovine morulae after culture and freezing. *Journal of Reproduction and Fertility* 96, 453–459.

Kinghorn, B.P., Smith, C. and Dekkers, J.C.M. (1991) Potential genetic gains in dairy cattle with gamete harvesting and *in vitro* fertilization. *Journal of Dairy Science* 74, 611–622.

Kinis, A. (1990) Studies related to *in vitro* fertilization and microsurgery of cattle ova. PhD Thesis, National University of Ireland, Dublin.

Kinis, A., Vergos, E., Gallagher, M. and Gordon, A. (1989a) Studies in nuclear transplantation using bovine oocytes and embryos produced by *in vitro* culture techniques. *Journal of Reproduction and Fertility*, Abstract Series No. 4, Abs. 39.

Kinis, A., Vergos, V., Gallagher, M. and Gordon, A.C. (1989b) Use of sodium citrate in the denudation of bovine oocytes prior to *in vitro* fertilization. *Proceedings of the Fifth Conference of the European Embryo Transfer Association* (Lyon), 162.

Kinis, A., Vergos, E., Gordon, I., Gordon, A., and Gallagher, M. (1990a) Studies in the production of chimaeric cattle embryos by aggregation of blastomeres from embryos derived from oocytes matured and fertilized *in vitro*. *Theriogenology* 33, 268.

Kinis, A., Vergos, E., Gallagher, M. and Gordon, I. (1990b) Factors affecting nuclear transfer in cattle using oocytes and embryos produced by *in vitro* culture. *Proceedings of the Third International Ruminant Reproduction Symposium* (Nice), Abs. 14.

Kinis, A., Vergos, E., Lonergan, P., Sharif, H., Gallagher, M. and Gordon, I. (1990c) The incidence of polyspermy after zona drilling and zona cutting in bovine *in vitro* fertilization. *Journal of Reproduction and Fertility*, Abstract Series No. 6, Abs. 69.

Kinis, A., Vergos, E. and Gordon, I. (1990d) Nuclear transplantation in mammalian eggs. *Proceedings of the Fifth Hellenic Veterinary Congress* (Thessaloniki), 42.

Kinis, A., Vergos, E. and Gordon, I. (1990e) Methods of producing chimaeric embryos. *Proceedings of the Fifth Hellenic Veterinary Congress* (Thessaloniki), 49–50.

Kinis, A., Vergos, E., Lonergan, P., Sharif, H. and Gordon, I. (1990f) Effect of zona drilling on the outcome of cattle IVF in oocytes exposed to a low sperm concentration. *Proceedings of the Fifth Hellenic Veterinary Congress* (Thessaloniki), 53.

Kinis, A., Vergos, E., Lonergan, P., Sharif, H., Gallagher, M. and Gordon, I. (1991a) The incidence of polyspermy and the developmental potential of bovine ova fertilized *in vitro* after zona drilling. *Theriogenology* 35, 226.

Kinis, A., Vergos, E., Gallagher, M. and Gordon, I. (1991b) Factors affecting nuclear transfer in cattle using oocytes and embryos produced by *in vitro* culture. *Journal of Reproduction and Fertility*, Suppl. 43, 261–262.

Kippax, I.S., Christie, W.B. and Rowan, T.G. (1991) Effects of method of splitting, stage of development and presence or absence of zona pellucida on foetal survival in commercial bovine embryo transfer of bisected embryos. *Theriogenology* 35, 25–35.

Kirkpatrick, B.W. and Monson, R.L. (1993) Sensitive sex determination assay applicable to bovine embryos derived from IVM and IVF. *Journal of Reproduction and Fertility* 98, 335–340.

Kishi, J., Noda, Y., Narimoto, K., Umaoka, Y. and Mori, T. (1991) Block to development in cultured rat 1-cell embryos is overcome using medium HECM-1. *Human Reproduction* 6, 1445–1448.

Kitiyanant, Y., Lhuangmahamongkol, S., Areekijseree, M., Tocharus, C., Thonabulsombat, C. and Pavasuthipaisit, K. (1993) Porcine oviductal cells support *in vitro* bovine embryo development. *Theriogenology* 39, 246.

Kitwood, S.E., Phillips, C.J.C. and Weise, M. (1993) Use of a vaginal mucus impedance meter to detect estrus in the cow. *Theriogenology* 40, 559–569.

Knight, P.G. (1990) Identification and purification of inhibin and inhibin-related proteins. *Proceedings of the Third International Ruminant Reproduction Symposium* (Nice), Abstract.

Knight, P.G. and Muttukrishna, S. (1993) Measurement of dimeric inhibin using a modified 2-site IRMA specific for oxidized (Met O) inhibin: comparison with total alpha- and beta-subunit immunoreactivity in bovine follicular fluid (bFF). *Journal of Reproduction and Fertility*, Abstract Series No. 11, 46 (Abs. 81).

Ko, J.C.H., Kastelic, J.P., Del Campo, M.R. and Ginther, O.J. (1991) Effects of a dominant follicle on ovarian follicular dynamics during the oestrous cycle in heifers. *Journal of Reproduction and Fertility* 91, 511–519.

Ko, Y. and Threlfall, W.R. (1988) The effects of 1,2-propanediol as a cryoprotectant on the freezing of mouse oocytes. *Theriogenology* 29, 987–995.

Kobayashi, K., Nagashima, H., Yamakawa, H., Kato, Y. and Ogawa, S. (1990) The survival of whole and bisected rabbit morulae after cryopreservation by the vitrification method. *Theriogenology* 33, 777–788.

Kobayashi, K., Takagi, Y., Hoshi, H. and Oikawa, T. (1991) A study of *in vitro* development of bovine embryos in serum-free conditioned media. *Biology of Reproduction* 44 (Suppl. 1), 56 (Abs. 15).

Kobayashi, K., Okuyama, M., Fujimoto, G., Rothman, C.M., Hill, D.L. and Ogawa, S. (1992a) Subzonal insemination with a single spermatozoon using manipulation assisted sperm adhesion onto the ooplasmic membrane in mouse ova. *Molecular Reproduction and Development* 31, 223–229.

Kobayashi, K., Kato, K., Saga, M., Yamane, M., Rothman, C.M. and Ogawa, S. (1992b) Subzonal insemination of a single mouse spermatozoon with a personal computer-controlled micromanipulation system. *Molecular Reproduction and Development* 33, 81–88.

Kobayashi, K., Hill, D.L., Rotham, C.M. and Underberger, C.F. and Ogawa, S. (1993) Co-culture of subzonally microinseminated human oocytes with cumulus cells. *Theriogenology* 39, 247.

Kobayashi, T., Oda, T., Yoshimura, Y., Takehara, Y., Natori, M. and Nozawa, S. (1991) Androstenedione and progesterone concentrations in preovulatory follicular fluid correlate with successful fertilization and cleavage of human oocytes *in vitro*. *Fertility and Sterility* 56, 301–305.

Kocoski, L., Douchi, O., Chan, H., del Sol, A.B., Ali Shah, K., Velasquez, R.R., Sapalli, L.E. and Popovski, K. (1991) The effect of different cryopreservation methods on the survival rate in bovine embryos. *Proceedings of the Seventh Meeting of the European Embryo Transfer Association* (Cambridge), 154.

Koike, M., Sugawara, S. and Masali, J. (1991) Changes in 6-phosphoglutonate dehydrogenase activity in rat embryos during preimplantation stages. *Japanese Journal of Animal Reproduction* 37, 281–283.

Kojima, T. (1991) Studies on cryopreservation of mammalian embryos – induction of ice formation with droplets of silver iodide immobilized in alginate gel. *Japanese Journal of Animal Reproduction* 37, 13p–24p.

Kojima, T., Onodera, K., Soma, T. and Oguri, N. (1988a) Survival after transfer of bovine embryos frozen using the seeding method by immobilized silver iodide. *Japanese Journal of Animal Reproduction* 34, 149–152.

Kojima, T., Soma, T. and Oguri, N. (1988b) Effect of ice nucleation by droplet of immobilized silver iodide on freezing of rabbit and bovine embryos. *Theriogenology* 30, 1199–1207.

Kokia, E., Hurwitz, A., Ricciarelli, E., Resnick, C.E. and Adashi, E.Y. (1992) Interleukin-1 stimulates ovarian prostaglandin biosynthesis: obligatory role for heterologous contact-independent cell–cell interaction. *Biology of Reproduction* 46 (Suppl. 1) 91 (Abs. 161).

Konig, I., Gotze, M., Torner, H., Kauffold, P., Alm, H. and Friedrich, M. (1990) Results of a pilot trial involving the collection of oocytes from the ovaries of slaughtered cows for *in vitro* maturation and fertilization. *Monatshefte für Veterinarmedizin* 45, 138–141.

Kono, T., Iwasaki, S. and Nakahara, T. (1989) Parthenogenetic activation by electric stimulus of bovine oocytes matured *in vitro*. *Theriogenology* 32, 569–576.

Kono, T., Kwon, O.Y. and Nakahara, T. (1991) Development of vitrified mouse oocytes after *in vitro* fertilization *Cryobiology* 28, 50-54.

Kono, T., Sotomoru, Y., Aono, F., Takahashi, T., Ogihara, K., Sekizawa, F., Arai, T. and Nakahara, T. (1993) Effect of ooplast activation on the development of nuclear transferred bovine embryos. *Theriogenology* 39, 248.

Koong, L.J., Anderson, G.B. and Garrett, W.N. (1982) Maternal energy status of beef cattle during single and twin pregnancy. *Journal of Animal Science* 54, 480–490.

Koopman, P., Gubbay, J., Vivian, N., Goodfellow, P. and Lovell-Badge, R. (1991) Male development of chromosomally female mice transgenic for Sry. *Nature* 351, 117–121.

Kopecny, V., Fakan, S., Pavlok, A., Pivko, J., Grafenau, P., Biggiogera, M., Leser, G. and Martin, T.E. (1991) Immunoelectron microscopic localization of small nuclear ribonucleoproteins during bovine early embryogenesis. *Molecular Reproduction and Development* 29, 209–219.

Kot, K., Tischner, M. and Allen, W.R. (1991) The effect of freezing on the survival of horse embryos. *Medycyna Weterynaryjna* 47(8), 371–373.

Kotsuji, F., Kamitani, N., Goto, K. and Tominaga, T. (1990) Bovine theca and granulosa cell interactions modulate their growth, morphology and function. *Biology of Reproduction* 43, 726–732.

Kotsuji, F., Kubo, M., Goto, K., Misawa, M., Taniguchi, M. and Tominaga, T. (1992) The role of granulosa and theca cell interactions in oocyte maturation. *Biology of Reproduction* 46 (Suppl. 1), 139 (Abs. 354).

Kovacs, A. and Foote, R.H. (1989) Chromosome preparation from bovine spermatozoa. *Theriogenology* 31, 213.

Koyama, H., Yang, X., Jiang, S., Suzuki, H. and Foote, R.H. (1993) Analysis of polarity of bovine and rabbit blastomeres by scanning electron microscopy. *Theriogenology* 39, 249.

Kozikova, L.V. and Sirotkin, A.V. (1992) Autoradiographical analysis of RNA synthesis in bovine oocytes and cumulus cells during *in vitro* maturation. *Journal of Reproduction and Fertility*, Abstract Series No. 9, 50 (Abs. 86).

Krebs, J. (1988) The effect of dose of the β-blocker Carazolol on the conception rate of artificially inseminated cattle (field trial). Thesis, Ludwig-Maximilians-Universitat, Munich, Germany, 44 pp.

Krimpenfort, P., Rademakers, A., Eyestone, W., Van de Schans, A., Van den Broek, S., Kooiman, P., Kootwijk, E., Platenburg, G., Pieper, F., Strijker, R. and de Boer, H. (1991) Generation of transgenic dairy cattle using *'in vitro'* embryo production. *Biotechnology* 9, 844–847.

Krisher, R.L., Petters, R.M., Johnson, B.H., Bavister, B.D. and Archibong, A.E. (1989a) Development of porcine embryos from the one-cell stage to blastocyst in mouse oviducts maintained in organ culture. *Journal of Experimental Zoology* 249, 235–239.

Krisher, R.L., Petters, R.M. and Johnson, B.H. (1989b) Effect of oviductal condition on the development of one-cell porcine embryos in mouse or rat oviducts maintained in organ culture. *Theriogenology* 32, 885–892.

Kroetsch, T.G. and Stubbings, R.B. (1992) Sire and insemination dose does affect *in vitro* fertilization of bovine oocytes. *Theriogenology* 37, 240.

Kronnie, G. te and Boer, P. de (1993) Embryonic development in the pig up to the 64-cell stage, with reference to DNA replication and cell-cycle times from the third cleavage division. *Theriogenology* 39, 919–928.

Kruip, Th.A.M. (1982) Macroscopic identification of tertiary follicles > 2 mm in the ovaries of cycling cows. In: *Factors Influencing Fertility in the Post-Partum Cow*, Vol. 20, *Current Topics in Veterinary Medicine and Animal Science*. Martinus Nijhoff, The Hague, pp. 95–101.

Kruip, Th.A.M. and Dieleman, S.J. (1982) Macroscopic classification of bovine follicles and its validation by micromorphological and steroid biochemical procedures. *Reproduction, Nutrition and Development* 22, 465–473.

Kruip, Th.A.M. and Dieleman, S.J. (1985) Steroid hormone concentration in the fluid of bovine follicles relative to size, quality and stage of the oestrous cycle. *Theriogenology* 24, 395–408.

Kruip, Th.A.M. and Dieleman, S.J. (1989) Intrinsic and extrinsic factors influencing steroid production *in vitro* by bovine follicles. *Theriogenology* 31, 531–544.

Kruip, Th.A.M. and Vernooy, B.T.M. (1982) Influence of temperature on the resumption meiosis of bovine oocytes *in vitro*. In: Rolland, R., Hall, E.V. van, Hillier, S.G., McNatty, K.P. and Schoemaker, J. (eds) *Follicular Maturation and Ovulation*. International Congress Series, pp. 282–288.

Kruip, Th.A.M., Dieleman, S.J. and Moor, R.M. (1979) Steroid production by bovine follicles *in vitro*: influence of size, stage of cycle and culture system. *Annales de Biologie animale Biochimie Biophysique* 19, 1537–1545.

Kruip, Th.A.M., Cran, D.G., Beneden, T.H. and Dieleman, S.J. (1983) Structural changes in bovine oocytes during final maturation *in vivo*. *Gamete Research* 8, 29–47.

Kruip, Th.A.M., Van Beneden, Th.H., Dieleman, S.J. and Bevers, M.M. (1988) The effect of oestradiol-17β on nuclear maturation of bovine oocytes. *Proceedings of the Eleventh International Congress on Animal Reproduction and Artificial Insemination* (Dublin) 3, 336 (3 pp.).

Kruip, Th.A.M., Pieterse, M.C., van Beneden, T.A.M., Vos, P., Wurth, Y.A. and Taverne, M.A.M. (1990) Increased success rate of IVM and IVF in the bovine after sonographic guided transvaginal collections of the oocytes. *Theriogenology* 33, 269.

Kruip, Th.A.M., Pieterse, M.C., van Beneden, T.H., Vos, P.L.A.M., Wurth, Y.A. and Taverne, M.A.M. (1991) A new method for bovine embryo production: a potential alternative to superovulation. *Veterinary Record* 128, 208–210.

Kruip, Th.A.M., Wurth, Y.A. and den Daas, N. (1992) Homologous oocyte penetration *in vitro*: a reliable parameter for the fertilizing potency of bull semen. *Proceedings of the Twelfth International Congress on Animal Reproduction* (The Hague) 2, 653–655.

Kruip, Th.A.M., Boni, R., Roelofsen, M.W.M., Wurth, Y.A. and Pieterse, M.C. (1993) Application of OPU for embryo production and breeding in cattle. *Theriogenology* 39, 251.

Krzyminska, U.B., Leung, P., O'Neill, C. and Pike, I.L. (1992) Microinsemination by sperm transfer. *Lancet* 339, 375.

Krzywinski, A. and Wierzchos, E. (1992) Reproductive biotechnology in deer farming. In: Lauria, A. and Gandolfi, F. (eds) *Embryonic Development and Manipulation in Animal Production*. Portland Press, London, pp. 249–254.

Kubajak, C.L., Boehm, S.K., Brantmeier, S.A., Bushmeyer, S.M. and Ax, R.L. (1984) Distribution and sizes of ovarian follicles harvested at various stages of the estrous cycle from slaughtered cows. *Proceedings of the Tenth International Congress on Animal Reproduction and Artificial Insemination* (Illinois) 3, 494 (3 pp.).

Kuehner, L.F., Rieger, D., Johnson, W.H. and Walton, J.S. (1993a) Dose-dependent increases in plasma growth hormone following clonidine injection in luteal phase heifers. *Theriogenology* 39, 252.

Kuehner, L.F., Rieger, D., Walton, J.S., Zhao, X. and Johnson, W.H. (1993b) The effect of a depot injection of recombinant bovine somatotropin on follicular development and embryo yield in superovulated Holstein heifers. *Theriogenology* 40, 1003–1013.

Kumar, J., Osborn, J.C., Cameron, A.W.N. and Trounson, A.O. (1991) Oocyte maturation. *Contributions to Obstetrics and Gynaecology* 2, 21–37.

Kurosaka, S., Takai, K., Kita, M., Tachiura, G. and Kanoh, J. (1990) Effect of heparin on bovine oocytes development in culture during and after *in vitro* fertilization. *Japanese Journal of Animal Reproduction* 36, 241–244.

Kusunoki, H., Sakaue, M., Kato, S. and Kanda, S. (1989) Induction of the acrosome reaction in ejaculated goat spermatozoa by preincubation in chemically defined medium. *Journal of Experimental Zoology* 249, 322–328.

Kuwayama, M., Shioya, Y., Iwasaki, S., Okuyama, Y., Fukushima, M. and Hanada, A. (1989) Effects of culture medium and time of transfer to the rabbit oviduct on the developmental capability of bovine oocytes matured and fertilized *in vitro*. *Japanese Journal of Animal Reproduction* 35, 1–6.

Kuwayama, M., Fujikawa, S. and Hamano, S. (1992a) Successful cryopreservation of bovine blastocysts by vitrification: the observation of vitrification by freeze replica technique. *Proceedings of the Twelfth International Congress on Animal Reproduction* (The Hague) 3, 1445–1447.

Kuwayama, M., Hamano, S. and Nagai, T. (1992b) Vitrification of bovine blastocysts obtained by *in vitro* culture of oocytes matured and fertilized *in vitro*. *Journal of Reproduction and Fertility* 96, 187–193.

Kuzan, F.B. and Seidel, G.E. (1986) Embryo transfer in animals. In: Gwatkin, R.B.L. (ed.) *Developmental Biology*, Vol. 4, *Manipulation of Mammalian Development*. Plenum Press, London, pp. 249–278.

Kuzan, F.B. and Wright, R.W. (1982) Observations on the development of bovine morulae on various cellular and non-cellular substrata. *Journal of Animal Science* 54, 811–816.

Kvasnicki, A.V. (1951) Interbreed ova transplantation. *Sovetsk. Zooteh*, pp. 36–42; *Animal Breeding Abstracts* 19, p. 224.

Lacalandra, G.M., Dell 'Aquila, M.E., Fusco, S., Sciorsci, R.L., Perrone P. and Minoia, P. (1992) *In vitro* fertility test of bull semen. *Proceedings of the Twelfth International Congress of Animal Reproduction* (The Hague) 2, 656–658.

Lacham-Kaplan, O. and Trounson, A. (1993) The effects of the sperm motility activators

2-deoxyadenosine and pentoxifylline used for sperm micro-injection on mouse and human embryo development. *Human Reproduction* 8, 945–952.

Lai, Y.M., Stein, D.E., Soong, Y.K., Tang, Y.X., Grifo, J., Malter, H.E., Talansky, B.E., Cohen, J., Liu, H.C. and Rosenwaks, Z. (1992a) Evaluation of Vero cell co-culture system for mouse embryos in various media. *Human Reproduction* 7, 276–280.

Lai, Y.M., Soong, Y.K., Chang, S.Y., Lee, C.L. and Wang, M.L. (1992b) Co-culture of human zygotes on Vero cells. *Proceedings of the Fourteenth World Congress on Fertility and Sterility* (Caracas), 85 (Abs. 188).

Laing, J.A. (1945) Observations on the survival time of the spermatozoa in the genital tract of the cow and its relation to fertility. *Journal of Agricultural Science (Cambridge)* 35, 72–83.

Laing, J.A. (1949) Infertility in cattle associated with death of ova at early stages after fertilization. *Journal of Comparative Pathology and Therapeutics* 59, 97–109.

Lajili, H., Humblot, P. and Thibier, M. (1991) Effect of PG F 2 alpha treatment on conception rates of dairy cows treated with a GnRH agonist 12 to 14 days after artificial insemination. *Theriogenology* 36, 335–347.

Lambert, R.D., Bernard, C., Rioux, J.E., Eeland, R., D'Amours, D. and Montreuil, A. (1983) Endoscopy in cattle by the paralumbar route: technique for ovarian examination and follicular aspiration. *Theriogenology* 20, 149–161.

Lambert, R.D., Sirard, M.A., Bernard, C. and Beland, R. (1984) The fertilization performance *in vitro* of bovine and rabbit spermatozoa capacitated *in vitro*. *Proceedings of the Tenth International Congress on Animal Reproduction and Artificial Insemination* (Illinois) 3, 292 (3 pp.).

Lambert, R.D., Sirard, M.A., Bernard, C., Beland, R., Rioux, J.E., Leclerc, P., Menard, D.P. and Bedoya, M. (1986) *In vitro* fertilization of bovine oocytes matured *in vivo* and collected at laparoscopy. *Theriogenology* 25, 117–133.

Lamming, G.E. and Mann, G.E. (1993) Progesterone concentration affects the development of the luteolytic mechanism in the cow. *Journal of Reproduction and Fertility*, Abstract Series No. 11, 30 (Abs. 8).

Lamming, G.E., Parkinson, T.J. and Flint, A.P.F. (1988) Evidence that bovine interferon will prolong luteal function in the ewe. *Journal of Reproduction and Fertility*, Series No. 1, Abs. 33.

Lancaster, R., Catt, J., Rhodes, S. and Polege, C. (1990) Bull and breed specific effects of heparin-capacitated bovine sperm on *in vitro* fertilization of *in vitro* matured follicular oocytes. *Proceedings of the Sixth European Congress of the Embryo Transfer Association* (Lyon), 166.

Landa, V. and Hajkova, M. (1989) Parthenogenetic activation of cattle oocytes by a single pulse of direct current. *Folia Biologica (Praha)* 35, 307–313.

Landa, V. and Riha, J. (1992) Directed aggregation chimaeras constructed from individual blastomeres of bovine 16-cell stage embryos. *Proceedings of the Twelfth International Congress on Animal Reproduction* (The Hague) 2, 703–705.

Landsverk, K., Refsdal, A.O. and Vatn, T. (1992a) Pregnancy rates with frozen bovine embryos using two different methods for dilution of glycerol. *Theriogenology* 37, 241.

Landsverk, K., Jaakma, U., Muursepp, I., Refsdal, A.O. and Valdemann, E. (1992b) A field experiment comparing pregnancy rates in the bovine after transfer of embryos stored at 4°C and frozen–thawed embryos. *Proceedings of the Twelfth International Congress on Animal Reproduction* (The Hague) 3, 1448–1450.

Lane, M. and Gardner, D.K. (1992) The effect of incubation volume and embryo density on the development and viability of mouse embryos *in vitro*. *Human Reproduction* 7, 558–562.

Lange, W., Rommel, P., Bottcher, M., Dopp, E. and Winkler, L. (1991) Generation of cattle

× sheep hybrid embryos. *Proceedings of the Seventh Meeting of the European Embryo Transfer Association* (Cambridge), 158.

Langhout, D.J. and Spicer, L.J. (1990) Serum-free culture of bovine granulosa cells (GC): effects of bovine growth hormone (GH) on proliferation and steroidogenesis *in vitro*. *Journal of Animal Science* 68 (Suppl. 1), 455.

Langhout, D.J., Spicer, L.J. and Geisert, R.D. (1991) Development of a culture system for bovine granulosa cells: effects of growth hormone, estradiol, and gonadotropins on cell proliferation, steroidogenesis, and protein synthesis. *Journal of Animal Science* 69, 3321–3334.

Lannou, D.L. and Blanchard, Y. (1988) Nuclear maturity and morphology of human spermatozoa selected by Percoll density gradient centrifugation or swim-up procedure. *Journal of Reproduction and Fertility* 84, 551–556.

Larocca, C., Carbo, A., Cuenca, L., Algorta, M., Caorsi, C., Ferdandez, A., Gonzalez, A., Alfonso, F., Bonneveaux, J., Bordaberry, S., Bottaro, R., Cibils, J., Stewart, S., Olivera, M.A. and Rodrigues, J.L. (1992) Bovine embryonic death after splitting and embryo transfer. *Proceedings of the Twelfth International Congress on Animal Reproduction* (The Hague) 2, 799–801.

Larocca, C., Kmaid, S. and Calvo, J. (1993) Effect of follicular fluid and estrus cow serum on maturation, fertilization and development of the bovine oocyte *in vitro*. *Theriogenology* 39, 253.

Larson, L.L. and Ball, P.J.H. (1992) Regulation of estrous cycles in dairy cattle: a review. *Theriogenology* 38, 255–267.

Larson, R.C., Ignotz, G.G. and Currie, W.B. (1989) Development of *in vitro* fertilized bovine embryos beyond the 8-cell block using completely defined medium containing fibronectin. *Journal of Reproduction and Fertility*, Abstract Series No. 3, 20.

Larson, R.C., Ignotz, G.G. and Currie, W.B. (1990) Defined medium containing TGF-β and bFGF permits development of bovine embryos beyond the '8-cell block'. *Journal of Reproduction and Fertility*, Abstract Series No. 5, Abs. 22.

Larson, R.C., Ignotz, G.G. and Currie, W.B. (1991) Platelet derived growth factor (PDGF) initiates completion of the fourth cell cycle of bovine embryo development. *Journal of Reproduction and Fertility*, Abstract Series No. 7, 6 (Abs. 1).

Larson, R.C., Ignotz, G.G. and Currie, W.B. (1992a) Effect of fibronectin on early embryo development in cows. *Journal of Reproduction and Fertility* 96, 289–297.

Larson, R.C., Ignotz, G.G. and Currie, W.B. (1992b) Transforming growth factor-beta and basic fibroblast growth factor synergistically promote early bovine embryo development during the fourth cell cycle. *Molecular Reproduction and Development* 33, 432–435.

Larson, S.F. and Parks, J.E. (1990) *In vitro* maturation and fertilization of bovine oocytes in defined medium. *Biology of Reproduction* 42 (Suppl. 1), 92.

Larson, R.C., Ignotz, G.G., Parks, J.E. and Currie, W.B. (1992c) Bioactivity of epidermal growth factor is enhanced by a binder present in fluid from small bovine follicles. *Journal of Reproduction and Fertility*, Abstract Series No. 9, 18 (Abs. 21).

Larsson, B., Shamsuddin, M. and Gustafsson, H. (1992) Post-thaw viability of bovine embryos obtained *in vitro*. *Proceedings of the Twelfth International Congress on Animal Reproduction* (The Hague) 3, 1451–1453.

Larsson, B., Shamsuddin, M. and Rodriguez-Martinez, H. (1993) *In vitro* oocyte maturation and fertilization in the bovine. *Reproduction in Domestic Animals* 28, 161–163.

Lassalle, B. and Testart, J. (1988) Human sperm injection into the perivitelline space (SI-PVS) of hamster oocytes: effect of sperm pretreatment by calcium-ionophore A23187 and freeze–thawing on the penetration rate and polyspermy. *Gamete Research* 20, 301–311.

Lassalle, B., Testart, J. and Renard, J.P. (1985) Human embryo features that influence the success of cryopreservation with the use of 1,2-propanediol. *Fertility and Sterility* 44, 645–651.

Lassalle, B., Courtot, A.M. and Testart, J. (1987) *In vitro* fertilization of hamster and human oocytes by microinjection of human sperm. *Gamete Research* 16, 69–78.

Lathe, R. (1985) Molecular tailoring of the farm animal germline. *ABRO Annual Report 1985*, 7–10.

Lathrop, W.F. and Foote, R.H. (1986) The swim-up test to predict fertility of bull spermatozoa. *Journal of Dairy Science* 69 (Suppl. 1), 239 (Abs.).

Laufer, N., Palanker, D., Shufaro, Y., Safran, A., Simon, A. and Lewis, A. (1993) The efficacy and safety of zona pellucida drilling by a 193-nm excimer laser. *Fertility and Sterility* 59, 889–895.

Lauria, A. and Gandolfi, F. (1993) Recent advances in sperm mediated gene transfer. *Molecular Reproduction and Development* 36, 255–257.

Laurincik, J., Picha, J., Pichova, D. and Oberfranc, M. (1991a) Timing of laparoscopic aspiration of preovulatory oocytes in heifers. *Theriogenology* 35, 415–423.

Laurincik, J., Picha, J., Pichova, D. and Pivko, J. (1991b) Effect of LH-RH on meiotic maturation of preovulatory oocytes in superovulated heifers. *Animal Reproduction Science* 25, 189–197.

Laurincik, J., Kroslak, P., Hyttel, P., Pivko, J. and Sirotkin, A.V. (1992) Bovine cumulus expansion and corona-oocyte disconnection during culture *in vitro*. *Reproduction, Nutrition and Development* 32, 151–161.

Laurincik, J., Oberfranc, M., Hyttel, P., Grafenau, P., Tomanek, M. and Pivko, J. (1993a) Characterization of the periovulatory period in superovulated heifers. *Theriogenology* 39, 537–544.

Laurincik, J., Grafenau, P., Hyttel, P. and Greve, T. (1993b) Character of preovulatory follicles and oocytes after different superovulatory treatments in heifers. *Theriogenology* 39, 545–551.

Lavitrano, M., Camaioni, A., Fazui, V.M., Dolci, S., Farace, M.G. and Spadafora, C. (1989) Sperm cells as vectors for introducing foreign DNA into eggs: genetic transformation of mice. *Cell* 57, 717–723.

Lavitrano, M., French, D., Zani, M., Frati, L. and Spadafora, C. (1992a) The interaction between exogenous DNA and sperm cells. *Molecular Reproduction and Development* 31, 161–169.

Lavitrano, M., Zani, M., Lulli, V., French, D., Sperandio, S., Malone, B., Francolini, M., Frati, L. and Spadafora, C. (1992b) Mechanism of sperm–DNA interaction: spermatozoa as vectors for introducing exogenous DNA into eggs. In: Lauria, A. and Gandolfi, F. (eds) *Embryonic Development and Manipulation in Animal Production*. Portland Press, London, pp. 165–174.

Lavoir, M.-C., Basrur, P.K. and Betteridge, K.J. (1992) Isolation and identification of bovine fetal germ cells. *Proceedings of the Twelfth International Congress on Animal Reproduction* (The Hague) 1, 351–353.

Lavoir, M.-C., Leibo, S.P. and Betteridge, K.J. (1993) Cryopreservation of bovine fetal germ cells: preliminary results. *Theriogenology* 39, 255.

Lavoir, M.-C., Basrur, P.K. and Betteridge, K.J. (1994) Isolation and identification of germ cells from fetal bovine ovaries. *Molecular Reproduction and Development* 37, 413–424.

Lawitts, J.A. and Biggers, J.D. (1991) Optimization of mouse embryo culture media using simplex methods. *Journal of Reproduction and Fertility* 91, 543–556.

Lawitts, J.A. and Biggers, J.D. (1992) Joint effects of sodium chloride, glutamine, and glucose in mouse preimplantation embryo culture media. *Molecular Reproduction and Development* 31, 189–194.

Lazar, L., Lenard, J. and Pokorny, J. (1992) Xenogenous fertilization of *in vitro* matured bovine oocytes in pseudopregnant rabbits. *Proceedings of the Twelfth International Congress on Animal Reproduction* (The Hague) 2, 659–661.

Lazzari, G. and Galli, C. (1992) *In vitro* embryo production from valuable cows slaughtered for reproductive failure or terminal illness. *Proceedings of the Eighth Conference of the European Embryo Transfer Association* (Lyon), 176.

Lazzari, G. and Galli, C. (1993a) *In vitro* embryo production from valuable cows slaughtered for reproductive failure or terminal illness. *Theriogenology* 39, 256.

Lazzari, G. and Galli, C. (1993b) Salvage of valuable germplasm of sterile cattle by *in vitro* technologies. *Proceedings of the Ninth Congress of the European Embryo Transfer Association* (Lyon), 87.

Lazzari, G. and Moor, R.M. (1990) Protein synthesis and oocyte–somatic cooperation during oocyte development in the pig. *Journal of Reproduction and Fertility*, Abstract Series No. 6, Abs. 20.

Lazzari, G., Reik, W. and Moor, R.M. (1991) Granulosa cell development involves major changes in DNA methylation and protein synthesis. *Journal of Reproduction and Fertility*, Abstract Series No. 7, Abs. 12.

Lechtzin, N., Garside, W., Heyner, S. and Hillman, N. (1991) Glass-bead column separation of motile and non-motile human spermatozoa. *Journal of In Vitro Fertilization and Embryo Transfer* 8, 96–100.

Leclerc, P., Langlais, J., Lambert, R.D., Sirard, M.A. and Chafouleas, J.G. (1989a) Effect of heparin on the expression of calmodulin-binding proteins in bull spermatozoa. *Journal of Reproduction and Fertility* 85, 615–622.

Leclerc, P., Sirard, M.-A., Do, F., Chafouleas, J.G. and Lambert, R.D. (1989b) Decrease of calmodulin levels during bovine sperm capacitation. *Biology of Reproduction* 40 (Suppl. 1), 147 (Abs. 300).

Leclerc, P., Sirard, M.A., Chafouleas, J.G. and Lambert, R.D. (1992) Decrease in calmodulin concentrations during heparin-induced capacitation in bovine spermatozoa. *Journal of Reproduction and Fertility* 94, 23–32.

Ledda, S., Loi, P., Cappai, P., Filia, F. and Naitana, S. (1991) Effect of glucose on early ovine embryos developed in simple serum free medium. *Journal of Reproduction and Fertility*, Abstract Series No. 7, 8 (Abs. 7).

Lee, B.E., Boone, W.R., Brackelsberg, P.O. and Carmichael, R.A. (1988) Development of screening systems for evaluation of materials used in mammalian embryo transfer. *Theriogenology* 30, 605–612.

Lee, C.N.and Ax, R.L. (1984) Concentrations and composition of glygosaminoglycans in the female bovine reproductive tract. *Journal of Dairy Science* 67, 2006–2009.

Lee, C.N., Lenz, R.W. and Ax, R.L. (1983) Bovine sperm undergo capacitation when exposed to glycosaminoglycans *in vitro*. *Journal of Animal Science* 57 (Suppl. 1), 530.

Lee, C.N., Handrow, R.R., Lenz, R.W. and Ax, R.L. (1985) Interactions of seminal plasma and glycosaminoglycans on acrosome reactions in bovine spermatozoa *in vitro*. *Gamete Research* 12, 345–355.

Lee, C.N., Clayton, M.K., Bushmeyer, S.M., First, N.L. and Ax, R.L. (1986) Glycosaminoglycans in ewe reproductive tract and their influence on acrosome reactions in bovine spermatozoa *in vitro*. *Journal of Animal Science* 63, 861–867.

Lee, C.N., Vincent, D.L., Nusser, K.D., Carpenter, J.R., Campbell, C.M. and Toma, W.Y. (1989) Chemical analyses of allantoic fluid from bovine embryos recovered from dams subjected to nutritional stress. *Biology of Reproduction* 40 (Suppl. 1), 65 (Abs. 54).

Lee, K.O. and Ng, S.C. (1991) Use of the growth hormone in gynaecology. *Contributions to Obstetrics and Gynaecology* 2, 55–61.

Lee, S. (1991) Biologist's involvement in the reproduction revolution. *Biologist* 38(3), 85–88.

Lee, S.H. and Ahuja, K.K. (1991) The distribution of fucosylated glycoconjugates on the oocyte surface after cortical granules release. *Journal of Reproduction and Fertility*, Abstract Series No. 7, 53 (Abs. 95).

Lee, S.H., Ko, T.H., Monaghan, P., Lonergan, P., Gallagher, M. and Gordon, I. (1992) Successful *in vitro* development of ethanol-activated bovine oocytes to the blastocyst stage following *in vitro* maturation. *Journal of Reproduction and Fertility*, Abstract Series No. 9, 56 (Abs. 97).

Lee, S.J., Boice, M.L, Li, X. and Day, B.N. (1992) Effects of follicular fluid on *in vitro* maturation and fertilization of pig oocytes. *Journal of Animal Science* 70 (Suppl. 1), 74 (Abs. 150).

Leese, H.J. (1988) The formation and function of oviduct fluid. *Journal of Reproduction and Fertility* 82, 843–856.

Leese, H.J. (1991) Metabolism of the preimplantation mammalian embryo. *Oxford Reviews of Reproductive Biology* 13, 35–72.

Leese, H.J. (1992) 5 450 000 000 and rising. *Human Reproduction* 7, 1183–1184.

Leeuw, A.M. de, Rall, W.F., den Daas, J.H.G. and Kruip, Th.A.M. (1991) Comparative studies on the efficacy of rapid cryopreservation methods for bovine embryos. *Proceedings of the Seventh Meeting of the European Embryo Transfer Association.* (Cambridge), 77.

Leeuw, A.M. de, Daas, J.H.G. den, Kruip, Th.A.M. and Rall, W.F. (1992) The relative efficacy of bovine embryo cryopreservation by vitrification and conventional slow freezing. *Proceedings of the Twelfth International Congress of Animal Reproduction* (The Hague) 3, 1398–1400.

LeFever, D.G., Holland, M.D., Greathouse, G.A., Schafer, D.W., Brinks, J.S. and Odde, K.G. (1991) Effect of GnRH treatment twelve days after timed insemination on pregnancy rate in beef cattle. *Journal of Animal Science* 69 (Suppl. 1), 397.

Le Gal, F., Gasqui, P. and Renard, J.-P. (1991) Modelling the osmotic behaviour of goat oocytes before and after maturation. *Journal of Reproduction and Fertility*, Abstract Series No. 8, 12 (Abs. 14).

Le Gal, F., Gall, L. and De Smedt, V. (1992) Changes in protein synthesis pattern during *in vitro* maturation of goat oocytes. *Molecular Reproduction and Development* 32, 1–8.

Le Gal, F., Baril, G., Vallet, J.C. and Leboeuf, B. (1993) *In vivo* and *in vitro* survival of goat embryos after freezing with ethylene glycol or glycerol. *Theriogenology* 40, 771–777.

Le Guienne, B. and Thibier, M. (1988) Premier blastocysts bovins obtenus en totalité *in vitro*: note préliminaire. *El. and Ins.* 224, 11–14.

Le Guienne, B., Gerard, M., Chupin, D. and Thibault, C. (1988) Dévéloppement *in vitro* de l'oeuf bovin après fécondation *in vitro*. *Proceedings of the Fourth Scientific Meeting of the European Embryo Transfer Association* (Lyon), 33.

Lehn-Jensen, H. (1983) Survival of cow blastocysts using cooling rates of 1°C/minute to −25°C before plunging. *Theriogenology* 19, 138.

Lehn-Jensen, H. (1984) Deep freezing of cattle embryos. *Proceedings of the Tenth International Congress on Animal Reproduction and Artificial Insemination* (Illinois) 4, 1–12.

Lehn-Jensen, H. and Ball, W.F. (1983) Cryomicroscopic obervations of cattle embryos during freezing and thawing. *Theriogenology* 19, 263–277.

Lehrer, A.R., Lewis, G.S. and Aizinbud, E. (1992) Oestrus detection in cattle: recent developments. *Animal Reproduction Science* 28, 355–361.

Lei, C., Wang, Z.K., Wei, P.H., Wang, J.Z., Kong, M.Q. and Wu, T.G. (1992) Comparison of different *in vitro* capacitation methods for pig sperm assessed by penetration of zona-free hamster eggs. *Theriogenology* 37, 243.

Leibfried, L. and First, N.L. (1979) Characterization of bovine follicular oocytes and their ability to mature *in vitro*. *Journal of Animal Science* 48, 76–86.

Leibfried, L. and First, N.L. (1980) Effect of bovine and porcine follicular fluid and granulosa cells on maturation of oocytes *in vitro*. *Biology of Reproduction* 23, 699–704.

Leibfried, M.L. and Bavister, B.D. (1980) The effects of taurine and hypotaurine on *in vitro* fertilization in the golden hamster. *Gamete Research* 4, 57–63.

Leibfried, M.L. and Bavister, B.D. (1982) Effects of epinephrine and hypotaurine on *in vitro* fertilization in the golden hamster. *Journal of Reproduction and Fertility* 66, 87–93.

Leibfried-Rutledge, M.L., Critser, E.S. and First, N.L. (1985) Fertilization potential of follicular oocytes classified by stage of cycle and size of follicle. *Theriogenology* 23, 753–759.

Leibfried-Rutledge, M.L., Critser, E.S. and First, N.L. (1986a) Effects of fetal calf serum and bovine serum albumin on *in vitro* maturation and fertilization of bovine and hamster cumulus–oocyte complexes. *Biology of Reproduction* 35, 850–857.

Leibfried-Rutledge, M.L., Critser, E.S., Eyestone, W.H., Northey, D.L. and First, N.L. (1986b) Developmental potential of bovine oocytes matured *in vitro* or *in vivo*. *Theriogenology* 25, 164.

Leibfried-Rutledge, M.L., Critser, E.S., Eyestone, W.H., Northey, D.L. and First, N.L. (1987) Development potential of bovine oocytes matured *in vitro* or *in vivo*. *Biology of Reproduction* 36, 376–383.

Leibfried-Rutledge, M.L., Critser, E.S., Parrish, J.J. and First, N.L. (1989a) *In vitro* maturation and fertilization of bovine oocytes. *Theriogenology* 31, 61–74.

Leibfried-Rutledge, M.L., Florman, H.M. and First, N.L. (1989b) The molecular biology of mammalian oocyte maturation. In: Schatten, I. (ed.), *The Molecular Biology of Fertilization*. Academic Press, New York, pp. 259–301.

Leibo, S.P. (1982a) A one-step method for direct non-surgical transfer of frozen–thawed bovine embryos. *Proceedings of an International Congress on Embryo Transfer in Mammals* (Annécy), 98 (Abs.).

Leibo, S.P. (1982b) Field trial of one-step frozen bovine embryos transferred non-surgically. *Theriogenology* 19, 139.

Leibo, S.P. (1984) A one-step method for direct non-surgical transfer of frozen–thawed bovine embryos. *Theriogenology* 21, 767–790.

Leibo, S.P. (1985) Field trial of one step diluted frozen thawed bovine embryos: an update. *Theriogenology* 23, 201.

Leibo, S.P. (1986a) Cryobiology: preservation of mammalian embryos. In: Evans, J.W. and Hollander, A. (eds) *Genetic Engineering of Animals*. Plenum Press, New York, pp. 251–272.

Leibo, S.P. (1986b) Commercial production of pregnancies from one-step diluted frozen–thawed bovine embryos. *Theriogenology* 25, 166.

Leibo, S.P. (1988) Cryopreservation of embryos. *Proceedings of the Eleventh International Congress on Animal Reproduction and Artificial Insemination* (Dublin) 5, 370–377.

Leibo, S.P. (1990) Ova and embryos for quality control in reproductive biology. *Theriogenology* 33, 67–76.

Leibo, S.P. and Loskutoff, N.M. (1993) Cryobiology of *in vitro*-derived bovine embryos. *Theriogenology* 39, 81–94.

Leibo, S.P. and Mazur, P. (1978) Methods for the preservation of mammalian embryos by freezing. In: Daniel, J.C. (ed.) *Methods in Mammalian Reproduction*. Academic Press, New York, pp. 179–201.

Leibo, S.P. and Rall, W.F. (1990) Prenatal diagnosis of sex in bovine fetuses by amniocentesis. *Theriogenology* 33, 531–552.

Leibo, S.P. and Winniger, D. (1986) Production of bovine pregnancies from embryos transported at 0°C by air. *Theriogenology* 25, 165.

Leibo, S.P., de Mayo, F.J. and O'Malley, B. (1991a) Production of transgenic mice from cryopreserved ova. *Serono Symposium on Preimplantion Embryo Development*, Abstract Vol., 62 (Abs. II–24).

Leibo, S.P., De Mayo, J.F. and O'Malley, B. (1991b) Production of transgenic mice from cryopreserved fertilized ova. *Molecular Reproduction and Development* 30, 313–319.

Leidl, W., Kato, H., Hollerrieder, J. and Braun, J. (1993) Analysis of sperm motility by computer assisted methods, with special consideration of *in vitro* fertilization. *Molecular Reproduction and Development* 36, 224–228.

Lengwinat, T., Blottner, S., Wersuhn, A. and Pitra, C. (1990) A comparison of the pregnancy rate, using fresh or frozen bull semen, after *in vitro* maturation and fertilization of non-ovulated cattle oocytes. *Monatshefte fur Veterinarmedizin* 45, 345–347.

Lenz, R.W., Ball, G.D., Lohse, J.K., First, N.L. and Ax, R.L. (1983a) Chondroitin sulfate facilitates an acrosome reaction in bovine spermatozoa as evidenced by light microscopy, electron microscopy and *in vitro* fertilization. *Biology of Reproduction* 28, 683–690.

Lenz, R.W., Ball, G.D., Leibfried, M.L., Ax, R.L. and First, N.L. (1983b) *In vitro* maturation and fertilization of bovine oocytes are temperature-dependent processes. *Biology of Reproduction* 29, 173–179.

Lenz, R.W., Martin, J.L. and Ax, R.L. (1986) Prediction of non-return rates of Holstein bulls by evaluating occurrence of acrosome reactions *in vitro*. *Journal of Dairy Science* 69 (Suppl. 1), 166 (Abs.).

Lenz, R.W., Martin, J.L., Bellin, M.E. and Ax, R.L. (1988) Predicting fertility of dairy bulls by inducing acrosome reactions in sperm with chondroitin sulfates. *Journal of Dairy Science* 71, 1073–1077.

Leonard, M., Kirszenbaum, M., Cotinot, C., Chesne, P., Heyman, Y., Stinnakre, M.G., Bishop, C., Delouis, C., Vaiman, M. and Fellows, M. (1987) Sexing bovine embryos using Y-chromosome specific DNA probe. *Theriogenology* 20, 311–319.

Lester, T.D., Gliedt, D.W., Miller, G.F., Johnson, Z.B. and Rorie, R.W. (1991) Evaluation of cryoprotectants for toxicity to bovine trophoblastic vesicles (bTV). *Journal of Animal Science* 69 (Suppl. 1), 48.

Letcher, R., Simmen, R.C.M., Bazer, F.W. and Simmen, F.A. (1989) Insulin-like growth factor-1 expression during early conceptus development in the pig. *Biology of Reproduction* 41, 1143–1151.

Levanduski, M.J. and Westhusin, M.E. (1990) Effect of cytoskeletal inhibitor on fusion and development of bovine nuclear transfer embryos. *Theriogenology* 33, 273.

Levesque, J.T. and Sirard, M.A. (1994) Proteins in oocytes from calves and adult cows before maturation: relationship with their development capacity. *Reproduction, Nutrition and Development* 34, 133–139.

Le Van Ty, Chupin, D. and Driancourt, M.A. (1989) Ovarian follicular populations in buffaloes and cows. *Animal Reproduction Science* 19, 171–178.

Levron, J., Stein, D.W., Brandes, J.M. and Itskovitz-Eldor, J. (1993) Presence of sperm in the perivitelline space predicts fertilization rate after partial zona dissection. *Fertility and Sterility* 59, 820–825.

Lewis, A.M., Kaye, P.L., Lising, R. and Cameron, R.D.A. (1992) Stimulation of protein synthesis and expansion of pig blastocysts by insulin *in vitro*. *Reproduction, Fertility and Development* 4, 119–123.

Leymaster, K.A. and Bennett, G.L. (1990) Models of litter size and their consequences. *Proceedings of the Fourth World Congress on Genetics Applied to Livestock Production* 16, 299–308.

Li, J. and Foote, R.H. (1992) Effect of taurine and superoxide dismutase upon cultured rabbit zygotes. *Biology of Reproduction* 46 (Suppl. 1), 121 (Abs. 281).

Li, J. and Foote, R.H. (1993) Culture of rabbit zygotes into blastocysts in protein-free medium with one to twenty per cent oxygen. *Journal of Reproduction and Fertility* 98, 163–167.

Li, J., Foote, R.H. and Simkin, M. (1993) Development of rabbit zygotes cultured in protein-free medium with catalase, taurine, or superoxide dismutase. *Biology of Reproduction* 48, 33–37.

Li, L.Y., Denniston, R.S. and Godke, R.A. (1993) Enhanced development of IVF-derived bovine embroys by culturing with mouse embroys. *Theriogenology* 39, 259.

Li, R. and Trounson, A. (1991) Rapid freezing of the mouse blastocyst: effects of cryoprotectants and time and temperature of exposure to cryoprotectant before direct plunging into liquid nitrogen. *Reproduction, Fertility and Development* 3, 175–183.

Li, R., Cameron, A.W.N., Batt, P.A. and Trounson, A.O. (1990) Maximum survival of frozen goat embryos is attained at the expanded, hatching and hatched blastocyst stages of development. *Reproduction, Fertility and Development* 2, 345–350.

Li, Y., Perezgrovas, R., Gazal, O., Schwabe, C. and Anderson, L.L. (1991a) Antiprogesterone, RU 486, facilitates parturition in cattle. *Biology of Reproduction* 44 (Suppl. 1), 166 (Abs. 455).

Li, Y., Perezgrovas, R., Gazal, O.S., Schwabe, C. and Anderson, L.L. (1991b) Antiprogesterone, RU486, facilitates parturition in cattle. *Endocrinology (Philadelphia)* 129, 765–770.

Liebrich, J. (1992) Capacity for *in vivo* development after transfer of cattle embroys produced *in vitro*. Thesis, Ludwig-Maximilians Universitat Munchen, 127 pp.

Leihman, P., Greve, T. and Xu, K.P. (1986a) Nuclear and cytoplasmic maturation of bovine oocytes. *Proceedings of the Animal Meeting of the Society for the Study of Fertility* (Bristol), Abs. 109.

Liehman, P., Greve, T. and Xu, K.P. (1986b) Nuclear and cytoplasmic maturation of bovine oocytes cultured with dbc AMP, FSH and HCG. *Acta Veterinaria Scandinavica* 27, 566–574.

Lim, J.M., Fukui, Y. and Ono, H. (1991) The post-thaw developmental capacity of frozen bovine oocytes following *in vitro* maturation and fertilization. *Theriogenology* 35, 1225–1235.

Lim, J.M., Fukui, Y. and Ono, H. (1992) Developmental competence of bovine oocytes frozen at various maturation stages followed by *in vitro* maturation and fertilization. *Theriogenology* 37, 351–361.

Lincoln, D.W. (1992) Human contraception: development of new scientific opportunities. *Journal of Reproduction and Fertility*, Suppl. 45, 175–192.

Lindemann, C.B. and Kanous, K. (1991) The cyclic nitroxide free radical, 4-hydroxy-2,2,6,6-tetramethylpiperine-1-oxyl (TEMPOL), is effective in prolonging the motility of bull sperm *in vitro*. *Biology of Reproduction* 44 (Suppl. 1), 117 (Abs. 259).

Lindner, G.M. and Ellis, D.E. (1985) Refrigeration of bovine embryos. *Theriogenology* 23, 202.

Lindner, G.M. and Wright, R.W. (1983) Bovine embryo morphology and evaluation. *Theriogenology* 20, 407–416.

Lindner, G.M., Anderson, G.B., BonDurant, R.H. and Cupps, P.T. (1983) Survival of bovine embryos stored at 4°C. *Theriogenology* 20, 311–319.

Lindsell, C.E., Rajkumar, K., Manning, A.W., Emery, S.K., Mapletoft, R.J. and Murphy, B.O. (1986) Variability in FSH : LH ratios among batches of commercially available gonadotropins. *Theriogenology* 25, 167.

Ling, Z.J. and Lu, K.H. (1990) Frequency of cleavage and development *in vitro* of bovine oocytes fertilized in different numbers in drops with different sperm concentrations. *Theriogenology* 33, 275.

Linnenbrink, B. (1990) Investigation of the fertilizing ability of AI bulls by *in vitro* fertilization tests using cattle oocytes. Thesis, Tierarztliche Hochschule, Hannover, Germany, 121 pp.

Lippi, J., Mortimer, D. and Jansen, R.P.S. (1993) Sub-zonal insemination for extreme male factor infertility. *Human Reproduction* 8, 908–915.

Liu, D.Y. and Baker, H.W.G. (1992) Tests of human sperm function and fertilization *in vitro*. *Fertility and Sterility* 58, 465–483.

Lobb, D.K. and Dorrington, J. (1992) Intraovarian regulation of follicular development. *Animal Reproduction Science* 28, 343–354.

Logue, D.N., Bax, J.A., Renton, J.P. and McNally, J. (1992) Use of a Norgestomet implant to reduce embryonic loss and resynchronize oestrus in dairy heifers. *Proceedings of the British Society for Animal Production* (Winter Meeting), Paper No. 119.

Loi, P., Gallus, M., Dattena, M., Cappai, P., Branca, A., Ledda, S. and Naitana, S. (1992) The use of IVM oocytes for nuclear transplantation in sheep embryos. *Journal of Reproduction and Fertility*, Abstract Series No. 10, 34 (Abs. 63).

Lonergan, P. (1990) Factors affecting *in vitro* maturation and fertilization of the bovine oocyte. M.Agr.Sc. Thesis, University College, Dublin.

Lonergan, P. (1992) Studies in the *in vitro* maturation, fertilization and culture of bovine follicular ooytes, PhD Thesis, National University of Ireland, Dublin.

Lonergan, P., Sharif, H., Monaghan, P., Wahid, H., Gallagher, M. and Gordon, I. (1991a) Effect of cooling bovine oocytes to room temperature before insemination on the subsequent fertilization rate and embryo yield. *Journal of Reproduction and Fertility*, Abstract Series No. 7, 53 (Abs. 96).

Lonergan, P., Sharif, H., Monaghan, P., Wahid, H., Gallagher, M. and Gordon, I. (1991b) The effect of follicle size on the type of bovine oocyte obtained for *in vitro* maturation. *Proceedings of the Seventh Meeting of the European Embryo Transfer Association* (Cambridge), 162.

Lonergan, P., Vergos, E., Kinis, A., Sharif, H., Gallagher, M. and Gordon, I. (1991c) The effect of recovery method on the type of bovine oocyte obtained for *in vitro* maturation. *Theriogenology* 35, 231.

Lonergan, P., Fair, T. and Gordon, I. (1992) Effect of time of transfer to granulosa cell monolayer and cell-stage at 48 hours post-insemination on bovine oocyte development following IVM/IVF/IVC. *Proceedings of the Eighth Conference of the European Embryo Transfer Association* (Lyon), 178.

Lonergan, P., Monaghan, P., Rizos, D., Bagusi, A., Boland, M. and Gordon, I. (1993a) Effect of pFSH administration prior to slaughter on follicle size distribution and *in vitro* development of bovine oocytes. *Theriogenology* 39, 260.

Lonergan, P., Hafne, A.L., Farstad, W. and Andresen, O. (1993b) Effect of semen extender on *in vitro* development of bovine embryos following IVF. *Reproduction in Domestic Animals* 28, 174–175.

Lonergan, P., Monaghan, P., Rizoz, D., Boland, M.P. and Gordon, I. (1994) Effect of follicle size on bovine oocyte quality and developmental competence following maturation, fertilization and culture *in vitro*. *Molecular Reproduction and Development* 37, 48–53.

Long, C., Levanduski, M., Duplantis, S. and Westhusin, M. (1991) Nuclear transfer in the bovine embryo: the effect of DC voltage and pulse duration on electrofusion and cell lysis. *Theriogenology* 35, 232.

Long, C.R., Chase, C.N., Balise, J.J., Duby, R.T. and Robl, J.M. (1993a) Effect of sperm removal time, sperm concentration and motility enhancers on fertilization parameters and development of bovine embryos *in vitro*. *Theriogenology* 39, 261.

Long, C.R., Pinto-Correia, C., Duby, R.T., Ponce de Leon, F.A., Boland, M.P., Roche, J.F. and Robl, J.M. (1993b) Chromatin and microtubule morphology during the first cell cycle in bovine zygotes. *Molecular Reproduction and Development* 36, 23–32.

Long, J.A., Dickey, J.F. and Richardson, M.E. (1992) Fractionated conditioned medium supports bovine embryonic development through the *in vitro* cell block. *Biology of Reproduction* 46 (Suppl. 1), 120 (Abs. 277).

Looney, C.R. and Bondioli, K.R. (1988) Bovine FSH produced by recombinant DNA technology. *Theriogenology* 29, 235.

Looney, C.R., Westhusin, M.E. and Bondioli, K.R. (1989) Effect of cooling temperatures on precompacted bovine embryos. *Theriogenology* 31, 218.

Looney, C.R., Lindsey, B.R., Gonseth, C.L. and Johnson, D.L. (1994) Commercial aspects of oocyte retrieval and *in vitro* fertilization (IVF) for embryo production in problem cows. *Theriogenology* 41, 67–72.

Loos, de F. and Kruip, Th.A.M. (1988) Ultrastructure of immature bovine oocytes. *Proceedings of the Eleventh Congress on Animal Reproduction and Artificial Insemination* (Dublin) 3, 323 (3 pp.).

Loos, de F., van Vliet, C., van Maurik, P. and Kruip, Th.A.M. (1989) Morphology of immature bovine oocytes. *Gamete Research* 24, 197–204.

Loos, de F., Kastrop, P., van Maurik, P., van Beneden, T.H. and Kruip, Th.A.M. (1991a) Heterologous cell contacts and metabolic coupling in bovine cumulus oocyte complexes. *Molecular Reproduction and Development* 28, 255–259.

Loos, de F.A.M., Bevers, M.M., Dieleman, S.J. and Kruip, T.A.M. (1991b) Morphology of preovulatory bovine follicles as related to oocyte maturation. *Theriogenology* 35, 527–535.

Loos, de F.A.M., Bevers, M.M., Dieleman, S.J. and Kruip, T.A.M. (1991c) Follicular and oocyte maturation in cows treated for superovulation. *Theriogenology* 35, 537–546.

Loos, de F.A.M., Maurik, P. van, Beneden, T. van and Kruip, T.A.M. (1992a) Structural aspects of bovine oocyte maturation *in vitro*. *Molecular Reproduction and Development* 31, 208–214.

Loos, de F.A.M., Jeuken, M., Zeinstra, E. and Bevers, M.M. (1992b) Maintenance of meiotic arrest in bovine oocytes cultured *in vitro*. *Proceedings of the Twelfth International Congress on Animal Reproduction* (The Hague) 1, 357–359.

Loos, de F.A.M., Ursem, P. and Bevers, M.M. (1993) The effect of growth factors on *in vitro* bovine oocyte maturation. *Journal of Reproduction and Fertility*, Abstract Series No. 11, 64 (Abs. 117).

Lopata, A. (1992) The neglected human blastocyst. *Journal of Assisted Reproduction and Genetics* 9, 508–512.

Lopes, R.F.F., Rodrigues, J.L., Chebel, R.J., Maia, H.M.M., Perseu, J.A.S., Termignoni, C. and Stoll Rial, G.M. (1992) Survival of bovine embryos after micromanipulation for sex determination using the polymerase chain reaction (PCR). *Proceedings of the Twelfth International Congress on Animal Reproduction* (The Hague) 2, 712–714.

Lorenzo, P., Illera, M.J., Sanchez, J., Silvan, G. and Illera, J.C. (1992) The effect of EGF on cumulus expansion and bovine oocyte maturation *in vitro*. *Theriogenology* 37, 250.

Lorenzo, P., Illera, M.J. and Sanchez, J. (1993) Bovine oocyte maturation and fertilization *in vitro* with growth factors and estrous cow serum. *Theriogenology* 39, 262.

Lorton, S.P. and First, N.L. (1979) Hyaluronidase does not disperse the cumulus oophorus surrounding bovine ova. *Biology of Reproduction* 21, 301–308.

Loskutoff, N.M., Pollard, J.W., Scodras, J.M., De la Fuente, R., King, W.A. and Betteridge, K.J. (1991a) Production of blastocysts from single blastomeres cultured after isolation from 4-cell bovine embryos generated *in vitro*. *Theriogenology* 35, 233.

Loskutoff, N.M., Xu, K.P. and Betteridge, K.J. (1991b) Effect of biopsy on the development of cleavage stage bovine embryos generated *in vitro*. *Journal of Reproduction and Fertility*, Abstract Series No. 7, 35 (Abs. 59).

Loskutoff, N.M., Johnson, W.H. and Betteridge, K.J. (1993) The developmental competence of bovine embryos with reduced cell numbers. *Theriogenology* 39, 95–107.

Loutradis, D., John, D. and Kiessling, A.A. (1987) Hypoxanthine causes a 2-cell block in random-bred mouse embryos. *Biology of Reproduction* 37, 311–316.

Lovell-Badge, R.H. (1985) Transgenic animals: new advances in the field. *Nature* 315, 628.

Lu, G. and Hsu, T.T. (1991) *In vitro* fertilization of oocytes in Taiwan water buffalo. *Proceedings of the Third World Buffalo Congress* (Varna, Bulgaria) 3, 604–609.

Lu, K.H. (1990) Studies related to the *in vitro* maturation and fertilization of ovarian oocytes in cattle. PhD Thesis, National University of Ireland, Dublin.

Lu, K.H. and Gordon, I. (1987) Effect of serum, hormones and cumulus cells on the *in vitro* maturation of bovine oocytes. *Proceedings of the Society for the Study of Fertility* (York), Abs. 81.

Lu, K.H. and Gordon, I. (1988) Effect of heparin on the capacitation of frozen–thawed bovine spermatozoa used in the *in vitro* fertilization of oocytes matured *in vitro*. *Proceedings of Eleventh International Congress on Animal Reproduction and Artificial Insemination* (Dublin) 3, 342 (3 pp.).

Lu, K.H. and Polge, C. (1991) Pregnancy and twinning rates after transfer of IVF embryos to the bred recipient. *Proceedings of the Seventh Congress of the European Embryo Transfer Association* (Cambridge), 164.

Lu, K.H. and Polge, C. (1992) A summary of two years' results in large scale *in vitro* bovine embryo production. *Proceedings of the Twelfth International Congress on Animal Reproduction* (The Hague) 3, 1315–1317.

Lu, K.H., Gordon, I., Chen, H.B. and McGovern, H. (1987a) *In vitro* culture of early bovine embryos derived from *in vitro* fertilization of follicular oocytes matured *in vitro*. *Proceedings of the Third Scientific Meeting of the European Embryo Transfer Association* (Lyon), 70.

Lu, K.H., Gordon, I., Gallagher, M. and McGovern, H. (1987b) Pregnancy established in cattle by transfer of embryos derived from *in vitro* fertilization of follicular oocytes matured *in vitro*. *Veterinary Record* 121, 159–260.

Lu, K.H., Gordon, I., Boland, M.P. and Crosby, T.F. (1987c) *In vitro* fertilization of bovine oocytes matured *in vitro*. *Proceedings of the British Society of Animal Production (Winter Meeting)* Paper 81.

Lu, K.H., Gordon, I., McGovern, H. and Gallagher, M. (1988a) Production of cattle embryos by *in vitro* maturation and fertilization of follicular oocytes and their subsequent culture *in vivo* in sheep. *Theriogenology* 29, 272.

Lu, K.H., McGovern, H., Gallagher, M., MacDonnell, H.F. and Gordon, I. (1988b) Birth of calves after *in vitro* maturation and fertilization of oocytes and the *in vivo/in vitro* culture of zygotes. *Proceedings of the Annual Meeting of the Irish Grassland and Animal Production Association*, 2 pp.

Lu, K.H., Gordon, I., Chen, H.B., Gallagher, M. and McGovern, H. (1988c) Birth of twins after transfer of cattle embryos produced by *in vitro* techniques. *Veterinary Record* 122, 539–540.

Lu, K.H., Shi, D.S., Jiang, H.S., Goulding, D., Boland, M.P. and Roche, J.F. (1991) Comparison of the developmental capacity of bovine oocytes from superovulated and non-stimulated heifers. *Theriogenology* 35, 234.

Lucas-Hahn, A. (1992) Status of cryopreserving micromanipulated bovine embryos. *Embryo Transfer Newsletter* 10(1), 18–23.

Lucas-Hahn, A. and Niemann, H. (1991a) Cryopreservation of isolated bovine blastomeres. *Theriogenology* 35, 235.

Lucas-Hahn, A. and Niemann, H. (1991b) Aspects of bovine embryo production *in vivo* and *in vitro*. *Proceedings of the Seventh Meeting of the European Embryo Transfer Association* (Cambridge), 63.

Luck, M.R., Rodgers, R.J. and Findlay, J.K. (1990) Secretion and gene expression of inhibin, oxytocin and steroid hormones during the *in vitro* differentiation of bovine granulosa cells. *Reproduction, Fertility and Development*, 2, 11–25.

Luck, M.R., Shale, J.A. and Payne, J.H. (1992) Direct stimulation of bovine granulosa cells by very low, non-toxic concentrations of bovine recombinant interferon. *Journal of Reproduction and Fertility*, Abstract Series No. 9, 25 (Abs. 36).

Lucke, J.N. (1991) Embryo transfer: putting welfare first. *Veterinary Record* 129, 474.

Lucy, M.C., Savio, J.D., Badinga, L. and Thatcher, W.W. (1991) Factors regulating ovarian follicular dynamics in cattle. *Journal of Animal Science* 69 (Suppl. 1), 438–439.

Lundin, K., Sjogren, A. and Hamberger, L. (1993) Fertilization of human oocytes with acrosomeless spermatozoa. *Journal of Reproduction and Fertility*, Abstract Series No. 12, 55.

Lussier, J.G., Matton, P. and Dufour, J.J. (1987) Growth rates of follicles in the ovary of the cow. *Journal of Reproduction and Fertility* 81, 301–307.

Lutterbach, A., Koll, R.A. and Brem, G. (1987) *In vitro* maturation of bovine oocytes in coculture with granulosa cells and their subsequent fertilization and development. *Zuchthygiene* 22, 145–150.

Luvoni, G. and Oliva, O. (1991) Effect of gonadotropins on maturation and *in vitro* fertilization of bovine oocytes. *Atti della Societa Italiana di Buiatria* 24, 571–574.

Mabruk, M., O'Sullivan, M.A., Sheahan, B.J. and Atkins, G.J. (1990) Infection of cultured bovine blastocysts with bovine viral diarrhoea and Border disease viruses. *Proceedings of the Eighth International Congress of Virology* (Berlin), 244.

McCaffrey, C., McEvoy, C., Gwazdauskas, F.C., Diskin, M.G. and Sreenan, J.M. (1990) Successful *in vitro* culture of recently fertilized cattle ova to the morula/blastocyst stage. *Proceedings of the Irish Grassland and Animal Production Association* (Sixteenth Annual Research Meeting), 53–54.

McCaffrey, C. and Sreenan, J.M. (1991a) The effect of cell to ovum contact on the development of 1–4 cell cattle ova during co-culture with oviductal cells. *Theriogenology* 35, 239.

McCaffrey, C. and Sreenan, J.M. (1991b) An *in vitro* culture system for cattle embryos. *Farm and Food Research*, July/September, 21–23.

McCaffrey, C., McEvoy, T.G., Diskin, M.G., Gwazdauskas, F.C., Kane, M.T. and Sreenan, J.M. (1991) Successful co-culture of 1–4 cell cattle ova to the morula or blastocyst stage. *Journal of Reproduction and Fertility* 91, 119–124.

McCaffrey, C., Diskin, M.G., Lu, K.H. and Sreenan, J.M. (1992a) The development of *in vitro* fertilized (IVF) cattle ova in an oviduct cell co-culture system. *Proceedings of the British Society of Animal Production* (Winter Meeting), Paper No. 41.

McCaffrey, C., Lu, K.H. and Sreenan, J.M. (1992b) Factors involved in the *in vitro* development of IVF cattle ova. *Proceedings of the Irish Grassland and Animal Production Association* (Eighteenth Annual Research Meeting), 33–34.

McCaffrey, C., Lu, K.H. and Sreenan, J.M. (1992c) Development of *in vivo* and *in vitro* fertilized (IVF) cattle ova during *in vitro* culture. *Proceedings of the Twelfth International Congress on Animal Reproduction* (The Hague) 1, 366–368.

Macaulay, A.S., Roussel, J.D. and Seybt, S.H. (1986) Cortisol response in heifers to artificial insemination, natural mating, and no mating at estrus. *Theriogenology* 26, 117–122.

McConnell, D.J. (1986) Methods and achievements of genetic engineering: prospects in agriculture: a review. *Livestock Production Science* 15, 109–131.

McCutcheon, G.A., Caffrey, P.J., Kelleher, D.L. and Brophy, P.O. (1991) Twinning in a suckler herd. 1. Effects on performance of cows and their calves. *Irish Journal of Agricultural Research* 30, 1–9.

MacDonnell, H.F., Boland, M.P. and Gordon, I. (1984) Plasma oestrogen levels throughout gestation in Friesian cows carrying single, twin and one set of triplet calves. *Proceedings of the Tenth International Congress of Animal Reproduction and Artificial Insemination* 2, 76 (3 pp.).

MacDonnell, H.F., Mullins, S. and Gordon, I. (1993) Foetal progress and onset of parturition monitored by plasma oestrogen levels in the cow. *Irish Journal of Medical Science* 162, 108–109.

McDougall, S., Burke, C.R., MacMillan, K.L. and Williamson, N.B. (1992) The effect of pretreatment with progesterone on the oestrous response to oestradiol-17-beta benzoate in the post-partum dairy cow. *Proceedings of the New Zealand Society of Animal Production* 52, 157–160.

McEvoy, J.D. (1992) Alteration of the sex ratio. *Animal Breeding Abstracts* 60, 97–111.

McEvoy, J.D., Mayne, C.S. and McCaughey, W.J. (1992a) Induced twinning in dairy cows using IVF embryos: effects on reproductive performance. *Proceedings of the Association of Veterinary Teachers and Research Workers* (Twenty-Ninth Winter Meeting), 4 (Abs.).

McEvoy, T.G. (1990) Studies related to gene transfer in cattle and salmon (*Salmo salar*). PhD Thesis, National University of Ireland, Dublin.

McEvoy, T.G. and Sreenan, J.M. (1990) The efficiency of production, centrifugation, microinjection and transfer of one- and two-cell bovine ova in a gene transfer program. *Theriogenology* 33, 819–828.

McEvoy, T.G., Stack, M., Barry, T., Keane, B., Gannon, F. and Sreenan, J.M. (1987) Direct gene transfer by microinjection. *Theriogenology* 27, 258.

McEvoy, T.G., Gannon, F. and Sreenan, J.M. (1990) Production, manipulation and transfer of recently fertilized bovine ova in a gene transfer programme. *Proceedings of the Third International Ruminant Reproduction Symposium* (Nice), Abs. 16.

McEvoy, T.G., Robinson, J.J. and Sreenan, J.M. (1992) A role for transgenic animals in food production? *Trends in Food Science and Technology* 3, 294–302.

McGaughhey, R.W. (1983) Regulation of oocyte maturation. *Oxford Review of Reproductive Biology* 5, 106–130.

McGinnis, L.K. and Youngs, C.R. (1991) Morphological evaluation versus DNA specific staining for assessment of *in vitro* development of ovine embryos. *Theriogenology* 35, 242.

McGinnis, L.K. and Youngs, C.R. (1992) *In vitro* development of ovine embryos in CZB medium. *Theriogenology* 37, 559–569.

McGinnis, L.K., Duplantis, S.C. and Youngs, C.R. (1993) Cryopreservation of sheep embryos using ethylene glycol. *Animal Reproduction Science* 30, 273–280.

McGrath, J. and Solter, D. (1983a) Nuclear transplantation in the mouse embryo by microsurgery and cell fusion. *Science* 220, 1300–1302.

McGrath, J. and Solter, D. (1983b) Nuclear transplantation in mouse embryos. *Journal of Experimental Zoology* 228, 355–362.

McGrath, J. and Solter, D. (1984a) Completion of mouse embryogenesis requires both the maternal and paternal genomes. *Cell* 37, 179–183.

McGrath, J. and Solter, D. (1984b) Inability of mouse blastomere nuclei transferred enucleated zygotes to support development *in vitro*. *Science* 226, 1317–1319.

McGrath, J. and Solter, D. (1986) Nuclear and cytoplasmic transfer in mammalian embryos. In: Gwatkin, R.B.L. (ed.) *Developmental Biology*, Vol. 4, *Manipulation of Mammalian Development*. Plenum Press, New York, pp. 37–55.

McGuirk, B. (1989) The relevance of MOET programmes to developing countries. *Theriogenology* 31, 29–40.

McGuirk, B. (1990) The development of a nucleus MOET breeding project in the United Kingdom. *Animal Science Papers and Reports (Warsaw)* 6, 57–65.

Machatkova, M., Petelikova, J., Horky, F. and Jokesova, E. (1988) The co-cultivation of the early bovine embryo with the cell line of trophoblast. *Proceedings of the Fourth Meeting of the European Embryo Transfer Association* (Lyon), 35.

Machatkova, M., Petelikova, J., Jokesova, E. and Horky, F. (1991a) The relation between IVM–IVF–IVC development of bovine embryos and oestral cycle stage of oocyte donors. *Proceedings of the Seventh Meeting of the European Embryo Transfer Association* (Cambridge), 170.

Machatkova, M., Petelikova, J., Jokesova, E. and Horky, F. (1991b) The relationship between the development of the IVM–IVF–IVC embryos and the stage of the oestrous cycle of the donors of oocytes. *In XV geneticke dny. Ceske* (Budejovice, Czechoslovakia), 16–18.

Machatkova, M., Petelikova, J. and Jokesova, E. (1993) Development of frozen IVM–IVF–IVC embryos *in vitro* and *in vivo*. *Proceedings of the Ninth Congress of the European Embryo Transfer Association* (Lyon), 232.

Machaty, Z., Paldi, A., Solti, L. and Vajta, G. (1992) Blastomere biopsy and sex determination in bovine IVF embryos. *Proceedings of the Twelfth International Congress on Animal Reproduction* (The Hague) 2, 715–717.

Machaty, Z., Paldi, A., Csaki, T., Varga, Z., Kiss, I., Barandi, Z. and Vajta, G. (1993) Biopsy and sex determination by PCR of IVF bovine embryos. *Journal of Reproduction and Fertility* 98, 467–470.

McIntosh, A. and Hazeleger, N.L. (1994) The use of ethylene glycol for freezing bovine embryos. *Theriogenology* 41, 253.

McKelvey, W.A.C. and Robinson, J.J. (1986) Embryo survival and growth in the ewe – recent studies on the effects of nutrition and on novel techniques for the recovery and transfer of embryos. *Annual Report of the Rowett Research Institute* 46, 9–24.

McKelvey, W.A.C., Robinson, J.J. and Aitken, R.P. (1985) A simplified technique for the transfer of ovine embryos by laparoscopy. *Veterinary Record* 117, 492–494.

Mackenzie, A.M., Macdonald, D.C., Scaife, J.R., Acamovic, T., Warren, M. and Paton, F.A.C. (1991) Effects of beta-carotene or polyunsaturated fatty acids and selenium supplementation on immune status of single and twin calves. *Proceedings of the British Society for Animal Production* (Winter Meeting), Paper No. 118.

MacKenzie, N.M. and Pinder, A.C. (1987) Flow cytometry and its applications in veterinary medicine. *Research in Veterinary Science* 42, 131–139.

McKiernan, S.H. and Bavister, B.D. (1990a) Analysis of energy substrate requirements of Hamster 1- and 2-cell embryos for development *in vitro* to the morula and blastocyst stages. *Biology of Reproduction* 42 (Suppl. 1), 57.

McKiernan, S.H. and Bavister, B.D. (1990b) Environmental variables influencing *in vitro* development of hamster 2-cell embryos to the blastocyst stage. *Biology of Reproduction* 43, 404–413.

McKiernan, S.H. and Bavister, B.D. (1993) Time is a critical parameter for assessment of embryo development *in vitro*. *Biology of Reproduction* 48 (Suppl. 1), 172.

McKiernan, S.H., Bavister, B.D. and Tasca, R.J. (1991) Energy substrate requirements for *in vitro* development of hamster 1- and 2-cell embryos to the blastocyst stage. *Human Reproduction* 6, 64–75.

McLaren, A. (1984a) Methods and success of nuclear transplantation in mammals. *Nature* 309, 671–672.

McLaren, A. (1984b) Chimeras and sexual differentiation. In: *Chimeras in Developmental Biology*. Academic Press, London, pp. 381–399.

McLaren, A. (1988) The developmental history of female germ cells in mammals. *Oxford Reviews of Reproductive Biology* 10, 162–179.

McLaughlin, K.J., Davies, L. and Seamark, R.F. (1990) *In vitro* embryo culture in the production of identical Merino lambs by nuclear transplantation. *Reproduction, Fertility and Development* 2, 619–622.

McLaughlin, K.J., Pugh, A.P., Logan, K. and Tervit, H.R. (1991) Assessment of oocyte source for ovine nuclear transfer. *Theriogenology* 35, 240.

McLaughlin, K.J., McClean, D.M., Lewis, P.A., Hicks, L., Stevens, G., Bartsch, B.D. and Seamark, R.F. (1992) *In vitro* culture of bovine nuclear transfer embryos in synthetic oviduct fluid medium (SOFM). *Theriogenology* 37, 255.

McLeod, B.J., Foulkes, J.A., Williams, M.E. and Weller, R.F. (1991) Predicting the time of ovulation in dairy cows using on-farm progesterone kits. *Animal Production* 52(1), 1–9.

McLeod, B.J., Hunter, M.G., Bleach, E.C.L., Glencross, R.G. and Wrathall, J.H.M. (1992) Preovulatory follicle development and luteal function in ewes immunized against a synthetic peptide sequence of the α subunit of bovine inhibin. *Journal of Endocrinology* 133, 413–419.

Macmillan, K.L, Taufa, V.K. and Day, A.M. (1986) Effects of an agonist of gonadotrophin releasing hormone (buserelin) in cattle. III. Pregnancy rates after a post-insemination injection during metoestrus or dioestrus. *Animal Reproduction* 11, 1–10.

Macmillan, K.L., Taufa, V.K., Barnes, D.R., Day, A.M. and Henry, R. (1988) Detecting estrus in synchronized heifers using tailpaint and an aerosol raddle. *Theriogenology* 30, 1099–1114.

Macmillan, K.L., Thatcher, W.W. and Drost, M. (1989) Recent developments in animal breeding programmes. *Proceedings of the New Zealand Society for Animal Production* 49, 91–96.

MacMillan, K.L., Washburn, S.P., Henderson, H.V., Petch, S.F. (1990a) Effects of varying the progesterone content of CIDR intravaginal devices and multiple CIDR treatments on plasma hormone concentrations and residual hormone content. *Proceedings of the New Zealand Society for Animal Production* 50, 473–475.

Macmillan, K.L., Taufa, V.K., Day, A.M. and Petersen, A.J. (1990b) Effects of supplemental progesterone on pregnancy rates in cattle. *Proceedings of the Third International Ruminant Reproduction Symposium* (Nice), Abs. 23.

Macmillan, K.L., Taufa, V.K., Barnes, D.R. and Day, A.M. (1991) Plasma progesterone concentrations in heifers and cows treated with a new intravaginal device. *Animal Reproduction Science* 26, 25–40.

Macmillan, W.H. and Macmillan, K.L. (1989) CIDR-B for managed reproduction in beef cows and heifers. *Proceedings of the New Zealand Society for Animal Production* 49, 85–89.

McNatty, K.P., Heath, D.A., Henderson, K.M., Lun, S., Hurst, P.R., Ellis, L.M., Montgomery, G.W., Morrison, L. and Thurley, D.C. (1984) Some aspects of thecal and granulosa cell function during follicular development in the bovine ovary. *Journal of Reproduction and Fertility* 72, 39–53.

McNutt, T.L. and Johnson, L.A. (1992) Influence of flow cytometric sorting on sperm membrane proteins. *Journal of Animal Science* 70 (Suppl. 1), 252 (Abs. 455).

McNutt, T., Rogowski, L., Vasilatos-Younken, R. and Killian, G. (1992) Adsorption of oviductal fluid proteins by the bovine sperm membrane during *in vitro* capacitation. *Molecular Reproduction and Development* 33, 313–323.

McWhir, J., O'Neill, G.T. and Wilmut, I. (1990) Transgenesis in farm animals. *Proceedings of the Third Symposium on Advanced Topics in Animal Reproduction* (FCAV-UNESP, Jaboticabal) 101–117.

Madan, M.L. (1990) Conservation of germplasm through embryo transfer in buffaloes. *Proceedings of the Twenty-Third International Dairy Congress* (Montreal) 1, 302–314.

Madan, M.L., Singla, S.K., Chauhan, M.B. and Manik, R.S. (1994) *In vitro* production and transfer of embryos in buffaloes. *Theriogenology* 41, 139–143.

Madison, V., Greve, T., Avery, B. and Wamberg, T. (1991a) The effect of endotoxin-contaminated medium on *in vitro* fertilization and development of bovine oocytes matured *in vitro*. *Reproduction, Nutrition and Development* 31, 159–165.

Madison, V., Avery, B. and Greve, T. (1991b) Effect of sperm concentration on *in vitro* penetration and development of bovine oocytes. *Journal of Animal Science* 69 (Suppl. 1), 461.

Madison, V., Avery, B. and Greve, T. (1992) Selection of immature bovine oocytes for developmental potential *in vitro*. *Animal Reproduction Science* 27, 1–11.

Mahadevan, M.M., Trounson, A.O., Wood, C. and Leeton, J.F. (1987) Effect of oocyte quality and sperm characteristics on the number of spermatozoa bound to the zona pellucida of human oocytes inseminated *in vitro*. *Journal of In Vitro Fertilization and Embryo Transfer* 4, 223–227.

Mahmoudzadeh, A.R., Van Soom, A., Van Vlaenderen, I. and De Kruif, A. (1993) A comparative study of the effect of one-step addition of different vitrification solutions on *in vitro* survival of vitrified bovine embryos. *Theriogenology* 39, 1291–1302.

Mahon, G.D. and Rawle, J.E. (1987) The export of deep-frozen bovine embryos. *Theriogenology* 27, 21–35.

Malayer, J.R. and Hansen, P.J. (1990) Effects of alanine and taurine on survival of preimplantation embryos during heat shock. *Journal of Animal Science* 68 (Suppl. 1), 419.

Malenko, G.P. (1990) Recovery and *in vitro* fertilization of oocytes for obtaining early embryos of cattle. From: Chpt 2, Borovsk, USSR, 127–128.

Mandelbaum, J., Junca, A.M., Tibi, C., Plachot, M., Alnot, M.O., Rim H., Salat-Baroux, J. and Cohen, J. (1988) Cryopreservation of immature and mature hamster and human oocytes. *Annals of the New York Academy of Sciences* 541, 550–561.

Mann, G.E., McNeilly, A.S. and Baird, D.T. (1992a) Hormone production *in vivo* and *in vitro* from follicles at different stages of the oestrous cycle in the sheep. *Journal of Endocrinology* 132, 225–234.

Mann, G.E., Campbell, B.K., McNeilly, A.S. and Baird, D.T. (1992b) The role of inhibin and oestradiol in the control of FSH secretion in the sheep. *Journal of Endocrinology* 133, 381–391.

Mann, J.R. (1988) Full term development of mouse eggs fertilized by a spermatozoan microinjected under the zona pellucida. *Biology of Reproduction* 38, 1077–1083.

Mannaerts, B.M.J.L. (1986) Cytological parameters for rating bovine embryo quality. *Current Topics in Veterinary Medicine and Animal Science* 34, 216–222.

Mannaerts, B.M.J.L. (1992) New developments in gonadotrophins. *Proceedings of the Fourteenth World Congress on Fertility and Sterility* (Caracas), Abstract Vol., 33.

Mapletoft, R.J. (1985) Embryo transfer in the cow: general procedures. *Rev. Sci. Tech. Off. Int. Epiz.* 4(4), 843–858.

Mapletoft, R.J. and Pierson, R.A. (1993) Factors affecting superovulation in the cow: practical considerations. *Embryo Transfer Newsletter* 11(3), 15–24.

Mapletoft, R.J., Wu, M. and Murphy, B.D. (1988) Superovulation of beef cattle with Follitrophin. *Proceedings of the Eleventh International Congress on Animal Reproduction and Artificial Insemination* (Dublin) 2, Brief Communications No. 173.

Marek, D.E., Pryor, J.H., Whitesell, T.H. and Looney, C.R. (1990) Nuclear transplantation in the bovine: effect of donor embryo age on subsequent embryo production. *Theriogenology* 33, 283.

Mariana, J.C., Monniaux, D., Driancourt, M.A. and Mauleon, P. (1991) Folliculogenesis. In: Cupps, P.T. (ed.) *Reproduction in Domestic Animals*, 4th edn. Academic Press, London, pp. 119–171.

Mariante, A. da Silva and Bem, A.R. de (1992) Animal genetic resources conservation programme in Brazil. *Animal Genetic Resources Information Bulletin (FAO, Rome)* 10, 9–32.

Marion, G.B. and Gier, H.T. (1971) Ovarian and uterine embryogenesis and morphology of the non-pregnant female mammal. *Journal of Animal Science* 32 (Suppl. 1), 24–47.

Marion, G.B. and Gier, H.T. and Choudary, J.B. (1968) Micromorphology of the bovine ovarian follicular system. *Journal of Animal Science* 27, 451–465.

Markert, C.L. (1983) Fertilization of mammalian eggs by sperm injection. *Journal of Experimental Zoology* 228, 195–201.

Marks, J.L. and Ax, R.L. (1985) Relationship of non-return rates of dairy bulls to binding affinity of heparin to sperm. *Journal of Dairy Science* 68, 2078–2082.

Marquant-Le Guienne, B. and Humblot, P. (1991) Evaluation of fertilizing ability of domestic animals by *in vitro* fertilization test. *Contraception, Fertilité et Sexualité* 19, 819–824.

Marquant-Le Guienne, B. and Humblot, P. (1992) Fertility testing of domestic animals by *in vitro* fertilization. *Annales de Zootechnie* 41, 361–370.

Marquant-Le Guienne, B., Gerard, M., Solari, A. and Thibault, C. (1989) *In vitro* culture of bovine eggs fertilized either *in vivo* or *in vitro*. *Reproduction, Nutrition and Development* 29, 559–568.

Marquant-Le Guienne, B., Humbolt, P., Thibier, M. and Thibault, C. (1990a) Evaluation of bull semen fertility by homologous *in vitro* fertilization tests. *Reproduction, Nutrition and Development* 30, 259–266.

Marquant-Le Guienne, B., Guyader, C., Humblot, P., Gerard, O. and Thibier, M. (1990b) *In vitro* testing of bull fertility by *in vitro* fertilization and embryonic development rates. *Proceedings of the Sixth Meeting of the European Embryo Transfer Association* (Lyon), 172.

Marquant-Le Guienne, B., Nibart, M., Esposito, L., Guyader, C., Kohen, G. and Thibier,M. (1991) Bovine *in vitro* fertilized embryos sexing by a male specific probe. *Proceedings of the Seventh Meeting of the European Embryo Transfer Association* (Cambridge), 172.

Marquant-Le Guienne, B., Nibart, M., Guyader, C., Kohen, G., Exposito, L., Thuard, J.M. and Thibier, M. (1992a) DNA probe sexing of young *in vitro* fertilized bovine embryos. *Theriogenology* 37, 253.

Marquant Le Guienne, B., Humblot, P., Guillon, N., Gerard, O. and Thibier, M. (1992b) *In vitro* fertilization as a tool to evaluate fertility in the bovine. *Proceedings of the Twelfth International Congress on Animal Reproduction* (The Hague) 2, 662–664.

Marquant-Le Guienne, B., Guillon, N. and Thibier, M. (1993a) The effect of glycine in capacitation medium on IVF and further development rates in the cow. *Theriogenology* 39, 264.

Marquant-Le Guienne, B., Humblot, P., Guillon, N. and Thibier, M. (1993b) Murine LIF improves the development of IVF cultured bovine morulae. *Journal of Reproduction and Fertility*, Abstract Series No. 12, 61.

Marrs, R.P. and Quinn, P. (1990) Oocyte and embryo cryopreservation techniques and results. In: *CRC Handbook of the Laboratory Diagnosis and Treatment of Infertility*. CRC Press, Boca Raton, Florida, pp. 345–352.

Marshall, L.K., Brinks, J.S., Golden, B.L. and Andersen, K.J. (1993) Heritability of gestation length in Limousin cattle. *Journal of Animal Science* 71 (Suppl. 1), 108.

Martin, J.M., Squires, E.L., Jasko, D.J. and Carney, N.J. (1991) Effect of storage of equine embryos at 5°C for 18 and 36 hours. *Theriogenology* 35, 238.

Martin, K.L. and Leese, H.J. (1993) The importance of glucose utilization in the mouse preimplantation embryo. *Journal of Reproduction and Fertility*, Abstract Series No. 11, 69 (Abs. 128).

Martin, R.H., Ko, E. and Radmaker, A. (1988) Human sperm chromosome complements after micro-injection of hamster eggs. *Journal of Reproduction and Fertility* 84, 179–186.

Martin, T.L., Fogwell, R.L. and Ireland, J.J. (1991) Concentrations of inhibins and steroids in follicular fluid during development of dominant follicles in heifers. *Biology of Reproduction* 44, 693–700.

Martinez, E., Vazquez, J.M., Matas, C., Roca, J., Coy, P. and Gadea, J. (1993) Evaluation of boar spermatozoa penetrating capacity using pig oocytes at the germinal vesicle stage. *Theriogenology* 40, 547–557.

Martino, A., Mogas, T., Palomo, M.J. and Paramio, M.T. (1992) Effect of method of recovery on the IVM of pre-pubertal goat oocytes. *Journal of Reproduction and Fertility*, Abstract Series No. 9, 53 (Abs. 91).

Martino, A., Mogas, T., Palomo, M.J. and Paramio, M.T. (1993) Effect of oocyte and granulosa cell source used during *in vitro* maturation on *in vitro* fertilization of goat oocytes. *Theriogenology* 39, 265.

Martinod, S., Maurer, R.R., Siegenthaler, B., Gerber, C. and Hansen, P.J. (1991) The effects of recombinant bovine interferon-alpha on fertility in ewes. *Theriogenology* 36, 231–239.

Maruo, T., Ladines-Llave, C.A., Samoto, T., Matsuo, H., Manalo, A.S., Ito, H. and Mochizuki, M. (1993) Expression of epidermal growth factor and its receptor in the human ovary during follicular growth and regression. *Endocrinology* 132, 924–931.

Marx, J.L. (1988) Cloning sheep and cattle embryos. *Science* 239, 463–464.

Massip, A., Van Der Zwalmen, P., Hanzen, C. and Ectors, F. (1982) Fast freezing of cow embryos in French straws with an automatic program. *Theriogenology* 18, 325–332.

Massip, A., Van Der Zwalmen, P., Scheffen, B. and Ectors, F. (1986) Pregnancies following transfer of cattle embryos preserved by vitrification. *Cryo-Letters* 7, 270–273.

Massip, A., Van Der Zwalmen, P. and Ectors, F. (1987) Recent progress in cryopreservation of cattle embryos. *Theriogenology* 27, 69–79.

Massip, A., Van Der Zwalmen, P., Scheffen, B. and Ectors, F. (1989) Some significant steps in the cryopreservation of mammalian embryos with a note on a vitrification procedure. *Animal Reproduction Science* 19, 117–129.

Massip, A., Mermillod, P., Wills, C. and Dessy, F. (1993a) Effects of dilution procedure and culture conditions after thawing on survival of frozen bovine blastocysts produced *in vitro*. *Journal of Reproduction and Fertility* 97, 65–69.

Massip, A., Van Langendonckt, A., Mermillod, P., Wills, C., Grisart, B. and Dessy, F. (1993b) Viability of bovine blastocysts produced by totally *in vitro* technique in serum-free oviduct conditioned media and transferred immediately or after cryopreservation. *Proceedings of the Ninth Congress of the European Embryo Transfer Association* (Lyon), 240.

Masuda, H. (1989) Electrophoretic separation of X- and Y-sperm. *Japanese Journal of Animal Reproduction* 35(5), 33–37.

Masuda, H., Kawakura, K., Takahashi, M., Sekino, M., Iwasaki, S. (1988) An attempt for separation of bovine X- and Y-bearing sperm by Percoll density gradient centrifugation. *Bulletin of the National Institute of the Animal Industry (Japan)* Summaries, No. 48, 4.

Mathialagan, N. and Roberts, R.M. (1991) Interleukin-6 expression in porcine, bovine and ovine preimplantation conceptuses. *Serono Symposium on Preimplantation Embryo Development*, Abstract Vol., 23 (Abs. II–26).

Matsuoka, K., Sakata, S., Ichino, K., ShiMaya, Y. and Suzuki, T. (1992) Effect of superovulated cow serum for culture of bovine oocytes to the blastocyst stage. *Theriogenology* 37, 254.

Matsuyama, K., Miyakoshi, H. and Fukui, Y. (1993) Effect of glucose levels during the *in vitro* culture in synthetic oviduct fluid medium on *in vitro* development of bovine oocytes matured and fertilized *in vitro*. *Theriogenology* 40, 595–605.

Mattioli, M. (1989) The role of follicular cells in maturation, fertilizability and developmental competence of pig oocytes. *Proceedings of a Symposium on In Vitro Approaches to Mammalian Gamete Maturation and Embryonic Development* (Milan), 19–28.

Mattioli, M. (1992) Biology of oocyte maturation. In: Lauria, A. and Gandolfi, F. (eds) *Embryonic Development and Manipulation in Animal Production*. Portland Press, London, pp. 17–28.

Mattioli, M., Galeati, G. and Seren, E. (1988a) Effect of follicle somatic cells during pig oocyte maturation on egg penetratability and male pronucleus formation. *Gamete Research* 20, 177–183.

Mattioli, M., Galeati, G., Bacci, M.L. and Seren, E. (1988b) Follicular factors influence oocyte fertilizability by modulating the intercellular cooperation between cumulus cells and oocyte. *Gamete Research* 21, 223–232.

Mattioli, M., Bacci, M.L., Galeati, G. and Seren, E. (1989) Developmental competence of pig oocytes matured and fertilized *in vitro*. *Theriogenology* 31, 1201–1207.

Mattioli, M., Barboni, B., Bacci, M.L. and Seren, E. (1990) Maturation of pig oocytes: observations on membrane potential. *Biology of Reproduction* 43, 318–322.

Mattioli, M., Bacci, M.L., Galeati, G. and Seren, E. (1991a) Effects of LH and FSH on the maturation of pig oocytes *in vitro*. *Theriogenology* 36, 95–105.

Mattioli, M., Galeati, G., Bacci, M.L. and Barboni, B. (1991b) Changes in maturation-promoting activity in the cytoplasm of pig oocytes throughout maturation. *Molecular Reproduction and Development* 30, 119–125.

Mattioli, M., Galeati, G., Barboni, B., Cappannari, C. and Seren, E. (1992) cAMP in pig oocytes: effects on meiotic resumption and intracellular levels throughout maturation *in vivo* and *in vitro*. *Journal of Reproduction and Fertility*, Abstract Series No. 9, 54 (Abs. 93).

Matton, P., Adelakoun, V., Couture, Y. and Dufour, J.J. (1981) Growth and replacement of the bovine ovarian follicles during the estrous cycle. *Journal of Animal Science* 52, 813–820.

Matzuk, M.M., Finegold, M.J., Su, J.-G.J., Hsueh, A.J.W. and Bradley, A. (1992) Alpha-inhibin is a tumour-suppressor gene with gonadal specificity in mice. *Nature* 360, 313–319.

Maurasse, C., Matton, P. and Dufour, J.J. (1985) Ovarian follicular populations at two stages of an estrous cycle in heifers given high energy diets. *Journal of Animal Science* 61, 1194–1200.

Maurel, M.C. (1991) Development of an ELISA kit for the determination of LH on farm. *Proceedings of the Seventh Meeting of the European Embryo Transfer Association* (Cambridge), 176.

Maurer, R.R. (1976) Storage of mammalian oocytes and embryos. *Canadian Journal of Animal Science* 51, 131–145.

May, J.V., Bridge, A.J. and Gangrade, B.K. (1991) Synergistic actions of epidermal growth factor (EGF) and platelet-derived growth factor (PDGF) upon thecal cell proliferation *in vitro*. *Biology of Reproduction* 44 (Suppl. 1), 84 (Abs. 128).

Mayne, C.S. and McCaughey, W.J. (1990a) An evaluation of the effects of twinning in dairy cows on animal production. *Proceedings of the British Society of Animal Production* (Winter Meeting), Paper 3.

Mayne, C.S. and McCaughey, W.J. (1990b) An evaluation of the effects of twinning in dairy cows on animal production. *Animal Production* 50, 545.

Mayne, C.S. and McEvoy, J.D. (1993a) IVF embryos – implications for the dairy herd. *Veterinary Annual* 33, 75–83.

Mayne, C.S. and McEvoy, J. (1993b) The effects of twinning by embryo transfer on dairy cow production and fertility and interaction with feeding level. *Sixty-fifth Annual Report of the Agricultural Research Institute of Northern Ireland (1991–92)*, 15 (Abs.).

Mayne, C.S., McCaughey, W.J. and McEvoy, J. (1991) Practical implications of embryo transfer on dairy herd management. *Proceedings of the British Cattle Veterinary Association* (Reading), 39–46.

Mayr, B., Schellander, K., Schleger, W. and Fuhrer, F. (1986) Bovine oocytes: an *in vitro* maturation and IVF study. *Proceedings of a Congress on Future Aspects in Human In Vitro Fertilization* (Vienna). *Journal of In Vitro Fertilization and Embryo Transfer* 3, 76–77.

Mazur, P. (1988) Stopping biological time: the freezing of living cells. *Annals of the New York Academy of Sciences* 541, 514–531.

Mebus, C.A. and Singh, E.L. (1991) Embryo transfer as a means of controlling the transmission of viral infections. XIII. Failure to transmit foot-and-mouth disease virus through the transfer of embryos from viremic donors. *Theriogenology* 35, 435–441.

Mee, J.F. (1991) Factors affecting the spontaneous twinning rate and the effect of twinning on calving problems in nine Irish dairy herds. *Irish Veterinary Journal* 44, 14–20.

Mehmood, A., Gordon, I., Monaghan, P., Lonergan, P., Wahid, H., Carolan, C. and Gallagher, M. (1992) Comparison of oocyte recovery procedures in sheep. *Proceedings of the Irish Grassland and Animal Production Association* (Eighteenth Annual Research Meeting), 85–86.

Meinecke, B. and Meinecke-Tillman, S. (1990) Assessment of the embryo quality in cattle and small ruminants. *Proceedings of the Sixth Conference of the European Embryo Transfer Association* (Lyon), 53–63.

Meinecke, B. and Meinecke-Tillmann, S. (1993) Effects of alpha-amanitin on nuclear maturation of porcine oocytes *in vitro*. *Journal of Reproduction and Fertility* 98, 195–201.

Meinecke-Tillmann, S. and Meinecke, B. (1991) Experiments on the establishment in culture of pluripotential cell lines from sheep, goat and pig embryos. *Proceedings of the Seventh Meeting of the European Embryo Transfer Association* (Cambridge), 178.

Meintjes, M., Bellow, M.S., Broussard, J.R., Paul, J.B. and Godke, R.A. (1993) Transvaginal aspiration of bovine oocytes from hormone-treated pregnant beef cattle for IVF. *Theriogenology* 39, 266.

Meintjes, M., Bellow, M.S., Broussard, J.R., Paccamonti, D., Eilts, B.E. and Godke, R.A. (1994) Repeated transvaginal ultrasound-guided oocyte retrieval from pregnant mares. *Theriogenology* 41, 255.

Meizel, S. and Turner, K.O. (1993) Effects of polyamine biosynthesis inhibitors on the progesterone-initiated increase in intracellular free Ca^{2+} and acrosome reactions in human sperm. *Molecular Reproduction and Development* 34, 457–465.

Menezo, Y.J.R., Guerin, J.F. and Czyba, J.-C. (1990) Improvement of human early embryo development *in vitro* by co-culture on monolayers of Vero cells. *Biology of Reproduction* 42, 301–306.

Menezo, Y., Hazout, A., Dumont, M., Herbaut, N. and Nicollet, B. (1992) Culture of embryos on Vero cells and transfer of blastocysts in humans. *Human Reproduction* 7, 101–106.

Menino, A.R., Williams, J.S., Gardiner, C.S. and John, T.L. (1989) Mouse embryo development in media containing growth factors. *Animal Reproduction Science* 18, 125–137.

Mermillod, P., Boccart, C. and Dessy, F. (1991) Use of rabbit or bovine oviduct epithelial cell monolayers for supporting early development of bovine embryos. *Theriogenology* 35, 241.

Mermillod, P., Boccart, C., Wils, C., Massip, A. and Dessy, F. (1992a) Effect of oviduct-conditioned medium and of cumulus cells on bovine embryo development *in vitro*. *Theriogenology* 37, 256.

Mermillod, P., Wils, C., Massip, A. and Dessy, F. (1992b) Oocyte collection and *in vitro* blastocyst production from individual slaughtered cows. *Proceedings of the Twelfth International Congress on Animal Reproduction* (The Hague) 3, 1321–1323.

Mermillod, P., Wils, C., Massip, A. and Dessy, F. (1992c) *In vitro* survival of frozen–thawed *in vitro* produced cattle blastocysts. *Proceedings of the Eighth Conference of the European Embryo Transfer Association* (Lyon), 194.

Mermillod, P., Wils, C., Massip, A. and Dessy, F. (1992d) Collection of oocytes and production of blastocysts *in vitro* from individual, slaughtered cows. *Journal of Reproduction and Fertility* 96, 717–723.

Mermillod, P., Massip, A. and Dessy, F. (1992e) *In vitro* production of cattle embryos: review and Belgian results. *International Journal of Developmental Biology* 36, 185–195.

Mermillod, P., Mourmeaux, J.L., Wils, C., Massip, A. and Dessy, F. (1992f) Protein-free oviduct conditioned medium for complete bovine embryo development. *Veterinary Record* 130, 13.

Mermillod, P., Vansteenbrugge, A., Wils, C., Massip, A. and Dessy, F. (1993a) Study of the bovine embryotrophic activity of serum free oviduct conditioned medium. *Theriogenology* 39, 267.

Mermillod, P., Vansteenbrugge, A., Wils, C., Mourmeaux, J.-L., Massip, A. and Dessy, F. (1993b) Characterization of the embryotrophic activity of exogenous protein-free oviduct-conditioned medium used in culture of cattle embryos. *Biology of Reproduction* 49, 582–587.

Merriman, J.A., Carroll, J. and Whittingham, D.G. (1993) Maturation and fertilization of oocytes grown *in vitro* and *in vivo*. *Journal of Reproduction and Fertility*, Abstract Series No. 11, 6 (Abs. 1).

Metezeau, P., Cotinot, C., Colas, G., Azoulay, M., Kiefer, H., Goldberg, M.E. and Kirszenbaum, M. (1991) Improvement of flow cytometry analysis and sorting of bull spermatozoa by optical monitoring of cell orientation as evaluated by DNA specific probing. *Molecular Reproduction and Development* 30, 250–257.

Mhawi, A.J., Kanka, J. and Motlik, J. (1991) Follicle and oocyte growth in early post-natal calves: cytochemical, autoradiographical and electron microscopical studies. *Reproduction, Nutrition and Development* 31, 115–126.

Miller, B.J., Graves, C.N. and Lodge, J.R. (1988) Effect of *in vitro* storage of bovine spermatozoa on acrosomal integrity, proteolytic activity, binding and initiation of penetration of oocytes. *Gamete Research* 20, 53–65.

Miller, D.J. and Ax, R.L. (1986) Seminal plasma inhibits the heparin-induced acrosome reaction in bovine sperm. *Biology of Reproduction* 34 (Suppl. 1), 352 (Abs.).

Miller, D.J. and Ax, R.L. (1988) An improved assay for measuring heparin binding to bull sperm. *Journal of Dairy Science* 71, 239–244.

Miller, D.J. and Ax, R.L. (1989a) Chemical *N*-desulfation of heparin negates its ability to capacitate bovine spermatozoa. *Gamete Research* 23, 451–465.

Miller, D.J. and Ax, R.L. (1989b) Heparin-binding proteins from seminal plasma potentiate heparin-induced capacitation of bovine spermatozoa *in vitro*. *Biology of Reproduction* 40 (Suppl. 1), 148 (Abs. 302).

Miller, D.J. and Ax, R.L. (1990) Carbohydrates and fertilization in animals. *Molecular Reproduction and Development* 26, 184–198.

Miller, D.J. and Hunter, A.G. (1987) Individual variation for *in vitro* fertilization success in dairy bulls. *Journal of Dairy Science* 70, 2150–2153.

Miller, D.J., Bellin, M.E., First, N.L. and Ax, R.L. (1987) Competition of follicular fluid glycosaminoglycans and desulfated heparin for heparin binding sites on bull sperm. *Journal of Animal Science* 65 (Suppl. 1), 363 (Abs.).

Miller, D.J., Macek, M.B. and Shur, B.D. (1992) Complementarity between sperm surface β-1,4-galactosyltransferase and egg coat ZP3 mediates sperm–egg binding. *Nature* 357, 589–593.

Miller, G.F., Lester, T.D., Gliedt, D.W., Rakes, J.M. and Rorie, R.W. (1991) Addition of trypsin inhibitor to hyaluronidase preparations used for oocyte cumulus cell removal prior to *in vitro* fertilization improves the subsequent embryo cleavage rate. *Journal of Animal Science* 69 (Suppl. 1), 47.

Miller, G.F., Gliedt, D.L., Lester, T.D., Pierson, J.N., Rakes, J.M. and Rorie, R.W. (1992) Addition of bovine oviductal epilelial cells (BOEC) and/or penicillamine, hypotaurine and epinephrine (PHE) to bovine *in vitro* fertilization (IVF) medium increases the subsequent embryo cleavage rate. *Theriogenology* 37, 259.

Miller, G.F., Gliedt, D.W., Rakes, J.M. and Rorie, R.W. (1994) Addition of penicillamine, hypotaurine and epinephrine (PHE) or bovine oviductal epithelial cells (BOEC) alone or in combination to bovine *in vitro* fertilization medium increases the subsequent embryo cleavage rate. *Theriogenology* 41, 689–696.

Miller, J.R. (1991) Embryo sexing – a broad survey. *Reproduction in Domestic Animals* 26, 54–57.

Miller, K.F. and Pursel, V.G. (1987) Absorption of compounds in medium by the oil covering microdrop cultures. *Gamete Research* 17, 57–61.

Miller, K.F., Bolt, D.J., Pursel, V.G., Hammer, R.E., Pinkert, C.A., Palmiter, R.D. and Brinster, R.L. (1986) Expression of human or bovine growth hormone gene with a mouse metallothionein-1 promoter in transgenic swine alters the secretion of porcine growth hormone and insulin-like growth factor-I. *Journal of Endocrinology* 120, 481–488.

Miller, K.F., Xie, S. and Pope, W.F. (1991) Immunoreactive inhibin in follicular fluid is related to meiotic stage of the oocyte during final maturation of the porcine follicle. *Molecular Reproduction and Development* 34, 35–39.

Miller, N.G.A., Howes, E.A., Cran, D.G. and Johnson, L.A. (1993) Studies on sorted X and Y enriched populations of bull spermatozoa. *British Society of Animal Production* (Winter Meeting), Paper 156 (2 pp.).

Minami, N., Bavister, B.D. and Iritani, A. (1988) Development of hamster two-cell embryos in the isolated mouse oviduct in organ culture system. *Gamete Research* 19, 235–240.

Minamihashi, A., Watson, A.J., Watson, P.H., Church, R.B. and Schultz, G.A. (1993) Bovine parthenogenetic blastocysts following *in vitro* maturation and oocyte activation with ethanol. *Theriogenology* 40, 63–76.

Minhas, B.S., Capehart, J.S., Bowen, M.J., Womack, J.E., McCrady, J.D., Harms, P.G., Wagner, T.E. and Kraemer, D.C. (1984) Visualization of pronuclei in living bovine zygotes. *Biology of Reproduction* 30, 687–691.

Minhas, B.S., Roudebush, W.E., Ricker, D.B. and Dodson, M.G. (1991) Treatment of sperm with high-ionic strength medium increases microsurgical fertilization rates of rabbit oocytes fertilized by subzonal placement of sperm. *Journal of In Vitro Fertilization and Embryo Transfer* 8, 111–115.

Minhas, B.S., Kim, H.N., Zhu, Y.P., Ripps, B.A. and Buster, J.E. (1993) Platelet activating factor treatment of spermatozoa enhances fertilization rates of rabbit oocytes utilizing subzonal insertion of sperm. *Theriogenology* 39, 269.

Minoia, P., Leopold, A., Dell-Aquila, M.E., Lacalandra, G.M., Fusco, S. and Maritato, F. (1991) Pregnancy in cattle with *in vitro*-produced embryos transported over a long distance. *Atti della Societa Italiana di Buiatria* 23, 133–138.

Misener, M., Pollard, J.W. and Metzger, K. (1991) *In vitro* culture of porcine embryos in CZB medium. *Theriogenology* 35, 244.

Misra, A.K., Joshi, B.V., Agrawala, P.L., Kasiraj, R., Sivaiah, S., Rangareddi, N.S. and Siddiqui, M.U. (1990) Multiple ovulation and embryo transfer in Indian buffalo (*Bubalus bubalis*). *Theriogenology* 33, 1131–1141.

Mitani, T., Utsumi, K. and Iritani, A. (1993) Developmental ability of enucleated bovine oocytes matured *in vitro* after fusion with single blastomeres of eight-cell embryos matured and fertilized *in vitro*. *Molecular Reproduction and Development* 34, 314–322.

Mittwoch, U. (1993) Blastocysts prepare for the race to be male. *Human Reproduction* 8, 1550–1555.

Miyake, M., Hosoi, Y., Utsumi, K., Kato, S. and Iritani, A. (1991) Effect of bovine serum albumin on *in vitro* fertilization in rats. *Science Reports of Faculty of Agriculture, Kobe University* 19, 149–156.

Miyano, T., Naohara, K., Kato, S. and Kanda, S. (1990a) Sex difference in X-linked glucose 6-phosphate dehydrogenase activity of mouse embryos. *Japanese Journal of Animal Reproduction* 36, 209–212.

Miyano, T., Yoshikawa, K., Kato, S., Harayama, H., Nanjo, I. and Kanda, S. (1990b) *In vitro* fertilization of *in vitro* and *in vivo* matured oocytes from prepubertal Meishan gilts. *Japanese Journal of Zootechnical Science* 61(11), 1011–1016.

Mizoshita, K., Brandes, T.L. and Youngs, C.R. (1993). Effect of antibiotic and heat inactivation of serum on development of bovine oocytes matured, fertilized and cultured *in vitro*. *Journal of Assisted Reproduction and Genetics* 10 (Suppl.), 229.

Mizuno, J., Nakada, K., Endo, K., Kobayashi, K., Nishikata, Y. and Miyazaki, S. (1994) Changes of intracellular calcium concentration at *in vitro* fertilization of bovine oocytes. *Theriogenology* 41, 259.

Mlodawska, W., Okolski, A. and Lietz, W. (1991) Recovery of equine oocytes, their developmental stage in relationship to ovarian size, and their maturation *in vitro*. *Acta Agraria et Silvestria, Series Zootechnica* 29, 59–65.

Mochizuki, H., Fukui, Y. and Ono, H. (1991a) Effect of number of granulosa cells added to culture medium for *in vitro* maturation, fertilization and development of bovine oocytes. *Theriogenology* 36, 973–986.

Mochizuki, H., Fukui, Y., Oosaki, K. and Ono, H. (1991b) Effect of granulosa cells added to culture medium for *in vitro* maturation and fertilization of bovine oocytes. *Japanese Journal of Animal Reproduction* 37, 263–271.

Modina, S., Brevini, T.A.L., Galli, C., Motta, M. and Gandolfi, F. (1989) Immunolocalization of SOP 92, a glycoprotein secreted by the sheep oviduct epithelium. *Journal of Reproduction and Fertility*, Abstract Series No. 4, Abs. 38.

Modlinski, J.A. and Smorag, Z. (1991) Preimplantation development of rabbit embryos after transfer of embryonic nuclei into different cytoplasmic environment. *Molecular Reproduction and Development* 28, 361–372.

Moffat, A.S. (1991) Transgenic animals may be down on the pharm. *Science* 245, 35–36.

Mogas, T., Martino, A., Palomo, M.J. and Paramio, M.T. (1992) Effect of method of recovery on the number and type of oocytes obtained for IVM. *Journal of Reproduction and Fertility* Abstract Series No. 9, 52 (Abs. 90).

Mogas, T., Palomo, M.J., Izquierdo, D., Martino, A. and Paramio, M.T. (1993) Effect of hormones on the *in vitro* maturation, fertilization and early cleavage of prepubertal goat oocytes. *Journal of Reproduction and Fertility*, Abstract Series No. 11, 67 (Abs. 124).

Mohr, L.R. and Trounson, A.O. (1981) Structural changes associated with freezing of bovine embryos. *Biology of Reproduction* 25, 1009–1025.

Monaghan, P. (1993) Studies related to the laboratory production of bovine embryos. PhD Thesis, National University of Ireland, Dublin.

Monaghan, P., Carolan, C., Lonergan, P., Sharif, H., Wahid, H. and Gordon, I. (1993) The effect of maturation time on the subsequent *in vitro* development of bovine oocytes. *Theriogenology* 39, 270.

Monget, P., Monniaux, D., Pisselet, C. and Durand, P. (1993) Changes in insulin-like growth factor-I (IGF-I), IGF-II, and their binding proteins during growth and atresia of ovine ovarian follicles. *Endocrinology* 132, 1438–1446.

Monis, H. and Bavister, B.D. (1990a) Development of four-cell hamster embryos to the blastocyst stage *in vitro* and its regulation by components of the culture milieu. *Reproduction, Fertility and Development* 2, 1–9.

Monis, H. and Bavister, B.D. (1990b) Analysis of the inhibitory effect of inorganic phosphate on development of four-cell hamster embryos *in vitro*. *Journal of Experimental Zoology* 256, 75–83.

Monk, M. (1987) Biochemical microassays for X-chromosome-linked enzymes HPRT and PGK. In: Monk, M. (ed.) *Mammalian Development – a Practical Approach* IRL Press, Oxford, pp. 139–161.

Monk, M. and Handyside, A.H. (1988) Sexing of preimplantation mouse embryos by measurement of X-linked gene dosage in a single blastomere. *Journal of Reproduction and Fertility* 82, 365–368.

Monniaux, D., Chupin, D. and Saumande, J. (1983) Superovulatory responses of cattle. *Theriogenology* 19, 55–81.

Monniaux, D., Monget, P., Mariana, J.C., Nicolle, A., Pisselet, C., Cognie, Y., Poulin, N., Fontaine, J. and Durand, P. (1993) Functional changes throughout follicular atresia in the ewe. *Reproduction, Nutrition and Development* 33, 82–83.

Monson, R., Northey, D.L., Gottfredson, R., Peschel, D.R., Rutledge, J.J. and Schaefer, D.M. (1992) Pregnancy rates of *in vitro* produced bovine embryos following non-surgical transfer. *Theriogenology* 37, 261.

Moodie, G.R. and Graham, E.F. (1989) The effect of incubating sheep ovaries for various times and temperatures on oocyte maturation rates *in vitro*. *Biology of Reproduction* 40 (Suppl. 1), 53 (Abs. 18).

Moodie, G.R., Hunter, A.G. and Graham, E.F. (1992) Studies on the cryopreservation of sheep germinal vesicle oocytes: effects of cryo-solution composition, concentration, exposure time and cumulus cells on subsequent *in vitro* maturation. *Biology of Reproduction* 46 (Suppl. 1), 64 (Abs. 56).

Moohan, J.M. and Lindsay, K.S. (1993) Human sperm selected by discontinuous Percoll density gradient exhibit enhanced motion characteristics. *Journal of Reproduction and Fertility*, Abstract Series No. 12, 53.

Moohan, J.M., Winston, R.M.L. and Lindsay, K.S. (1993) Variability of human sperm response to immediate and prolonged exposure to pentoxifylline. *Human Reproduction* 8, 1696–1700.

Moor, R.M. (1978) Role of steroids in the maturation of ovine oocytes. *Annales de Biologie animale Biochimie Biophysique* 18, 477–482.

Moor, R.M. (1988) Regulation of the meiotic cycle in oocytes of domestic mammals. *Annals of the New York Academy of Science* 541, 248–258.

Moor, R.M. and Crosby, I.M. (1985) Temperature-induced abnormalities in sheep oocytes during maturation. *Journal of Reproduction and Fertility* 75, 467–473.

Moor, R.M. and Gandolfi, F. (1987) Molecular and cellular changes associated with maturation and early development of sheep eggs. *Journal of Reproduction and Fertility*, Suppl. 34, 55–69.

Moor, R.M. and Seamark, R.F. (1986) Cell signaling, permeability and microvasculatory changes during antral follicle development in mammals. *Journal of Dairy Science* 69, 927–943.

Moor, R.M. and Trounson, A.O. (1977) Hormonal and follicular factors affecting maturation of sheep oocytes *in vitro* and their subsequent developmental capacity. *Journal of Reproduction and Fertility* 49, 101–109.

Moor, R.M. and Warnes, G.M. (1978) Effect of oocyte maturation in mammals. In: Crighton, D.B., Foxcroft, G.R., Haynes, N.B. and Lamming, G.E. (eds) *Control of Ovulation*. Butterworths, London, pp. 159–176.

Moor, R.M., Hay, M.F., Dott, H.M. and Cran, D.G. (1978) Macroscopic identification and steroidogenoic function of atretic follicles in sheep. *Journal of Endocrinology* 77, 309–318.

Moor, R.M., Kruip, Th.A.M. and Green, D. (1984) Intraovarian control of folliculogenesis: limits to superovulation. *Theriogenology* 21, 103–116.

Moor, R.M., Mattioli, M., Ding, J. and Nagai, T. (1990) Maturation of pig oocytes *in vivo* and *in vitro*. *Journal of Reproduction and Fertility*, Suppl. 40, 197–210.

Moor, R.M., Sun, F.Z. and Galli, C. (1992a) Reconstruction of ungulate embryos by nuclear transplantation. *Animal Reproduction Science* 28, 423–431.

Moor, R., Hutchings, A., Hawkins, C. and Jung, T. (1992b) Towards a molecular understanding of meiosis in domestic animals. In: Laura, A. and Gandolfi, F. (eds), *Embryonic Development and Manipulation in Animal Production*. Portland Press, London, pp. 1–15.

Moore, A.S. and Spahr, S.L. (1991) Activity monitor and an enzyme immunoassay for milk progesterone to aid in the detection of estrus. *Journal of Dairy Science* 74, 3857–3862.

Moore, K. and Bondioli, K.R. (1990) Altering metabolic substrates of culture media enhances bovine IVM/IVF embryo development beyond the eight cell block. *Biology of Reproduction* 42 (Suppl. 1), 55.

Moore, K. and Bondioli, K.R. (1993) Glycine and alanine supplementation of culture medium enhances development of *in vitro* matured and fertilized cattle embryos. *Biology of Reproduction* 48, 833–840.

Morbeck, D.E., Esbenshade, K.L., Flowers, W.L. and Britt, J.H. (1991) Morphological characteristics and growth rate of preantral and antral follicles in the prepuberal gilt. *Biology of Reproduction* 44 (Suppl. 1), 152 (Abs. 399).

Mordel, N., Ohad, S., Zentner, B., Schenker, J.G., Gordon, J. and Laufer, N. (1992) Enhancing *in vitro* fertilization of mouse oocytes by partial zona pellucida digestion. *Journal of Assisted Reproduction and Genetics* 9, 128–132.

Moreno, J.F. and Westhusin, M. (1993) A comparison of two systems for culture of bovine zygotes *in vitro*. *Biology of Reproduction* 48 (Suppl. 1), 169.

Moreno, J.F., Turczynski, C.J., Flores-Foxworth, G. and Kraemer, D.C. (1992) Influence of parity of donor and presence of a CL on quantity and quality of bovine oocytes from ovarian follicles aspirated post-mortem. *Biology of Reproduction* 46 (Suppl. 1), 65 (Abs. 60).

Moreno, J.F., Flores-Foxworth, G., Westhusin, M. and Kraemer, D.C. (1993) Influence of pregnancy and presence of a CL on quantity and quality of bovine oocytes obtained from ovarian follicles aspirated post-mortem. *Theriogenology* 39, 271.

Morgan, P.M. and Kane, M.T. (1993) Protein content of rabbit embryos: one cell to peri-implantation blastocyst. *Journal of Reproduction and Fertility* 97, 101–106.

Mori, K., Daitoh, T., Kmada, M., Maeda, N., Maegawa, M., Hirano, K., Irahara, M. and Aono, T. (1993) Blocking of human fertilization by carbohydrates. *Human Reproduction* 8, 1729–1732.

Morrell, J.M. (1991) Applications of flow cytometry to artificial insemination: a review. *Veterinary Record* 129, 375–378.

Morrell, J.M. and Dresser, D.W. (1989) Offspring from inseminations with mammalian sperm stained with Hoechst 33342, either with or without flow cytometry. *Mutation Research* 224, 177–183.

Morrell, J.M. and Dresser, D.W. (1992) Preparation of living sperm for DNA analysis by flow cytometry. *Proceedings of the Twelfth International Congress on Animal Reproduction* (The Hague) 1, 496–498.

Morrell, J.M., Keeler, K.D., Noakes, D.E., Mackenzie, N.M. and Dresser, D.W. (1988) Sexing of sperm by flow cytometry. *Veterinary Record* 122, 322–324.

Morris, B.A., Rhind, S.M., Clayton, J., Price, C.R. and Webb, R. (1988) Passive immunisation of cows with ovine testosterone anti-serum to increase ovulation rates. *Proceedings of the Eleventh International Congress on Animal Reproduction and Artificial Insemination* (Dublin) 4, 498 (3 pp.).

Morris, C.A. (1990) Theoretical and realised responses to selection for reproductive rate. *Pro-*

ceedings of the Fourth World Congress on Genetics Applied to Livestock Production 16, 309–318.

Morris, C.A. (1991) Screening herds for cows with a history of twin calving. *Proceedings of the New Zealand Society of Animal Production* 51, 447–451.

Morris, C.A., Day, A.M. and Macmillan, K.L. (1988) Studies of ovulation rate, embryo survival and calving rate in a twin breeding experiment. *Proceedings of the Eleventh International Congress on Animal Reproduction and Artificial Insemination* (Dublin) 4, 484 (3 pp.).

Morris, C.A., Day, A.M., Amyes, N.C. and Hurford, A.P. (1992) Ovulation and calving data from a herd selected for twin calving. *New Zealand Journal of Agricultural Reseacrh* 35, 379–391.

Morris, D.G., Diskin, M.G. and Sreenan, J.M. (1992a) Immunization against inhibin increases ovulation and twin-calving rate in cows. *Proceedings of the Irish Grassland and Animal Production Association* (Eighteenth Annual Research Meeting), 29–30.

Morris, D.G., Diskin, M.G. and Sreenan, J.M. (1992b) Increased ovulation and twin-calving rates in heifers immunized against bovine inhibin peptides. *Proceedings of the British Society of Animal Production* (Winter Meeting), Paper No. 43.

Morris, D.G., Diskin, M.G. and Sreenan, J.M. (1992c) Immunization of heifers against inhibin peptides increases ovulation rate and twin-calving rate. *Proceedings of the Twelfth International Congress on Animal Reproduction* (The Hague) 3, 1216–1218.

Morris, D.G., McDermott, M.G., Diskin, M.G., Morrison, C.A., Swift, P.J. and Sreenan, J.M. (1993) Effect of immunization against synthetic peptide sequences of bovine inhibin α-subunit on ovulation rate and twin-calving rate in heifers. *Journal of Reproduction and Fertility* 97, 255–261.

Morstin, J. and Katska, L. (1986) Effect of temperature on the ultrastructure of bovine oocytes in culture. *Animal Reproduction Science* 12, 13–19.

Mortimer, R.G., Seidel, G.E., Amann, R.P., Grotelueschen, D.M. and Olson, S.E. (1991) Use of *in vitro* fertilization to evaluate spermatozoa with proximal droplets in young beef bulls. *Theriogenology* 35, 247.

Motlik, J. (1989) Cytoplasmic aspects of oocyte growth and maturation in mammals. *Journal of Reproduction and Fertility*, Suppl. 38, 17–25.

Motlik, J. and Fulka, J. (1986) Factors affecting meiotic competence in pig oocytes. *Theriogenology* 25, 87–96.

Motlik, J. and Kubelka, M. (1990) Cell-cycle aspects of growth and maturation of mammalian oocytes. *Molecular Reproduction and Development* 27, 366–375.

Motlik, J., Koefoed-Johnsen, H.H. and Fulka, J. (1985) Breakdown of the germinal vesicle in bovine oocytes cultivated *in vitro*. *Journal of Experimental Zoology* 205, 377–384.

Motlik, J., Fulka, J. and Flechon, J.E. (1986) Changes in intercellular coupling between pig oocytes and cumulus cells during maturation *in vivo* and *in vitro*. *Journal of Reproduction and Fertility* 76, 31–37.

Motlik, J., Byonglye, E. and Shioya, Y. (1990) Two sensitivity levels of cattle oocytes to puromycin. *Biology of Reproduction* 43, 994–998.

Motlik, J., Kalous, J., Sutovsky, P. and Rimkevicova, Z. (1991) Pig membrana granulosa cells prevent resumption of meoisis in cattle and mouse oocytes. *Serono Symposium on Preimplantation Embryo Development*, Abstract Vol. 40 (Abs. I-32).

Mukhopadhyay, A.K., Holstein, K., Szkudlinski, M., Brunswig-Spickenheier, B. and Leidenberger, F.A. (1991) The relationship between prorenin levels in follicular fluid and follicular atresia in bovine ovaries. *Endocrinology* 129, 2367–2375.

Mulheron, G.W. and Schomberg, D.W. (1990) Rat granulosa cells express transforming growth factor-β type 2 messenger ribonucleic acid which is regulatable by follicle-stimulating hormone *in vitro*. *Endocrinology* 126, 1777–1779.

Muller, M. (1990) State of the art, limitations and prospective of gene transfer in domestic animals. *Proceedings of the Sixth Conference of the European Embryo Transfer Association* (Lyon), 97–103.

Munar, C.J. and Hasler, J.F. (1989) Results of the first frozen bovine embryos exported from the USA to Argentina. *Theriogenology* 31, 230.

Munkittrick, T.W., Nebel, R.L. and Sacke, R.G. (1992) Accessory sperm numbers for cattle inseminated with protamine sulfate microcapsules. *Journal of Dairy Science* 75, 725–731.

Munne, S., Lee, A., Rosenwaks, Z., Grifo, J. and Cohen, J. (1993) Diagnosis of major chromosome aneuploidies in human preimplantation embryos. *Human Reproduction* 8, 2185–2191.

Munro, R.K. and Bertram, J. (1990) Progesterone administration after insemination did not affect the fertility of cattle following a controlled breeding program. *Australian Journal of Experimental Agriculture* 30, 179–181.

Munson, L., Ellington, J.E. and Schlafer, D.H. (1991) Bovine trophoblastic cell vesicle attachment to polarized endometrial epithelial cells *in vitro*. *In Vitro Cellular and Developmental Biology* 27A(1), 31–38.

Munson, L.E., Wilkinson, J.E. and Bechtel, M. (1992) Transforming growth factor-β (TGF-β) and platelet-derived growth factor (PDGF) are synergistic mitogens for bovine trophoblastic and endometrial epithelial cells. *Biology of Reproduction* 46 (Suppl. 1), 67 (Abs. 66).

Murase, T., Okuda, K. and Niwa, K. (1992) Effects of estrous bovine cervical mucus on capacitation and the acrosome reaction of frozen–thawed spermatozoa *in vitro*. *Journal of Reproduction and Development* 38, 279–284.

Murdoch, W.J. (1992) Comparative morphometry and steroidogenic function of antral ovine follicles destined for ovulation or atresia. *Domestic Animal Endocrinology* 9, 219–224.

Murray, A.W. (1992) Creative blocks: cell-cycle check points and feedback controls. *Nature* 359, 599–604.

Murray, M.K. (1990) The *in vitro* synthesis of proteins in the sheep oviduct. *Biology of Reproduction* 42 (Suppl. 1), 74.

Murray, R.D. and Ward, W.R. (1993) Welfare implications of modern artificial breeding techniques for dairy cattle and sheep. *Veterinary Record* 133, 283–286.

Musse, P. (1990) Effect of highly purified gonadotrophins and serum from oestrous sows on the *in vitro* cumulus expansion and nuclear maturation of porcine oocytes. Thesis, Tierarztliche Hochschule Hannover, Germany, 115 pp.

Myers, M.W., Zhang, L., Stroud, B.K. and Godke, R.A. (1992) Development of *in vitro* matured bovine oocytes fertilized either *in vitro* or *in vivo*. *Proceedings of the Twelfth International Congress on Animal Reproduction* (The Hague) 2, 668–670.

Mylne, J., McKelvey, B., Broadbent, P. and MacMillan, D. (1992) Salvage of bovine oocytes. *Veterinary Record* 130, 59.

Nadir, S., Saacke, R.G., Bame, J., Mullins, J. and Degelos, S. (1992) Effect of semen freezing and sperm dosage on accessory sperm number and embryo quality in artificially inseminated cows. *Journal of Animal Science* 70 (Suppl. 1), 256 (Abs. 473).

Nadir, S., Saacke, R.G., Bame, J., Mullins, J. and Degelos, S. (1993) Effect of freezing semen and dosage of sperm on number of accessory sperm, fertility, and embryo quality in artificially inseminated cattle. *Journal of Animal Science* 71(1), 199–204.

Nagai, T. (1992) Development of bovine *in vitro*-matured follicular oocytes activated with ethanol. *Theriogenology* 37, 869–875.

Nagai, T. (1994) Current status and perspectives in IVM–IVF of porcine oocytes. *Theriogenology* 41, 73–78.

Nagai, T. and Moor, R.M. (1990) Effect of oviduct cells on the incidence of polyspermy in pig eggs fertilized *in vitro*. *Molecular Reproduction and Development* 26, 377–382.

Nagai, T., Ding, J.C. and Moor, R.M. (1988) Factors affecting sperm penetration and pronuclear formation in pig oocytes matured *in vitro*. *Journal of Reproduction and Fertility*, Abstract Series No. 2, Abs. 68.

Nagai, T., Takahashi, T., Shioya, Y. and Oguri, N. (1990) Maturation and fertilization of pig follicular oocytes cultured in pig amniotic fluid. *Theriogenology* 34, 195–204.

Nagao, Y., Saeki, K. and Kainuma, H. (1991) Development of bovine *in vitro* matured and fertilized oocytes co-cultured with oviductal tissue in serum-free medium. *Theriogenology* 35, 249.

Nagao, Y., Saeki, K., Hoshi, M. and Kainuma, H. (1993) Effect of oxygen concentration on the development of *in vitro* matured and fertilized bovine oocytes cultured in a protein-free medium. *Theriogenology* 39, 273.

Nagashima, H., Kato, Y., Yamakawa, H., Matsumoto, T. and Ogawa, S. (1989) Changes in freezing tolerance of pig blastocysts in peri-hatching stage. *Japanese Journal of Animal Reproduction* 35, 130–134.

Nagashima, H., Yamakawa, H. and Niemann, H. (1992) Freezability of porcine blastocysts at different peri-hatching stages. *Theriogenology* 37, 839–850.

Nagashima, H., Kashiwazaki, N., Ashman, R., Grupen, C., Seamark, R.F. and Nottle, M. (1994) Recent advances in cryopreservation of porcine embryos. *Theriogenology* 41, 113–118.

Nagy, H., Barany, I., Papp, D., Koppany, A., Meszaros, J., Holdas, S. and Becze, J. (1986) Induction of twinning in beef cattle. *Proceedings of the Thirty-Seventh Annual Meeting of the European Association for Animal Production* (Budapest) 1, 183 (Abs.).

Nagy, Z.P., Joris, H., Liu, J., Devroey, P. and Van Steirteghem, A.C. (1993a) Sequential nuclear changes in human oocytes at fertilization after intracytoplasmic sperm injection. *Journal of Reproduction and Fertility*, Abstract Series No. 12, 25.

Nagy, Z.P., Joris, H., Liu, J., Staessen, C., Devroey, P. and Van Steirteghem, A.C. (1993b) Intracytoplasmic single sperm injection of 1-day-old unfertilized human oocytes. *Human Reproduction* 8, 2180–2184.

Naitana, S., Loi, P., Ledda, S. and Cappai, P. (1992) Embryo transfer and new technology in sheep reproduction. In: Lauria, A. and Gandolfi, F. (eds) *Embryonic Development and Manipulation in Animal Production*. Portland Press, London, pp. 183–194.

Naito, K. and Toyoda, Y. (1992) Effects of microinjection of glutathione on male pronucleus formation in porcine oocytes matured *in vitro*. *Journal of Reproduction and Development* 38, 173–178.

Naito, K., Fukuda, Y. and Toyoda, Y. (1988) Effects of follicular fluid on male pronucleus formation in porcine oocytes matured *in vitro*. *Gamete Research* 21, 289–295.

Naito, K., Kosaka, M., Fukuda, Y., Ishibashi, I. and Toyoda, Y. (1990) Analysis of the factor(s) present in follicular fluids promoting male pronucleus formation ability of porcine follicular oocytes. *Japanese Journal of Animal Reproduction* 36, 213–218.

Naito, K., Toyoda, Y. and Yanagimachi, R. (1992) Production of normal mice from oocytes fertilized and developed without zonae pellucidae. *Human Reproduction* 7, 281–285.

Nakagata, N. (1989) High survival rate of unfertilized mouse oocytes after vitrification. *Journal of Reproduction and Fertility* 87, 479–483.

Nakagawa, A., Suzuki, T., Suga, T., Inoue, T., Takahashi, Y. and Kanagawa, H. (1991) Morphology and size of embryos recovered from superovulated cows. *Journal of Veterinary Medical Science* 53, 287–290.

Nakahara, T., Taninoki, S., Bozono, M. and Iwasaki, H. (1990) Relation of the position of bovine ova and zona fracture in the straw during freezing and thawing. *Japanese Journal of Animal Reproduction* 36, 184–187.

Nakao, H. and Nakatsuji, N. (1990) Effects of co-culture, medium components and gas phase on *in vitro* culture of *in vitro* matured and *in vitro* fertilized bovine embryos. *Theriogenology* 33, 591–600.

Nakatsuka, M., Yoshida, N. and Kudo, T. (1992) Platelet activating factor in culture media as an indicator of human embryonic development after *in vitro* fertilization. *Human Reproduction* 7, 1435–1439.

Nancarrow, C.D., Marshall, J.T.A., Clarkson, J.L., Murray, J.D., Shanahan, C.M., Wynn, P.C. and Ward, K.A. (1990) Expression and physiology of performance regulating genes in sheep. *Proceedings of the Third International Ruminant Reproduction Symposium* (Nice), Abstract.

Nass, S.J. and Ax, R.L. (1989) Heparin-binding proteins from bovine accessory sex glands bind to cauda epididymal spermatozoa. *Biology of Reproduction* 40 (Suppl. 1), 149 (Abs. 304).

Nasser, L.F., Adams, G.P., Bo, G.A. and Mapletoft, R.J. (1993) Ovarian superstimulatory response relative to follicular wave emergence in heifers. *Theriogenology* 40, 713–724.

Navara, C.S., Sims, M.M. and First, N.L. (1992) Timing of polarization in bovine embryos and developmental potential of polarised blastomeres. *Biology of Reproduction* 46 (Suppl. 1), 71 (Abs. 82).

Nayudu, P.L. and Osborn, S.M. (1992) Factors influencing the rate of preantral and antral growth of mouse ovarian follicles *in vitro*. *Journal of Reproduction and Fertility* 95, 349–362.

Nayudu, P.L., Osborn, S.M. and Gosden, R.G. (1990) *In vitro* preantral and antral growth of isolated mouse ovarian follicles. *Journal of Reproduction and Fertility*, Abstract Series No. 6, Abs. 19.

Nayudu, P.L., Nowshari, M.A. and Hodges, J.K. (1992) *In vitro* follicular development and oocyte maturation in the mouse and common marmoset. *Journal of Reproduction and Fertility*, Abstract Series No. 9, 54 (Abs. 94).

Nebel, R.L. (1988) On-farm milk progesterone tests. *Journal of Dairy Science* 71, 1682–1690.

Nebel, R.L., Altemose, D.L., Munkittrick, T.W., Sprecher, D.J. and McGilliard, M.L. (1989) Comparisons of eight commercial on-farm milk progesterone tests. *Theriogenology* 31, 753–764.

Neev, J., Tadir, Y., Ho, P., Berns, M.W., Asch, R.H. and Ord, T. (1992) Microscope-delivered ultraviolet laser zona dissection: principles and practices. *Journal of Assisted Reproduction and Genetics* 9, 513–523.

Neev, J., Gonzalez, A., Licciardi, F., Alikani, M., Tadir, Y., Berns, M. and Cohen, J. (1993) Opening of the mouse zona pellucida by laser without a micromanipulator. *Human Reproduction* 8, 939–944.

Nelson, C.F. and Nelson, L.D. (1988) Cryopreservation of 7- to 9-day bovine embryos. *Theriogenology* 29, 281.

Nephew, K.P., McClure, K.E. and Pope, W.F. (1989) Embryonic migration relative to maternal recognition of pregnancy in sheep. *Journal of Animal Science* 67, 999–1005.

Nephew, K.P., McClure, K.E., Xie, S., Parrett, N.A. and Pope, W.F. (1990) The role of estradiol in intrauterine migration of ovine embryos. *Journal of Animal Science* 68 (Suppl. 1), 130.

Nephew, K.P., Xie, S., Broermann-Ridder, D.M., McClure, K.E. and Pope, W.F. (1992) Influence of the embryo on intrauterine migration in sheep. *Journal of Animal Science* 70, 1911–1915.

Neugebauer, G. (1992) Effect of the origin (donor/ovaries) of cattle oocytes on the results of *in vitro* fertilization. Thesis, Tierarztliche Hochschule, Hannover, Germany, 85 pp.

Newcomb, R., Christie, W.B. and Rowson, L.E.A. (1978) Birth of calves after *in vivo* fertilization of oocytes removed from follicles and matured *in vitro*. *Veterinary Record* 102, 461–462.

Newcomb, R., Christie, W.B. and Rowson, L.E.A. (1980) Fetal survival rate after the surgical transfer of two bovine embryos. *Journal of Reproduction and Fertility* 59, 31–36.

Ng, S.C., Bongso, T.A., Liow, S.L., Edirisinghe, R., Tok, V. and Ratnam, S.S. (1992a) Controversies in microinjection. *Journal of Assisted Reproduction and Genetics* 9, 186–189.

Ng, S.C., Liow, S.L., Bongso, T.A., Montag, M., Tok, V., Sathananthan, A.H. and Ratnma, S.S. (1992b) Micromanipulation and possible new approaches to help the severe male-factor patient. *Journal of Assisted Reproduction and Genetics* 9, 191–196.

Nibart, M. (1992) Practical application of two advanced biotechnologies to bovine embryo transfer: splitting and sexing. In: Lauria, A. and Gandolfi, F. (eds) *Embryonic Development and Manipulation in Animal Production*. Portland Press, London, pp. 215–224.

Nibart, M., Sripongpun, S., Cedden, F., Mechekour, F., Le Guienne, B. and Thibier, M. (1988a) Histological study of bovine intact and demi-embryos. *Theriogenology* 29, 283.

Nibart, M., Mechekour, F., Schwartz, J.L., Procureur, R., Chesne, P., Chupin, D. and Thibier, M. (1988b) Taux de gestation après transfer cervical d'embryons bovins 'biopsés' maintenus 24 heures en survie à 0°C. *Proceedings of the Fourth European Embryo Transfer Association Meeting* (Lyon), 105.

Nibart, M., Mechekour, F., Procureur, R., Chesne, B., Chupin, D. and Thibier, M. (1989) Pregnancy rates after cervical transfer of biopsied bovine embryos. *Theriogenology* 31, 234.

Nibart, M., Kohen, G., Epositol, L., Baudi, P., Thuard, J.M., Desmettre, P. and Thibier, M. (1992) Rapid bovine embryo sexing by DNA probe: field results. *Proceedings of the Twelfth International Congress of Animal Reproduction* (The Hague) 2, 727–729.

Nibart, M., Thuard, J.M., Esposito, L., Herpe, P., Hascoet, J., Marrec, C., Lebrun, D. and Rohou, A. (1993) Bovine embryo sexing and modelization in an attempt to produce female calves by embryo transfer. *Proceedings of the Ninth Congress of the European Embryo Transfer Association* (Lyon), 244.

Nicholas, J.S. (1993) Development of transplanted rat eggs. *Proceedings of the Society of Experimental Biology and Medicine* 30, 1111–1113.

Nicholas, F.W. (1979) The genetic implications of multiple ovulation and embryo transfer in small dairy herds. *Proceedings of the Thirtieth Annual Meeting of the European Association for Animal Production* (Harrogate), CG1.11.y.

Nicholas, F.W. (1985) The implications of developments in reproductive biology for livestock improvement programmes. *Proceedings of the Australian Association of Animal Breeding and Genetics Fifth Conference* (Sydney), 77–82.

Nicholas, F.W. and Smith, C. (1983) Increased rates of genetic change in dairy cattle by embryo transfer and splitting. *Animal Production* 36, 341–353.

Niemann, H. (1985) Freezing of bovine embryos: effects of a one-step addition of 1.4 M glycerol. *Theriogenology* 23, 369–379.

Niemann, H. (1989) To South America in a thermos flask: first experience of international transportation of pig embryos. *Tierzuchter* 41(8), 336–337.

Niemann, H. (1991) Cryopreservation of ova and embryos from livestock: current status and research needs. *Theriogenology* 35, 109–124.

Niemann, H. and Elsaesser, F. (1986) Evidence for estrogen-dependent blastocyst formation in the pig. *Biology of Reproduction* 35, 10–16.

Niemann, H. and Strelchenko, N. (1994) Isolation and maintenance of rabbit embryonic stem (ES) cell like cells. *Theriogenology* 41, 265.

Niemann, H., Schilling, E., Sacher, B. and Smidt, D. (1982) Effect of bovine embryo quality on survival rate after fast freezing and thawing. *Proceedings of an International Congress on Embryo Transfer in Mammals* (Annecy), 101.

Niemann, H., Freitag, M. and Elsaesser, F. (1989) The role of oestrogens in early embryonic development. *Journal of Reproduction and Fertility*, Suppl. 38, 73–83.

Niemann, H., Lucas-Hahn, A. and Stoffregen, C. (1993) Cryopreservation of bovine oocytes and embryos following microsurgical operations. *Molecular Reproduction and Development* 36, 232–235.

Nijs, M., Vanderzwalmen, P., Segal-Bertin, G., Geerts, L., Van Roosendaal, E., Segal, L.,

Schoysman-Deboeck, A. and Schoysman, R. (1993) A monozygotic twin pregnancy after application of zona rubbing on a frozen–thawed blastocyst. *Human Reproduction* 8, 127–129.

Nivot, A. and Liegeois, L. (1990) Bovine embryo sexing with a rapid Y chromosome detection assay. *Proceedings of the Sixth Conference of the European Embryo Transfer Association* (Lyon), 176.

Nivot, A. and Liegeois, L. (1991) One year's results using a bovine embryo sexing process in France and in Europe. *Proceedings of the Seventh Meeting of the European Embryo Transfer Association* (Cambridge), 180.

Niwa, K. and Ohgoda, O. (1988) Synergistic effect of caffeine and heparin on *in vitro* fertilization of cattle oocytes matured in culture. *Theriogenology* 30, 733–741.

Niwa, K., Ohgoda, O. and Yuhara, M. (1988) Effects of caffeine in media for pretreatment of frozen–thawed sperm on *in vitro* penetration of cattle oocytes. *Proceedings of the Eleventh International Congress on Animal Reproduction and Artificial Insemination* (Dublin) 3, 346 (3 pp.).

Niwa, K., Park, C.K. and Okuda, K. (1991) Penetration *in vitro* of bovine oocytes during maturation by frozen–thawed spermatozoa. *Journal of Reproduction and Fertility* 91, 329–336.

Niwa, K., Ohgoda, O. and Park, C.K. (1992) Effect of glucose in medium with caffeine and/or heparin on *in vitro* penetration of bovine oocytes matured in culture. *Proceedings of the Twelfth International Congress of Animal Reproduction* (The Hague) 2, 671–673.

Noda, Y., Matsumoto, H., Umaoka, Y., Tatsumi, K., Kishi, J. and Mori, T. (1991) Involvement of superoxide radicals in the mouse two-cell block. *Molecular Reproduction and Development* 28, 356–360.

Nonogaki, T., Noda, Y., Narimoto, K., Umaoka, Y. and Mori, T. (1992) Effects of superoxide dismutase on mouse *in vitro* fertilization and embryo culture system. *Journal of Assisted Reproduction and Genetics* 9, 274–280.

Northey, D.L., Nuttleman, P.R. and Rosenkrans, G.F. (1991) Removal of bovine oocyte cytoplasm prior to fertilization reduces cell number in embryos. *Biology of Reproduction* 44 (Suppl. 1), 156 (Abs. 413).

Notarianni, E. and Laurie, S. (1992) Embryonic stem cells from domestic animals: establishment and potential applications. In: Lauria, A. and Gandolfi, F. (eds) *Embryonic Development and Manipulation in Animal Production*. Portland Press, London, pp. 175–182.

Notarianni, E., Galli, C., Laurie, S., Moor, R.M. and Evens, M.J. (1990) Derivation of pluripotent, embryonic cell lines from porcine and ovine blastocysts. *Proceedings of the Fourth World Congress on Genetics Applied to Livestock Production*, 58–64.

Nowshari, M.A. and Holtz, W. (1991) *In vitro* and *in vivo* development of zona pellucida free goat demi-embryos. *Serono Symposium on Preimplantation Embryo Development*, Abstract Vol., 65 (Abs. II-29).

Nowshari, M.A., Nayudu, P.L. and Hodges, J.K. (1993) Cryopreservation of mouse oocytes by ultra-rapid versus conventional freezing methods. *Journal of Reproduction and Fertility* Abstract Series No. 11, 65 (Abs. 119).

Nuttinck, F., Mermillod, P., Massip, A. and Dessy, F. (1993) Characterization of *in vitro* growth of bovine preantral ovarian follicles: a preliminary study. *Theriogenology* 39, 811–821.

O'Brien, J.K., Rhodes, S.L., Maxwell, W.M.C. and Evans, G. (1994) Hormonal requirements for *in vitro* maturation of sheep oocytes. *Theriogenology* 41, 266.

Oda, K., Gibbons, W.E. and Leibo, S.P. (1992) Osmotic shock of fertilized mouse ova. *Journal of Reproduction and Fertility* 95, 737–747.

Odde, K.G. (1990) A review of synchronization of estrus in post-partum. *Journal of Animal Science* 68, 817–830.

O'Farrell, K., Mee, J., Murphy, J. and Reitsma, P. (1990) Induced twinning in dairy cows. *Farm and Food Research* 21(2), 25–27.

O'Farrell, K., Mee, J., Murphy, J. and Reitsma, P. (1991) Induced twinning in dairy cows. *Irish Veterinary News* 13(3), 10–13.

Ogawa, S., Yamakawa, H., Yamanoi, J., Nishida, S., Takeshima, T.,Matumoto, T., Taughi, Y., Nagashima, H. and Kanou, Y. (1987) Detection of fluorescent body in spermatozoa of bull, boar, dog, rabbit, rat and mouse. *Japanese Journal of Animal Reproduction* 33, 140–144.

O'Gorman, D.P.J. (1970) Culture and attempted fertilization of the bovine follicular oocyte. M.Agr.Sc. Thesis, University College, Dublin.

Ogura, A. and Yanagimachi, R. (1993) Round spermatid nuclei injected into hamster oocytes form pronuclei and participate in syngamy. *Biology of Reproduction* 48, 219–225.

Oguri, N. and Tsutsumi, Y. (1974) *Journal of Reproduction and Fertility*, 41, 313–320.

Ohgoda, O., Niwa, K., Yuhara, M., Takahashi, S. and Kanoya, K. (1987) Variations in penetration rates *in vitro* of bovine follicular oocytes do not reflect conception rates after artificial insemination using frozen semen from different bulls. *Theriogenology* 29, 1375–1381.

Oikawa, T. (1978) A simple method for the isolation of a large number of ova from pig ovaries. *Gamete Research* 1, 265–267.

Oikawa, T., Yogo, K.I. and Sugawara, S. (1991) A study to establish the objective criteria for the selection of excellent ovarian oocytes. *Serono Symposium on Preimplantation Embryo Development*, Abstract Vol., 41 (Abs. 33).

Okolski, A. and Sionina, D. (1992) *In vitro* maturation of equine oocytes in a coculture with granulosa and theca internal cells. *Proceedings of the Third International Symposium on Equine Embryo Transfer* (Buenos Aires), 40–41.

Okolski, A., Bezard, J., Magistrini, M. and Palmer, E. (1990) Experiment on culture and *in vitro* fertilization of oocytes from normal and atretic equine ovarian follicles. *Proceedings of the Fifth International Symposium on Equine Reproduction* (Deauville), Abstracts, 126.

Okolski, A., Bezard, J., Magistrini, M. and Palmer, E. (1991) Maturation of oocytes from normal and atretic equine ovarian follicles as affected by steroid concentrations. *Journal of Reproduction and Fertility*, Suppl. 44, 385–392.

Olar, T.T. and Potts, A.S. (1993) Effect of fetal bovine serum of different gestational ages on mouse embryo growth and development. *Journal of Assisted Reproduction and Genetics* 10, 236–238.

Oliveira, E.B. de, Catt, J.W., Garcia, J.M. and Esper, C.R. (1991a) Some factors affecting '*in vitro*' fertilization rate in zebu cattle (*Bos indicus*). *Proceedings of the Seventh Congress of the European Embryo Transfer Association* (Cambridge), 184.

Oliveira, E.B. de, Catt, J.W., Garcia, J.M. and Esper, C.R. (1991b) Totally '*in vitro*' fertilization and development of *Bos indicus* embryos, up to the blastocyst stage. *Proceedings of the Seventh Congress of the European Embryo Transfer Association* (Cambridge), 186.

Oliveira, E.B. de, Catt, J., Garcia, J.M. and Esper, C.R. (1992) '*In vitro*' fertilization in zebu cattle (*Bos indicus*): some peculiarities affecting early development. *Proceedings of the Twelfth International Congress on Animal Reproduction* (The Hague) 3, 1330–1332.

Oliveira, E.B. de, Watanabe, Y.F. and Garcia, J.M. (1994) Establishment of an IVF program for zebu cattle (*Bos indicus*) in Brazil. *Theriogenology* 41, 188.

Olson, S.E., Romero, A., Thomas, W.K. and Seidel, G.E. (1990) Effects of FSH and heparin on *in vitro* maturation, fertilization and development of bovine oocytes. *Theriogenology* 33, 293.

Olson, S.E., Thomas, W.K. and Seidel, G.E. (1991) Effects of gonadotropins during *in vitro* maturation of bovine oocytes on subsequent embryonic development. *Theriogenology* 35, 250.

Olson, S.E., Curtin, R. and Seidel, G.E. (1993) Development of bovine embryos in a chemically defined medium at two oxygen tensions and with EDTA or DTPA. *Biology of Reproduction* 48 (Suppl. 1), 169.

Omran, S.N., Ayliffe, T.R. and Boyd, J.S. (1988) Preliminary observations of bovine ovarian structures using B-mode real time ultrasound. *Veterinary Record* 122, 465–466.

O'Neill, C. (1991) A consideration of the factors which influence and control the viability and developmental potential of the preimplantation embryo. *Baillière's Clinical Obstetrics and Gynaecology* 5, 159–178.

Onishi, A. and Youngs, C.R. (1993) Comparison of immunosurgery with calcium ionophore treatment for isolation of inner cell masses of porcine blastocysts. *Theriogenology* 39, 274.

Ooba, T., Sricharoen, P., Areekijseree, M. and Pavasuthipaisit, K. (1991) Evaluation of acrosome reaction in bovine sperm by a triple-staining technique. *Biology of Reproduction* 44 (Suppl. 1), 137 (Abs. 339).

Ooe, M., Yamamoto, M., Takagi, and Suzuki, T. (1993) Direct transfer of frozen–thawed bovine embryos derived from *in vivo* or *in vitro* fertilization using ethylene glycol. *Journal of Reproduction and Development* 19(5), 11–15.

Osborn, J.C. and Moor, R.M. (1985) Oocyte maturation and developmental competence. In: Thompson, W., Joyce, I. and Newton, J. (eds) *In Vitro Fertilization and Donor Insemination: Proceedings of the Twelfth Study Group of the Royal College of Obstetries and Gynaecology*. pp. 101–114.

O'Shea, T., Hillard, M.A., Anderson, S.T., Bindon, B.M., Findlay, J.K., Tsonis, C.G. and Wilkins, J.F. (1994) Inhibin immunization for increasing ovulation rate and superovulation. *Theriogenology* 41, 3–17.

Otoi, T., Tachikawa, S., Kondo, S. and Suzuki, T. (1992) Developmental capacity of bovine oocytes cryopreserved after maturation *in vitro* and of frozen–thawed bovine embryos derived from frozen mature oocytes. *Theriogenology* 38, 711–719.

Otoi, T., Ogura, T., Tachikawa, S., Kitamura, S. and Suzuki, T. (1993a) Deep freezing of bovine oocytes using different cryoprotectants. *Theriogenology* 39, 275.

Otoi, T., Tachikawa, S., Kondo, S. and Suzuki, T. (1993b) Effects of different lots of semen from the same bull on *in vitro* development of bovine oocytes fertilized *in vitro*. *Theriogenology* 39, 713–718.

Otoi, T., Tachikawa, S., Kondo, S. and Suzuki, T. (1993c) Developmental capacity of bovine oocytes frozen in different cryoprotectants. *Theriogenology* 40, 801–807.

Ott, T.L., Van Heeke, G., Fliss, F.M.V., Johnson, H.M. and Bazer, F.W. (1991) Antiluteolytic and antiviral activities of recombinant ovine trophoblast protein-1 (roTP-1) overproduced from a synthetic gene in *S. cerevisiae*. *Biology of Reproduction* 44 (Suppl. 1), 89 (Abs. 148).

Ott, T.L., Van Heeke, G., Hostetler, C.E., Schalue, T.K., Olmsted, J.J., Johnson, H.M. and Bazer, F.W. (1993) Intrauterine injection of recombinant ovine interferon-tau extends the interestrous interval in sheep. *Theriogenology* 40, 757–769.

Ouhibi, N., Menezo, Y, Benet, G. and Nicollet, B. (1988) Culture of epithelial cells derived from the oviduct of different species. *Proceedings of the Fourth Meeting of the European Embryo Transfer Association* (Lyon), 37.

Ouhibi, N., Hamidi, J., Guillaud, J. and Menezo, Y. (1990) Co-culture of 1-cell mouse embryos on different cell supports. *Human Reproduction* 5, 737–743.

Overstrom, E.W. (1992) Manipulation of early embryonic development. *Animal Reproduction Science* 28, 277–285.

Overstrom, E.W., Paquin-Platts, D., Toner, M. and Cravalho, E.G. (1990) Cryoprotectant and thermal effects on cytoskeletal organization and *in vitro* fertilization rates of mouse oocytes. *Biology of Reproduction* 42 (Suppl. 1), 175.

Owens, J.L., Edey, T.N., Bindon, B.M. and Piper, L.R. (1985) Parturient behaviour and calf survival in a herd selected for twinning. *Applied Animal Behaviour Science* 13, 321–333.

Ozil, J.P. (1992) Embryo cloning. *Human Reproduction* 7 (Suppl. 2), 73.

Pabon, J.E., Findley, W.E. and Gibbons, W.E. (1989) The toxic effect of short exposures to the atmospheric oxygen concentration on early mouse embryonic development. *Fertility and Sterility* 51, 896–900.

Paisley, L.G., Mickelsen, W.D. and Anderson, P.B. (1986) Mechanisms and therapy for retained fetal membranes and uterine infections of cows: a review. *Theriogenology* 25, 353–381.

Palermo, G. and Van Steirteghem, A. (1991) Enhancement of acrosome reaction and subzonal insemination of a single spermatozoon in mouse eggs. *Molecular Reproduction and Development* 30, 339–345.

Palermo, G., Joris, H., Devroey, P. and Steirteghem, C.A. van (1992a) Induction of acrosome reaction in human spermatozoa used for subzonal insemination. *Human Reproduction* 7, 248–254.

Palermo, G., Joris, H., Camus, M., Devroey, P. and Van Steirteghem, A.C. (1992b) Assisted fertilization by direct intracytoplasmic sperm injection to treat severe male factor infertility. *Human Reproduction* 7 (Suppl. 2), 83.

Palma, G.A. (1994) Effect of FSH and estradiol-19 β for maturation of calf oocytes on *in vitro* development to blastocysts. *Theriogenology* 41, 267.

Palma, G.A., Clement-Sengewald, A., Berg, U. and Brem, G. (1992) Role of the embryo number in the development of *in vitro* produced bovine embryos. *Theriogenology* 37, 271.

Palma, G.A., Clement-Sengewald, A. and Krefft, H. (1993a) *In vitro* production of cattle embryos from calf oocytes. *Theriogenology* 39, 278.

Palma, G.A., Clement-Sengewald, A., Krefft, H. and Brem, G. (1993b) Effect of insulin-like growth factor (IGF-I) on development of *in vitro* produced bovine embryos in two culture densities. *Proceedings of the Ninth Congress of the European Embryo Transfer Association* (Lyon), 248.

Palma, G.A., Krefft, H., Clement-Sengewald, A., Zhu, J. and Brem, G. (1993c) Results of large-scale *in vitro* bovine embryo production. *Proceedings of the Ninth Congress of the European Embryo Transfer Association* (Lyon), 250.

Palmer, E., Bezard, J., Magistrini, M. and Duchamp, G. (1990) *In vitro* fertilization in the horse: a retrospective study. *Proceedings of the Fifth International Symposium on Equine Reproduction* (Deauville), Abstracts, 124–125.

Palmer, E., Bezard, J., Magistrini, M. and Duchamp, G. (1991) *In vitro* fertilization in the horse: a retrospective study. *Journal of Reproduction and Fertility*, Suppl. 44, 375–384.

Palomo, M.J., Martino, A., Mogas, T. and Paramio, M.T. (1992) Effect of culture type on the *in vitro* maturation of prepubertal goat oocytes. *Journal of Reproduction and Fertility*, Abstract Series No. 9, 53 (Abs. 92).

Pampiglione, J.S., Tan, S.-L. and Campbell, S. (1993) The use of the stimulated acrosome reaction test as a test of fertilizing ability in human spermatozoa. *Fertility and Sterility* 59, 1280–1284.

Panter-Brick, C., Lotstein, D.S. and Ellison, P.T. (1993) Seasonality of reproductive function and weight loss in rural Nepali women. *Human Reproduction* 8, 684–690.

Papaioannou, V.E. and Dieterlen-Lievre, F.F. (1984) Making chimeras. In: *Chimeras in Developmental Biology*. Academic Press, London, pp. 3–37.

Papaioannou, V.E. and Ebert, K.M. (1986) Development of fertilized embryos transferred to oviducts of immature mice. *Journal of Reproduction and Fertility* 76, 603–608.

Papstein, H.J., Bunge, O., Ender, K. and Matthes, H.D. (1993) Twin calves from crossbred beef cows using embryo transfer. *Neue Landwirtschaft* 4, 63–65.

Parinaud, J., Labal, B. and Vieitez, G. (1992) High progesterone concentrations induce acrosome reaction with a low cytotoxic effect. *Fertility and Sterility* 58, 599–602.

Park, C.S. and Pursel, V.G. (1991) *In vitro* fertilization of pig ova. *Journal of Animal Science* 69 (Suppl. 1), 404.

Park, C.-K., Ohgoda, O. and Niwa, K. (1989) Penetration of bovine follicular oocytes by frozen–thawed spermatozoa in the presence of caffeine and heparin. *Journal of Reproduction and Fertility* 86, 577–582.

Park, R.L., Powell, K.L., Andrus, D.D., Bingham, W.W. and Wallentine, M.V. (1991) Conception rate and pregnancies per flush by herdsmen using embryo transfer in a large university herd: a six-year study. *Journal of Dairy Science* 74 (Suppl. 1), 199 (Abs. P171).

Park, S.B., Park, H.K. and Iritani, A. (1990a) Maturation and fertilization of porcine follicular oocytes cocultured with granulosa cells during *in vitro* culture. *Korean Journal of Animal Sciences* 32, 15–19.

Park, S.B., Park, H.K. and Iritani, A. (1990b) *In vitro* fertilization in the pig. *Korean Journal of Animal Sciences* 32, 20–26.

Park, S.B., Yeon, S.H. and Thak, T.Y. (1991) Influence of cumulus cells on gonadotrophin-induced maturation of pig oocytes *in vitro*. *Research Reports of the Rural Development Administration: Livestock* 33, 43–46.

Park, Y.S. and Lin, Y.C. (1993) Effect of epidermal growth factor (EGF) and defined simple media on *in vitro* bovine oocyte maturation and early embryonic development. *Theriogenology* 39, 475–484.

Parkinson, T.J. (1991) Early establishment of a new pregnancy. *Proceedings of the British Cattle Veterinary Association* (Reading), 5–14.

Parkinson, T.J., Lamming, G.E., Flint, A.P.F. and Jenner, L.J. (1992) Administration of recombinant bovine interferon-α_1 at the time of maternal recognition of pregnancy inhibits prostaglandin $F_{2\alpha}$ secretion and causes luteal maintenance in cyclic ewes. *Journal of Reproduction and Fertility* 94, 489–500.

Parks, J.E. and Hough, S.R. (1990) Effects of platelet activating factor on the motility and acrosome reaction of bovine spermatozoa. *Theriogenology* 34, 903–912.

Parks, J.E. and Ruffing, N.A. (1992) Factors affecting low temperature survival of mammalian oocytes. *Theriogenology* 37, 59–73.

Parlevliet, J.M., de Loos, F., Fazeli, A.R., Bevers, M., Vos, P.L.A.M., Pieterse, M., Taverne, M. and Colenbrander, B. (1993) Transvaginal ultrasound-guided ovum pick up in the horse. *Journal of Reproduction and Fertility*, Abstract Series No. 12, 40.

Parrish, J.J., Parrish, J.L. and First, N.L. (1984) Effect of swim-up separation and heparin pretreatment of frozen–thawed spermatozoa on *in vitro* fertilization of bovine oocytes. *Biology of Reproduction* 30 (Suppl. 1), 112 (Abs.).

Parrish, J.J., Susko-Parrish, J.L. and First, N.L. (1985a) Role of heparin in bovine sperm capacitation. *Biology of Reproduction* 32 (Suppl. 1), 211 (Abs.).

Parrish, J.J., Susko-Parrish, J.L. and First, N.L. (1985b) *In vitro* fertilization of bovine oocytes using heparin treated and swim-up separated frozen–thawed bovine semen is repeatable and results in high frequencies of fertilization. *Theriogenology* 23, 216.

Parrish, J.J., Susko-Parrish, J.L. and First, N.L. (1985c) Effect of heparin and chondroitin sulfate on the acrosome reaction and fertility of bovine sperm *in vitro*. *Theriogenology* 24, 537–549.

Parrish, J.J., Susko-Parrish, J.L., Critser, E., Leibfried, L. and First, N.L. (1986a) Bovine *in vitro* fertilization. *Proceedings of the Fourth World Conference on Human In Vitro Fertilization* (Melbourne), *Journal of In Vitro Fertilization and Embryo Transfer* 3, 202 (Abs.).

Parrish, J.J., Susko-Parrish, J.L., Leibfried-Rutledge, M.L., Critser, E.S., Eyestone, W.H. and First, N.L. (1986b) Bovine *in vitro* fertilization with frozen–thawed semen. *Theriogenology* 25, 591–600.

Parrish, J.J., Susko-Parrish, J.L., Winer, M.A. and First, N.L. (1988) Capacitation of bovine sperm by heparin. *Biology of Reproduction* 38, 1171–1180.

Parrish, J.J., Susko-Parrish, J.L., Handrow, R.R., Sims, M.M. and First, N.L. (1989a) Capacitation of bovine spermatozoa by oviduct fluid. *Biology of Reproduction* 40, 1020–1025.

Parrish, J.J., Susko-Parrish, J.L. and First, N.L. (1989b) Capacitation of bovine sperm by heparin: inhibitory effect of glucose and role of intracellular pH. *Biology of Reproduction* 41, 683–699.

Parrish, J.J., Kim, C.I. and Bae, I.H. (1992) Current concepts of cell-cycle regulation and its relationship to oocyte maturation, fertilization and embryo development. *Theriogenology* 38, 277–296.

Patterson, D.C. (1993a) Comparative growth of artificially reared twin and single born calves. *Sixty-fifth Annual Report of the Agricultural Research Institute of Northern Ireland (1991–92)*, 14 (Abs.).

Patterson, D.C. (1993b) The effect of pre-and post-natal growth rate with beef calves. *Sixty-sixth Annual Report of the Agricultural Research Institute of Northern Ireland*, 20–28.

Patterson, D.C., Steen, R.W.J. and Kilpatrick, D.J. (1993) A comparison of growth, feed efficiency and carcass characteristics of single and twin beef calves derived by embryo transfer. *Animal Production* 57, 81–90.

Pavasuthipaisit, K., Kitiyanant, Y., Thonabulsombat, C., Tocharus, C., Sriurairratna, S. and White, K.L. (1992) *In vitro* maturation and fertilization of swamp buffalo oocytes and their subsequent development. *Theriogenology* 38, 545–555.

Pavasuthipaisit, K., Tocharus, C., Thonabulsombat, C. and Kitiyanant, Y. (1993) The viability testing of frozen–thawed bovine embryos produced *in vitro*. *Theriogenology* 39, 280.

Pavasuthipaisit, K., Kitiyanant, Y. and Tocharus, C. (1994) Embryonic development of bovine oocytes fertilized by sperm microinjection: comparison between subzonal and ooplasmic injection. *Theriogenology* 41, 270.

Pavlok, A., Motlik, J., Kanka, J. and Fulka, J. (1989) *In vitro* techniques of bovine oocyte maturation, fertilization and embryo culture resulting in the birth of a calf. *Reproduction, Nutrition and Development* 29, 611–616.

Pavlok, A., Lucas-Hahn, A. and Niemann, H. (1992) Fertilization and developmental competence of bovine oocytes derived from different categories of antral follicles. *Molecular Reproduction and Development* 31, 63–67.

Pavlok, A., Kopecny, V., Lucas-Hahn, A. and Niemann, H. (1993) Transcriptional activity and nuclear ultrastructure of 8-cell bovine embryos developed by *in vitro* maturation and fertilization of oocytes from different growth categories of antral follicles. *Molecular Reproduction and Development* 35, 233–243.

Pawshe, C.H. and Totey, S.M. (1993) Effect of cumulus cell monolayer on the *in vitro* maturation of the denuded oocytes of the buffalo (*Bubalis bubalis*). *Theriogenology* 39, 281.

Payas, A.J., Broadbent, P.J., Dolman, D.F. and Christie, W.B. (1989) Factors affecting pregnancy rate in embryo transfer recipients with reference to plasma progesterone. *Theriogenology* 31, 238.

Payne, D., McLaughlin, K.J., Depypere, H.T., Kirby, C.A., Warnes, G.M. and Matthews, C.D. (1991) Experience with zona drilling and zona cutting to improve fertilization rates of human oocytes *in vitro*. *Human Reproduction* 6, 423–431.

Peippo, J. and Bredbacka, P. (1992) Sex related growth rate differences in preimplantation stage mouse embryos *in vivo* vs. *in vitro*. *Proceedings of the Nordic A.I. Virtanen Institute Symposium on the In Vitro Culture of Domestic Animal Embryos*, 28–29.

Peli, J. and Schellander, K. (1993) Effect of LH and estradiol-17β concentration in estrus cow serum on *in vitro* produced bovine embryos. *Theriogenology* 39, 282.

Peluso, J.J. and Hirschel, M.D. (1987) Role of gonadotropins and insulin in controlling steroidogenesis and growth of antral bovine follicles in perifusion culture. *Theriogenology* 28, 503–512.

Peluso, J.J. and Hirschel, M.D. (1988) Factors controlling growth of bovine primary and preantral follicles in perifusion culture. *Theriogenology* 30, 613–627.

Persson, E., Rodriguez-Martinez, H. and Dantzer, V. (1993) Immunocytochemical localization of IGF-I, PDGF-A and PDGF receptors in the porcine and bovine female reproductive tract. *Reproduction in Domestic Animals* 28, 205–206.

Peter, A.T., Jones, P.P. and Robinson, J.P. (1993) Fractionation of bovine spermatozoa for sex selection: a rapid immunomagnetic technique to remove spermatozoa that contain the H-Y antigen. *Theriogenology* 40, 1177–1185.

Peters, A.R. (1985) Hormonal control of the bovine oestrous cycle. I. The natural cycle. *British Veterinary Journal* 141, 564–575.

Petersen, B.A. and Holm, P. (1993) Embryo transfer in sheep. *Tidsskrift for Dansk Fareavl* 58, 3–5.

Peterson, A.J., Schofield, K.M., McLaughlin, R.R. (1990) Adsorption of DNA to the sperm of rams, bulls and fallow bucks. *Proceedings of the New Zealand Society of Animal Production* 50, 407–410.

Petr, J., Zetova, L., Fulka, J. and Jilek, F. (1989) Quantitative inhibitory influence of porcine cumulus cells upon the maturation of pig and cattle oocytes *in vitro*. *Reproduction, Nutrition and Development* 29, 541–550.

Petr, J., Tomanek, J., Fulka, J., Mika, J. and Jilek, F. (1991) Effects of GnRH on preovulatory endocrinology and oocyte maturation in PMSG-superovulated cows. *Animal Reproduction Science* 24, 37–52.

Petters, R.M. (1986) Recombinant DNA, gene transfer and the future of animal agriculture. *Journal of Animal Science* 62, 1759–1768.

Petters, R.M. (1992) Embryo development *in vitro* to the blastocyst stage in cattle, pigs and sheep. *Animal Reproduction Science* 28, 415–421.

Petters, R.M. and Reed, M.L. (1991) Addition of taurine or hypotaurine to culture medium improves development of one and two-cell pig embryos *in vitro*. *Theriogenology* 35, 253.

Pettit, W.H. (1985) Commercial freezing of bovine embryos in glass ampules. *Theriogenology* 23, 13–16.

Peura, T.T. (1990) Effects of acetate, citrate and malate on development of *in vitro* fertilized bovine embryos. *Theriogenology* 33, 296.

Peura, T.T. (1993) *In vitro* culture of mammalian preimplantation embryos. *Reproduction in Domestic Animals* 28, 170–173.

Peura, T. and Aalto, J. (1989) Pregnancies resulting from bovine oocytes matured and fertilized *in vitro*. *Suomen Elainlaakarilehti* 95, 69–73.

Peura, T., Hyttinen, J.-M., Janne, J. and Aalto, J. (1991a) Transfer of *in vitro* matured, fertilized and cultured bovine embryos of predetermined sex. *Theriogenology* 35, 254.

Peura, T., Hyttinen, J.M., Turunen, M. and Janne, J. (1991b) A reliable sex determination assay for bovine preimplantation embryos using the polymerase chain reaction. *Theriogenology* 35, 547–555.

Peura, T.T., Hyttinen, J.-M., Tolvanen, M., Aalto, J., Raino, V. and Janne, J. (1992a) Pregnancies resulting from *in vitro* produced, pronucleus-injected and bisected bovine embryos. *Proceedings of the Twelfth International Congress on Animal Reproduction* (The Hague) 2, 733–735.

Peura, T., Hyttinen, J.M., Aalto, J., Kosenen, R. and Janne, J. (1992b) *In vitro* maturation and fertilization, centrifugation, pronucleus-injection, *in vitro* culture and bisection do have detrimental effects on bovine embryo viability. *Proceedings of the Nordic A.I. Virtanen Institute Symposium In Vitro Culture of Domestic Animal Embryos*, 36–39.

Peura, T.T., Kosonen, R. and Utriainen, M. (1993a) Effects of centrifugation, microscopic evaluation and pronucleus-injection on the development of bovine zygotes. *Theriogenology* 39, 283.

Peura, T.T., Tolvanen, M. and Janne, J. (1993b) Development of bovine zygotes after short-term storage in manipulation chamber: effects of temperature and storage. *Journal of Reproduction and Fertility*, Abstract Series No. 11, 30 (Abs. 50).

Phillips, C.J.C. (1993) *Cattle Behaviour*. Farming Press, Ipswich, 212 pp.

Phocas, I., Sarandakou, A., Rizos, D., Dimitriadou, F., Mantzavinos, T. and Zourlas, P.A. (1992) Secretion of α-immunoreactive inhibin by human pre-embryos cultured *in vitro*. *Human Reproduction* 7, 545–549.

Picard, L. and Betteridge, K.J. (1989) The micromanipulation of farm animal embryos. In: Babiuk, L.A. and Phillips, J.P. (eds) *Animal Biotechnology*. Pergamon Press, Oxford, 141–178.

Picard, L., Schneider, U., Betteridge, K.J. and King, W.A. (1988) Effects of the zona pellucida, agar embedding and culture on the survival of micromanipulated bovine embryos after freezing and thawing. *Journal of In Vitro Fertilization and Embryo Transfer* 5, 268–274.

Picard, L., Chartrain, I., King, W.A. and Betteridge, K.J. (1990) Production of chimaeric bovine embryos and calves by aggregation of inner cell masses with morulae. *Molecular Reproduction and Development* 27, 295–304.

Pickering, S.J., Braude, P.R., Johnson, M.H., Cant, A. and Currie, J. (1990) Transient cooling to room temperature can cause irreversible disruption of the meiotic spindle in the human oocyte. *Fertility and Sterility* 54, 102–108.

Piedrahita, J.A., Guillespie, L. and Maeda, N. (1990) Establishing an embryonic stem (ES) cell system utilizing hamster embryos. *Biology of Reproduction* 42 (Suppl. 1), 175.

Pierson, R.A. and Ginther, O.J. (1986) Ultrasonic evaluation of the ovaries in cycling and pregnant heifers. *Journal of Animal Science* 63 (Suppl. 1), 336 (Abs.).

Pierson, R.A. and Ginther, O.J. (1987a) Intraovarian effect of the corpus luteum on ovarian follicles during early pregnancy in heifers. *Animal Reproduction Science* 15, 53–60.

Pierson, R.A. and Ginther, O.J. (1987b) Reliability of diagnostic ultrasonography for identification and measurement of follicles and detecting the corpus luteum in heifers. *Theriogenology* 28, 929–936.

Pierson, R.A. and Ginther, O.J. (1988) Ultrasonic imaging of the ovaries and uterus in cattle. *Theriogenology* 29, 21–37.

Pierson, R.A., Kastelic, J.P. and Ginther, O.J. (1988) Basic principles and techniques for transsrectal ultrasonography in cattle and horses. *Theriogenology* 29, 3–20.

Pieterse, M.C. (1990) Clinical use of ultrasound in bovine reproduction. Thesis, Rijksuniversiteit, Utrecht, The Netherlands, 171 pp.

Pieterse, M.C., Taverne, M.A.M. and Kruip, Th.A.M. (1988a) Collection of bovine oocytes by means of a transvaginal ultrasound guided puncture technique. *Proceedings of the Eleventh International Congress on Animal Reproduction and Artificial Insemination* (Dublin) 3, 348 (3 pp.).

Pieterse, M.C., Kappen, K.A., Kruip, Th.A.M. and Taverne, M.A.M. (1988b) Aspiration of bovine oocytes during transvaginal ultrasound scanning of the ovaries. *Theriogenology* 30, 751–762.

Pieterse, M.C., Vos, P.L.A.M., Kruip, T.A.M., Wurth, Y.A., Beneden, T.H. van, Willemse, A.H. and Taverne, M.A.M. (1991a) Transvaginal ultrasound guided follicular aspiration of bovine oocytes. *Theriogenology* 35, 19–24.

Pieterse, M.C., Vos, P.L.A.M., Kruip, T.A.M., Willemse, A.H. and Taverne, M.A.M. (1991b) Characteristics of bovine estrous cycles during repeated transvaginal, ultrasound-guided puncturing of follicles for ovum pick-up. *Theriogenology* 35, 401–413.

Pieterse, M.C., Vos, P.L.A.M., Kruip, Th.A.M., Wurth, Y.A., Van Beneden, Th.H., Willemse, A.H. and Taverne, M.A.M. (1991c) Transvaginal ultrasound guided follicular aspiration of bovine oocytes. *Theriogenology* 35, 857–862.

Pincus, G. and Enzmann, E.V. (1935) The comparative behaviour of mammalian eggs *in vivo* and *in vitro*. *Journal of Experimental Medicine* 62, 665–675.

Pincus, G. and Saunders, B. (1939) The comparative behaviour of mammalian eggs *in vivo* and *in vitro*. VI. The maturation of human ovarian ova. *Anatomical Record* 75, 737–741.

Pinkel, D., Garner, D.L., Gledhill, B.L., Lake, S., Stephenson, D. and Johnson, L.A. (1985) Flow cytometric determination of the proportions of X- and Y-chromosome bearing sperm in samples of purportedly separated bull sperm. *Journal of Animal Science* 60, 1303–1307.

Pinkert, C.A., Kooyman, D.L., Baumgartner, A. and Keisler, D.H. (1989) *In vitro* development of zygotes from superovulated prepubertal and mature gilts. *Journal of Reproduction and Fertility* 87, 63–66.

Pinto-Correia, C., Collas, P., Able Ponce De Leon F. and Robl, J.M. (1993) Chromatin and microtubule organization in the first cell cycle in rabbit parthenotes and nuclear transplant embryos. *Molecular Reproduction and Development* 34, 33–42.

Pinyopummintr, T. (1990) Bovine embryos develop to the morula/blastocyst stage *in vitro* in a chemically defined protein-free medium. *Biology of Reproduction* 42 (Suppl. 1), 61.

Pinyopummintr, T. and Bavister, B.D. (1991) *In vitro*-matured/*in vitro*-fertilized bovine oocytes can develop into morulae/blastocysts in chemically defined, protein-free culture media. *Biology of Reproduction* 45, 736–742.

Pinyopummintr, T. and Bavister, B.D. (1993) Effect of timing of addition and types of serum supplementation on bovine embryo development *in vitro*. *Biology of Reproduction* 48 (Suppl. 1), 174.

Pinyopummintr, T. and Bavister, B.D. (1994) Effect of gaseous atmosphere on *in vitro* maturation and *in vitro* fertilization of bovine oocytes. *Theriogenology* 41, 276.

Pipkin, J.L., Prien, S.D., Pipkin, K.E. and Clark, J.R. (1991) Culture and co-culture of porcine embryos. *Journal Animal Science* 69 (Suppl. 1), 417.

Plachot, M., Beaudelot, F. and Salat-Baroux, J. (1992) Granulosa cells improve human embryo development *in vitro*. *Human Reproduction* 7 (Suppl. 2), 140.

Plachot, M., Antoine, J.M., Alvarez, S., Firmin, C., Pfister, A., Mandelbaum, J., Junca, A.-M. and Salat-Baroux, J. (1993) Granulosa cells improve human embryo development *in vitro*. *Human Reproduction* 8, 2133–2140.

Plante, C., Bousquet, D.L., Guay, P. and Goff, A.K. (1987) Luteotropohic activity of bovine embryos. *Theriogenology* 28, 801–813.

Plante, C., Hansen, P. and Thatcher, W.W. (1988) Prolongation of luteal lifespan in cows by intra-uterine infusion of recombinant bovine alpha-interferon. *Endocrinology* 122, 2342–2344.

Plante, C., Hansen, P.J., Martinod, S., Siegenthaler, B., Thatcher, W.W., Pollard, J.W. and Leslie, M.V. (1990) Effect of intrauterine and intramuscular administration of recombinant bovine interferon α on luteal lifespan in cattle. *Journal of Dairy Science* 72, 1859–1865.

Plante, C., Pollard, J.W., Kobayashi, S. and Leibo, S.P. (1993) Chilling sensitivity of porcine morulae. *Theriogenology* 39, 285.

Plante, L. and King, W.A. (1993) Developmental potential of spontaneously activated bovine oocytes. *Theriogenology* 39, 286.

Plante, Y., Schmutz, S.M. and Lang, K.D.M. (1992) Restriction fragment length polymorphism in the mitochondrial DNA of cloned cattle. *Theriogenology* 38, 897–904.

Polak de Fried, E., Blanco, L., Lancuba, S. and Asch, R.H. (1992) Immunosuppression in *in vitro* fertilization (IVF) patients. *Proceedings of the Fourteenth World Congress on Fertility and Sterility* (Caracas), 133 (Abs. 356).

Polak de Fried, E., Blanco, L., Lancuba, S. and Asch, R.H. (1993) Improvement of clinical pregnancy rate and implantation rate of *in vitro* fertilization-embryo transfer patients by using methylprednisone. *Human Reproduction* 8, 393–395.

Polge, C. and Rowson, L.E.A. (1952) Fertilizing capacity of bull spermatozoa after freezing at −7°C. *Nature (London)* 169, 626–627.

Polge, C. and Rowson, L.E.A. (1975) Recent progress in techniques for increasing reproductive potential in farm animals. In: Reid, R.L. (ed.) *Proceedings of the Third World Conference on Animal Production*. Sydney University Press, Sydney. pp. 633–643.

Polge, C., Adams, C.E. and Baker, R.D. (1972) Development and survival of pig embryos in the rabbit oviduct. *Proceedings of the Seventh International Congress on Animal Reproduction and Artificial Insemination* (Munich), 513–517.

Polge, C., Wilmut, I. and Rowson, L.E.A. (1974) The low temperature preservation of cow, sheep and pig embryos. *Cryobiology* 11, 560.

Polge, C., Broadbent, P. and Lu, K.H. (1991) Pregnancy rates following transfer of fresh or frozen bovine embryos produced *in vitro*. *International Symposium on Animal Biotechnology* (Kyoto, Japan), 15–17 October, 5.

Pollard, J.W. and Leibo, S.P. (1993) Comparative cryobiology of *in vitro* and *in vivo* derived bovine embryos. *Theriogenology* 39, 287.

Pollard, J.W. and Leibo, S.P. (1994) Chilling sensitivity of mammalian embryos. *Theriogenology* 41, 101–106.

Pollard, J.W., Xu, K.P., Rorie, R., King, W.A. and Betteridge, K.J. (1989) Influence of various oviductal epithelial cell culture on the development of early cleavage bovine embryos *in vitro*. *Theriogenology* 31, 239.

Pollard, J.W., Scodras, J.M., Plante, L., King, W.A. and Betteridge, K.J. (1990) Partial characterization of the embryotrophic nature of bovine endosalpingeal epithelial cell conditioned media. *Biology of Reproduction* 42 (Suppl. 1), 168.

Pollard, J.W., Scodras, J.M., Plante, L., King, W.A. and Betteridge, K.J. (1991a) Definitions of the cleavage stage(s) at which oviductal epithelial cells enable bovine embryos to pass through the *in vitro* 8–16 cell block. *Theriogenology* 35, 256.

Pollard, J.W., Plante, C., King, W.A., Hansen, P.J., Betteridge, K.J. and Suarez, S.S. (1991b) Fertilizing capacity of bovine sperm may be maintained by binding to oviductal epithelial cells. *Biology of Reproduction* 44, 102–107.

Pollard, J.W., Scodras, J.M., Madison, V., Greve, T., Hyttel, P., King, W.A. and Betteridge, K.J. (1992) Hyperactivation of bovine sperm: incidence, induction and physiological significance. *Proceedings of the Twelfth International Congress on Animal Reproduction* (The Hague) 2, 677–679.

Pomeroy, K.O., Dodds, J.F. and Seidel, G.E., Jr (1988) Caffeine promotes *in vitro* fertilization of mouse ova within 15 minutes. *Journal of Experimental Zoology* 284, 207–212.

Pontbriand, D., Goff, A.K., Xu, K.P. and King, W.A. (1989) Effect of steroids on protein synthesis in mature and immature bovine oocytes. *Theriogenology* 31, 240.

Pool, S.H., Rorie, R.W., Pendleton, R.J., Menino, A.R. and Godke, R.A. (1988a) Culture of early-stage bovine embryos inside day-13 and day-14 precultured trophoblastic vesicles. *Annals of the New York Academy of Sciences* 541, 407–418.

Pool, S.H., Wiemer, K.E., Rorie, R.W. and Godke, R.A. (1988b) The use of trophoblastic vesicles and fetal uterine monolayer cells for the culture of precompaction stage bovine embryos. *Proceedings of the Eleventh International Congress on Animal Reproduction and Artificial Insemination* (Dublin) 4, 479 (3 pp.).

Pope, W.F. (1988) Uterine asynchrony: a cause of embryonic loss. *Biology of Reproduction* 39, 999–1003.

Pope, W.F. (1992) Embryogenesis recapitulates oogenesis in swine. *Proceedings of the Society of Experimental Biology and Medicine* 199, 273–281.

Pope, W.F., Xie, S., Broermann, D.M. and Nephew, K.P. (1990) Causes and consequences of early embryonic diversity in pigs. *Journal of Reproduction and Fertility*, Suppl. 40, 251–260.

Potts, M. and Rosenfield, A. (1990a) The fifth freedom revisited: I. Background and existing programmes. *Lancet* 336, 1227–1231.

Potts, M. and Rosenfield, A. (1990b) The fifth freedom revisited: II. The way forward. *Lancet* 336, 1293–1295.

Powell, D.J., Galli, C. and Moor, R.M. (1992) The fate of DNA injected into mammalian oocytes and zygotes at different stages of the cell cycle. *Journal of Reproduction and Fertility* 95, 211–220.

Powell, R. and Barnes, F.L. (1992) The kinetics of oocyte activation and polar body formation in bovine embryo clones. *Molecular Reproduction and Development* 33, 53–58.

Prado, R., Rhind, S.M., Wright, I.A., Russel, A.J.F., McMillen, S.M., Smith, A.L. and McNeilly, A.S. (1989) Ovarian follicular populations, steroidogenic capacity and physiological status in post-partum suckling beef cows in high and low body condition. *British Society of Animal Production* (Winter Meeting), Paper No. 149.

Prather, R.S. and First, N.L. (1988) A review of early mouse embryogenesis and its applications to domestic species. *Journal of Animal Science* 66, 2626–2635.

Prather, R.S. and First, N.L. (1990a) Cloning of embryos. *Journal of Reproduction and Fertility*, Suppl. 40, 227–234.

Prather, R.S. and First, N.L. (1990b) Nuclear transfer in mammalian embryos. *International Review of Cytology* 120, 169–190.

Prather, R.S. and First, N.L. (1993) Cell-to-cell coupling in early stage bovine embryos: a preliminary report. *Theriogenology* 39, 561–567.

Prather, R.S., Barnes, F.L., Robl, J.M. and First, N.L. (1986) Multiplication of bovine embryos. *Biology of Reproduction* 34 (Suppl. 1), 285 (Abs.).

Prather, R.S., Barnes, F.L., Sims, M.M., Robl, J.M., Eyestone, W.H. and First, N.L. (1987a) Nuclear transplantation in the bovine embryo: assessment of donor nuclei and recipient oocyte. *Biology of Reproduction* 37, 859–866.

Prather, R.S., Spire, M.F. and Schalles, R.R. (1987b) Evaluation of cryopreservation techniques for bovine embryos. *Theriogenology* 28, 195–204.

Prather, R.S., Sims, M.M., Maul, G.G., First, N.L. and Schatten, G. (1989a) Nuclear lamin antigens are developmentally regulated during porcine and bovine embryogenesis. *Biology of Reproduction* 40, 123–132.

Prather, R.S., Sims, M.M. and First, N.L. (1989b) Nuclear transplantation in early pig embryos. *Biology of Reproduction* 41, 414–418.

Prather, R.S., Sims, M.M. and First, N.L. (1990) Nuclear transplantation in the pig embryo: nuclear swelling. *Journal of Experimental Zoology* 255, 355–358.

Prather, R.S., Eichen, P.A., Nicks, D.K. and Peters, M.S. (1991a) Activation of porcine oocytes matured *in vitro*. *Theriogenology* 35, 258.

Prather, R.S., Eichen, P.A., Nicks, D.K. and Peters, M.S. (1991b) Artificial activation of porcine oocytes matured *in vitro*. *Molecular Reproduction and Development* 28, 405–409.

Prather, R.S., Sims, M.M. and First, N.L. (1991c) Culture of porcine embryos from the one- and two-cell stage to the blastocyst stage in sheep oviducts. *Theriogenology* 35, 1147–1151.

Prather, R.S., Stumpf, T.T. and Rickords, L.F. (1992) Reprogramming the nucleus and synchronizing it with the cytoplasm. *Proceedings of a Symposium on Cloning Mammals by Nuclear Transplantation* (Fort Collins), 26–27.

Preinberg, G.A., Tirmanis, I.Y. and Turka, L.K. (1989) Quantitative and qualitative evaluation of isolated oocytes in cows. *Referativnyi Zhurnal* 2.58.349.

Presicce, G.A. and Yang, X. (1993) Dynamics of activation of *in vitro* matured bovine follicular oocytes following combined ethanol and cycloheximide treatment. *Theriogenology* 39, 290.

Presicce, G.A. and Yang, X. (1994) Nuclear dynamics of parthenogenesis of bovine oocytes matured *in vitro* for 20 and 40 hours and activated with combined ethanol and cycloheximide treatment. *Molecular Reproduction and Development* 37, 61–68.

Price, C.A. (1991) The control of FSH secretion in the larger domestic species. *Journal of Endocrinology* 131, 177–184.

Price, C.A., Carriere, P.D., Bhatia, B. and Groome, N.P. (1993) Changes in inhibin-beta production, and follicular dimeric inhibin and oestradiol concentrations during growth and atresia of bovine follicles. *Journal of Reproduction and Fertility*, Abstract Series No. 11, 46 (Abs. 82).

Pritchard, J.F., Pool, S.H., Blakewood, E.G. and Godke, R.A. (1990) Culture of early-stage caprine embryos using goat oviductal and uterine cell monolayers. *Theriogenology* 33, 300.

Pritchard, J.F., Pool, S.H., Blakewood, E.G., Menezo, Y. and Godke, R.A. (1991) Culture of early stage caprine embryos using goat oviductal cell monolayers. *Theriogenology* 35, 259.

Pritchard, J.F., Thibodeaux, J.K., Pool, S.H., Blakewood, E.G., Menezo, Y. and Godke, R.A. (1992) *In vitro* co-culture of early stage caprine embryos with oviduct and uterine epithelial cells. *Human Reproduction* 7, 553–557.

Prochazka, R., Durnford, R., Fiser, P.S. and Marcus, G.J. (1993) Parthenogenetic development of activated *in vitro* matured bovine oocytes. *Theriogenology* 39, 1025–1032.

Prokofiev, M.I., Ernst, L.K., Suraeva, N.M., Lagutina, I.S, Udavlennikova, N.N., Kesyan, A.Z. and Dolgohatskiy, A.I. (1992) Bovine oocyte maturation, fertilization and further development *in vitro* and after transfer into recipients. *Theriogenology* 38, 461–469.

Pugh, P.A., Tervit, H.R. and Thompson, J.G.E. (1991a) Parthenogenetic activation of sheep oocytes by electric current. *Theriogenology* 35, 260.

Pugh, P.A., Fukui, Y., Tervit, H.R. and Thompson, J.G. (1991b) Developmental ability of *in vitro* matured sheep oocytes collected during the non-breeding season and fertilized *in vitro* with frozen ram semen. *Theriogenology* 36, 771–778.

Pullar, D. and Collins, C.A. (1993) Relationships between observable oestrous behaviour and number of heifers in oestrus. *British Society of Animal Production* (Winter Meeting), Paper 56 (2 pp.).

Pulvermacher, R.J. (1991) Effectiveness of an automated estrus detection system for dairy cows. *Journal of Dairy Science* 74 (Suppl. 1), 193 (Abs. P154).

Pursel, V.G. and Rexroad, C.E. (1993) Recent progress in the transgenic modification of swine and sheep. *Molecular Reproduction and Development* 36, 251–254.

Pursel, V.G., Pinkert, C.A., Miller, K.F., Bolt, D.J., Campbell, R.G., Palmiter, R.D., Brinster, R.L. and Hammer, R.E. (1989) Genetic engineering of livestock. *Science* 244, 1281–1288.

Pursel, V.G., Bolt, D.J., Miller, K.F., Pinkert, C.A., Hammer, R.E., Palmiter, R.D. and Brinster, R.L. (1990) Expression and performance in transgenic pigs. *Journal of Reproduction and Fertility*, Suppl. 40, 235–245.

Purvis, K. and Egdetveit, I. (1993) Factors affecting sperm yield during swim-up. *Journal of Assisted Reproduction and Genetics* 10, 145–150.

Purwantara, B., Schmidt, M. and Greve, T. (1991) Intraovarian changes and embryo recovery rate in two different superovulation regimen in cattle. *Proceedings of the Seventh Meeting of the European Embryo Transfer Association* (Cambridge), 194.

Putney, D.J., Malayer, J.R., Gross, T.S., Hansen, P.J., Thatcher, W.W. and Drost, M. (1987) Heat shock induced alterations in prostaglandin and protein secretion by bovine conceptuses and uterine endometrium. *Biology of Reproduction* 36 (Suppl. 1), 151 (Abs.).

Putney, D.J., Drost, M. and Thatcher, W. (1989) Influence of summer heat stress on pregnancy rates of lactating dairy cattle following embryo transfer or artificial insemination. *Theriogenology* 31, 765–778.

Putney, D.J., Mullins, S., Thatcher, W.W., Drost, M. and Gross, T.S. (1989b) Embryonic development in superovulated dairy cattle exposed to elevated ambient temperatures between the onset of estrus and insemination. *Animal Reproduction Science* 19, 37–51.

Pye, J., Holmes, M.C. and Crawford, M. (1981) A simplified piece of equipment for controlled cooling and freezing of mouse embryos. In: Zeilmaker, G.H. (ed.) *Frozen Storage of Laboratory Animals*. Gustav Fischer Verlag, New York, pp. 61–66.

Quinn, P., Hirayama, T. and Marrs, R.P. (1991) Cooperative interaction among mouse zygotes cultured in protein-free medium: blastocyst development and hatching. *Serono Symposium on Preimplantation Embryo Development*, Abstract Vol., 68 (Abs. II-35).

Quirke, J.F. (1968) Culture and attempted fertilization of the sheep follicular oocyte. M.Agr.Sc. Thesis, University College, Dublin.

Qvist, R., Blackwell, L.F., Bourne, H. and Brown, J.B. (1990) Development of mouse ovarian follicles from primary to preovulatory stages *in vitro*. *Journal of Reproduction and Fertility* 89, 169–180.

Rabahi, F., Monniaux, D., Pisselet, C., Chupin, D. and Durand, P. (1991) Qualitative and quantitative changes in protein synthesis of bovine follicular cells during the preovulatory period. *Molecular Reproduction and Development* 30, 265–274.

Rabahi, F., Monniaux, D., Pisselet, C. and Durand, P.H. (1992) Control of *in vitro* maturation of cumulus-oocyte complexes by preovulatory bovine granulosa cells. *Journal of Reproduction and Fertility*, Abstract Series No. 9, 51 (Abs. 87).

Rabahi, F., Monniaux, D., Pisselet, C. and Durand, P. (1993) Control of *in vitro* maturation of bovine cumulus-oocyte complex by preovulatory granulosa cells. *Molecular Reproduction and Development* 34, 431–442.

Radford, A.J., Hodgson, A.L.M., Rothel, J.S. and Wood, P.R. (1991) Structure of ovine interferon. *Australian Veterinary Journal* 68, 82–86.

Rajakoski, E. (1960) The ovarian follicular system in sexually mature heifers with special reference to seasonal, cyclical and left-right variations. *Acta Endocrinologica* 52, 1–68.

Rajamahendran, R. and Sianangama, P.C. (1992) Effect of human chorionic gonadotrophin on dominant follicles in cows: formation of accessory corpora lutea, progesterone production and pregnancy rates. *Journal of Reproduction and Fertility* 95, 577–584.

Rajamahendran, R., Canseco, R.S., Gwazadauskas, F.C. and Vinson, W.E. (1985) Observations on *in vitro* development of bovine morulae in Hams F-10 and Dulbecco's phosphate buffered saline supplemented with normal steer serum. *Theriogenology* 24, 369–374.

Rall, W.F. (1992) Cryopreservation of oocytes and embryos: methods and applications. *Animal Reproduction Science* 28, 237–245.

Rall, W.F. and Fahy, G.M. (1985) Ice-free cryopreservation of mouse embryos at −196°C by vitrification. *Nature (London)* 313, 573–575.

Rall, W.F. and Meyer, T.K. (1989) Zona fracture damage and its avoidance during the cryopreservation of mammalian embryos. *Theriogenology* 31, 683–692.

Rana, N., Jeyendran, R.S., Holmgren, W.J., Rotman, C. and Zaneveld, J.D. (1989) Glass wool-filtered spermatozoa and their oocyte penetrating capacity. *Journal of In Vitro Fertilization and Embryo Transfer* 6, 280–284.

Randall, G.W., O'Connor, E.F. and Gantt, P.A. (1991) Synergy between tumour necrosis factor and endotoxin decreases early embryo development *in vitro*. *Journal of In Vitro Fertilization and Embryo Transfer* 8, 304–307.

Rao, V.H., Sarmah, B.C. and Bhattacharyya, N.K. (1984) Xenogenous fertilization of goat ova in the rabbit oviduct. *Journal of Reproduction and Fertility* 71, 377–379.

Rao, V.H., Heyman, Y., Chesne, P. and Renard, J.P. (1992) Multiplication of selected rabbit embryos by recloning of frozen stored nuclear transfer embryos. *Proceedings of the Eighth Conference of the European Embryo Transfer Association* (Lyon), 202.

Rao, V.H., Chesne, P., Renard, J.P. and Heyman, Y. (1993) Influence of age of *in vitro* matured oocytes and electrical stimulations on the development of cattle nuclear transfer embryos. *Theriogenology* 39, 292.

Rath, D. (1993) Current status of ultrasound-guided retrieval of bovine oocytes. *Embryo Transfer Newsletter* 11(2), 10–15.

Rath, D., Johnson, L.A. and Welch, G.R. (1993) *In vitro* culture of porcine embryos: development to blastocysts after *in vitro* fertilization (IVF) with flow cytometrically sorted and unsorted semen. *Theriogenology* 39, 293.

Rath, D., Johnson, L.A., Welch, G.R. and Neimann, H. (1994) Successful gamete intrafallopian transfer (GIFT) in the porcine. *Theriogenology* 41, 1173–1179.

Raychoudhury, S.S. and Suarez, S.S. (1991) Porcine sperm binding to oviductal explants in culture. *Theriogenology* 36, 1059–1070.

Reed, M.L. and Petters, R.M. (1991) An evaluation of hamster, rat, and mouse sperm-cell motility in media formulated with water of different qualities. *Journal of In Vitro Fertilization and Embryo Transfer* 8, 41–47.

Reed, M.L., Estrada, J.L. and Petters, R.M. (1991) Effects of epidermal growth factor, insulin-like growth factor-1, and porcine follicular fluid on porcine oocyte nuclear maturation *in vitro*. *Serono Symposium on Preimplantation Embryo Development*, Abstract Vol., 42 (Abs. I-36).

Reed, M.L., Illera, M.J. and Petters, R.M.(1992) *In vitro* culture of pig embryos. *Theriogenology* 37, 95–109.

Rees, W.H.G. (1988) The international animal disease situation and its effect on trade in cattle, bovine semen and embryos. *British Cattle Breeders' Club Digest* 43, 20–25.

Reeves, J.J., Chang, C.F., Deavila, D.M., Grieger, D.M., Johnson, H.E. and Roberts, A.J. (1989) Vaccines against endogenous hormones: a possible future tool in animal production. *Journal of Dairy Science* 72, 3363–3371.

Refsdal, A.O., Kjaestad, H. and Vatn, T. (1988) Transfer of refrigerated bovine embryos. *Proceedings of the Eleventh Congress on Animal Reproduction and Artificial Insemination* (Dublin) 2, 186 (3 pp.).

Rehman, N., Collins, A.R. and Wright, R.W. (1993) Effect of sperm exposure time on *in vitro* fertilization and embryo development of bovine oocytes. *Theriogenology* 39, 294.

Reichenbach, H.D., Liebrich, J., Berg, U. and Brem, G. (1990) Pregnancy rates and births following transfer of *in vitro* produced IVM–IVF bovine embryos to synchronous or asynchronous recipients. *Proceedings of the Sixth Conference of the European Embryo Transfer Association* (Lyon), 182.

Reichenbach, H.D., Liebrich, J., Berg, U. and Brem, G. (1992) Pregnancy rates and births after unilateral or bilateral transfer of bovine embryos produced *in vitro*. *Journal of Reproduction and Fertility* 95, 363–370.

Reichenbach, H.D., Wiebke, N.H., Besenfelder, U.H., Modl, J. and Brem, G. (1993a) Transvaginal laparoscopic guided aspiration of bovine follicular oocytes: preliminary results. *Theriogenology* 39, 295.

Reichenbach, H.-D., Modl, J. and Brem, G. (1993b) Piglets born after transcervical transfer of embryos into recipient gilts. *Veterinary Record* 133, 36–39.

Reichenbach, H.-D., Wiebke, N.H., Modl, J. and Brem, G. (1994) Bovine follicular oocytes collected by laparoscopic guided transvaginal aspiration. *Theriogenology* 41, 283.

Reid, J.P., Wilton, J.W. and Walton, J.S. (1986) Comparative productivity of cows receiving two embryos at transfer. *Canadian Journal of Animal Science* 66, 373–380.

Reintjes, C.V. (1991) Experiments on the co-culture of cattle oocytes, matured and fertilized *in vitro*. Thesis, Tierarztliche Hochschule, Hannover, Germany, 94 pp.

Renard, J.P. (1990) Cloning of embryos in mammals. *Proceedings of the Sixth Conference of the European Embryo Transfer Association* (Lyon), 105–112.

Renard, J.P., Heyman, Y. and Du Mesnil du Buisson, F. (1977) Unilateral and bilateral cervical transfer of bovine embryos at the blastocyst stage. *Theriogenology* 7, 189–194.

Renard, J.P., Menezo, Y., Saumande, J. and Heyman, Y. (1978) Attempts to predict the viability of cattle embryos produced by superovulation. In: Sreenan, J.M. (ed.) *Current Topics in Veterinary Medicine*, Vol. 1. Martinus Nijhoff, The Hague, pp. 398–417.

Renard, J.-P., Ozil, J.P. and Heyman, Y. (1979) The use of embryo transfer in the field for increased calf crops in beef and dairy cattle. *Animal Reproduction Science* 2, 353–361.

Renard, J.-P., Philippon, A. and Menezo, Y. (1980) *In vitro* uptake of glucose by bovine blastocysts. *Journal of Reproduction and Fertility* 58, 161–164.

Renard, J.P., Heyman, Y., Leymonie, P. and Plat, J.C. (1983) Sucrose dilution: a technique for field transfer of bovine embryos frozen in the straw. *Theriogenology* 19, 145.

Resnick, J.L., Bixler, L.S., Cheng, L. and Donovan, P.J. (1992) Long-term proliferation of mouse primordial germ cells in culture. *Nature* 359, 550–551.

Rettmer, I., Stevenson, J.S. and Corah, L.R. (1992a) Pregnancy rates in beef cattle after administering a GnRH agonist 11 to 14 days after insemination. *Journal of Animal Science* 70, 7–12.

Rettmer, I., Stevenson, J.S. and Corah, L.R. (1992b) Endocrine responses and ovarian changes in inseminated dairy heifers after an injection of a GnRH agonist 11 to 13 days after estrus. *Journal of Animal Science* 70, 508–517.

Revel, F., Mermillod, P., Peynot, N., Renard, J.P. and Heyman, Y. (1993) *In vitro* production of blastocysts from calf ovaries. *Proceedings of the Ninth Congress of the European Embryo Transfer Association* (Lyon), 270.

Rexroad, C.E., Jr and Powell, A.M. (1986) Co-culture of sheep ova and cells from the sheep oviduct. *Theriogenology* 25, 187.

Rexroad, C.E., Jr and Powell, A.M. (1987) Co-culture of single-cell sheep embryos on monolayers of oviductal, uterine and kidney cells. *Journal of Animal Science* 65 (Suppl. 1), 403 (Abs.).

Rexroad, C.E., Jr and Powell, A.M. (1988a) Co-culture of ovine eggs with oviductal cells in Medium 199. *Journal of Animal Science* 66, 947–953.

Rexroad, C.E., Jr and Powell, A.M. (1988b) Co-culture of ovine eggs with oviductal cells and trophoblastic vesicles. *Theriogenology* 29, 295.

Rexroad, C.E. and Powell, A.M. (1989) Asynchronous transfer improves development of sheep embryos co-cultured in serum-free medium. *Biology of Reproduction* 40 (Suppl. 1), 137 (Abs. 268).

Rexroad, C.E., Jr and Powell, A.M. (1990) Feeder layers for coculture of sheep embryos. *Journal of Animal Science* 68 (Suppl. 1), 430.

Rexroad, C.E. and Powell, A.M. (1991a) Development of transferred ovine embryos that were co-cultured on oviductal cells or on STO cells. *Journal of Animal Science* 69 (Suppl. 1), 403.

Rexroad, C.E. and Powell, A.M. (1991b) Effect of serum-free co-culture and synchrony of recipients on development of cultured sheep embryos to fetuses. *Journal of Animal Science* 69, 2066–2072.

Rexroad, C.E., Jr and Wall, R.J. (1987) Development of one-cell fertilized sheep ova following microinjection into pronuclei. *Theriogenology* 27, 611–619.

Rexroad, C.E., Jr, Powell, A.M., Rohan, R. and Wall, R.J. (1990) Evaluation of co-culture as a method for selecting viable microinjected sheep embryos for transfer. *Animal Biotechnology* 1, 1–10.

Rexroad, C.E., Mayo, K., Bolt, D.J., Elsasser, T.H., Miller, K.F., Behringer, R.R., Palmiter, R.D. and Brinster, R.L. (1991) Transferrin- and albumin-directed expression of growth-related peptides in transgenic sheep. *Journal of Animal Science* 69, 2995–3004.

Reynolds, P. and Street, P. (1991) *Efficiency in Beef Production: Twinning in Suckler Cows*. University of Reading Report, Reading, 25 pp.

Rhind, S.M., Schemm, S.R. and Schanbacher, B.D. (1991) Effects of active immunization against inhibin and androstenedione, separately or together, on hormone profiles, ovarian follicle populations and ovulation rate in heifers. *Proceedings of the British Society for Animal Production* (Winter Meeting), Paper No. 17.

Rice, C. and McGaughey, R.W. (1981) Effect of testosterone and dibutyryl cAMP on the spontaneous maturation of pig oocytes. *Journal of Reproduction and Fertility* 62, 245–256.

Richard, F. and Sirard, M.A. (1993) The effect of co-culture of follicular components with bovine oocytes on meiotic resumption. *Biology of Reproduction* 48 (Suppl. 1), 123.

Richards, D.W., Sikes, J.D. and Murphy, C.N. (1988) Non-surgical transfer and survival of frozen–thawed bovine embryos supplemented with raffinose. *Theriogenology* 21, 138–149.

Richardson, M.E., Peltz, J.M., Long, J.A. and Dickey, J.F. (1992) *In vitro* culture of goat (*Capra hircus*) embryos. *Biology of Reproduction* 46 (Suppl. 1), 121 (Abs. 282).

Riddell, K.P., Stringfellow, D.A., Gray, B.W., Riddell, M.G. and Worley, S.D. (1993a) Antimicrobial treatment of bovine embryos. *Theriogenology* 39, 297.

Riddell, K.P., Stringfellow, D.A., Gray, B.W., Riddell, M.G., Wright, J.C. and Galik, P.K. (1993b) Structural and viral association comparisons of bovine zonae pellucidae from follicular oocytes, day-7 embryos and day-7 degenerated ova. *Theriogenology* 40, 1281–1291.

Rider, S. (1991) ET doesn't pay. *Dairy Farmer*, August, 53.

Rieger, D. (1984) The measurement of metabolic activity as an approach to evaluating viability and diagnosing sex in early embryos. *Theriogenology* 21, 138–149.

Rieger, D. (1992) Relationships between energy metabolism and development of early mammalian embryos. *Theriogenology* 37, 75–93.

Rieger, D. and Guay, P. (1988) Measurement of the metabolism of energy substrates in individual bovine blastocysts. *Journal of Reproduction and Fertility* 83, 585–591.

Rieger, D. and Loskutoff, N.M. (1993) Changes in the metabolism of glucose, pyruvate and glutamine during *in vitro* maturation of bovine oocytes. *Theriogenology* 39, 298.

Rieger, D., Palmer, E., Langeau, D. and Tiffin, G. (1989) Simultaneous measurement of the metabolism of two energy substrates in individual horse and cow embryos. *Theriogenology* 31, 249.

Rieger, D., Bruyas, J.F., Lagneaux, D., Bezard, J. and Palmer, E. (1990) The effect of cryopreservation on the metabolic activity of day 6.5 horse embryos. *Proceedings of the Fifth International Symposium on Equine Reproduction* (Deauville), 132–133.

Rieger, D., Bruyas, J.-F., Lagneaux, D., Bezard, J. and Palmer, E. (1991a) The effect of cryopreservation on the metabolic activity of day-6.5 horse embryos. *Journal of Reproduction and Fertility*, Suppl. 44, 411–417.

Rieger, D., Walton, J.S., Goodwin, M.L. and Johnson, W.H. (1991b) The effect of co-treatment with recombinant bovine somatotrophin on plasma progesterone concentration and number of embryos collected from superovulated Holstein heifers. *Theriogenology* 35, 863–868.

Rieger, D., Loskutoff, N.M. and Betteridge, K.J. (1991c) Changes in glucose and glutamine metabolism during development of bovine embryos produced *in vitro*. *Serono Symposium on Preimplantation Embryo Development*, Abstract Vol., 44 (Abs. I-40).

Rieger, D., Loskutoff, N.M. and Betteridge, K.J. (1992) Developmentally related changes in the metabolism of glucose and glutamine by cattle embryos produced and co-cultured *in vitro*. *Journal of Reproduction and Fertility* 95, 585–595.

Rieger, D. and Loskutoff, N.M. (1994) Changes in the metabolism of glucose, pyruvate, glutamine and glycine during maturation of cattle oocytes *in vitro*. *Journal of Reproduction and Fertility* 100, 257–262.

Rigby, I. (1989) Once bred heifer systems. *British Society of Animal Production* (Winter Meeting), Paper No. 8.

Riha, J. and Petelikova, J. (1990) Some experiences with induced twinning in cattle in the Czech Republic. *Nas Chov* 50, 489–491.

Riha, J. and Polasek, M. (1987) The production of twin calves with embryo transfer to inseminated recipients. *Zivocisna Vyroba* 32, 489–497.

Riha, J., Polasek, M., Rozsival, A., Fulka, J., Pavlok, A. and Motlik, J. (1984) Transfer of embryos to inseminated cows for producing twins in cattle. *Zivocisna Vyroba* 29, 1–9.

Riha, J., Polasek, M., Vertelar, J., Jilek, L. and Banas, P. (1986) Non-surgical transfer of embryos to inseminated cows to induce twin pregnancy in cattle. *Zivocisna Vyroba* 31, 107–116.

Riha, J., Kneissl, J., Rehorka, V. and Polasek, M. (1989) The effect of recipient parity on their pregnancies after non-surgical embryo transfer. *Zivocisna Vyroba* 34, 19–24.

Riha, J., Sramek, J., Pozdisek, J. and Netopil, J. (1990) The course of pregnancy and parturition in heifers used at a young age for embryo transfer-twin production. *Zivocisna Vyroba* 35, 1–8.

Riha, J., Landa, V., Kneissl, J., Jindra, M. and Kloucek, Z. (1991) Vitrification of cattle embryos by direct dropping into liquid nitrogen and embryo survival after non-surgical transfer. *Zivocisna Vyroba* 36, 113–119.

Riha, J., Sramek, J. and Drimaj, M. (1992) The production of twins under commercial and experimental conditions. *Vyzkum v Chovu Skotu* 34(4), 1–5.

Rinehart, J.S., Bavister, B.D. and Gerrity, M. (1988) Quality control in the *in vitro* laboratory: comparison of bioassay systems for water quality. *Journal of In Vitro Fertilization and Embryo Transfer* 5, 335–342.

Risopatron, J.M., Campo, M.R. Del and Campo, C.H. Del (1991) Effect of storage temperature on the maturation, fertilization and development of cattle oocytes. *Archivos de Medicina Veterinaria* 23, 143–150.

Roberts, A.J. and Schlitz, S.M. (1991) Regulation of bovine thecal cell steroidogenesis by transforming growth factors alpha (TGF-alpha) and beta (TGF-beta). *Biology of Reproduction* 44 (Suppl. 1), 85 (Abs. 129).

Roberts, A.J. and Skinner, M.K. (1990) Estrogen regulation of thecal cell steroidogenesis and differentiation: thecal cell–granulosa cell interactions. *Endocrinology (Philadelphia)* 127, 2918–2929.

Roberts, A.J. and Skinner, M.K. (1991) Transforming growth factor alpha and beta differentially regulate growth and steroidogenesis of bovine thecal cells during antral follicle development. *Endocrinology (Philadelphia)* 129, 2041–2048.

Roberts, R.M., Schalue-Francis, T., Francis, H. and Keisler, D. (1990a) Maternal recognition of pregnancy and embryonic loss. *Theriogenology* 33, 175.

Roberts, R.M., Klemann, S., Li, J. and Farin, C.E. (1990b) The structure and properties of oTP-1, bTP-1 and their mRNA. *Proceedings of the Third International Ruminant Reproduction Symposium* (Nice), Abstract.

Roberts, R.M., Leaman, D.W. and Cross, J.C. (1992) Role of interferon in maternal recognition of pregnancy in ruminants. *Proceedings of the Society of Experimental Biology and Medicine* 200, 7–18.

Robertson, E.J. (1991) Using embryonic stem cells to introduce mutations into the mouse germ line. *Biology of Reproduction* 44, 238–245.

Robertson, J.L., Minhas, B.S., Palmer, T.V., Ansari, T.S. and Dodson, M.G. (1988) The absorption of ultraviolet light by the addition of para-amino-benzoic acid to nutrient media used in *in vitro* fertilization. *Theriogenology* 29, 298.

Robinson, J.J. (1990) The pastoral animal industries in the 21st century. *Proceedings of the New Zealand Society of Animal Production* 50, 345–359.

Robinson, J., Wallace, J. and Aitken, R. (1989) Significance of improved embryo transfer techniques for sheep breeding. *Agricultural and Food Research Council News* (Royal Show Supplement), 16–17.

Robinson, N.A., Leslie, K.E. and Walton, J.S. (1989) Effect of treatment with progesterone on pregnancy rate and plasma concentrations of progesterone in Holstein cows. *Journal of Dairy Science* 72, 202–207.

Robl, J.M. and Stice, S.L. (1989) Prospects for the commercial cloning of animals by nuclear transplantation. *Theriogenology* 31, 75–84.

Robl, J.M., Gilligan, B., Critser, E.S. and First, N.L. (1986a) Nuclear transplantation in mouse embryos: assessment of recipient cell stage. *Biology of Reproduction* 34, 733–739.

Robl, J.M., Prather, R., Eyestone, W.H., Barnes, F., Northey, D., Gilligan, B. and First, N.L. (1986b) Nuclear transplantation in bovine embryos. *Theriogenology* 25, 189.

Robl, J.M., Prather, R., Barnes, F., Eyestone, W.H., Northey, D., Gilligan, B. and First, N.L. (1987) Nuclear transplantation in bovine embryos. *Journal of Animal Science* 64, 642–647.

Robl, J.M., Fissore, R., Collas, P. and Duby, R.T. (1992) Cell fusion and oocyte activation. *Proceedings of a Symposium on Cloning Mammals by Nuclear Transplantation* (Fort Collins), 24–25.

Roche, J.M. (1989) New techniques in hormonal manipulation of cattle production. In: Phillips, C.J.C. (ed.) *New Techniques in Cattle Production*. Butterworths, London, pp. 48–60.

Rodriguez, H.F., Wiemer, K.W., Denniston, R.S. and Godke, R.A. (1990) A bilayered fetal-cell co-culture system for culturing bovine embryos. *Theriogenology* 33, 309.

Rodriguez, H.F., Denniston, R.S. and Godke, R.A. (1991) Co-culture of bovine embryos using a bilayer of bovine oviductal and granulosa cells. *Theriogenology* 35, 264.

Rodriguez, H.F., Denniston, R.S. and Godke, R.A. (1992) Co-culture of IVF-derived bovine embryos using bilayers of oviduct and granulosa cells. *Journal of Dairy Science* 75 (Suppl. 1), 302 (Abs. 9 So).

Rodriguez, J.L., Lopes, R.F.F., Nibart, M. and Thibier, M. (1991) Pregnancy rates obtained with frozen bovine embryos imported from Europe. *Proceedings of the Seventh Meeting of the European Embryo Transfer Association* (Cambridge), 202.

Roelofsen, M.W.M., Van Dillen, J., Aarts, T., Rodewijk and Van Der Streek, G. (1993) The collection of bovine oocytes by ovum pick up (OPU). *Proceedings of the Ninth Congress of the European Embryo Transfer Association* (Lyon), 274.

Rogers, J.B. (1978) Mammalian sperm capacitation and fertilization *in vitro*: a critique of methodology. *Gamete Research* 1, 165–223.

Roldan, E.R.S. and Gomendio, M. (1992) Morphological, functional and biochemical changes underlying the preparation and selection of fertilizing spermatozoa '*in vivo*'. *Animal Reproduction Science* 28, 69–78.

Romero, A., Thomas, W.K., Olson, S.E. and Seidel, G.E. (1990) Follicular fluid obtained after the LH surge stimulates meiotic maturation of bovine oocytes. *Theriogenology* 33, 310.

Romero, A., Olson, S.E., Thomas, W.K. and Seidel, G.E., Jr (1991) Maturation of bovine oocytes in follicular fluid collected after the LH surge enhances subsequent embryonic development. *Biology of Reproduction* 44 (Suppl. 1), 149 (Abs. 385).

Ron-El, R., Joris, H., Liu, J., Nagy, P. and Van Steirteghem, A. (1993) Intracytoplasmic sperm injection in the mouse. *Human Reproduction* 8 (Suppl. 1), 128.

Rorie, R.W., Pendleton, R.J., Pool, S.H., White, K.L. and Godke, R.A. (1987a) Cryopreservation of bovine trophoblastic vesicles. *Theriogenology* 27, 272.

Rorie, R.W., Pool, S.H., Pendleton, R.J. and Godke, R.A. (1987b) Co-culture of intact and half embryos of cattle with bovine trophoblastic vesicles. *Journal of Animal Science* 65 (Suppl. 1), 403–404.

Rorie, R.W., Gliedt, D.W., Miller, G.F. and Lester, T.D. (1991a) Development of *in vitro* produced bovine embryos in synthetic oviductal fluid (SOF) medium supplemented with different lots of fraction V bovine serum albumin (BSA). *Journal of Animal Science* 69 (Suppl. 1), 47–48.

Rorie, R.W., Lester, T.D. and Miller, F.G. (1991b) *In vitro* development of bovine embryos cultured in synthetic oviductal fluid (SOF) medium supplemented with either fetal calf serum (FCS) or bovine serum albumin (BSA) and the presence or absence of oviductal cells. *Journal of Animal Science* 69 (Suppl. 1), 402.

Roschlau, D., Roschlau, K., Roselius, R., Dexne, U., Michaelis, U., Strehl, R. and Unicki, P. (1992) Experiences in sexing of bovine embryos commercial programmes. *Proceedings of the Eighth Conference of the European Embryo Transfer Association* (Lyon), 204.

Roschlau, K., Rommel, P., Andreewa, L., Zackel, M., Roschlau, D., Zackel, B., Schwerin, M., Huhn, R. and Gazarjan, K.G. (1989) Gene transfer experiments in cattle. *Journal of Reproduction and Fertility*, Suppl. 38, 153–160.

Rose, E.P. De and Wilton, J.W. (1991) Productivity and profitability of twin-births in beef cattle. *Journal of Animal Science* 69, 3085–3093.

Rose, T.A. and Bavister, B.D. (1992) Effect of oocyte maturation medium on *in vitro* development of *in vitro* fertilized bovine embryos. *Molecular Reproduction and Development* 31, 72–77.

Rose, T.A., Fischer, B. and Sheffield, L.G. (1991) Role of EGF in oocyte–somatic cell interactions in the bovine follicle. *Biology of Reproduction* 44 (Suppl. 1), 148 (Abs. 381).

Rosenkrans, C.F., Jr and First, N.L. (1991) Culture of bovine zygotes to the blastocyst stage: effects of amino acids and vitamins. *Theriogenology* 35, 266.

Rosenkrans, C.F., Jr, Zeng, G.Z., Schoff, P.K. and First, N.L. (1990) A simple medium for *in vitro* development of bovine embryos. *Journal of Animal Science* 68 (Suppl. 1), 430.

Rosenkrans, C.F., Bovy, B.T., Sims, M.M. and First, N.L. (1991) Development of bovine embryos *in vitro* as affected by atmosphere, osmolarity and mannitol. *Journal of Animal Science* 69 (Suppl. 1), 404.

Rosenkrans, C.F., Zeng, G.Q., McNamara, G.T., Schoff, P.K. and First, N.L. (1993) Development of bovine embryos *in vitro* as affected by energy substrates. *Biology of Reproduction* 49, 459–462.

Rosenkranz, C. (1993) Low temperature (20°C) during transport of porcine ovaries improves *in vitro* maturation of COCs. *Journal of Reproduction and Fertility*, Abstract Series No. 11, 64 (Abs. 118).

Roudebush, W.E., Fukuda, A.I. and Minhas, B.S. (1993) Enhanced embryo development of rabbit oocytes fertilized *in vitro* with platelet activating factor (PAF)-treated spermatozoa. *Journal of Assisted Reproduction and Genetics* 10, 91–94.

Rowson, L.E.A. (1971) The role of reproductive research in animal production. *Journal of Reproduction and Fertility* 26, 113–126.

Rowson, L.E.A., Lamming, G.E. and Fry, R.M. (1953) The relationship between ovarian hormones and uterine infection. *Veterinary Record* 65, 335–340.

Rowson, L.E.A., Moor, R.M. and Lawson, R.A.S. (1969) Fertility following egg transfer in the cow: effect of method, medium and synchronization of oestrus. *Journal of Reproduction and Fertility* 18, 517–523.

Rowson, L.E.A., Lawson, R.A.S. and Moor, R.M. (1971) Production of twins in cattle by egg transfer. *Journal of Reproduction and Fertility* 25, 261–268.

Roy, S.K. and Greenwald, G.S. (1989) Hormonal requirements for the growth and differentiation of hamster preantral follicles in long-term culture. *Journal of Reproduction and Fertility* 87, 103–114.

Roy, S.K. and Treacy, B.J. (1993) Isolation and long-term culture of human preantral follicles. *Fertility and Sterility* 59, 783–790.

Ruane, J. (1988) Review of the use of embryo transfer in the genetic improvement of dairy cattle. *Animal Breeding Abstracts* 56, 439–446.

Ruane, J. (1993) Documenting the world's domestic animal resources. *Animal Genetic Resources Information* (FAO/UNEP) 11, 13–21.

Rubin, S.E., Magnoni, G.E. and Boatman, D.E. (1993) Oviductal modulation of sperm and egg interactions. *Biology of Reproduction* 48 (Suppl. 1), 167.

Rubinsky, B., Arav, A. and Devries, A.L. (1991) Cryopreservation of oocytes using directional cooling and antifreeze glycoproteins. *Cryo-Letters* 12, 93–106.

Rubinstein, S. and Breitbart, H. (1991) Role of spermine in mammalian sperm capacitation and acrosome reaction. *Biochemical Journal* 278, 25–28.

Ruiz, F.J., Oltenacu, P.A. and Smith, R.D. (1992) Cost-benefit evaluation of on-farm milk progesterone testing to monitor return to cyclicity and to classify ovarian cysts. *Journal of Dairy Science* 75, 1036–1043.

Ruknudin, A. (1989) Cytochemical study of intracellular calcium in hamster spermatozoa during the acrosome reaction. *Gamete Research* 22, 375–384.

Ruknudin, A., Dakoune, J.P. and Silver, I.A. (1988) Intracellular calcium in hamster spermatozoa in testis, epididymis and during the acrosome reaction. *Animal Reproduction Science* 16, 145–153.

Rush, L., Tibbitts, F.D. and Foote, W.D. (1973) Hormonal control of oocyte maturation *in vitro*. *Proceedings of the Western Section of the American Society of Animal Science* 24, 251–253.

Russe, I. (1983) Oogenesis in cattle and sheep. *Bibliotheca Anatomica* 24, 77–92.

Rust, J., Ritchie, W.A., McWhir, J. and Wilmut, I. (1993) Developmental potential of nuclei from sheep embryonic ectoderm. *Theriogenology* 39, 300.

Ryan, D. (1993a) Ultrasound for non-pregnancy diagnosis. *Irish Farmers' Journal* 45(32), 25.

Ryan, D. (1993b) Research in dairy herd reproduction. *Irish Farmers Journal* 45(50), 22–23.

Ryan, D.P., Blakewood, E.G., Swanson, W.F., Rodrigues, H. and Godke, R.A. (1990a) The use of follicle stimulating hormone (FSH) to stimulate follicle development for *in vitro* fertilization during the first trimester of pregnancy in cattle. *Theriogenology* 33, 315.

Ryan, D.P., Blakewood, E.G., Lynn, J.W., Munyakazi, L. and Godke, R.A. (1990b) The effect of heat-shock treatment on the *in vitro* development of early stage bovine embryos. *Biology of Reproduction* 42 (Suppl. 1), 56.

Ryan, D.P., Kopel, E., Boland, M.P. and Godke, R.A. (1991) Pregnancy rates in dairy cows following the administration of a GnRH analogue at the time of artificial insemination or at mid-cycle post-insemination. *Theriogenology* 36, 367–377.

Ryan, D.P., Condon, T., Grealy, M., Sreenan, J. and O'Farrell, K.O. (1992a) Physiological responses and pregnancy rates in dairy cows following the administration of gonadotrophin-releasing hormone at the time of artificial insemination or at mid-cycle post-insemination. *Proceedings of the Association of Veterinary Teachers and Research Workers* (Twenty-ninth Winter Meeting) 2 (Abs.).

Ryan, D.P., Blakewood, E.G., Lynn, J.W., Munyakazi, L. and Godke, R.A. (1992b) Effect of heat-stress on bovine embryo development *in vitro*. *Journal of Animal Science* 70, 3490–3497.

Ryan, D.P., Prichard, J.F., Kopel, E. and Godke, R.A. (1993a) Comparing early embryo mortality in dairy cows during hot and cool seasons of the year. *Theriogenology* 39, 719–737.

Ryan, D.P., Blakewood, E.G., Swanson, W.F., Rodriguez, H. and Godke, R.A. (1993b) Using hormone-treated pregnant cows as a potential source of oocytes for *in vitro* fertilization. *Theriogenology* 40, 1039–1055.

Ryan, D.P., D'Hoore, L., Snijders, S. and O'Farrell, K.J. (1994) Intrauterine transfer of bovine trophoblastic vesicles during dioestrus after breeding to increase pregnancy rates in dairy cows. *Theriogenology* 41, 287.

Ryan, J.P., O'Neill, C. and Wales, R.G. (1990a) Oxidative metabolism of energy substrates by preimplantation mouse embryos in the presence of platelet-activating factor. *Journal of Reproduction and Fertility* 89, 301–307.

Ryan, J.P., Spinks, N.R., O'Neill, C. and Wales, R.G. (1990b) Implantation potential and fetal viability of mouse embryos cultured in media supplemented with platelet-activating factor. *Journal of Reproduction and Fertility* 89, 309–315.

Ryan, J.P., O'Neill, C., Ammit, A.J. and Roberts, C.G. (1992) Metabolic and developmental responses of preimplantation embryos to platelet activating factor (PAF). *Reproduction, Fertility and Development* 4, 387–398.

Saeki, K., Nagao, Y., Utaka, K. and Ishimori, H. (1990a) Maturation of bovine oocytes by co-culture with rabbit preovulatory follicle cells. *Japanese Journal of Zootechnical Science* 61(1), 89–90.

Saeki, K., Leibfried-Rutledge, M.L. and First, N.L. (1990b) Are fetal calf serum and hormones necessary during *in vitro* maturation of cattle oocytes for subsequent development. *Theriogenology* 33, 316.

Saeki, K., Hoshi, M. and Leibfried-Rutledge, M.L. (1990c) Effects of gonadotrophins and estradiol-17β (E) during *in vitro* maturation of bovine oocytes in serum-free medium on subsequent fertilization and development. *Biology of Reproduction* 42 (Suppl. 1), 92.

Saeki, K., Hoshi, M., Liebfried-Rutledge, M.L. and First, N.L. (1990d) *In vitro* fertilization and development of bovine oocytes matured with commercially available follicle stimulating hormone. *Theriogenology* 34, 1035–1039.

Saeki, K., Hoshi, M., Leibfried-Rutledge, M.L. and First, N.L. (1991a) *In vitro* fertilization and development of bovine oocytes matured in serum-free medium. *Biology of Reproduction* 44, 256–260.

Saeki, K., Kato, H., Hosoi, Y., Miyake, M., Utsumi, K. and Iritani, A. (1991b) Early morphological events of *in vitro* fertilized bovine oocytes with frozen–thawed spermatozoa. *Theriogenology* 35, 1051–1058.

Saeki, K., Nagao, Y., Hoshi, M. and Kainuma, H. (1993) *In vitro* fertilization of bovine oocytes in a protein-free medium and subsequent development to the blastocyst stage. *Theriogenology* 39, 302.

Saiki, R.K., Gelfand, D.H., Stoffel, S., Scharf, S.J., Higuchi, R., Horn, G.T., Mullis, K.B. and Ehrlich, H.A. (1988) Primer directed enzymatic amplification of DNA with a thermostable DNA polymerase. *Science* 239, 487–494.

Saito, H., Totsukawa, K. and Kayaba, T. (1992) A study on the detection of the fluorescent body (F body) in mammalian spermatozoa. *Journal of the Yamagata Agriculture and Forestry Society* 48, 27–31.

Saito, S., Choi, Y.H. and Oguri, N. (1993) The parthenogenetic development of porcine oocytes matured *in vitro* after electrical, ionophore A and ethanol stimulation. *Theriogenology* 39, 303.

Saito, N., Imai, K. and Tomizawa, M. (1994) Effect of sugars-addition on the survival of vitrified bovine blastocysts produced *in vitro*. *Theriogenology* 41, 1053–1060.

Sakkas, D. and Trounson, A.O. (1990) Co-culture of mouse embryos with oviduct and uterine cells prepared from mice at different days of pseudopregnancy. *Journal of Reproduction and Fertility* 90, 109–118.

Sakkas, D., Batt, P.A. and Cameron, A.W.N. (1989) Development of preimplantation goat (*Capra hircus*) embryos *in vivo* and *in vitro*. *Journal of Reproduction and Fertility* 87, 359–365.

Sakkas, D., Lacham, O., Gianaroli, L. and Trounson, A.O. (1992) Subzonal sperm microinjection in cases of severe male factor infertility and repeated *in vitro* fertilization failure. *Fertility and Sterility* 57, 1279–1288.

Sakul, H., Bradford, G.E., BonDurant, R.H., Anderson, G.B. and Donahue, S.E. (1992) Cryopreservation of embryos as a means of germ plasm conservation in sheep. *Journal of Animal Science* 70 (Suppl. 1), 143.

Salhab, S.A. and Merilan, C.P. (1991) Some effects of collection equipment, glycerolation, and post-thaw re-equilibration times on the motility and survival of bovine spermatozoa. *Animal Reproduction Science* 24, 53–61.

Sanbuissho, A. (1989) Influence of serum and gonadotrophins on *in vitro* bovine oocyte maturation and fertilization. *Dissertation Abstracts International B, Sciences and Engineering* 49, 3046.

Sanbuissho, A. and Threlfall, W.R. (1988) The influence of serum and gonadotropins on bovine oocyte maturation *in vitro*. *Theriogenology* 29, 301.

Sanbuissho, A. and Threlfall, W.R. (1989) The effects of estrous cow serum on the *in vitro* maturation and fertilization of the bovine follicular oocyte *in vitro*. *Theriogenology* 31, 300–309.

Sanbuissho, A., Coskun, S. and Lin, Y.C. (1990a) Stimulatory action of epidermal growth factor (EGF) on bovine oocyte maturation in serum free defined medium. *Theriogenology* 33, 318.

Sanbuissho, A., Coskun, S. and Lin, Y.C. (1990b) Role of cyclic adenosine monophosphate (cAMP) in maturation process of *in vitro* cultured bovine follicular oocytes. *Biology of Reproduction* 42 (Suppl. 1), 93.

Sapienza, F., Verheyen, G., Tournaye, H., Janssens, R., Pletinex, I., Derde, M. and Van Steirteghem, A. (1993) An auto-controlled study in *in vitro* fertilization reveals the benefit of Percoll centrifugation to swim-up in the preparation of poor-quality semen. *Human Reproduction* 8, 1856–1862.

Sathananthan, A.H., Ng, S.C., Trounson, A.O., Bongso, A., Ratnam, S.S., Ho, J., Mok, H. and Lee, M.N. (1988) The effects of ultrarapid freezing on meiotic and mitotic spindles of mouse oocytes and embryos. *Gamete Research* 21, 385–401.

Sato, E., Iritani, A. and Nishikawa, Y. (1977) Factors involving *in vitro* maturation of pig and cattle follicular oocytes cultured *in vitro*. *Japanese Journal of Animal Reproduction* 23, 12–18.

Sato, E., Matsuo, M. and Miyamoto, H. (1990) Meiotic maturation of bovine oocytes *in vitro*: improvement of meiotic competence by bibutyryl cyclic adenosine 3′,5′-monophosphate. *Journal of Animal Science* 68(4), 1182–1187.

Savio, J.D., Bongers, H., Drost, M., Lucy, M.C. and Thatcher, W.W. (1991) Follicular dynamics and superovulatory response in Holstein cows treated with FSH-P in different endocrine states. *Theriogenology* 35, 915–929.

Savio, J.D., Thatcher, W.W., Morris, G.R., Entwistle, K., Drost, M. and Mattiacci, M.R. (1993) Effects of induction of low plasma progesterone concentrations with a progesterone-releasing intravaginal device on follicular turnover and fertility in cattle. *Journal of Reproduction and Fertility* 98, 77–84.

Scanlon, A.R., Boland, M.P., Ireland, J.J., Martin, T.L., O'Callaghan, D. and Roche, J.F. (1990) Effect of active immunization against an inhibin fragment on follicular dymamics and ovulation rate in beef heifers. *Journal of Reproduction and Fertility*, Abstract Series No. 6, Abs. 29.

Scanlon, P.F. (1969) Studies in reproduction in the cow. PhD Thesis, National University of Ireland, Dublin.

Schally, A.V. (1992) LH-RH analogues in medicine. *Proceedings of the Fourteenth World Congress on Fertility and Sterility* (Caracas), Abstract Vol., 3.

Schalue-Francis, T.K., Cross, J.C., Keisler, D., Sikes, J.D. and Roberts, R.M. (1989) Bovine interferon alpha-1 increases pregnancy rates in ewes. *Biology of Reproduction* 40 (Suppl. 1), 85 (Abs. 114).

Schalue-Francis, T.K., Keisler, D.H., Farin, P.W., Cross, J.C. and Roberts, R.M. (1990) Bovine interferon alpha-1 increases the lambing rate in ewes. *Journal of Animal Science* 68 (Suppl. 1), 125.

Schalue-Francis, T.K., Farin, P.W., Cross, J.C., Keisler, D. and Roberts, R.M. (1991) Effect of injected bovine interferon-α 1 on oestrous cycle length and pregnancy success in sheep. *Journal of Reproduction and Fertility* 91, 347–356.

Scheffen, B., Van Der Zwalmen, P. and Massip, A. (1986) A simple and efficient procedure for preservation of mouse embryos by vitrification. *Cryo-Letters* 7, 260–269.

Schellander, K., Brackett, B.G., Fuhrer, F. and Schleger, W. (1988) *In vitro* fertilization of frozen–thawed cattle oocytes. *Proceedings of the Eleventh International Congress on Animal Reproduction and Artificial Insemination* (Dublin) 3, 349 (3 pp.).

Schellander, K., Hassan-Hauser, C., Fuhrer, F., Korb, H. and Scheger, W. (1989a) Culture of bovine embryos for stem cell production. *Theriogenology* 31, 254.

Schellander, K., Brackett, B.G., Keefer, C.L. and Fayrer-Hosken, R.A. (1989b) Testing capacitation of bull sperm with zona-free hamster ova. *Animal Reproduction Science* 18, 95–104.

Schellander, K., Fayrer-Hosken, R.A., Keefer, C.L., Brown, L.M., Malter, H., McBride, C.E. and Brackett, B.G. (1989c) *In vitro* fertilization of bovine follicular oocytes recovered by laparoscopy. *Theriogenology* 31, 927–934.

Schellander, K., Hassan-Hauser, C., Fuhrer, F., Kalat, M., Mayr, B. and Korb, H. (1989d) Classification of cattle oocyte populations from antral follicles according to their *in vitro* maturation, fertilization and cleavage ability. *Wiener Tierarztliche Monatsschrift* 76, 320–331.

Schellander, K., Schellander, E., Fuhrer, F., Hauser, C., Schleger, W., Seregi, J., Peli, J., Treuer, A., Solti, L., Haraszti, J., Szasz, F., Benyei, B. and Torok, M. (1989e) *In vitro* fertilization of bovine oocytes. *Magyar Allatorvosok Lapja* 44, 561–563.

Schellander, K., Fuhrer, F.,Brackett, B.G., Korb, H. and Schleger, W. (1990a) *In vitro* fertilization and cleavage of bovine oocytes matured in medium supplemented with estrous cow serum. *Theriogenology* 33, 477–485.

Schellander, K., Mayr, B., Kalat, M., Fuhrer, F., Keefer, C.L. and Brackett, B.G. (1990b) A DNA-specific technique for pig and cattle oocytes using counterstain-enhanced fluorescence. *Caryologia* 43, 215–221.

Schenk, S.L. (1880) Das Saugethierei kunstlick befrucht ausserhalb des mutterthieres. *Mitt. Embr. Inst. K.K. Univ. Wien* 1, 107.

Schiewe, M.C., Schmidt, P.M., Pontbriand, D. and Wildt, D.E. (1988) Toxicity potential of residual ethylene oxide on fresh or frozen embryos maintained in plastic straws. *Gamete Research* 19, 31–39.

Schiewe, M.C., Schmidt, P.M., Wildt, D.E. and Rall, W.F. (1990a) Quality control measures in an embryo research program. *Theriogenology* 33, 9–22.

Schiewe, M.C., Rall, W.F., Stuart, L.D. and Wildt, D.E. (1990b) *In situ* straw dilution of ovine embryos cryopreserved by conventional freezing or vitrification. *Theriogenology* 33, 321.

Schiewe, M.C., Rall, W.F., Stuart, L.D. and Wildt, D.E. (1991) Analysis of cryoprotectant, cooling rate and *in situ* dilution using conventional freezing or vitrification for cryopreserving sheep embryos. *Theriogenology* 36, 279–293.

Schilling, E. and Doepke, H.H. (1978) A rapid diagnostic test for the viability of early cattle and rabbit embryos using diacetyl-fluorescein. *Naturawissenschaften* 65, 658–664.

Schilling, E., Smidt, D., Sacher, G., Petac, D. and El Kaschab, S. (1979) Diagnosis of the viability of early bovine embryos by fluoresence microscopy. *Annales de Biologie animale Biochimie Biophysique* 19, 1625–1629.

Schilling, E., Niemann, H. and Smidt, D. (1982) Evaluation of fresh and frozen cattle embryos by fluoresence microscopy. *Journal of In Vitro Fertilization and Embryo Transfer* 28, 349–355.

Schleger, W., Schellander, K., Mayr, B. and Pichler, W. (1990) Factors affecting cleavage in bovine oocytes fertilized *in vitro*. *Wiener Tieraerztliche Monatsschrift* 77, 39–43.

Schmidt, M., Avery, B. and Smith, S. (1991) Culture of frozen/thawed bovine demi-embryos with bovine oviduct epithelial cells. *Reproduction in Domestic Animals* 26, 88–90.

Schmidt, M., Smith, S.D., Avery, B., Purwantara, B. and Greve, T. (1992a) Coculture of bovine demi-embryos prior to freezing. *Acta Veterinaria Scandinavica* 33, 237–243.

Schmidt, M., Avery, B., Smith, S.D., Purwantara, B. and Greve, T. (1992b) The freezability of biopsied bovine embryos. *Theriogenology* 38, 615–621.

Schmidt, M., Smith, S.D., Avery, B., Purwantara, B. and Greve, T. (1992c) Blastomere content of cultured/frozen bovine demiembryos. *Acta Veterinaria Scandinavica* 33, 363–367.

Schmidt, M., Hyttel, P., Greve, T. and Avery, B. (1993) Ultrastructure of frozen/thawed bovine *in vitro* matured oocytes. *Theriogenology* 39, 304.

Schneider, U. and Mazur, P. (1986) Implications and applications of long-term preservation by freezing. In: Morrow, D.A. (ed.) *Current Therapy in Theriogenology*, 2nd edn. W.B. Saunders, Philadelphia, pp. 81–83.

Schrick, F.N., Inskeep, E.K. and Butcher, R.L. (1993) Pregnancy rates for embryos transferred from early postpartum beef cows into recipients with normal estrous cycles. *Biology of Reproduction* 49, 617–621.

Schroder, A., Miller, J.R., Thomsen, P.D., Roschlau, K., Avery, B., Poulsen, P.H., Schmidt, M. and Schwerin, M. (1990) Sex determination of bovine embryos using the polymerase chain reaction. *Animal Biotechnology* 1(2), 121–133.

Schroeder, A., Schultz, R.M., Kopf, G.S., Taylor, F.R., Becker, R.B. and Eppig, J.J. (1990) Fetuin inhibits zona pellucida hardening and conversion of ZP2 to $ZP2_f$ during spontaneous mouse oocyte maturation *in vitro* in the absence of serum. *Biology of Reproduction* 43, 891–897.

Schroeder, A.C., Johnston, D. and Eppig, J.J. (1991) Reversal of postmortem degeneration of mouse oocytes during meiotic maturation *in vitro*. *Journal of Experimental Zoology* 258, 240–245.

Schulman, J.D. (1993) Ethical issues in gender selection by X/Y sperm separation. *Human Reproduction* 8, 1541.

Schultz, G. (1993) Growth factor gene expression in preimplantation mammalian embryos. *Journal of Reproduction and Fertility*, Abstract Series No. 11, 3 (Abs. S2).

Schultz, R.M. and Dardik, A. (1991) Stimulatory effect of TGF-alpha/EGF on the rate of mouse blastocoel expansion. *Biology of Reproduction* 44 (Suppl. 1), 88 (Abs. 144).

Schumacher, A. and Fischer, B. (1988) Influence of visible light and room temperature on cell proliferation in preimplantation rabbit embryos. *Journal of Reproduction and Fertility* 84, 197–204.

Schumacher, A., Hegele-Hartung, C., Mootz, U., Fischer, B. and Beier, H.M. (1991a) Detrimental effects of visible light on embryoblast development in preimplantation rabbit embryos. *Journal of Reproduction and Fertility*, Abstract Series No. 7, 8 (Abs. 5).

Schumacher, A., Hegele-Hartung, C., Mootz, U., Fischer, B. and Beier, H.M. (1991b) Development of rabbit preimplantation embryos injured by prolonged exposure to visible light or room temperature. *Serono Symposium on Preimplantation Embryo Development*, Abstract Vol., 68 (Abs. II-36).

Schweigert, F.J. and Zucker, H. (1988) Concentration of vitamin A, β-carotene and vitamin E in individual bovine follicles of different quality. *Journal of Reproduction and Fertility* 82, 575–579.

Schweigert, F.J., Rambeck, W.A. and Zucker, H. (1987) β-Carotene and vitamin A in the follicular development of the bovine species. In: Roche, J.F. and O'Callaghan, D. (eds) *Follicular Growth and Ovulation Rate in Farm Animals*. Martinus Nijhoff, Dordrecht, The Netherlands pp. 55–62.

Schweigert, F.J., Wierich, M., Rambeck, W.A. and Zucker, H. (1988) Carotene cleavage activity in bovine ovarian follicles. *Theriogenology* 30, 923–930.

Schwerin, M., Blottner, S., Thomsen, P.D., Roschlau, D. and Brockmann, G. (1991) Quantification of Y chromosome bearing spermatozoa of cattle using *in situ* hybridization. *Molecular Reproduction and Development* 30, 39–43.

Schwiderski, H., Seifert, F. and Genz, S. (1990a) Splitting of cattle embryos: international position and own results. *Archiv für Tierzucht* 33, 139–148.

Schwiderski, H., Kauffold, P., Jura, J. and Duschinski, U. (1990b) Effect of microinjection of motile bovine spermatozoa into the perivitelline space of oocytes. *Reproduction in Domestic Animals* 25(6), 277–282.

Scodras, J.M., Pollard, J.W., Plante, L. and Betteridge, K.J. (1990) Establishment of a bovine trophoblast cell line supporting the development of *in vitro* matured and fertilized oocytes. *Biology of Reproduction* 42 (Suppl. 1), 167.

Scodras, J.M., Pollard, J.W. and Betteridge, K.J. (1991) Effect of somatic cell type on bovine embryonic development in co-culture. *Theriogenology* 35, 269.

Scott, S.J. (1993) Observations on the use of ultrasound to predict bovine embryo recipient suitability. *Proceedings of the Ninth Congress of the European Embryo Transfer Association* (Lyon), 276.

Scott, C.A., Robertson, L., de Moura, R.T.D., Paterson, C. and Boyd, J.S. (1994) Technical aspects of transvaginal ultrasound-guided follicular aspiration in cow. *Veterinary Record* 134, 440–443.

Seamark, R.F., Hadjisavas, M. and Robertson, S.A. (1992) Influence of the immune system on reproductive function. *Animal Reproduction Science* 28, 171–178.

Seaton, A.D., Catt, J.W., Rhodes, S.L., McDonald, M.F. and Welch, R.A.S. (1991) The use of unfrozen semen for *in vitro* fertilization of *in vitro* matured bovine oocytes. *Proceedings of the New Zealand Society of Animal Production* 51, 67–71.

Seidel, G.E., Jr (1983) Applications of microsurgery to mammalian embryos. *Theriogenology* 17, 1–12.

Seidel, G.E., Jr (1984) Applications of embryo transfer and related technologies to cattle. *Journal of Dairy Science* 67, 2786–2796.

Seidel, G.E., Jr (1986) Genetic engineering. In: Morrow, D.A. (ed.) *Current Therapy in Theriogenology*, 2nd edn. W.B. Saunders, Philadelphia, pp. 88–92.

Seidel, G.E., Jr (1988) Sexing mammalian sperm and embryos. *Proceedings of the Eleventh International Congress on Animal Reproduction and Artificial Insemination* (Dublin) 5, 137–144.

Seidel, G.E. (1989) Methods and comparative aspects of embryo cryopreservation in domestic animals. *Equine Veterinary Journal*, Suppl. 8, 77–79.

Seidel, G.E., Jr (1991) Embryo transfer: the next 100 years. *Theriogenology* 35, 171–180.

Seidel, G.E., Jr (1992) Overview of cloning mammals by nuclear transplantation. *Proceedings of a Symposium on Cloning Mammals by Nuclear Transplantation* (Fort Collins), 1–4.

Seidel, G.E., Jr, Elsden, R.P., Takeda, T. and Ferrand, G.D. (1983) Field trials with cryopreserved bovine embryos. In: Beir, H.M. and Lindner, H.R. (eds) *Fertilization of the Human Egg In Vitro*. Springer-Verlag, Berlin, pp. 343–352.

Seidel, G.E., Elsden, R.P. and Brink, Z. (1990) Cryopreservation of bovine embryos in media with chemically defined macromolecules. *Theriogenology* 33, 322.

Seidel, G.E., Jr, Glass, T. and Olson, S.E. (1991a) Culture of 1-cell bovine embryos to blastocysts in chemically defined media. *Biology of Reproduction* 44 (Suppl. 1), 155 (Abs. 412).

Seidel, G.E., Nauta, W. and Olson, S.E. (1991b) Effects of myoinositol, transferrin and insulin on culture of bovine embryos. *Journal of Animal Science* 69 (Suppl. 1), 403.

Seike, N., Saeki, K., Utaka, K., Sakai, M., Takakura, R., Nagao, Y. and Kanagawa, H. (1989) Production of bovine identical twins via transfer of demi-embryos without zonae pellucidae. *Theriogenology* 32, 211–220.

Sekine, J., Sakurada, T. and Oura, R. (1992) Optimum temperature of ovary transportation for *in vitro* fertilization of bovine oocytes. *Veterinary Record* 131, 372.

Selgrath, J.P. and Ebert, K.M. (1990) Fertilization of porcine oocytes in the mouse oviduct. *Biology of Reproduction* 42 (Suppl. 1), 129.

Selgrath, J.P., Memon, M.A., Smith, T.E. and Ebert, K.M. (1990) Collection and transfer of microinjectable embryos from dairy goats. *Theriogenology* 34, 1195–1205.

Seller, M.J. and Perkins-Cole, K.J. (1987) Sex difference in mouse embryonic development at neurulation. *Journal of Reproduction and Fertility* 79, 159–161.

Semple, E., Loskutoff, N., Leibo, S.P. and Betteridge, K.J. (1993) Effects of culture medium and maturation time on *in vitro* development of bovine oocytes into blastocysts. *Theriogenology* 39, 307.

Senger, P.L. (1993) The heat detection problem – new concepts, technologies and possibilities. *Journal of Dairy Science* 76 (Suppl. 1), 311.

Sergeev, N.I. (1986) Obtaining twins by embryo transfer to inseminated recipient cows. *Doklady Vsesoyuzoni Akademii Sel'skokhozyaistvennydh Nauk* 4, 26–28.

Sergejew, N. and Baibekow, F. (1992) *In vitro* maturation and *in vitro* fertilization of bovine oocytes. *Reproduction in Domestic Animals* 27, 44–45.

Serta, R.T., Seibel, M.M. and Kiessling, A.A. (1991) Maturation and fertilization of mouse oocytes in serum-free media. *Serono Symposium on Preimplantation Embryo Development*, Abstract Vol., 45 (Abs. I-42).

Sertich, P.L., Love, L.B., Hodgson, M.R. and Kenney, R.M. (1988) 24-hour cooled storage of equine embryos. *Theriogenology* 30, 947–952.

Seshagiri, P.B. (1990) Evidence for glycolytic inhibition of respiration and the oxidation of tricarboxylic acid cycle intermediates in cultured hamster 8-cell embryos. *Biology of Reproduction* 42 (Suppl. 1), 133.

Seshagiri, P.B. and Bavister, B.D. (1990a) Production of high quality viable blastocysts from hamster 8-cell embryos using hamster embryo culture medium-2. *Theriogenology* 33, 323.

Seshagiri, P.B. and Bavister, B.D. (1990b) Assessment of hamster blastocysts derived from eight-cell embryos cultured in hamster embryo culture medium-2 (HECM-2): cell numbers and viability following embryo transfer. *Journal of In Vitro Fertilization and Embryo Transfer* 7, 229–235.

Seshagiri, P.B. and Bavister, B.D. (1991a) Relative developmental abilities of hamster 2 and 8-cell embryos cultured in hamster embryo culture medium-1 and -2. *Journal of Experimental Zoology* 257, 51–57.

Seshagiri, P.B. and Bavister, B.D. (1991b) Glucose and phosphate inhibit respiration and oxidative metabolism in cultured hamster eight-cell embryos: evidence for the 'Crabtree effect'. *Molecular Reproduction and Development* 30, 105–111.

Setiabudi, R. and Gustavsson, I. (1991) Establishment of embryo sexing techniques in Sweden. *Reproduction in Domestic Animals* 26, 78–81.

Shabpareh, V., Squires, E.L., Seidel, G.E. and Jasko, D.J. (1993) Methods for collecting and maturing equine oocytes *in vitro*. *Theriogenology* 40, 1161–1175.

Shalgi, R., Smith, T.T. and Yanagimachi, R. (1992) A quantitative comparison of the passage of capacitated and uncapacitated hamster spermatozoa through the uterotubal junction. *Biology of Reproduction* 46, 419–424.

Shamsuddin, M. and Rodriguez-Martinez, H. (1993) Use of hyaluronic acid-selected spermatozoa for *in vitro* fertilization in the bovine. *Journal of Reproduction and Fertility*, Abstract Series No. 11, 66 (Abs. 122).

Shamsuddin, M., Larsson, B., Gustafsson, H., Gustari, S., Bartolome, J. and Rodriguez-Martinez, H. (1992a) Comparative morphological evaluation of *in vivo* and *in vitro* produced bovine embryos. *Proceedings of the Twelfth International Congress of Animal Reproduction* (The Hague) 3, 1333–1335.

Shamsuddin, M., Larsson, B. and Rodriguez-Martinez, H. (1992b) Use of hyaluronic acid for the selection of spermatozoa for *in vitro* fertilization in bovine. *Proceedings of the*

Nordic A.I. Virtanen Institute Symposium on the In Vitro Culture of Domestic Animal Embryos, 17–18.

Shamsuddin, M., Larsson, B., Gustafsson, H. and Rodriguez-Martinez, H. (1992c) Number of cells in the pre-implantation bovine embryos obtained from different methods of *in vitro* culture of IVM–IVF oocytes. *Proceedings of the Nordic A.I. Virtanen Institute Symposium on the In Vitro Culture of Domestic Animal Embryos*, 30–31.

Shamsuddin, M., Larsson, B., Gustafsson, H. and Rodriguez-Martinez, H. (1993a) *In vitro* development up to hatching of bovine *in vitro*-matured and fertilized oocytes with or without support from somatic cells. *Theriogenology* 39, 1067–1079.

Shamsuddin, M., Larsson, B. and Rodriguez-Martinez, H. (1993b) Culture of bovine IVM/IVF embryos up to blastocyst stage in defined medium using insulin, transferrin and selenium or growth factors. *Reproduction in Domestic Animals* 28, 209–210.

Sharif, H. (1992) The use of the isolated mouse oviduct as an *in vitro* culture system for cattle oocytes and embryos. PhD Thesis, National University of Ireland, Dublin.

Sharif, H., Vergos, E., Lonergan, P., Gallagher, M., Kinis, A. and Gordon, I. (1991a) Development of early bovine embryos in mouse oviducts maintained in organ culture. *Theriogenology* 35, 270.

Sharif, H., Lonergan, P., Monaghan, P., Wahid, H., Gallagher, M. and Gordon, I. (1991b) Culture of bovine embryos in the isolated mouse oviduct. *Proceedings of the Irish Grassland and Animal Production Association* (Seventeenth Annual Research Meeting), 29–30.

Sharif, H., Lonergan, P., Monaghan, P., Wahid, H.A., Gallagher, M. and Gordon, I. (1991c) Effect of oviductal condition on the development of early bovine embryos in cultured mouse oviducts. *Journal of Reproduction and Fertility*, Abstract Series No. 7, 57 (Abs. 103).

Sharif, H., Lonergan, P., Monaghan, P., Wahid, H., Gallagher, M. and Gordon, I. (1991d) Early bovine embryonic development in cultured mouse oviduct obtained at various stages of the oestrous cycle. *Proceedings of the Seventh Meeting of the European Embryo Transfer Association* (Cambridge), 206.

Sharif, H., Zamri-Saad, M., Lonergan, P., Monaghan, P., Gallagher, M. and Gordon, I. (1992a) Histological evaluation of the isolated mouse oviduct as used for an *in vitro* culture system for early bovine embryos. *Proceedings of the Irish Grassland and Animal Production Association* (*Eighteenth* Annual Research Meeting), 83–84.

Sharif, H., Lonergan, P., Monaghan, P. and Gordon, I. (1992b) The influence of culture temperature on bovine development in the isolated mouse oviduct system. *Journal of Reproduction and Fertility*, Abstract Series No. 9, 61 (Abs. 108).

Sharif, H., Gordon, I., Lonergan, P. and Monaghan, P. (1992c) Fertilization and subsequent development of *in vitro* matured bovine oocytes in the isolated mouse oviduct system. *Proceedings of the Eighth Conference of the European Embryo Transfer Association* (Lyon), 214.

Sharif, H., Lonergan, P., Monaghan, P., Gallagher, M. and Gordon, I. (1992d) Comparison of bovine embryonic development in an isolated mouse oviduct culture system with bovine oviductal epithelial cell co-culture. *Proceedings of the Twelfth International Congress of Animal Reproduction* (The Hague) 3, 1336–1338.

Sharif, H., Monaghan, P. and Gordon, I. (1993) Influence of amino-acids and vitamins on the development of early bovine embryos in the isolated mouse oviduct system. *Theriogenology* 39, 308.

Sharma, D., Teneja, V.K. and Majumdar, A.C. (1991) Buffalo bull fertility test using dead oocyte penetration technique. *Indian Journal of Animal Sciences* 61, 836–837.

Shaw, J.M. and Trounson, A.O. (1989) Parthenogenetic activation of unfertilized mouse oocytes by exposure to 1,2-propanediol is influenced by temperature, oocyte age and cumulus removal. *Gamete Research* 24, 269–279.

Shaw, P.W., Fuller, B.J., Bernard, A. and Shaw, R.W. (1991) Vitrification of mouse oocytes: improved rates of survival, fertilization and development to blastocysts. *Molecular Reproduction and Development* 29, 373–378.

Shaw, R.C. (1989) Into the future with cattle breeding organisations. *British Cattle Breeders Conference*, Digest No. 44, 38–40.

Shea, B.F. (1981) Evaluating the bovine embryo. *Theriogenology* 15, 31–42.

Shea, B.F., Janzen, R.E. and McAlister, R. (1983) Recovery and fertilization of bovine follicular oocytes. *Theriogenology* 19, 385–390.

Sheldon, I.M. and Dobson, H. (1993) Effects of gonadotrophin releasing hormone administered 11 days after insemination on the pregnancy rates of cattle to first and later services. *Veterinary Record* 133, 160–163.

Shelton, J.N. (1987) Prospects for the use of embryos in the control of disease and the transport of genotypes. *Australian Veterinary Journal* 64, 6–10.

Shelton, J.N. (1988) Embryo manipulation in research and animal production. *Australian Journal of Biological Science* 41, 117–132.

Shelton, J.N. (1992) Factors affecting viability of fresh and frozen-thawed sheep demi-embryos. *Theriogenology* 37, 713–721.

Shi, D.S. (1991) Studies related to factors affecting the yield of bovine embryos produced by *in vitro* techniques. PhD Thesis, National University of Ireland, Dublin.

Shi, D.S., Lu, K.H. and Gordon, I. (1990) Effects of bulls on fertilization of bovine oocytes and their subsequent development *in vitro*. *Theriogenology* 33, 324.

Shi, D.S., Lu, K.H., Gordon, I. and Polge, C. (1991) Variation in cleavage and embryonic development of bovine oocytes *in vitro* fertilized with different bull ejaculates. *Theriogenology* 35, 271.

Shi, Z., Jiang, S. and Yang, X. (1993) Synergistic effect of A23187 and cycloheximide allows effective activation of freshly matured bovine oocytes. *Theriogenology* 39, 309.

Shim, C., Lee, D.K., Lee, C.C., Cho, W.K. and Kim, K. (1992) Inhibitory effect of purines in meiotic maturation of denuded mouse oviducts. *Molecular Reproduction and Development* 31, 280–286.

Shioya, Y., Kuwayama, M., Ueda, S., Saitou, S., Oota, H. and Hanada, A. (1988a) Effect of the time between slaughter and aspiration of follicles on the developmental capability of bovine oocytes matured and fertilized *in vitro*. *Japanese Journal of Animal Reproduction* 34, 39–44.

Shioya, Y., Kuwayama, M., Fukushima, M., Iwasaki, S. and Hanada, A. (1988b) *In vitro* fertilization and cleavage capability of bovine follicular oocytes classified by cumulus cells and matured *in vitro*. *Theriogenology* 30, 489–496.

Shisong, C. and Wrathall, A.E. (1989) The importance of the zona pellucida for disease control in livestock by embryo transfer. *British Veterinary Journal* 145, 129–140.

Shorgan, B., Suolian, Z., Xiaoxian, X., Yiefei, P., Hongwu, C., Xihe, L., Qinglan, H. and Qin, S. (1990) *In vitro* development of ovine oocytes matured and fertilized *in vitro* and lambing after embryo transfer. *Japanese Journal of Animal Reproduction* 36, 225–229.

Shull, M.M., Ormsby, I., Kier, A.B., Pawlowski, S., Diebold, R.J., Yin, M., Allen, R., Sidman, C., Proetzel, G., Calvin, D., Annunziata, N. and Doetschman, T. (1992) Targeted disruption of the mouse transforming growth factor-β1 gene results in multifocal inflammatory disease. *Nature* 359, 693–699.

Shushan, A. and Schenker, J.G. (1993) Prenatal sex determination and selection. *Human Reproduction* 8, 1545–1549.

Shyu, J.J., Graves, C.N. and Wagner, W.C. (1990) A morphological study of porcine oocyte maturation *in vitro*. *Biology of Reproduction* 42 (Suppl. 1), 93.

Sianangama, P.C. and Rajamahendran, R. (1992) Effect of human chorionic gonadotrophin administered at specific times following breeding on milk progesterone and pregnancy in cows. *Theriogenology* 38, 85–96.

Sikes, J.D., Hernandez-Ledezma, J.J., Villanueva, C. and Roberts, R.M. (1993) Effects of CZB versus Medium 199 and of conditioning culture media with either bovine oviductal epithelial cells or buffalo rat liver cells on development of bovine zygotes derived by IVM–IVF procedures. *Theriogenology* 39, 310.

Silvan, G., Illera, J.C., Illera, M.J., Lorenzo, P. and Illera, M. (1993) Percentage of atretic and non-atretic follicles in different follicular size of heifers and cows. *Theriogenology* 39, 311.

Silvia, W.J., Lewis, G.S., McCracken, J.A., Thatcher, W.W. and Wilson, L. (1991) Hormonal regulation of uterine secretion of prostaglandin $F_{2\alpha}$ during luteolysis in ruminants. *Biology of Reproduction* 45, 655–663.

Simm, G., Steane, D.E. and Wray, N.R. (1990) Developments in beef cattle breeding programmes in Europe. *Proceedings of the Fourth World Congress on Genetics Applied to Livestock Production*, 231–243.

Simmen, F.A. and Simmen, R.C.M. (1991) Peptide growth factors and proto-oncogenes in mammalian conceptus development. *Biology of Reproduction* 44, 1–5.

Simmen, R.C.M., Ko, Y. and Simmen, F.A. (1993) Insulin-like growth factors and blastocyst development. *Theriogenology* 39, 163–175.

Simmons, C.R. and Graves, C.N. (1990) A comparison of embryo cryopreservation techniques – controlled freezing versus vitrification. *Journal of Animal Science* 68 (Suppl. 1), 116–117.

Simon, L., Bungartz, L., Rath, D. and Niemann, H. (1993) Repeated bovine oocyte collection by means of a permanently rinsed ultrasound guided aspiration unit. *Theriogenology* 39, 312.

Simon, M., Jilek, F. and Fulka, J. (1989) Effect of cycloheximide upon maturation of bovine oocytes. *Reproduction, Nutrition, Development* 29, 533–540.

Simons, J.P. and Land, R.B. (1987) Transgenic livestock. *Journal of Reproduction and Fertility*, Suppl. 34, 237–250.

Simons, J.P., McClenaghan, M. and Clark, A.J. (1987) Alteration of the quality of milk by expression of sheep β-lactoglobulin in transgenic mice. *Nature* 328, 530–532.

Simons, J.P., Wilmut, I., Clark, A.J., Archibald, A.L., Bishop, J.O. and Lathe, R. (1988) Gene transfer into sheep. *Biotechnology* 6, 179–183.

Simons, J.P., Ali, S., Archibald, A.L., Brown, P., Harris, S., McClenaghan, M., Whitelaw, C.B.A., Wilmut, I. and Clark, A.J. (1990) Modifications of milk composition by gene transfer. *Proceedings of the Fourth World Congress on Genetics Applied to Livestock Production* 14, 132–135.

Simpson, A.M. and White, I.G. (1987) Interrelationships between motility, cAMP, respiration and calcium uptake of ram and boar sperm. *Animal Reproduction Science* 15, 189–207.

Sims, M.M. and First, N.L. (1993) Production of fetuses from totipotent cultured bovine inner cell mass cells. *Theriogenology* 39, 313.

Sims, M.M., Rosenkrans, C.F. and First, N.L. (1991) Development *in vitro* of bovine embryos derived from nuclear transfer. *Theriogenology* 35, 272.

Sinclair, K.D., Broadbent, P.J., Dolman, D.F. and McNally, J.R. (1992) The effect of a Norgestomet ear implant (Crestar) and prostaglandin regime on pregnancy rate and re-synchronisation of oestrus in embryo transfer recipients. *Proceedings of the Twelfth International Congress on Animal Reproduction* (The Hague) 2, 814–816.

Sinclair, K.D., Broadbent, P.J., Dolman, D.F., Taylor, A.G. and Reaper, J.F. (1994) *In vitro* produced embryos as a means of achieving pregnancy and improving productivity in beef cows. *Theriogenology* 41, 294.

Singh, B. and Armstrong, D.T. (1992) Factors influencing oocyte maturation and cumulus cell expansion in pigs *in vitro*. *Biology of Reproduction* 46 (Suppl. 1), 138 (Abs. 351).

Singh, B. and Armstrong, D.T. (1994) Localization of epidermal growth factor and its receptor in the porcine ovarian follicle, and its effects on *in vitro* maturation (IVM) and fertilization (IVF) of porcine oocytes. *Theriogenology* 41, 295.

Singh, B., Zhang, X. and Armstrong, D.T. (1993a) Porcine oocyte release cumulus expansion-enabling activity even though porcine cumulus expansion *in vitro* is independent of the oocyte. *Endocrinology* 132, 1860–1862.

Singh, B., Barbe, G.J. and Armstrong, D.T. (1993b) Factors influencing resumption of meiotic maturation and cumulus expansion of porcine oocyte–cumulus cell complexes *in vitro*. *Molecular Reproduction and Development* 36, 113–119.

Singh, E.L. (1987) Disease transmission and embryo transfer. *Proceedings of a Symposium on the Application of Egg and Embryo Technologies to Domestic Animals* (Copenhagen), 16–18.

Singh, E.L. (1988) The potential of semen and embryos for introducing pathogens into the uterus. *Proceedings of the Eleventh International Congress on Animal Reproduction and Artificial Insemination* (Dublin) 5, 72–79.

Singh, E.L. and Hare, W.C.D. (1986) Embryo–pathogen interactions in relation to disease transmission. In: Morrow, D.A. (ed.) *Current Therapy in Theriogenology*, 2nd edn. W.B. Saunders, Philadelphia, pp. 84–87.

Singh, R. and Majumdar, A.C. (1992) Chronological changes of buffalo follicular oocytes maturation *in vitro*. *Indian Journal of Animal Sciences* 82, 205–209.

Siracusa, G., Felici, M.D. and Salustri, A. (1990) Meiotic maturation of the mammalian oocyte. In: Asch, R.H., Balmaceda, J.P. and Johnson, I. (eds) *Gamete Physiology: Serona Symposia USA, Norwell, Massachusetts*, pp. 129–144.

Sirard, M.A. (1989a) Temporary inhibition of *in vitro* meiotic resumption by adenylate cyclase stimulation in immature bovine oocytes. *Theriogenology* 31, 257.

Sirard, M.A. (1989b) Practical aspects of *in vitro* fertilization in cattle. *Journal of Reproduction and Fertility*, Suppl. 38, 127–134.

Sirard, M.A. (1990) Temporary inhibition of meiosis resumption *in vitro* by adenylate cyclase stimulation in immature bovine oocytes. *Theriogenology* 33, 757–767.

Sirard, M.A. and Bilodeau, S. (1990a) Live granulosa cells can maintain meiotic inhibition of bovine oocytes *in vitro*. *Biology of Reproduction* 42 (Suppl. 1), 90.

Sirard, M.A. and Bilodeau, S. (1990b) Granulosa cells inhibit the resumption of meiosis in bovine oocytes *in vitro*. *Biology of Reproduction* 43, 777–783.

Sirard, M.A. and Coenen, K. (1993a) The use of a defined medium to evaluate meiotic resumption of immature bovine oocytes. *Theriogenology* 39, 314.

Sirard, M.A. and Coenen, K. (1993b) The co-culture of cumulus-enclosed bovine oocytes and hemi-sections of follicles: effects on meiotic resumption. *Theriogenology* 40, 933–942.

Sirard, M.A. and First, N.L. (1988) *In vitro* inhibition of oocyte nuclear maturation in the bovine. *Biology of Reproduction* 39, 229–234.

Sirard, M.A. and Lambert, R.D. (1985) *In vitro* fertilization of follicular oocytes obtained by laparoscopy. *Biology of Reproduction* 33, 487–494.

Sirard, M.A. and Lambert, R.D. (1986) Birth of calves after *in vitro* fertilization using laparoscopy and rabbit oviduct incubation of zygotes. *Veterinary Record* 119, 167–169.

Sirard, M.A. and Lambert, R.D. (1988) *In vitro* inhibition of oocyte nuclear maturation in the bovine. *Biology of Reproduction* 39, 229–234.

Sirard, M.A., Lambert, R.D., Guay, P., Menard, D.P. and Bedoya, M. (1985a) *In vivo* and *in vitro* development of *in vitro* fertilized bovine follicular oocytes obtained by laparoscopy. *Theriogenology* 23, 230.

Sirard, M.A., Lambert, R.D., Beland, R. and Bernard, C. (1985b) The effects of repeated laparoscopic surgery used for ovarian examination and follicular aspiration in cows. *Animal Reproduction Science* 9, 25–30.

Sirard, M.A., Lambert, R.D. and Gray, P. (1986a) *In vitro* development of *in vitro* fertilized bovine follicular oocytes obtained by laparoscopy. *Animal Reproduction Science* 12, 21–29.

Sirard, M.A., Fortier, M., Lambert, R.D. and Leclerc, P. (1986b) Enhanced motility of bovine spermatozoa treated with supernatant of bovine ovarian cell culture. *Theriogenology* 25, 197.

Sirard, M.A., Lambert, R.D., Menard, D.P. and Bedoya, M. (1986c) *In vitro* fertilization in the cow: 6 calves are born from surgical or non-surgical uterine transfer to heifers. *Theriogenology* 25, 198.

Sirard, M.A., Florman, H., Barnes, F. and First, N.L. (1988a) *In vitro* maturation of bovine oocytes: nuclear changes and protein requirement. *Theriogenology* 29, 307.

Sirard, M.A., Parrish, J.J., Ware, C.B., Leibfried-Rutledge, M.L. and First, N.L. (1988b) The culture of bovine oocytes to obtain developmentally competent embryos. *Biology of Reproduction* 39, 546–552.

Sirard, M.A., Florman, H.M., Leibfried-Rutledge, M.L., Barnes, F.L., Sims, M.L. and First, N.L. (1989) Timing for nuclear progression and protein synthesis necessary for meiotic maturation of bovine oocytes. *Biology of Reproduction* 40, 1257–1263.

Sirard, M.A., Coenen, K. and Bilodeau, S. (1992) Effect of fresh or cultured follicular fractions on meiotic resumption in bovine oocytes. *Theriogenology* 37, 39–57.

Sirois, J. and Fortune, J.E. (1990) Lengthening the bovine estrous cycle with low levels of exogenous progesterone: a model for studying ovarian follicular dominance. *Endocrinology* 127, 916–925.

Sirotkin, A.V. and Nitray, J. (1992) Effects of growth hormone on bovine granulosa cells and oocytes in culture. *Journal of Reproduction and Fertility*, Abstract Series No. 9, 49 (Abs. 84).

Sivakumaran, K., Calder, M. and Rajamahendran, R. (1991) The influence of reproductive status on oocyte number and maturation *in vitro* and the effect of co-culture systems on fertilization and subsequent development. *Journal of Animal Science* 69 (Suppl. 1), 438.

Sjogren, A., Holmes, P.V. and Hillensjo, T. (1991) Interleukin-1-alpha modulates luteinizing hormone stimulated cyclic AMP and progesterone release from human granulosa cells *in vitro*. *Human Reproduction* 6, 910–913.

Sjogren, A., Hamberger, L., Lundin, K. and Soderlund, B. (1992) Avoidance of polyspermia in microsurgical fertilization techniques. *Human Reproduction* 7 (Suppl. 2), 138.

Skrzyszowska, M. and Smorag, Z. (1989) Cell loss in bisected mouse, sheep and cow embryos. *Theriogenology* 32, 115–122.

Skyer, D.M., Garverick, H.A., Youngquist, R.S. and Krause, G.F. (1987) Ovarian follicular populations and *in vitro* steroidogenesis on three different days of the estrous cycle. *Journal of Animal Science* 64, 1710–1716.

Slade, N.P., Takeda, T., Squires, E.L., Elsden, R.P. and Seidel, G.E. (1985) A new procedure for the cryopreservation of equine embryos. *Theriogenology* 24, 54–58.

Slavik, T. and Fulka, J. (1991) Pregnancies after transfer of sheep embryos produced from oocytes matured and fertilized *in vitro*. *Folia Biologica-Praha* 37, 94–100.

Slavik, T. and Fulka, J. (1992) *In vitro* fertilization of intact sheep and cattle oocytes with goat spermatozoa. *Theriogenology* 38, 721–726.

Slavik, T., Pavlok, A. and Fulka, J. (1990) Penetration of intact bovine ova with ram spermatozoa *in vitro*. *Molecular Reproduction and Development* 25, 345–347.

Slavik, T., Fulka, J. and Goll, I. (1992) Pregnancy rate after the transfer of sheep embryos originated from randomly chosen oocytes matured and fertilized *in vitro*. *Theriogenology* 38, 749–756.

Slaweta, R. and Laskowska, T. (1987) The effect of glutathione on the motility and fertility of frozen bull sperm. *Animal Reproduction Science* 13, 249–253.

Smirnova, T.A. and Galieva, L.D. (1989) State of the oocyte nucleus of cows in relation to development of the microcirculation of follicles. *Byulleten, Vsesoyuznyi Nauchno-Issledovatel'skii Institut Razvedeniya i Genetiki Sel'skokhozyaistvennykh Zhivotnykh* 115, 21–22.

Smith, B.J., Spire, M.F., Davis, D.L. and Schalles, R.R. (1986) Bovine embryo development in Dulbecco's phosphate buffered saline and Ham's F-10 medium with Hepes buffer. *Theriogenology* 25, 199.

Smith, C. (1988) Genetic improvement of livestock, using nucleus breeding units. *World Animal Review* 65, 2–10.

Smith, C. (1989) Cloning and genetic improvement of beef cattle. *Animal Production* 49, 49–62.

Smith, C. (1990) Breeding strategies to capitalize on new technologies. *Proceedings of the Twenty-third International Dairy Congress* (Montreal) 1, 579–584.

Smith, C., Meuwissin, T.H.E. and Gibson, J.P. (1987) On the use of transgenes in livestock improvement. *Animal Breeding Abstracts* 55, 1–10.

Smith, L.C. (1990) Cloning mammals by nuclear transplantation. *Proceedings of the Third Symposium on Advanced Topics in Animal Reproduction* (FCAV-UNESP, Jaboticabal), 119–141.

Smith, L.C. (1993) Membrane and intracellular effects of ultraviolet irradiation with Hoechst 33342 on bovine secondary oocytes matured *in vitro*. *Journal of Reproduction and Fertility* 99, 39–44.

Smith, L.C. and Wilmut, I. (1988) Factors influencing nuclear transplantation in sheep embryos. *Journal of Reproduction and Fertility,* Abstract Series No. 1, Abs. 10.

Smith, L.C. and Wilmut, I. (1989) Influence of nuclear and cytoplasmic activity on the development *in vivo* of sheep embryos after nuclear transplantation. *Biology of Reproduction* 40, 1027–1035.

Smith, L.C. and Wilmut, I. (1990) Factors affecting the viability of nuclear transplanted embryos. *Theriogenology* 33, 153–164.

Smith, L.C., France Desbiens and Goff, A.K. (1989) Protein synthesis in bovine oocytes after ovulation and fertilization. *Journal of Reproduction and Fertility,* Abstract Series No. 4, Abs. 14.

Smith, S., Schmidt, M., Purwantara, B. and Greve, T. (1993) Oviduct epithelial cell co-culture of early porcine embryos. *Acta Veterinaria Scandinavica* 33, 349–355.

Smith, S.D., Soloy, E., Avery, B., Holm, P., Callesen, H. and Greve, T. (1994) Nucleus remodelling in reconstructed cattle embryos. *Theriogenology* 41, 298.

Solano, R., de Armas, R., Pupo, C.A. and Castro, F.O. (1994) Short term preservation of intrafollicular oocytes at 4°C. *Theriogenology* 41, 299.

Solti, L., Machaty, Z., Barandi, Zs., Torok, M. and Vajta, G. (1992) IVF embryos of known parental origin from the endangered Hungarian Grey cattle breed. *Theriogenology* 37, 301.

Soong, Y.K., Lai, Y.M., Chang, S.Y., Wang, M.L., Chang, M.Y. and Lee, C.L. (1993) A successful pregnancy after subzonal insertion with epididymal sperm and coculture on Vero cell monolayer. *Fertility and Sterility* 59, 1308–1310.

Sparks, A.E.T., Gwazdauskas, F.C. and McGilliard, M.L. (1992a) Culture of one-cell bovine embryos in explanted mouse oviduct and bovine oviductal epithelial cells. *Theriogenology* 37, 587–594.

Sparks, A.E.T., Canseco, R.S., Russell, C.G., Moll, H.D., Velander, W.H., Johnson, J.L. and Gwazdauskas, F.C. (1992b) Utilization of pronuclear DNA microinjection, *in vitro* culture, microsurgery, and polymerase chain reaction technique for production and selection of transgenic bovine preimplantation embryos. *Biology of Reproduction* 46 (Suppl. 1), 71 (Abs. 83).

Spedding, A.W. (1989) The potential for increasing slaughter weight of suckled calves. *British Society Animal Production* (Winter Meeting), Paper No. 9.

Spicer, L.J. and Echternkamp, S.E.(1986) Ovarian follicular growth, function and turnover in cattle: a review. *Journal of Animal Science* 62, 428–451.

Spicer, L.J. and Geisert, R.D. (1992) Concentrations of insulin-like growth factor-I, estradiol and progesterone in follicular fluid of ovarian follicles during early pregnancy in cattle. *Theriogenology* 37, 749–760.

Spicer, L.J. and Zavy, M.T. (1992) Concentrations of insulin-like growth factor-I in serum of sheep with different ovulation rates: changes during the estrous cycle. *Theriogenology* 37, 395–405.

Spicer, L.J., Tucker, H.A., Convey, E.M. and Echternkamp, S.E. (1987) Comparison of surface diameters and dissected diameters of bovine ovarian follicles. *Journal of Animal Science* 64, 226–230.

Spicer, L.J., Enright, W.J., Murphy, M.G. and Roche, J.F. (1991) Effect of dietary intake on concentrations of insulin-like growth factor-I in plasma and follicular fluid, and ovarian function in heifers. *Domestic Animal Endocrinology* 8, 431–437.

Spicer, L.J., Crowe, M.A., Prendiville, D.F., Goulding, D. and Enright, W.J. (1992) Systemic but not intraovarian concentrations of insulin-like growth factor-I are affected by short-term fasting. *Biology of Reproduction* 46, 920–925.

Spiropoulos, J. and Long, S.E. (1989) Female meiosis in dairy cattle. *Journal of Reproduction and Fertility*, Abstract Series No. 3, 48.

Squires, E.L. (1989) Commercial equine embryo transfer: synopsis of panel discussion. *Equine Veterinary Journal*, Suppl. 8, 75.

Squires, E.L. and Tarr, S.F. (1991) Advances in the transfer and manipulation of equine embryos. *Veterinary Medicine* 86, 626–631.

Squires, E.L., Seidel, G.E. and McKinnon, A.O. (1989) Transfer of cryopreserved equine embryos to progestin-treated ovariectomized mares in winter. *Equine Veterinary Journal*, Suppl. 8, 89–91.

Sreenan, J.M. (1968) *In vivo* and *in vitro* culture of cattle eggs. *Proceedings of the Sixth International Congress on Animal Reproduction and Artificial Insemination* (Paris) 1, 577–580.

Sreenan, J.M. (1969) Studies related to ovum transfer in cattle. PhD Thesis, National University of Ireland, Dublin.

Sreenan, J.M. (1970) *In vitro* maturation and attempted fertilization of cattle follicular oocytes. *Journal of Agricultural Science* (Cambridge) 75, 393–396.

Sreenan, J.M. (1988) Embryo transfer: its uses and recent developments. *Veterinary Record* 122, 624–629.

Sreenan, J.M. and Beehan, D. (1974) Egg transfer in the cow: pregnancy rate and egg survival. *Journal of Reproduction and Fertility* 41, 497–499.

Sreenan, J.M. and Beehan, D. (1976) Embryonic survival and development at various stages of gestation after bilateral egg transfer in the cow. *Journal of Reproduction and Fertility* 47, 127–128.

Sreenan, J.M. and Diskin, M.G. (1986) The extent and timing of embryonic mortality in the cow. In: Sreenan, J.M. and Diskin, M.G. (eds) *Embryonic Mortality in Farm Animals*. Martinus Nijhoff, The Hague, The Netherlands, pp. 1–11.

Sreenan, J.M. and Diskin, M.G. (1987) Factors affecting pregnancy rate following embryo transfer in the cow. *Theriogenology* 27, 99–113.

Sreenan, J.M. and Diskin, M.G. (1988a) The induction of unilateral and bilateral twin-pregnancy in the cow. *Proceedings of the Eleventh International Congress on Animal Reproduction and Artificial Insemination* (Dublin) 2, 109 (3 pp.).

Sreenan, J.M. and Diskin, M.G. (1988b) The national calf crop: increasing its size and beef potential. *Farm and Food Research* 19(4), 28–30.

Sreenan, J.M. and Diskin, M.G. (1989) Effect of a unilateral or bilateral twin embryo distribution on twinning and embryo survival rate in the cow. *Journal of Reproduction and Fertility* 87, 657–664.

Sreenan, J.M. and Diskin, M.G. (1992) *Breeding the Dairy Herd*. Teagasc Publication, Dublin, 112 pp.

Sreenan, J.M. and Diskin, M.G. (1994) Factors affecting herd conception rate. *Irish Farmers' Journal* 46(18), 30–31.

Sreenan, J.M. and McDonagh, T. (1979a) Embryo transfer in cattle is now a practical proposition. *Farm and Food Research* 10(2), 53–55.

Sreenan, J.M. and McDonagh, T. (1979b) Comparison of the embryo survival rate in heifers following artificial insemination, non-surgical blastocyst transfer or both. *Journal of Reproduction and Fertility* 56, 281–284.

Sreenan, J.M. and McEvoy, T.G. (1989a) Developments in cattle breeding technology. *Farm and Food Research* 20(5), 4–7.

Sreenan, J.M. and McEvoy, T.G. (1989b) Methodology of gene transfer in farm animals and fish. *Proceedings of the Fortieth Annual Meeting of the European Association of Animal Production*. (Dublin), G 4.1 (7 pp.).

Sreenan, J.M. and Scanlon, P.F. (1968) Continued cleavage of fertilized bovine ova in the rabbit. *Nature* 217, 867.

Sreenan, J.M., Scanlon, P.F. and Gordon, I. (1968) Culture of fertilized cattle eggs. *Journal of Agricultural Science (Cambridge)* 70, 183–185.

Sreenan, J.M., Scanlon, P.F. and Gordon, I. (1970) Storage of fertilized cattle ova *in vitro*. *Journal of Agricultural Science (Cambridge)* 74, 593–594.

Sreenan, J.M., Beehan, D. and Mulvihill, P. (1975) Egg transfer in the cow: factors affecting pregnancy and twinning rates following bilateral transfers. *Journal of Reproduction and Fertility* 44, 77–85.

Sreenan, J.M., Diskin, M.G. and McDonagh, T. (1981) Induction of twin-calving by non-surgical embryo transfer: a field trial. *Veterinary Record* 109, 77–80.

Sreenan, J.M., Morris, D., Tait, A. and Diskin, M.G. (1987) Manipulation of the immune system to increase ovulation rate in the cow. In: Roche, J.F. and O'Callaghan, D. (eds) *Follicular Growth and Ovulation Rate in Farm Animals*. Martinus Nijhoff, Dordrecht, The Netherlands, pp. 73–86.

Sreenan, J.M., O'Farrell, K.J. and Diskin, M.G. (1991) Reproductive waste in cows. *Farm and Food*, July/September, 28–30.

Staigmiller, R.B. and England, B.G. (1982) Folliculogenesis in the bovine. *Theriogenology* 17, 43–52.

Staigmiller, R.B. and Moor, R.M. (1984a) Effect of follicle cells on the maturation and developmental competence of ovine oocytes matured outside the follicle. *Gamete Research* 9, 221–229.

Staigmiller, R.B. and Moor, R.M. (1984b) *In vitro* maturation of extrafollicular ovine oocytes. *Proceedings of the Tenth International Congress on Animal Reproduction and Artificial Insemination* (Illinois) 2, 43.

Staigmiller, R.B., Bellows, R.A., Anderson, G.B., Seidel, G.E., Foote, W.D., Menino, A.R. and Wright, R.W. (1992) Superovulation of cattle with equine pituitary extract and porcine FSH. *Theriogenology* 37, 1091–1099.

Stancic, B., Laurincik, J., Sahinovic, R., Kubavicova, E. and Oberfranc, M. (1992) The recovery, numbers and morphology of follicular oocytes in pigs. *Biotechnologija u Stocarstvu* 8(1–2), 23–31.

Stanger, J.D. (1983) The effect of catecholamines and their antagonists on the fertilization of cumulus free mouse ova *in vitro* at a suboptimal sperm density. *Gamete Research* 7, 111–122.

Staples, T.R. and Page, R.D. (1987) Freezing bovine embryos with a portable liquid nitrogen freezer requiring no external seeding. *Theriogenology* 28, 647–659.

Steptoe, P.C. and Edwards, R.G. (1978) Birth after the reimplantation of a human embryo. *Lancet* ii, 366.

Stevenson, J.S. and Mee, M.O. (1991a) Fertility of lactating Holstein cows after postinsemination treatment with a progesterone-releasing intravaginal device (PRID). *Journal of Dairy Science* 74 (Suppl. 1), 196 (Abs. P162).

Stevenson, J.S. and Mee, M.O. (1991b) Pregnancy rates of Holstein cows after post-insemination treatment with a progesterone-releasing intravaginal device. *Journal of Dairy Science* 74, 3849–3856.

Stevenson, J.S. and Phatak, A.P. (1992) Post-insemination treatment of Holstein cows with a GnRH agonist. *Journal of Dairy Science* 75 (Suppl. 1), 241 (Abs. P290).

Stevenson, J.S., Phatak, A.P., Rettmer, I. and Stewart, R.E. (1992) Hormonal, follicular, and luteal characteristics in lactating cows after administration of a GnRH agonist on day-12 after estrus. *Journal of Dairy Science* 75 (Suppl. 1), 241 (Abs. P289).

Stevenson, K.R., Gilmour, R.S. and Wathes, D.C. (1992) IGF-I synthesis and localization in the ovine uterus during the oestrous cycle and early pregnancy. *Journal of Reproduction and Fertility*, Abstract Series No. 10, 11 (Abs. 17).

Stewart, C.L., Kaspar, P., Brunet, L.J., Bhatt, H., Gadi, I., Kontgen, F. and Abbondanzo, S.J. (1992) Blastocyst implantation depends on maternal expression of leukaemia inhibitory factor. *Nature* 359, 76–79.

Stewart, M., Broadbent, P.J. and Dolman, D.F. (1991) An evaluation of *in vitro* fertilized (IVF) beef embryos. *Proceedings of the British Society of Animal Production* (Winter Meeting), Paper No. 170.

Stewart-Savage, J. and Bavister, B.D. (1988) Deterioration of stored culture media as monitored by a sperm motility bioassay. *Journal of In Vitro Fertilization and Embryo Transfer* 5, 76–80.

Stice, S.L. (1992) Multiple generation bovine embryo cloning. *Proceedings of a Symposium on Cloning Mammals by Nuclear Transplantation* (Fort Collins), 28–31.

Stice, S.L. and Keefer, C. (1992) Improved developmental rates for bovine nucleus transfer embryos using cold shock activated oocytes. *Biology of Reproduction* 46 (Suppl. 1), 166 (Abs. 462).

Stice, S.L. and Keefer, C.L. (1993) Multiple generational bovine embryo cloning. *Biology of Reproduction* 48, 715–719.

Stice, S.L. and Robl, J.M. (1988) Nuclear reprogramming in nuclear transplant rabbit embryos. *Biology of Reproduction* 39, 657–664.

Stice, S.L., Keefer, C.L., Maki-Laurila, M. and Phillips, P.E. (1991) Producing multiple generations of bovine nuclear transplant embryos. *Theriogenology* 35, 273.

Stice, S.L., Keefer, C.L., Maki-Laurila, M. and Matthews, L. (1993) Donor blastomere cell cycle stage affects developmental competence of bovine nuclear transfer embryos. *Theriogenology* 39, 318.

Stice, S., Strelchenko, N., Betthauser, J., Scott, B., Jurgella, G., Jackson, J., David, V., Keefer, C. and Matthews, L. (1994) Bovine pluripotent embryonic cells contribute to nuclear transfer and chimeric fetuses. *Theriogenology* 41, 301.

Stock, A.E. and Fortune, J.E. (1993) Ovarian follicular dominance in cattle: relationship between prolonged growth of the ovulatory follicle and endocrine parameters. *Endocrinology* 132, 1108–1114.

Stock, A.E. and Hansel, W. (1992) Assay of embryo-derived platelet activating factor (EDPAF) by an equine platelet aggregation assay: preliminary data concerning its presence in bovine embryo culture media. *Theriogenology* 38, 757–768.

Stock, A.E., Ellington, J.E. and Fortune, J.E. (1993) Follicular dominance in cattle: superovulatory responses in the presence vs absence of a dominant follicle. *Biology of Reproduction* 48 (Suppl. 1), 189.

Stojanov, T., Robinson, S.J., Rhodes, S.L., O'Brien, J.K., Evans, G. and Maxwell, W.M.C. (1994) *In vitro* fertilization with chilled-stored ram spermatozoa. *Theriogenology* 41, 302.

Stolzenburg, U. and Schonmuth, G. (1990a) Selection experiment on twinning in dairy cattle. *Journal of Animal Breeding and Genetics* 107, 16–31.

Stolzenburg, U. and Schonmuth, G. (1990b) Selection of bulls for twinning. *Journal of Animal Breeding and Genetics* 107, 32–42.

Storey, B.T., Hourani, C. and Kim, J.B. (1992) A transient rise in intracellular Ca^{2+} is a precursor reaction to the zona pellucida-induced acrosome reaction in mouse sperm and is blocked by the induced acrosome reaction inhibitor 3-quinuclidinyl benzilate. *Molecular Reproduction and Development* 32, 41–50.

Strelchenko, N. and Stice, S. (1994) Bovine embryonic pluripotent cell lines derived from morula stage embryos. *Theriogenology* 41, 304.

Strelchenko, N., Saito, S. and Niemann, H. (1991) Towards the establishment of bovine embryonic stem cells. *Theriogenology* 35, 274.

Strickland, J.D., Dorgan, W.J. and Moody, E.L. (1976) Recovery of bovine oocytes from whole ovaries. *Proceedings of the Western Section of the American Society of Animal Science* 27, 215–216.

Strickland, J.D., Stallings, C.M., Dargan, W.J. and Moody, E.L. (1977) Recovery rate, size and viability of trypsin liberated bovine oocytes. *Proceedings of the Western Section of the American Society of Animal Science* 28, 158–159.

Stringfellow, D.A. (1985) The potential of bovine embryo transfer for infectious disease control. *Rev. Sci. Tech. Off. Int. Epiz.* 4, 859–866.

Stringfellow, D.A. and Seidel, S.M. (eds) (1990) *Manual of the International Embryo Transfer Society: a Procedural Guide and General Information for the Use of Embryo Transfer Technology, Emphasizing Sanitary Precautions*. International Embryo Transfer Society, Champaign, p. 17.

Stringfellow, D.A., Wolfe, D.F., McGuire, J.A., Lauerman, L.H., Gray, B.W. and Sparling, P.H. (1986) Effects of embryo freezing and thawing techniques on the survivability of *Brucella abortus*. *Theriogenology* 26, 553–559.

Stringfellow, D.A., Panangala, V.S. and Galik, P.A. (1988) Recovery and culture of ova from *Brucella abortus* infected cows. *Theriogenology* 27, 280.

Stringfellow, D.A., Lauerman, L.H., Nasti, K.B. and Galik, P.K. (1990a) Trypsin treatment of bovine embryos after *in vitro* exposure to infectious bovine rhinotracheitis virus or bovine herpesvirus-4. *Theriogenology* 33, 331.

Stringfellow, D.A., Lauerman, L.H., Nasti, K.B. and Galik, P.K. (1990b) Trypsin treatment of bovine embryos after *in vitro* exposure to infectious bovine rhinotracheitis virus or bovine herpesvirus-4. *Theriogenology* 34, 427–434.

Stringfellow, D.A., Gray, B.W., Toivio-Kinnucan, M., Galik, P., Riddell, K.P., Brock, K.V. and Kemppainen, R.J. (1991a) A continuous cell line established from a preimplantation bovine embryo. *Theriogenology* 35, 275.

Stringfellow, D.A., Riddell, K.P. and Zurovac, O. (1991b) The potential of embryo transfer for infectious disease control in livestock. *New Zealand Veterinary Journal* 39, 8–17.

Stringfellow, D.A., Riddell, M.G., Riddell, K.P., Carson, R.L., Smith, R.C. and Gray, B.W. (1992) *In vitro* fertilization can help salvage genetic potential of cattle. *Highlights of Agricultural Research (Alabama Agricultural Experiment Station)* 39(4), 15.

Stringfellow, D., Riddell, M., Riddell, K., Carson, R., Smith, R. and Gray, B. (1993) Use of *in vitro* fertilization for production of calves from involuntary cull cows. *Theriogenology* 39, 320.

Stroud, B.K. and Myers, M.W. (1993) Clinical results in a commercial IVF facility. *Ars Veterinaria* 9(2), 105–113.

Stubbings, R.B. (1989) Factors affecting maturation and development of bovine oocytes. *Dissertation Abstracts International B, Sciences and Engineering* 50, 34.

Stubbings, R.B. (1990) Bovine *in vitro* fertilization: present achievements and future challenges. *Revista Brasileira de Reproducao Animal*, Suplemento 2, 62–71.

Stubbings, R.B. and Wosik, C.P. (1991) Glass wool versus swim-up separation of bovine spermatozoa for *in vitro* fertilization. *Theriogenology* 35, 276.

Stubbings, R.B., Ohashi, O. and Basrur, P.K. (1987) A method for the dissociation of cumulus cells from bovine oocytes. *Theriogenology* 27, 280.

Stubbings, R.B., Armstrong, D.T., Berialt, R.A. and Basrur, P.K. (1988a) A method for aspirating bovine oocytes from small vesicular follicles: preliminary results. *Theriogenology* 29, 312.

Stubbings, R.B., Betteridge, K.J. and Basrur, P.K. (1988b) Investigations of culture requirements for bovine oocytes from small vesicular follicles: preliminary results. *Theriogenology* 29, 313.

Stubbings, R.B., Walton, J.S., Armstrong, D.T. and Basrur, P.K. (1990a) Recovery of bovine oocytes from small vesicular follicles for *in vitro* maturation and fertilization. *Veterinary Research Communications* 14, 71–81.

Stubbings, R.B., Liptrap, R.M. and Basrur, P.K. (1990b) The assessment of follicular parameters for the selection of oocytes recovered from superovulated heifers. *Animal Reproduction Science* 23, 181–195.

Stubbings, R.B., Liptrap, R.M., Betteridge, K.J., Walton, J.S., Armstrong, D.J. and Basrur, P.K. (1990c) Requirements for bovine oocyte maturation *in vitro*. *Reproduction in Domestic Animals* 25, 158–166.

Stubbings, R.B., Wosik, C. and Armstrong, D.T. (1993) Ovarian response in calves to multiple versus a single subcutaneous injection of Folltropin. *Theriogenology* 39, 321.

Stumpf, T.T., Terlouw, S.L., Funahashi, H., Prather, R.S. and Day, B.D. (1993) Processing of nuclei transplanted into *in vitro* matured porcine oocytes. *Theriogenology* 39, 322.

Suarez, S.S. (1988) Hamster sperm motility transformation during development of hyperactivation *in vitro* and epididymal maturation. *Gamete Research* 19, 51–65.

Suarez, S.S. (1992) Control of sperm transport in the fallopian tubes. *Journal of Reproduction and Fertility*, Abstract Series No. 10, 1 (Abs. S1).

Subramaniam, A., Devarajan, K.P. and Mohanan, M. (1990) Incidence and avoidance of zona fracture in cryopreserved bovine ova and embryos. *New Zealand Veterinary Journal* 38, 156–157.

Sueldo, C., Subias, E. and Oehninger, S. (1992) Progesterone enhancement of human sperm fertilizing capacity. *Proceedings of the Fourteenth World Congress on Fertility and Sterility* (Caracas), 119 (Abs. 311).

Sugawara, S., Hamonon, K., Miyamoto, A., Kameyama, K., Horiuchi, T. and Masaki, J. (1984) *In vitro* fertilization and subsequent development of bovine oocytes precultured in synthetic media. *Proceedings of the Tenth International Congress on Animal Reproduction and Artificial Insemination* (Illinois) 3, 382 (3 pp.).

Suh, T.K., Rehman, N., Collins, A. and Wright, R.W. (1993) Effects of granulosa cells from different size follicles in co-culture with bovine oocytes on maturation, fertilization and cleavage. *Theriogenology* 39, 323.

Sun, F.Z. and Moor, R.M. (1992) Nuclear transplantation in oocytes and eggs of domestic animals. *Institute of Animal Physiology and Genetics Research Babraham and Roslin Biennial Report (1990–1991)*, 58–59.

Sun, F.Z., Hoyland, J., Huang, X., Mason, W.T. and Moor, R.M. (1990) Ionic, morphological and biochemical changes that occur to porcine eggs during activation induced by electrical pulses. *Journal of Reproduction and Fertility*, Abstract Series No. 6, Abs. 21.

Sun,F.Z., Bradshaw, J.P., Galli, C. and Moor, R.M. (1993) Transient intracellular calcium concentration ([Ca^{2+}]i) rises in bovine oocytes following sperm penetration, and the role of IP3-induced Ca^{2+} release on fertilization. *Journal of Reproduction and Fertility*, Abstract Series No. 11, 66 (Abs. 121).

Sun, Q.Y., Qin, P.C., Li, G.P., Liu, G.Y. and Yang, Q.Z. (1994) Chronological events and ultrastructure of bovine *in vitro* fertilization. *Theriogenology* 41, 305.

Surrey, E.S. and Quinn, P.J. (1990) Successful ultrarapid freezing of unfertilized oocytes. *Journal of In Vitro Fertilization and Embryo Transfer* 7, 262–266.

Susko-Parrish, J.L. and Parrish, J.J. (1988) Glucose and cAMP modulators affect capacitation of bovine sperm by heparin. *Biology of Reproduction* 38 (Suppl. 1), 136 (Abs.).

Susko-Parrish, J.L., Wheeler, M.B., Ax, R.L., First, N.L. and Parrish, J.J. (1990) The effect of penicillamine, hypotaurine, epinephrine and sodium metabisulfite on bovine *in vitro* fertilization. *Theriogenology* 33, 333.

Susko-Parrish, J., Nuttleman, P. and Leibfried-Rutledge, M.L. (1991) Effect of bovine oocyte aging *in vitro* on development. *Biology of Reproduction* 44 (Suppl. 1), 156 (Abs. 414).

Suss, U. and Madison, V. (1983) Morphology and meiotic development of bovine oocytes cultured *in vitro*. *Archives of Andrology* 11, 217–218.

Suss, U. and Wuthrich, K. (1985) Stages of the first meiotic division in bovine oocytes matured *in vitro*. *Theriogenology* 23, 281.

Suss, U., Wuthrich, K. and Stranzinger, G. (1988a) Chromosome configuration and time sequence of the first meiotic division in bovine oocytes matured *in vitro*. *Biology of Reproduction* 38, 871–880.

Suss,U., Wuthrich, K. and Stranzinger, G. (1988b) Ultrastructure of the inter-cellular coupling of cumulus cells before and after *in vitro* maturation of bovine oocytes. *Theriogenology* 29, 314.

Sutovsky, P., Flechon, J.E., Flechon, B., Motlik, J., Peynot, N., Chesne, P. and Heyman, Y. (1993) Dynamic changes of gap junctions and cytoskeleton during *in vitro* culture of cattle oocyte cumulus complexes. *Biology of Reproduction* 49, 1277–1287.

Suzuki, H., Yang, X. and Foote, R.H. (1994) Surface characteristics and size changes of immature, *in vitro* matured and *in vitro* fertilized bovine oocytes. *Theriogenology* 41, 307.

Suzuki, O., Geshi, M., Yonai, M. and Sakaguchi, M. (1994) Effects of method of embryo production and transfer on pregnancy rate, embryo survival rate, abortion and calf production in beef cows. *Theriogenology* 41, 309.

Suzuki, T. (1989) Sexing of bovine sperm by discontinuous Percoll density gradient. *Japanese Journal of Animal Reproduction* 35, 29–32.

Suzuki, T. (1993) Bovine embryo transfer and related techniques. *Molecular Reproduction and Development* 36, 236–237.

Suzuki, T. and Hayashi, T. (1991) Detrimental effects of the position of straws during deep freezing of cattle zonae pellucidae and oocytes. *Bulletin of the Faculty of Agriculture, Yamaguchi University* 39, 1–6.

Suzuki, T., Ishida, T. and Sakai, Y. (1988) Twinning of ipsilateral and bilateral transfer of bovine frozen embryos. *Proceedings of the Seventy-fourth Meeting of the Japanese Society of Animal Reproduction*, 83 (abstract in Japanese).

Suzuki, T., Ishida, T., Sakai, Y. and Shimohira, I. (1989a) Induction of twinning in dairy heifers by ipsilateral or bilateral transfer of two frozen embryos. *Theriogenology* 31, 261.

Suzuki, T., Ishida, T. and Sakai, Y. (1989b) Cryopreservation of bovine embryos in the medium with glycerol (1.4M) and sucrose. *Japanese Journal of Animal Reproduction* 35, 125–129.

Suzuki, T., Yamamoto, M., Ooe, M., Sakata, A. and Matsuoka, K. (1990a) Comparison of one-step sucrose dilution and direct transfer of frozen bovine embryos in glycerol and 1,2-propanediol. *Theriogenology* 33, 334.

Suzuki, T., Guo, Z., Chen, J. and Kuo, T. (1990b) Removal examination of cryoprotectants from frozen–thawed bovine embryos with glycerol and 1,2-propanediol. *Japanese Journal of Animal Reproduction* 36, 105–109.

Suzuki, T., Yamamoto, M., Ooe, M., Sakata, A., Matsuoka, M., Nishikata, Y. and Okamoto, K. (1990c) Effect of sucrose concentration used for one-step dilution upon *in vitro* and *in vivo* survival of bovine embryos refrigerated in glycerol and 1,2-propanediol. *Theriogenology* 34, 1051–1057.

Suzuki, T., Sakai, Y., Ishida, T. and Kanouchi, T. (1991a) Production of identical twins from bovine embryos split pre- or post-freezing. *Japanese Journal of Animal Reproduction* 37, 237–242.

Suzuki, T., Singla, S.K., Sujata, J. and Madan, M.L. (1991b) Cleavage capability of water buffalo follicular oocytes classified by cumulus cells and fertilized *in vitro*. *Journal of Veterinary Medical Science* 53, 475–478.

Suzuki, T., Singla, S.K., Sujata, J. and Madan, M.L. (1992) *In vitro* fertilization of water buffalo follicular oocytes and their ability to cleave *in vitro*. *Theriogenology* 38, 1187–1194.

Suzuki, T., Takagi, M., Yamamoto, M., Saha, S., Boediono, A. and Oe, M. (1993a) Pregnancy rate and survival in culture of bovine IVF embryos frozen in various cryoprotectants and thawed in a one step system. *Theriogenology* 39, 324.

Suzuki, T., Takagi, M., Yamamoto, M., Boediono, A., Saha, S., Sakakibara, H. and Oe, M. (1993b) Pregnancy rate and survival in culture of *in vitro* fertilized bovine embryos frozen in various cryoprotectants and thawed using a one-step system. *Theriogenology* 40, 651–659.

Svitoyus, A., Vashkas, G. and Ukhotskene, D. (1990) Transfer of two embryos for obtaining twins. *Referativnyi Zhurnal*, 3.58.327.

Swartz, H.A., Johnson, M.E. and Ellersieck, M. (1990) Effect of twinning on efficiency of beef production. *Journal of Animal Science* 68 (Suppl. 1), 77.

Sweeney, A. and Tomkins, P.T. (1993a) Accessibility of acrosomal markers in bovine spermatozoa following Baekonization. *Irish Chemical News* 8, 23 (Abs.).

Sweeney, A. and Tomkins, P.T. (1993b) Optimisation of the electrically-induced sperm acrosome reaction: advantages of Baekonization. *Irish Chemical News* 8, 24 (Abs.).

Szell, A. and Hudson, R.H.H. (1991) Factors affecting the survival of bisected sheep embryos *in vivo*. *Theriogenology* 36, 379–387.

Szell, A., Zhang, J. and Hudson, R. (1990) Rapid cryopreservation of sheep embryos by direct transfer into liquid nitrogen vapour at −180°C. *Reproduction, Fertility and Development* 2, 613–618.

Szollosi, D., Desmedt, V., Crozet, N. and Brender, C. (1988) *In vitro* maturation of sheep ovarian oocytes. *Reproduction, Nutrition and Development* 28(4B), 1047–1080.

Tabibzadeh, S. and Sun, X.Z. (1992) Cytokine expression in human endometrium throughout the menstrual cycle. *Human Reproduction* 7, 1214–1221.

Tachikawa, S., Otoi, T., Kondo, S., Machida, T. and Kasai, M. (1993) Successful vitrification of bovine blastocysts, derived by *in vitro* maturation and fertilization. *Molecular Reproduction and Development* 34, 266–271.

Tada, T. and Takagi, N. (1992) Early development and X-chromosome inactivation in mouse parthenogenetic embryos. *Molecular Reproduction and Development* 31, 20–27.

Tadir, Y., Neev, J. and Berns, M.W. (1992) Laser in assisted reproduction and genetics. *Journal of Assisted Reproduction and Genetics* 9, 303–305.

Tadir, Y., Neev, J., Ho, P. and Berns, M.W. (1993) Lasers for gamete micromanipulation: basic concepts. *Journal of Assisted Reproduction and Genetics* 10, 121–125.

Taft, R., King, R., Anderson, S. and Killian, G. (1992) *In vitro* capacitation of sperm from fertile and subfertile bulls using oviductal fluid. *Journal of Animal Science* 70 (Suppl. 1) 256 (Abs. 474).

Taha, T.A. and Schellander, K. (1992) Isolation and culture of primary follicles from cattle ovaries. *Proceedings of the Twelfth International Congress of Animal Reproduction* (The Hague) 1, 275–277.

Tajik, P., Niwa, K. and Murase, T. (1993) Effects of different protein supplements in fertilization medium on *in vitro* penetration of cumulus-intact and cumulus-free bovine oocytes matured in culture. *Theriogenology* 40, 949–958.

Takada, N., Ohisa, N., Numabe, T. and Ishikawa, Y. (1991) Production of twin calves by transfer of embryos produced *in vitro*. *Veterinary Record* 128, 307.

Takagi, Y., Mori, K., Tomizawa, M., Takahashi, T., Sugawara, S. and Masaki, J. (1991) Development of bovine oocytes matured, fertilized and cultured in a serum-free, chemically defined medium. *Theriogenology* 35, 1197–1207.

Takagi, Y., Mori, K., Takahashi, T., Sugawara, S. and Masaki, J. (1992) Differences in development of bovine oocytes recovered by aspiration or by mincing. *Journal of Animal Science* 70, 1923–1927.

Takagi, M., Saha, S., Boediono, A. and Suzuki, T. (1993) Survival rate of frozen–thawed IVF bovine embryos in relation to equilibration time using various cryoprotectants. *Theriogenology* 39, 325.

Takahashi, Y. and First, N.L. (1992) *In vitro* development of bovine one-cell embryos: influence of glucose, lactate, pyruvate, amino-acids and vitamins. *Theriogenology* 37, 963–978.

Takahashi, Y. and First, N.L. (1993) *In vitro* culture of bovine one-cell embryos fertilized *in vitro* using synthetic oviduct fluid medium with and without glucose and supplemented with fetal calf serum. *Animal Reproduction Science* 31, 33–47.

Takahashi, Y. and Hanada, A. (1984) Penetration of zona-free hamster eggs *in vitro* by ejaculated bull sperm after treatment with ionophore A23187. *Japanese Journal of Animal Reproduction* 30(1), 30–38.

Takahashi, M., Nagai, T., Okamura, N. and Okano, A. (1992) Effect of 2-mercaptoethanol on development of bovine IVF embryos. *Proceedings of the Twelfth International Congress on Animal Reproduction* (The Hague) 3, 1342–1344.

Takahashi, M., Nagai, T., Hamano, S., Kuwayama, M., Okamura, N. and Okano, A. (1993a) Effect of cysteamine on *in vitro* development and intracellular glutathione content of bovine embryos. *Theriogenology* 39, 326.

Takahashi, M., Nagai, T., Hamano, S., Kuyayama, M., Okamura, N. and Okano, A. (1993b) Effect of thiol compounds on *in vitro* development and intracellular glutathione content of bovine embryos. *Biology of Reproduction* 49, 228–232.

Takahashi, S., Tokunaga, T. and Imai, H. (1994) Co-culture system using mouse embryonic fibroblast cells for *in vitro* fertilized and nuclear transplanted bovine embryos. *Theriogenology* 41, 311.

Takano, H., Koyama, K., Kozai, C., Kato, Y. and Tsunoda, Y. (1993) Effect of aging of recipient oocytes on the development of bovine nuclear transfer embryos *in vitro*. *Theriogenology* 39, 909–917.

Takeda, T. (1987) Effect of thawing procedures on damage to zonae pellucidae of bovine ova frozen in plastic straws. *Theriogenology* 23, 284.

Takeda, T., Elsden, R.P. and Seidel, G.E., Jr. (1985) Survival of cryopreserved bovine embryos cooled at 0.5 or 1°C/minute. *Theriogenology* 23, 232.

Takeda, T., Henderson, W.B. and Hasler, J.F. (1987) Deep freezing of split and intact bovine embryos. *Theriogenology* 27, 285.

Takeuchi, K., Yamamoto, S., Mori, A. and Nagata, Y. (1989) The effect of oxygen on the development of the early mouse embryo *in vitro*. *Japanese Journal of Fertility and Sterility* 34, 26–28.

Talbot, N., Powell, A., Pursel, V. and Rexroad, C.E. (1991) Efficient isolation of homogenous cells from the inner cell mass of pig and sheep blastocysts. *Serono Symposium on Preimplantation Embryo Development*, Abstract Vol., 90 (Abs. II–40).

Tan, J.H., Yang, Z.M., Qin, P.C. and Pashen, P.L. (1988) Light and electron microscope studies on follicular oocytes of Chinese Yellow cattle prior to *in vitro* maturation. *Theriogenology* 29, 317.

Tan, L., Liao, H., Zhang, Y., Liu, R. and Yan, Z. (1991) The survival of quartered bovine embryos with and without zonae pellucidae. *Theriogenology* 35, 279.

Tan, L.Y., Liao, H.M., Zhang, Y., Liu, R.H., Yan, Z.Q., Zhou, G.Q. and Li, H.R. (1992) A study of simplified quartering of cattle embryos. *Acta Veterinaria et Zootechnica Sinica* 23, 289–294.

Tan, S.J. and Lu, K.H. (1990) Effects of different oestrous cycle stages of ovaries and sizes of follicles on generation of IVF early bovine embryos. *Theriogenology* 33, 335.

Tao, T., Shie, J.,Yanqing, H., Changheng, G., Wenzhu, M., Tiezhu, A. and Yuding, Z. (1991) The survival of zona-free bovine demi- and quarter embryos. *Theriogenology* 35, 281.

Tateno, H. and Mikamo, K. (1987) A chromosomal method to distinguish between X- and Y-bearing spermatozoa of the bull in zona-free hamster ova. *Journal of Reproduction and Fertility* 81, 119–125.

Taylor, C.T., Ankers, L., Smullen, L.A. and Kingsland, C.R. (1992a) Effect of antioxidants on human preimplantation embryo development *in vitro*. *Human Reproduction* 7 (Suppl. 2), 144.

Taylor, C.T., Lawrence, Y.M., Kingsland, C.R. and Cuthbertson, K. (1992b) Oscillations in intracellular free calcium in human oocytes activated by sperm. *Human Reproduction* 7 (Suppl. 2), 173.

Taylor, C.T., Lawrence, Y.M., Kingsland, C.R., Biljan, M.M. and Cuthbertson, K.S.R. (1993) Oscillatious in intra cellular free calcium induced by spermatozoa in human oocytes at fertilization. *Human Reproduction* 8, 2174–2179.

Taylor, L.C. and Headon, D.R. (1993) Chemical approaches to enhancing animal reproduction. *Irish Chemical News* 8(3), 28–30.

Taylor, St.C.S., Moore, A.J., Thiessen, R.B. and Bailey, C.M. (1985) Efficiency of food utilization in traditional and sex-controlled systems of beef production. *Animal Production* 40, 401–440.

Techakumphu, M., Lohachit, C., Chantaraprateep, P., Prateep, P. and Kobayashi, G. (1989) Preliminary report on cryopreservation of Thai swamp buffalo embryos: manual and automatic methods. *Buffalo Bulletin* 8(2), 29–36.

Techakumphu, M., Adenot, P., Chesne, P. and Rao, V.H. (1993) Viability of bovine blastomeres after metaphase arrest with Nocodazole. *Theriogenology* 39, 328.

Telford, N.A., Watson, A.J. and Schultz, G.A. (1990) Transition from maternal to embryonic control in early mammalian development: a comparison of several species. *Molecular Reproduction and Development* 26, 90–100.

Terqui, M., Berthelot, F. and Driancourt, M.A. (1988a) Morphological and functional features of ovine follicles maintained in perfusion with pulsatile hormone delivery. *Journal of Reproduction and Fertility*, Abstract Series No. 1, Abs. 99.

Terqui, M., Berthelot, F. and Driancourt, M.A. (1988b) Morphological and functional features of ovine follicles in perifusion with pulsatile hormone delivery. *Reproduction, Nutrition and Develoment* 28(4B), 1093–1104.

Terriou, P., Giorgetti, C., Hans, E., Spach, J.L. and Roulier, R. (1993) Partial zona dissection enhances the fertilization rate in cases of sperm defect only when total failure has occurred in previous IVF fertilization attempts. *Human Reproduction* 8, 722–726.

Tervit, H.R. (1990) Potential use of embryo technology in gene transfer in dairy cattle. *Proceedings of the Twenty-third International Dairy Congress* (Montreal) 1, 566–578.

Tervit, H.R., Whittingham, D.G. and Rowson, L.E.A. (1972) Successful culture *in vitro* of sheep and cattle ova. *Journal of Reproduction and Fertility* 30, 493–497.

Tervit, H.R., Pugh, P.A., McGowan, L.T., Bell, A.C.S. and Wells, R.W. (1994) The freezability of sheep embryos is affected by culture system and source (*in vivo* or *in vitro* derived). *Theriogenology* 41, 315.

Tesarik, J. (1992) Control of the fertilization process by the egg coat: how does it work in humans? *Journal of Assisted Reproduction and Genetics* 9, 313–317.

Testart, J. and Du Mesnil du Buisson, F. (1966) Etude biométrique des placentomes dans les gestations simples ou gémellaires des bovins. *Annales de Biologie animale Biochimie Biophysique* 6, 483–493.

Testart, J., Lassalle, B., Belaisch-Allart, J., Hazout, A., Forman, R., Rainhorn, J.D. and Frydman, R. (1986) High pregnancy rate after early human embryo freezing. *Fertility and Sterility* 46, 268–272.

Thatcher, S.S., Restrepo, U., Lavy, G. and DeCherney, A.H. (1989) *In vitro* fertilization and sex ratio. *Lancet* i, 6 May, 1025–1026.

Thatcher, W.W., Hansen, P.J., Plante, C., Badinga, L., Cleeff, J. Van, Danet-Desnoyers, G., Savio, J.D., Mirando, M.A. and Bazer, F.W. (1990) Understanding and exploiting the physiology and endocrinology of reproduction to enhance reproductive efficiency in cattle. *Proceedings of the New Zealand Society of Animal Production* 50, 109–121.

Thibault, C. (1977) Hammond Memorial Lecture. Are follicular maturation and oocyte maturation independent processes? *Journal of Reproduction and Fertility* 51, 1–15.

Thibault, C., Szollosi, D. and Gerard, M. (1987) Mammalian oocyte maturation. *Reproduction, Nutrition and Development* 27(5), 865–896.

Thibier, M. (1990) Transfer of bovine embryos: the safest means of exchanging genetic material. *Proceedings of the Sixth Conference of the European Embryo Transfer Association* (Lyon), 197–207.

Thibier, M. (1992) Tentative prospective and health regulations of the bovine embryo transfer industry. In: Lauria, A. and Gandolfi, F. (eds), *Embryonic Development and Manipulation in Animal Production*. Portland Press, London, pp. 265–271.

Thibier, M. and Nibart, M. (1992) Clinical aspects of embryo transfer in some domestic animals. *Animal Reproduction Science* 28, 139–148.

Thibodeaux, J.K., Goodeaux, L.L., Roussel, J.D., Amborski, G.F., Moreau, J.D. and Godke, R.A. (1991) Stage of the bovine estrous cycle and *in vitro* characteristics of uterine and oviductal epithelial cells. *Journal of Dairy Science* 74 (Suppl. 1), 295 (Abs. So8).

Thibodeaux, J.K., Del Vecchio, R.P. and Hansel, W. (1992a) The role of platelet derived growth factor (PDGF) in development of early bovine embryos. *Biology of Reproduction* 46 (Suppl. 1), 76 (Abs. 103).

Thibodeaux, J.K., Menezo, Y., Roussel, J.D., Hansel, W., Goodeaux, L.L., Thompson, D.L. and Godke, R.A. (1992b) Coculture of *in vitro* fertilized bovine embryos with oviductal epithelial cells originating from different stages of the estrous cycle. *Journal of Dairy Science* 75, 1448–1455.

Thibodeaux, J.K., Del Vecchio, R.P., Broussard, J.R., Dickey, J.F. and Hansel, W. (1993) Stimulation of development of *in vitro* matured and *in vitro* fertilized bovine embryos by platelets. *Journal of Animal Science* 71, 1910–1916.

Thomas, W.K. and Seidel, G.E. (1991) *In vitro* methods for producing transgenic bovine embryos. *Journal of Animal Science* 69 (Suppl. 1), 401–402.

Thomas, W.K., Schnieke, A. and Seidel, G.E. (1993) Methods for producing transgenic bovine embryos from *in vitro* matured and fertilized oocytes. *Theriogenology* 40, 679–688.

Thompson, J.G.E. and Wales, R.G. (1987) Observation on the loss of the cellular vestment surrounding superovulated sheep oocytes. *Animal Reproduction Science* 15, 169–175.

Thompson, J.G.E., Parton, G.A.J., Cruikshank, G.W., Smith, J.F. and Wales, R.G. (1989a) Development of sheep preimplantation embryos in media supplemented with glucose and acetate. *Theriogenology* 32, 323–330.

Thompson, J.G.E., Petersen, A.J. and Tervit, H.R. (1989b) Development of cattle with a modified genotype: embryological techniques needed for the efficient production of transgenics. *Proceedings of the New Zealand Society of Animal Production* 49, 65–70.

Thompson, J.G., Simpson, A.C., Pugh, P.A. and Tervit, H.R. (1991) An examination of the requirement for glucose in preimplantation sheep embryos during *in vitro* culture. *Serono Symposium on Preimplantation Embryo Development*, Abstract Vol., 47 (Abs.I-46).

Thompson, J.G., Simpson, A.C., Pugh, P.A. and Tervit, H.R. (1992a) Requirement for glucose during *in vitro* culture of sheep preimplantation embryos. *Molecular Reproduction and Development* 31, 253–257.

Thompson, J.G., Pugh, P.A., Tervit, H.R. and Berg, D.L. (1992b) *In vitro* production of domestic animal embryos: advances made at Ruakura, New Zealand. *Australasian Biotechnology* 5, 280–287.

Thompson, J.G., Simpson, A.C., Pugh, P.A., McGowan, L.T., James, R.W., Berg, D.K., Payne, S.R. and Tervit, H.R. (1992c) Effect of glucose level in culture medium on survival of *in vitro* cultured sheep embryos following transfer to recipient ewes. *Proceedings of the New Zealand Society of Animal Production* 52, 255–256.

Thompson, J.G., Simpson, A.C., Pugh, P.A. and Tervit, H.R. (1992d) *In vitro* development of early sheep embryos is superior in medium supplemented with human serum compared with sheep serum or human serum albumin. *Animal Reproduction Science* 29, 61–68.

Thomson, M.S., Bird, R.C., Stringfellow, D.A., Rossi, C.R. and Lauerman, L.H. (1989) Chemically-induced viral resistance in preimplantation bovine embryos. *Theriogenology* 31, 267.

Thonabulsombat, C., Pavasuthipaisit, K., Chongthammakum, S., Kitiyanant, Y., Sanitwongse, B. and Tocharus, C. (1990) Successful culture of bovine embryos from oocytes matured and fertilized *in vitro* to the blastocyst stage. *Thai Journal of Veterinary Medicine* 20, 275–290.

Thonon, F., Ectors, F.J., Delval, A., Fontes, R.S., Touati, K. and Beckers, J.F. (1993) *In vitro* maturation, fertilization and development rates of bovine oocytes connected with the reproductive status of the donor. *Theriogenology* 39, 330.

Tiffin, G., Rieger, D., Betteridge, K.J., Yadav, B.R. and King, W.A. (1990) Measurement of the activity of the pentose phosphate pathway in sexed bovine embryos. *Theriogenology* 33, 339.

Tiffin, G.J., Rieger, D., Betteridge, K.J., Yadav, B.R. and King, W.A. (1991) Glucose and glutamine metabolism in pre-attachment cattle embryos in relation to sex and stage of development. *Journal of Reproduction and Fertility* 93, 125–132.

Tobback, C., Hough, S. and Foote, R.H. (1991) A procedure for cryopreservation of hamster oocytes yielding highly conserved oocytes suitable for sperm penetration tests. *Fertility and Sterility* 55, 184–188.

Todorov, J., Todorova, T., Zimmermann, U., Arnold, W.M., Leiding, C., Hahn, R. and Hahn, J. (1992) The influence of a prolonged *in vitro* maturation time and of a repeated field-pulse sequence upon the development rates of the oocytes in bovine nuclear transfer. *Proceedings of the Twelfth International Congress on Animal Reproduction* (The Hague) 2, 754–756.

Tomkins, P.T. and Houghton, J.A. (1988) The rapid induction of the acrosome reaction of human spermatozoa by elecropermeabilization. *Fertility and Sterility* 50, 329–336.

Topps, J.H., Islam, M.N., Broadbent, P.J. and Paterson, G.F.M. (1989) Effects of pre-calving nutrition on the performance of twin-bearing cows and their calves. *British Society of Animal Production* (Winter Meeting), Paper No. 1.

Topps, J.H., Broadbent, P.J., Methu, J.N. and Xaba, B.B. (1990) The effect of different distributions of the same total dietary between late pregnancy and early lactation on the performance of twin-bearing suckler cows. *British Society of Animal Production* (Winter Meeting), Paper No. 2.

Tornell, J., Billig, H. and Hillensjo, T. (1991) Regulation of oocyte maturation by changes in ovarian levels of cyclic nucleotides. *Human Reproduction* 6, 411–422.

Torner, H., Kauffold, P., Gotze, M., Konig, I., Dautzenberg, H. and Loth, F. (1986) Microcapsulation for immobilisation of mammalian embryos prior to implantation. *Archiv für Experimentalle Veterinar Medizin (Leipzig)* 49, 541–547.

Torner, H., Alm, H. and Goristanov, I. (1992) IVM/IVF of calf oocytes. *Proceedings of the Twelfth International Congress on Animal Reproduction* (The Hague) 1, 381–383.

Torrance, C., Telfer, E. and Gosden, R.G. (1989) Quantitative study of the development of isolated mouse pre-antral follicles in collagen gel culture. *Journal of Reproduction and Fertility* 87, 367–374.

Totey, S.M., Singh, G., Taneja, M. and Talwar, G.P. (1991a) *In vitro* maturation and fertilization of follicular oocytes from buffalo. *Theriogenology* 35, 284.

Totey, S.M., Taneja, M., Pawshe, C.H., Singh, G. and Talwar, G.P. (1991b) *In vitro* maturation, fertilization and development of buffalo oocytes. *Proceedings of the Third World Buffalo Congress* (Varna, Bulgaria) 3, 610–617.

Totey, S.M., Singh, G., Taneja, M., Pawshe, C.H. and Talwar, G.P. (1992) *In vitro* maturation, fertilization and development of follicular oocytes from buffalo (*Bubalis bubalis*). *Journal of Reproduction and Fertility* 95, 597–607.

Totey, S.M., Pawshe, C.H. and Singh, G.P. (1993a) Effects of bull and heparin and sperm concentration on *in vitro* fertilization of buffalo (*Bubalis bubalis*) oocytes matured *in vitro*. *Theriogenology* 39, 887–898.

Totey, S.M., Pawshe, C.H. and Singh, G.P. (1993b) *In vitro* maturation and fertilization of buffalo oocytes (*Bubalis bubalis*): effects of media, hormones and sera. *Theriogenology* 39, 1153–1171.

Touati, K., Beckers, J.F. and Ectors, F. (1991) Hormonal control of folliculogenesis in the bovine: better superovulatory responses after pure FSH administration preceding the classical treatment. *Theriogenology* 35, 285.

Tournaye, H., Van der Linden, M., Van den Abbeel, E., Devroey, P. and Van Steirteghem, A.C. (1993) Effect of pentoxifylline on mouse *in vitro* fertilization and subsequent embryonic development. *Journal of Reproduction and Fertility*, Abstract Series No. 12, 55.

Traub, A.I., Earnshaw, J.C., Brannigan, P.D. and Thompson, W. (1982) A critical assessment of the response to caffeine of human sperm motility. *Fertility and Sterility* 37, 436–437.

Tregaskes, L.D., Broadbent, P.J., Dolman, D.F. and Grimmer, S.P. (1992) Evaluation of Crestar for synchronizing oestrus in maiden heifers used as embryo transfer recipients. *Proceedings of the British Society of Animal Production* (Winter Meeting), Paper No. 45.

Tregaskes, L.D., Broadbent, P.J., Dolman, D.F., Grimmer, S.P. and Franklin, M.F. (1994) Evaluation of Crestar, a synthetic progestogen regime, for synchronizing oestrus in maiden heifers used as recipients of embryo transfers. *Veterinary Record* 134, 92–94.

Tribulo, H., Bo, G.A., Jofre, F., Carcedo, J., Alonso, A. and Mapletoft, R.J. (1991) The effect of LH concentration in a porcine pituitary extract and season on superovulatory response of *Bos indicus* heifers. *Theriogenology* 35, 286.

Trounson, A. (1986a) Preservation of human eggs and embryos. *Fertility and Sterility* 46, 1–12.

Trounson, A. (1986b) Recent progress in human *in vitro* fertilization and embryo transfer. In: Gwatkin, R.B.L. (ed.) *Developmental Biology*, Vol. 4, *Manipulation of Mammalian Development*. Plenum Press, London, pp. 149–194.

Trounson, A. (1992) The production of ruminant embryos *in vitro*. *Animal Reproduction Science* 28, 125–137.

Trounson, A.O. and Mohr, L. (1983) Human pregnancy following cryopreservation, thawing and transfer of an eight-cell embryo. *Nature* 305, 707–709.

Trounson, A.O. and Shaw, J. (1992) Fertilization and *in vitro* development of frozen–thawed MII mouse oocytes. *Journal of Reproduction and Fertility*, Abstract Series No. 9, 55 (Abs. 96).

Trounson, A.O., Willadsen, S.M. and Rowson, L.E.A. (1976a) The storage of cows eggs at room temperature and at low temperatures. *Journal of Reproduction and Fertility* 46, 173–178.

Trounson, A.O., Willadsen, S.M. and Rowson, L.E.A. (1976b) The influence of *in vitro* culture and cooling on the survival and development of cow embryos. *Journal of Reproduction and Fertility* 47, 367–370.

Trounson, A.O., Willadsen, S.M. and Rowson, L.E.A. (1977) Fertilization and development capability of bovine follicular oocytes matured *in vitro* and *in vivo* and transferred to the oviducts of rabbits and cows. *Journal of Reproduction and Fertility* 51, 321–327.

Trounson, A.O., Shea, B.F., Ollis, G.W. and Jacobson, M.E. (1978) Frozen storage and transfer of bovine embryos. *Journal of Animal Science* 47, 677–681.

Trounson, A.O., Pugh, A., Aarta, M.H. and Fielden, E.D. (1980) The development of bovine embryos after freezing and international transport. *Proceedings of the Twelfth Annual Conference of the Society of Reproductive Biology* (Armidale, New South Wales), Abs. 39.

Trounson, A., Pushett, D., Maclellan, L.J., Lewis, I. and Gardner, D.K. (1994) Current status of IVM/IVF and embryo culture in humans and farm animals. *Theriogenology* 41, 57–66.

Truelson, S.L., Mortimer, R.G. and Graham, J.K. (1992) The *in vitro* penetration of lipid treated bull sperm into bovine and zona-free hamster eggs. *Journal of Animal Science* 70 (Suppl. 1), 103 (Abs. 38).

Tsai, F.C.H. and Gardner, D.K. (1994) Nicotinamide, a component of complex culture media, inhibits mouse embryo development *in vitro* and reduces subsequent developmental potential after transfer. *Fertility and Sterility* 61, 376–382.

Tsujii, H. (1988) Xenogenous fertilization of bovine follicular oocytes in rat uterus. *Japanese Journal of Fertility and Sterility* 33, 155–159.

Tsunoda, Y., Todunaga, T. and Sugie, T. (1985) Altered sex ratio of live young after transfer of fast and slow-developing mouse embryos. *Gamete Research* 12, 301–304.

Tsuzuki, Y., Go, J., Fujihara, N. and Ashizawa, K. (1992) Ultrastructural studies of bovine blastocysts fertilized *in vitro* associated with freezing and thawing. *Proceedings of the Twelfth International Congress on Animal Reproduction* (The Hague) 3, 1487–1489.

Tucker, E.M., Moor, R.M. and Rowson, L.E.A. (1974) Tetraparental sheep chimeras induced by blastomere transplantation. *Immunology* 26, 613–621.

Tucker, M., Wiker, S. and Massey, J. (1993) Rational approach to assisted fertilization. *Human Reproduction* 8, 1778–1785.

Tuli, R.K., Schmidt-Baulain, R. and Holz, W. (1992) Computer-assisted motility assessment of spermatozoa from fresh and frozen–thawed semen of the bull, boar and goat. *Theriogenology* 38, 487–490.

Turzillo, A.M. and Fortune, J.E. (1993) Effects of suppressing plasma FSH on ovarian follicular dominance in cattle. *Journal of Reproduction and Fertility* 98, 113–119.

Twagiramungu, H., Guilbault, L.A., Prouix, J. and Dufour, J.J. (1992) Synchronization of estrus and fertility in beef cattle with two injections of buserelin and prostaglandin. *Theriogenology* 38, 1131–1144.

Ueda, S., Kuwayama, M., Shuioya, Y. and Hanada, A. (1988) Culture of *in vitro* fertilized bovine embryos in a chemically defined medium (BO-medium). *Japanese Journal of Animal Reproduction* 34, 56–60.

Uguz, C. and Parrish, J.J. (1993) Rp-adenosine-3′,5′-cyclic phosphorothioate (Rp) inhibits capacitation of bovine sperm by heparin. *Theriogenology* 39, 332.

Ulbricht, T.L.V. and Southgate, D.A.T. (1991) Coronary heart disease: seven dietary factors. *Lancet* 338, 985–992.

Umaoka, Y., Noda, Y., Narimoto, K. and Mori, T. (1991) Developmental potentiality of embryos cultured under low oxygen tension with superoxide dismutase. *Journal of In Vitro Fertilization and Embryo Transfer* 8, 245–249.

Umaoka, Y., Noda, Y., Narimoto, K. and Mori, T. (1992a) Effects of oxygen toxicity on early development of mouse embryos. *Molecular Reproduction and Development* 31, 28–33.

Umaoka, Y., Noda, Y., Nakayama, T., Narimoto, K., Mori, T.and Iritani, A. (1992b) Effect of visual light on *in vitro* embryonic development in the hamster. *Theriogenology* 38, 1043–1054.

Umaoka, Y., Noda, Y., Nakayama, T., Narimoto, K., Mori, T. and Iritani, A. (1993) Development of hamster one-cell embryos recovered under different conditions to the blastocyst stage. *Theriogenology* 39, 485–498.

Unsworth, E.F. and Mayne, C.S. (1991) The energy balance of non lactating dairy cows bearing single or twin calves during late gestation. *Proceedings of the Twelfth European Association of Animal Production Symposium on Energy Metabolism of Farm Animals* 58, 280–283.

Upreti, G.C., Riches, P.C. and Johnson, L.A. (1988) Attempted sexing of bovine spermatozoa by fractionation on a Percoll density gradient. *Gamete Research* 20, 83–92.

Ushijima, H. and Eto, T. (1993) Bovine nuclear transfer using parthenogenetically activated oocyte as recipient cytoplast. *Theriogenology* 39, 333.

Ushijima, H., Etoh, T. and Tsunoda, Y. (1991) Establishment of pregnancy in cattle by reconstituted eggs produced by fusion of an *in vitro* matured oocyte and a blastomere from a 16-cell embryo produced *in vivo*. *Theriogenology* 35, 287.

Ushijima, M. Okuda, T., Nakayama, A., Moji, K., Ishida, K., Murata, H., Iguchi, A. and Etoh, T. (1988) Relationship between the cell number and quality of day-8 bovine blastocysts. *Proceedings of the Third Meeting of the Japanese Society of Animal Embryo Transfer* 9, 37–38 (in Japanese).

Utsumi, K. and Iritani, A. (1990) Production of cattle identical twins by splitting blastocysts using a metal microblade. *Theriogenology* 33, 341.

Utsumi, K. and Iritani, A. (1993) Embryo sexing by male specific antibody and by PCR using male specific (SRY) primer. *Molecular Reproduction and Development* 36, 238–241.

Utsumi, K., Katoh, H. and Iritani, A. (1988) Developmental ability of bovine follicular oocytes matured in culture and fertilized *in vitro*. *Theriogenology* 29, 320.

Utsumi, K., Kato, H. and Iritani, A. (1991) Full-term development of bovine follicular oocytes matured in culture and fertilized *in vitro*. *Theriogenology* 35, 695–703.

Utsumi, K., Kawamoto, T., Kim, J.H., Takeda, H. and Iritani, A. (1992) Sexing of bovine embryos by polymerase chain reaction using Y-specific DNA as primer. *Proceedings of the Twelfth International Congress on Animal Reproduction* (The Hague) 2, 757–759.

Utsumi, K., Hayashi, M., Takakura, R., Utaka, K. and Iritani, A. (1993) Embryo sex selection by a rat male-specific antibody and the cytogenetic and developmental confirmation in cattle embryos. *Molecular Reproduction and Development* 34, 25–32.

Vachal, J. and Sereda, L. (1990) MOET-breeding system in Czechoslovakia. *Animal Science Papers and Reports (Warsaw)* 6, 99–101.

Vaillancourt, D., Bierschwal, C.J., Ogwu, D., Elmore, R.G., Martin, C.E., Sharp, A.J. and Youngquist, R.S. (1979) Correlation between pregnancy diagnosis by membrane slip and embryonic mortality. *Journal of American Veterinary Medicine Association* 175, 466–468.

Vajta, G., Machaty, Z., Barandi, Z., Soos, A. and Solti, L. (1991) Transfer of *in vitro* fertilized and cultivated swine embryos. *Theriogenology* 35, 289.

Vajta, G., Machaty, Z., Barandi, Zs. and Varga, Zs. (1992) Embryos derived from the *in vitro* fertilization of oocytes of pregnant cows. *Theriogenology* 37, 811–815.

Valdez, C.A., Mazni, O.A., Takahashi, Y., Hishinuma, M. and Kanagawa, H. (1990) Effects of equilibration time, precooling and developmental stage on the survival of mouse embryos cryopreserved by vitrification. *Theriogenology* 33, 627–636.

Valdez, C.A., Mazni, O.A., Takahashi, Y., Fujikawa, S. and Kanagawa, H. (1992) Successful crypopreservation of mouse blastocysts using a new vitrification solution. *Journal of Reproduction and Fertility* 96, 793–802.

Valdivia, R.P.A., Kunieda, T., Azuma, S. and Toyoda, Y. (1993) PCR sexing and developmental rate differences in preimplantation mouse embryos fertilized and cultured *in vitro*. *Molecular Reproduction and Development* 35, 121–126.

Vallet, J.C., Baril, G. and Loysel, C. (1989) Efficiency of laparoscopic embryo transfer in goats. *Proceedings of the Fifth Congress of the European Embryo Transfer Association* (Lyon), 186.

Van Blerkom, J. (1991) Microtubule mediation of cytoplasmic and nuclear maturation during the early stages of resumed meiosis in cultured mouse oocytes. *Proceedings of the National Academy of Sciences (USA)* 88, 5031–5035.

Van Blerkom, J. (1993) Development of human embryos to the hatched blastocyst stage in the presence or absence of a monolayer of Vero cells. *Human Reproduction* 8, 1525–1539.

Van Blerkom, J., Bell, H. and Weipz, D. (1990) Cellular and developmental biological aspects of bovine meiotic maturation, fertilization, and preimplantation embryogenesis *in vitro*. *Journal of Electron Microscopy Technique* 16, 298–323.

Van Cleeff, J., Drost, M. and Thatcher, W.W. (1991) Effects of postinsemination progesterone supplementation on fertility and subsequent estrous responses of dairy heifers. *Theriogenology* 36, 795–807.

Van den Hurk, R. and Dijkstra, G. (1992) An immunocytochemical study of bovine ovaries, with special reference to atypical granulosa cells. *Development and Function of the Reproductive Organs: Serona Symposium* (Peebles), 62.

Van den Hurk, R., Spek, E.R., Dijkstra, G., Hulshof, S.C.J. and Van Vorstenbosch, C.J.A.H.V. (1991) The effect of PMSG on theca cell penetration in the membrana granulosa of antral bovine follicles. *Journal of Reproduction and Fertility*, Abstract Series No. 7, 56 (Abs. 101).

Vandeplassche, M., Bufaye, R. and Bouters, R. (1979) The twin capacity of the uterus in heifers and cows. *Deutsche Tierärztliche Wochenschrift* 86, 470–473.

Van der Auwera, I., Cornillie, F., Pijnenborg, R. and Koninckx, P.R. (1992) The age of the pronucleate mouse ova influences their development *in vitro* and survival after freezing. *Human Reproduction* 7, 660–664.

Van der Elst, J., Nerinckx, S. and Van Steirteghem, A.C. (1992) *In vitro* maturation of mouse germinal vesicle-stage oocytes following cooling, exposure to cryoprotectants and ultrarapid freezing: limited effect on the morphology of the second meiotic spindle. *Human Reproduction* 7, 1440–1446.

Vanderhyden, B.C. (1989) Studies on the regulation of oocyte maturation in the rat with assessment by *in vitro* fertilization and fetal development. *Dissertation Abstracts International B, Sciences and Engineering* 49, 3622–3623.

Vanderhyden, B.C. (1993) Species differences in the regulation of cumulus expansion by an oocyte-secreted factor(s). *Journal of Reproduction and Fertility* 98, 219–227.

Vanderhyden, B.C. and Armstrong, D.T. (1989) Role of cumulus cells and serum on the *in vitro* maturation, fertilization, and subsequent development of rat oocytes. *Biology of Reproduction* 40, 720–728.

Vanderhyden, B.C., McLaughlin, K.J., Rutledge, J.M. and Armstrong, D.T. (1989) Zona drilling increases the penetrability of rat oocytes matured *in vitro*. *Biology of Reproduction* 40, 953–960.

Vanderhyden, B.C., Caron, P.J. and Eppig, J.J. (1990) The role of the oocyte in promoting granulosa cell differentiation and proliferation. *Biology of Reproduction* 42 (Suppl. 1), 174.

Vanderhyden, B.C., Telfer, E.E. and Eppig, J.J. (1991) Murine oocytes modulate granulosa cell proliferation in preantral and antral follicles. *Biology of Reproduction* 44 (Suppl. 1), 153 (Abs. 401).

Van der Schans, A., Vander Westerlaken, L.A.J., de Wit, A.A.C., Eyestone, W.H. and de Boer, H.A. (1991) Ultrasound-guided transvaginal collection of oocytes in the cow. *Theriogenology* 35, 288.

Van der Schans, A., van Rens, B.T.T.M., van der Westerlaken, L.A.J. and de Wit, A.A.C. (1992) Bovine embryo production by repeated transvaginal oocyte collection and *in vitro* fertilization. *Proceedings of the Twelfth International Congress on Animal Reproduction* (The Hague) 3, 1366–1368.

Van Inzen, W.G., Wurth, Y.A., Verstrate-Voolstra, C., Kruip, Th.A.M. and Weima, S.M. (1992) Culture of bovine embryos to the blastocyst stage in the presence of embryotrophic factors secreted by buffalo rat liver (BRL) cells. *Proceedings of the Twelfth International Congress on Animal Reproduction* (The Hague) 3, 1360–1362.

Vankelecom, A., Joris, H., Liu, J., Nagy, Z., Desmet, B., Tournaye, H., Devroey, P. and Van Steirteghem, A.C. (1993) Intracytoplasmic sperm injection and subzonal insemination in patients with extreme oligospermia. *Human Reproduction* 8 (Suppl. 1), 21.

Van Lieshout, J.A.H. (1987) The development of embryo transfer in the Netherlands. *Proceedings of the Third Conference of the European Embryo Transfer Association*, 65–76.

Van Soom, A. and Kruif, A. de (1992) A comparative study of *in vivo* and *in vitro* derived bovine embryos. *Proceedings of the Twelfth International Congress on Animal Reproduction* (The Hague) 3, 1363–1365.

Van Soom, A., Vlaenderen, I. Van and de Kruif, A. (1991) Effect of modification of heparin dose on bovine embryonic development after IVF. *ARTA* 2, 116–117.

Van Soom, A., Van Vlanenderen, V., Mahmoudzadeh, A.R., Deluyker, H. and de Kruif, A. (1992) Compaction rate of *in vitro* fertilized bovine embryos related to the interval from insemination to first cleavage. *Theriogenology* 38, 905–919.

Van Soom, A., Mahmoudzadeh, A.R. and De Kruif, A. (1993) Timing of bovine blastocyst formation and cell number of early blastocysts derived from *in vitro* techniques. *Journal of Reproduction and Fertility*, Abstract Series No. 12, 64.

Van Soom, A., Van Langendonckt, A., Mahmoudzadeh, A.R., Deluyker, H., Dessy, F.and de Kruif, A. (1994) Effect of oil quality on *in vitro* embryonic development in the bovine. *Theriogenology* 41, 325.

Vansteenbrugge, A., Van Langendonckt, A., Massip, A. and Dessy, F. (1993) Development of bovine embryos in bovine oviduct-conditioned or BRL-conditioned media. *Journal of Reproduction and Fertility*, Abstract Series No. 12, 63.

Van Steirteghem, A.C., Liu, J., Joris, H., Nagy, Z., Janssenswillen, C., Tournaye, H., Derde, M.-P., Van Assche, E. and Devroey, P. (1993a) Higher success rate by intracytoplasmic sperm injection than by subzonal insemination: report of a second series of 300 consecutive treatment cycles. *Human Reproduction* 8, 1055–1060.

Van Steirteghem, A.C., Nagy, Z., Joris, H., Liu, J., Staessen, C., Smitz, J., Wisanto, A. and Devroey, P. (1993b) High fertilization and implantation rates after intracytoplasmic sperm injection. *Human Reproduction* 8, 1061–1066.

Van Steirteghem, A., Liu, J., Nagy, Z., Joris, H., Tournaye, H., Liebaers, I. and Devroey, P. (1993c) Use of assisted fertilization. *Human Reproduction* 8, 1784–1785.

Van Stekelenburg-Hamers, A.E.P., van Inzen, W.G., van Achterberg, T.A.E., Kruip, Th.A.M. and Weima, S.M. (1992) Nucleus transfer in bovine IVM/IVF embryos and IVM oocytes: establishment of electrofusion and activation conditions. *Proceedings of the Twelfth International Congress on Animal Reproduction* (The Hague) 2, 760–762.

Van Stekelenburg-Hamers, A.E.P., Kruip, Th.A.M. and Weima, S.M. (1993a) Age dependence of activation of *in vitro* matured bovine oocytes after maturation in different media. *Theriogenology* 39, 316.

Van Stekelenburg-Hamers, A.E.P., Van Inzen, W.G., Van Achterberg, T.A.E., Kruip, Th.A.M., De Latt, S. and Weima, S.M. (1993b) Nuclear transfer and electrofusion in bovine *in vitro*-matured/*in vitro* fertilized embryos: effect of media and electrical fusion parameters. *Molecular Reproduction and Development* 36, 307–312.

Van Stekelenburg-Hamers, A.E.P., Rebel, H.G., Van Inzen, W.G., De Loos, F.A.M., Drost, M., Mummery, C.L., Weima, S.M. and Trounson, A.O. (1994) Stage-specific appearance of the mouse antigen TEC-3 in normal and nuclear transfer bovine embryos: re-expression after nuclear transfer. *Molecular Reproduction and Development* 37, 27–33.

Van Vleck, L.D., Gregory, K.E. and Echternkamp, S.E. (1990) Genetic correlation between ovulation rate and twinning rate. *Journal of Animal Science* 68 (Suppl. 1), 66–67.

Van Vliet, R.A., Verrinder Gibbins, A.M. and Walton, J.S. (1989) Livestock embryo sexing: a review of current methods,with emphasis on Y-specific DNA probes. *Theriogenology* 32, 421–438.

Van Werven, T., Schukken, Y.H., Lloyd, J., Brand, A., Heeringa, H.Tj. and Shea, M. (1992) The effects of duration of retained placenta on reproduction, milk production, postpartum disease and culling rate. *Theriogenology* 37, 1191–1203.

Veeck, L.L. (1991) *Atlas of the Human Oocyte and Early Conceptus*, Vol. 2. Williams and Wilkins, Baltimore, Maryland.

Veeck, L.L., Amundson, C.H., Brothman, L.J., DeScisciolo, C., Maloney, M.K., Muasher, S.J. and Jones, H.W. (1993) Significantly enhanced pregnancy rates per cycle through cryopreservation and thaw of pronuclear stage oocytes. *Fertility and Sterility* 59, 1202–1207.

Vergos, E. (1990) *In vitro* fertilization and embryo culture in cattle. PhD Thesis, National University of Ireland, Dublin.

Vergos, E., Gordon, A., Gallagher, M. and Gordon, I. (1989a) *In vitro* culture of embryos produced by *in vitro* maturation and fertilization of bovine follicular oocytes. *British Society of Animal Production* (Winter Meeting), Paper No. 14.

Vergos, E., Kinis, A., Gallagher, M. and Gordon, A. (1989b) Development of bovine embryos on oviductal monolayer in relation to cell stage at 48 h after *in vitro* fertilization. *Journal of Reproduction and Fertility*, Abstract Series No. 4, Abs. 37.

Vergos, E., Kinis, A., Lonergan, P., Sharif, H., Gallagher, M. and Gordon, I. (1991) The effect of culture system on the *in vitro* development of bovine oocytes matured and fertilized *in vitro*. *Theriogenology* 35, 290.

Vergos, E., Alifakiotis, T., Konstandidou, M. and Gordon, A.C. (1993) Embryo production by *in vitro* techniques in dairy cattle in Greece. *Veterinary Record* 133, 70.

Veselinovic, S., Popovski, K., Veselinovic, S., Micic, R., Medic, D., Ivkov, V., Nivot, A. and Carmichael, R.A. (1992) Results of transplantation of embryos imported both from USA and France. *Proceedings of the Eighth Conference of the European Embryo Transfer Association* (The Hague), 226.

Vidal, F., Moragas, M., Catala, V., Torello, M.J., Santalo, J., Calderon, G., Gimenez, C., Barri, P.N., Egozcue, J. and Veiga, A. (1993) Sephadex filtration and human serum albumin gradients do not select spermatozoa by sex chromosome: a fluorescent *in-situ* hybridization study. *Human Reproduction* 8, 1740–1743.

Vincent, C., Heyman, Y., Garnier, V., Smorag, Z. and Renard, J.P. (1985) *In vitro* survival of early-stage rabbit and cow embryos directly frozen to intermediate temperature (−25 to −30°C) before plunging in liquid nitrogen. *Theriogenology* 23, 234.

Vincent, C., Garnier, V., Heyman, Y. and Renard, J.P. (1989) Solvent effects on cytoskeletal organization and *in-vivo* survival after freezing of rabbit oocytes. *Journal of Reproduction and Fertility* 87, 809–820.

Vines, G. (1992) Guess what's coming to dinner? *New Scientist* 136 (1847), 13–14.

Vines, G. (1993) Shots in the dark for infertility. *New Scientist* 140 (1901), 13–14.

Vitale, N.J., Barry, D.M., Meintjes, M. and Godke, R.A. (1994) *In vitro* development of refrozen bovine embryos. *Theriogenology* 41, 327.

Viuff, D., Madison, V., Hyttel, P., Avery, B. and Greve, T. (1991) Fluorescent intravital staining of bovine oocytes and zygotes. *Theriogenology*, 35, 291.

Viuff, D., Hyttel, P. and Alexandersen, S. (1994) Expression and localization of growth factor mRNA in the bovine oviduct. *Theriogenology* 41, 328.

Vivanco, H.W., Rangel, R., Lynch, P. and Rhodes, A. (1991) Large-scale commercial application of bisection of sheep embryos. *Theriogenology* 35, 292.

Vlaenderen, I. Van, Soom, A. Van and Kruife, A. de (1991) Bull specific effect on *in vitro* penetration/fertilization of bovine oocytes using various levels of heparin. *ARTA* 2, 105–106.

Voelkel, S.A. and Hu, Y.X. (1992a) Direct transfer of frozen-thawed bovine embryos. *Theriogenology* 37, 23–37.

Voelkel, S.A. and Hu, Y.X. (1992b) Use of ethylene glycol as a cryoprotectant for bovine embryos allowing direct transfer of frozen-thawed embryos to recipient females. *Theriogenology* 37, 687–697.

Voelkel, S.A. and Hu, Y.X. (1992c) Effect of gas atmosphere on the development of one-cell bovine embryos in two culture systems. *Theriogenology* 37, 1117–1131.

Voelkel, S.A., Vider, S.D., Lanbeth, V.A. and Godke, R.A. (1984) A new approach to freezing bovine embryos without the use of liquid cooling agents. *Theriogenology* 21, 272.

Volpe, A., Coukos, G., Barreca, A., Giordano, G., Artini, P.G. and Genazzani, A.R. (1991) Clinical use of growth hormone-releasing factor for induction of superovulation. *Human Reproduction* 6, 1228–1232.

Volpe, A., Artini, P.G., Barreca, A., Minuto, F., Coukos, G. and Genazzani, A.R. (1992) Effects of growth hormone administration in addition to gonadotrophins in normally ovulating women and polycystic ovary syndrome (PCO) patients. *Human Reproduction* 7, 1347–1352.

Vos, P.L.A.M., de Loos, F.A.M., Pieterse, M.C., Bevers, M.M., Taverne, M.A.M. and Dieleman, S.J. (1990) Transvaginal ultra-sound-guided follicle puncture as a technique to collect individual combinations of oocyte–cumulus complexes and follicular fluids at consecutive times relative to the preovulatory LH surge in PMSG/PG-treated cows. *Proceedings of the Sixth Conference of the European Embryo Transfer Association* (Lyon), 83–96.

Voss, A.K. and Fortune, J.E. (1991) Oxytocin secretion by bovine granulosa cells: effects of stage of follicular development, gonadotrophins and co-culture with theca interna. *Endocrinology* 128, 1991–1999.

Voss, A.K. and Fortune, J.E. (1992) Expression of the oxytocin/neurophysin I gene in bovine granulosa cells increases after the LH surge. *Biology of Reproduction* 46 (Suppl. 1), 80 Abs 118.

Wachtel, S., Nakamura, D., Wachtel, G., Felton, W., Kent, M. and Jaswaney, V. (1988) Sex selection with monoclonal H-Y antibody. *Fertility and Sterility* 50, 355–360.

Wade, J., Kelly, M.F., Hayden, T.J. and Gordon, I. (1987) Non-surgical recovery and transfer of equine embryos. *Theriogenology* 27, 290.

Wade, J., Gallagher, M. and Gordon, I. (1989) Equine embryo transfer in Ireland: from research into commercial practice. *Equine Veterinary Journal*, Suppl. 8, 76 (abs.).

Wagner, H.G.R. (1987) Present status of embryo transfer in cattle. *World Animal Review* 64, 2–11.

Wagner, T.E., Murray, F.A., Minhas, B. and Draemer, D.C. (1984) The possibility of transgenic livestock. *Theriogenology* 21, 29–44.

Wahid, H. (1993) Studies related to the *in vitro* production of sheep embryos. PhD Thesis, National University of Ireland, Dublin.

Wahid, H., Gordon, I., Sharif, H., Lonergan, P., Monaghan, P. and Gallagher, M. (1991) Development of ovine blastocysts following maturation, fertilization and culture of oocytes *in vitro*. *Proceedings of the Seventh Meeting of the European Embryo Transfer Association* (Cambridge), 214.

Wahid, H., Monaghan, P. and Gordon, I. (1992) *In vitro* maturation (IVM) of the sheep follicular oocyte. *Journal of Reproduction and Fertility*, Abstract Series No. 9, 52 (Abs. 89).

Waldmann, U.R. (1991) Results of cattle oocyte *in vitro* fertilization trials with reference to their development potential after transfer to recipients. Thesis, Tierarztliche Hochschule, Hannover, Germany, 89 pp.

Wales, R.G., Khurana, N.K., Edirisinghe, W.R. and Pike, I.L. (1985) Metabolism of glucose by preimplantation mouse embryos in the presence of glucagon, insulin, epinephrine, cAMP, theophylline and caffeine. *Australian Journal of Biological Science* 38, 421–428.

Walker, S.K., Warnes, G.M., Quinn, P., Seamark, R.F. and Smith, D.H. (1985) Laparoscopic technique for the transfer of embryos in sheep. *Australian Veterinary Journal* 62, 105–106.

Walker, S.K., Heard, T.M., Verma, P.J., Bawden, C.S., Sivaprasad, A.V., Rogers, G.E. and Seamark, R.F. (1990a) *In vitro* culture of sheep zygotes following pronuclear microinjection. *Theriogenology* 33, 342.

Walker, S.K., Heard, T.M., Verma, P.J., Rogers, G.E., Bawden, C.S., Sivaprasad, A.V., Mclaughlin, K.J. and Seamark, R.F. (1990b) *In vitro* assessment of the viability of sheep zygotes after pronuclear microinjection. *Reproduction, Fertility and Development* 2, 633–640.

Walker, S.K., Heard, T.M. and Seamark, R.F. (1992a) *In vitro* culture of sheep embryos without co-culture: successes and perspectives. *Theriogenology* 37, 111–126.

Walker, S.K., Heard, T.M., Bee, C.A., Frensham, A.B., Warnes, D.M. and Seamark, R.F. (1992b) Culture of embryos of farm animals. In: Lauria, A. and Gandolfi, F. (eds) *Embryonic Development and Manipulation in Animal Production*. Portland Press, London, 77–92.

Walker, S.K., Hill, J.L., Bee, C.A. and Warnes, D.M. (1994) Improving the rate of production of sheep embryos using *in vitro* maturation and fertilization. *Theriogenology* 41, 330.

Wall, R.J. and Hawk, H.W. (1988) Development of centrifuged cow zygotes cultured in rabbit oviducts. *Journal of Reproduction and Fertility* 82, 673–680.

Wall, R.J. and Seidel, G.E. (1992) Transgenic farm animals – a critical review. *Theriogenology* 38, 337–357.

Wall, R.J., Hawk, H.W. and Nel, N. (1992) Making transgenic livestock: genetic engineering on a large scale. *Journal of Cellular Biochemistry* 49, 113–120.

Walton, J.S., Goodwin, M.L. and Leslie, K.E. (1991) Gonadotrophin-induced accessory corpora lutea in the cow – an equine philosophy for the support of bovine pregnancy? *Journal of Dairy Science* 74 (Suppl. 1), 196 (Abs. P163).

Walton, J.S., Christie, K.A. and Stubbings, R.B. (1993) Evaluation of frequency of ultrasonically guided follicle aspiration of bovine ovarian dynamics. *Theriogenology* 39, 336.

Wang, W.H., Niwa, K. and Okuda, K. (1991) *In vitro* penetration of pig oocytes matured in culture by frozen-thawed ejaculated spermatozoa. *Journal of Reproduction and Fertility* 93, 491–496.

Wang, W.L. (1991) Factors affecting metabolic requirements and *in vitro* culture of bovine embryos. PhD Thesis, National University of Ireland, Dublin.

Wang, W.L., Jiang, H.S., Lu, K.H., McCarthy, D. and Gordon, I. (1989) The effect of media and sera on the *in vitro* development of early bovine embryos. *Journal of Reproduction and Fertility*, Abstract Series No. 3, 50.

Wang, W.L., Jiang, H.S., Lu, K.H. and Gordon, I. (1990) Effect of condition medium and glucose concentration on the *in vitro* development of early bovine embryos. *Theriogenology* 33, 343.

Wang, W.L., Jiang, H.S., Shi, D.S., Lu, K.H., Gordon, I. and Polge, C. (1991) The effect of temperature on the *in vitro* maturation, fertilization and development of bovine oocytes. *Proceedings of the Seventh Congress of the European Embryo Transfer Association* (Cambridge), 216.

Wang, W.L., Jiang, H.S., Lu, K.H., Gordon, I. and Polge, C. (1992) The effect of gas phase on the *in vitro* development of bovine embryos derived from *in vitro* maturation and fertilization of ovarian oocytes. *Theriogenology* 37, 320.

Wang, Z.K., Wei, P.H., Wang, J.Z., Lei, C. and Kou, M.Q. (1992) Maturation and fertilization of porcine oocytes *in vitro*. *Theriogenology* 37, 733–739.

Ward, K.A. and Nancarrow, C.D. (1991) The genetic engineering of production traits in domestic animals. *Experientia* 47, 913–922.

Ward, K.A., Murray, J.D. and Nancarrow, C.D. (1989) Current research programmes in animal biotechnology in Australia. *AgBiotech News and Information* 1, 19–22.

Ward, W.S. and Coffey, D.S. (1991) DNA packaging and organization in mammalian spermatozoa: comparison with somatic cells. *Biology of Reproduction* 44, 569–574.

Ware, C.B. (1986) Studies in embryo micromanipulation and freezing. PhD Thesis, National University of Ireland, Dublin.

Ware, C.B. and First, N.L. (1988) Development of embryonic stem cells from farm animals. *Biology of Reproduction* 38 (Suppl. 1), 241 (Abs).

Ware, C.B., Barnes, F.L., Maiki-Laurila, M. and First, N.L. (1989) Age dependence of bovine oocyte activation. *Gamete Research* 22, 265–275.

Warwick, B.L. and Berry, R.O. (1949) Inter-genetic specific embryo transfers. *Journal of Heredity* 40, 297–303.

Warwick, B.L., Berry, R.O. and Horlacher, W.R. (1934) Results of mating rams to angora female goats. *Proceedings of the American Society of Animal Production, Annual Meeting*, pp. 225–227.

Warikoo, P.K. (1988) Hypoxanthine and cAMP maintain meiotic arrest of rhesus monkey oocytes *in vitro*. *Biology of Reproduction* 38 (Suppl. 1), 63 (Abs).

Warikoo, P.K. and Bavister, B.D. (1989) Hypoxanthine and cyclic adenosine 5-monophosphate maintain meiotic arrest of rhesus monkey oocytes *in vitro*. *Fertility and Sterility* 51, 886–889.

Warnes, G.M., Moor, R.M. and Johnson, M.H. (1977) Changes in protein synthesis during maturation of sheep oocytes *in vivo* and *in vitro*. *Journal of Reproduction and Fertility* 49, 331–335.

Washburn, S.P., Howard, H.G., Jochle, W. and Macmillan, K.L. (1989) Control of estrous cycles in mature dairy heifers with a progesterone-releasing device. *Journal of Animal Science* 67 (Suppl. 1), 382 (Abs. 927).

Wassarman, P.M. (1988a) Fertilization in mammals. *Scientific American*, December, 52–58.

Wassarman, P.M. (1988b) The mammalian ovum. In: Knobil, E., Neil, J. *et al.* (eds), *The Physiology of Reproduction*. Raven Press, New York, pp. 69–102.

Wassarman, P.M. (1994) Gamete interactions during mammalian fertilization. *Theriogenology* 41, 31–44.

Watanabe, Y.F. and Oliveira Filho, E.B. (1992) Vitrification of bovine oocytes before and after maturation *in vitro*. *Proceedings of the Twelfth International Congress on Animal Reproduction* (The Hague) 3, 1499–1501.

Waterman, R.A., Hawk, H.W. and Wall, R.J. (1991) *In vitro* maturation, fertilization and culture of bovie oocytes using one semi-defined basal medium. *Biology of Reproduction* 44 (Suppl. 1), 74 (Abs. 86).

Wathes, D.C., Denning Kendall, P.A., Perks, C. and Brown, D. (1992a) Effects of stage of the cycle and estradiol-17β on oxytocin synthesis by ovine granulosa and luteal cells. *Endocrinology*, 130, 1009–1016.

Wathes, D.C., Ayad, V.J., Kumari, M., Matthews, E.L. and Wathes, C.M. (1992b) A comparison of plasma vasopressin and oxytocin concentrations during the oestrous cycle of the ewe. *Journal of Endocrinology* 134, 107–113.

Watson, A.J. (1992) The cell biology of blastocyst development. *Molecular Reproduction and Development* 33, 492–504.

Watson, A.J., Hogan, A., Hahnel, A., Wiemer, K.E. and Schultz, G.A. (1992a) Expression of growth factor ligand and receptor genes in the preimplantation bovine embryo. *Molecular Reproduction and Development* 31, 87–95.

Watson, A.J., Wiemer, K.E., Arcellana-Panlilio, M. and Schultz, G.A. (1992b) U2 small nuclear RNA localization and expression during bovine preimplantation development. *Molecular Reproduction and Development* 31, 231–240.

Watson, A.J., Hogan, A., Hahnel, A. and Schultz, G.A. (1993) Activation of the embryonic genome: comparisons between mouse and bovine development. In: Bavister, B.D. (ed.) *Preimplantation Embryo Development: Serona Symposium*. Springer-Verlag, New York, pp 115–130.

Watson, P.F., Jones, P.S. and Plummer, J.M. (1991) A quantitative comparison of the spontaneous and ionophore-induced acrosome reaction in ejaculated ram spermatozoa: the effects of temperature, time and individual. *Animal Reproduction Science* 24, 93–108.

Way, A.L. and Henault, M.A. (1992) Comparison of staining techniques used for analysis of bovine sperm viability and acrosomal integrity. *Biology of Reproduction* 46 (Suppl. 1), 93 (Abs. 172).

Wazzan, W.C., Gwatkin, R.B.L. and Thomas, A.J. Jr (1990) Zona drilling enhances fertilization by mouse caput epididymal sperm. *Molecular Reproduction and Development* 27, 332–336.

Webb, R., Gong, J.G., Law, A.S. and Rusbridge, S.M. (1992) Control of ovarian function in cattle. *Journal of Reproduction and Fertility* Suppl. 45, 141–156.

Webb, R., Gong, J.G. and Bramley, T.A. (1994) Role of growth hormone and intrafollicular peptides in follicle development in cattle. *Theriogenology* 41, 25–30.

Weber, J.A., Woods, G.L., Freeman, D.A. and Vanderwall, D.K. (1993) Oviductal and uterine influence on the development of day-2 equine embryos *in vivo* and *in vitro*. *Theriogenology* 40, 689–698.

Wegner, C.C. and Killian, G.J. (1992) Origin of oestrus-associated glycoproteins in bovine oviductal fluid. *Journal of Reproduction and Fertility* 95, 841–854.

Wehrman, M.E., Roberson, M.S., Cupp, A.S., Kojima, F.N., Stumpf, T.T., Werth, L.A., Wolfe, M.W., Kittok, R.J. and Kinder, J.E. (1993) Increasing exogenous progesterone during synchronization of estrus decreases endogenous 17-beta estradiol and increases conception in cows. *Biology of Reproduction* 49, 214–220.

Weima, S.M., Van Achterberg, T.A.E., Kruip, Th.A.M. and Mummery, C.L. (1991) Isolation of cell lines from bovine blastocysts using different feeder layers. *Serono Symposium on Preimplantation Embryo Development*, Abstract Vol., 11 (Abs. II–42).

Weima, S.M., Achterberg, T.A.E. van, Rebel, H.G., Kruip, Th.A.M. and Mummery, C.L.M. (1992) Isolation and characterization of permanent cell lines derived from bovine blastocysts. *Proceedings of the Twelfth International Congress on Animal Reproduction* (The Hague) 2, 766–768.

Wells, D.G., Thompson, J.G.E., Tervit, H.R., James, R.W. and Udy, G.B. (1990) Experiences in the application of embryo bisection in sheep MOET programmes. *Proceedings of the New Zealand Society of Animal Production* 50, 431–435.

Wells, D.N., McWhir, J., Hooper, M.L. and Wilmut, I. (1991) Factors influencing the isolation of murine embryonic stem cells. *Theriogenology* 35, 293.

Werb, Z. (1990) Expression of EGF and TGF-alpha genes in early mammalian development. *Molecular Reproduction and Development* 27, 10–15.

Wert, S.E. and Larsen, W.J. (1989) Meiotic resumption and gap junction modulation in the cultured rat cumulus oocyte complex. *Gamete Research* 22, 143–162.

West, J.D., Angell, R.R., Thatcher, S.S., Gosden, J.R., Hastie, N.D., Glasier, A.F. and Baird, D.T. (1987) Sexing the human pre-embryo by DNA–DNA *in situ* hybridisation. *Lancet* i, 1345–1347.

West, J.D., Gosden, J.R., Angell, R.R., West, K.M., Glasier, A.F., Thatcher, S.S. and Baird, D.T. (1988) Sexing whole human pre-embryos by *in-situ* hybridization with a Y-chromosome specific DNA probe. *Human Reproduction* 3, 1010–1019.

Westhusin, M.E., Slapak, J.R., Fuller, D.T. and Kraemer, D.C. (1989) Culture of agar-embedded one and two cell bovine embryos and embryos produced by nuclear transfer in the sheep and rabbit oviduct. *Theriogenology* 31, 291.

Westhusin, M.E., Levanduski, M.J., Scarborough, R., Looney, C.R. and Bondioli, K.R. (1990) Utilization of fluorescent staining to identify enucleated demi-oocytes for utilization in bovine nuclear transfer. *Biology of Reproduction*, 42 (Suppl. 1), 176.

Westhusin, M.E., Pryor, J.H. and Bondioli, K.R. (1991) Nuclear transfer in the bovine embryo: a comparison of 5-day, 6-day, frozen-thawed, and nuclear transfer donor embryos. *Molecular Reproduction and Development* 28, 119–123.

Westhusin, M.E., Levanduski, M.J., Scarborough, R., Looney, C.R. and Bondioli, K.R. (1992) Viable embryos and normal calves after nuclear transfer into Hoechst stained enucleated demi-oocytes of cows. *Journal of Reproduction and Fertility* 95, 475–480.

Wheeler, M.B. and Seidel, G.E. Jr (1986) Time course of *in vitro* capacitation of frozen and unfrozen bovine spermatozoa. *Theriogenology* 25, 216.

Wheeler, M.B. and Seidel, G.E. Jr. (1987) Zona pellucida penetration assay for capacitation of bovine sperm. *Gamete Research* 237–250.

Wheeler, M.B. and Seidel, G.E. Jr.(1988a) Capacitation of bull sperm by lysophospholipids and trypsin. *Theriogenology* 29, 325.

Wheeler, M.B. and Seidel, G.E. Jr. (1988b) Capacitation of bovine spermatozoa by lysophospholipids and trypsin. *Gamete Research* 22, 193–204.

White, I.G., Belanger, L., Hough, S., Ellington, J. and Foote, R.H. (1992) Biochemical changes in bull spermatozoa during capacitation *in vitro*. *Theriogenology* 37, 571–578.

White, K.L. (1989) Embryo and gamete sex selection. In: Babiuk, L.A. and Philips, J.P. (eds) *Animal Biotechnology*. Pergamon Press, Oxford, pp. 179–202.

White, K.L., Anderson, G.B. and BonDurant, R.H. (1987a) Expression of a male-specific factor on various stages of preimplantation bovine embryos. *Biology of Reproduction* 37, 867–873.

White, K.L., Anderson, G.B., BonDurant, R.H., Donahue, S. and Pashen, R.L. (1987b) Viability of bisected bovine embryos after detection of H-Y antigen. *Theriogenology* 27, 293.

White, K.L., Rickords, L.F., Southern, L., Hehnke, K., Thompson, D.L. and Wood, T.C. (1988)

Early embryonic development *in vitro* by co-culture with oviductal epithelial cells in the porcine. *Biology of Reproduction* 38 (Suppl. 1), 421 (Abs.).
White, K.L, Hehnke, K., Rickords, L.F., Southern, L.L. Thompson, D.L. and Wood, T.C. (1989) Early embryonic development *in vitro* by co-culture with oviductal epithelial cells in pigs. *Biology of Reproduction* 41, 425–430.
White, K.L., Bunch, T.C., Reed, W.A., Wang, S. and Yue, C. (1993) Bovine oocyte activation is mediated by IP3-sensitive intracellular calcium pools. *Biology of Reproduction* 48 (Suppl. 1), 145.
White, L.M., Beal, W.E., Barne, J.H., Saacke, R.G. and Marshall, C.E. (1984) Characteristics of bovine spermatozoa after migration through bovine serum albumin gradient. *Journal of Animal Science* 59, 454–459.
Whitfield, C.H. and Parkinson, T.J. (1992) Relationship between fertility of bovine semen and *in vitro* induction of acrosome reactions by heparin. *Theriogenology* 38, 11- 20.
Whittingham, D.G. (1971) Survival of mouse embryos after freezing and thawing. *Nature (London)* 233, 125.
Whittingham, D.G. (1975). Survival of rat embryos after freezing and thawing. *Journal of Reproduction and Fertility* 43, 575–578.
Whittingham, D.G. (1980) Parthenogenesis in mammals. In: Finn, C.A. (ed.) *Oxford Reviews of Reproductive Biology*, Vol. 2. Clarendon Press, Oxford, pp. 205–231.
Wideman, D., Dorn, C.G. and Kraemer, D.C. (1989) Sex detection of the bovine fetus using linear array real-time ultrasonography. *Theriogenology* 31, 272.
Wiebke, N.H., Reichenbach, H.D., Wenigerkind, H. and Brem, G. (1993) Repeated transvaginal laparoscopic guided aspiration of bovine follicular oocytes at different intervals. *Proceedings of the Ninth Congress of the European Embryo Transfer Association* (Lyon), 288.
Wiemer, K.E., Casey, P.L., Mitchell, P.S. and Godke, R.A. (1989) Pregnancies following 24-hour co-culture of equine embryos on foetal bovine uterine monolayer cells. *Equine Veterinary Journal*, Suppl. 8, 117–122.
Wiemer, K.E., Watson, A.J., Polanski, V., McKenna, A.I., Schultz, G.A. and Willadsen, S. (1991a) Effects of maturation and co-culture treatments on the developmental capacity of early bovine embryos. *Biology of Reproduction* 44 (Suppl. 1), 97 (Abs. 178).
Wiemer, K.E., Watson, A.J., Polanski, V., McKenna, A.I., Fick, G.H. and Schultz, G.A. (1991b) Effects of maturation and co- culture treatments on the developmental capacity of early bovine embryos. *Molecular Reproduction and Development* 30, 330–338.
Wiemer, K.E., Hoffman, D.I., Maxson, W.S., Eager, S., Muhlberger, B., Fiore, I. and Cuervo, M. (1993) Embryonic morphology and rate of implantation of human embryos following co-culture on bovine oviductal epithelial cells. *Human Reproduction* 8, 97–101.
Wilkinson, J.I.D. & Tarrant, M.E. (1991) Fertility of cows receiving somidobove in European studies. *Journal of Dairy Science* 74 (Suppl. 1), 192 (Abs. P151).
Willadsen, S.M. (1977) Factors affecting the survival of sheep and cattle embryos during deep-freezing and thawing. In: *The Freezing of Mammalian Embryos.* Ciba Foundation Symposium 52, Excerpta Medica, Amsterdam, pp. 175–194.
Willadsen, S.M. (1979) A method for culture of micromanipulated sheep embryos and its use to produce monozygotic twins. *Nature* 277, 298–300.
Willadsen, S.M. (1981) The developmental capacity of blastomeres from 4- and 8-cell sheep embryos. *Journal of Embryology and Experimental Morphology* 65, 165–172.
Willadsen, S.M. (1986a) Nuclear transplantation in sheep embryos. *Nature* 320, 63–66.
Willadsen, S.M. (1986b) Micromanipulation of embryos of the large domestic species. *Journal of the Royal Agricultural Society of England* 146, 160–171.
Willadsen, S.M. (1989) Cloning of sheep and cow embryos. *Genome* 31, 956–962.
Willadsen, S.M. and Fehilly, C.B. (1983) The developmental potential and regulatory capacity

of blastomeres from two, four and eight-cell sheep embryos. In: Beier, H.M. and Lindner, H.R. (eds) *Fertilization of the Human Egg In Vitro*. Springer-Verlag, Berlin, pp. 353–357.

Willadsen, S.M. & Polge, C. (1981) Attempts to produce monozygotic quadruplets in cattle by blastomere separation. *Veterinary Record* 108, 211–213.

Willadsen, S.M., Polge, C., Rowson, L.E.A. and Moor, R.M. (1974) Preservation of sheep embryos in liquid nitrogen. *Cryobiology* 11, 560.

Willadsen, S.M., Polge, C., Rowson, L.E.A. & Moor, R.M. (1976) Deep freezing of sheep embryos. *Journal of Reproduction and Fertility* 46, 151–154.

Willadsen, S.M., Polge, C. and Rowson, L.E.A. (1978) *In vitro* storage of cattle embryos. In: Sreenan, J.M. (ed.) *Control of Reproduction in the Cow*. Martinus Nijhoff, The Hague, pp. 427–437.

Willadsen, S.M., Janzen, R.E., McAlister, R.J., Shea, B.F., Hamilton, G. and McDermand, D. (1991) The viability of late morulae and blastocysts produced by nuclear transplantation in cattle. *Theriogenology* 35, 161–170.

Willett, E.L., Black, W.G., Casida, L.E., Stone, W.H. and Buckner, P.G. (1951) Successful transplantation of a fertilized bovine ovum. *Science* 113, 247.

Williams, B.L., Knight, J.W., Canseco, R.S., Sparks, A.E.T., Gwazdauskas, F.C., Kornegay, E.T., Velander, W.H. and Wilkins, T.D. (1990) Comparison of synchronization procedures for the collection of one-cell porcine embryos. *Journal of Animal Science* 68 (Suppl. 1) 431–432.

Williams, B.L., Sparks, A.E.T., Canseco, R.S., Knight, J.W., Johnson, J.L., Velander, W.H., Page, R.L.,Drohan, W.N., Young J.M., Pearson, R.E., Wilkins, T.D. and Gwazdauskas, F.C. (1992) *In vitro* development of zygotes from prepubertal gilts after microinjection of DNA. *Journal of Animal Science* 70, 2207–2211.

Williams, D.O. and Evans, C. (1985) The production of twin calves by non-surgical transfer of two embryos to cows in a suckler herd. *British Society of Animal Production* (Winter Meeting), Paper No. 126 (2 pp.).

Williams, G.L. (1990) Suckling as a regulator of postpartum rebreeding in cattle: a review. *Journal of Animal Science* 68, 831–852.

Williams, J.E. and Butler, J.E. (1991) Expression of insulin-like growth factor-I (IGF-I) gene by ovine oviductal cells. *Journal of Animal Science* 69 (Suppl. 1), 406.

Williams, M.E. (1991) The strategic use of milk progesterone tests to improve herd ovulation detection. *Proceedings of the British Cattle Veterinary Association* (Reading), 123–134.

Williams, M.E. and Esslemont, R.J. (1993) A decision support system using milk progesterone tests to improve fertility in commercial dairy herds. *Veterinary Record* 132, 503–506.

Williams, P.E.V., Broadbent, P.J., MacDearmid, A. and Mollison, G.S. (1989) The energy metabolism of twin bearing cows. *British Society of Animal Production* (Winter Meeting), Paper No. 67.

Williams, R.F. and Hodgen, G.D. (1989) Follicular recruitment, selection and dominance in spontaneous versus induced cycles. *Serona Symposia Publications* 63, 69–79.

Williams, R.M., Graham, J.K. and Hammerstedt, R.H. (1991) Determination of the capacity of ram epididymal and ejaculated sperm to undergo the acrosome reaction and penetrate ova. *Biology of Reproduction* 44, 1080–1091.

Williams, T.J. (1988) A technique for sexing mouse embryos by a visual colorimetric assay of the X-linked enzyme, glucose-6-phosphate dehydrogenase. *Theriogenology* 25, 733–739.

Williams, T.J. and Moore, L.(1988) Quick-splitting of bovine embryos. *Theriogenology* 29, 477–484.

Willis, P., Caudle, A.B. and Fayrer-Hosken, R.A. (1991a) Effects of time and hormones on *in vitro* maturation of equine oocytes. *Theriogenology* 35, 294.

Willis, P., Caudle, A.B. and Fayrer-Hosken, R.A. (1991b) Equine oocyte *in vitro* maturation: influences of sera, time and hormones. *Molecular Reproduction and Development* 30, 360–368.

Willis, P., Caudle, A.B. and Fayrer-Hosken, R.A. (1994) Fine structure of equine oocytes matured *in vitro* for 15 hours. *Molecular Reproduction and Development* 37, 87–92.

Wilmut, I. (1986) Cryopreservation of mammalian eggs and embryos. In: Gwatkin, R.B.L. (ed.) *Developmental Biology*, Vol. 4, *Manipulation of Mammalian Development*. Plenum Press, London, pp. 217–247.

Wilmut, I. (1991) Genetic manipulation in cattle breeding and production. *Proceedings of the British Cattle Veterinary Association* (Reading), 26–37.

Wilmut, I. and Campbell, K. (1992) Embryo multiplication in livestock: present procedures and the potential for improvement. In: Lauria, A. and Gandolfi, F. (eds) *Embryonic Development and Manipulation in Animal Production*. Portland Press, London, 135–145.

Wilmut, I. and Rowson, L.E.A. (1973). Experiments on the low-temperature preservation of cow embryos. *Veterinary Record* 92, 686–690.

Wilmut, I. and Smith, L.C. (1988) Biotechnology and the bovine embryo: at present and in the future. *Proceedings of the Fourth Scientific Meeting of the European Embryo Transfer Association* (Lyon), 19–31.

Wilmut, I., Archibald, A.L., Harris, S., McClenaghan, M., Simons, J.P., Whitelaw, C.B.A. and Clark, A.J. (1990a) Methods of gene transfer and their potential use to modify milk composition. *Theriogenology* 33, 113–123.

Wilmut, I., Haley, C.S., Simons, J.P. and Webb, R. (1990b) Prospects for application of molecular genetic manipulation to improve reproduction. *Proceedings of the Fourth World Congress on Genetics Applied to Livestock Production* 16, 281–291.

Wilmut, I., Archibald, A.L., McClenaghan, M., Simons, J.P., Whitelaw, C.B.A. and Clark, A.J. (1991) Production of pharmaceutical proteins in milk. *Experientia* 47, 905–912.

Wilmut, I., Campbell, K.H.S. and O'Neill, G.T. (1992a) Sources of totipotent nuclei including embryonic stem cells. *Proceedings of a Symposium on Cloning Mammals by Nuclear Transplantation* (Fort Collins), 8–16.

Wilmut, I., Haley, C.S. and Woolliams, J.A. (1992b) Impact of biotechnology on animal breeding. *Animal Reproduction Science* 28, 149–162.

Wilmut, I., and Haley, C.S., Simons, J.P. and Webb, R. (1992c) The potential role of molecular genetic manipulation in the improvement of reproductive performance. *Journal of Reproduction and Fertility*, Suppl. 45, 157–173.

Wilson, J.M. and Zalesky, D.D. (1988) Early pregnancy determination in the bovine utilising ultrasonography. *Theriogenology* 29, 330.

Wilson, J.M., Zalesky, D.D., Looney, C.R., Bondioli, K.R. and Magness, R.R. (1992) Hormone secretion by preimplantation embryos in a dynamic *in vitro* culture system. *Biology of Reproduction* 46, 295–300.

Wiltbank, M.C., Diskin, M.G. and Niswender, G.D. (1990) Differential actions of second messenger systems in the corpus luteum of sheep. *Proceedings of the Third International Ruminant Reproduction Symposium* (Nice), Abstract.

Wiltbank, M.C., Wiepz, G.J., Knickerbocker, J.J., Belfiore, C.J. and Niswender, G.D. (1992) Proteins secreted from the early ovine conceptus block the action of prostaglandin F2 alpha on large luteal cells. *Biology of Reproduction* 46, 475–482.

Winston, R.M.L. and Handyside, A.H. (1993) New challenges in human *in vitro* fertilization. *Science* 260, 932–936.

Wintenberger, S., Dauzier, L. and Thibault, C. (1953) Le développement *in vitro* de l'oeuf de la brebis et de celui de la chevre. *C.R. Seanc. Soc. Biol.* 147, 1971.

Wintenberger-Torres, S. (1956) Les rapports entre l'oeuf en segmentation et le tractus maternel chez la brebis. *Proceedings of the Third International Congress on Animal Reproduction and Artificial Insemination* (Cambridge), 63–64.

Wise, T., Vernon, M.W. and Maurer, R.R. (1986) Oxytocin, prostaglandins E and F, estradiol, progesterone, sodium and potassium in preovulatory bovine follicles either developed normally or stimulated by follicle stimulating hormone. *Theriogenology* 26, 757–778.

Wittemer, C., Moreau, L., Dellenbach, P., Gerlinger, P., Le Meur, M.-A. and Chambon, P. (1991) A case of human pregnancy after microinjection of capacitated sperm into the perivitelline space. *Journal of In Vitro Fertilization and Embryo Transfer* 8, 222–224.

Wolfe, B.A. and Kraemer, D.C. (1992) Methods in bovine nuclear transfer. *Theriogenology* 37, 5–15.

Wolfe, D.F., Riddell, M.G., Mysinger, P.W., Stringfellow, D.A., Carson, R.L. and Garrett, P.D. (1990) A caudal flank approach for the collection of oviductal-stage bovine embryos. *Theriogenology* 34, 167–174.

Womack, J.E. (1987) Genetic engineering in agriculture: animal genetics and development. *Trends in Genetics* 3, 65–68.

Wood, M.J., Carroll, J. and Whittingham, D.G. (1989) The addition of serum to the medium limits zona hardening in frozen oocytes. *Journal of Reproduction and Fertility,* Abstract Series No. 3, 33.

Wood, S.A. and Kaye, P.L. (1989) Effects of epidermal growth factor on preimplantation mouse embryos. *Journal of Reproduction and Fertility* 85, 575–582.

Woodruff, T.K., Lyon, R.J., Hansen, S.E., Rice, G.C. and Mather, J.P. (1990) Inhibin and activin locally regulate rat ovarian folliculogenesis. *Endocrinology (Philadelphia)* 127, 3196–3205.

Woolliams, J.A. (1990) Strategies to maximise selection progress in dairy cattle. *Proceedings of the Fourth World Congress on Genetics Applied to Livestock Production,* 15–24.

Woolliams, J.A. and Wilmut, I. (1989) Embryo manipulation in cattle breeding and production. *Animal Production* 48, 3–30.

Worsfold, A.I., Williams, G.L. and Williams, D.O. (1989) Oestrone sulphate measurement in bovine serum during late pregnancy and its relationship with the number of calves born. *British Veterinary Journal* 145, 46–49.

Wrathall, A.E. (1992) Embryo transfer and slow viruses. *Embryo Transfer Newsletter* 10(3), 7–16.

Wrathall, J.H.M. and Knight, P.G. (1991) Effects of exogenous steroids on immunoactive (ir) inhibin secretion by bovine granulosa cells *in vitro*. *Journal of Reproduction and Fertility,* Abstract Series No. 7, 66 (Abs. 122).

Wrathall, J.H.M. and Knight, P.G. (1993) Effects of exogenous oestradiol, LH and various ovarian peptides on androstenedione secretion by bovine theca cells *in vitro*. *Journal of Reproduction and Fertility,* Abstract Series No. 11, 45 (Abs. 79).

Wrathall, J.H.M., McLeod, B.J., Glencross, R.G. and Knight, P.G. (1992) Effects of active immunization against a synthetic peptide sequence of the inhibin α-subunit on plasma gonadotrophin concentrations, ovulation rate and lambing rate in ewes. *Journal of Reproduction and Fertility* 95, 175–182.

Wright, G., Carver, A., Cottom, D., Reeves, D., Scott, A., Simons, P., Wilmut, I., Garner, I. and Colman, A. (1991) High level expression of active human alpha-1-antitrypsin in the milk of transgenic sheep. *Bio/Technology* 9, 830–834.

Wright, I.A. and Peters, A.R. (1991) Nutrition: reproduction interactions in cows. *Proceedings of the British Society of Animal Production* (Winter Meeting), Paper No. 35.

Wright, I.A., White, I.R., Russel, A.J.F., Whyte, T.K. and McBean, A.J. (1988) Prediction of calving date in beef cows by real-time ultrasonic scanning. *Veterinary Record* 123, 228–229.

Wright, I.A., Rhind, S.M. and Whyte, T.K. (1992a) A note on the effects of pattern of food intake and body condition on the duration of the post-partum anoestrous period and LH profiles in beef cows. *Animal Production* 54, 143–146.

Wright, I.A., Rhind, S.M., Smith, A.J. and Whyte, T.K. (1992b) The effect of pheromones from cows in oestrus on the duration of the post-partum anoestrous period in beef cows. *Proceedings of the British Society of Animal Production* (Winter Meeting), Paper No. 66.

Wright, J.M. (1985) Commercial freezing of bovine embryos in straws. *Theriogenology* 23, 17–29.

Wright, R.W. (1986) Culture and short-term storage of embryos. In: Morrow, D.A. (ed.) *Current Therapy in Theriogenology*, 2nd edn. W.B. Saunders, Philadelphia, pp. 79–80.

Wright, R.W. and Bondioli, K.R. (1981) Aspects of *in vitro* fertilization and embryo culture in domestic animals. *Journal of Animal Science* 53, 702–729.

Wright, R.W., Anderson, G.B., Cupps, P.T. and Drost, M. (1976a) Successful culture *in vitro* of bovine embryos to the blastocyst stage. *Biology of Reproduction* 14, 157–162.

Wright, R.W., Anderson, G.B., Cupps, P.T. and Drost, M. (1976b) Blastocyst expansion and hatching of bovine ova cultured *in vitro*. *Journal of Animal Science* 43, 170–174.

Wu, S.C., Cheng, S.P., Liu, P.S., Chiou, C.M., Chu, C.C. (1991a) *In vitro* maturation of pig oocytes and their subsequent development capacity: I. Effects of the integrity of the oocyte complex and the addition of hormones to the oocyte maturation medium. *Journal of the Chinese Society of Animal Science* 20, 455–468.

Wu, S.C., Cheng, S.P., Liu, P.S., Chiou, C.M., Chu, C.C. (1991b) *In vitro* maturation of pig oocytes and their subsequent developmental capacity: II. Effects of follicle size and addition of pig follicular fluid to the oocyte maturation medium. *Journal of the Chinese Society of Animal Science* 20, 469–480.

Wubishet, A., Kesler, D.J., Graves, C.N., Spahr, S.L. and Favero, R.J. (1991) Preovulatory LH profiles of superovulated cows and progesterone concentrations at embryo recovery. *Theriogenology* 35, 451–457.

Wurth, Y.A. and Kruip, Th.A.M. (1992) Bovine embryo production *in vitro* after selection of the follicles and oocytes. *Proceedings of the Twelfth International Congress on Animal Reproduction* (The Hague) 1, 387–389.

Xia, P. and Armstrong, D.T. (1994) Embryonic development of *in vitro* matured and fertilized porcine oocytes after co-culture with estrogen-treated oviductal epithelial cells. *Theriogenology* 41, 340.

Xia, P. and Qin, P. (1992) Studies on polyspermy *in vivo* in pigs. *Biology of Reproduction* 46, 68 (Abs. 71).

Xiaoxia, Z., Zhanqun, J., Shie, J., Jianmin, L., Yuding, Z. (1991) *In vitro* co-culture of *in vitro* fertilized (IVF) bovine embryos with different cell monolayers. *Theriogenology* 35, 295.

Xie, S., Broermann, D.M., Nephew, K.P., Geisert, R.D. and Pope, W.F. (1990a) Ovulation and early embryogenesis in swine. *Biology of Reproduction* 43, 236–240.

Xie, S., Broermann, D.M., Nephew, K.P., Bishop, M.D. and Pope, W.F. (1990b) Relationship between oocyte maturation and fertilization on zygotic diversity in swine. *Journal of Animal Science* 68, 2027–2033.

Xie, S., Broermann, D.M., Nephew, K.P., Ottobre, J.S., Day, M.L. and Pope, W.F. (1990c) Changes in follicular endocrinology during final maturation of porcine oocytes. *Domestic Animal Endocrinology* 7, 75–82.

Xu, K.P. and Greve, T. (1987) Early events of *in vitro* fertilization of bovine follicular oocytes. *Proceedings of a Symposium on the Application of Egg and Embryo Technologies to Domestic Animals* (Copenhagen), 36–40.

Xu, K.P. and Greve, T. (1988) A detailed analysis of early events during *in vitro* fertilization of bovine follicular oocytes. *Journal of Reproduction and Fertility* 82, 127–134.

Xu, K.P. and King, W.A. (1990a) The biology of mammalian fertilization and embryo development. *AgBiotech News and Information* 2(1), 25–28.

Xu, K.P. and King, W.A. (1990b) Effects of oviductal cells and heparin on bovine sperm capacitation *in vitro*. *Biology of Reproduction* 42 (Suppl. 1), 89.

Xu, K.P., Greve, T., Smith, S., Liehman, P., Callesen, H. and Hyttel, P. (1986a) Parthenogenetic activation of cattle oocytes matured *in vitro* and cultured in rabbit oviducts. *Theriogenology* 25, 218.

Xu, K.P., Greve, T., Smith, S. and Hyttel, P. (1986b) Chronological changes of bovine follicular oocyte maturation *in vitro*. *Acta Veterinaria Scandinavica* 27(4), 505–519.

Xu, K.P., Heier, R. and Greve, T. (1987a) Dynamic changes of estradiol concentration in two culture systems for bovine oocyte maturation *in vitro*. *Theriogenology* 27, 297.

Xu, K.P., Greve, T., Callesen, H. and Hyttel, P. (1987b) Pregnancy resulting from cattle oocytes matured and fertilized *in vitro*. *Journal of Reproduction and Fertility* 81, 501–504.

Xu, K.P., Greve, T., Callesen, H. and Hyttel, P. (1988a) Birth of a calf following *in vitro* fertilization of *in vitro* matured oocytes. *Journal of Reproduction and Fertility*, Abstract Series No. 1, Abs. 18.

Xu, K.P., Heier, R. and Greve, T. (1988b) Dynamic changes of estradiol and progesterone concentrations during *in vitro* maturation in cattle. *Theriogenology* 30, 245–255.

Xu, K.P., Picard, L., King, W.A. and Goff, A.K. (1989) Sexing and bisecting bovine embryos produced by *in vitro* maturation and fertilization. *Theriogenology* 31, 274.

Xu, K.P., Loskutoff, N.M., Betteridge, K.J. and King, W.A. (1990) Immature zona-free bovine oocytes can be matured, fertilized and cultured to the blastocyst stage *in vitro*. *Journal of Reproduction and Fertility*, Abstract Series No. 5, Abs. 103.

Xu, K.P., Loskutoff, N.M. and Betteridge, K.J. (1991a) Pregnancies resulting from bovine embryos derived from *in vitro* culture of zona-free zygotes produced by *in vitro* maturation and fertilization of follicular oocytes. *Theriogenology* 35, 296.

Xu, K.P., Yadav, B.R., King, W.A. and Betteridge, K.J. (1991b) Sex related differences in the development of bovine embryos produced *in vitro*. *Biology of Reproduction* 44 (Suppl. 1), 97 (Abs. 177).

Xu, K.P., Hill, B. and Betteridge, K.J. (1992a) Application of *in vitro* fertilization techniques to obtain calves from valuable cows after slaughter. *Veterinary Record* 130, 204–206.

Xu, K.P., Yadav, D.R., King, W.A. and Betteridge, K.J. (1992b) Sex-related differences in developmental rates of bovine embryos produced and cultured *in vitro*. *Molecular Reproduction and Development* 31, 249–252.

Xu, K.P., Yadav, B.R., Rorie, R.W., Plante, L., Betteridge, K.J. and King, W.A. (1992c) Development and viability of bovine embryos derived from oocytes matured and fertilized *in vitro* and co-cultured with bovine oviducal epithelial cells. *Journal of Reproduction and Fertility* 94, 33–43.

Xu, X.J., Yang, Q.Z., Qian, R.C., Nakahara, Z., Niwa, K. and Qin, P.C. (1992d) Development of *in vitro* fertilized bovine ova in two co-culture systems. *Proceedings of the Twelfth International Congress on Animal Reproduction* (The Hague) 3, 1372–1374.

Yadav, B.R., King, W.A., Xu, K.P., Picard, L. and Betteridge, K.J. (1990) Sex ratio of bovine embryos produced by fertilization *in vivo* and *in vitro*. *Theriogenology* 33, 356.

Yadav, B.R., King, W.A. and Betteridge, K.J. (1993) Relationships between the completion of first cleavage and the chromosomal complement, sex, and developmental rates of bovine embryos generated *in vitro*. *Molecular Reproduction and Development* 36, 434–439.

Yamamoto, Y., Oguri, N., Tsutsumi, Y. and Hachinohe, Y. (1982). Experiments in the freezing and storage of equine embryos. *Journal of Reproduction and Fertility*, Suppl. 32, 399–403.

Yamashima, H. (1989) Practical studies on bovine embryo transfer. *Japanese Journal of Animal Reproduction* 35, 20–23.

Yamazaki, Y. and Ishibashi, I. (1990) *In vitro* fertilizing and developmental capacity of mouse follicular oocytes matured in TYH medium supplemented with fetal calf serum for various periods of culture. *Japanese Journal of Animal Reproduction* 36, 189–194.

Yanagida, K.,Yanagimachi, R., Perreault, S.D. and Kleinfeld, R.G. (1991) Thermostability of sperm nuclei assessed by microinjection into hamster oocytes. *Biology of Reproduction* 44, 440–447.

Yanagimachi, R. (1990) Capacitation and the acrosome reaction. In: Asch, R.H., Balmaceda, J.P. and Johnston, I. (eds) *Gamete Physiology: Serona Symposia*, USA, Norwell, Massachusetts, pp. 31–42.

Yang, B.K., Yang, X. and Foote, R.H. (1993a) The effect of growth factors on development of IVM/IVF bovine embryos. *Theriogenology* 39, 343.

Yang, B.K., Yang, X. and Foote, R.H. (1993b) Effect of growth factors on morula and blastocyst development of *in vitro* matured and *in vitro* fertilized bovine oocytes. *Theriogenology* 40, 521–530.

Yang, B.S., Oh, S.J., Lee, M.S. and Kim, H.S. (1989) A triple-staining technique for evaluating normal acrosome reactions of cattle spermatozoa. *Research Reports of the Rural Development Administration: Livestock* 31, 7–10.

Yang, L., Wu, F. and Zhang, Z. (1992) Application of PMSG and its antiserum to artificial induction of twinning in beef cows. *Proceedings of the Twelfth International Congress on Animal Reproduction* (The Hague) 1, 292–294.

Yang, N.S. (1991) Studies related to the freezing and storage of bovine embryos produced by *in vitro* culture methods. PhD Thesis, National University of Ireland, Dublin.

Yang, N.S., Lu, K.H. and Gordon, I. (1990) *In vitro* fertilization (IVF) and culture (IVC) of bovine oocytes from stored ovaries. *Theriogenology* 33, 352.

Yang, N.S., Duff, R., Lu, K.H., Gordon, I. and Polge, C. (1991) Effect of storage temperature and time on the viability of bovine embryos produced *in vitro*. *Theriogenology* 35, 297.

Yang, N.S., Lu, K.H., Gordon, I. and Polge, C. (1992) Vitrification of bovine blastocysts produced *in vitro*. *Theriogenology* 37, 326.

Yang, Q.Z., Xia, P., Xu, X.J., Feng, H.L., Qian, R.C. and Qin, P.C. (1992) *In vitro* maturation, *in vitro* fertilization and *in vitro* development of cattle oocytes. *Proceedings of the Twelfth International Congress on Animal Reproduction* (The Hague) 1, 553–555.

Yang, X. (1991) Embryo cloning by nuclear transfer in cattle and rabbits. *Embryo Transfer Newsletter* 9(4), 10–22.

Yang, X. and Anderson, G.B. (1992) Micromanipulation of mammalian embryos: principles, progress and future possibilities. *Theriogenology* 38, 315–335.

Yang, X. and Foote, R.H. (1990) Pregnancy rate in the rabbit model following hyperosmotic sucrose treatment of embryos. *Theriogenology* 33, 354.

Yang, X., Chen, J. and Foote, R.H. (1988) Blastocyst development from rabbit ova fertilized by injected sperm. *Journal of Reproduction and Fertility*, Abstract Series No. 1, Abs. 16.

Yang, X., Chen, J., Chen, Y.Q. and Foote, R.H. (1990) Improved developmental potential of rabbit oocytes fertilized by sperm microinjection into the perivitelline space enlarged by hypertonic media. *Journal of Experimental Zoology* 255, 114–119.

Yang, X., Jiang, S., Kovacs, A. and Foote, R.H. (1991) Age dependent activation, enucleation and nuclear transfer of bovine oocytes matured *in vitro* and *in vivo*. *Biology of Reproduction* 44 (Suppl. 1), 141 (Abs. 355).

Yang, X., Jiang, S.-E., Kovacs, A. and Foote, R.H. (1992) Nuclear totipotency of cultured rabbit morulae to support full-term development following nuclear transfer. *Biology of Reproduction* 47, 636–643.

Yang, X., Jiang, S. and Foote, R.H. (1993a) Bovine oocyte development following different oocyte maturation and sperm capacitation procedures. *Molecular Reproduction and Development* 34, 94–100.

Yang, X., Jiang, S., Farrell, P., Foote, R.H. and McGrath, A.B. (1993) Nuclear transfer in cattle: effect of nuclear donor cells, cytoplast age, co-culture, and embryo transfer. *Molecular Reproduction and Development* 35, 29–36.

Yang, Y.B. and Lu, K.H. (1990) The influence of bovine oocyte type on the *in vitro* fertilization and subsequent *in vitro* development. *Theriogenology* 33, 355.

Yang, Z. and Smith, L.C. (1993) Parthenogenetic development *in vitro* of bovine secondary oocytes exposed to a single electric pulse. *Theriogenology* 39, 344.

Yates, C.A. and De Kretser, D.M. (1987) Male-factor infertility and *in vitro* fertilization. *Journal of In Vitro Fertilization and Embryo Transfer* 4, 141–147.

Yeung, W.S.B., Ho, P.C., Lau, E.Y.L. and Chan, S.T.H. (1992) Improved development of human embryos *in vitro* by a human oviductal cell co-culture system. *Human Reproduction* 7, 1144–1149.

Yong, Z., Jianchen, W., Jufen, Q. and Zhiming, H. (1991) Nuclear transplantation in goats. *Theriogenology* 35, 299.

Yoon, S.H., Park, S.P., Park, T.G., Ko, D.H. and Chung, K.S. (1989) Studies on *in vitro* maturation of bovine follicular oocytes. 2. Effects of oestrous bovine serum, bovine follicular fluid and matured cumulus cells on *in vitro* maturation and fertilization of follicular oocytes. *Korean Journal of Dairy Science* 11, 139–146.

Yoshida, M. (1993) Role of glutathione in the maturation and fertilization of pig oocytes *in vitro*. *Molecular Reproduction and Development* 35, 76–81.

Yoshida, M. and Ishigaki, K. (1991) Effect of maturation media on male pronucleus formation of pig oocytes matured and fertilized *in vitro*. *Serono Symposium on Preimplantation Embryo Development*, Abstract Vol., 48 (Abs. I-48).

Yoshida, M., Bamba, K. and Kojima, Y. (1989) Effects of gonadotrophins and estradiol-17β on the timing of nuclear maturation and cumulus mass expansion in pig oocytes cultured *in vitro*. *Japanese Journal of Animal Reproduction* 35, 86–91.

Yoshida, M., Ishizaki, Y. and Kawagishi, H. (1990) Blastocyst formation by pig embryos resulting from *in vitro* fertilization of oocytes matured *in vitro*. *Journal of Reproduction and Fertility* 88, 1–8.

Yoshida, M., Ishigaki, K. and Pursel, V.G. (1992a) Effect of maturation media on male pronucleus formation in pig oocytes matured *in vitro*. *Molecular Reproduction and Development* 31, 68–71.

Yoshida, M., Ishizaki, Y., Kawagishi, H., Bamba, K. and Kojima, Y. (1992b) Effects of pig follicular fluid on maturation of pig oocytes *in vitro* and on their subsequent fertilizing and developmental capacity *in vitro*. *Journal of Reproduction and Fertility* 95, 481–488.

Yoshida, M., Mizoguchi, Y., Ishigaki, K., Kojima, T. and Nagai, T. (1993a) Birth of piglets derived from *in vitro* fertilization of pig oocytes matured *in vitro*. *Theriogenology* 39, 1303–1311.

Yoshida, M., Ishigaki, K., Nagai, T., Chikyu, M. and Pursel, V.G. (1993b) Glutathione concentration during maturation and after fertilization in pig oocytes: relevance to the ability of oocytes to form male pronucleus. *Biology of Reproduction* 49, 89–94.

Yoshimura, Y., Hosoi, Y., Iritani, A., Nakamura, Y., Atlas, S.J. and Wallach, E.E. (1989) Developmental potential of rabbit oocyte matured *in vitro*: the possible contribution of prolactin. *Biology of Reproduction* 40, 26–33.

Yoshimura, Y., Nakamura, Y., Oda, T., Yamada, H., Nanno, T., Ando, M., Ubukata, Y. and Suzuki, M. (1990) Effects of gonadotrophin-releasing hormone agonists on meiotic maturation of follicle-enclosed oocytes in rabbits. *Biology of Reproduction* 43, 1012–1018.

Yoshimura, Y., Nakamura, Y., Yamada, H., Nanno, T., Ubukata, Y., Ando, M. and Suzuki,

M. (1991) Gonadotropin-releasing hormone agonists induce meiotic maturation and degeneration of oocytes in the *in vitro* perfused rabbit ovary. *Fertility and Sterility* 55, 177–183.

Yoshimura, Y., Nakamura, Y., Oda, T., Ando, M., Ubukata, Y., Karube, M., Koyama, N. and Yamada, H. (1992) Induction of meiotic maturation of follicle-enclosed oocytes of rabbits by a transient increase followed by an abrupt decrease in cyclic AMP concentration. *Journal of Reproduction and Fertility* 95, 803–812.

Youngs, C.R., Pendleton, R.J., Rorie, R.W., Casey, P.L. and Godke, R.A. (1988) A preliminary study on the use of a computerized image analysis system for the characterization of dairy goat embryos. *Proceedings of the Eleventh International Congress on Animal Reproduction and Artificial Insemination* (Dublin) 2, 200 (3 pp.)

Youngs, C.R., McGinnis, L.K. and Weber, P.K. (1992) Glutamine supports *in vitro* development of sheep embryos in medium lacking pyruvate, lactate and glucose. *Biology of Reproduction* 46 (Suppl. 1), 120 (Abs. 279).

Younis, A.I. and Brackett, B.G. (1991) Importance of cumulus cells and insemination intervals for development of bovine oocytes into morulae and blastocysts *in vitro*. *Theriogenology* 36, 11–21.

Younis, A.I. and Brackett, B.G. (1992) Thyroid stimulating hormone enhancement of bovine oocyte maturation *in vitro*. *Molecular Reproduction and Development* 31, 144–151.

Younis, A.I., Keefer, C.L. and Brackett, B.G. (1989a) Fertilization of bovine oocytes by sperm injection. *Theriogenology* 31, 276.

Younis, A.I., Brackett, B.G. and Fayrer-Hosken, R.A. (1989b) Influence of serum and hormones on bovine oocyte maturation and fertilization *in vitro*. *Gamete Research* 23, 189–201.

Younis, A.I., Zuelke, K.A., Harper, K.M., Oliveira, M.A.L. and Brackett, B.G. (1990) Pregnancies after embryo transfer of *in vitro* fertilized goat oocytes. *Biology of Reproduction* 42 (Suppl. 1), 128.

Younis, A.I., Zuelke, K.A., Harper, K.M., Oliveira, M.A.L. and Brackett, B.G. (1991) *In vitro* fertilization of goat oocytes. *Biology of Reproduction* 44, 1177–1182.

Younis, J.S., Simon, A., Koren, R., Dorembus, D., Schenker, J.G. and Laufer, N. (1992) The effect of growth hormone supplementation on *in vitro* fertilization outcome: a prospective randomized placebo-controlled double-blind study. *Fertility and Sterility* 58, 575–580.

Yovich, J.L. (1993) Pentoxifylline: actions and applications in assisted reproduction. *Human Reproduction* 8, 1786–1791.

Yuswiati, E. and Holtz, W. (1990) Successful transfer of vitrified goat embryos. *Theriogenology* 34, 629–632.

Zamboni, L. (1972) Comparative studies on the ultrastructure of mammalian oocytes. In: Biggers, J.D. and Schuetz, A.W. (eds), *Oogenesis*. Butterworths, London, pp. 5–45.

Zaneveld, L.J.D., Jonge, C.J. De, Anderson, R.A. and Mack, S.R. (1991) Human sperm capacitation and the acrosome reaction. *Human Reproduction* 6, 1265–1274.

Zarutskie, P.W., Dixon, L.L. and Hiller, S.L. (1992) Identifying sources of bacterial endotoxin contamination in an *in vitro* fertilization (IVF) culture environment. *Journal of Assisted Reproduction and Genetics* 9, 77–80.

Zeng, Y.T., Zhang, M.L., Chen, M.J., Zhou, X.D., Huang, Y., Ren, Z. R. and Huang, S.Z. (1992) Sexing bovine embryos using PCR amplification of bovine SRY sequence. *Proceedings of the Twelfth International Congress on Animal Reproduction* (The Hague) 2, 773–775.

Zetova, L., Mardesic, T., Mikova, M., Muller, P. and Stroufova, A. (1992) Cocultures of human embryos improve IVF+ET results. *Human Reproduction* 7 (Suppl. 2), 159.

Zetova, L., Mardesic, T., Mikova, M., Muller, P. and Stroufova, A. (1993) Improved development of human embryos cultured on a Vero cell monolayer. *Journal of Assisted Reproduction and Genetics* 10, 234–236.

Zhang, J.J., Muzs, L.Z. and Boyle, M.S. (1990) *In vitro* fertilization of horse follicular oocytes matured *in vitro*. *Molecular Reproduction and Development* 26, 361–365.

Zhang, L., Blakewood, E.G., Denniston, R.S. and Godke, R.A. (1991a) The effect of insulin on maturation and development of *in vitro* fertilized bovine oocytes. *Theriogenology* 35, 301.

Zhang, L., Denniston, R.S., Bunch, T.D. and Godke, R.A. (1991b) Cows vs. heifers for the production of *in vitro* matured, *in vitro* fertilized (IVF) embryos. *Journal of Animal Science* 69 (Suppl. 1), 49.

Zhang, L., Denniston, R.S. and Godke, R.A. (1991c) The effect of insulin on the development of *in vitro* fertilized porcine oocytes. *Journal of Animal Science* 69 (Suppl. 1), III.

Zhang, X., Rutledge, J. and Armstrong, D.T. (1991) Studies on zona hardening in rat oocytes that are matured *in vitro* in a serum-free medium. *Molecular Reproduction and Development* 28, 292–296.

Zhang, Y., Wang, J.C., Qian, J.F. and Hao, Z.M. (1991) Nuclear transplantation in goat embryos. *Scientia Agricultura Sinica* 24(5), 1–6.

Zheng, Y.S. and Sirard, M.A. (1992) The effect of sera, bovine serum albumin and follicular cells on *in vitro* maturation and fertilization of porcine oocytes. *Theriogenology* 37, 779–790.

Zheng, Y.S., Fiser, P. and Sirard, M.A. (1992) The use of ejaculated boar semen after freezing in 2 or 6% glycerol for *in vitro* fertilization of porcine oocytes matured *in vitro*. *Theriogenology* 38, 1065–1075.

Zolti, M., Ben-Rafael, Z., Meirom, R., Shemesh, M., Bider, D., Mashiach, S. and Apte, R.N. (1991) Cytokine involvement in oocytes and early embryos. *Fertility and Sterility* 56, 265–272.

Zuelke, K.A. and Brackett, B.G. (1990) Luteinizing hormone-enhanced *in vitro* maturation of bovine oocytes with and without protein supplementation. *Biology of Reproduction* 43, 784–787.

Zuelke, K.A. and Brackett, B.G. (1992) Effects of luteinizing hormone on glucose metabolism in cumulus enclosed bovine oocytes matured *in vitro*. *Biology of Reproduction* 46 (Suppl. 1), 117 (Abs. 267).

Zuelke, K.A. and Brackett, B.G. (1993) Increased glutamine metabolism in bovine cumulus cell-enclosed and denuded oocytes after *in vitro* maturation with luteinizing hormone. *Biology of Reproduction* 48, 815–820.

Zuelke, K.A., Keith, C.H. and Brackett, B.G. (1991) LH induction of calcium oscillations in immature bovine oocytes. *Biology of Reproduction* 44 (Suppl. 1), 141 (Abs. 354).

Index